V. Ravichandran · Atul Kumar Razdan

# Fundamental Discrete Structures

V. Ravichandran
Department of Mathematics
National Institute of Technology
Tiruchirappalli, Tamil Nadu, India

Atul Kumar Razdan
Department of Mathematics
Inderprastha Engineering College
Ghaziabad, Uttar Pradesh, India

ISSN 2731-9318 ISSN 2731-9326 (electronic)
University Texts in the Mathematical Sciences
ISBN 978-981-96-0071-7 ISBN 978-981-96-0069-4 (eBook)
https://doi.org/10.1007/978-981-96-0069-4

Mathematics Subject Classification: 97E30, 15A06, 05C50, 03G05, 03G10, 94C10, 20B05, 15A21, 68R10, 03D05, 94B05, 68R01, 97E60, 68R05, 20A05, 15A03, 15A04, 15A18



This Springer imprint is published by the registered company Springer Nature Singapore Pte Ltd.
The registered company address is: 152 Beach Road, #21-01/04 Gateway East, Singapore 189721, Singapore

# University Texts in the Mathematical Sciences

Textbooks in this series cover a wide variety of courses in mathematics, statistics and computational methods. Ranging across undergraduate and graduate levels, books may focus on theoretical or applied aspects. All texts include frequent examples and exercises of varying complexity. Illustrations, projects, historical remarks, program code and real-world examples may offer additional opportunities for engagement. Texts may be used as a primary or supplemental resource for coursework and are often suitable for independent study.

*To Prof. Dato' Indera Rosihan M. Ali and Prof. Mini Shaji Thomas*

*—V. Ravichandran*

*To my wife Anju and children, Anuja and Aman*

*—Atul Kumar Razdan*

# Preface

*Mathematics makes a nice distinction between the usually synonymous terms elementary and simple, with elementary taken to mean that not very much mathematical knowledge is needed to read the work and simple to mean that not very much mathematical ability is needed to understand it.*

*Julian Havel (1854–1912)*

Discrete mathematics is about operations and structures that constitute the mathematical foundation to configurations and process involving variables changing only *in steps*. It provides tools and valuable insight to model problems related to *data structures, algorithms, programming, languages, automata theory, software testing*, and, in general, most systematic procedures concerning computability. It is thus imperative for every budding computer scientist to have a good grasp over the basic concepts of discrete mathematics. In most situations, a major concern is how to deal with the "complexity", which studies the resources used by different types of computational procedures.

The topics covered in this book are good enough to model and solve computation problems in a variety of subject areas. For example, a graph is a tool to study structures defined in terms of connections among objects. It also suggests ways how to store and retrieve data efficiently. Moreover, some graph-based combinatorial techniques are used extensively in the analysis and design of algorithms, and also in related issues concerning data structures. We also discuss applications of the core concepts to computer science, and also to its allied subjects, such as related to *data science, machine learning, communication systems, network security, and discrete optimization.* The emphasis throughout the text is on helping the reader master combinatorial techniques that play a significant role in solving many different types of technical problems.

The contents presented in this book may be used to design different levels of *12-week courses* catering to the need of undergraduates pursuing a degree course in computer science or some allied branch. The authors have been teaching some such semester courses to students in various engineering colleges in India. Most of the ideas introduced in early chapters are easy to grasp. The presentation is sufficiently interactive to help readers have an active learning experience. Exercises provided at the end of each chapter are specifically arranged to allow the reader to master the basic stuff before moving on to the next chapter. This book is also suitable for readers who intend to learn the subject on their own. We would **very much** appreciate being told of any corrections or possible improvements **however minor**.

We are also thankful to Dr. Shamim Ahmad, Publishing Editor, Mathematics, Springer India, for his continued support. We also thank Mr. Prasanth Anandan, Production Editor, Springer Nature, and his team for their wonderful work in getting the book into various formats for publication.

Tiruchirappalli, India
Ghaziabad, India
September 2024

V. Ravichandran
Atul Kumar Razdan

# Introduction

*Pure mathematics is, in its way, the poetry of logical ideas.*
*Albert Einstein (1879–1955)*

A study was conducted in 1980 to understand the mathematical needs of a budding computer scientist [1]. The findings led to a review of the whole elementary mathematics curriculum. In the process, *discrete mathematics* took on a new significance. Since the topics covered in a typical discrete mathematics course provide a rich variety of tools needed to deal with problems in various allied subjects, it forms the backbone of the computer science ([2–5]). In addition to concepts and various types of computation tools, it also provides the fundamentals of the mathematical reasoning. As we know, to *think* about a scientific problem, we need a relevant and precise language. In the case of mathematics and computer science, sets, relations, and functions provide a universal language.

For example, a "dictionary" for data structures is a set, together with operations such as *insert, delete*, and *lookup*. Also, a set in *Java* is an object that holds other objects such as a *Collection, List, Hashset, and Tree*. Moreover, a particular type of set of strings over an alphabet is called a *language*. In general, set theory provides various fundamental tools to model and solve computation problems ([2–5]). Listed below are some important concepts based on sets that most computer science graduates would come across at some point in time during their academic pursuit:

1. A *data structure* is about implementation of sets, with a special emphasis on efficient implementations of related operations;
2. A *relational database* uses relations among sets. It doesn't matter when *records* have more than two elements. For example, a triplet $(a, b, c)$ can be viewed as the tuple $(a, (b, c))$, and so on.
3. A *formal language* is a set of strings, and the related *programming* uses usual set operations, concatenation, and Kleen's star operation. Simple illustrations include programs used to handle design, testing, and, maintenance of code for various types of software.

4. In a programming language semantics, the *semantic domain* are sets that have (typically) a complete partial order structure.
5. Axiomatic set theory finds applications in certain parts of *computability theory*, and also in the study of logic used in advance computability.

A yet another good reason why set theory is considered as a natural language of computer science is the fact that each *relation* (and hence each *function*) can be represented as sets, which also facilitates the operation of *composition* of relations and functions. In this book, we discuss some key concepts such as related to mathematical logic, sets, relations, functions, lattices, Boolean algebras, algebraic structures, discrete probability, and graphs.

In Chap. 1, we discuss various forms of *deductive reasoning* related to "if... then ..." statements, as developed by Boole, De Morgan, Frege, Russel, Wittgenstein, and Post. The main topics covered in this chapter are truth tables, equivalences, implication, predicate logic, quantifiers, and also the theory of inference. A brief about the philosophical form of reasoning mainly due to ancient Greek philosophers Aristotle and Chrysippus is also included.

In Chap. 2, we discuss different ways to prove an implication $p \Rightarrow q$, considering how the *hypotheses p* is formulated and how the *conclusion q* is to be drawn. There are mainly two types of *methods of proof*. In majority cases, we first apply the *direct method* of proof, which derives the *conclusion q* by using related axioms, some known logical equivalences, and rules of inference. On the other hand, the *indirect method* of proof (or *proof by contradiction*) takes $\neg q$ as an *additional premise* and, subsequently, attempts to find a contradiction. Dirichlet's *pigeon-hole principle* is an interesting special case of this method. This principle is used in group theory to show that every element of a finite group has an order. A typical proof of the fact that every finite integral domain is a field stems from similar logic. Sometimes, we also need to use *proof by cases* and the *principle of mathematical induction*.

In Chap. 3, we introduce sets, relations, and related operations. We also discuss some important properties of binary relations on a set. Applications of the digraph and the matrix of a binary relation are also illustrated. We discuss important theorems related to equivalence relations and partitions of a set. In the last section of the chapter, we introduce the fundamental concept of a *partially order set* (or simply a *poset*). We also use *Hasse diagrams* to study the properties of some important posets such *divisor posets*.

In Chap. 4, we introduce some important functions. The concept of a function is fundamental to everything we do in mathematics and computer science. Dirichlet defined a function in 1837 as a "formula or an expression" that gives a unique value for each assigned value. In 1888, Dedekind defined a function as a *rule* $f$ that assigns to each element $x$ of a nonempty set $X$ a unique element $f(x)$ of another nonempty set $Y$. We prove important theorems related to properties of functions. We also discuss the concept of the cardinality of an infinite set. Some applications to problems in computer science dealing with *hashing* and *growth* are also included. We have also included here a brief about recursive functions.

In Chap. 5, we discuss some interesting combinatorial methods based on certain basic *counting techniques* related to combination, arrangements of objects, and *partitions*. The main focus is to discuss the work of nineteenth- and twentieth-century mathematicians such as Pascal, Jacob Bernoulli, and Euler, who developed fundamental ideas related to topics such graph theory, coding theory, and discrete probability. In general, combinatorics is a branch of mathematics that deals with some important *enumeration problems* concerning certain engineering problems. Many exciting examples are included to illustrate the importance of "counting" in many different types of applications. Some other problems that are discussed in the chapter concern the existence of certain types of combinatorial objects subject to various constraints. The classic *stars and bars problem* provides key ideas how to count the number of integer solutions to equations of the form

$$x_1 + x_2 + \cdots + x_n = k,$$

subject to some *constraints* on $x_i$'s. We also discuss two methods of finding an explicit formula for the $n$th term of an unknown sequence $(a_n)$ of real (or complex) numbers that satisfies a linear *recurrence relation*, i.e., $a_n$ is defined in terms of some of its previous terms, say $a_{n-1}, a_{n-2}, \ldots, a_{n-k}$. These are the method of *characteristic equation* and the *generating function method*.

In Chap. 6, we discuss some basic concepts of discrete probability theory inasmuch as needed to discuss some significant applications related to the interface of real world with computer science. The probability is a measure of "likelihood" of occurrence of an unpredictable event, and the probability theory deals with the analysis of such events. We mainly discuss basic aspects of Kolmogorov's probability model and prove certain core results involving the *expectation* and *variance* of a random variable. The central idea is to use *average behavior* of problems to analyze, even when the associated distributions are not known. We conclude the chapter with some useful concentration inequalities and limit theorems. The topics covered in this chapter find applications to problems concerning randomized algorithms, machine learning, computer vision, error-correction coding, cryptography, optimization, robotics, and also to problems concerning computational linguistics. Further, in the case of performance modeling, the probabilistic ideas help predict delays, schedule jobs and resources, and also to manage provision capacity.

In Chap. 7, we discuss some interesting properties of lattices and Boolean algebras that find applications to several scientific problems, including related to computer science. We prove some important theorems related to lattices. A poset $(L, \preceq)$ is called a *lattice* if, for every pair of elements $a, b \in L$, both supremum $\sup\{a, b\}$ and infimum $\inf\{a, b\}$ exist. Lattices play a significant role in dealing with some important engineering problems such as dealing with *structural configuration, hierarchy*, and *scheduling*. A distributive complemented lattice $(L, \preceq, \vee, \wedge, 0, 1)$ is called a *Boolean algebra*. We discuss some properties of Boolean algebras and application of *Karnaugh map* to digital circuits. More generally, Boolean algebras and related concepts provide mathematical foundation to *switching theory*.

In Chap. 8, we discuss basic algebraic structures such as groups, rings, and fields. A machine-executable computation uses a "systematic procedure" involving strings of symbols and operations. In abstract sense, such a procedure is a "function" that takes strings as inputs. So, one may obtain a machine solution of a problem by using the basic properties of these types of algebraic structures. Majority of concepts introduced in this chapter were first discovered while formulating a unified "theory of numbers", which also include the *p-adic numbers* and *Hamilton's quaternions*. The origin of group theory is attributed to Galois' work on *field extensions* and Cayley's on *group of substitutions*. On the other hand, Richard Dedekind introduced the core concepts of ring theory while editing Dirichlet's lectures on number theory. Some applications of the concepts developed in this chapter are discussed in later chapters.

Chapter 9 is an introduction to basic linear algebra. The fundamental concept of a *vector space* is a tool to study *linear combinations*, i.e., finite sums of scalar multiple of vectors, in the most generalized sense. The related algebraic methods play a significant role in many applications, including machine learning. In general, the topics covered in this chapter are suitable for a senior-level undergraduate pursuing an engineering degree in computer science or some allied branch, or a course related to a part of operations research. We will also discuss some illustrations of historical significance.

In Chap. 10, we discuss *graph theory*. Graph is a fascinating modern tool of computation. It consists of "nodes" and "links". The main emphasis of any *graph model* is always on how nodes of diverse forms are linked together. A major role it plays in the development of graph algorithms such as used to find the *shortest path, minimum spanning tree, an element in a data, graph planarity*, and so on. The number of practical problems concerning physical, biological, and social systems, that can be analyzed using a graph algorithm is enormous. Graph-based concepts presented here concern some fundamental processes and configurations related to computer science. In this chapter, we have also included applications of trees to network problems related to data organization, computational devices, communication, etc.

In Chap. 11, we introduce basics of *finite fields, error-correction coding*, and *cryptography*. We discuss some basic error-detection and error-correction encoding schemes. The main emphasis in various examples discussed in this chapter will be on linear codes. We next introduce some basics of three fundamental types of cryptosystems, with emphasis on describing how simple mathematical tools help design efficient encryption and decryption algorithms. The prefix "crypt" means *hidden* and the suffix "graphy" means *writing*. So, the main purpose of cryptography is to secure the data, communicated over public channel or in a computer system with shared resources. The undergraduates pursuing a course in communication systems are the intended audience for this chapter.

In Chap. 12, we introduce some basic stuff related to certain interesting *discrete optimization* problems that are significant to both theoreticians and working professionals. As a special case, combinatorial problems study the ways discrete structures can be selected, arranged, or combined. Such problems arise in a number of processes related to computer science such as *memory allocation, scheduling*, and also in case of *data packet routing*. Also, they involve finding ways to efficiently allocate resources in mathematically formulated problems such as related to graph coloring, finding shortest paths, partitioning, packing, and covering sets. Some of the active research areas related to *discrete optimization* are such as *combinatorial optimization, integer programming*, and *constraint programming*. The first involves graphs, matroids, and some other types of discrete structures. Integer programming-based algorithms find applications in a number of practical problems concerning computer science. Some such tools also find applications in computer science areas such as *error-correction coding, computational geometry*, and many others. In this chapter, we will not dwell on data structures, implementation details, or more sophisticated approaches. The graduates pursuing a course in computer science, operation research, and some other similar subject areas, are the intended audience for this chapter.

A good grasp over the foundational material covered in the first four chapters is important to have an active learning experience throughout the book. A larger part of this book is accessible to anyone who has good hold on concepts taught at lower level undergraduate mathematics. Solving all exercises provided at the end of a chapter would help have an appropriate mathematical maturity to understand the concepts presented in subsequent chapters. As most would agree, while introducing a mathematical concept, a learner feels less motivated if the discussion starts with well-crafted definitions, followed by proving related theorems. In our experience, learning is more enriching when concepts are introduced starting with its origin, followed by a discussion about the subsequent developments.

## References

1. Howson, A.G., Kahane, J.P., Lauginie, P., and de Turckheim, E.: Mathematics as a Service Subject, Cambridge University Press, (1988)
2. Cameron, P.J.: Combinatorics: Topics, Techniques, Algorithms, Cambridge University Press, (1994)
3. Gallier, J.: Discrete Mathematics, Universitext, Springer, (2011)
4. Graham, R.L., Knuth, D.E. and Patashnik, O.: Concrete Mathematics: A Foundation for Computer Science (2/e), Addison-Wesley, (1994)
5. Lewis, H. and Zax, R.: Essential Discrete Mathematics for Computer Science, Princeton University Press, (2019)

# Contents

# List of Figures

# Chapter 1
# Mathematical Logic

*For hundreds of pages the closely-reasoned arguments unroll, axioms and theorems interlock. And what remains with us in the end? A general sense that the world can be expressed in closely-reasoned arguments, in interlocking axioms and theorems.*

***Micheal Frayan*** *(*1933*–)*

This chapter introduces some core topics related to mathematical logic. We begin our discussion with *propositional calculus*. It involves statement formulas consisting of atomic propositions connected by *logical operators* such as "*conjunction*", "*disjunction*", or "*negations*". The focus is on developing tools to study certain fundamental *equivalences* and *implications*. Next, we discuss *first-order logic* involving *predicates* and *quantifiers*. Our final aim is to discuss *syllogisms* that are used to arrive at a *valid conclusion* from a set of given premises by way of following a stepwise procedure, each known as a *valid argument*. In all that follows, *syllogism* refers to an argument that is used as a *rule of inference*.

Since ideas introduced in this chapter form the fundamentals of some important related topics such as *natural language processing*, it is strongly advised the reader pursuing computer science to grasp most of it. Also, it is equally important to attempt all the practice problems provided at the end of the chapter. The interested reader may like to read some advance concepts from excellent texts such as [1–4].



V. Ravichandran and A. K. Razdan, *Fundamental Discrete Structures*,
University Texts in the Mathematical Sciences,
https://doi.org/10.1007/978-981-96-0069-4_1

## 1.1 Early Work

Since the time of Greek philosophers *Aristotle* (384–322 B.C.E.), and even before in the work of *Chrysippus of Soli* (278–206 B.C.E.), mathematicians and philosophers had been familiar with *syllogisms* as forms of valid deductive reasoning. As such, in traditional terms, logic refers to a philosophical type of reasoning that has been applied to legitimatize a conclusion drawn from a set of premises. For many centuries, the studies related to logic lacked formal machinery, and all the *argument forms* were verbal. Most of it had been about drawing interpretations of the works of Aristotle and Chrysippus's *syllogisms*.

The central theme of the present chapter, namely, the *deductive reasoning* has always been subject to two differing views: *Use some syllogisms to draw a conclusion from a finite number of premises or, like ancient Greek philosophers, use a speculative statement as a conclusion*. However, the main problem that we need to study here is about how to assign a *truth value* to an implication of the form "if ..., then ...", where "if" part is also called *the antecedent* and "then" part is also called *a consequent*.

We are not too ambitious to make a detour into the history of deductive reasoning. The sole purpose is to familiarize the reader how modern logic emerged in over more than two millennia. We begin with a brief about historical development focusing mainly on writings of the two most influential Greek philosophers Aristotle and Chrysippus, who together developed the basics of what we now call as *propositional calculus*. A more elaborate discussion based on logical calculus of deduction is provided in subsequent sections of the chapter. For a modern treatment, we will also glance through the work of Boole, Russell, and Whitehead.

Greeks treated the problem of assigning the *truth value* to "if-then" types of statements. The *syllogisms* that are of central importance to theory of logic are attributed to Chrysippus. Unlike an *inductive argument* that moves from specific to general, the term *syllogism* refers to a deductive argument that moves from general to specific, where the truth value of two (or more) *premises* implies a conclusion that is consistent with known facts [5]. The five *indemonstrable forms* (or *argument forms*) that are attributed to Chrysippus are listed as given below [6–8]:

1. *If the first, then the second. The first. Therefore, the second.* For example, if it is day, there is light. It is day. Therefore, there is light.
2. *If the first, then the second. Not the second. Therefore, not the first.* For example, if it is day, there is light. There is not light. Therefore, it is not day.
3. *Not both the first and the second. The first. Therefore, not the second.* For example, it is not the case both that it is day and it is night. It is day. Therefore, it is not night.
4. *Either the first or the second. The first. Therefore, not the second.* Notice that no example of this rule is provided in citations given above.
5. *Either the first or the second. Not the first. Therefore, the second.* For example, either it is day or it is night. It is not night. Therefore, it is day. Notice that, in this example, the first and second are switched from the general rule above.

Notice that, in each one of the above five syllogisms, there are two types of statements taken as a major and a minor premises, followed by a conclusion. Greeks called the first two "if-then" types of "syllogisms" as *hypothetical propositions* [9]. The matter related to what truth value be assigned to a hypothetical proposition was debated in antiquity [10]. Further, the third syllogism has a major "and" type of premise, whereas the fourth syllogism has a major "or" type of premise.

Aristotle had a good understanding about the role a language plays in theory of logic, and so he focused on statements that are either true or false. The main idea was to strengthen dialectics to assure correctness of arguments used in deductive reasoning. Syllogisms helped him to arrange premises leading to a valid *argument*. He often applied syllogisms to verbal statements such as given below:

All humans are mortal.

All mortals will die.

Therefore, all humans will die.

Aristotle discussed various types of syllogisms, starting with the *first figure* of term arrangements: "If $r$ is predicated of all $q$, and $q$ of all $p$, $r$ must be predicated of all $p$". In what follows, the above is called "Aristotle's first figure", while $p$, $q$, and $r$ are called terms. A property or quality $r$ predicated of all $q$ means that every $q$ has property $r$ (or that every $q$ is an $r$). For example, in the *first figure* "Having four sides is predicated of every rectangle, and being a rectangle is predicated of every square", we have

$p$ : A figure is a square;

$q$ : A figure is a rectangle;

$r$ : A figure has four sides,

which are more convenient to find if the above statement is rewritten in the active voice. Notice that, however, a syllogism may be valid without being true or true without being valid. As discussed in Chap. 2, we need both to describe logical legitimacy of a *proof* of a mathematical statement. Some later writers categorized these *syllogisms* and made these famous by using certain interesting names.

We conclude our discussion here by making some important remarks about intellectual efforts of various mathematicians that rendered mathematical logic its present-day *symbolic logic* form. It is an interesting fact to know that more than two millennia passed before Russell and Whitehead were able to define an implication in terms of a statement of equivalent deductive power. Therefore, we are now taking a journey from verbal rules of deduction stated by the ancient Greeks to a present-day discussion of such rules via truth tables.

The development of modern theory of symbolic logic is attributed to *George Boole* (1815–1864), *Augustus De Morgan* (1806–1871), *Charles Sanders Peirce* (1839–1914), *Ernst Schröder* (1841–1902), *Giuseppe Peano* (1858–1932), and *Gottlob*

*Frege* (1848–1925). All their work culminated into magnum opus *Principia Mathematica*[1] due to *Bertrand Russell* (1872–1970) and *Alfred North Whitehead* (1861–1947) [11]. The related work is now used to communicate modern mathematics in terms of sets and functions [3].

A more streamlined notation to display the *truth values* of a compound proposition was introduced independently by *Ludwig Wittgenstein* (1889–1951) and *Emil Post* (1897–1954), who used tabular formats to reduce questions in propositional calculus to simple computations. Wittgenstein coined the term *schemata* for such tables, while the more familiar *truth table* terminology is due to Post [12]. In Sect. 1.2, we shall use truth tables to discuss some related concepts such as equivalences, implications, and also the theory of inference that is based on Aristotle's syllogisms.

Boole's monumental treatise [13, 14] studies rendering a verbal statements to a symbolic form, by using for the first time the arithmetic symbols $\times$, $+$, and $-$ to replace, respectively, the *connectives* "and", "or", and "not". On the other hand, the German mathematician and philosopher Frege introduced a different system of notation for logic, called a *concept-script* [15, 16], by using a special symbol known as the *condition stroke*. It carries a truth value recognized today as being very close to the "if-then" statements appearing in Chrysippus's first two rules of inference, as discussed later. Boole's work ushered in a revolution in logic, which was advanced further by De Morgan, Peirce, Schröder, and Peano. The next key step in this revolution in logic was made by the great German mathematician and philosopher Frege, who created an original symbolic system of logic with a view to develop foundations for whole of mathematics on the basis of formal logic.

By the early twentieth century, the stage was set for *Bertrand Russell* (1872–1970) and *Alfred North Whitehead* (1861–1947) to give a modern account of logic and the foundations of mathematics in their influential treatise *Principia Mathematica* [11]. Their joint work develops the doctrine of logicism, and seeks to place on firm foundations emerging ideas in set theory. In part I of *Principia Mathematica*, the theme of deduction is studied in detail with this notion expressed collectively as "the principles by which conclusions are inferred from premisses" [11, p. 90]. A crucial notion in this study is the implicative function stated above.

In his first publication *Logische-philosophische Abhandlung* [17], translated as *Tractatus Logico-Philosophicus* [18, 19], the Austrian-born philosopher Wittgenstein's introduced a simple and unifying principle to evaluate the truth-possibilities of symbolic statements as *schemata*, by resolving compound statements into elementary propositions. The way musical notation is a "picture" of music, he speaks of symbolic logic as a "picture of reality". His *Tractatus* consists of seven main statements, with sub-statements used to modify such statements in turn. Independently of Wittgenstein, the Poland-born Post developed a highly efficient method to represent the truth values of compound statements involving the connectives "and", "or",

[1] The most influential treatise was published in three volumes between 1910 and 1913. *Principia* is a culmination of work that had been done on logic and the foundations of mathematics in the preceding century, and it had a tremendous influence on further development of the subject in the twentieth century.

"not", and "if-then", which he called as "truth tables" in his doctoral dissertation "Introduction to a General Theory of Elementary Propositions" [12].

## 1.2 Propositional Logic

As said earlier, *propositional logic* helps in analyzing patterns in mathematical communications, defining rules of reasoning, and also in describing how a proof works. In general, a broader classification of the kind of sentences we use in common conversation is given as follows:

1. A statement such as "*The capital of India is New Delhi*" is a *declarative sentence* (or a declaration).
2. An *interrogative sentence* (or simply a question) such as "*What is the height of Himalayas?*" is used to request for an information.
3. An *exclamatory sentence* (or simply an exclamation) is generally a more emphatic form of statement such as "*What a wonderful joke this is!*".
4. An *imperative sentence* (or simply a command) is ordinarily used to make a demand or request such as "*Submit your discrete mathematics homework by due date*".

Notice that it is only a declarative sentence that we can be assigned the *truth value*[2] T (for true) or F (for false), but not both. That is, we can say with certainty whether it is a *true* or a *false* statement. The other three types of sentences listed above do not have any such characteristic property. The following is an excerpt from *On Interpretation* [20]:

> Every sentence has meaning, not as being the natural means by which a physical faculty is realized, but, ... by convention. Yet every sentence is not a proposition; only such are propositions as have in them either truth or falsity. Thus, a prayer is a sentence, but is neither true nor false.
>
> Let us therefore dismiss all other types of sentences but the proposition, for this last concerns our present inquiry .... Every proposition must contain a verb or the tense of a verb. ...
>
> In the case of that which is or which has taken place, propositions ... must be true or false.

**Definition 1.1** A sentence is called a **proposition** if it takes a definite **truth value** as TRUE or FALSE, respectively, denoted by T or F.

For example, based on Aristotle's description, one of the sentences listed below is a *proposition*:

1. What is your name?
2. Close the door of the class room.
3. That peacock is very beautiful!
4. I am a liar.

---

[2] In some applications, it is more natural to use 1 or 0, respectively, for T or F.

**Example 1.1** All the propositions listed below have the truth value T:

1. New Delhi is the capital of India.
2. Dog is an animal.
3. The sum of an odd integer and an even integer is an odd integer.
4. The product of an odd integer with an even integer is an even integer.
5. The triplet (3, 4, 5) constitutes an integer solution to the equation $x^2 + y^2 = z^2$.

**Example 1.2** All the propositions listed below have the truth value F:

1. $2 + 4 = 7$.
2. The sum of an odd integer and an even integer is an even integer.
3. There is no integral solution to the equation $x^2 + y^2 = z^2$.
4. For every real number $x$, we have $x^2 < 0$.

## *Standard Logical Connectives*

We usually write *proposition variables* as $p, q, r, \ldots$. A *compound statement* is formed from two or more propositions by using standard *logical connectives* "**and**" and "**or**", respectively, called the *conjunction* and *disjunction*. In some cases, we also use the unary operation "negation". For a proposition $p$, the negation of the $p$ is called "**not**" $p$.

**Definition 1.2** Let $p$ and $q$ be two propositions.

1. The *conjunction* of $p$ and $q$, denoted by $p \wedge q$, is the proposition with truth value T if and only if both $p$ and $q$ have the truth value T.
2. The *disjunction* of $p$ and $q$, denoted by $p \vee q$, is the proposition whose truth value is F if and only if both $p$ and $q$ have the truth value F.
3. The *negation* of a proposition $p$, denoted by $\neg p$, is the proposition whose truth value is opposite to the truth value of $p$. That is, if a proposition $p$ has the truth value F, then $\neg p$ has the truth value T. Similarly, if $p$ has the truth value T, then $\neg p$ has the truth value F.

Also, for two proposition variables $p$ and $q$, the conjunction as the *product* is denoted by $p \cdot q$, and the disjunction as the *sum* is denoted by $p + q$.

The truth tables of conjunction, disjunction, and negation are given below.

| $p$ | $q$ | $p \wedge q$ |
|---|---|---|
| F | F | F |
| F | T | F |
| T | F | F |
| T | T | T |

| $p$ | $q$ | $p \vee q$ |
|---|---|---|
| F | F | F |
| F | T | T |
| T | F | T |
| T | T | T |

| $p$ | $\neg p$ |
|---|---|
| F | T |
| T | F |

**Example 1.3** Consider the propositions $p$ and $q$ as given below:

$p$ :"Raman passed the examination",
$q$ :"Samy passed the examination".

Then the proposition $p \wedge q$ is given by

$p \wedge q$ :"Raman and Samy passed the examination."

Consider the propositions $p$ and $q$ as given below:

$p$ :"Raman passed the examination",
$q$ :"Samy passed the examination".

Then the proposition $p \vee q$ is given by

$p \vee q$ :"Raman or Samy passed the examination."

Also, if $p$ is the proposition "to enter into a country, you need to have a passport or a voter registration card", then "*not*" $p$ is the proposition "It is not the case that you need a passport or a voter registration card to enter into a country".

We may also use *conditional connectives* such as "*if* $\cdots$, *then* $\cdots$" and "$\cdots$ *if and only if* $\cdots$". Notice that a *simple proposition* is also called as an *atomic proposition* in the sense that it involves no connectives. A compound proposition involving $n$ atomic propositions $p_1, \ldots, p_n$ is usually denoted by $P(p_1, \ldots, p_n)$.

**Example 1.4** Notice that all the propositions given in Examples 1.1 through 1.2 are atomic. However, on the other hand, each sentence listed below is a compound proposition:

1. Ramanujan is an Indian mathematician and Hardy is a British mathematician.
2. Either it is already raining outside or it may at least be cloudy.
3. If there is no light in the classroom, then no class will be held today.
4. If a real-valued function $f$ is differentiable at a point $x = a$, it is continuous at that point.

**Definition 1.3** The *conditional connective* "if $p$ then $q$" of two propositions $p, q$, denoted by $p \to q$, is the proposition whose truth value is F if and only if the truth value of $p$ is T and the truth value of $q$ is F. In this case, the proposition $p$ is called a *hypothesis* and $q$ is called a *conclusion*. The *bi-conditional connective* "$p$ if and only if $q$", denoted by $p \leftrightarrow q$, is the proposition whose truth value is T if and only if both $p$ and $q$ have the same truth value.

The truth table of the compound proposition formed by two conditional and bi-conditional connectives is, respectively, as given below:

| $p$ | $q$ | $p \rightarrow q$ |
|---|---|---|
| F | F | T |
| F | T | T |
| T | F | F |
| T | T | T |

| $p$ | $q$ | $p \leftrightarrow q$ |
|---|---|---|
| F | F | T |
| F | T | F |
| T | F | F |
| T | T | T |

**Example 1.5** If $p$ and $q$ denote, respectively, the propositions "Water is heated to 100 °C" and "Water becomes vapor", then $p \rightarrow q$ is the proposition " If Water is heated to 100 °C, then "Water becomes vapor". If $p$ and $q$ denote, respectively, the propositions "a number is a prime number", "the number has exactly two positive divisors", then $p \leftrightarrow q$ is the proposition "a number is a prime number if and only if the number has exactly two positive divisors".

The propositions $p \rightarrow q$ and $p \leftrightarrow q$ are usually expressed as follows:

| $p \rightarrow q$ | $p \leftrightarrow q$ |
|---|---|
| if $p$, (then) $q$ | $p$ if and only if $q$ |
| $p$ implies $q$ | $p$ is necessary and sufficient for $q$ |
| $p$ only if $q$ | If $p$, (then) $q$ and conversely |
| $q$ is necessary for $p$ | |
| $p$ is sufficient for $q$ | |

**Example 1.6** Let $p$ and $q$ denote the propositions "Ramanujan is Indian" and "Ramanujan is a Mathematician", respectively. Then $p \vee q$ is the proposition "Ramanujan is Indian or he is a Mathematician"; $p \wedge q$ is the proposition "Ramanujan is an Indian Mathematician"; $p \rightarrow q$ is the proposition "If Ramanujan is an Indian, then Ramanujan is a Mathematician". Similarly, $p \leftrightarrow q$ is the proposition "Ramanujan is Indian if and only if Ramanujan is a mathematician".

**Example 1.7** Let us consider the following compound propositions:

1. Saravanan is studying well and he will pass the examination.
2. Saravanan is not studying well and he will pass the examination.
3. Saravanan is studying well or he will not pass the examination.
4. If Saravanan is studying well, then he will join university next year.

Let $p, q, r$ denote the atomic propositions "Saravanan is studying well", "Saravanan will pass the examination", "Saravanan will join university next year", respectively, then the above four propositions can be written as $p \wedge q$, $\neg p \wedge q$, $p \vee \neg q$, $p \rightarrow q$, respectively.

**Definition 1.4** If $p$, $q$ are two propositions, then their **exclusive disjunction** or **exclusive or** denoted by $p \overline{\vee} q$ is the proposition whose truth value is T if and only if both $p$ and $q$ have opposite truth values. The truth table for this connective is as given by the left side table below:

| $p$ | $q$ | $p \overline{\vee} q$ |
|---|---|---|
| F | F | F |
| F | T | T |
| T | F | T |
| T | T | F |

| $p$ | $q$ | $p \uparrow q$ |
|---|---|---|
| F | F | T |
| F | T | T |
| T | F | T |
| T | T | F |

**Definition 1.5** The **not and** or **NAND** of the propositions $p$, $q$, denoted by $p \uparrow q$, is the proposition which is the negation of $p \wedge q$:

$$p \uparrow q \text{ is } \neg(p \wedge q).$$

The truth table for NAND is as given by the right side table above.

**Definition 1.6** The **not or** or **NOR** of $p$ and $q$ denoted by $p \downarrow q$ is the proposition which is the negation of $(p \vee q)$:

$$p \downarrow q \text{ is } \neg(p \vee q).$$

The truth table for NOR is given below:

| $p$ | $q$ | $p \downarrow q$ |
|---|---|---|
| F | F | T |
| F | T | F |
| T | F | F |
| T | T | F |

In general, a *statement formula* $P(p_1, \ldots, p_n)$ has no truth values unless specific truth values are substituted for the propositions variable $p_1, p_2, p_3, \ldots$. Therefore, a truth value of a statement formula depends on the truth values assigned to the constituting atomic propositions. Notice that, since the truth value of each atomic propositions $p_1, \ldots, p_n$ is T or F, the proposition $P(p_1, \ldots, p_n)$ takes $2^n$ values, where each is T or F.

**Remark 1.1** A usual approach to list all $2^n$ combinations of truth values for the $n$-tuple $(p_1, \ldots, p_n)$ is to write

$$\frac{1}{2} \times 2^n \text{ F's followed by } \frac{1}{2} \times 2^n \text{ T's}$$

in the first column (under $p_1$) of the truth table. Next, we write

$$\frac{1}{4} \times 2^n \text{ F's followed by } \frac{1}{4} \times 2^n \text{ T's}$$

in the second column (under $p_2$), and so on. Finally, in the column under $p_n$, we write F and T alternatively, in that order. Accordingly, $2^n$ truth values of $P(p_1, \ldots, p_n)$ are written in the last column of the truth table for the proposition $P(p_1, \ldots, p_n)$.

**Definition 1.7** A statement formula is called a **well-formed formula** (or simply a **formula**) if it is obtained by using the following recursive procedure:

1. Each propositional variable is a formula. Also, both symbols T, F are formulas.
2. If $p$ and $q$ are formulas, then so are their disjunction $(p \vee q)$ and conjunction $(p \wedge q)$.
3. If $p$ is formula, then so is $\neg p$.
4. Any statement formula obtained by applying the above three rules (any number of times) is a formula.

In view of (2) and (3), we also have $(p \rightarrow q)$ and $(p \leftrightarrow q)$ are formulas. Also,

$$\big((p \vee q) \rightarrow r\big) \quad \text{and} \quad \big((p \wedge q) \leftrightarrow (\neg p)\big) \tag{1.2.1}$$

are well-formed formulas; the statement formulas

$$\begin{aligned} \neg((\neg p) \vee (q \vee r)) \quad &\text{and} \quad \neg(p \vee q) \vee \neg(\neg p \vee \neg q); \\ (p \vee q) \rightarrow (p \wedge \neg q) \quad &\text{and} \quad (p \rightarrow q) \rightarrow (q \rightarrow p) \end{aligned}$$

are well-formed formulas in propositional variables $p, q, r$. As in (1.2.1), we can omit some of the parenthesis and write a formula in a simpler form. For example, by removing the outermost parenthesis, the two formulas in (1.2.1) may be written simply as

$$(p \vee q) \rightarrow r \quad \text{and} \quad (p \wedge q) \leftrightarrow \neg p. \tag{1.2.2}$$

In general, to reduce the number of parentheses used, we follow the convention that the negation affects as little as possible. For example, in $\neg p \vee (q \vee r)$, the negation is applied only to $p$. In case we wish to apply the negation to all the terms, we need to write $\neg(p \vee (q \vee r))$.

## *Tautology and Contradiction*

A propositional formula is called a *tautology* if it is *universally true*, i.e., it has truth value T, for every input truth values of the constituting propositions. It is common to write a tautology as $\mathscr{T}$. Therefore, a formula is a tautology if the last column of its truth table comprises only $T$. In next example, we give some simple formulas that are tautologies.

**Example 1.8** As the proposition $p \vee \neg p$ has the truth value T, for both choices $(F, T)$ and $(T, F)$, it follows that $p \vee \neg p$ and is a tautology. Also, it follows from the truth table given below:

| $p$ | $q$ | $p \wedge q$ | $\neg(p \wedge q)$ | $p \vee \neg(p \wedge q)$ |
|---|---|---|---|---|
| F | F | F | T | T |
| F | T | F | T | T |
| T | F | F | T | T |
| T | T | T | F | T |

that the proposition $p \vee \neg(p \wedge q)$ is universally true, i.e., its truth value is T, for all input truth values of $p$ and $q$. Hence, it is a tautology.

A propositional formula is called a *contradiction* if it is *universally false*, i.e., it has truth value F, for every input truth values of its constituting propositions. It is common to write a contradiction as $\mathscr{F}$. Therefore, a formula is a contradiction if the last column of its truth table comprises only "F". In next example, we give some simple formulas that are contradictions.

**Example 1.9** The proposition $p \wedge \neg p$ has the truth value F for all choices of truth values of $p$ and $q$:

| $p$ | $\neg p$ | $p \wedge \neg p$ |
|---|---|---|
| F | T | F |
| T | F | F |

Therefore, $p \wedge \neg p$ is a contradiction. Also, the truth table given below shows that $q \wedge (p \wedge \neg q)$ is a contradiction:

| $p$ | $q$ | $\neg q$ | $p \wedge \neg q$ | $q \wedge (p \wedge \neg q)$ |
|---|---|---|---|---|
| F | F | T | F | F |
| F | T | F | F | F |
| T | F | T | T | F |
| T | T | F | F | F |

The truth value of $q \wedge (p \wedge \neg q)$ is F for all choices of the truth values of $p$ and $q$ and hence it is a contradiction.

A formula that is neither a tautology nor a contradiction is called a *satisfiable formula*.

**Definition 1.8** A formula $B$ obtained from another formula $A$ by substituting formulas for some variables in the proposition is called a *substitution instance* of $A$ if the variable is replaced by the same formula in all places wherever it occurs.

**Example 1.10** Some substitution instances of the formula $(q \vee p) \rightarrow (q \vee \neg p)$ are

1. $((q \vee \neg q) \vee p) \rightarrow ((q \vee \neg q) \vee \neg p)$,
2. $(p \vee p) \rightarrow (p \vee \neg p)$,
3. $(p \vee q) \rightarrow (p \vee \neg q)$,
4. $((p \rightarrow q) \vee (s \vee t)) \rightarrow ((p \rightarrow q) \vee \neg(s \vee t))$.

The formula $(q \vee (p \vee q)) \rightarrow (q \vee \neg p)$ is not a substitution instance of $(q \vee p) \rightarrow (q \vee \neg p)$ because $p$ is not replaced by $p \vee q$ in all the places.

**Table 1.1** Propositional algebra

Standard equivalences

| | | | |
|---|---|---|---|
| Idempotent laws | $p \wedge p \equiv p$<br>$p \vee p \equiv p$ | Identity laws | $p \wedge \mathcal{T} \equiv p$<br>$p \vee \mathcal{F} \equiv p$ |
| Associative laws | $(p \wedge q) \wedge r \equiv p \wedge (q \wedge r)$<br>$(p \vee q) \vee r \equiv p \vee (q \vee r)$ | Absorption laws | $p \vee (p \wedge q) \equiv p$<br>$p \wedge (p \vee q) \equiv p$ |
| Commutative laws | $p \wedge q \equiv q \wedge p$<br>$p \vee q \equiv q \vee p$ | Double negation laws | $\neg(\neg p) \equiv p$ |
| De Morgan's laws | $\neg(p \wedge q) \equiv (\neg p \vee \neg q)$<br>$\neg(p \vee q) \equiv (\neg p \wedge \neg q)$ | Contrapositive laws | $p \to q \equiv \neg q \to \neg p$ |
| Distributive laws | $p \wedge (q \vee r) \equiv (p \wedge q) \vee (p \wedge r)$<br>$p \vee (q \wedge r) \equiv (p \vee q) \wedge (p \vee r)$ | Conditional as disjunction | $p \to q \equiv \neg p \vee q$ |
| Complement laws | $p \wedge \neg p \equiv \mathcal{F}$<br>$p \vee \neg p \equiv \mathcal{T}$ | Bi-conditional as conditional | $p \leftrightarrow q \equiv (p \to q) \wedge (q \to p)$ |
| Dominance laws | $p \vee \mathcal{T} \equiv \mathcal{T}$<br>$p \wedge \mathcal{F} \equiv \mathcal{F}$ | Exportation laws | $p \to (q \to r) \equiv (p \wedge q) \to r$ |

Notice that the substitution instance of a tautology is again a tautology; the substitution instance of a contradiction is again a contradiction.

**Example 1.11** By replacing $p$ by $p \vee q$ in the tautology $p \vee \neg p$, we obtain the tautology $(p \vee q) \vee \neg(p \vee q)$. It so happens because the truth value of the proposition $(p \vee q) \vee \neg(p \vee q)$ does not depend on the truth values of $p \vee q$. Similarly, when $p$ is replaced by $r \vee (q \to r)$ in the contradiction $p \wedge \neg p$, we obtain a yet another contradiction $(r \vee (q \to r)) \vee \neg(r \vee (q \to r))$.

## *Equivalence and Implications*

We say propositions $P(p_1, \ldots, p_n)$ and $Q(p_1, \ldots, p_n)$ are *logically equivalent* (or simply *equivalent*[3]) if the two have the same truth value for every choice of $2^n$ truth values for the $n$-tuple $(p_1, \ldots, p_n)$. In this case, we write $P \equiv Q$. Notice that $P \leftrightarrow Q$ is a tautology whenever $P \equiv Q$. We have listed some important equivalences in Table 1.1. The reader may use truth tables to prove these equivalences. For example, since the truth tables for $p \vee \mathcal{T} \leftrightarrow \mathcal{T}$ and $p \wedge \mathcal{F} \leftrightarrow \mathcal{F}$ are, respectively, given by

| $p$ | $\mathcal{T}$ | $p \vee \mathcal{T}$ | $p \vee \mathcal{T} \leftrightarrow \mathcal{T}$ |
|---|---|---|---|
| F | T | T | T |
| T | T | T | T |

| $p$ | $\mathcal{F}$ | $p \wedge \mathcal{F}$ | $p \wedge \mathcal{F} \leftrightarrow \mathcal{F}$ |
|---|---|---|---|
| F | F | F | T |
| T | F | F | T |

Therefore, the two propositions $p \vee \mathcal{T} \leftrightarrow \mathcal{T}$ and $p \wedge \mathcal{F} \leftrightarrow \mathcal{F}$ are tautologies, and hence we have $p \vee \mathcal{T} \Leftrightarrow \mathcal{T}$ and $p \wedge \mathcal{F} \Leftrightarrow \mathcal{F}$. This proves the *Dominance Laws*.

---

[3] This indeed defines an equivalence relation on the class of propositions.

Other important equivalences follow from the standard equivalences listed in Table 1.1.

**Example 1.12** We prove here the equivalence:

$$(p \vee q) \wedge \neg p \equiv \neg p \wedge q. \tag{1.2.3}$$

We have

$$\begin{aligned} (p \vee q) \wedge \neg p &\equiv \neg p \wedge (p \vee q) p && \text{Commutative law} \\ &\equiv (\neg p \vee p) \wedge (\neg p \vee q) && \text{Distributive Law} \\ &\equiv T \wedge (\neg p \vee q) && \neg p \vee p \Leftrightarrow T \\ &\equiv (\neg p \vee q) && \text{Identity law} \end{aligned}$$

This proves the equivalence (1.2.3).

**Example 1.13** We prove here the equivalence:

$$\neg(p \vee (\neg p \wedge q)) \equiv \neg p \wedge \neg q. \tag{1.2.4}$$

In this case, we have

$$\begin{aligned} \neg(p \vee (\neg p \wedge q)) &\equiv \neg p \wedge \neg(\neg p \wedge q) && \text{De Morgan's law} \\ &\equiv \neg p \wedge (\neg\neg p \vee \neg q) && \text{De Morgan's law} \\ &\equiv \neg p \wedge (p \vee \neg q) && \text{Double negation law} \\ &\equiv (\neg p \wedge p) \vee (\neg p \wedge \neg q) && \text{Distributive law} \\ &\equiv T \vee (\neg p \wedge \neg q) && p \wedge \neg \Leftrightarrow T \\ &\equiv (\neg p \wedge \neg q)) && \text{Identity law} \end{aligned}$$

The above proves the equivalence (1.2.4).

**Example 1.14** We prove here the equivalence:

$$\neg(\neg((p \vee q) \wedge r) \vee \neg q) \equiv q \wedge r. \tag{1.2.5}$$

In this case, we have

$$\begin{aligned} &\neg(\neg((p \vee q) \wedge r) \vee \neg q) \\ &\equiv \neg\neg((p \vee q) \wedge r) \wedge \neg\neg q && \text{De Morgan's law} \\ &\equiv ((p \vee q) \wedge r) \wedge q && \text{Double negation law} \\ &\equiv (p \vee q) \wedge (r \wedge q) && \text{Associative law} \\ &\equiv (p \vee q) \wedge (q \wedge r) && \text{Commutative law} \\ &\equiv ((p \vee q) \wedge q) \wedge r && \text{Associative law} \\ &\equiv q \wedge r && \text{Absorption law} \end{aligned}$$

The above proves the equivalence (1.2.5).

**Example 1.15** We prove here the equivalence:

$$(p \vee q) \wedge (\neg p \wedge (\neg p \wedge q)) \equiv \neg p \wedge q. \tag{1.2.6}$$

In this case, we have

$$\begin{aligned}
&(p \vee q) \wedge (\neg p \wedge (\neg p \wedge q)) && \\
&\equiv (p \vee q) \wedge ((\neg p \wedge \neg p) \wedge q) && \text{Associative law} \\
&\equiv (p \vee q) \wedge (\neg p \wedge q) && \text{Idempotent law} \\
&\equiv (p \vee q) \wedge (q \wedge \neg p) && \text{Commutative law} \\
&\equiv ((p \vee q) \wedge q) \wedge \neg p && \text{Associative law} \\
&\equiv (q \wedge (p \vee q)) \wedge \neg p && \text{Commutative law} \\
&\equiv q \wedge \neg p && \text{Absorption law} \\
&\equiv \neg p \wedge q && \text{Commutative law}
\end{aligned}$$

The above proves the equivalence (1.2.6).

**Definition 1.9** Let $p$ and $q$ be two propositions. We say $p$ *logically implies* $q$ if $p \to q$ is a tautology. In this case, we write $p \Rightarrow q$. We write $p \Leftrightarrow q$ if and only if $p \Rightarrow q$ and $q \Rightarrow p$.

The presently accepted definition of an implication, stated by Bertrand Russell (1872–1970) and Alfred N. Whitehead (1861–1947) in Principia Mathematica [6], is in terms of a certain "or" statement of equivalent deductive power to Chrysippus's first argument form, which he gave in terms of an "if-then" statement. The definitions of Russell and Whitehead can be used to settle the logical equivalence of Chrysippus's rules.

**Example 1.16** It is easy to see that

$$p \vee q \equiv \neg(\neg p) \vee q \equiv \neg p \to q.$$

Similarly, $\neg(p \to q) \equiv \neg(\neg p \vee q) \equiv p \wedge \neg q$, which shows

$$\neg(p \to q) \Rightarrow p \quad \text{and} \quad \neg(p \to q) \Rightarrow \neg q.$$

We also have

$$\begin{aligned}
\neg p \to (p \to q) &\Leftrightarrow \neg(\neg p) \vee (p \to q) \\
&\equiv p \vee (\neg p \vee q) \\
&\equiv (p \vee \neg q) \vee q \\
&\equiv T \vee q \Leftrightarrow T.
\end{aligned}$$

Therefore, we have $\neg p \Rightarrow p \to q$.

We have listed some important implications in Table 1.2. Once again, the reader may

**Table 1.2** Logical reasoning for an argument

| Fundamental logical implications | |
|---|---|
| Detachment (or Modus Ponens) | $(p \to q) \wedge p \Rightarrow q$ |
| Contrapositive (or Modus Tollens) | $(p \to q) \wedge \neg q \Rightarrow \neg p$ |
| Disjunctive addition | $p \Rightarrow (p \vee q)$ |
| Conjunctive simplification | $(p \wedge q) \Rightarrow p$ |
| Disjunctive syllogism | $(p \vee q) \wedge \neg p \Rightarrow q$ |
| Hypothetical syllogism | $(p \to q) \wedge (q \to r) \Rightarrow (p \to r)$ |

use truth tables to prove these implications. For example, since the truth table for $(p \to q) \wedge p \Rightarrow q$ is given by

| $p$ | $q$ | $p \to q$ | $(p \to q) \wedge p$ | $(p \to q) \wedge p \to q$ |
|---|---|---|---|---|
| F | F | T | F | T |
| F | T | T | F | T |
| T | F | F | F | T |
| T | T | T | T | T |

it follows that $(p \to q) \wedge p \to q$ is a tautology, and hence $(p \to q) \wedge p \Rightarrow q$. This proves the *detachment law* (also known as *modus ponens*) (see Table 1.2). In the next examples, wc prove two other implications listed in Table 1.2.

**Example 1.17** We prove here *modus tollens*:

$$(p \to q) \wedge \neg q \quad \Rightarrow \quad \neg p. \tag{1.2.7}$$

For, since the truth table of the proposition $((p \to q) \wedge \neg q) \to \neg p$ is given by

| $p$ | $q$ | $p \to q$ | $\neg q$ | $(p \to q) \wedge \neg q$ | $\neg p$ | $((p \to q) \wedge \neg q) \to \neg p$ |
|---|---|---|---|---|---|---|
| F | F | T | T | T | T | T |
| F | T | T | F | F | T | T |
| T | F | F | T | F | F | T |
| T | T | T | F | F | F | T |

it follows that $((p \to q) \wedge \neg q) \to \neg p$ is a tautology, and hence (1.2.7) holds.

**Example 1.18** We prove here the *hypothetical syllogism*[4]:

$$(p \to q) \wedge (q \to r) \quad \Rightarrow \quad (p \to r). \tag{1.2.8}$$

[4] It is also known as the *chain rule*.

Since the truth table for the proposition

$$S_1: \quad (p \to q) \wedge (q \to r) \quad \to \quad (p \to r)$$

is given by

| $p$ | $q$ | $r$ | $p \to q$ | $q \to r$ | $(p \to q) \wedge (q \to r)$ | $p \to r$ | $S_1$ |
|---|---|---|---|---|---|---|---|
| F | F | F | T | T | T | T | T |
| F | F | T | T | T | T | T | T |
| F | T | F | T | F | F | T | T |
| F | T | T | T | T | T | T | T |
| T | F | F | F | T | F | F | T |
| T | F | T | F | T | F | T | T |
| T | T | F | T | F | F | F | T |
| T | T | T | T | T | T | T | T |

it follows that the proposition $S_1$ is a tautology, and hence (1.2.8) holds.

**Example 1.19** We prove here the implication: $p \Rightarrow (p \vee q)$. Notice that it is enough to show that $p \to (p \vee q)$ is a tautology. Indeed, we have

$$\begin{aligned} p \to (p \vee q) &\equiv \neg p \vee (p \vee q) && \text{Conditional as Disjunction} \\ &\equiv (\neg p \vee p) \vee q && \text{Associative Law} \\ &\equiv T \vee q \\ &\equiv T. \end{aligned}$$

In previous two examples, we used truth tables to prove some implications. However, we can also use known equivalences to do the same.

**Example 1.20** We prove here the implication: $q \Rightarrow (p \to q)$. We have

$$\begin{aligned} q \ &\Rightarrow \neg p \vee q && \text{Disjunctive addition} \\ &\equiv p \to q && \text{Conditional as Disjunction} \end{aligned}$$

Notice that, alternatively, a direct proof may be given by showing that $q \to (p \to q)$ is a tautology. In this case, we have

$$\begin{aligned} q \to (p \to q) &\equiv \neg q \vee (\neg p \vee q) && \text{Conditional as Disjunction} \\ &\equiv \neg q \vee (q \vee \neg p) && \text{Commutative law} \\ &\equiv (\neg q \vee q) \vee \neg p && \text{Associative law} \\ &\equiv T \vee \neg p \\ &\equiv T. \end{aligned}$$

**Example 1.21** We prove here the implication: $(p \wedge q) \Rightarrow p$. Notice that it is enough to show that the proposition $(p \wedge q) \rightarrow p$ is a tautology. We have

$$\begin{aligned}
(p \wedge q) \rightarrow p &\equiv \neg(p \wedge q) \vee p && \text{Conditional as Disjunction} \\
&\equiv (\neg p \vee \neg q) \vee p && \text{De Morgan Law} \\
&\equiv p \vee (\neg p \vee \neg q) && \text{Commutative Law} \\
&\equiv (p \vee \neg p) \vee \neg q && \text{Associative Law} \\
&\equiv T \vee \neg q \\
&\equiv T.
\end{aligned}$$

**Definition 1.10** For a given proposition of the form $p \rightarrow q$, the proposition $q \rightarrow p$ is said to be its *converse*; the proposition $\neg p \rightarrow \neg q$ is said to be its *inverse*; and the proposition $\neg q \rightarrow \neg p$ is said to be its *contrapositive*.

Notice that, for any two propositions $p$ and $q$, we have $p \rightarrow q \equiv \neg q \rightarrow \neg p$. This holds for the converse and the inverse. The first assertion is true because we have

$$\begin{aligned}
\neg q \rightarrow \neg p &\equiv \neg\neg q \vee \neg p \\
&\equiv q \vee \neg p \\
&\equiv \neg p \vee q \\
&\equiv p \rightarrow q,
\end{aligned}$$

and also $\neg p \rightarrow \neg q \equiv q \rightarrow p$.

## *Functionally Complete Set of Connectives*

We defined earlier the connectives $\neg$, $\vee$, $\wedge$, $\rightarrow$, $\leftrightarrow$, $\overline{\vee}$, $\uparrow$, and $\downarrow$. However, in actual practice, not all connectives are required. For example, by DeMorgan's law, we have $\neg(p \vee q) \equiv (\neg p \wedge \neg q)$, which on taking negation gives

$$p \vee q \equiv \neg(\neg p \wedge \neg q).$$

The above shows that any formula involving the connective $\vee$ can be written in an equivalent form involving only the connectives $\neg$ and $\wedge$.

**Definition 1.11** A set of connectives is said to be a *functionally complete* provided it does not contain any redundant connectives, i.e., every formula can be written in an equivalent form by using such connectives.

Notice that, to prove that a set of connective is functionally complete, we need to show that any formula involving the connectives such as listed above can be written in an equivalent form with the connectives in the given set. In each case, it is enough to show that a connective not in the given set can be written in an equivalent form using connectives from the set. In what follows, we show that the sets of connectives given below are functionally complete:

$$\{\neg, \vee\}, \quad \{\neg, \wedge\}, \quad \{\neg, \rightarrow\}, \quad \{\uparrow\}, \quad \{\downarrow\}.$$

Let us start with the first set in above list.

**Theorem 1.1** *The set of connectives* $\{\neg, \vee\}$ *is a functionally complete.*

***Proof*** As said above, to prove the assertion, we need to write the connectives $p \wedge q$, $p \rightarrow q$, $p \leftrightarrow q$, $p \overline{\vee} q$, $p \uparrow q$, and $p \downarrow q$ in an equivalent form involving the connectives $\neg$, $\vee$. For instance,

1. by using De Morgan's law and the equivalence $\neg\neg p \equiv p$, we have

$$(p \wedge q) \equiv \neg\neg(p \wedge q) \equiv \neg(\neg p \vee \neg q);$$

2. $p \rightarrow q \equiv \neg p \vee q$;
3. by writing bi-conditional in terms of conditionals, followed by writing conditionals as disjunction, and then writing conjunction in terms of disjunction, we obtain

$$\begin{aligned} p \leftrightarrow q &\equiv (p \rightarrow q) \wedge (q \rightarrow p) \\ &\equiv (\neg p \vee q) \wedge (\neg q \vee p) \\ &\equiv \neg(\neg(\neg p \vee q) \vee \neg(\neg q \vee p)). \end{aligned}$$

4. The reader may supply a proof of the equivalence $p \overline{\vee} q \equiv \neg(p \leftrightarrow q)$. Therefore,

$$p \overline{\vee} q \equiv \neg(p \leftrightarrow q) \equiv \neg(\neg p \vee q) \vee \neg(\neg q \vee p).$$

5. By the definition of $p \uparrow q$, we have

$$p \uparrow q \equiv \neg(p \wedge q) \equiv (\neg p \vee \neg q).$$

6. By the definition of $p \downarrow q$, we have $p \downarrow q \equiv \neg(p \vee q)$.

We have thus shown that the formulas $p \wedge q$, $p \rightarrow q$, $p \leftrightarrow q$, $p \overline{\vee} q$, $p \uparrow q$, and $p \downarrow q$ can be written in an equivalent form involving the connectives $\neg$, $\vee$. Also $\{\neg, \vee\}$ does not contain any redundant connectives. Therefore the set $\{\neg, \vee\}$ is functionally complete. □

**Theorem 1.2** *The set of connectives* $\{\neg, \wedge\}$ *is functionally complete.*

***Proof*** In this case, we need to write the connectives $p \vee q$, $p \rightarrow q$, $p \leftrightarrow q$, $p \overline{\vee} q$, $p \uparrow q$, and $p \downarrow q$ in terms of the two connectives $\neg$ and $\wedge$. Notice that

1. $p \vee q \equiv \neg\neg(p \vee q) \equiv \neg(\neg p \wedge \neg q)$;
2. $p \rightarrow q \equiv \neg p \vee q \equiv \neg\neg(\neg p \vee q) \equiv \neg(p \vee \neg q)$;
3. $p \leftrightarrow q \equiv (\neg p \vee q) \wedge (p \vee \neg q) \equiv \neg(p \wedge \neg q) \wedge \neg(\neg p \wedge q)$;
4. $p \overline{\vee} q \equiv \neg(p \leftrightarrow q) \equiv \neg(\neg(p \wedge \neg q) \wedge \neg(\neg p \wedge q))$;
5. $p \uparrow q \equiv \neg(p \wedge q)$;
6. $p \downarrow q \equiv \neg(p \vee q) \equiv \neg p \wedge \neg q$.

The above equivalences prove that the set of connectives $\{\neg, \wedge\}$ is functionally complete. □

**Theorem 1.3** *The single connective* $\{\uparrow\}$ *is functionally complete.*

***Proof*** The following equivalences prove that the set of connective $\{\uparrow\}$ is functionally complete:

1. $\neg p \Leftrightarrow \neg(p \wedge p) \Leftrightarrow p \uparrow p$,
2. $p \vee q \Leftrightarrow \neg(\neg p \wedge \neg q) \Leftrightarrow (\neg p) \uparrow (\neg q)$
   $\Leftrightarrow (p \uparrow p) \uparrow (q \uparrow q)$,
3. $p \wedge q \Leftrightarrow \neg\neg(p \wedge q) \Leftrightarrow \neg(p \uparrow q) \Leftrightarrow (p \uparrow q) \uparrow (p \uparrow q)$,
4. $p \rightarrow q \Leftrightarrow (\neg p \vee q) \Leftrightarrow \neg(p \wedge \neg q) \Leftrightarrow p \uparrow (\neg q)$
   $\Leftrightarrow p \uparrow (q \uparrow q)$,
5. $p \leftrightarrow q \Leftrightarrow (p \rightarrow q) \wedge (q \rightarrow p)$ which on writing conditionals and conjunctions in terms of $\uparrow$ gives

$$
\begin{aligned}
p \leftrightarrow q \Leftrightarrow & ((p \uparrow (q \uparrow q)) \uparrow (q \uparrow (p \uparrow p))) \\
& \uparrow ((p \uparrow (q \uparrow q)) \uparrow (q \uparrow (p \uparrow p))).
\end{aligned}
$$

6. Since $p \overline{\vee} q \Leftrightarrow \neg(p \leftrightarrow q)$ and expressions involving $\neg$ and $\leftrightarrow$ can be written in terms of $\uparrow$, we can write $p \overline{\vee} q$ in terms of $\uparrow$. □

The *replacement process* refers to the procedure such that a part of a formula $S_1$ is replaced by a proposition that itself is a formula, say $S_2$. Further, if it is assumed that the part of the formula is replaced by an equivalent formula, then $S_1$ and $S_2$ are equivalent. For example, replacing $p$ in $p \vee \neg q$ by $p \rightarrow q$, we obtain the formula $(p \rightarrow q) \vee \neg q$. Similarly, the formula $(\neg p \rightarrow p) \vee \neg q$ is obtained from $p \vee \neg q$ by the replacement process. Also, since

$$
\neg p \rightarrow p \;\equiv\; \neg\neg p \vee p \;\equiv\; p \vee p \Leftrightarrow p,
$$

we have $p \vee \neg q \equiv (\neg p \rightarrow p) \vee \neg q$.

## *Duality*

Since any statement formula can be written in an equivalent form involving only the connectives $\neg, \vee$, in this section, we consider statement formulas involving connectives $\neg$, $\vee$, and $\wedge$. A statement formula $A$ involving the statement variables $p_1, p_2, \ldots, p_n$ is written as $A(p_1, p_2, \ldots, p_n)$. The dual of this statement formula $A$, denoted by $A^*$, is the proposition obtained by replacing $\vee$ by $\wedge$, $\wedge$ by $\vee$, T by F, and F by T, respectively.

**Example 1.22** The dual of $(p \vee q) \wedge \neg r$ is the formula $(p \wedge q) \vee \neg r$. Similarly the dual of $(p \vee T) \vee (F \wedge q)$ is the formula $(p \wedge F) \wedge (T \vee q)$.

**Theorem 1.4** (Duality Law) *If $A(p_1, p_2, \ldots, p_n)$, $B(p_1, p_2, \ldots, p_n)$ are statement formulas involving the statement variables $p_1, p_2, \ldots, p_n$, then*

$$A^*(p_1, p_2, \ldots, p_n) \Leftrightarrow \neg A(\neg p_1, \neg p_2, \ldots, \neg p_n),$$

*and $A \Leftrightarrow B$ if and only $A^* \Leftrightarrow B^*$.*

***Proof*** Note that if two propositions are equivalent, then they have the same dual. Hence, it is enough to verify for statement formula in their normal form. But then the result follows easily by using De Morgan's laws. The second part follows from the first easily. □

**Example 1.23** De Morgan's laws given below are dual to each other:

$$\neg(p \wedge q) \Leftrightarrow (\neg p \vee \neg q), \quad \neg(p \vee q) \Leftrightarrow (\neg p \wedge \neg q).$$

Note that the second law follows from the first by duality law.

**Example 1.24** Since $p \to q \Leftrightarrow \neg p \vee q$, the dual of $p \to q$ is $\neg p \wedge q \Leftrightarrow \neg(p \wedge \neg q) \Leftrightarrow \neg(q \to p)$.

## *Normal Forms*

In all that follows, the term *product* refers to the conjunction and a *sum* to the disjunction, of formulas. The product of variables and their negation is called an *elementary product*, and the sum of variables and their negation is called an *elementary sum.*

**Example 1.25** The expressions such as

$$p \wedge q, \quad p \wedge \neg q, \quad p \wedge \neg q \wedge r, \quad \neg p \wedge \neg q \wedge r$$

are elementary products, and the expressions such as

$$p \vee q, \quad p \vee \neg q, \quad p \vee \neg q \vee r, \quad \neg p \vee \neg q \vee r$$

are elementary sums. On the other hand, expressions such as

$$(p \wedge q) \vee (p \wedge \neg q) \vee (p \wedge \neg q \wedge r), \qquad p \vee (q \wedge r)$$

are sum of elementary products, and the expressions such as

$$p \wedge q, \qquad p \wedge (p \vee \neg q) \wedge (p \vee \neg q \vee r)$$

are product of elementary sums.

**Definition 1.12** The *sum of products* normal form of a given formula is the proposition that is written as the sum of elementary products. The *product of sums* normal form of a given formula is the proposition that is written as the product of elementary sums.

**Example 1.26** We find here the sum of products normal form of the proposition $p \wedge (q \to r)$. The proposition $p \wedge (q \to r)$ is equivalent to $p \wedge (\neg q \vee r)$. Since $p \wedge (\neg q \vee r)$ is the product of $p$ and the elementary sum$(\neg q \vee r)$, it follows that $p \wedge (\neg q \vee r)$ is the product of sums normal form for $p \wedge (q \to r)$. By using distributive law, we have

$$p \wedge (q \to r) \equiv p \wedge (\neg q \vee r) \equiv (p \wedge \neg q) \vee (p \wedge r).$$

The last term in the above equivalence is the sum of products normal form for the proposition $p \wedge (q \to r)$.

In general, given $n$ propositions $p_1, p_2, \ldots, p_n$, there are several elementary products and elementary sums possible. For example,

$$p_1, \qquad p_1 \vee \neg p_3, \qquad p_1 \vee p_2 \vee \cdots \vee p_n$$

are all elementary sums. Among the above, the elementary sum of the form

$$B_1 \vee B_2 \vee \cdots \vee B_n,$$

with each $B_i$ being a $p_i$ or $\neg p_i$, is called a *maxterm*. Similarly, the elementary product

$$B_1 \wedge B_2 \wedge \cdots \wedge B_n,$$

with each $B_i$ being a $p_i$ or $\neg p_i$, is called a *minterm*. Notice that, in each of the above two situations, all the variables appear as it is or as negation, but not as both.

**Definition 1.13** A proposition is said to be in *principal conjunctive normal form* (or simply PCNF) if it is equivalent to a product of maxterms. A proposition is said to be in *principal disjunctive normal form* (or simply PDNF) if it is equivalent to a sum of minterms.

**Remark 1.2** As said earlier, if only two propositional variables are involved in a given proposition, then the truth table contains exactly 4 rows. In general, if a proposition involves $n$ variables, then the truth table contains $2^n$ rows. Notice that, since the truth values of two propositional variables $p$ and $q$ are FF, FT, TF, and TT, we may assign the binary numbers from 0 to 3 with 0 replaced by F and 1 replaced by T. In general, if there are $n$ variables, then the truth values assigned to these $n$ variables will be the binary numbers from 0 to $2^n - 1$ with 0 replaced by $F$ and 1 replaced by $T$.

We first consider minterms and maxterms in terms of $p$ and $q$. Since

$$\begin{aligned} T &\equiv p \vee \neg p \\ &\equiv (p \wedge T) \vee (\neg p \wedge T) \\ &\equiv (p \wedge (q \vee \neg q)) \vee (\neg p \wedge (q \vee \neg q)) \\ &\equiv (p \wedge q) \vee (p \wedge \neg q) \vee (\neg p \wedge q) \vee (\neg p \wedge \neg q) \end{aligned}$$

and by taking negotiation, we obtain

$$T \equiv (p \vee q) \wedge (p \vee \neg q) \wedge (\neg p \vee q) \wedge (\neg p \vee \neg q).$$

In this case (and in general too), the PDNF of a tautology contains all the minterms and the PDNF of a contradiction contains all the maxterms. Also it is trivial that PCNF of a tautology has no maxterms (empty product) while the PDNF has no minterms (empty sum).

Notice that, for any formula $S$, the minterm normal form of the negation of $S$ consists of all minterms which are not in the minterm normal form of $S$. Similarly, the maxterm normal form of the negation of $S$ consists of all the maxterms which are not in the maxterm normal form of $S$. The above observation helps while obtaining the one normal form from the other.

**Example 1.27** We find here the PDNF of the formula $S : \ (p \vee \neg q) \wedge (\neg p \vee q)$. Notice that, from the given maxterm normal form of $S$, it follows that the maxterm normal form $\neg S$ consists of the remaining two maxterms. We thus have

$$\neg S \ \equiv \ (p \vee q) \wedge (\neg p \vee \neg q),$$

which on taking negation gives

$$S \equiv \neg(p \vee q) \vee \neg(\neg p \vee \neg q) \ \equiv \ (\neg p \wedge \neg q) \vee (p \wedge q),$$

by using DeMorgan's law. Therefore, the PDNF of $S$ is obtained as

$$(\neg p \wedge \neg q) \vee (p \wedge q).$$

**Example 1.28** We find here the normal forms of $p \to q$. Notice that $p \to q$ is equivalent to $\neg p \vee q$. Since $(\neg p \vee q)$ is a elementary product (consisting of a single term) of sums, it is a product of elementary sums normal form of $p \to q$. Since $(\neg p \vee q)$ is a maxterm, the normal form is also canonical. Since $\neg p \vee q$ is a sum of elementary product $\neg p$ and $q$, it is the sum of products normal form of $p \to q$. But this normal form is not canonical. Since

$$\begin{aligned} p \to q \Leftrightarrow \neg p \vee q \\ &\equiv (\neg p \wedge T) \vee (T \wedge q) \\ &\equiv (\neg p \wedge (q \vee \neg q)) \vee ((p \vee \neg p) \wedge q) \\ &\equiv ((\neg p \wedge q) \vee (\neg p \wedge \neg q)) \vee ((p \wedge q) \vee (\neg p \wedge q)) \\ &\equiv (\neg p \wedge q) \vee (\neg p \wedge \neg q) \vee (p \wedge q), \end{aligned}$$

we see that the canonical sum of products normal form of $p \to q$ is

$$(\neg p \wedge q) \vee (\neg p \wedge \neg q) \vee (p \wedge q).$$

Of course, we can also derive this normal form by using the normal form of $\neg(p \to q)$. Since $p \to q \equiv (\neg p \vee q)$, we see that

$$\neg(p \to q) \quad \equiv \quad (p \vee q) \wedge (p \vee \neg q) \wedge (\neg p \vee \neg q).$$

Notice that, by taking negation and using De Morgan's law, we have

$$\begin{aligned} p \to q \quad &\equiv \neg(p \vee q) \vee \neg(p \vee \neg q) \vee \neg(\neg p \vee \neg q) \\ &\equiv (\neg p \wedge \neg q) \vee (\neg p \wedge q) \vee (p \wedge q). \end{aligned}$$

**Example 1.29** We find here *sum of products* and *product of sums* normal forms of $p \leftrightarrow q$. Recall that

$$p \leftrightarrow q \quad \equiv (p \to q) \wedge (q \to p) \quad \equiv \quad (\neg p \vee q) \wedge (\neg q \vee p).$$

Therefore, $(\neg p \vee q) \wedge (\neg q \vee p)$ is the product of sums normal form of $p \leftrightarrow q$. Notice that

$$\begin{aligned} p \leftrightarrow q \quad &\equiv \quad (p \to q) \wedge (q \to p) \\ &\equiv \quad (\neg p \vee q) \wedge (\neg q \vee p) \\ &\equiv \quad ((\neg p \vee q) \wedge \neg q) \vee ((\neg p \vee q) \wedge p) \\ &\equiv \quad ((\neg p \wedge \neg q) \vee (q \wedge \neg q)) \vee ((\neg q \wedge p) \vee (q \wedge p)) \\ &\equiv ((\neg p \wedge \neg q) \vee \mathscr{F}) \vee (\mathscr{F} \vee (q \wedge p)) \\ &\equiv \quad (\neg p \wedge \neg q) \vee (q \wedge p). \end{aligned}$$

Therefore $(\neg p \wedge \neg q) \vee (q \wedge p)$ is the sum of products normal form of $p \leftrightarrow q$. We can also obtain the sum of product normal form of $p \leftrightarrow q$ by using the product of sums normal form. For, since

$$p \leftrightarrow q \quad \equiv \quad (\neg p \vee q) \wedge (\neg q \vee p),$$

it follows that

$$\neg(p \leftrightarrow q) \quad \equiv \quad (p \vee q) \wedge (\neg p \vee \neg q),$$

which on taking negation, and using De Morgan's laws, gives

$$p \leftrightarrow q \quad \equiv \quad \neg(p \vee q) \vee \neg(\neg p \vee \neg q) \Leftrightarrow (\neg p \wedge \neg q) \vee (p \wedge q).$$

## *PCNF and PDNF Using Truth Tables*

Notice that if we have only two variables $p$ and $q$, then the minterms are $p \wedge q$, $p \wedge \neg q$, $\neg p \wedge q$, $\neg p \wedge \neg q$. Their sum is a tautology since

$$\begin{aligned}
&(p \wedge q) \vee (p \wedge \neg q) \vee (\neg p \wedge q) \vee (\neg p \wedge \neg q)\\
&\Leftrightarrow (p \wedge (q \vee \neg q)) \vee (\neg p \wedge (q \vee \neg q))\\
&\Leftrightarrow (p \wedge T) \vee (\neg p \wedge T)\\
&\Leftrightarrow p \vee \neg p\\
&\Leftrightarrow T.
\end{aligned}$$

Also, the sum of the minterms is a tautology since for any of choice of truth values of the individual proposition, one of the minterms will have truth value T and hence the sum will have the truth value T. This shows that the minterm normal form of a tautology consists of all minterms. The minterm normal form of a contradiction contains no term. This will force us to define empty sum of minterms to be F. In general, for a proposition in the variables $p_1$, $p_2$, ..., $p_n$, we can easily write the minterm normal form of a given proposition by using the following algorithm:

1. construct the truth table for $p$;
2. for each line in the truth table in which $p$ is true, form a minterm $B_1 \wedge B_2 \wedge \cdots \wedge B_n$ where $B_i$ is $p_i$ if $p_i$ is true on that line, otherwise let $B_i = \neg p_i$;
3. form the product of all minterms obtained in the previous step; this is the required minterm normal form of the given proposition.

We obtain the maxterm normal form of a given proposition by using the following algorithm:

1. construct the truth table for $p$;
2. for each line in the truth table in which $p$ is false, form a maxterm $B_1 \vee B_2 \vee \cdots \vee B_n$ where $B_i$ is $p_i$ if $p_i$ is false on that line, otherwise let $B_i = \neg p_i$;

3. form the product of all maxterms obtained in the previous step; this is the required maxterm normal form of the given proposition.

In the examples, we find the maxterm/minterm normal form for certain propositions to illustrate the above algorithms.

**Example 1.30** We find here the PCNF and PDNF of the proposition $p \wedge (p \vee \neg q)$. The truth table for $p \wedge (p \vee \neg q)$ is given by

| $p$ | $q$ | $\neg q$ | $p \vee \neg q$ | $p \wedge (p \vee \neg q)$ |
|---|---|---|---|---|
| F | F | T | T | F |
| F | T | F | F | F |
| T | F | T | T | T |
| T | T | F | T | T |

The minterm normal form of $p \wedge (p \vee \neg q)$ is

$$(p \wedge \neg q) \vee (p \wedge q).$$

Also the maxterm normal form is

$$(p \vee q) \wedge (p \vee \neg q).$$

**Example 1.31** We find here the PCNF and the PDNF of the proposition $p \wedge (q \rightarrow r)$. The truth table for the proposition $p \wedge (q \rightarrow r)$ is given by

| $p$ | $q$ | $r$ | $q \rightarrow r$ | $p \wedge (q \rightarrow r)$ |
|---|---|---|---|---|
| F | F | F | T | F |
| F | F | T | T | F |
| F | T | F | F | F |
| F | T | T | T | F |
| T | F | F | T | T |
| T | F | T | T | T |
| T | T | F | F | F |
| T | T | T | T | T |

The minterm normal form of $p \wedge (q \rightarrow r)$ is

$$(p \wedge \neg q \wedge \neg r) \vee (p \wedge \neg q \wedge r) \vee (p \wedge q \wedge r).$$

Also the maxterm normal form is given by

$$(\neg p \vee \neg q \vee \neg r) \wedge (\neg p \vee \neg q \vee r) \wedge (\neg p \vee q \vee \neg r)$$
$$\wedge (\neg p \vee q \vee r) \wedge (\neg p \vee \neg q \vee r).$$

## *Rules of Inference*

We discuss here standard nine *rules of inferences* of propositional calculus. In an axiomatic system for a theory, a proof is a chain of logically valid statements consisting of certain axioms of the theory or a tautology or some previously known result. Each is called a *logical argument* of the proof. Mostly, it is possible to use certain axioms to derive a *direct proof* of an implication $p \Rightarrow q$. For example, assuming the existence of the set of natural numbers $\mathbb{N}$ based on Peano's axioms, we can construct the set of real numbers $\mathbb{R}$, by using basic rules of logic and deductive reasoning.

In general, the following **rules of inferences** are applied to show that a valid conclusion follows from a given set of premises:

1. Rule-P: A hypothesis or premise can be introduced at any point in the derivation.
2. Rule-T: A formula $r$ can be introduced at any point in the derivation if $r$ is implied by one or more of preceding formulas in the derivation.
3. Rule-CP: If we can derive $s$ from $r$ and a set of premises $p_1, \ldots, p_n$, then we can derive $r \to s$ from the set of premises $p_1, \ldots, p_n$ alone.

Notice that the last rule says that if we have to derive a conditional conclusion $r \to s$ from premises $p_1, \ldots, p_n$, then it is enough to show that the hypothesis $r$ of the conditional conclusion and the premises $p_1, \ldots, p_n$ implies $s$. This follows because

$$\begin{aligned} p \to (r \to s) \quad &\leftrightarrow \quad \neg p \vee (\neg r \vee s) \\ &\leftrightarrow \quad (\neg p \vee \neg r) \vee s \\ &\leftrightarrow \quad \neg(p \wedge r) \vee s \\ &\leftrightarrow \quad (p \wedge r) \to s. \end{aligned}$$

The procedure that is used to validating a conclusion, starting from the specified premises and applying the rules of inferences, is known as the *method of direct proof*. Further, since

$$\begin{aligned} (p \wedge \neg q) \to \mathbf{F} \quad &\leftrightarrow \quad \neg(p \wedge \neg q) \vee \mathbf{F} \\ &\leftrightarrow \quad \neg p \vee q \\ &\leftrightarrow \quad p \to q, \end{aligned}$$

we see that $p \wedge \neg q \Rightarrow \mathbf{F}$ is equivalent to $p \Rightarrow q$. This gives another method of proof, called *indirect method* or the *method of contradiction* (or *reductio-ad-absurdum*): To prove that $p_1, \ldots, p_n$ implies $q$, show that $p_1, \ldots, p_n, \neg q$ implies a contradiction. In such case, the propositions $p_1, \ldots, p_n, \neg q$ are *inconsistent*.

**Definition 1.14** We say premises $p_1, \ldots, p_n$ are **inconsistent** if $p_1 \wedge \cdots \wedge p_n \to \mathbf{F}$.

Notice that $p \wedge \neg p \to \mathbf{F}$, always. For an illustration, we show that $p \to q, q \to r$, and $p$ implies $r$. A direct proof is as given below:

| | | |
|---|---|---|
| (1) | $p \to q$ | Rule P |
| (2) | $q \to r$ | Rule P |
| (3) | $p \to r$ | (1), (2), chain rule |
| (4) | $p$ | Rule P |
| (5) | $r$ | modus ponens, (3), (4) |

An indirect proof of the above implication may be given as follows:

| | | |
|---|---|---|
| (1) | $\neg r$ | negated conclusion |
| (2) | $q \to r$ | Rule P |
| (3) | $\neg q$ | modus tollens, (1), (2) |
| (4) | $p \to q$ | Rule P |
| (5) | $\neg p$ | modus tollens, (3), (4) |
| (6) | $p$ | Rule P |
| (7) | $p \wedge \neg p$ | (5), (6), $p \wedge \neg p$ is a contradiction |

We now show that $p \to q$ and $\neg q$ implies $\neg p$. In fact, the result is the *modus tollens*. A direct proof of it is as given below:

| | | |
|---|---|---|
| (1) | $p \to q$ | Rule P |
| (2) | $\neg q \to \neg p$ | (1), $p \to q \leftrightarrow \neg q \to \neg p$ |
| (3) | $\neg q$ | Rule P |
| (4) | $\neg p$ | modus ponens, (2), (3) |

An indirect proof of the above implication may be given as follows:

| | | |
|---|---|---|
| (1) | $p$ | negated conclusion |
| (2) | $p \to q$ | Rule P |
| (3) | $q$ | modus ponens, (1), (2) |
| (4) | $\neg q$ | premise |
| (5) | $q \wedge \neg q$ | (3), (4), $q \wedge \neg q$ is a contradiction |

Similarly, it can be shown that $\neg(p \wedge \neg q)$, $\neg q \vee r$, and $\neg r$ implies $\neg p$. In the next examples, we give proof of an implication when a conclusion is a conditional.

**Example 1.32** We show that $p \to q$, $\neg p \to r$, $r \to s$ implies $\neg q \to s$. We first derive the conditional conclusion directly.

| | | |
|---|---|---|
| (1) | $\neg p \to r$ | premise |
| (2) | $r \to s$ | premise |
| (3) | $\neg p \to s$ | chain rule, (1), (2) |
| (4) | $p \to q$ | premise |
| (5) | $\neg q \to \neg p$ | contrapositive, (4) |
| (6) | $\neg q \to s$ | chain rule, (5), (3) |

The following is a proof when the premise of the conditional conclusion is used as additional premise and we have derived a conclusion of the conditional conclusion.

| | | |
|---|---|---|
| (1) | $p \to q$ | premise |
| (2) | $\neg q$ | premise |
| (3) | $\neg p$ | modus tollens, (1), (2) |
| (4) | $\neg p \to r$ | premise |
| (5) | $r$ | modus ponens, (3), (4) |
| (6) | $r \to s$ | premise |
| (7) | $s$ | modus ponens, (5), (6) |

The next example is an illustration of propositions that are inconsistent.

**Example 1.33** We show that the premises $p \to q$, $p \to r$, $q \to \neg r$, $p$ are inconsistent. We prove this by showing that $p \to q$, $p \to r$, $q \to \neg r$, $p$ implies a contradiction $p \wedge \neg p$.

| | | |
|---|---|---|
| (1) | $p$ | premise |
| (2) | $p \to q$ | premise |
| (3) | $q$ | (1), (2), modus ponens |
| (4) | $q \to \neg r$ | premise |
| (5) | $\neg r$ | (3), (4), modus ponens |
| (6) | $p \to r$ | premise |
| (7) | $\neg r \to \neg p$ | (6), contrapositive |
| (8) | $\neg p$ | (5), (7), modus ponens |
| (9) | $p \wedge \neg p$ | (1), (8) |

Recall that, for two simple propositions $p$ and $q$, we say $p \Rightarrow q$ if $p \to q$ is a tautology. When there are more than one premises, say $p_1, p_2, \ldots, p_n$, we say these premises imply a conclusion $q$ or that $q$ is a valid conclusion[5] drawn from the premises $p_1, \ldots, p_n$ if we have

$$\left(p_1 \wedge p_2 \wedge \cdots \wedge p_n\right) \quad \Rightarrow \quad q.$$

The next theorem proves an important rule, known as the *rule CP* (conditional proof).

**Theorem 1.5** *If $h_1, h_2, \ldots, h_n$ and $p$ together imply $q$, then $h_1, h_2, \ldots, h_n$ imply $p \to q$.*

***Proof*** As $h_1, h_2, \ldots, h_n$ and $p$ together imply $q$, we have

$$\left(h_1 \wedge h_2 \wedge \cdots \wedge h_n \wedge p\right) \quad \Rightarrow \quad q.$$

[5] We also say $q$ follows logically from premises $p_1, p_2, \ldots, p_n$.

It thus follows that

$$(h_1 \wedge h_2 \wedge \cdots \wedge h_n \wedge p) \quad \to \quad q$$

is a tautology. By Exercise 1.13 (b), we know that

$$(p \wedge q) \quad \to \quad r \quad \equiv \quad p \to (q \to r).$$

Therefore, we have

$$(h_1 \wedge h_2 \wedge \cdots \wedge h_n) \quad \to \quad (p \to q)$$

is a tautology, which proves

$$(h_1 \wedge h_2 \wedge \cdots \wedge h_n) \quad \Rightarrow \quad (p \to q).$$

□

To show that $q$ follows from $p_1, \ldots, p_n$, we use truth table for

$$(p_1 \wedge p_2 \wedge \cdots \wedge p_n) \quad \to \quad q.$$

But the method of truth table will be tedious if the number of variables in these propositions is large. By using the following rules of inferences, we can show that a conclusion follows from a set of premises:

1. Rule-P A hypothesis or premise can be introduced at any point in the derivation.
2. Rule-T A formula $r$ can be introduced at any point in the derivation if $r$ is implied by one or more of preceding formulas in the derivation.
3. Rule-CP If we can derive $s$ from $r$ and a set of premises $p_1, \ldots, p_n$, then we can derive $r \to s$ from the set of premises $p_1, \ldots, p_n$ alone (Theorem 1.5).

As said earlier, Rule-CP says that if we want to derive a conditional conclusion $r \to s$ from a set of premises $p_1, \ldots, p_n$, then it is enough to show that the hypothesis $r$ of the conditional conclusion and the premises $p_1, \ldots, p_n$ implies $s$. Such a procedure is known as the method of *direct proof*. Further, since

$$(p \wedge \neg q) \to \mathscr{F} \equiv \neg(p \wedge \neg q) \vee \mathscr{F} \equiv \neg p \vee q \Leftrightarrow p \to q,$$

we see that $p \wedge \neg q \Rightarrow \mathscr{F}$ is equivalent to $p \Rightarrow q$, which is known as the method of *indirect proof*.[6] That is, to prove that premises $p_1, \ldots, p_n$ imply a conclusion $q$ is equivalent to showing that $p_1, \ldots, p_n, \neg q$ leads to a contradiction. In latter situation, we also say that premises $p_1, \ldots, p_n, \neg q$ are *inconsistent*.

**Definition 1.15** The propositions $p_1, p_2, \ldots, p_n$ are **inconsistent** if $p_1, \ldots, p_n$ implies a contradiction.

---

[6] This is also known as the *method of contradiction* or *reductio-ad-absurdum*.

Notice that, as $p \wedge \neg p$ is always a contradiction, we have $p, .\neg p$ are inconsistent, for any proposition. In the next illustration, we consider a verbal problem.

**Example 1.34** Consider the statement: "If the traffic is heavy, then traveling is difficult. If customers arrived on time, then travelling was not difficult. They arrived on time. Therefore, the traffic was not heavy". We check whether the above constitutes a valid argument. For, we may write

$p$ : There are meeting;
$q$ : Traveling was difficult;
$r$ : They arrived on time,

so that the symbolic form of above statement is given by

$$p \to q, \; r \to \neg q, \; r \;\; \Rightarrow \;\; \neg p.$$

That the argument is valid can be shown as follows:

| | | |
|---|---|---|
| (1) | $r$ | premise |
| (2) | $r \to \neg q$ | premise |
| (3) | $\neg q$ | (1), (2), modus ponens |
| (4) | $p \to q$ | premise |
| (5) | $\neg p$ | (3), (4), modus tollens |

**Example 1.35** We check here that the below statements constitute a valid argument: "If $A$ works hard, then either $B$ or $C$ will enjoy themselves, If $B$ enjoy himself, then $A$ will not work hard. If $D$ enjoy himself, then $C$ will not. Therefore, if $A$ works hard, then $D$ will not enjoy himself". For, we take

$p$ : $A$ works hard;
$q$ : $B$ enjoy himself;
$r$ : $C$ enjoy himself;
$s$ : $D$ enjoy himself.

Then the above statement in symbolic form can be written as

$$p \to (q \vee r), \; q \to \neg p, \; s \to \neg r \;\; \Rightarrow \;\; p \to \neg s.$$

The argument is valid may be shown as follows: We use premise of a conclusion as additional hypothesis and obtain a conclusion from a conclusion, namely,

$$p \to (q \vee r), q \to \neg p, s \to \neg r, p \Rightarrow \neg s.$$

| | | |
|---|---|---|
| (1) | $p$ | premise (assumed) |
| (2) | $p \to (q \vee r)$ | premise |
| (3) | $q \vee r$ | (1), (2), modus pollens |
| (4) | $q \to \neg p$ | premise |
| (5) | $p \to \neg q$ | (4), contrapositive |

| | | |
|---|---|---|
| (6) | $\neg q$ | (1), (5), modus pollens |
| (7) | $r$ | (3), (6), disjunctive syllogism |
| (8) | $s \to \neg r$ | premise |
| (9) | $r \to \neg s$ | (8), contrapositive |
| (10) | $\neg s$ | (7), (9), modus pollens |

**Example 1.36** We prove here that $p \to q$, $q \to r$, and $p$ implies $r$. A direct proof is given as follows:

| | | |
|---|---|---|
| (1) | $p \to q$ | Rule P |
| (2) | $q \to r$ | Rule P |
| (3) | $p \to r$ | (1), (2), chain rule |
| (4) | $p$ | Rule P |
| (5) | $r$ | modus ponens, (3), (4) |

Next, we give an indirect proof of this implication:

| | | |
|---|---|---|
| (1) | $\neg r$ | negated conclusion |
| (2) | $q \to r$ | Rule P |
| (3) | $\neg q$ | modus tollens, (1), (2) |
| (4) | $p \to q$ | Rule P |
| (5) | $\neg p$ | modus tollens, (3), (4) |
| (6) | $p$ | Rule P |
| (7) | $p \wedge \neg p$ | (5), (6), $p \wedge \neg p$ is a contradiction |

**Example 1.37** We show here that $p \Leftrightarrow p \vee (p \wedge q)$. We first notice that

$$p \equiv p \wedge T \equiv p \wedge (q \vee \neg q) \equiv (p \wedge q) \vee (p \wedge \neg q)$$

so that using the above, we obtain

$$\begin{aligned} p \vee (p \wedge q) &\equiv (p \wedge q) \vee (p \wedge q) \vee (p \wedge \neg q) \\ &\equiv (p \wedge q) \vee (p \wedge \neg q) \\ &\equiv p. \end{aligned}$$

**Example 1.38** We show here that $\neg(p \wedge \neg q)$, $\neg q \vee r$, and $\neg r$ implies $\neg p$. We first give a direct proof as follows:

| | | |
|---|---|---|
| (1) | $\neg r$ | Rule P |
| (2) | $\neg q \vee r$ | Rule P |
| (3) | $\neg q$ | (1), (2), disjunctive syllogism |
| (4) | $\neg(p \wedge \neg q)$ | Rule P |
| (5) | $\neg p \vee \neg\neg q$ | (4), De Morgan's law |
| (6) | $\neg p \vee q$ | (5), Double negation law |
| (7) | $\neg p$ | (3), (6), disjunctive syllogism |

Next, an indirect proof of the implication may be given as follows:

| | | |
|---|---|---|
| (1) | $\neg(p \wedge \neg q)$ | Rule P |
| (2) | $\neg p \vee \neg\neg q$ | (1), De Morgan's law |
| (3) | $\neg p \vee q$ | (2), Double negation law |
| (4) | $p \to q$ | (3), conditional as disjunction |
| (5) | $p$ | negated conclusion |
| (6) | $\neg(\neg q)$ | (4), (5), modus ponens |
| (7) | $\neg q \vee r$ | Rule P |
| (8) | $r$ | (6), (7),disjunctive syllogism |
| (9) | $\neg r$ | Rule P |
| (10) | $r \wedge \neg r$ | (8), (9), $r \wedge \neg r$ is a contradiction |

**Example 1.39** We show here that $\neg p \vee q$, $\neg r \to \neg q$, and $\neg r$ implies $\neg p$. We can give a direct proof as follows:

| | | |
|---|---|---|
| (1) | $\neg p \vee q$ | Rule P |
| (2) | $p \to q$ | $p \to q \equiv \neg p \vee q$, (1) |
| (3) | $\neg r \to \neg q$ | Rule P |
| (4) | $q \to r$ | contrapositive, (3) |
| (5) | $p \to r$ | chain rule, (2), (4) |
| (6) | $\neg r$ | Rule P |
| (7) | $\neg p$ | modus tollens, (5), (6) |

We may also give an indirect proof of this implication as follows:

| | | |
|---|---|---|
| (1) | $p$ | negated conclusion |
| (2) | $\neg p \vee q$ | Rule P |
| (3) | $q$ | disjunctive syllogism, (1), (2) |
| (4) | $\neg r \to \neg q$ | Rule P |
| (5) | $q \to r$ | contrapositive, (3) |
| (6) | $r$ | modus ponens, (3), (5) |
| (7) | $\neg r$ | Rule P |
| (8) | $r \wedge \neg r$ | (6), (7); $r \wedge \neg r$ is a contradiction |

**Example 1.40** We show here that

$$p \to q, (\neg q \vee r) \wedge \neg r, \neg(\neg p \wedge s) \Rightarrow \neg s.$$

A direct proof may be given as follows:

| | | |
|---|---|---|
| (1) | $p \to q$ | Rule P |
| (2) | $\neg q \to \neg p$ | (1), contrapositive law |
| (3) | $(\neg q \vee r) \wedge \neg r$ | premise |
| (4) | $\neg q$ | (3), modus tollens |
| (5) | $\neg p$ | (2), (4), modus ponens |
| (6) | $\neg(\neg p \wedge s)$ | Rule P |
| (7) | $p \vee \neg s$ | (6), De Morgan's law |
| (8) | $\neg s$ | (5), (7), disjunctive syllogism |

**Example 1.41** We prove here that $r \wedge (p \vee q)$ is a valid conclusion from the premises $p \vee q$, $q \rightarrow r$, $p \rightarrow m$, and $\neg m$. A direct proof may be given as follows:

| | | |
|---|---|---|
| (1) | $\neg m$ | Rule P |
| (2) | $p \rightarrow m$ | Rule P |
| (3) | $\neg p$ | (1), (2), modus tollens |
| (4) | $p \vee q$ | Rule P |
| (5) | $q$ | (3), (4), disjunctive syllogism |
| (6) | $q \rightarrow r$ | Rule P |
| (7) | $r$ | (5), (6), modus ponens |
| (8) | $r \wedge (p \vee q)$ | (4), (7) |

An indirect proof may be given as follows:

| | | |
|---|---|---|
| (1) | $\neg(r \wedge (p \vee q))$ | negated conclusion |
| (2) | $\neg r \vee \neg(p \vee q)$ | (1), De Morgan's law |
| (3) | $(p \vee q) \rightarrow \neg r$ | (2), conditional as disjunction |
| (4) | $(p \vee q)$ | premise |
| (5) | $\neg r$ | (3), (4), modus ponens |
| (6) | $q \rightarrow r$ | Rule P |
| (7) | $\neg q$ | (5), (6), modus tollens |
| (8) | $p$ | (4), (7), disjunctive syllogism |
| (9) | $p \rightarrow m$ | Rule P |
| (10) | $m$ | (8), (9), modus ponens |
| (11) | $\neg m$ | Rule P |
| (12) | $m \wedge \neg m$ | (10), (11), a contradiction |

In next examples, we give proof of an implication where a conclusion is a conditional.

**Example 1.42** We show here that $p \rightarrow q, \neg p \rightarrow r, r \rightarrow s$ implies $\neg q \rightarrow s$. Let us first derive the conditional conclusion directly as follows:

| | | |
|---|---|---|
| (1) | $\neg p \rightarrow r$ | premise |
| (2) | $r \rightarrow s$ | premise |
| (3) | $\neg p \rightarrow s$ | chain rule, (1), (2) |
| (4) | $p \rightarrow q$ | premise |
| (5) | $\neg q \rightarrow \neg p$ | contrapositive, (4) |
| (6) | $\neg q \rightarrow s$ | chain rule, (5), (3) |

The following is proof where the premise of the conditional conclusion is used as additional premise and we have derived a conclusion of the conditional conclusion.

| | | |
|---|---|---|
| (1) | $p \rightarrow q$ | premise |
| (2) | $\neg q$ | premise |
| (3) | $\neg p$ | modus tollens, (1), (2) |
| (4) | $\neg p \rightarrow r$ | premise |
| (5) | $r$ | modus ponens, (3), (4) |
| (6) | $r \rightarrow s$ | premise |
| (7) | $s$ | modus ponens, (5), (6) |

**Example 1.43** Prove that $p \to (q \to s)$, $r \to p$, and $q$ implies $r \to s$. Let us first give a direct derivation of the conditional conclusion as follows:

| | | |
|---|---|---|
| (1) | $r \to p$ | premise |
| (2) | $p \to (q \to s)$ | premise |
| (3) | $r \to (q \to s)$ | chain rule, (1), (2) |
| (4) | $(r \wedge q) \to s$ | exportation law, (3) |
| (5) | $(q \wedge r) \to s$ | commutative law, (4) |
| (6) | $q \to (r \to s)$ | exportation law, (5) |
| (7) | $q$ | premise |
| (8) | $r \to s$ | modus ponens, (6), (7) |

We now give a derivation of a conclusion of the conditional conclusion by adding the premise of the conditional conclusion as additional premise.

| | | |
|---|---|---|
| (1) | $r$ | premise of conclusion |
| (2) | $r \to p$ | premise |
| (3) | $p$ | modus ponens, (1), (2) |
| (4) | $p \to (q \to s)$ | premise |
| (5) | $q \to s$ | modus ponens, (3), (4) |
| (6) | $q$ | premise |
| (8) | $s$ | modus ponens, (6), (5) |

Our final example gives an example of propositions which are inconsistent.

**Example 1.44** We show here that the premises $p \to q$, $p \to r$, $q \to \neg r$, $p$ are inconsistent. We prove this by showing that $p \to q$, $p \to r$, $q \to \neg r$, $p$ implies a contradiction $p \wedge \neg p$.

| | | |
|---|---|---|
| (1) | $p$ | premise |
| (2) | $p \to q$ | premise |
| (3) | $q$ | (1), (2), modus ponens |
| (4) | $q \to \neg r$ | premise |
| (5) | $\neg r$ | (3), (4), modus ponens |
| (6) | $p \to r$ | premise |
| (7) | $\neg r \to \neg p$ | (6), contrapositive |
| (8) | $\neg p$ | (5), (7), modus ponens |
| (9) | $p \wedge \neg p$ | (1), (8) |

## 1.3 Predicate Logic

The *sentential logic* we discussed in the previous section is not adequate to express everything we do in mathematics or to process *natural language* sentences. The main point is, for example, commonly used sentences such as "*Some students are intelligent*" or "*Rahul likes playing football*" are not propositions. We thus need a more general form of logic that offers tools to overcome limitations of propositional calculus.

We introduce here *predicate logic*, also known as the *first-order logic*, which has a much wider applications to problems that include concerning computer science. First-order logic is yet another way to represent knowledge in artificial intelligence. In addition, it allows enough expressive power to represent statements of natural language in a concise way. More generally, predicate logic helps develops information about the many different types of subjects in a more easy way and can also in expressing the relationship among these. Such subjects could be symbols, people, numbers, colors, theories, etc., and a relation may be unary such as intelligent, red, even integer, greater than, an Indian, etc. In the context of natural language, first-order logic also has two important parts *syntax* and *semantics*, where the syntax determines which collection of symbols is a logical expression in first-order logic and semantics refers to symbols.

Let us start with the simplest case.

**Definition 1.16** A *predicate statement* (or simply a *predicate*) $P$ expresses a property of a *predicate variable* (or a *subject*) $x$. In this case, we denote a predicate by $P(x)$, which is read as "$x$ is $P$". A set of *values* that the variable $x$ may be assigned is called the *domain of discourse* (or *universe*) of the predicate. By a *predicate constant* we mean a number, name of a person, animal, etc.

Since $P(x)$ involves a single variable, we also say that $P(x)$ is a *single place predicate*. For example, if we take $D$ symbolize the statement "*is divisible by* 3", then we have

$$D(x) \; : \; x \text{ is divisible by } 3$$

is a predicate with the set of integers $\mathbb{Z}$ as a universe. Clearly, a predicate $P(x)$ is not a proposition unless the predicate variable $x$ is assigned a *particular value* from the specified domain. For example, with $D$ as above, we have $D(6),\ D(12)$, and $\neg D(5)$ are, respectively, the propositions "6 is divisible by 3", "12 is divisible by 3", and "5 is not divisible by 3".

More generally, if $s$ is a member of related universe, then the proposition $A(s)$ is called a *substitution instance* (or simply *instantiation*) of the predicate $A(x)$ for $x = s$. For example, if $H$ is the predicate statement "is a mortal" so that

$$H(x) \; : \; x \text{ is a mortal},$$

with the set of "*all human beings*" as a universe, then $H(a)$ and $H(b)$ are the substitution instances of the predicate statement $H(x)$, where $a$ and $b$ are names "Aryabhatta" and "Baskara". Notice that both $H(a)$ and $H(b)$ are propositions with truth value $T$.

**Example 1.45** Let $M$ stand for the statement "*is an Indian mathematician*" so that the related single place predicate can be written as

$$M(x) \; : \; x \text{ is an Indian mathematician},$$

where a universe for the subject $x$ is the set of "*all mathematician*". Let $b$, $p$, and $r$, respectively, denote the *predicate constants* "R. Balasubramanian", "K. R. Parthasarathy", and "S. Ramanujan". Then $M(b)$, $M(p)$, and $M(r)$ are, respectively, the statements "*R. Balasubramanian is an Indian mathematician*", "*K. R. Parthasarathy is an Indian mathematician*", and "*S. Ramanujan is an Indian mathematician*". Notice that all these statements have the truth value $T$. However, if $h$ is the predicate constant "G. H. Hardy", then the statement $M(h)$ has the truth value $F$.

**Example 1.46** Consider the predicate $G(x)\ :\ x > 2$ with the set $\mathbb{Z}^+$ as a universe. Notice that, in particular, the proposition $G(1)$ is false and $G(3)$ is true. Further, for the predicate

$$P(x)\ :\ x \text{ is a prime number},$$

with the set $\mathbb{Z}^+$ as universe, it is easy to find the truth values of the propositions $P(2)$, $P(5)$, and $P(1729)$. The first two are true and the third is false.

**Definition 1.17** An expression or a relation (mathematical or otherwise) consisting of a predicate symbol $P$, and $n$ *predicate variables* (or simply *variables*) $x_1, \ldots, x_n$, is called an *$n$-place statement function* of $n$ variables. It is denoted by $P(x_1, \ldots, x_n)$. The proposition obtained by replacing each of these $n$ variables by some members of a specified universe is called a *substitution instance* of the statement function $P(x_1, \ldots, x_n)$.

The most basic statement functions are formed from a predicate symbol followed by a parenthesis with a sequence of variables. For example,

$$O(x, y)\ :\ x \text{ is older than } y$$

is a *two-place* atomic statement function, with the set of all human beings as a universe. We also write the above statement function as "Older(x,y)". Notice that *Older(Ravi, Atul)* is a substitution instance of this statement function. Now, let $h$ and $r$ be, respectively, the predicate constants "G. H. Hardy" and "S. Ramanujan". Then both $O(h, r)$ and $O(r, h)$ are substitution instances of the statement function $O(x, y)$. The truth value of $O(h, r)$ is $T$, while the truth value of $O(r, h)$ is $F$. Further,

$$S(x, y) :\ x = y + 2;$$
$$T(x, y, z) :\ x + y = z$$

are, respectively, the *two-place* and *three-place* atomic statement functions. As said earlier, none of the above two statement functions is a proposition. In particular, we have the proposition $S(2, 0)$ is true and the proposition $T(3, 0, 2)$ is false.

**Remark 1.3** In technical terms, statement functions are used to show that valid inputs to a computer program always produce the desired output. The statements that describe valid inputs are known as *preconditions*, and *postconditions* refer to conditions that the output must satisfy when a computer program is run.

### *Statement Functions*

As in the case of propositional calculus, more complex statement functions are formed using standard logical connectives. Let us start our discussion with some simple illustrations. For instance, consider the statement functions given by

$$\begin{aligned} D(x) &: \ x \text{ is divisible by } 3; \\ G(x, y) &: \ x \text{ is greater than } y. \end{aligned}$$

Then the statement "$x$ is divisible by 3 and $x$ is greater than $y$" can be expressed as $D(x) \wedge G(x, y)$. Notice that the proposition

$$\neg D(5) \ : \ \text{the number 5 is not divisible by 3}$$

can also be expressed as "*it is not true that the number* 5 *is divisible by* 3". Next, let $M$ stand for the predicate "*is a man*" and $N$ for the predicate "*is an Indian mathematician*" so that we can write statement functions $M(x)$ and $N(x)$, respectively, as

$$\begin{aligned} M(x) &: \ x \text{ is a man}; \\ N(x) &: \ x \text{ is an Indian mathematician}. \end{aligned}$$

Now, if $h$ and $r$, respectively, stand for the *predicate constants* such as names "G. H. Hardy" and "S. Ramanujan", then the compound statements "*Ramanujan is an Indian mathematician*" and "*Hardy is not an Indian mathematician*" can be expressed, respectively, as

$$M(r) \wedge N(r) \quad \text{and} \quad M(h) \wedge \neg N(h).$$

In general, the statements such as

$$M(x) \wedge N(x) \qquad \text{and} \qquad M(x) \wedge \neg N(x)$$

are called *statement formulas*.

**Definition 1.18** An expression or a relation consisting of simple statement functions (of one or more variables) connected by standard logical connectives is called a *predicate formula*. An $n$*-place predicate formula* in $n$-variables $x_1, \ldots, x_n$ is denoted by $P(x_1, \ldots, x_n)$.

In the next example, we give some illustrations of some 2*-place* and 3*-place* predicate formulas.

**Example 1.47** Let $P$ denote the predicate "are parents of", and

$$P(x, y, z) \ : \ x \text{ and } y \text{ are parents of } z.$$

Then the statement "Surya and Kunti are parents of Karna" is symbolized as $P$(Surya, Kunti, Karna). Next, let $T$ be the predicate "played tennis against", and

$$T(x, y) \; : \; x \text{ played tennis against } y.$$

Let $p$ and $s$ be the predicate constants "Paes and Bhupathy" and "Sampras and Agassi". Then $T(p, s)$ symbolizes the statement "Paes and Bhupathi played tennis against Sampras and Agassi". Also, the predicate $G$, $M$ Exercise 1.76 take two subjects, and so it is a special case of a 2-place predicates. Notice that the predicate $B$ in Exercise 1.75 takes just one subject, and so it is a special case of a 1-place predicate.

## *Quantifiers*

A *quantifier* allows to express the scope of involved variable in a logical expression. It is a language element that is used to specify the quantity of specimen in the universe. Some common quantifiers used in everyday English language are phrases such as *for all, everyone, everything, for some, for many, for none, for at least one*, and *a few*. In general, the two most commonly used *quantifiers* in first-order logic are as follows:

1. The *existential quantifier* specifies a statement function to be true for at least one substitution instance.
2. The *universal quantifier* specifies a statement function to be true for every substitution instances.

In what follows, we shall discuss part of the first-order logic that helps analyze complex statement formulas in terms of atomic statement functions, standard logical connectives, and quantifiers.

**Definition 1.19** Let $P(x)$ be a predicate with domain $U$.

1. The *existential quantification* of $P(x)$ is the proposition such that $P(a)$ is true for at least one subject $a$ in $U$. It is denoted by $(\exists x)P(x)$, where the symbol $\exists$ is read as "*there exists*" or "*there is*". Notice that the main connective for existential quantifier $\exists$ is the connective disjunction $\wedge$.
2. The *universal quantification* of $P(x)$ is the proposition such that $P(a)$ is true for all subjects $a$ in $U$. It is denoted by $(\forall x)\, P(x)$ (or simply by $(x)\, P(x)$), where the symbol $\forall$ is read as "*for all*", "*for each*", or "*for every*". Notice that the main connective for universal quantifier $\forall$ is the conditional connective $\rightarrow$.

**Example 1.48** (*Existential Quantification*) Consider the statement "Some real numbers are rational". We can rewrite it as "For some x, where x is a real number, and x is a rational number". So, for

$$R(x) \; : \; x \text{ is a real number;}$$
$$Q(x) \; : \; x \text{ is a rational number.}$$

we can symbolize the given statement as $(\exists x)(R(x) \wedge Q(x))$. Also, since the statement "Some students are hard working" can be rewritten as "For some x, x is a student, and x is hard working", we shall use the predicates

$$B(x) : x \text{ is a student;}$$
$$H(x) : x \text{ is hard working,}$$

so that, in this case, we can symbolize the statement as $(\exists x)(B(x) \wedge H(x))$. Also, for the predicates

$$student(x) : x \text{ is a student;}$$
$$intelligent(x) : x \text{ is intelligent.}$$

Symbolic statement $(\exists x)student(x) \wedge intelligent(x)$ is read as "Some students are intelligent".

**Example 1.49** (*Universal Quantification*) Consider the statement "All students are hard working". Since it can also be written as "For every x, if x is a student, then x is hard working", we symbolize it as $(x)[B(x) \rightarrow H(x)]$, where

$$B(x) : x \text{ is a student;}$$
$$H(x) : x \text{ is hard working.}$$

Also, the statement "All men drink tea" can symbolize as $(x)man(x) \rightarrow drink(x, tea)$, where we have

$$man(x) : x \text{ is a man;}$$
$$drink(x, tea) : x \text{ drinks tea.}$$

Similarly, the statement "All birds can fly" can symbolize as $(x)bird(x) \rightarrow fly(x)$, where we have

$$bird(x) : x \text{ is a bird;}$$
$$fly(x) : x \text{ can fly.}$$

Further, since the statement "Every Monday I go to library" can also be written as "For every x, if x is a Monday, then x is when I go to library", we may use the following predicates:

$$M(x) : x \text{ is a Monday;}$$
$$L(x) : x \text{ is when I go to library.}$$

Then $(x)(M(x) \rightarrow L(x))$ is the symbolic form of the given statement.

### Statement Formulas

We start our discussion here with some examples of an $n$-place *statement formula* $P(x_1, \ldots, x_n)$, for $n \geq 2$. Let us begin with the 2-place formula $\forall x\, man(x) \to respect(x, parent)$ that is symbolic form of the statement "Every man respects his parents". Also, the statement "Some boys play cricket" has 2-place formula given by $\exists x boy(x) \to play(x, cricket)$. Further, the statement "Not all students like both Mathematics and Science" has 2-place formula given by

$$\neg\forall x\big[student(x) \to like(x, mathematics) \wedge like(x, science)\big].$$

It is little more *nested* to express statements such as "Only one student failed in mathematics" symbolically. In this case, taking

$$S(x) : \ x \text{ is a student;}$$
$$F(x, y) : \ \text{student}\, x \text{ failed in subject } y,$$

the given statement can be expressed symbolically as

$$\exists x\big[S(x) \to F(x, mathematics) \wedge \forall y\big[x\neg(x = y) \wedge S(y) \to \neg F(x, mathematics)\big]\big].$$

**Definition 1.20** A statement formula (or simply *formula*) is called a *well-formed formula* if it is obtained by applying the following procedure finite number of times:

1. Each atomic formula is a well-formed formula.
2. If $A$ and $B$ are well-formed formulas, then so are $(\neg A)$, $(A \vee B)$, $(A \wedge B)$, $(A \to B)$, and $A \leftrightarrow B)$.
3. If $A$ is a well-formed formula, and $x$ is any variable, then $(x)A$ and $(\exists x)A$ are also well-formed formulas.

A simple illustration of well-formed formula is a statement such as $(x)\big[(P(x) \vee Q(x)) \wedge R(x)\big]$. Notice this is not a statement formula in first-order logic sense. On the other hand, $(x)\big[P(x) \wedge Q(x)\big] \wedge (\exists x)\big[S(x)\big]$ is a well-formed formula, and also a statement formula in first-order logic sense. For the next definition, recall that the scope of a quantifier is the predicate to which it applies.

**Definition 1.21** A variable $x$ appearing in a statement formula is said to be a *bound variable* if it occurs within the scope of the quantifier $(x)$ or $(\exists x)$. Otherwise, we say $x$ is called a *free variable*.

For example, in the statement formula $(\exists x)(P(x) \vee Q(x)) \wedge R(x)$, the scope of the existential quantifier is the formula $P(x) \vee Q(x)$ and the occurrence of $x$ in $P(x) \vee Q(x)$ is a bound occurrence. However, the $x$ in $R(x)$ is a free variable. Also, in the statement formula $(x)P(x) \vee (Q(x) \wedge R(x))$, the scope the universal quantifier is $P(x)$ and $x$ in $Q(x) \wedge R(x)$ is a free variable. Thus the occurrence of the variable $x$ in $P(x)$ has no relationship to the free variable $x$ of $Q(x) \wedge R(x)$. In fact, we can write the statement formula $(x)(P(x) \vee (Q(x) \wedge R(x))$ as

$$(y)(P(y) \vee Q(y)) \wedge (Q(x) \wedge R(x)),$$

where $y$ is a bound variable while $x$ is a free variable.

**Remark 1.4** In general, while dealing with double application of universal quantifiers, we need to remember that $\forall x \forall y$ is similar to $\forall y \forall x$. Also, while dealing with double application of existential quantifiers, we have $\exists x \exists y$ is similar to $\exists y \exists x$. However, notice that $\exists x \forall y$ is *not* similar to $\forall y \exists x$.

## 1.4 Theory of Inference

A proposition in FOL (first-order logic) quantified with an existential quantifier is analogs to the disjunction in usual propositional logic sense. Also, a set $U$ is called a *domain of discourse* (or simply a *domain*) if the subjects are taken only from the set $U$. If $U = \{x_1, \ldots, x_n\}$ is a finite set, then the predicate $(\exists x)P(x)$ is true if and only if $P(x_1) \vee P(x_2) \vee \cdots \vee P(x_n)$ is true. It thus follows that

$$(\exists x)P(x) \quad \Leftrightarrow \quad P(x_1) \vee P(x_2) \vee \cdots \vee P(x_n). \tag{1.4.1}$$

Similarly, a proposition in FOL quantified with an universal quantifier is analogs to the conjunctions in propositional logic sense. That is, $(x)P(x)$ is true if and only if $P(x_1) \wedge P(x_2) \wedge \cdots \wedge P(x_n)$ is true. It thus follows that

$$(x)P(x) \quad \Leftrightarrow \quad P(x_1) \wedge P(x_2) \wedge \cdots \wedge P(x_n). \tag{1.4.2}$$

For example, suppose the universe of the discourse be the set $\{a, b, c, d\}$. Therefore, a formula $(x)P(x)$ is same as $P(a) \wedge P(b) \wedge P(c) \wedge P(d)$. Similarly, the formula $(x)(\neg P(x)) \vee (x)P(x)$ is the formula

$$[\neg P(a) \wedge \neg P(b) \wedge \neg P(c) \wedge \neg P(d)] \vee [P(a) \wedge P(b) \wedge P(c) \wedge P(d)].$$

**Theorem 1.6** *Let $P$ and $Q$ be predicates with a finite domain. Then we have*

$$\neg[(\exists x)P(x)] \quad \Leftrightarrow \quad (x)[\neg P(x)]. \tag{1.4.3}$$

***Proof*** Let the finite domain be given by $U = \{x_1, \ldots, x_n\}$. By De Morgan's law of propositional calculus, we have

$$\begin{aligned}\neg[(\exists x)P(x)] &\Leftrightarrow \neg[P(x_1) \vee P(x_2) \vee \cdots \vee P(x_n)]\\ &\Leftrightarrow \neg P(x_1) \wedge \neg P(x_2) \wedge \cdots \wedge \neg P(x_n)\\ &\Leftrightarrow (x)[\neg P(x)].\end{aligned}$$

□

**Theorem 1.7** *Let P and Q be predicates with a finite domain. Then we have*

$$\neg[(x)P(x)] \quad \Leftrightarrow \quad (\exists x)[\neg P(x)]. \tag{1.4.4}$$

***Proof*** Let the finite domain be given by $U = \{x_1, \ldots, x_n\}$. By using De Morgan's law of propositional calculus, we have

$$\begin{aligned}\neg[(x)P(x)] &\Leftrightarrow \neg[P(x_1) \wedge P(x_2) \wedge \cdots \wedge P(x_n)]\\ &\Leftrightarrow \neg P(x_1) \vee \neg P(x_2) \vee \cdots \vee \neg P(x_n)\\ &\Leftrightarrow (\exists x)[\neg P(x)].\end{aligned}$$

□

Recall that a statement formula $P(x_1, \ldots, x_n)$ is not be a proposition unless subjects are predicate constants or the variables $x_1, \ldots, x_n$ are assigned "*values*" from the specified domain. As said earlier, the proposition so obtained is called a *substitution instance* of the statement formula. Therefore, most rules of inference of sentential logic apply. In addition, we also use the following four rules of inference of FOL:

**Rule-UG** When $P(x)$ is true for all subjects in the specified domain, we have $(x)P(x)$ is true. It thus follows that we have $P(x) \Rightarrow (y)P(y)$. This implication is known as the rule of *universal generalization* (or simply UG). Clearly, it can be applied only when $P(x)$ is true for all $x$ in the specified domain.

**Rule-US** Since $(x)P(x)$ is true if and only if $P(x)$ is true for every $x$ in the specified domain, we have $(x)P(x) \Rightarrow P(y)$, where $P(y)$ is a proposition, and $y$ is any subject from the domain. This implication is known as the rule of *universal specification* (or simply US).

**Rule-EG** If $P(x)$ is true for at least one subject $a$ in the specified domain, then we have a conclusion: $P(a) \Rightarrow (\exists x)P(x)$. This is known as the rule of *existential generalization* (or simply EG).

**Rule-ES** When $(\exists x)P(x)$ is true, $P(x)$ is true for some $a$ in the specified domain, and we obtain the implication: $(\exists x)P(x) \Rightarrow P(a)$. This rule is known as the rule of *existential specification* (or simply ES). Notice that a variable cannot be generalized universally if it was obtained by an application of the Rule-ES.

**Example 1.50** We show here

$$(\exists x)(P(x) \wedge Q(x)) \quad \Rightarrow \quad (\exists x)P(x).$$

We use the derivation given below:

| | | |
|---|---|---|
| (1) | $(\exists x)(P(x) \wedge Q(x))$ | premise |
| (2) | $P(y) \wedge Q(y)$ | ES (1) |
| (3) | $P(y)$ | (2), $P \wedge Q \Rightarrow P$ |
| (4) | $(\exists x)P(x)$ | (3), EG |

It thus follows that $(\exists x)(P(x) \wedge Q(x)) \Rightarrow (\exists x)P(x)$.

**Example 1.51** We show here the predicate $(\exists x)(M(x))$ follows logically from the premises $(x)(H(x) \to M(x))$ and $(\exists x)(H(x))$. We use the derivation as given below:

| | | |
|---|---|---|
| (1) | $(\exists x)H(x)$ | premise |
| (2) | $H(a)$ | (1), ES |
| (3) | $(x)(H(x) \to M(x))$ | premise |
| (4) | $H(a) \to M(a)$ | (5), US |
| (5) | $M(a)$ | (2), (4), modus ponens |
| (6) | $(\exists x)(M(x)$ | (5), EG |

It thus follows that $(\exists x)(M(x)$ follows logically from the premises $(x)(H(x) \to M(x))$ and $(\exists x)(H(x))$.

We now provide an account of what it means to be an FOL truth, an FOL consequence, or an FOL equivalence. The core idea is FOL validities (or consequences, or equivalences) are truths (or consequences, or equivalences) solely by virtue of the truth functional connectives, the quantifiers, and the identity symbol. That is, to determine whether a statement is an FOL valid (or an argument a case of FOL consequence, or a pair of statements FOL equivalent) we ignore the meanings of the names and predicates they contain. The focus here is on *logical equivalence* of statement function involving quantifiers. As we shall see, the main argument is a simple extension of the concept introduced earlier while discussing similar types of problems involving statement formulas.

**Definition 1.22** Let $S$ and $T$ be two statement functions involving predicates and quantifiers. We say $S$ and $T$ are **FOL equivalent**, denoted by $S \equiv T$, if they have the same truth value no matter which predicates are substituted for these statements, and which domain of discourse is used for the involved predicate variables.

**Theorem 1.8** *Let $P$ and $Q$ be predicates with a finite domain. Then we have*

$$(\forall x)[P(x) \wedge Q(x)] \quad \Leftrightarrow \quad (\forall x)P(x) \wedge (\forall x)Q(x).$$

*The above equivalence fails if the* conjunction *is replaced by* disjunction.

***Proof*** Suppose $P$ and $Q$ are two predicates, with a common universe, such that the nested statement $\forall x \left[P(x) \wedge Q(x)\right]$ holds. Then, $P(a) \wedge Q(a)$ is true, for any $a$ in the universe. Therefore, for all $a$ in the universe, $P(a)$ is true, and also $Q(a)$ is true, for all $a$ in the universe. Hence, nested statement $\forall x\ P(x) \wedge \forall x\ Q(x)$ holds. The converse can be proven on similar lines. As a counter-example for the second part, let

$$P(x)\ :\ x \text{ is even} \qquad \text{and} \qquad Q(x)\ :\ x \text{ is odd},$$

with the set of integers $\mathbb{Z}$ as a universe for $x$. Now, the nested statement $\forall x \left[P(x) \vee Q(x)\right]$ reads as "every integer $x$ is either even or odd", which is a valid statement. On the other hand, the nested statement $\forall x\ P(x) \vee \forall x\ Q(x)$ is absurd. □

**Theorem 1.9** *Let P and Q be predicates with a finite domain. Then we have*

$$(\exists x)[P(x) \vee Q(x)] \quad \Leftrightarrow \quad (\exists x)P(x) \vee (\exists x)Q(x).$$

*The above equivalence fails if the* disjunction *is replaced by* conjunction.

***Proof*** Suppose $P$ and $Q$ are two predicates, with a common universe, such that the nested statement $\exists x \left[P(x) \vee Q(x)\right]$ holds. Then, for some $a$ in the universe, $P(a)$ is true or $Q(a)$ is true. Therefore, for some $a$ in the universe, $P(a)$ is true; or, for some $a$ in the universe, $Q(a)$ is true. Hence, nested statement $\exists x\ P(x) \vee \exists x\ Q(x)$ holds. The converse can be proven on similar lines. As a counter-example for the second part, let $P(x)$ and $Q(x)$ as in the proof of Theorem 1.8. In this case, the nested statement $\left(\exists x\ P(x)\right) \wedge \left(\exists y\ Q(y)\right)$ reads as "there is an even integer $x$ and there is an odd integer $y$", which is true. On the other hand, the nested statement $\exists x \left[P(x) \wedge Q(x)\right]$ reads as "some integer is both even and odd", which is absurd. □

**Theorem 1.10** *Let P and Q be predicates with a finite domain. Then we have*

$$(\exists x)(P(x) \wedge Q(x)) \quad \Rightarrow \quad (\exists x)P(x) \wedge (\exists(x))Q(x).$$

*However, the converse is not true.*

***Proof*** We may use the derivation as given below:

| | | |
|---|---|---|
| (1) | $(\exists x)(P(x) \wedge Q(x))$ | premise |
| (2) | $P(a) \wedge Q(a)$ | (1), ES |
| (3) | $P(a)$ | (2), conjunctive simplification |
| (4) | $Q(a)$ | (2) conjunctive simplification |
| (5) | $(\exists x)P(x)$ | (3), EG |
| (6) | $(\exists x)Q(x)$ | (4), EG |
| (7) | $(\exists x)P(x) \wedge (\exists x)Q(x)$ | (5), (6) |

This proves that

$$(\exists x)(P(x) \wedge Q(x)) \Rightarrow (\exists x)P(x) \wedge \exists(x)Q(x).$$

Next we prove the converse may not hold in general. We can argue as follows:

| | | |
|---|---|---|
| (1) | $(\exists x)P(x) \wedge (\exists x)Q(x)$ | premise |
| (2) | $(\exists x)P(x)$ | (1) |
| (3) | $P(a)$ | (2), ES |
| (4) | $(\exists x)Q(x)$ | (1) |
| (5) | $Q(b)$ | (4), ES |

We cannot use the same constant in steps (3) and (5) so that it is not possible to conclude the desired conclusion. □

## Exercises 1

1.1. Symbolize the statement: "If Uma attends classes regularly and if either she is attentive in the class or studies well, then she gets the top grade".

1.2. Translate the following sentence into a logical expression: "You can access the internet from campus only if you are a computer science student or you are not a freshman". Also negate the sentence.

1.3. Translate the following sentence into a logical expression: "If tigers have wings, then the earth travels round the sun". Also, state its truth value.

1.4. Construct the truth table for the propositions given below:

a. $p \wedge (p \vee q)$ and $p \vee (p \wedge q)$.
b. $(p \rightarrow q) \rightarrow p$ and $(p \wedge \neg q) \vee p$.
c. $(q \wedge (p \rightarrow q)) \rightarrow p$.
d. $p \leftrightarrow q$ and $(p \rightarrow q) \wedge (q \rightarrow p)$.
e. $\neg(\neg p \vee \neg q)$ and $p \wedge q$.
f. $(p \wedge q) \vee (\neg p \wedge q) \vee (p \wedge \neg q)$.
g. $(\neg p \wedge (\neg q \wedge r)) \vee ((q \wedge r) \vee (p \wedge r))$.

1.5. Construct the truth table for the propositions given below:

a. $p \wedge (q \wedge p)$ and $\neg(\neg p \wedge \neg q)$.
b. $(p \vee \neg q) \wedge (\neg p \vee q)$ and $(p \wedge \neg q) \vee (\neg p \wedge q)$.
c. $(p \rightarrow q) \rightarrow (q \rightarrow p)$ and $(p \rightarrow \neg q) \wedge (q \rightarrow \neg p)$.
d. $(p \rightarrow q) \wedge (q \rightarrow p)$ and $(p \rightarrow q) \leftrightarrow (\neg p \vee q)$.
e. $(p \vee q) \wedge (\neg p \wedge \neg q)$ and $(p \vee q) \vee (\neg p \wedge \neg q)$.
f. $(p \leftrightarrow q) \leftrightarrow ((p \rightarrow q) \wedge (q \rightarrow r))$ and $p \leftrightarrow ((p \rightarrow q) \vee r)$.
g. $\neg p \wedge (\neg q \wedge r)) \vee (q \wedge r) \vee (p \wedge r)$.

What can you infer from these tables about these propositions?

1.6. Prove that the following are tautologies:

a. $(p \wedge q) \to (p \vee q)$.
b. $q \vee (p \wedge \neg q) \vee (\neg p \wedge \neg q)$.
c. $((p \vee q) \wedge \neg(\neg p \wedge (\neg q \vee \neg r))) \vee (\neg p \wedge \neg q) \vee (\neg q \wedge \neg r)$.

1.7. Use truth table to show that $\vee$ is commutative. That is, $(p \vee q) \leftrightarrow (q \vee p)$ is a tautology.
1.8. Use truth table to show that $\neg p \to (p \to q)$ is a tautology.
1.9. Is the formula $q \vee (p \wedge \neg q) \vee (\neg p \wedge \neg q)$ a tautology?
1.10. Is the proposition $((p \to q) \to r) \vee \neg p$ a tautology?
1.11. Use truth table to show that $(p \wedge q) \wedge \neg(p \vee q)$ is a contradiction.
1.12. Prove the following equivalences:

a. $\neg(p \vee (\neg p \wedge q)) \equiv \neg p \wedge \neg q$.
b. $\neg(p \leftrightarrow q) \equiv \neg p \leftrightarrow q$.
c. $p \to (q \to p) \equiv \neg p \to (p \to q)$.
d. $\neg(p \wedge q) \to (\neg p \vee (\neg p \vee q)) \equiv (\neg p \vee q)$.
e. $p \to (q \vee r) \equiv (p \to q) \vee (p \to r)$.
f. $p \to q \equiv \neg p \vee q \equiv \neg q \to \neg p$.
g. $(p \leftrightarrow q) \equiv (p \wedge q) \vee (\neg p \wedge \neg q)$.
h. $(p \wedge q) \vee (p \wedge \neg q) \equiv (p \vee q) \wedge (p \vee \neg q) \equiv p$.

1.13. Prove the following equivalences:

a. $(p \to q) \wedge (p \to r) \equiv p \to (q \vee r)$.

b. $p \to (q \to r) \equiv (p \wedge q) \to r$.

c. $p \to (q \to p) \equiv \neg p \to (p \to q)$.

1.14. Prove the *Involution Law*: $p \equiv \neg\neg p$.
1.15. Prove the *Identity Laws*: $p \wedge \mathscr{T} \equiv p$ and $p \vee \mathscr{F} \equiv p$.
1.16. Prove the *Idempotent Laws*: $p \wedge p \equiv p$ and $p \vee p \equiv p$.
1.17. Prove the *Absorption Laws*: $p \vee (p \wedge q) \equiv$ and $p \wedge (p \vee q) \equiv p$.
1.18. Prove the following implications:

a. $p \Rightarrow p$.
b. $(p \to q) \to p \Rightarrow p$.
c. $p \Rightarrow (q \to p)$.
d. $p \to (q \to r) \Rightarrow (p \to q) \to (p \to r)$.

1.19. Prove *De Morgan's Laws*:

$$\neg(p \wedge q) \Leftrightarrow (\neg p \vee \neg q) \quad \text{and} \quad \neg(p \vee q) \Leftrightarrow (\neg p \wedge \neg q).$$

1.20. Prove the *Distributive Laws*:

$$p \wedge (q \vee r) \equiv (p \wedge q) \vee (p \wedge r) \quad \text{and} \quad p \vee (q \wedge r) \equiv (p \vee q) \wedge (p \vee r).$$

1.21. Prove the *Exportation Law*: $p \to (q \to r) \equiv (p \wedge q) \to r$.
1.22. Show that $(\neg p \wedge (\neg q \wedge r)) \vee (q \wedge r) \vee (p \wedge r) \equiv r$.
1.23. Show that $((p \vee q) \wedge \neg(\neg p \wedge (\neg q \vee \neg r))) \vee (\neg p \wedge \neg q) \vee (\neg p \wedge \neg r)$ is a tautology.
1.24. Prove the equivalence: $p \to q \equiv \neg q \to \neg p$.
1.25. Prove the equivalence: $p \to q \equiv p \to (p \wedge q)$.
1.26. Prove the equivalence: $(p \to r) \wedge (q \to r) \Leftrightarrow (p \vee q) \to r$.
1.27. Prove the equivalence: $((p \vee \neg q) \to q) \to ((p \vee \neg p) \to r) \Leftrightarrow q \to r$.
1.28. Show that $(p \to q) \to q \Leftrightarrow (p \vee q)$.
1.29. Show that $p \to (q \to r) \Leftrightarrow (p \to q) \to (p \to r)$.
1.30. Show that $(p \to q) \wedge (\neg p \to q) \equiv q$.
1.31. Prove the equivalence: $\neg(p \wedge q) \to (\neg p \vee (\neg p \vee q)) \equiv \neg p \wedge q$.
1.32. Prove that $(p \wedge q) \Rightarrow (p \vee q)$.
1.33. Prove the implications: $\neg(p \to q) \Rightarrow p$ and $\neg(p \to q) \Rightarrow \neg q$.
1.34. Prove the implication: $(p \to q) \wedge p \Rightarrow q$.
1.35. Prove the implication: $(p \to q) \wedge \neg q \Rightarrow \neg p$.
1.36. Show that $negp \Rightarrow p \to q$.
1.37. Show that any formula in two variables is equivalent to one of the following 16 formulas: $F$, $p \wedge q$, $p \wedge \neg q$, $p$, $\neg \wedge q$, $q$, $\neg(p \leftrightarrow q)$, $p \vee q$, $\neg(p \vee q)$, $\neg(p \overline{\vee} q)$, $\neg q$, $p \vee \neg q$, $\neg p$, $p \to q$, $\neg p \vee \neg q$, $T$.
1.38. Show that $p, q \Rightarrow (p \to q)$.
1.39. Show that $p \vee q$, $p \to r$ and $q \to r$ implies $r$. Or, equivalently, show that

$$[((p \vee q) \wedge (p \to r) \wedge (q \to r))] \to r$$

is a tautology.
1.40. Show that the set of connectives $\{\neg, \to\}$ is functionally complete.
1.41. Show that the single connective $\{\downarrow\}$ is functionally complete.
1.42. Show that the number of non-equivalent formulas in $n$ variables is $2^{2^n}$ which is equal to the number of connectives that can be defined.
1.43. Write the duals of following:

a. $(p \wedge q) \vee r$.
b. $(p \wedge q) \vee T$.
c. $\neg(p \wedge q) \vee (\neg p \vee \neg(q \wedge \neg s))$.

1.44. Find the duals of following:

a. $(p \vee q) \to \neg r$.
b. $(p \wedge q) \leftrightarrow (\neg(p \vee \neg r))$.

1.45. Show that $\{\vee, \wedge\}$, $\{\neg\}$, $\{\vee\}$, $\{\wedge\}$ are not functionally complete.

1.46. Obtain the PDNF and the PCNF of the following:

a. $(\neg p \vee \neg q) \rightarrow (p \wedge \neg q)$.
b. $(p \vee \neg q) \wedge q$.
c. $(q \rightarrow p) \wedge (\neg p \wedge q)$.
d. $(p \wedge q) \vee (\neg p \wedge q \wedge r)$.
e. $\neg p \vee q$.
f. $(p \wedge q) \vee (\neg p \wedge r) \vee (q \wedge r)$.
g. $(p \vee q) \wedge (\neg p \vee r) \wedge (q \vee r)$.
h. $\neg p \rightarrow r) \wedge (q \leftrightarrow p)$.
i. $p \rightarrow ((p \rightarrow q) \wedge \neg(\neg q \vee \neg p))$.
j. $p \rightarrow ((p \rightarrow q) \wedge \neg(\neg q \vee \neg p))$.
k. $p \wedge (p \rightarrow q)$.
l. $\neg(p \vee q) \leftrightarrow (p \wedge q)$.
m. $(\neg p \vee \neg q) \rightarrow (p \leftrightarrow \neg q)$.
n. $(p \wedge q) \vee (\neg p \wedge q) \vee (q \wedge r)$.

1.47. Find the PDNF of $q \vee (p \wedge \neg q) \vee (\neg p \wedge \neg q)$, and hence show that it is a tautology.
1.48. Using PDNF, show that $p \vee (p \wedge q) \equiv p$.
1.49. Use PDNF to show that $p \vee (\neg p \wedge q) \equiv p \vee q$.
1.50. Obtain the product of sums canonical form for

$$(p \wedge q \wedge r) \vee (\neg p \wedge r \wedge q) \vee (\neg p \wedge \neg q \wedge \neg r).$$

1.51. Obtain the PCNF and PDNF of $S : (p \wedge q) \vee (\neg p \wedge q \wedge r)$.
1.52. Obtain the PCNF and PDNF of $S : (\neg p \rightarrow r) \wedge (p \leftrightarrow q)$.
1.53. Find the PDNF and PCNF of $S : p \vee (\neg p \rightarrow (q \vee (\neg q \rightarrow r)))$.
1.54. Find the PCNF and PDNF of $S : (\neg p \rightarrow r) \wedge (q \leftrightarrow p)$.
1.55. Find the PDNF and PCNF of $S : \neg((p \vee q) \wedge r)) \wedge (p \vee r)$.
1.56. Find the PCNF and PDNF of $(p \rightarrow (q \wedge r)) \wedge (\neg p \rightarrow (\neg q \wedge \neg r))$.
1.57. Find the PCNF and the PDNF of $(p \wedge r) \vee (p \wedge \neg q)$.
1.58. Use truth tables to determine the validity of a conclusion $q$ drawn from premises $p \rightarrow q$, $\neg p$.
1.59. Use truth tables to determine the validity of a conclusion $\neg p$ drawn from premises $p \rightarrow q$, $\neg(p \wedge q)$.
1.60. Show that the following premises are inconsistent: "If Raman is irregular to school, then he fails in the school. If Raman fails in the school, then he is uneducated. If Raman studied by himself, then he is not uneducated. Raman is regular to the school and studies by himself".
1.61. Show that $\neg p \wedge \neg q$ implies $\neg(p \wedge q)$.
1.62. Show that $p \rightarrow q$ and $\neg q$ implies $\neg p$.
1.63. Prove that

$$p \rightarrow q, q \rightarrow \neg r, r, p \vee s \Rightarrow s.$$

1.64. Show that

$$(p \vee q) \wedge (p \to r) \wedge (q \to s) \Leftrightarrow s \vee r.$$

1.65. Show that

$$(p \wedge q) \to r, \neg r \vee s, \neg s \Rightarrow \neg p \vee \neg q.$$

1.66. Give a direct and an indirect proofs of the implication:

$$p \to q, q \to r, \neg(p \wedge r), p \vee r \Rightarrow r.$$

1.67. Prove that

$$\neg p \to (q \vee r), (s \vee t) \to \neg p, s \vee t \Rightarrow q \vee r.$$

1.68. Show that $r \vee s$ logically follows from the premises

$$c \vee d, (c \vee d) \to \neg h, \neg h \to (a \wedge \neg b), (a \wedge \neg b) \to (r \vee s).$$

1.69. Show that $(p \to q) \to r, p \wedge s, q \wedge t \Rightarrow r$.

1.70. Prove that

$$\neg q, p \to q, p \vee t \Rightarrow t.$$

1.71. Prove that

$$p, p \to (q \to (r \wedge s)) \Rightarrow q \to s.$$

1.72. Show that

$$p \to (q \to r), q \to (r \to s) \Rightarrow p \to (q \to s).$$

1.73. Show the following:

a. $p \to q, q \to r, p \Rightarrow r$.
b. $p \vee q,\ p \vee q \to \neg r,\ \neg r \to (s \wedge \neg t),\ (s \wedge \neg t) \to u \vee s \Rightarrow u \vee s$.
c. $p \vee q,\ q \to r,\ p \to s,\ \neg s \Rightarrow r \wedge (p \vee q)$.
d. $\neg p \wedge \neg q \Rightarrow \neg(p \wedge q)$.
e. $p \vee q,\ p \to r,\ q \to s \Rightarrow s \vee r$.
f. $p \to (q \to s),\ \neg r \vee p,\ q \Rightarrow r \to s$.

1.74. Write the following statements as predicate formulas:

a. At least one integer is even.
b. There exists a positive integer that is odd.
c. If $x$ is even, then $x$ is not divisible by 3.
d. No even integer is divisible by 7.

e. There exists an even integer divisible by 11.
f. If $x$ is even and $x$ is a perfect cube, then $x$ is divisible by 8.

You may take the set of integers as the specified domain.

1.75. Write each of the following statements as a predicate formula:

a. Thirunelveli is a town in India which is situated on the bank of the river Thamiraparani.
b. Madurai is a town in India which is situated on the bank of the river Vaigai.
c. Tiruchirapalli is a town in India which is situated on the bank of the river Cauveri.
d. Rajamundri is a town in India which is situated on the bank of the river Godavari.
e. Calcutta is a town in India which is situated on the bank of the river Hoogly.
f. Guwahati is a town in India which is situated on the bank of the river Bramaputra.
g. Banarus is a town in India which is situated on the bank of the river Ganga.
h. Ahmedabad is a town in India which is on the bank of the river Sabarmathi.
i. The number 6 is greater than 4.
j. The number 8 is greater than 1.
k. The number 5 is not greater than 8.

1.76. Write the statements "Indira Gandhi is the mother of Rajiv Gandhi" and "Kunti is the mother of Karna" as predicate formulas.

1.77. Express the following statements using variables, predicates, and quantifiers:

a. Every apple is red.
b. All Indians are intelligent.
c. There exists an integer which is a square root of an integer.
d. Some integers are not square of any integers.

1.78. Symbolize the following statement with and without using the set of integers as the specified domain: "Given any positive integer, there is a greater positive integer".

1.79. Symbolize the following statement: For every $x$, there is an $y$ such that $x^2 + y^2 \geq 100$.

1.80. Determine the truth values of $(\exists x)(x^3 + 8 = 0)$ and $(\forall x)(x^3 + 8 = 0)$, where the specified domain is the set of all real numbers.

1.81. Let $q(x, y, z)$ denote "$x + y = z$". What are the truth values of the quantifications: $(x)(y)(z)Q(x, y, z)$ and $(\exists z)(x)(y)Q(x, y, z)$.

1.82. Let $q(x, y)$ denote "$x + y = 0$". What are the truth values of the quantifications: $(\exists y)(x)Q(x, y)$ and $(y)(\exists x)Q(x, y)$.

1.83. Translate the following as a sentence in English language:

$$\forall x \forall y((x < 0) \wedge (y < 0) \longrightarrow (xy > 0)).$$

You may assume that the domain is the set of all real numbers.

1.84. Let $P(x)$ denote the predicate formula "$x$ is a person", $F(x, y)$ the formula "$x$ is the father of $y$", and $M(x, y)$ the formula "$x$ is the mother of $y$". Using these formulas, write the predicate "$x$ is a father of mother of $y$".

1.85. Write symbolically the statement "All the world loves a lover".

1.86. Indicate the variables that are free and bound and also determine the scope of the quantifiers in the following statement formula:

$$(x)(P(x) \wedge Q(x)) \rightarrow (x)P(x) \wedge Q(x).$$

1.87. If the specified domain is $\{a, b, c, d\}$, eliminate the quantifier in the following formula $(x)P(x)$.

1.88. If the specified domain is $\{a, b, c\}$, eliminate the quantifier in the following formula:

$$(\exists x)(P(x) \wedge Q(x)).$$

1.89. If the specified domain is $\{a, b, c\}$, eliminate the quantifier in the following formula:

$$(\exists x)P(x) \rightarrow (x)Q(x).$$

1.90. Show the following:

a. $(x)(P(x) \vee Q(x)) \Rightarrow (x)P(x) \vee (\exists x)Q(x)$.
b. $(\exists x)(P(x) \wedge Q(x)) \Rightarrow (\exists x)P(x) \wedge (\exists x)Q(x)$.
c. $(\exists x)(P(x) \wedge Q(x)) \rightarrow (y)(R(y) \rightarrow S(y)), \quad (\exists y)(R(y) \wedge \neg S(y)) \quad \Rightarrow$ $(x)S(x) \rightarrow \neg S(x)$.

1.91. Verify the following implications:

a. $(\exists z)Q(z)$ follows from $(x)(P(x) \rightarrow Q(x))$, $(\exists y)P(y)$.
b. $(x)P(x)$ follows from $(\exists)(P(x) \wedge Q(x))$.
c. $(\exists x)(P(x) \wedge Q(x))$ follows from $(\exists x)P(x)$, $(\exists x)Q(x)$.
d. The conditional conclusion $(x)(P(x) \rightarrow Q(x))$ follows from $(\exists x)P(x) \rightarrow$ $(x)Q(x)$.

1.92. Demonstrate the following implication:

$$(x)(P(x) \vee Q(x)), (x)(\neg P(x)) \quad \Rightarrow \quad (x)Q(x).$$

1.93. Show that

a. $(x)(P(x) \rightarrow Q(x)), (x)(R(x) \rightarrow \neg Q(x)) \Rightarrow (x)(R(x) \rightarrow \neg P(x))$.
b. $(x)(P(x) \vee Q(x)) \Rightarrow (x)P(x) \vee (\exists x)Q(x)$.
c. $\neg P(a, b)$ follows logically from

$$(x)\ (y)\ (P(x, y) \rightarrow W(x, y)) \text{ and } \neg W(a, b).$$

d. $(\exists x)(F(x) \rightarrow \neg S(x))$ follows from

$$(\exists x)(F(x) \wedge S(x)) \rightarrow (y)(M(y) \rightarrow W(y)), (\exists y)(M(y) \wedge \neg W(y)).$$

1.94. Prove the validity of the following argument: All animals are mortal. All human beings are animals. Therefore all human beings are mortal.

## References

1. Mattson, H.F. (Jr.): Discrete Mathematics. Wiley (1993)
2. Russell, B., Whitehead, A.N.: Principia Mathematica. Cambridge University Press, Cambridge, England (1997)
3. Sfard, A.: Thinking as Communicating: Human Development, The Growth of Discourse, and Mathematizing. Cambridge University Press, Cambridge, England (2008)
4. Sipser, M.: Introduction to the Theory of Computation (3/e). Cengage Learning (2012)
5. Smith, R.: Aristotle's Logic. Stanford Encyclopedia of Philosophy. http://plato.stanford.edu
6. Laertius, D.: Lives of Eminent Philosophers, Hicks, R. D. (translator). Harvard University Press, Cambridge, Massachusetts (1965)
7. Gould, J.B.: The Philosophy of Chrysippus. SUNY Press, Albany, NY (1970)
8. Long, A.A., Sedley, D.N.: The Hellenistic Philosophers, vol. I. Cambridge University Press, Cambridge, England (1987)
9. Empiricus, S.: Against the Logicians, Bury, R. G. (translator). Harvard University Press, Cambridge Massachusetts (1967)
10. Gould, J.B.: Chrysippus: On the Criteria for the Truth of a Conditional Proposition. Phronesis **12**, 152–161 (1967)
11. Russell, B., Whitehead, A.N.: Principia Mathematica, vol. 1. Cambridge University Press, Cambridge, England (1910); (vol. 2, 1912); (vol. 3, 1913)
12. Post, E.L.: Introduction to a general theory of elementary propositions. Amer. J. Math. **43**, 163–185 (1921)
13. Boole, G.: An Investigation of the Laws of Thought. Walton and Maberly, London (1854)
14. Boole, G.: An Investigation of the Laws of Thought. Dover Publications Inc, New York (1958)
15. Frege, G.: Grundgesetze der Arithmetik: Begriffschriftlich abgeleitet, (photographic reproduction) Georg Olms, Hildesheim (1962)
16. Frege, G.: The Basic Laws of Arithmetic: Exposition of the System, Furth, M. (translator). University of California Press, Los Angeles (1964)
17. Wittgenstein, L.: Logische-philosophische Abhandlung. Annalen der Naturphilosophie **14**, 198–262 (1921)
18. Wittgenstein, L.: Tractatus Logico-Philosophicus, Ramsey, F., Ogden, C.K. (translators). Kegan Paul, London (1922)
19. Wittgenstein, L.: Tractatus Logico-Philosophicus, Pears, D.F., McGuinness, B.F. (translators). Routledge & Kegan Paul, London (1961)
20. Aristotle: On Interpretation, in [14]

# Chapter 2
# Methods of Proof

*Mathematics is not a deductive science—that's a cliché. When you try to prove a theorem, you don't just list the hypotheses, and then start to reason. What you do is trial and error, experimentation, guesswork.*

***Paul Halmos** (1916–2006)*

An *axiom* is a statement that is accepted to be true on the basis of sufficient explanatory depth. A *conjecture* is a statement that is largely believed to be true, whereas a *theorem* is a statement that is known to be true. A *proof* of a theorem of the form $P \Rightarrow Q$ is an exploration such that the conclusion $Q$ is derived from the premise $P$ by putting together relevant definitions, and also some known facts. There are mainly two methods to prove a theorem. We may use a *direct proof* or an *indirect proof* to prove a theorem. The related ideas discussed in this chapter find applications to most concepts introduced in later parts of the book. The primary focus is to develop certain crucial skills that the reader finds useful to write and communicate a concise, clear, and complete proof of a theorem. The interested reader may read excellent texts such as [1, 2], or [3], for further details.



V. Ravichandran and A. K. Razdan, *Fundamental Discrete Structures*,
University Texts in the Mathematical Sciences,
https://doi.org/10.1007/978-981-96-0069-4_2

## 2.1 Direct Proof

The sixth-century Greece philosopher *Thales of Miletus* is recognized as the first to understand the importance of what and how we know something is true. According to Aristotle, he knew that the *deductive reasoning* is central to a *proof* of a theorem.[1] As shown in the previous chapter, deductive reasoning is applied to draw a *valid conclusion* from a set of premises in finite number of steps, where each step that is based on Aristotle's syllogisms is called a *valid argument*. Therefore, a mathematically correct argument uses logical reasoning based on *Aristotle's syllogisms* and certain *rules of inferences*.

**Definition 2.1** A **direct proof** of an implication $P \Rightarrow Q$ is a chain of "coherent" logical deductions that starts with certain well-established *fundamental axioms*, makes use of the related definitions, and also some known results.

The real art in constructing a *direct proof* of a theorem depends on how one finds a correct balance between the logical deductions employed and the arguments used. It is always a good idea to have a tolerant openness and awareness while writing a *good proof*. We also need to apply some strictness such as in Euclid's proof of the fundamental result: *There are infinitely many primes* (Theorem 2.12). There are enough evidences to accept the fact that in general the above approach has been very productive in the advancement of knowledge related to developments of certain fundamental theories.

**Definition 2.2** An implication $P \Rightarrow Q$ is said to be *trivially true* if conclusion $Q$ holds regardless of the truth value of the hypotheses $P$. That is, we have

$$Q \rightarrow (P \rightarrow Q)$$

is a tautology.

When hypothesis $P$ is a conjunction of a number of premises $P_1, \ldots, P_n$ such that one or more $P_i$ is false, the falsity of $P$ implies that the implication $P \Rightarrow Q$ is *vacuously true*. That is, the implication holds regardless of the truth value of the statement $Q$. In this case, we have

$$\neg P \rightarrow (P \rightarrow Q)$$

is a tautology. In all other situations, our first preference is to find a direct proof. We start with a well-known fact as stated in the next theorem. Recall that two integers are said to have the same *parity* if both are odd or both are even. Otherwise, we say integers have *opposite parity*.

**Theorem 2.1** *The sum of any two integers of the same parity is an even integer, and the sum of any two integers of opposite parity is odd.*

[1] The theorems such as Theorems 8.45 and 2.13 were known to Pythagoreans.

***Proof*** We first suppose $a$ and $b$ are odd integers so that we can write

$$a = 2m + 1 \quad \text{and} \quad b = 2n + 1, \quad \text{for some} \quad m, n \in \mathbb{Z}.$$

Then, by associativity and commutativity of addition, we have

$$a + b = (2m + 1) + (2n + 1) = 2(m + n + 1),$$

which is an even integer. We can prove other two assertions by similar argument. □

Notice that we can also write the statement of Theorem 2.1 as a *quantified statement* involving three single-variable predicates, each having the set of integers $\mathbb{Z}$ as universe of discourse. Also, in the above proof, we mainly used the definitions when an integer to said to be odd and even. In general, a *definition* is a universally accepted statement about a mathematical concept.

**Theorem 2.2** *For any odd integers n, we have*

$$\left\lfloor \frac{n^2}{4} \right\rfloor = \left(\frac{n-1}{2}\right)\left(\frac{n+1}{2}\right),$$

*where* $\lfloor\ \rfloor$ *denotes the* floor function.

***Proof*** A *direct proof* may be given as follows: Write $n = 2k + 1$, for some integer $k$, so that

$$n^2 = 4k^2 + 4k + 1 \quad \Rightarrow \quad \left\lfloor \frac{n^2}{4} \right\rfloor = \left\lfloor \frac{4k^2 + 4k + 1}{4} \right\rfloor = k^2 + k.$$

We also have

$$\frac{n-1}{2} = k \ \text{ and } \ \frac{n+1}{2} = k + 1 \quad \Rightarrow \quad \left(\frac{n-1}{2}\right)\left(\frac{n+1}{2}\right) = k^2 + k.$$

Hence, the equality, as asserted. □

**Remark 2.1** In general, an axiomatic mathematical theory is about formulation of relevant axioms, definitions, and also theorems derived as logical consequences of the previous two. As in this book, we develop elementary aspect of theories related to fundamental concepts such as *sets, functions, groups, rings, lattices*, and *graphs*. As has been the case since antiquity, the focus is to develop a suitable mathematical theory that is good enough to prove certain fundamental theorems.[2] A good example to quote is Euclid's deductive reasoning approach to geometry, as codified in *Elements*.

---

[2] It is though true that many mathematical theories evolved by way of proving certain conjectures, but the main emphasis has always been to study various types of natural processes.

It is convenient to write a direct proof of an implication $P \Rightarrow Q$ especially when the hypothesis $P$ can be translated into an expression involving equations or inequalities so as to facilitate some further algebraic manipulations. The *principle of mathematical induction* is a typical method that helps verify the truth of a universal statement involving a set of non-negative integers. Other interesting method of a direct proof is known as the *proof by cases*.

## Principle of Induction

In this section, our main concern is to discuss a powerful direct method of proof, known as the **proof by induction**. It is also known as the *principle of induction* (or simply POI). In both its weak and strong forms, it helps to prove that a "conjectured statement" $p(n)$ holds for all $n \in K \subseteq \mathbb{N}_0$. Some simple illustrations of a statement $p(n)$ are as given below:

(i) (*Bernoulli Inequality*) $(1+a)^n \geq 1+na$, with $a \geq -1$, for $n \geq 0$ (see Example 2.1).
(ii) $1-na < (1-a)^n < 1/(1+na)$, with $0 < a < 1$, for $n \geq 2$.
(iii) $2n < n^2$, for $n \geq 3$;
(iv) the *closed form* of the *Fibonacci sequence* $\langle f_n \rangle$ defined *recursively* as $f_1 = f_2 = 1,\ f_{n+2} = f_n + f_{n+1},\ n \in \mathbb{N}$, is given by

$$f_n = \frac{1}{\sqrt{5}}\left(\varphi^n + (-1)^{n-1}\frac{1}{\varphi^n}\right), \quad n \in \mathbb{N}, \tag{2.1.1}$$

where $\varphi = (1+\sqrt{5})/2$ is known as the *golden ratio* (see Example 2.1).

## Principle of Induction: Weak Form

For a simple demonstration of how a *proof by induction* works, we consider the following well-known identity:

$$1+2+3+\cdots+n = \frac{n(n+1)}{2}, \quad \text{for all} \quad n \geq 1. \tag{2.1.2}$$

For instance, let $p(n)$ denote the statement

$$1+2+3+\cdots+n = \frac{n(n+1)}{2}.$$

In particular, $p(1)$ is the statement $1 = 1(1+1)/2$, which is true. This is called the *base step* of mathematical induction. Next, we assume that $p(n)$ holds for $n = k$, with $k \geq 1$. That is, we have

$$1 + 2 + 3 + \cdots + k = \frac{k(k+1)}{2}. \tag{2.1.3}$$

This is known as the *induction hypothesis*. Now, to complete the proof of the assertion (2.1.2), we have to show that $p(n)$ also holds for $n = k + 1$. For, adding $k + 1$ to both sides of equation (2.1.3), we obtain

$$\begin{aligned} 1 + 2 + 3 + \cdots + k + (k+1) &= \frac{k(k+1)}{2} + (k+1) \\ &= \frac{(k+1)(k+2)}{2}, \end{aligned}$$

which proves $p(k + 1)$ is also true. We have thus shown that

$$p(k) \Rightarrow p(k+1), \quad \text{for} \quad k \geq 1. \tag{2.1.4}$$

In particular, since $p(1)$ holds, $p(2)$ holds, and so $p(3)$ holds. Continuing in this manner, it follows that $p(n)$ is true, for all $n \in \mathbb{N}$. Therefore, the implication in (2.1.4) is the central idea of the mathematical induction. This is called the *induction step*. A formal statement is as given below.

**Principle of Mathematical Induction:** *If* $p(n)$ *is a statement, where n is a positive integer, such that we have*

1. **(Base Step)** $p(n)$ is true for some $n = n_0$ and
2. **(Induction Step)** $p(k) \Rightarrow p(k + 1)$, for $k \geq n_0$.

*Then* $p(n)$ *holds, for all* $n \geq n_0$.

The *weak form* of the principle of mathematical induction is the case when $n_0 = 1$. As shown later in Theorem A.2, we can use the *Well-ordering principle*[3] to prove the weak form of the mathematical induction.

**Remark 2.2** It is important to verify both (1) and (2) for an assertion $p(n)$ to hold for all $n \in \mathbb{N}$. For example, (2) holds if $p(n) : 2|2n - 1$, but $p(n)$ is not true for any $n \in \mathbb{N}$. On the other hand, $p(n) : n^2 = n$ is true only for $n = 1$. Also, the conjectured statement $p(n)$ that $n^2 - n + 41$ *is a prime* is true for $1 \leq n \leq 40$, but $p(41)$ fails to hold. However, we are free to start from any *base value*, other than the 1.

We now give some applications of principle of induction. We first consider some important inequalities. To start with, we show that

$$n! > 2^n, \qquad \text{for all} \quad n \geq 4. \tag{2.1.5}$$

[3] The **well-ordering principle** states that every nonempty subset of $\mathbb{N}$ contains a least element.

Notice that the base value in this case is $n_0 = 4$. Now, since we have

$$4! = 24 > 16 = 2^4,$$

it follows that the *base step* holds, i.e., the inequality (2.1.5) holds for $n_0 = 4$. We next assume that the inequality holds for $n = m \geq 4$. This is our *induction hypothesis*. That is, we have assumed that the inequality $m! > 2^m$ holds. To complete the *induction step*, it remains to show that the inequality (2.1.5) also holds for $n = m + 1$. For example, we have

$$(m+1)! = (m+1)m! > (m+1)2^m,$$

by induction hypothesis. We also have used the fact that $m + 1 > 2$, for $m \geq 4$. Therefore,

$$(m+1)2^m > (2)2^m = 2^{m+1} \quad \Rightarrow \quad (m+1)! > 2^{m+1}.$$

It thus follows that $n! > 2^n$, for all $n \geq 4$.

**Example 2.1** We prove **Bernoulli inequality**: For $h > -1$,

$$1 + nh \leq \left(1 + nh\right)^n, \quad \text{for all non-negative integers} \quad n. \tag{2.1.6}$$

Notice that, in this case, the base value is $n_0 = 1$. Since $1 + h \leq (1 + h)^1$, the assertion holds for $n = 1$. Next, as induction hypothesis, suppose (2.1.6) holds for $n = m$. That is,

$$1 + mh \leq (1 + h)^m, \quad \text{for} \quad m \geq 1.$$

Now, to complete the induction step, we prove that (2.1.6) also holds for $n = m + 1$. For, we have

$$\begin{aligned}
(1+h)^{m+1} &= (1+h)^m(1+h) \\
&\geq (1+mh)(1+h) \\
&= 1 + mh + h + mh^2 \\
&= 1 + (m+1)h + mh^2 \\
&\geq 1 + (m+1)h.
\end{aligned}$$

Therefore, (2.1.6) holds for all non-negative integers $n$.

The next theorem generalizes the usual *triangle inequality*.

**Theorem 2.3** *If $a_1, a_2, a_3, \ldots, a_n$ are real numbers ($n \geq 1$), then we have*

$$|a_1 + \cdots + a_n| \leq |a_1| + \cdots + |a_n|.$$

***Proof*** The assertion holds trivially for the base value $n = 1$. To apply the principle of induction, suppose that $n \geq 1$, and assume that the inequality holds for $n = m$.

That is,

$$|a_1 + \cdots + a_m| \leq |a_1| + \cdots + |a_m| .$$

We need to show that

$$\left|\sum_{r=1}^{m+1} a_r\right| \leq \sum_{r=1}^{m+1} |a_r| .$$

Now, since we have

$$\begin{aligned}\left|\sum_{r=1}^{m+1} a_r\right| &= \left|\sum_{r=1}^{m} a_r + a_{m+1}\right| \\ &\leq \left|\sum_{r=1}^{m} a_r\right| + |a_{m+1}| \\ &\leq \sum_{r=1}^{m} |a_r| + |a_{m+1}| \\ &= \sum_{r=1}^{m+1} |a_r| ,\end{aligned}$$

it follows that

$$|a_1 + \cdots + a_{m+1}| \leq |a_1| + \cdots + |a_{m+1}| .$$

Notice that the first inequality in above argument follows from the standard triangle inequality $|x + y| \leq |x| + |y|$, while the second follows by our induction hypothesis. Hence, the inequality holds for all $n \geq 1$. □

**Example 2.2** We show that

$$\sum_{i=1}^{n} i^3 < \frac{(n+1)^4}{4} < \sum_{i=1}^{n+1} i^3, \quad \text{for all} \quad n \geq 1.$$

For $n = 1$, the above inequality becomes $1 < 4 < 9$, which is true. Next, suppose that the inequality holds for $n = m$. That is,

$$\sum_{i=1}^{m} i^3 < \frac{(m+1)^4}{4} < \sum_{i=1}^{m+1} i^3.$$

Now, for $n = m + 1$, we have

$$\begin{aligned}\sum_{i=1}^{m+1} i^3 &= \sum_{i=1}^{m} i^3 + (m+1)^3 \\ &< \frac{(m+1)^4}{4} + (m+1)^3 \\ &= \frac{(m+1)^4 + 4(m+1)^3}{4} \\ &< \frac{(m+2)^4}{4},\end{aligned}$$

provided $(m+2)^4 > (m+1)^4 + 4(m+1)^3$, which holds because

$$\begin{aligned}(m+2)^4 &= [(m+1)+1]^4 \\ &= (m+1)^4 + 4(m+1)^3 + 6(m+1)^2 + 4(m+1) + 1.\end{aligned}$$

Similarly, we have

$$\begin{aligned}\sum_{i=1}^{m+2} i^3 &= \sum_{i=1}^{m+1} i^3 + (m+2)^3 \\ &> \frac{(m+1)^4}{4} + (m+2)^3 \\ &= \frac{(m+1)^4 + 4(m+2)^3}{4} \\ &> \frac{(m+2)^4}{4}\end{aligned}$$

provided $(m+1)^4 + 4(m+2)^3 > (m+2)^4$ or $6m^2 + 20m + 17 > 0$, which is obvious. Therefore, we have shown that

$$\sum_{i=1}^{m+1} i^3 < \frac{(m+2)^4}{4} < \sum_{i=1}^{m+2} i^3.$$

Hence, by the principle of induction, it follows that

$$\sum_{i=1}^{n} i^3 < \frac{(n+1)^4}{4} < \sum_{i=1}^{n+1} i^3$$

is true for all $n \geq 1$.

**Definition 2.3** The sequence $\langle H_n \rangle$ given by

$$H_n = 1 + \frac{1}{2} + \frac{1}{3} + \cdots + \frac{1}{n}, \quad \text{for} \quad n \geq 1 \tag{2.1.7}$$

is known as sequence of *harmonic numbers*.

**Example 2.3** We use principle of induction to show that

$$H_{2^n} \geq 1 + \frac{n}{2}, \tag{2.1.8}$$

for any integer $n \geq 0$. Since we have

$$H_{2^0} = H_1 = 1 \geq 1 + \frac{0}{2},$$

the assertion holds for $n = 0$. Next, suppose that

$$H_{2^m} \geq 1 + \frac{m}{2}.$$

To complete the induction, we need to show that

$$H_{2^{m+1}} \geq 1 + \frac{m+1}{2}.$$

Now, as we have

$$\begin{aligned}
H_{2^{m+1}} &= 1 + \frac{1}{2} + \frac{1}{3} + \cdots + \frac{1}{2^m} + \frac{1}{2^m + 1} + \cdots + \frac{1}{2^{m+1}} \\
&= H_{2^m} + \frac{1}{2^m + 1} + \cdots + \frac{1}{2^{m+1}} \\
&\geq \left(1 + \frac{m}{2}\right) + \frac{1}{2^m + 1} + \cdots + \frac{1}{2^{m+1}} \\
&\geq \left(1 + \frac{m}{2}\right) + 2^m \frac{1}{2^{m+1}} \\
&\geq \left(1 + \frac{m}{2}\right) + \frac{1}{2} \\
&= 1 + \frac{m+1}{2},
\end{aligned}$$

the inequality (2.1.8) holds for all positive integer $n$.

**Example 2.4** With the sequence $\langle H_n \rangle$ as defined above, we show here that

$$H_1^2 + H_2^2 + \cdots + H_n^2 = (n+1)H_n^2 - (2n+1)H_n + 2n.$$

For, notice that the above equality holds trivially for $n = 1$. Next, suppose that it holds for $n = k$. That is, we have

$$H_1^2 + H_2^2 + \cdots + H_k^2 = (k+1)H_k^2 - (2k+1)H_k + 2k.$$

Now, for $n = k + 1$, we have to show that

$$H_1^2 + H_2^2 + \cdots + H_k^2 + H_{k+1}^2 = (k+2)H_{k+1}^2 - (2k+3)H_{k+1} + 2(k+1).$$

For, by using the equality

$$H_{k+1} = H_k + \frac{1}{k+1},$$

we see that

$$\begin{aligned}
&(k+2)H_{k+1}^2 - (2k+3)H_{k+1} + 2(k+1) \\
&= (k+1)H_k^2 + 2H_k + \frac{1}{k+1} + H_{k+1}^2 \\
&\quad - (2k+1)H_k - (2k+1)\frac{1}{k+1} - 2H_k - \frac{2}{k+1} + 2k + 2 \\
&= H_{k+1}^2 + (k+1)H_k^2 - (2k+1)H_k + 2k \\
&= H_1^2 + H_2^2 + \cdots + H_{k+1}^2.
\end{aligned}$$

This proves the assertion.

We next use principle of induction to prove some interesting *combinatorial identities*. We may recall that binomial theorem for exponents 2 and 3 states that

$$\begin{aligned}
(a+b)^2 &= a^2 + 2ab = b^2; \\
(a+b)^3 &= a^3 + 3a^2b + 3ab^2 + b^3.
\end{aligned}$$

We prove it for arbitrary integer $n \in \mathbb{N}$. In 1654, the French mathematician *Blaise Pascal* (1623–1662) used *proof by induction* to prove tabular presentation for *binomial coefficients*, known as **Pascal's triangle**.

**Theorem 2.4** (Binomial Theorem) *For any integer $n \geq 0$, we have*

$$(a+b)^n = \sum_{i=0}^{n} \binom{n}{i} a^{n-i} b^i.$$

***Proof*** The assertion holds trivially for $n = 1$. We now assume that the assertion holds for $n = k$. That is, we have

$$(a+b)^k = \sum_{i=0}^{k} \binom{k}{i} a^{k-i} b^i.$$

Now we have to show that

$$(a+b)^{k+1} = \sum_{i=0}^{k+1} \binom{k+1}{i} a^{k+1-i} b^i.$$

Since

$$(a+b)^{k+1} = (a+b)^k (a+b),$$

we have

$$\begin{aligned}
(a+b)^{k+1} &= \sum_{i=0}^{k} \binom{k}{i} a^{k-i} b^{i+1} + \sum_{i=0}^{k} \binom{k}{i} a^{k+1-i} b^i \\
&= \sum_{i=1}^{k+1} \binom{k}{i-1} a^{k+1-i} b^i + \sum_{i=0}^{k} \binom{k}{i} a^{k+1-i} b^i \\
&= a^{k+1} + \sum_{i=1}^{k} \left( \binom{k}{i-1} + \binom{k}{i} \right) a^{k+1-i} b^i + b^{k+1} \\
&= \sum_{i=0}^{k+1} \binom{k+1}{i} a^{k+1-i} b^i.
\end{aligned}$$

The last equality follows since

$$\binom{k}{i-1} + \binom{k}{i} = \binom{k+1}{i},$$

which we shall prove in later chapter. □

**Example 2.5** We will use here principle of induction to show that the sum of $k$th powers of first $n$ natural numbers, denoted by $S_k := S_k(n)$, is a polynomial in $n$ of degree $k + 1$, without constant term. Since we have

$$S_k = \sum_{i=1}^{n} i^k = 1^k + 2^k + 3^k + \cdots + n^k,$$

we know that $S_1 = n(n+1)/2$. We can also show that

$$S_2 = \frac{n(n+1)(2n+1)}{6} \quad \text{and} \quad S_3 = \left(\frac{n(n+1)}{2}\right)^2.$$

Notice that $S_1$, $S_2$, and $S_3$ are, respectively, polynomials in $n$ of degree 2, 3, and 4, with no constant term. By the previous example, we also have

$$(i+1)^{k+1} - i^{k+1} = \binom{k+1}{1} i^k + \binom{k+1}{2} i^{k-1} + \cdots + \binom{k+1}{k+1}. \tag{2.1.9}$$

By adding $n$ equations obtained from (2.1.9) by taking $i = 1, 2, \ldots, n$, we get

$$(n+1)^{k+1} - 1 = \binom{k+1}{1} S_k(n) + \binom{k+1}{2} S_{k-1}(n) + \cdots + \binom{k+1}{k+1} S_0(n).$$

Notice that $S_0(n) = n$, and

$$S_k(n) = \frac{1}{k+1}\Bigg((n+1)^{k+1} - 1 - \left(\binom{k+1}{2} S_{k-1}(n) + \cdots + \binom{k+1}{k+1} S_0(n)\right)\Bigg).$$

Therefore, it follows from the principle of induction that $S_k(n)$ is a polynomial in $n$ of degree $k+1$, without constant term. Also, the coefficient of $n^{k+1}$ in the polynomial $S_k(n)$ is $1/(k+1)$.

**Example 2.6** We use induction to show that, for any natural number $n > 0$, we have

$$\left(1 - \frac{1}{4}\right)\left(1 - \frac{1}{9}\right) \cdots \left(1 - \frac{1}{(n+1)^2}\right) = \frac{n+2}{2(n+1)}.$$

Since we have

$$1 - \frac{1}{4} = \frac{1+2}{2(1+1)},$$

the assertion holds for $n = 1$. Next, suppose that for $k > 1$

$$\left(1 - \frac{1}{4}\right)\left(1 - \frac{1}{9}\right) \cdots \left(1 - \frac{1}{(k+1)^2}\right) = \frac{k+2}{2(k+1)}.$$

Then, by the induction hypothesis,

$$\begin{aligned}
&\left(1-\frac{1}{4}\right)\left(1-\frac{1}{9}\right)\cdots\left(1-\frac{1}{(k+2)^2}\right)\\
&=\frac{k+2}{2(k+1)}\left(1-\frac{1}{(k+2)^2}\right)\\
&=\frac{k+2}{2(k+1)}\left(\frac{k^2+4k+3}{(k+2)^2}\right)\\
&=\frac{k+2}{2(k+1)}\left(\frac{(k+1)(k+3)}{(k+2)^2}\right)\\
&=\frac{k+3}{2(k+2)},
\end{aligned}$$

establishing the truth of the assertion for $k+1$.

We next use the principle of induction to prove some *divisibility statements*. We may recall that a nonzero integer $a$ is said to *divide* an integer $b$ if $b$ is an integer multiple of $a$. In this case, we also write $a|b$. We start with the next example.

**Example 2.7** We show that $11^{n+2}+12^{2n+1}$ is divisible by 133, for any integer $n \geq 0$. First, since we have

$$11^{n+2}+12^{2n+1}=11^2+12^1=133,$$

it follows that the assertion holds for $n=0$. Next, suppose that $11^{k+2}+12^{2k+1}$ is divisible by 133. Now, since we have

$$\begin{aligned}
11^{k+3}+12^{2k+3}&=11\times 11^{k+2}+144\times 12^{2k+1}\\
&=11\times(11^{k+2}+12^{2k+1})+133\times 12^{2k+1},
\end{aligned}$$

it follows that $11^{k+3}+12^{2k+3}$ is divisible by 133. Hence, by principle of induction, the assertion holds for all integer $n \geq 0$.

**Example 2.8** We show that $n^3-n$ is divisible by 6, for every positive integer $n$. Since $1^3-1=0$ is divisible by 6, the assertion holds for $n=1$. Next, suppose $m^3-m$ is divisible by 6 so that we have $m^3-m=6\ell$, for some $\ell \geq 1$. Now, since $m$ or $m+1$ is even, $m(m+1)/2$ is an integer. Therefore, we have

$$\begin{aligned}
(m+1)^3-(m+1)&=(m^3+3m^2+3m+1)-(m+1)\\
&=(m^3-m)+(3m^2+3m)\\
&=6\ell+3(m^2+m)\\
&=6k,
\end{aligned}$$

where $k = \ell + m(m+1)/2$ is an integer. This proves that $(m+1)^3 - (m+1)$ is divisible by 6. Hence, the assertion holds for $n = m+1$ as well.

**Example 2.9** We show that $n^3 + (n+1)^3 + (n+2)^3$ is divisible by 9, for every integer $n \geq 0$. When $n = 0$, $0^3 + (0+1)^3 + (0+2)^3 = 9$. So the statement is true in this case. We assume the truth of the result when $n = m \geq 0$ and examine the result for $n = m+1$. We find that

$$\begin{aligned}
&(m+1)^3 + (m+2)^3 + (m+3)^3 \\
&= (m+1)^3 + (m+2)^3 + (m^3 + 9m^2 + 27m + 27) \\
&= (m^3 + (m+1)^3 + (m+2)^3) + (9(m^2 + 3m + 3)),
\end{aligned}$$

where the first summand $m^3 + (m+1)^3 + (m+2)^3$ is divisible by 9 because of induction hypothesis. Consequently, since the result is true for $n = 0$, and since the truth at $n = m$ ($\geq 0$) implies the truth for $n = m+1$, it follows from the principle of induction that the statement is true for all integers $n \geq 0$.

**Example 2.10** We show that $x^{2n-1} + y^{2n-1}$ is divisible by $x + y$, for all integers $n \geq 0$. For $n = 1$, the claim becomes $(x+y)|(x+y)$ and this is obvious. Now, we assume that the result is true for $n = m$:

$$(x+y)|\left(x^{2m-1} + y^{2m-1}\right)$$

and we have to prove that

$$(x+y)|\left(x^{2m+1} + y^{2m+1}\right).$$

Since $(x^{2m-1} + y^{2m-1})$ and $(x^2 - y^2)$ are divisible by $x + y$, it follows that

$$x^{2m+1} + y^{2m+1} = (x^{2m-1} + y^{2m-1})x^2 - y^{2m-1}(x^2 - y^2)$$

is divisible by $x + y$, as asserted.

We can also use principle of induction to verify algorithms. To illustrate, recall that this algorithm evaluates a polynomial.

$$P(x) = a_n x^n + a_{n-1} x^{n-1} + \cdots + a_1 x + a_0.$$

We follow the steps as given below:

**Step-1.** Let $S = a_0$ and $m = 1$.
**Step-2.** If $m \leq n$, then replace $S$ by $S + a_m x^m$ and $m$ by $m + 1$.
**Step-3.** $P(x) = S$.

Let S(n) be a statement: After completing step 2 of the algorithm $n$ times, we have

$$S = a_n x^n + a_{n-1} x^{n-1} + \cdots + a_1 x + a_0.$$

We will prove that $S(n)$ holds for all integers $n \geq 0$.

(a) For $n = 0$ the value of $S$ is the value $a_0$ given in Step-1. Since this equality is the statement $S(0)$, we have $S(0)$ is true.
(b) To perform the inductive step, we assume that $S(n)$ is true for integer $n = m$, and show that $S(n)$ is also true for $n = m + 1$. According to our inductive hypothesis, we have

$$S = a_m x^m + a_{m-1} x^{m-1} + \cdots + a_1 x + a_0,$$

and this is the value that $S$ has after step 2 has been completed $m$ times. It thus follows that, after completing Step-2 one more time ($m + 1$ times in all), the value of $S$ is given by

$$\begin{aligned} S + a_{m+1} x^{m+1} &= (a_m x^m + a_{m-1} x^{m-1} + \cdots + a_1 x + a_0) + a_{m+1} x^{m+1} \\ &= a_{m+1} x^{m+1} + a_m x^m + a_{m-1} x^{m-1} + \cdots + a_1 x + a_0. \end{aligned}$$

Therefore, $S(m + 1)$ is true. This completes the proof of the inductive step. Since $(a)$ and $(b)$ both hold, the principle of induction guarantees that $S(n)$ is the statement that $P(x) = S$. The proof given above also shows that the value of $S$ obtained after competing Step-2 of the algorithm $m$ times is given by

$$S = a_m x^m + a_{m-1} x^{m-1} + \cdots + a_1 x + a_0.$$

Since this relation holds for all iterations of Step-2, it is called a loop invariant.

## *Principle of Induction: Strong Form*

While giving an inductive proof of a statement, a common difficulty is to find an appropriate *induction hypothesis*. The illustrations given below show why it is important to strengthen our notion of "induction hypothesis" to include all non-negative integers $\geq n_0$ (base value) up to $n = m$.

We now give a formal statement of "strong form" of principle of induction.

**Strong Form of Mathematical Induction:** *If* $p(n)$ *is a statement about non-negative integers* $n \geq n_0$ *such that*

$$p(n_0),\ p(n_0 + 1),\ \ldots,\ p(n) \quad \Rightarrow \quad p(n + 1),$$

*then* $p(n)$ *is true, for all* $n \geq n_0$.

As a first application, we prove the "fundamental theorem of arithmetic". Recall that an integer $n > 1$ is prime if its only divisors are 1 and itself; otherwise, it is called a composite number.

**Theorem 2.5** *An integer $n > 1$ is a prime or a product of the form:*

$$n = p_1^{k_1} p_2^{k_2} \cdots p_t^{k_t}, \tag{2.1.10}$$

*where each $p_i$ is a prime divisor of $n$, and the integers $k_i \geq 0$, for $i = 1, \ldots, t$. Moreover, subject to the arrangement $p_1 < p_2 < \cdots < p_t$, the representation (5.5.1) is unique.*

***Proof*** The assertion holds trivially for the prime $n = 2$. Therefore, to complete the proof by using *strong form* principle of induction, we may assume that it holds for every integer $2 \leq k < n + 1$ that is a prime or a product of primes. Now, if $n + 1$ is a prime number, then we are done. Otherwise, suppose we have $n + 1 = ab$, where $a$ and $b$ are integers satisfying the condition $2 \leq a, b < n + 1$. Therefore, by the induction hypothesis, integers $a$ and $b$ are either primes or a product of primes. Hence, in all possible combinations, the product $ab = n + 1$ is again of the desired form. The proof of uniqueness part is left for the reader as an exercise. □

**Definition 2.4** The **Fibonacci Sequence** $\langle F_n \rangle$ is a casual sequence given by the recurrence relation:

$$F_{n+2} = F_n + F_{n+1}, \quad \text{with} \quad F_0 = 0 \text{ and } F_1 = 1. \tag{2.1.11}$$

**Example 2.11** Consider the two real numbers $\alpha$ and $\beta$, respectively, given by

$$\alpha = \frac{1 + \sqrt{5}}{2} \quad \text{and} \quad \beta = \frac{1 - \sqrt{5}}{2},$$

where $\alpha$ is also called the *golden ratio*. We show that an explicit formula for the *Fibonacci sequence* is given by

$$F_n = \frac{1}{\sqrt{5}}\Big(\alpha^n - \beta^n\Big), \qquad \text{for all} \quad n \geq 0. \tag{2.1.12}$$

For instance, since $\beta = -\alpha^{-1}$, we have

$$\alpha^2 = 1 + \alpha \quad \text{and} \quad \beta^2 = 1 + \beta. \tag{2.1.13}$$

Now, since (2.1.14) holds for $n = 0$, 1, we may assume that it holds for all integers $1 \leq n \leq k + 1$. That is, we have

$$F_k = \frac{1}{\sqrt{5}}\big(\alpha^k - \beta^k\big) \text{ and } F_{k+1} = \frac{1}{\sqrt{5}}\big(\alpha^{k+1} - \beta^{k+1}\big). \tag{2.1.14}$$

To complete the proof by strong form of principle of induction, we need to show that the relation (2.1.14) also holds for $n = k + 2$. Notice that it follows from (2.1.13) that

$$\alpha^{k+2} = \alpha^k\big(1+\alpha\big) \quad \text{and} \quad \beta^{k+2} = \beta^k\big(1+\beta\big).$$

We thus have

$$\begin{aligned} F_k + F_{k+1} &= \frac{1}{\sqrt{5}}\big(\alpha^k(1+\alpha) - \beta^k(1+\beta)\big) \\ &= \frac{}{\sqrt{5}}\big(\alpha^{k+2} - \beta^{k+2}\big) \\ &= F_{k+2}. \end{aligned}$$

Hence, (2.1.14) holds for all $n \geq 0$, by the principle of strong induction.

**Example 2.12** We show here that

$$F_{n+k} = F_k F_{n+1} + F_{k-1} F_n, \quad \text{for} \quad k \geq 0 \ \text{and} \ n \geq 1.$$

Clearly, the assertion holds for $n = 0$. Next, assume that it also holds for $n = 1, 2, \ldots, m$. We proceed to prove the assertion for $n = m + 1$. By our (strong) induction hypothesis, we have

$$\begin{aligned} F_{m+k-1} &= F_k F_m + F_{k-1} F_{m-1}, \\ F_{m+k} &= F_k F_{m+1} + F_{k-1} F_m. \end{aligned}$$

We have to show that

$$F_{m+k+1} = F_k F_{m+2} + F_{k-1} F_{m+1}.$$

Now, since we have

$$F_{m+2} = F_m + F_{m+1} \quad \text{and} \quad F_{m+1} = F_{m-1} + F_m,$$

it follows that

$$\begin{aligned} F_k F_{m+2} + F_{k-1} F_{m+1} &= F_k F_{m+1} + F_m F_k + F_{k-1} F_m + F_{k-1} F_{m-1} \\ &= F_{m+k} + F_{m-1+k} \\ &= F_{m+k+1}. \end{aligned}$$

## *Proof by Cases*

This method of *direct proof* of an implication $P \Rightarrow Q$ is based on the idea that if $P = P_1 \vee \cdots \vee P_n$ (exhaustively), then the conclusion $Q$ follows once all the implications $P_i \Rightarrow Q$ are proved separately. It is also known as the *proof by exhaustion*. The name *proof by cases* is justified because each implication $P_i \Rightarrow Q$ is a *case*. The validity of a proof by cases is guaranteed by the logical equivalence as given below:

$$\left[\left(P_1 \vee \cdots \vee P_n\right) \rightarrow Q\right] \quad \Leftrightarrow \quad \left[\left(p_1 \rightarrow Q\right) \vee \cdots \vee \left(p_n \rightarrow Q\right)\right].$$

Clearly, this method applies only when the number of cases to be dealt with are relatively manageable (also see Remark 2.3). In some complex situations, we may also have to use the principle of induction or even a *computer-aided proof*.[4] We start with some simple illustrations as given below.

**Theorem 2.6** *For a real number $x$, we have*

$$\frac{x^2 - 1}{x + 2} > 0 \quad \Rightarrow \quad x > 1 \quad or \quad -2 < x < -1.$$

***Proof*** Clearly, for the quotient in above statement to be positive, we must have that either all the three numbers $x + 1$, $x - 1$, and $x + 2$ be positive or any two of these are negative with the third being positive. Therefore, we need to consider the following four *exhaustive cases*:

**Case-1**: Suppose $x + 1 > 0$, $x - 1 > 0$, and $x + 2 > 0$ so that we have $x > -1$, $x > 1$, and $x > -2$. Therefore, in this case, we obtain $x > 1$.

**Case-2**: Suppose $x + 1 < 0$, $x - 1 < 0$, and $x + 2 > 0$ so that we have $x < -1$, $x < 1$, and $x > -2$. Therefore, in this case, we obtain $-2 < x < -1$.

**Case-3**: Suppose $x + 1 > 0$, $x - 1 < 0$, and $x + 2 < 0$ so that we have $x > -1$, $x < 1$, and $x < -2$. Clearly, in this case, there is no real number $x$ satisfying the above inequalities.

**Case-4**: Suppose $x + 1 < 0$, $x - 1 > 0$, and $x + 2 < 0$ so that we have $x < -1$, $x > 1$, and $x < -2$. Again, in this case, there is no real number $x$ satisfying the above inequalities.

Hence, it follows that the quotient is positive only when $x > 1$ or $-2 < x < -1$. □

**Theorem 2.7** *For all non-negative integers $m > n$, we have*

$$m^2 - n^2 \not\equiv 2 \pmod{4}.$$

---

[4] In artificial intelligence, an **expert system** refers to a program specifically designed to emulate the decision-making ability of a human on the basis of reasoning through bodies of knowledge, which help solve complex problems represented as an implication.

***Proof*** For, since we have

$$m^2 - n^2 = \big(m+n\big)\big(m-n\big),$$

we need to consider the following three *exhaustive cases*:

**Case-1**: When both $m$ and $n$ are even, so $m^2 - n^2$ as the product of two even integers $(m+n)$ and $(m-n)$. Therefore, in this case, division by 4 gives remainder 0.

**Case-2**: When both $m$ and $n$ are odd. In this case, as $(m+n)$ and $(m-n)$ are even integers, $m^2 - n^2$ is also an even integers. Therefore, division by 4 gives remainder 0.

**Case-3**: When $m$ and $n$ have opposite parity. In this case, both $(m+n)$ and $(m-n)$ are odd integers, so that their product $m^2 - n^2$ is also an odd integer. Therefore, on division by 4, it cannot give remainder 2.

Hence, we have shown that assertion holds in all cases. □

**Theorem 2.8** *For any integer n, the* cube number $n^3$ *is a multiple of* 9 *or one more than a multiple of* 9 *or one less than a multiple of* 9.

***Proof*** In this case, we cannot use parity of an integer. However, using the fact that every integer is a multiple of 3 or one more than a multiple of 3 or one less than a multiple of 3, we only need to consider the following three *mutually exhaustive* cases:

**Case-1**: When both $n = 3k$, for some integer $k$. In this case, we have $n^3 = 27k^2$ is a multiple of 9.

**Case-2**: When both $n = 3k + 1$, for some integer $k$. In this case, we have

$$n^3 = 27k^3 + 27k^2 + 9k + 1 = 9\big(3k^3 + 3k^2 + k\big) + 1,$$

which is 1 more than a multiple of 9.

**Case-3**: When both $n = 3k - 1$, for some integer $k$. In this case, we have

$$n^3 = 27k^3 - 27k^2 + 9k - 1 = 9\big(3k^3 - 3k^2 + k\big) - 1,$$

which is 1 less than a multiple of 9.

Hence, we have shown that assertion holds in all cases. □

**Theorem 2.9** *For any* $n \in \mathbb{N}$, *the number*

$$1 + (-1)^n\big(2n - 1\big)$$

*is a multiple of* 4.

It may be little tedious in some extreme situations to apply *proof by cases*. For example, suppose we use it to prove that $n^2 - n + 41$ is a prime number, for every natural number $n < 41$. Notice that, since $n^2 - n + 41$ is not a prime number for $n = 41$, there cannot be an argument that works in general. As $n$ increases, the procedure gets difficult in the sense that it may not be easy to *test the primality* of the number $n^2 - n + 41$. For example, when $n = 40$, the simplest way to test the primality of the number 1601 is to show that no prime $\leq \sqrt{1601}$ divides 1601.

**Remark 2.3** In addition to being less elegant, there is no upper limit on the number of cases that one may need while using *proof by cases*. However, for some important theorems such as related to *order of finite projective plane, classification of finite simple groups, Boolean Pythagorean triples problem*, no alternative proof is known till date. In some texts such as [4], an *exhaustive proof* is treated as a special type of *proof by cases*. However, no such distinction is made in this book.

## 2.2 Indirect Proof

We can also prove an implication $P \Rightarrow Q$ by using a method of *indirect proof* such as the *method of contrapositive* or *reductio-ad-absurdum* meaning "reduction to absurdity". In the former case, we prove the contrapositive statement $\neg Q \Rightarrow \neg P$ by applying some method of direct proof. The latter method is used when we fail to find a direct proof of a statement or its contrapositive. In this case, we take $P$ and $\neg Q$ both to be true, and try to derive a contradiction. It is also known as the method of *proof by contradiction*. A yet another interesting special case of the method of *non-constructive* indirect proof is known as *pigeon-hole principle*, which we shall discuss later in Sect. 2.3.

### *Proof by Contrapositive*

In general, when we don't find a direct proof of an implication $P \Rightarrow Q$, a *proof by contrapositive* is a preferred choice. The validity of the method is guaranteed by the following equivalence:

$$P \Rightarrow Q \equiv \neg Q \Rightarrow \neg P.$$

There are cases when it is more convenient to prove the contrapositive of an implication. To start with, we prove a statement that is analogous to Theorem 2.1.

**Theorem 2.10** *For any two integers $a$ and $b$, the product $ab$ is even implies $a$ or $b$ is even.*

***Proof*** We prove the *contrapositive statement*: If $a$ and $b$ are two odd, then the product $ab$ is odd. For instance, we can write

$$a = 2m + 1 \quad \text{and} \quad b = 2n + 1, \quad \text{for some} \quad m, n \in \mathbb{Z}.$$

It follows from the associative and commutative of addition, and also the distributive property, that

$$ab = \big(2m + 1\big)\big(2n + 1\big) = 2\big(2mn + m + n\big) + 1,$$

which is an odd integer. This completes the proof. □

By using an argument as above, we can show that if $x^2$ is an even integer then so is $x$. Another simple illustration is as given in the next theorem.

**Theorem 2.11** *For an integer $n$, we have $n^2$ is even implies $n$ is even.*

## *Proof by Contradiction*

A proof of an implication $P \Rightarrow Q$ by contradiction is an interesting method of *indirect proof*. In this case, we assume $\neg Q$ is also true, and try to derive a contradiction to $P$ or some other previously known theorem. The validity of *proof by contradiction* follows from the following equivalence:

$$\neg(P \wedge \neg Q) \equiv \big(P \Rightarrow Q\big).$$

More precisely, if we show that $P \wedge \neg Q$ is false, then $\neg(P \wedge \neg Q)$ is true so that the proposition $P \Rightarrow Q$ is true too. It is also known as the method of *reduction to absurdity* (or *reductio-ad-absurdum*, in Latin). For example, the indirect proof given in the next theorem is called the *method of contradiction*. In some cases, the only known proof is by way of contradiction. We first use this method to prove the fundamental result of Euclid given in next theorem is a classical application of *proof by contradiction*.

**Theorem 2.12** (Euclid) *There are infinitely many prime numbers.*

***Proof*** On the contrary, suppose there are only finitely many prime numbers, say $p_1, p_2, \ldots, p_n$. Consider the integer

$$N = p_1 p_2 \cdots p_n + 1.$$

Clearly, $N > 1$. However, none of $n$ prime numbers $p_i$ divides $N$, which contradicts Theorem 8.45. This completes the proof. □

We next prove the well-known fact that $\sqrt{2}$ is an irrational number.

**Theorem 2.13** *$\sqrt{2}$ is not a rational number.*

***Proof*** If possible, suppose $\sqrt{2}$ is a rational number. That is, we may assume that

$$\sqrt{2} = \frac{p}{q}, \quad \text{with } p, q \in \mathbb{Z},\ q \neq 0,$$

such that $p$ and $q$ have no common factor. Squaring both sides of the above equation, we obtain $p^2 = 2q^2$. Thus, $p^2$ is even, and so $p$ is even (Theorem 2.11). Let $p = 2k$, for some integer $k$. Therefore,

$$p^2 = 2q^2 \quad \Rightarrow \quad 4k^2 = 2q^2 \quad \Rightarrow \quad q^2 = 2k^2.$$

Once again, we can conclude that $q$ is even, which contradicts the assumption that $p$ and $q$ have no common factor. Hence, $\sqrt{2}$ is an irrational number. □

## 2.3 Pigeon-Hole Principle

A typical idea that comes very handy in many situations is called the *pigeon-hole principle*, which is also known as the *Dirichlet box principle*.

**Theorem 2.14** (Pigeon-hole Principle) *If $n + 1$ identical objects (pigeons) are to be placed into $n$ identical boxes (pigeon holes), then there will be at least one box that contains more than one object.*

***Proof*** It must be intuitively obvious that *if there are more pigeons than the number of pigeon holes then at least one pigeon hole must accommodate more than one pigeon*. However, for a contradiction, suppose no box contains more than one object. That is, each box has zero or one object in it. Therefore, the total number of objects in $n$ boxes is at most $n$. However, as we are given that $n + 1$ objects are placed in $n$ boxes, we got our contradiction. Hence, at least one box contains more than one object. □

More generally, if $nq + 1$ objects are placed into $n$ boxes, then at least one box contains more than $q$ objects. The next theorem is a straightforward generalization of Theorem 2.14.

**Theorem 2.15** (Generalized Pigeon-hole Principle) *Let $m$ and $n$ be positive integers such that $m > n$. If $m$ objects are placed into $n$ boxes then at least one box contains at least $\lceil m/n \rceil$ objects, where $\lceil x \rceil$ denotes the smallest integer $\geq x$.*

***Proof*** By *division algorithm*, we can find positive integers $q, r$ such that

$$m = qn + r, \qquad \text{with } 0 \leq r < n.$$

Suppose no box contains $\lceil m/n \rceil$ objects, and let us consider the following two cases:

1. When $r = 0$ so that $\lceil m/n \rceil = m/n = q$, we have that every box contains less than $q$ objects and hence there are less than $nq = m$ objects in total in the $n$ boxes, which is a contradiction.
2. When $1 \leq r < n$ so that $\lceil m/n \rceil = q + 1$, every box contains at most $q$ objects, and hence $nq = m - r < m$ objects in total, which again give a contradiction. □

**Example 2.13** This example illustrates the principle in three different situations.

1. Consider the 7 days of a week as boxes and people as objects. Therefore, by pigeon-hole principle, there should be at least two persons among a group of eight who have their birthdays on the same day of the week.
2. By the same argument, it follows that among a group of 22 persons at least 4 have their birthdays on the same day of the week. It can be shown by using Theorem 2.15 with $m = 22$ and $n = 7$.
3. We next find the least number students in a class so that 6 students obtain the same grade A, B, C, D, or F. For example, consider the grades as boxes and students as objects. Therefore, for $n = 5$, we need to find the smallest $m$ such that $\lceil m/5 \rceil = 6$. Clearly, such a lowest possible value of $m$ is 26. Hence, if there are 26 students, then 6 of them will get the same grade.

It is also possible to formulate the *pigeon-hole principle* in terms of sets and functions. For, recall that the *cardinality* of a set $A$ is the number of elements the set $A$ contains, which we denote by $|A|$. We need the following well-known implications to prove the next theorem:

$$\begin{aligned} f : S \to T \quad \text{is one-one} \quad &\Rightarrow \quad |S| \leq |T|; \\ f : S \to T \quad \text{is onto} \quad &\Rightarrow \quad |S| \geq |T|. \end{aligned}$$

**Theorem 2.16** *Let $S$ and $T$ be finite sets. For a function $f : S \to T$, if any of the following two conditions hold then so does the third:*

1. *$f$ is one-to-one.*
2. *$f$ is onto.*
3. *$|S| = |T|$.*

***Proof*** We first take both (1) and (2) hold so that a function $f : S \to T$ is both one-one and onto. It then follows from the above remark that $|S| \leq |T|$ and $|S| \geq |T|$. That is, $|S| = |T|$, which proves (3). Next, suppose both (1) and (3) hold so that a function $f : S \to T$ is one-one, and we also have $|S| = |T|$. Now, using the fact that the range $R(f)$ of $f$ is a subset of the set $T$, it follows that

$$|T| = |S| = |R(f)| \leq |T|.$$

Notice that the second equality holds because $f$ is one-one. Therefore, we have $|R(f)| = |T|$, which proves that $R(f) = T$ because $R(f) \subset T$. Hence, $f$ is onto, which proves (2). Finally, we show that if both (2) and (3) hold, then (1) holds. For, let $f$ be onto, and $|S| = |T|$. If $f$ is not one-one, then we have

$$|S| > |R(f)| \quad \Rightarrow \quad |R(f)| < |T| \quad \Rightarrow \quad f \text{ is not onto.}$$

This contradiction proves that $f$ has to one-one. □

We discuss now some interesting combinatorial situations illustrating applications of the above-stated versions of the pigeon-hole principle. The next three uses the concept of *divisibility*.

**Example 2.14** We show here, in any set of $n$ consecutive positive integers, there is an integer divisible by $n$. To use Theorem 2.16, consider the remainders of the $n$ consecutive integers when divided by $n$, which are necessarily one among the integers $0, 1, \ldots, n-1$, in some order. Now, as we have $n$ positive integers and the number of distinct remainders is also $n$, some integer must has the remainder 0. Therefore, we obtained the integer that is divisible by $n$.

**Example 2.15** We show here, in any set of $n+1$ positive integers, there are two integers such that their difference is divisible by $n$. For, consider the remainders of the $n+1$ numbers when divided by $n$. Since there are only $n$ possible remainders (namely, $0, 1, \ldots, n-1$), at least two of the $n+1$ numbers must have the same remainder, and hence their difference is divisible by $n$. In this example, the remainders are the pigeon holes and the given $n+1$ numbers are the pigeons.

**Example 2.16** We show here, in any set of $n+2$ positive integers, there are two integers such that their sum or difference is divisible by $2n$. If any two of the n + 2 given integers have the same remainder when divided by $2n$, then their difference is divisible by $2n$ and the result follows. If not, all the numbers have distinct remainders when divided by $2n$. Group these remainders into $n+1$ classes:

$$\{0\}, \{1, 2n-1\}, \{2, 2n-2\}, \ldots, \{n-1, n+1\}, \{n\}.$$

Since there are $n+2$ numbers, at least one of these classes must contain two remainders, and so the two numbers involved must sum to a multiple of $2n$.

**Example 2.17** We show here, in any set of $3n$ consecutive natural numbers, any subset of $n+1$ of these numbers has two members that differ by at most 2. Let the $3n$ consecutive numbers be $a, a+1, \ldots, a+3n-1$. Each of the $n+1$ numbers in the given subset belongs to one of the $n$ sets

$$\{a, a+1, a+2\}, \{a+3, a+4, a+5\}, \{a+6, a+7, a+8\},$$

$$\ldots, \{a+3n-3, a+3n-2, a+3n-1\}.$$

Consider $n$ boxes each one of them labeled by the three numbers in the $n$ sets above. Consider the $n + 1$ numbers as pigeons; place them into the boxes labeled by the remainder that you get when dividing the number by $n$. By the pigeon-hole principle, one of these sets has two members of the subset and these two numbers differ by at most 2.

**Example 2.18** We show here, if $a_1, a_2, \ldots, a_n$ are integers, then $n$ divides

$$a_i + a_{i+1} + \cdots + a_{i+k}, \qquad \text{for some} \quad i \geq 1 \quad \text{and} \quad k \geq 0.$$

For instance, consider the remainders of the $n$ numbers

$$a_1, a_1 + a_2, a_1 + a_2 + a_3, \ldots, a_1 + a_2 + a_3 + \cdots + a_n$$

when divided by $n$. If any one of them has remainder zero, the result follows with $i = 1$ and for some $k$. If not, there are only $n - 1$ possible remainders (namely, 1, 2 $\ldots, n - 1$). Since there are $n$ numbers and $n - 1$ remainders, there are at least two of the $n$ numbers

$$a_1, a_1 + a_2, a_1 + a_2 + a_3, \ldots, a_1 + a_2 + a_3 + \cdots + a_n$$

having the same remainder, and hence their difference is divisible by $n$. Thus

$$a_i + a_{i+1} + \cdots + a_{i+k}$$

is divisible by $n$ for some $i \geq 1$ and $k \geq 0$.

**Example 2.19** We show here, for any arbitrary positive integer $N$, there is a multiple of $N$ that contains only the given digits $d$ and 0. Before proving the general statement, let us see an illustration. For instance, take $d = 7$ and $N = 4$, and consider the integers $7, 77, 777, 7777$. These 4 integers when divided by 4 give the remainders 3, 1, 1, 1, respectively. If any one of these had remainder 0, then the same could have been the integer we wanted (with only digit 7). Now there are three integers (though two is enough) having the same remainders. The difference of 777 and 77 is 700. Therefore, the number 700 is the required integer. Notice that it is a multiple of 4, and it only has digits 7 and 0. By similar reasoning, the other two solutions are 7700 and 7000. We now turn to the proof of the general case. For instance, consider $1, 2, 3, \ldots, N$ digit integers containing only the digit $d$. If we divide these $N$ integers by $N$, then either there is an integer that is divisible by $N$ or there are two whose difference is divisible by $N$. In the former case, the integer that is divisible by $N$ is the required integer (and it contains only the digit $d$). In the later case, the difference is the required integer and it has digits $d$ and 0 only and is also a multiple of $N$.

We may recall that a *sequence* of real numbers is a function $x : \mathbb{N} \to \mathbb{R}$. We may write $x(n) = x_n$, for $n \in \mathbb{N}$, so that a sequence $x$ may also be written as

$$\langle x_1, x_2, \dots \rangle \quad \text{or simply} \quad \langle x_n \rangle.$$

We say $\langle x_n \rangle$ is called a *non-decreasing* sequence if

$$x_1 \le x_2 \le x_3 \le \dots$$

Similarly, $\langle x_n \rangle$ is called a *non-increasing* sequence if

$$x_1 \ge x_2 \ge x_3 \ge \dots$$

Some simple examples of non-decreasing subsequences of length 4 of the sequence $3, 7, 2, 5, 8, 4, 6, 1, 9$ are the sequences (i) $2, 5, 8, 9$; (ii) $3, 7, 8, 9$; (iii) $2, 4, 6, 9$. A sequence obtained by picking some terms of a sequence $\langle x_n \rangle$ in the same order is called a *subsequence* of the sequence. In particular, for $[n] = \{1, 2, \dots, n\}$, a function $f : [n] \to \mathbb{R}$ is called a *finite sequence* of *length* $n$. A sequence of length $n$ is usually written as $x_1, x_2, \dots, x_n$, where $x_i$ is called $i$th term of the sequence. Some subsequences of the finite sequence $a, b, c, d, e, f, g$ are as given below: (i) $b, e, f, g$; (ii) $a, c, d, e, g$; (iii) $a, d, g$. The next theorem is due to *Paul Erdős* (1913–1996) and *George Szekeres* (1911–2005) [5].

**Theorem 2.17** (Erdős-Szekeres, 1935) *Every sequence of $n^2 + 1$ distinct real numbers contains a non-decreasing or non-increasing subsequence of length $n + 1$.*

***Proof*** Given a sequence $a_1$, ..., $a_{n^2+1}$, for each $a_j$ let $d_j$ and $i_j$ be the lengths of the longest non-increasing and non-decreasing subsequences beginning with $a_j$. This gives a sequence of $n^2 + 1$ ordered pairs $(d_j, i_j)$. If there were no non-decreasing or non-increasing subsequence of length $n + 1$, then there are only $n^2$ possible ordered pairs $(d_j, i_j)$, since $1 \le d_j \le n$ and $1 \le i_j \le n$. By the pigeon-hole principle, at least two ordered pairs must be identical. Hence, there are $p$ and $q$ such that $d_p = d_q$ and $i_p = i_q$. If $a_p < a_q$, then the sequence $a_p$ followed by the non-decreasing subsequence starting at $a_q$ gives a non-decreasing subsequence of length greater than $i_q$, which is a contradiction. A similar contradiction on the choice of $d_p$ follows if $a_q < a_p$. Hence, a non-increasing or non-decreasing subsequence of length $n + 1$ must exist. □

**Theorem 2.18** (General Pigeon-hole Principle) *If $p_1 + p_2 + \dots + p_n + 1$ or more pigeons are put into $n$ pigeon holes numbered 1 through $n$, then for some $r$, pigeon hole $r$ has $p_r + 1$ or more pigeons.*

***Proof*** If there is no pigeon hole with the required property, then, for every $r$, the pigeon hole $r$ has $p_r$ or fewer pigeons in it. Thus there are no more than $p_1 + p_2 + \dots + p_n$ pigeons in the $n$ pigeon holes, a contradiction. □

**Example 2.20** Given any three positive integers whose sum is 16, we show that one of them should be at least 6. Let $a$, $b$, and $c$ be the three positive integers such that $a + b + c = 16$. Consider three pigeon holes and place 16 pigeons in it; place $a$ pigeon in first, $b$ in the second, and $c$ in the third pigeon hole. By pigeon-hole principle, there is a pigeon hole with at least $\lceil 16/3 \rceil = 6$ pigeons it. In other words, one of the three numbers is at least 6.

**Example 2.21** The circumference of a *roulette wheel* is divided into 36 sectors such that each sector is assigned numbers from $1, 2, \ldots, 36$ in some order. We show that there are three consecutive sectors such that the sum of their assigned numbers is at least 56. Let $a_i$ $(i = 1, 2, \ldots, 36)$ be the numbers assigned to the sectors of the roulette wheel. Then $1 \leq a_i \leq 36$ and

$$a_1 + a_2 + \cdots + a_{36} = 1 + 2 + \cdots + 36 = 36 \times 37/2 = 666.$$

The following are the sum of the numbers assigned to three consecutive sectors: $a_1 + a_2 + a_3$, $a_2 + a_3 + a_4$, $\ldots$, $a_{35} + a_{36} + a_1$, $a_{36} + a_1 + a_2$. There are 36 such sums and the sum of all these sums is $3 \times 666 = 1998$. We need to show that one of these sums is at least 56. Assume that each one of the sums $a_1 + a_2 + a_3$, $a_2 + a_3 + a_4$, $\ldots$, $a_{35} + a_{36} + a_1$, $a_{36} + a_1 + a_2$ is a pigeon hole and that we need to distribute 1998 pigeons in them. Then there is one pigeon hole with at least $\lceil 1998/36 \rceil = 56$ pigeons. In other words, there is one sum whose value is at least 56.

A *constructive proof* of an existential statement of the form $\exists x\big[P(x)\big]$, for some $x$ in the universe of discourse. For example, to show that there exists a triplet of positive integers[5] $(a, b, c)$ such that $a^2 + b^2 = c^2$, a constructive approach is to prove the statement by taking $a = 3$, $b = 4$, and $c = 5$. In general, a constructive proof amounts to finding some explicit values in the universe of discourse for which a given statement holds.

**Remark 2.4** In some cases, it is difficult to construct an example to support an existential statement. Therefore, one may also try to give a proof by contradiction. That is, we assume no such an example exists, and try to arrive at a contradiction. If not successful, the existence is established without having to produce any specific example.

## Exercises 2

2.1 Show that, for any even integer $a$, we have $a^2 - 6a + 5$ is an odd integer.
2.2 Let $x, y \in \mathbb{R}^+$ such that $x \leq y$. Show that $\sqrt{x} \leq \sqrt{y}$.
2.3 Let $x, y \in \mathbb{R}^+$. Show that $2\sqrt{xy} \leq x + y$.

[5] Such triplet of positive integers is known as the *Pythagorean numbers*.

2.4 Let $p$ be a prime number, and $k$ be an integer such that $0 < k < p$. Show that $p|\binom{p}{k}$.

2.5 Let $2 \leq n \in \mathbb{N}$. Show that $(n-1)$ numbers given by

$$n! + 2, \quad n! + 3, \quad n! + 4, \ \ldots, n! + n$$

are composite numbers.

2.6 Show that every odd integer is a difference of two squares.

2.7 Let $S_k$ denote the sum of $k$th powers of first $n$ natural numbers. Prove that each one of the following statements holds for all $n \geq 1$:

a. $S_2 = \dfrac{n(n+1)(n+2)}{6}$.

b. $S_3 = \left[\dfrac{n(n+1)}{2}\right]^2$.

c. $S_4 = \dfrac{n(n+1)(2n+1)(3n^2+3n-1)}{30}$.

2.8 Show that, for any non-negative integer $n$ and $a, b \in \mathbb{R}^+$, we have

$$a^n + b^n \geq \left(\frac{a+b}{2}\right)^n.$$

2.9 Show that $\sum_{i=1}^{n} i^2 \leq n^3$, for all positive integer $n$.

2.10 Show that, whenever $n$ is a positive integer greater than 1, we have

$$1 + \frac{1}{4} + \frac{1}{9} + \cdots + \frac{1}{n^2} < 2 - \frac{1}{n}.$$

2.11 Prove the following inequalities:

a. $2^n > n^3$ for $n \geq 10$ (Hint: $2 > (1 + 1/10)^3 \geq (1 + 1/k)^3$ for $k \geq 10$.).
b. $2^n > n^2$ for $n \geq 4$.
c. $n^3 < 3^n$ for $n \geq 4$.
d. $2^n \geq 3n^2 + 5$ for $n \geq 8$.
e. $(1 + 1/3)^n \geq (1 + n/3)$ for $n \geq 0$.
f. $2^n < n!$ for $n \geq 4$.
g. $3^n < n!$ for $n \geq 6$.
h. $n! < n^n$ for $n \geq 1$.
i. $F_n < 2^n$ for $n \geq 0$ where $F_n$ is the $n$th Fibonacci number.
j. $F_n \geq (3/2)^{n-2}$, $n \geq 2$.
k. $F_n = F_r F_{n-r-1} + F_{r+1} F_{n-r}$ for $0 \leq r < n$.
l. $H_{2^n} \geq 1 + \frac{n}{2}$ for $n \geq 1$
m. $H_{2^n} \leq 1 + n$ for $n \geq 1$.

2.12 Show that

$$\sqrt{\frac{\frac{5}{4}}{4n+1}} \leq \frac{1\cdot 3\cdot 5\cdots(2n-1)}{2\cdot 4\cdot 6\cdots 2n} \leq \sqrt{\frac{\frac{3}{4}}{2n+1}}, \quad \text{for} \quad n \geq 1. \tag{2.3.1}$$

2.13 Prove the following when $n \geq 1$:

a. $1 + a + a^2 + \cdots + a^n = \frac{a^{n+1}-1}{a-1}$ where $a \neq 1$.
b. $1.1! + 2.2! + 3.3! + \cdots + n.n! = (n+1)! - 1$.
c. $1.2 + 2.3 + 3.4 + \cdots + n(n+1) = \frac{n(n+1)(n+2)}{3}$.
d. $1.2.3 + 2.3.4 + \cdots + n(n+1)(n+2) = \frac{n(n+1)(n+2)(n+3)}{4}$
e. $\frac{k!}{0!} + \frac{(k+1)!}{1!} + \frac{(k+2)!}{2!} + \cdots + \frac{(n+k-1)!}{(n-1)!} = \frac{(n+k)!}{(n-1)!(k+1)}$.
f. $1^2 - 2^2 + 3^2 - 4^2 + \cdots + (-1)^n n^2 = (-)^{n-1}\frac{n(n+1)}{2}$.
g. $1^2 + 3^2 + 5^2 + \cdots + (2n-1)^2 = \frac{n(2n-1)(2n+1)}{3}$.
h. $\frac{1^2}{1.3} + \frac{2^2}{3.5} + \cdots + \frac{n^2}{(2n-1)(2n+1)} = \frac{n(n+1)}{2(2n+1)}$.
i. $\frac{1}{1.2} + \frac{1}{2.3} + \cdots + \frac{1}{n(n+1)} = \frac{n}{n+1}$
j. $\frac{1}{1.3} + \frac{1}{3.5} + \cdots + \frac{1}{(2n-1)(2n+1)} = \frac{n}{2n+1}$.
k. $\sum_{i=1}^{n} \frac{1}{[k(i-1)+1][ki+1]} = \frac{n}{kn+1}$.
l. $\sum_{i=1}^{n} i2^i = (n-1)2^{n+1} + 2$.

2.14 For every positive integer $n$, prove that

$$1 + 3 + 5 + \cdots + (2n-1) = n^2.$$

2.15 Let $\langle H_n \rangle$ be the sequence of harmonic numbers. Show that

a. $H_1 + H_2 + \cdots + H_n = (n+1)H_n - n$;
b. $H_1^3 + H_2^3 + \cdots + H_n^3 = (n+1)H_n^3 - 3(n+\frac{1}{2})H_n^2 + 3(2n+1)H_n - 6n + \frac{1}{2}(\frac{1}{1^2} + \frac{1}{2^2} + \cdots + \frac{1}{n^2})$.

2.16 Suppose we have weights of units $1, 3, 3^2, \ldots, 3^{n-1}$. Show that, using a beam balance only once, we can weigh from 1 to $(3^n - 1)/2$ by using the given weights.
2.17 Prove that the sum of interior angles of a convex polygon with $n$ sides ($n \geq 3$) is $(n-2)\pi$.
2.18 Show that every number greater than or equal to 6 is a sum of a non-negative integral multiple of 3 and a non-negative integral multiple of 4.
2.19 Prove that, for all $n \geq 1$, we have

$$3 + 33 + \cdots + 33\cdots 3\,(n \text{ times}) = \frac{10^{n+1} - 9n - 10}{27}.$$

2.20 Prove that, for all $n \geq 1$, we have

$$\frac{1}{1 \cdot 3} + \frac{1}{3 \cdot 5} + \cdots + \frac{1}{(2n-1)(2n+1)} = \frac{n}{2n+1}.$$

2.21 Prove that, for all $n \geq 1$, we have

$$\frac{n^3}{3} + \frac{n^5}{5} + \frac{7n}{15} \quad \text{is a natural number.}$$

2.22 Prove the following when $n \geq 0$:

a. 2 divides $n^2 + n$.
b. 3 divides (i) $n^3 + 2n$, (ii) $n^4 - 4n^2$, (iii) $2^{2n} - 1$.
c. 4 divides $6^n - 2^n$.
d. 5 divides (i) $3^{2n} + 4^{n+1}$, (ii) $n^5 - n$.
e. 6 divides $n^3 - n$.
f. 7 divides $2^{n+2} + 3^{2n+1}$.
g. 8 divides $(2n-1)^2 - 1$ for $n \geq 1$.
h. 9 divides $n^3 + (n+1)^3 + (n+2)^3$.
i. 21 divides $4^{n+1} + 5^{2n-1}$.
j. 24 divides $5^{2n} - 1$.

2.23 For every integer $n \geq 0$, the number $4^{2n+1} + 3^{n+2}$ is a multiple of 13.
2.24 Prove the following identities satisfied by Fibonacci numbers:

a. $\sum_{i=0}^{n} F_i = F_{n+2} - 1 \quad (n \geq 0)$.
b. $\sum_{i=0}^{n} (-1)^i F_i = (-1)^n F_{n-1} - 1 \quad (n \geq 1)$.
c. $\sum_{i=1}^{n} F_{2i-1} = F_{2n} \quad (n \geq 1)$.
d. $\sum_{i=0}^{n} F_{2i} = F_{2n+1} - 1 \quad (n \geq 0)$.
e. $\sum_{i=0}^{n} F_i^2 = F_n F_{n+1} \quad (n \geq 0)$.
f. $\sum_{i=1}^{2n-1} F_i F_{i+1} = F_{2n}^2 \quad (n \geq 1)$.
g. $\sum_{i=1}^{2n} F_i F_{i+1} = F_{2n+1}^2 - 1 \quad (n \geq 1)$.
h. $F_{n+1} F_{n-1} - F_n^2 = (-1)^n \quad (n \geq 1)$.
i. $F_{n+1}^2 + F_n^2 = F_{2n+1} \quad (n \geq 0)$.

2.25 For every integer $n \geq 0$, prove that $2^{2n+1} + 1$ is divisible by 3.
2.26 Show that

$$\frac{n^7}{7} + \frac{n^3}{3} + \frac{11^n}{21}$$

is an integer, for every integer $n \geq 1$.
2.27 Show that 57 divides $7^{n+2} + 8^{2n+1}$, for $n \in \mathbb{N}$.
2.28 Prove that $x^n - y^n$ is divisible by $x - y$, for all integers $n \geq 0$.
2.29 Show that the number of strings over the set $\{0, 1\}$ of length $n$ that contain an even number of ones is $2^{n-1}$.

2.30 Suppose each one of $n > 1$ persons in a room shook hand with every other person. Show that $n(n-1)/2$ hand shakes took place.
2.31 Suppose $n \geq 3$ lines in plane are such that no two of which are parallel, and no three of which pass through a point. Prove that such $n$ lines cut the plane into $(n^2 + n + 2)$ regions.
2.32 Show that a $n \times n$ chessboard with one square removed can be completely covered using $L$-shaped pieces if $n > 5$ is odd and is not divisible by 3.
2.33 Prove that the number of subsets of a finite set with $n$ elements is $2^n$.
2.34 Define the disjunction and conjunction of $n$ propositions. State and prove the corresponding De Morgan's Laws.
2.35 Prove that the $n$th derivative of $\sin x$ is $\sin(n\pi/2 + x)$.
2.36 Given any 17 positive integers, show that there are five whose sum is divisible by 5.
2.37 If 51 numbers are chosen from $1, 2, 3, ..., 99, 100$, show that there are two which do not have any common prime divisor.
2.38 Show that among any nine distinct real numbers, there are two, say $a$ and $b$, such that

$$0 < (a-b)/(1+ab) < \sqrt{2} - 1.$$

2.39 A circle is divided into 10 sectors and these are numbered by the integers 1 to 10 in random order. Show that there are three consecutive sectors whose assigned numbers add to at least 17.
2.40 A chess player wants to prepare for a championship match by playing some practice games in 30 days. He wants to play at least one game a day but no more than 45 games altogether. Show that no matter how he schedules the games there is period of consecutive days within which he plays exactly 14 games.
2.41 Suppose a player practices on $d$ consecutive days, playing at least one game a day and a total of no more than $b$ games where $d < b < 2d$. Prove that, for each $i < 2d - b - 1$, there is a block of consecutive days on which, in total, exactly $i$ games are played.
2.42 Consider an equilateral triangle whose sides have length 6. Show that, for any given 10 points inside the triangle, there are at least two of them within a distance of 2 from each other.
2.43 Among six persons, either there are three mutual friends or there are three mutual strangers.
2.44 Show that every sequence of $mn + 1$ distinct real numbers must have an increasing of length $m + 1$ or a decreasing subsequence of length $n + 1$.
2.45 Consider a square of side length 1 and choose 51 points from this square at random. Show that three of these points can be covered by a circle of radius 1/7.
2.46 Suppose a triangle can be placed inside a square of unit area in such a way that the center of the square is not inside the triangle. Show that one side of the triangle has length less than 1.
2.47 In every 16-digit number, show that there is a string of one or more consecutive digits such that the product of these digits is a perfect square.

2.48 Show that, in any set of $n + 1$ positive integers with each being $\leq 2n$, there are at least two such integers that one is a multiple of the other.

2.49 A man hiked for 10 h and covered a total distance of 45 km. It is known that he hiked 6 km in the first hour and only 3 km in the last hour. Show that he must have hiked at least 9 km within a certain period of 2 consecutive hours.

2.50 A chess player wants to prepare for a championship match by playing some practice games in 77 days. He wants to play at least one game a day but no more than 132 games altogether. Show that no matter how he schedules the games there is a period of consecutive days within which he plays exactly 21 games.

2.51 Let $m$ be an odd positive integer. Prove that there exists a positive integer $n$ such that $m$ divides $2^n - 1$.

2.52 Given five points inside a square of side length 2, show that there are two of them within a distance of $\sqrt{2}$.

2.53 Given five points in an equilateral triangle of side length 1, show that there are two points whose distance apart is at most 1/2.

2.54 Given $n^2 + 1$ points in an equilateral triangle of side length 1, show that there are two points whose distance apart is at most $1/n$.

2.55 Given 27 distinct positive odd integers all less than 100, show that there is a pair of integers whose sum is 102.

## References

1. Devlin, K.: Sets, Functions, and Logic: An Introduction to Abstract Mathematics (3/e). Chapman & Hall/CRC (2003)
2. Nissanke, N.: Introductory Logic and Sets for Computer Scientists. Addison-Wesley (1999)
3. Stanat, D.F., McAllister, D.F.: Discrete Mathematics in Computer Science. Prentice-Hall (1977)
4. Rosen, K.H.: Discrete Mathematics and its Applications with Combinatorics and Graph Theory (8/e). McGraw Hill, UK (2018)
5. Erdős, P., Szekeres, G.: A combinatorial problem in geometry. Compos. Math. **2**, 463–470 (1935)

# Chapter 3
# Sets and Relations

*No one shall expel us from the paradise which Cantor created for us.*

***David Hilbert*** *(1862–1943)*

The concept of a set is one of the major scientific developments of the twentieth century. The level of rigor provided by the *theory of sets* ultimately led the modern mathematicians to discover fundamental notions such as the *abstract groups* and the *topological spaces*. A majority of ideas introduced in this chapter provide mathematical foundations to study important aspects of computer science such as related to *data structures, algorithms*, and *programming*. The main emphasis is to discuss some foundational tools needed to study certain important "configurations" and "processes" concerning the theory of computation. We also prove some useful theorems. The notations introduced in this chapter are used throughout the book. The interested reader may refer [1–3] for further details.

## 3.1 Sets and Operations

The **axiomatic set theory** was formulated by the German mathematician and logician *Ernst Zermelo* (1871–1953), in (1908), and also by the Israeli mathematician *Abraham Fraenkel* (1891–1965), in (1922). It was further enriched by the Norwegian mathematician *Thoralf Skolem* (1887–1963) and by the Hungarian-American mathematician *John von Neumann* (1903–1957). The *axiomatic set theory* is a large and complicated subject in its own right [4]. However, in this book, we follow



V. Ravichandran and A. K. Razdan, *Fundamental Discrete Structures*,
University Texts in the Mathematical Sciences,
https://doi.org/10.1007/978-981-96-0069-4_3

"naive" approach to the set theory, which was originally developed by *George Cantor* (1845–1918) and *Richard Dedekind* (1831–1916).

**Definition 3.1** A *set* is an unordered collection of objects, which is usually denoted by $A, B, C, \dots, X, Y, Z$, and we write $a, b, c, \dots, x, y, z$ for the objects it contains. In general, a set may contain many diverse types of objects.

In axiomatic set theory, both "collection" and "object" are treated as undefined entities. A set is written showing its objects implicitly or explicitly inside curly braces. As long as a set contains small number of objects, we can list its objects explicitly. For example, the set of lower case English letters is expressed as

$$E = \{a, b, c, \dots, x, y, z\},$$

where *ellipsis* ($\dots$) stand for the missing letters. The set $D$ of digits is expressed as

$$D = \{0, 1, 2, 3, 4, 5, 6, 7, 8, 9\}.$$

Similarly, the set $V$ of *five vowels* is expressed as

$$V = \{a,\ e,\ i,\ o,\ u\}.$$

However, not every set can be expressed as in above illustrations. In some well-known cases, it may be enough to list some objects of the set. For example, the set of *natural numbers* $\mathbb{N}$, the set of *non-negative integers* $\mathbb{N}_0 := \mathbb{Z}^+$, and the set of *integers* $\mathbb{Z}$ are, respectively, expressed as

$$\begin{aligned}\mathbb{N} &:= \{1, 2, 3, \dots\};\\ \mathbb{N}_0 &:= \{0, 1, 2, 3, \dots\};\\ \mathbb{Z} &:= \{\dots, -2, -1, 0, 1, 2, \dots\}.\end{aligned}$$

Alternatively, one may specify a set by mentioning a *property* common to all objects it contains. For example, we can express the set of *rational numbers* $\mathbb{Q}$ as

$$\mathbb{Q} := \{p/q \mid p, q \in \mathbb{Z} \text{ and } q \neq 0\}.$$

Also, the set $P$ of all perfect squares $1, 4, 9, \dots$ may be expressed as

$$P := \{n^2 \mid n \text{ is a positive integer}\},$$

and the set of *even integers* $2\mathbb{Z}$ may be expressed as

$$2\mathbb{Z} := \{2n \mid n \in \mathbb{Z}\}.$$

In most applications, a set under consideration may contain mathematical entities such as the *numbers, vectors, matrices, polynomials, functions*, or the digital stuff like *text strings, images, audio or video files, hardware parts of a network*, etc. The core idea of the set theory is to treat a set as a *single entity*, uniquely determined by the objects it contains.

**Definition 3.2** (*Membership Relation*) The statement that "$x$ is an object of a set $X$" is expressed symbolically as $x \in X$, which we read as $x$ is an **element of** (or a **member of**) the set $X$. Also, the statement that the object $x$ does not belongs to the set $X$ is expressed symbolically as $x \notin X$.

For example, with $X = \big\{\{1\}, 2, 3\big\}$, we have

$$\{1\} \in X, \qquad 1 \notin X, \qquad 2 \in X, \qquad 4 \notin X.$$

It thus makes sense to say that a collection $X$ constitutes a set if, given an object $x$, the predicate $x \in X$ has the truth value $T$ or $F$.

**Definition 3.3** (*Empty Set*) A set with no element is called the *empty set* (or the *null set*). It is denoted by $\emptyset$ (or simply as $\{\,\}$). In abstract terms, we also write

$$\emptyset := \big\{x \mid x \neq x\big\}. \tag{3.1.1}$$

The concept introduced in the next definition is of fundamental importance in various applications. It also helps make precise the meaning of the *equality of sets* in terms of *inclusion relation*.

**Definition 3.4** (*Subset*) Let $A$ and $B$ be two sets. We say $A$ is a *subset of* $B$ if every *element of* the set $A$ is also an element of the set $B$. In this case, we write $A \subseteq B$. Symbolically,

$$A \subseteq B \quad \text{if and only if} \quad x \in A \Rightarrow x \in B. \tag{3.1.2}$$

We also say that $A$ is *contained in* $B$. For example, with $A = \{\{1\}, 2, 3\}$, we have

$$\{2, 3\} \subseteq A, \quad \{2\} \subseteq A, \quad \{\{1\}, 3\} \subseteq A, \quad \{\{1\}\} \subseteq A.$$

However, $\{1\} \nsubseteq A$ because $\{1\} \in A$.

**Definition 3.5** (*Power Set*) The **power set** of a set $A$, denoted by $\mathscr{P}(A)$, is the set of all subsets of the set $A$. When $A = \emptyset$, we have $\mathscr{P}(A) = \{\emptyset\}$.

For $A = \{a\}$, we have $\mathscr{P}(A) = \{\emptyset, A\}$; for $A = \{a, b\}$, we have

$$\mathscr{P}(A) = \big\{\emptyset, \{a\}, \{b\}, A\big\};$$

for $A = \{a, b, c\}$, we have

$$\mathscr{P}(A) = \big\{\emptyset, \{a\}, \{b\}, \{c\}, \{a, b\}, \{a, c\}, \{b, c\}, A\big\};$$

and so on. We will prove shortly that the *power set* of a set with $n$ elements has $2^n$ elements (Theorem 3.1). In fact, this assertion also holds for infinite sets.

**Definition 3.6** (*Set Equality*) Two sets $A$ and $B$ are said to be **equal** if they contain the same elements. Equivalently, in view of (3.1.2), we have

$$A = B \quad \text{if and only if} \quad A \subseteq B \;\text{ and }\; B \subseteq A. \tag{3.1.3}$$

In particular, a set $A$ is called a **proper subset** of a set $B$ if $A \subseteq B$, but $A \neq B$. In this case, we write $A \subset B$.

For example, if $A = \{\{1\}, 2, 3\}$ and $B = \{\{1\}, 3, 2\}$, then $A = B$. However, for $C = \{1, 2, 3\}$, we have $A \neq C$. We also have

$$\mathbb{N} \subset \mathbb{Z} \subset \mathbb{Q} \subset \mathbb{R}.$$

Given certain types of nonempty sets, we can always find a larger set that contains all the sets as subsets. We call this larger set a *universal set* of the given sets. The next definition makes the notion precise.

**Definition 3.7** (*Universal Set*) A set $U$ is a **universal set** of sets $X_i$ if a property $p(x)$ holds only for elements $x \in X_i$, and not for any other element of $U$. That is, we can specify each set $X_i$ as

$$X = \big\{x \in U \mid p(x)\big\}. \tag{3.1.4}$$

Said differently, a set $U$ is called a *universal set* of a collection of sets if each contains the same types of objects as in $U$. For example, for any number set, the set $\mathbb{R}$ is a universal set.

**Definition 3.8** (*Cardinality*) The number of elements a set $A$ contains is called the **cardinality** of the set $A$. It is denoted by $|A|$ (or $\#(A)$). We say $A$ is a **finite set** if we have $|A| < \infty$. Otherwise, $A$ is called an *infinite set*.

The only set of cardinality 0 is the empty set $\emptyset$. More precise definitions of a finite and an infinite set are given later in Sect. 4.3. The next theorem proves an interesting fact about the cardinality of the power set of a finite set.

**Theorem 3.1** *For a finite set $A$ with $n$ elements, we have*

$$\big|\mathscr{P}(A)\big| = 2^n. \tag{3.1.5}$$

***Proof*** In general, a subset is formed by choosing $0 \le k \le n$ elements of the set $A$. It thus follows that there are $\binom{n}{k}$ sets containing $k$ elements of the set $A$, for $0 \le k \le n$. That is, $A$ has

$$\binom{n}{0} = 1 \quad \text{subset containing no element;}$$

$$\binom{n}{1} = n \quad \text{subsets containing 1 element;}$$

$$\binom{n}{2} = n(n-1)/2 \quad \text{subsets containing 2 elements;}$$

$$\vdots$$

$$\binom{n}{n} = 1 \quad \text{subset containing } n \text{ elements.}$$

Therefore, in total, the set $A$ has the following number of subsets:

$$\binom{n}{0} + \binom{n}{1} + \binom{n}{2} + \cdots + \binom{n}{n} = (1+1)^n = 2^n.$$

Hence, (3.1.5) holds. Alternatively, think about two boxes such that each contains an element of the set $A$. That is, some elements of $A$ are in the first box, and the remaining in the second box. Each way of placing elements of $A$ in two boxes determines a subset of the set $A$. Since each element of $A$ can be placed uniquely in either of the two boxes in 2 ways, it follows that the number of ways of placing an element into the two boxes is $2^n$. □

**Definition 3.9** A **Venn diagram** is a graphical representation that depicts a universal set $U$ as the interior of a rectangle, and a subset of $U$ is shown as the interior of closed regions such as circles lying inside it.

That is, a *Venn diagram* is a pictorial tool to visualize two or more sets by circular region that may or may not overlap. For example, Fig. 3.1 shows a Venn diagram of a universal set $U$, with three *overlapping subsets*. In a situation when a universal set $U$ is clear from the context, we may avoid drawing a rectangle as the graphical representation of $U$ (see Fig. 3.2).

## *Set Operations*

We now introduce some important operations on sets such as *union, intersection*, and the *complementation*. We also define the *Cartesian product* of sets. We can also define these operations for an *indexed family* of sets. We start with some *algebraic*

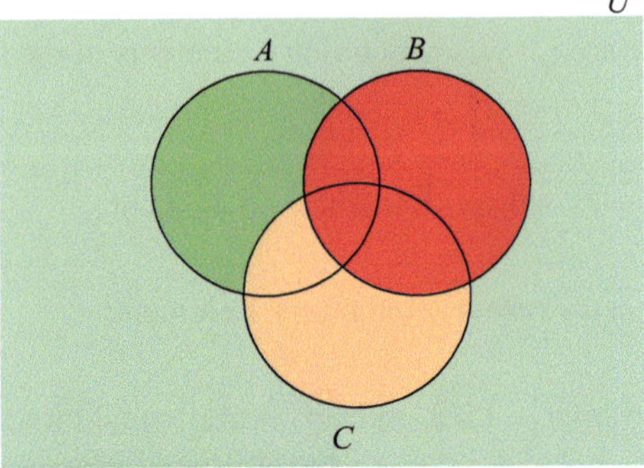


**Fig. 3.1** A Venn diagram of a universal set $U$ with three subsets

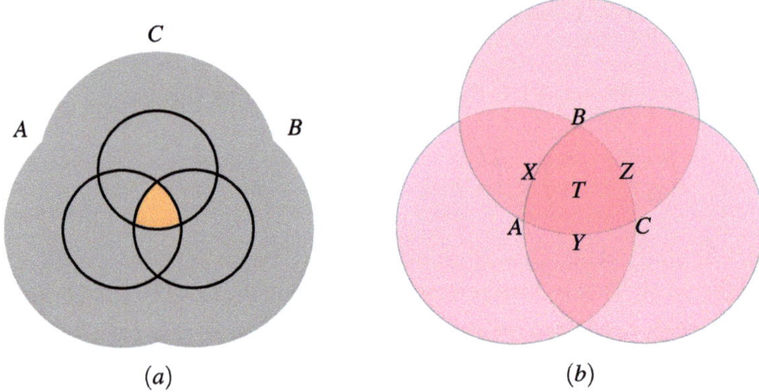


**Fig. 3.2** Venn diagrams of three overlapping sets

*operations*. Let $A$ and $B$ be two sets, and $U$ be a universal set. Given an object $x \in U$, we consider the following four cases:

(a) $x \in A$ or $x \in B$. In this case, the set given by

$$\{x \mid x \in A \text{ or } x \in B\} \tag{3.1.6}$$

is the *union* of the two sets $A$ and $B$. More precisely, the **union** $A \cup B$ of two sets $A$ and $B$ is the set containing elements of both the sets, without repetition. It is denoted by $A \cup B$. The Venn diagram of the union of two sets is shown in Fig. 3.3a. In particular, it follows that we always have

$$A \subseteq A \cup B \quad \text{and} \quad B \subseteq A \cup B.$$

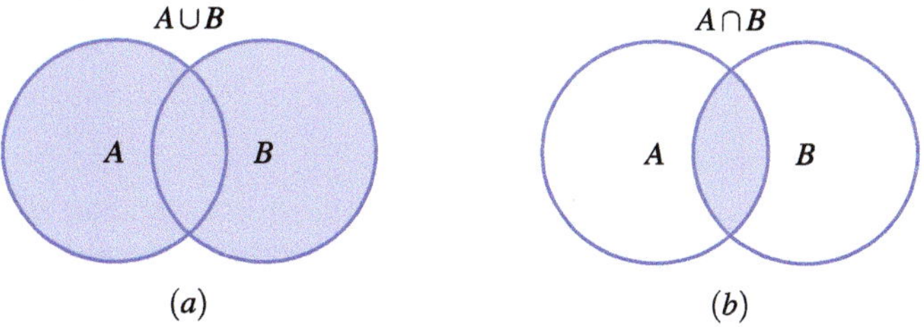


**Fig. 3.3** Venn diagram for the *union* ∪ and *intersection* ∩

(b) $x \in A$ and $x \in B$. In this case, the set

$$\{x \mid x \in A \text{ and } x \in B\} \tag{3.1.7}$$

is the *intersection* of the sets $A$ and $B$. More precisely, the **intersection** of two sets $A$ and $B$ is the set that only contains elements *common* to both the sets $A$ and $B$. It is denoted by $A \cap B$. A Venn diagram for intersection of two sets is as shown on right side of Fig. 3.3, where the overlapping region represents the intersection $A \cap B$ of the sets $A$ and $B$. In particular, we have

$$A \cap B \subseteq A \quad \text{and} \quad A \cap B \subseteq B.$$

Two sets $A$ and $B$ are said to be *disjoint sets* if $A \cap B = \emptyset$.

(c) $x \in A$, but $x \notin B$; or, $x \notin A$, but $x \in B$. In this case, we, respectively, write

$$A \setminus B = \{x \mid x \in A \text{ but } x \notin B\}; \tag{3.1.8a}$$
$$B \setminus A = \{x \mid x \in B \text{ but } x \notin A\}. \tag{3.1.8b}$$

Both sets are *relative complement* with respect to each other (see Fig. 3.4). More precisely, the set $A \setminus B$ given by (3.1.8a) is called the **relative complement** of $A$ modulo the set $B$. Similarly, the set $B \setminus A$ given by (3.1.8b) is called the **relative complement** of $B$ modulo the set $A$. If $A$ is a set with a universal set $U$, the relative complement $U \setminus A$ is called the **complement** of the set $A$. It is denoted by $A'$ (or $A^c$ or $\bar{A}$).

(d) $x \notin A$ and $x \notin B$ so that $x \in A^c$ and $x \in B^c$. In this case, $x \in A^c \cap B^c$ implies that $x \notin (A \cup B)$, by *De Morgan's law*.

For example, if $A = \{a, b, c, d\}$ and $B = \{b, d, e, f\}$, then we have

$$A \cup B = \{a, b, c, d, e, f\};$$
$$A \cap B = \{b, d\};$$
$$A \setminus B = \{a, c\} \quad \text{and} \quad B \setminus A = \{e, f\}.$$

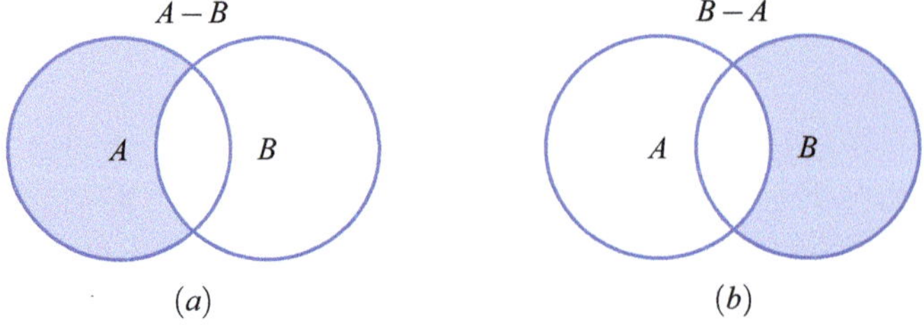


**Fig. 3.4** Venn diagrams for the *relative complements*

Also, if we take

$$E = \{a, b, c, \ldots, x, y, z\} \quad \text{and} \quad V = \{a, e, i, o, u\},$$

then the complement $V^c$ is the set of *consonants*. That is,

$$V^c = \{b, c, d, f, g, h, j, k, l, m, n, p, q, r, s, t, v, w, x, y, z\}.$$

Further, for the sets given by

$$A = \{a, b, c, d, e\}, \quad B = \{b, d, f, g\}, \quad C = \{a, b, d\}, \quad \text{and} \quad D = \{a, b, c\},$$

we have $A \subset E,\ B \subset E,\ C \subset A$, and

$$\begin{aligned} A \cup B &= \{a, b, c, d, e, f, g\}; \\ A \cap B &= \{b, d\}; \\ A \setminus B &= \{a, c, e\} \quad \text{and} \quad B \setminus A = \{f, g\}; \\ A^c &= \{f, g, h, \ldots, x, y, z\}. \end{aligned}$$

We use the concept of relative complementation to introduce an interesting set operation, denoted by $\oplus$ (or sometimes $\triangle$). For $A, B \subset X$, we define

$$A \oplus B = (A \setminus B) \cup (B \setminus A) = \{x \in X \mid \ x \in (A \cup B) \setminus (A \cap B)\}. \tag{3.1.9}$$

It is called the *symmetric difference* of the sets $A$ and $B$. For example, if $A = \{a, b, c, d, e\}$ and $B = \{b, d, f, g\}$, then

$$A \oplus B = \{a, c, e\} \cup \{f, g\} = \{a, c, e, f, g\}.$$

Also, notice that

$$\begin{aligned}(A \cup B) \setminus (A \cap B) &= \{a, b, c, d, e, f, g\} - \{b, d\} \\ &= \{a, c, e, f, g\} \\ &= A \oplus B.\end{aligned}$$

The above property holds for any two sets $A$ and $B$. We also have

$$A \oplus B = \big(A \cap B^c\big) \cup \big(A \cap B^c\big).$$

It follows from (3.1.9) that the operation $\oplus$ satisfies the properties as given below:

$$\begin{gathered}A \oplus B = B \oplus A, \qquad A \oplus A = \emptyset, \\ A \oplus \emptyset = A, \qquad A \oplus U = A^c, \qquad A \oplus A^c = U.\end{gathered}$$

What can be said about the sets $A$ and $B$ when $A \oplus B = A$?

**Definition 3.10** Let $\mathscr{A}$ be a family of sets, with universal set $U$. Then the *union of sets* in $\mathscr{A}$, denoted by $\cup_{A\in\mathscr{A}} A$, is the set given by

$$\bigcup_{A\in\mathscr{A}} A = \big\{x \in U \mid x \in A \text{ for at least one } A \in \mathscr{A}\big\}; \tag{3.1.10}$$

and the *intersection of sets* in $\mathscr{A}$, denoted by $\cap_{A\in\mathscr{A}} A$, is the set given by

$$\bigcap_{A\in\mathscr{A}} A = \big\{x \in U \mid x \in A \text{ for all sets } A \in \mathscr{A}\big\}. \tag{3.1.11}$$

In some situations, we are also interested to know what are union and intersection when $\mathscr{A} = \emptyset$. The answer is that the union is $\emptyset$, and the intersection is the universal set $U$. In general, let $I$ be a nonempty set, and $\mathscr{A}$ be a collection of sets such that

1. For each $i \in I$, there is a set $A_i \in \mathscr{A}$ and
2. for $i, j \in I$, $A_i = A_j$ if and only if $i = j$.

In this case, we call $\mathscr{A}$ an *indexed family of set*, with *indexing set* $I$. We also say that the family of sets in $\mathscr{A}$ is indexed by the set $I$, and write

$$\mathscr{A}^I = \big\{A_i \mid A_i \subseteq U \quad \text{and} \quad i \in I\big\}.$$

The *union* and *intersection* of the indexed family $\mathscr{A}^I$ are, respectively, given by

$$\bigcup_{i\in I} A_i = \big\{x \in U \mid x \in A_i \text{ for some } i \in I\big\}; \tag{3.1.12a}$$

$$\bigcap_{i\in I} A_i = \big\{x \in U \mid x \in A_i \text{ for all } i \in I\big\}. \tag{3.1.12b}$$

In particular, with $I = [n] := \{1, 2, \ldots, n\}$, we write

$$\mathscr{A} = \{A_k \mid k \in [n]\} = \{A_1, A_2, \ldots, A_n\},$$

we write the union and intersection of sets in $\mathscr{A}$, respectively, as

$$\bigcup_{k=1}^{n} A_k = \{x \in U \mid x \in A_k \text{ for some } k = 1, \ldots, n\}; \tag{3.1.13a}$$

$$\bigcap_{k=1}^{n} A_k = \{x \in U \mid x \in A_k \text{ for all } k = 1, \ldots, n\}. \tag{3.1.13b}$$

**Theorem 3.2** *Let $\{A_i : i \in I\}$ be an indexed set, and $B$ be any set. Then we have*

$$B \cup \left(\cap_{i\in I} A_i\right) = \cap_{i\in I}\left(B \cup A_i\right) \quad \textit{and} \quad B \cap \left(\cup_{i\in I} A_i\right) = \cup_{i\in I}\left(B \cap A_i\right).$$

***Proof*** We just prove one of them and other follows in the similar way. Let

$$\begin{aligned}
& x \in B \bigcup \left(\bigcap\nolimits_{i\in I} A_i\right) \\
\Leftrightarrow\quad & x \in B \text{ or } x \in \left(\bigcap\nolimits_{i\in I} A_i\right) \\
\Leftrightarrow\quad & x \in B \text{ or } \left(x \in A_i, \text{ for all } i \in I\right) \\
\Leftrightarrow\quad & \left(x \in B \text{ or } x \in A_i\right), \text{ for all } i \in I \\
\Leftrightarrow\quad & \left(x \in B \bigcup A_i\right), \text{ for all } i \in I \\
\Leftrightarrow\quad & x \in \bigcap\nolimits_{i\in I} \left(B \bigcup A_i\right).
\end{aligned}$$

This show that

$$B \cup \left(\cap_{i\in I} A_i\right) = \cap_{i\in I}\left(B \cup A_i\right).$$

Similarly, we can prove the identity as given below:

$$B \cap \left(\cup_{i\in I} A_i\right) = \cup_{i\in I}\left(B \cap A_i\right).$$

□

The three types of set operations defined above satisfy the following fundamental properties, which makes the power set $\mathscr{P}(X)$ of a set $X$ an important object in mathematics. Some of the basic identities for sets are given in Table 3.1. These identities and several others are left for the reader as exercises.

Similar to the normal form for statement formulas, we have *normal forms* for sets. In particular, for any three sets $A$, $B$, and $C$, we have

$$(A \cap B \cap C),\ (A \cap B \cap C^c),\ (A \cap B^c \cap C),\ (A \cap B^c \cap C^c),$$

$$(A^c \cap B \cap C),\ (A^c \cap B \cap C^c),\ (A^c \cap B^c \cap C),\ (A^c \cap B^c \cap C^c)$$

**Table 3.1** Basic set identities

| Fundamental set identities | |
|---|---|
| Idempotent laws | $A \cap A = A, \quad A \cap A = A$ |
| Associative laws | $(A \cap B) \cap C = A \cap (B \cap C)$, $(A \cap B) \cap C = A \cap (B \cap C)$ |
| Commutative laws | $A \cap B = B \cap A, \quad A \cap B = B \cap A$ |
| De Morgan's laws | $(A \cap B)^c = (A^c \cap B^c)$, $(A \cap B)^c = (A^c \cap B^c)$ |
| Distributive laws | $A \cap (B \cap C) = (A \cap B) \cap (A \cap C)$, $A \cap (B \cap C) = (A \cap B) \cap (A \cap C)$ |
| Complement laws | $A \cap A^c = \emptyset, \quad A \cup A^c = U$ |
| Dominance laws | $A \cup U = U, \quad A \cap \emptyset = \emptyset$ |
| Identity laws | $A \cap U = A, \quad A \cup \emptyset = A$ |
| Absorption laws | $A \cup (A \cap B) = A$, $A \cap (A \cup B) = A$ |
| Involution law | $(A^c)^c = A$ |

are *minterms* generated by sets $A$, $B$, and $C$. A set that can be written as union of minterms is said to be in *minterm normal form*. Similarly, we can define *maxterms* and *maxterm normal form*.

**Example 3.1** We write here the set $B$ in terms of the minterms generated by $A$, $B$, and $C$. For instance, since $A \cup A^c = U$ and $U \cap A = A$, we have

$$\begin{aligned} B &= (A \cup A^c) \cap B \\ &= (A \cap B) \cup (A^c \cap B) \qquad \text{(by distributive laws)} \\ &= ((A \cap B) \cap (C \cup C^c)) \cup ((A^c \cap B) \cap (C \cup C^c)) \\ &= (A \cap B \cap C) \cup (A \cap B \cap C^c) \\ &\quad \cup (A^c \cap B \cap C) \cup (A^c \cap B \cap C^c). \end{aligned}$$

By replacing $B$ in the above equation by $B^c$, we obtain

$$B' = (A \cap B^c \cap C) \cup (A \cap B^c \cap C^c) \cup (A^c \cap B^c \cap C) \cup (A^c \cap B^c \cap C^c).$$

Notice that the minterm normal form of the complement of a set $S$ contains all the minterms which are not present in the minterm normal form of $S$. Also, the union of all minterms is $E$ and the intersection of all minterms is $\emptyset$. Further, if two sets $A$ and $B$ have $U$ as the universal set, the Venn diagram is drawn inside a larger rectangular region representing the set $U$. The region representing the universe $U$ is partitioned into 4 mutually exclusive regions as indicated in the above diagram. The universe $U$ is divided into 8 regions each representing the following sets:

| Region | Set | Region | Set |
|---|---|---|---|
| 1 | $A \cap B^c \cap C^c$ | 2 | $A \cap B \cap C^c$ |
| 3 | $A^c \cap B \cap C^c$ | 4 | $A \cap B^c \cap C$ |
| 5 | $A \cap B \cap C$ | 6 | $A^c \cap B \cap C$ |
| 7 | $A^c \cap B^c \cap C$ | 8 | $A^c \cap B^c \cap C^c$ |

The statement of the next theorem is known as the *inclusion-exclusion principle* (for two sets). In fact, it holds for the union of any number of finite sets.

**Theorem 3.3** (Inclusion-Exclusion Principle) *For finite sets $A$ and $B$, we have*

$$\big|A \cup B\big| = \big|A\big| + \big|B\big| - \big|A \cap B\big|. \tag{3.1.14}$$

***Proof*** The statement holds because, while finding the cardinality of the set $A \cup B$, we count elements in both the sets $A$ and $B$. But, we only need to count the common elements in $A$ and $B$ only once. Therefore, the proof of the equality follows from the fact that there are $|A \cap B|$ elements common to both $A$ and $B$. □

We next define *Cartesian product* (or simply the **product**) of two or more *nonempty sets*. The concept is a simple abstraction of the familiar concept in analytical geometry of the "Cartesian coordinates" of a point. We know that, for $x, y, z \in \mathbb{R}$, an ordered pair $(x, y)$ represents a point in the *plane* $\mathbb{R}^2$; a triplet $(x, y, z)$ is a point in the *space* $\mathbb{R}^3$; and so on. To begin with, we first consider the case of two sets.

**Definition 3.11** Let $A$ and $B$ be two nonempty sets. The **Cartesian product** of sets $A$ and $B$ (in that order) is the set of **ordered pairs**[1] $(a, b)$, with $a \in A$ and $b \in B$. It is denoted by $A \times B$. The elements $a \in A$ and $b \in B$ are, respectively, called the *first* and the *second* coordinates of the *ordered pairs* $(a, b)$.

That is, the product $A \times B$ is the set given by

$$A \times B := \big\{(a, b) \mid a \in A \text{ and } b \in B\big\}. \tag{3.1.15}$$

In particular, when $|A| = m$ and $|B| = n$, each element of the set $A$ contributes $n$ elements to the product $A \times B$. Therefore, we have $|A \times B| = m\,n$. In particular, for $A = \{a, b, c\}$ and $B = \{0, 1\}$, the product $A \times B$ given by

$$A \times B = \big\{(a, 0), (a, 1), (b, 0), (b, 1), (c, 0), (c, 1)\big\}.$$

has six elements. Also, since the product $B \times A$ is the set given by

$$B \times A = \big\{(0, a), (0, b), (0, c), (1, a), (1, b), (1, c)\big\},$$

it follows that $A \times B \neq B \times A$, in general. The two simple applications given in the next example have interesting generalizations that are significant to certain more complex problems related to computer science.

[1] According to *Kazimierz Kuratowski* (1896–1980), we also write $(a, b) = \{\{a\}, \{a, b\}\}$.

**Example 3.2** We first consider a simple case of the *allocation problem*. Suppose $S$ is the set of students of an Institution, and $C$ is the set of science courses they can opt for in a semester. The statement "a student $s$ has opted for the course $c$" may be written symbolically as the element $(s, c) \in S \times C$. A similar application concerns the *transportation problem*. Let $B$ be the set of buses available at a *state bus terminus*, and $C$ be the set of cities in the adjoining states. In this case, the statement that the "bus $b$ has the city $c$ as destination" may be expressed symbolically as $(b, c) \in B \times C$. In general, the problem is about finding an optimal *transportation schedule* in a given network.

We next define the product of a finite collection of nonempty sets.

**Definition 3.12** The **product** of $n$ nonempty sets $A_1, \ldots, A_n$, denoted by

$$\prod_{i=1}^{n} A_i := A_1 \times \cdots \times A_n, \tag{3.1.16}$$

is the set of **n-tuples** $\big(a_1, \ldots, a_n\big)$, with $a_i \in A_i$ for $i = 1, \ldots, n$. The element $a_i \in A_i$ is called the **ith coordinate** of the *n-tuple* $\big(a_1, \ldots, a_n\big)$.

That is, the product $\prod_{i=1}^{n} A_i$ is the set given by

$$\prod_{i=1}^{n} A_i := \big\{\big(a_1, \ldots, a_n\big) \mid a_i \in A_i, \text{ for } 1 \le i \le n\big\}. \tag{3.1.17}$$

When each $A_i$ equals some nonempty set $A$, we write the product (3.1.16) as

$$A^n := A \times \cdots \times A \ \ (n\text{-copies}). \tag{3.1.18}$$

For example, the product $\mathbb{R}^2 = \mathbb{R} \times \mathbb{R}$ is the usual plane; the product $\mathbb{R}^3 = \mathbb{R} \times \mathbb{R} \times \mathbb{R}$ is the usual three-dimensional space, and so on. In general, we say

$$\big(a_1, \ldots, a_n\big) = \big(b_1, \ldots, b_n\big), \quad \text{with } \ a_i,\, b_i \in A_i, \tag{3.1.19}$$

if and only if $a_i = b_i$, for each $i = 1, \ldots, n$. In the next theorem, we prove some simple *set identities* related to product of sets.

**Theorem 3.4** *Let A, B, and C be three nonempty sets. Then we have*

1. $A \times \big(B \cap C\big) = \big(A \times B\big) \cap \big(A \times C\big)$;
2. $\big(A \cap B\big) \times \big(C \cap D\big) = \big(A \times C\big) \cap \big(B \times D\big)$.

***Proof*** To prove (1), we have

$$\begin{aligned} A \times (B \cap C) &= \{(x, y) \mid x \in A \text{ and } (y \in B \cap C)\} \\ &= \{(x, y) \mid x \in A \text{ and } (y \in B \text{ and } y \in C)\} \\ &= \{(x, y) \mid (x \in A \text{ and } y \in B) \text{ and } (x \in A \text{ and } y \in C)\} \\ &= \{(x, y) \mid (x, y) \in A \times B \text{ and } (x, y) \in A \times C\} \\ &= (A \times B) \cap (A \times C). \end{aligned}$$

To prove (2), we have

$$\begin{aligned} &(A \cap B) \times (C \cap D) \\ &= \{(x, y) \mid (x \in A \cap B) \text{ and } (y \in C \cap D)\} \\ &= \{(x, y) \mid (x \in A \text{ and } x \in B) \text{ and } (y \in C \text{ and } y \in D)\} \\ &= \{(x, y) \mid (x \in A \text{ and } y \in C) \text{ and } (x \in B \text{ and } y \in D)\} \\ &= \{(x, y) \mid (x, y) \in A \times C \text{ and } (x, y) \in B \times D\} \\ &= (A \times C) \cap (B \times D). \end{aligned}$$

□

**Remark 3.1** It is bit more technical to define the product of an infinite collection of nonempty sets. It uses the concept of *choice function* (Definition 4.1.14). In this case, the existence of the product is equivalent to the **axiom of choice**.

We conclude giving the construction of an important infinite set, called the **set of strings** over an alphabet. It finds applications in numerous subject areas such as *automata theory, compilers, and programming languages*.

**Example 3.3** (*Strings*) Let $\Sigma = \{a_1, \ldots, a_n\}$ be a finite set of symbols, which is called an *alphabet*. By a **string** $w$ over $\Sigma$ we mean a finite sequence of symbols $x_1, x_2, \ldots, x_m \in \Sigma$ written as a "juxtaposed expression" of the form:

$$w := x_1 x_2 \cdots x_m, \qquad \text{where each} \quad x_i \in \Sigma. \tag{3.1.20}$$

The symbols $x_i$ in the string $w$ don't have to be distinct. A string without any symbol is called the *empty string*, which is usually denoted by $\varepsilon$. The *length* of a string is the number of symbols it contains. In particular, the *length* of the string $w$ in (3.1.20) is $m$. More generally, we can form strings over $\Sigma$ of different lengths by using symbols $a_i$ in any manner, possibly with repetitions. For example,

$$a_1a_1a_4a_3a_1 = a_1^2a_4a_3a_1 \quad \text{and} \quad a_5a_2a_2a_2a_3a_1a_1 = a_5a_2^3a_3a_1^2,$$

are, respectively, the strings of length 5 and 7. For $m \geq 0$, let $\Sigma^m$ denote the set of strings of length $m$ over alphabet $\Sigma$. Notice that, since the only string of length zero is the *empty string* $\varepsilon$, we have $\Sigma^0 := \{\varepsilon\}$. The *set of strings* of all possible finite lengths over an alphabet $\Sigma$ is denoted by $\Sigma^*$. That is, we have

$$\Sigma^* := \Sigma^0 \cup \Sigma^1 \cup \Sigma^2 \cup \cdots \cup \Sigma^n \cup \cdots . \tag{3.1.21}$$

The set $\Sigma^*$ is an interesting object of study. For example, most information such as *numbers, names, pictures, and audio* are processed as **bit strings**, i.e., as strings over the alphabet $\Sigma = \{0, 1\}$. Also, a string over the alphabet $\Sigma = \{a, b, \dots\}$ is called a **character string**. Therefore, the study of a *formal language* as a set of strings is central to computer science. We will talk more on this in later chapters.

## 3.2 Relations and Operations

Various types of computation models introduced in the later chapters use relations on sets, and related operations. Our focus in this section is to discuss properties of some special type of binary relations. We first introduce the concept of a relation between two nonempty sets.

**Definition 3.13** Let $A$ and $B$ be two nonempty sets. A *relation* $R$ from $A$ to $B$, denoted by $R : A \to B$, is a subset of the product $A \times B$. When $(a, b) \in R \subseteq A \times B$, we say *$a \in A$ is related to $b \in B$* with respect to the relation $R$. In this case, we also write $a\,R\,b$ or $a \sim_R b$. On the other hand, when $(a, b) \notin R$, we say that $a$ and $b$ *are not related* with respect to relation $R$, and write $a \not R b$ (or $a \not\sim_R b$).

The next theorem is a very useful fact.

**Theorem 3.5** *Let $A$ and $B$ be two sets, with $|A| = m$ and $|B| = n$. Then the total number of relations from $A$ to $B$ is $2^{mn}$.*

***Proof*** Clearly, the product set $A \times B$ has $mn$ elements. Therefore, the cardinality of the power set of the set $A \times B$ is $2^{mn}$. Hence, by Definition 3.13, we have that the total number of relations $R : A \to B$ is $2^{mn}$. □

For a relation $R : A \to B$, the two sets given by

$$\begin{aligned} \mathrm{Dom}(R) &:= \{a \in A \mid (a, b) \in R, \quad \text{for some } b \in B\}; \\ \mathrm{Ran}(R) &:= \{b \in B \mid (a, b) \in R, \quad \text{for some } a \in A\} \end{aligned}$$

are, respectively, called the *domain* and *range* of the relation $R$. Notice that a relation $S$ from $B$ to $A$, denoted by $S : B \to A$, is a subset of the product $B \times A$.

**Example 3.4** Taking $A = \{a, b, c\}$ and $B = \{0, 1\}$, we have

$$A \times B = \{(a, 0), (a, 1), (b, 0), (b, 1), (c, 0), (c, 1)\}.$$

Therefore, by Theorem 3.5, there are $2^6 = 64$ relations from $A$ to $B$. Five such relations are as listed below:

$$R_1 = \{(a,0),(b,0),(c,1)\}, \qquad R_2 = \{(a,1)\},$$
$$R_3 = \{(a,0),(a,1)\}, \qquad R_4 = A \times B, \qquad R_5 = \emptyset.$$

Notice that we have

$$\begin{aligned} \mathrm{Dom}(R_1) &= \{a,b,c\} & \mathrm{Ran}(R_1) &= \{0,1\}; \\ \mathrm{Dom}(R_2) &= \{a\} & \mathrm{Ran}(R_2) &= \{1\}; \\ \mathrm{Dom}(R_3) &= \{a\} & \mathrm{Ran}(R_3) &= \{0,1\}; \\ \mathrm{Dom}(R_4) &= A & \mathrm{Ran}(R_4) &= B. \end{aligned}$$

We can use set operations to form new relations. For example, the *complement* of a relation $R : A \to B$ is the relation $R^c : A \to B$ given by

$$R^c := \{(a,b) \in A \times B \mid (a,b) \notin R\}.$$

We also write $R^c$ as $\sim R$. For example, with $A = \{a,b,c,d\}$ and $B = \{0,1,2\}$, consider the relation $R$ given by

$$R = \{(a,0),(a,2),(b,1),(b,2),(c,2),(d,0),(d,1),(d,2)\}.$$

We then have

$$R^c = \{(a,1),(b,0),(c,0),(c,1)\}.$$

Also, the *inverse* (or the *converse*) of a relation $R : A \to B$ is the relation $R^{-1} : B \to A$ given by

$$R^{-1} := \{(b,a) \in B \times A \mid aRb\}.$$

Next, if $R,\ S : A \to B$ are two relations, then the *union* $R \cup S$ and the *intersection* $R \cap S$ are, respectively, given by

$$R \cup S := \{(a,b) \in A \times B \mid aRb \text{ or } aSb\};$$
$$R \cap S := \{(a,b) \in A \times B \mid aRb \text{ and } aSb\}.$$

Further, the *difference relations* $R - S,\ S - R : A \to B$ are the relations given by

$$R - S := \{(a,b) \in A \times B :\ aRb \text{ and } a \not S\, b\};$$
$$S - R := \{(a,b) \in A \times B :\ aSb \text{ and } a \not R\, b\}.$$

Therefore, the *symmetric difference* $R \oplus S : A \to B$ of two relations $R, S : A \to B$ is the relation given by

$$R \oplus S := \{(a, b) \in A \times B : \text{ either } aRb \text{ or } aS\,b, \text{ but not both}\}.$$

**Example 3.5** Let $A = \{a, b, c, d\}$ and $B = \{0, 1, 2\}$, and consider the two relations $R, S : A \to B$ given by

$$R = \{(a, 0), (a, 2), (b, 1), (b, 2), (c, 2), (d, 0), (d, 1), (d, 2)\},$$
$$S = \{(a, 1), (b, 1), (b, 2), (c, 1), (c, 2), (d, 2)\}.$$

Then their union and intersection are, respectively, given by

$$R \cup S = \{(a, 0), (a, 1), (a, 2), (b, 1), (b, 2), (c, 1), (c, 2), (d, 0), (d, 1), (d, 2)\};$$
$$R \cap S = \{(b, 1), (b, 2), (c, 2), (d, 2)\}.$$

Next, for the relations $R, S : A \to B$ given by

$$R = \{(a, 0), (a, 2), (b, 1), (b, 2), (c, 2), (d, 0), (d, 1), (d, 2)\},$$
$$S = \{(a, 1), (b, 1), (b, 2), (c, 1), (c, 2), (d, 2)\},$$

we have that

$$R - S = \{(a, 0), (a, 2), (d, 0), (d, 1)\}, \qquad S - R = \{(a, 1), (c, 1)\},$$
$$R \oplus S = \{(a, 0), (a, 1), (a, 2), (c, 1), (d, 0), (d, 1)\}.$$

**Definition 3.14** (*Composition of Relations*) Let $A$, $B$, and $C$ be three nonempty sets, and consider two relations $R : A \to B$ and $S : B \to C$. Then the *composition* of relations $R$ and $S$ (in that order) is the relation denoted by $R \circ S : A \to C$, which is given by

$$R \circ S := \{(a, c) \in A \times C : (a, b) \in R \text{ and } (b, c) \in S, \text{ for some } b \in B\}.$$

That is, $a(R \circ S)c$ if and only if there is an element $b \in B$ such that $aRb$ and $bSc$.

**Example 3.6** Let $A = \{a, b, c, d\}$, $B = \{0, 1, 2\}$, and $C = \{x, y, z\}$. Consider the relations $R : A \to B$ and $S : B \to C$ given by

$$R = \{(a, 0), (a, 2), (b, 1), (b, 2), (c, 2), (d, 0), (d, 1), (d, 2)\};$$
$$S = \{(a, 1), (b, 1), (b, 2), (c, 1), (c, 2), (d, 2)\};$$
$$T = \{(0, x), (0, y), (1, z), (2, x), (2, y)\}.$$

For the above three relations, we have

$$\begin{aligned} R^{-1} &= \{(1,a),(0,b),(0,c),(1,c)\} \\ T^{-1} &= \{(x,0),(y,0),(z,1),(x,2),(y,2)\} \\ R \circ S &= \{(a,x),(a,y),(b,x),(b,y),(b,z),(c,x),(c,y),(d,x),(d,y),(d,z)\}. \end{aligned}$$

**Definition 3.15** (*Power Relations*) Let $R$ be a relation on a set $A$. The **nth power** $R^n$ of $R$ is defined inductively as

$$R^1 := R, \quad \text{and} \quad R^{n+1} := R^n \circ R, \qquad \text{for} \quad n \geq 1. \tag{3.2.1}$$

It is convenient to take $R^0 := \emptyset$.

It thus follows that we have

$$R^n := R \circ \cdots \circ R \quad (n \text{ times}), \qquad \text{for} \quad n \geq 1. \tag{3.2.2}$$

## 3.3 Properties of Relations

Let $A$ be a nonempty set. We may recall that a *binary relation* on $A$ is a subset of the set $A \times A$. In particular, the set $\emptyset$ is called the **null relation**, and the set $A \times A$ is called the *full relation*. The **diagonal relation** on $A$, denoted by $\Delta_A$, is given by

$$\Delta_A := \big\{(a,a) \mid a \in A\big\}. \tag{3.3.1}$$

The *full relation* and the *diagonal relation* on $A = \{a, b, c\}$ are, respectively, given by

$$\begin{aligned} A \times A &= \big\{(a,a),(a,b),(a,c),(b,a),(b,b),(b,c),(c,a),(c,b),(c,c)\big\}; \\ \Delta_A &:= \big\{(a,a),(b,b),(c,c)\big\}. \end{aligned}$$

Some other binary relations on $A = \{a, b, c\}$ are as given below:

$$\begin{aligned} R_6 &= \big\{(a,a),(b,b),(c,c)\big\}; \\ R_7 &= \big\{(a,a),(a,b),(b,a),(b,b),(c,c)\big\}; \\ R_8 &= \big\{(a,a),(a,b),(b,b),(c,c)\big\}; \\ R_9 &= \big\{(a,a),(a,b),(b,a),(c,a),(a,c)\big\}. \end{aligned} \tag{3.3.2}$$

Let $X$ be a nonempty set, and $A = \mathscr{P}(X)$, the *power set* of the set $X$. Then the *inclusion relation* $R$ on the set $A$ is the set given by

$$R := \{(U, V) \in A \times A \mid U \subseteq U\}.$$

We may also write the above relation as $\subseteq$. Further, the usual *order relation* $<$ on the set $\mathbb{R}$ contains all ordered pairs $(x, y)$, with $x \neq y \in \mathbb{R}$. That is, we have

$$<= \{(x, y) \mid x, y \in \mathbb{R}, \text{ with } x < y\},$$

which is read as *less than*. Similarly, *less than or equal to* $\leq$ is a binary relation on $\mathbb{R}$ given by

$$\leq= \{(x, y) \mid x, y \in \mathbb{R}, \text{ with } x \leq y\}.$$

The *greater than* $>$ and *greater than or equal to* $\geq$ are also binary relations on $\mathbb{R}$ that are obtained by reversing the order in the above two relations, respectively.

In general, some binary relations on a set have certain special properties. For example, each one of the relations $R_6$, $R_7$, $R_8$, and $R_9$ in (3.3.2) contains all the ordered pairs $(a, a)$, for $a \in A$. That is, each is a *reflexive relation*. A formal definition is given below.

**Definition 3.16** (*Reflexive Relation*) A relation $R$ on a nonempty set $A$ is called a **reflexive relation** if we have

$$(a, a) \in R, \quad \text{for all} \quad a \in A. \tag{3.3.3}$$

On the other hand, $R$ is called an **irreflexive relation** if we have

$$(a, a) \notin R, \quad \text{for any} \quad a \in A. \tag{3.3.4}$$

Notice that $R$ is a *reflexive relation* if and only if we have $\Delta_A \subseteq R$, and $R$ is an *irreflexive relation* if and only if we have $\Delta_A \cap R = \emptyset$. Also, a binary relation $R$ on $A$ is reflexive if and only if the complement set $R^c$ is an irreflexive relation. Notice that a non-reflexive relation is not same as an irreflexive, and vice versa.

**Example 3.7** On a collection of sets, the proper inclusion relation $\subset$ is irreflexive, and so is the strict order $<$ on any subset of $\mathbb{R}$. Let $A = \{1, 2, 3\}$, and consider the relations $R = \{(1, 1)\}$ and $S = \{(1, 1), (3, 3)\}$. These two relations are not reflexive because $(2, 2), (3, 3) \notin R$ and $(2, 2) \notin S$. They are also not irreflexive because $(1, 1) \in R$ and $(1, 1), (3, 3) \in S$. Further, let $A = \{a, b, c\}$, and consider the relations

$$\begin{aligned}
R_1 &= \{(a, a), (b, b), (c, c)\} \\
R_2 &= \{(a, a), (b, b), (c, c), (a, b), (b, a)\} \\
R_3 &= \{(a, b), (b, c), (c, a)\} \\
R_4 &= \{(a, b)\}
\end{aligned}$$

$$R_5 = \{(a, b), (b, a), (b, c)\}$$
$$R_6 = \{(a, a), (b, b), (c, c), (a, b), (b, a), (b, c), (c, a)\}$$
$$R_7 = \{(a, a), (b, b), (a, b), (b, a), (b, c), (c, a)\}.$$

As the relations $R_1$, $R_2$, and $R_6$ contain all the three order pairs $(a, a)$, $(b, b)$, and $(c, c)$, it follows that only these three are reflexive relations. However, as no ordered pair of the form $(x, x)$ belongs to relations $R_3$, $R_4$, and $R_5$, these three are irreflexive relations. Notice that the relation $R_7$ is neither reflexive (as $(c, c) \notin R_7$) nor irreflexive (as $(a, a) \in R_7$).

We also have that the relations $R_6$, $R_7$, $R_9$, and $R_{10}$ in (3.3.2) are such that each one contains both the pairs $(y, x)$ and $(x, y)$, for some distinct elements $x, y \in A$. That is, each is a *symmetric relation* on the set $A$. The next definition formalizes the concept.

**Definition 3.17 (Symmetric Relation)** A relation $R$ on a nonempty set $A$ is called a **symmetric relation** if we have

$$(a, b) \in R \quad \Rightarrow \quad (b, a) \in R. \tag{3.3.5}$$

Also, we say $R$ is an **antisymmetric relation** if we have

$$(a, b) \in R \text{ and } (b, a) \in R \quad \Rightarrow \quad a = b. \tag{3.3.6}$$

Equivalently, $R$ is an *antisymmetric relation* if $(a, b) \in R$ and $a \neq b$, then $(b, a) \notin R$.

Consider the seven relation in Example 3.7 defined on the set $A = \{a, b, c\}$. Notice that the relation $R_1$ is both symmetric and antisymmetric; the relation $R_2$ is symmetric, and since $(a, b), (b, a) \in R_2$, it is not antisymmetric; the relations $R_3$, $R_4$ are antisymmetric, but not symmetric; the relations $R_5$, $R_6$, $R_7$ are neither symmetric nor antisymmetric. Further, let $A = \{1, 2, 3, 4\}$, and consider the relations $R = \{(1, 1), (2, 2), (3, 3, ), (4, 4), (5, 5)\}$ and $S = \{(1, 1), (4, 4), (5, 5)\}$. These two relations are symmetric as well as antisymmetric. In general, the property introduced in the next definition is of central significance to many applications, as discussed in the later parts of the book.

**Definition 3.18** (*Transitive Relation*) A relation $R$ on a nonempty set $A$ is called a **transitive relation** if, for $a, b, c \in A$, we have

$$(a, b) \text{ and } (b, c) \in R \quad \Rightarrow \quad (a, c) \in R. \tag{3.3.7}$$

In Example 3.7, each $R_1$, $R_2$, and $R_4$ is a *transitive relation*, whereas all $R_3$, $R_5$, $R_6$, and $R_7$ are non-transitive relations. Also, on the set $A = \{1, 2, 3, 4, 5\}$, the relation $R$ given by

$$R = \{(1, 1), (1, 3), (2, 2), (2, 3), (3, 1), (3, 2), (3, 3, ), (4, 4), (5, 5)\}$$

is reflexive because $(a, a) \in R$, for all $a \in A$ it is symmetric, because for any $(a, b) \in R$ we also have $(b, a) \in R$. However, $R$ is not antisymmetric because both $(2, 3)$ and $(3, 2)$ belong to $R$. Also, $R$ is not transitive because $(2, 3), (3, 1) \in R$, but $(2, 1) \notin R$. Next, on $A = \{1, 2, 3, \ldots, 10\}$, the relation $R$ given by

$$R = \big\{(x, y) \mid x, y \in A, \text{ and } x + y = 10\big\}$$

can be written explicitly as

$$R = \big\{(1, 9), (2, 8), (3, 7), (4, 6), (5, 5), (6, 4), (7, 3), (8, 2), (9, 1)\big\}.$$

Clearly, it is neither reflexive (as $(1, 1) \notin R$) nor irreflexive (as $(5, 5) \in R$). However, it is symmetric, but not antisymmetric (as both $(1, 9)$ and $(9, 1)$ belong to $R$). Also, it not a transitive relation because $(1, 9), (9, 1) \in R$, but $(1, 1) \notin R$. On the set $\mathbb{R}$, consider the relation

$$R = \big\{(x, y) \mid x, y \in \mathbb{R}, x < y\big\}.$$

It is not a reflexive relation because $1 \not< 1$; it is an irreflexive relation because $a \not< a$, for all $a \in \mathbb{R}$; it is not a symmetric relation because $1 < 2$, but 2 is not less than 1; it is antisymmetric relation because if $a < b$ then $b \not< a$; it is transitive relation because $a < b$ and $b < c$ implies that $a < c$.

**Example 3.8** Let $P$ be a collection of subsets of some nonempty set $A$, say $P = \mathscr{P}(A)$, the power set of $A$. Let $R$ be the usual *inclusion relation* $\subseteq$. That is, for $S, T \in P$, we have

$$(S, T) \in R \iff S \subseteq T.$$

Since $S \subseteq S$, for any $S \in P$, the inclusion relation is reflexive. Now, by the definition of equality of sets, we have

$$S \subseteq T \text{ and } T \subseteq S \quad \Rightarrow \quad S = T.$$

Therefore, the relation $R$ is *antisymmetric*. Finally, as

$$S \subseteq T \quad \text{and} \quad T \subseteq U \quad \Rightarrow \quad S \subseteq U,$$

we conclude that the relation $R$ is transitive. Hence, the inclusion relation $\subseteq$ on the set $P$ is reflexive, antisymmetric, and transitive. As said earlier in Example 3.22, the relation $R$ on any subset of the set $\mathbb{N}$ given by

$$(a, b) \in R \quad \Leftrightarrow \quad a \mid b \tag{3.3.8}$$

is reflexive, antisymmetric, and transitive.

## 3.4 Graphical and Matrix Representations

Let $A$ be a finite set, and $R$ be a binary relation $R$ on $A$. The *digraph* of $R$ is a graphically representation, denoted by $D_R(A)$ (or simply $D_R$). A formal definition is as given below.

**Definition 3.19** Let $A = \{a_1, \ldots, a_n\}$, and $R$ be a binary relation $R$ on $A$. The **directed graph** (or simply a **digraph**) $D_R$ of $R$ has $a_i$ as vertices (usually plotted as points in the plane), and there is a directed edge (usually drawn as an arrow) from $a_i \in A$ to $a_j \in A$, for every pair $(a_i, a_j) \in R$.

Suppose both $(a_i, a_j)$ and $(a_j, a_i)$ belong to the relation $R$, for some pair $a_i \neq a_j \in A$. In this case, the two directed edges from $a_i$ to $a_j$ and from $a_j$ to $a_i$ are called *symmetric edges*. An arc from a vertex $a_i \in A$ to itself is called a **loop** at the vertex $a_i$. Some familiar properties of a relation $R$ in terms of its digraph $D_R$ are as given below:

1. $R$ is *reflexive* if and only if each vertex of the digraph $D_R$ has a *loop*.
2. $R$ is *irreflexive* if and only if no vertex of the digraph $D_R$ has a *loop*.
3. $R$ is *symmetric* if and only if the digraph $D_R$ has *symmetric edges* only.
4. $R$ is *antisymmetric* if and only if the digraph $D_R$ has no *symmetric edges*.

No simple graphical interpretations of the transitivity property of a relation $R$ can be given in terms of its digraph $D_R$. Very roughly, a relation $R$ is *transitive* if the digraph $D_R$ has no *open triangles*. Some illustrations given later may help visualize the geometry of transitivity property of a relation on a set.

**Example 3.9** The relation $R$ on the set $A = \{a, b, c, d, e\}$ given by

$$R = \Delta_A \bigcup \big\{(a, b), (a, c), (b, c), (b, d), (c, d), (c, e), (d, e), (d, a), (e, a), (e, b)\big\}$$

has the digraph as shown in Fig. 3.5, where $\Delta_A$ denotes the *diagonal relation* on the set $A$. Also, the digraph of the poset $(D_{60}, \mid)$ introduced in Example 3.22 is as shown in Fig. 3.6.

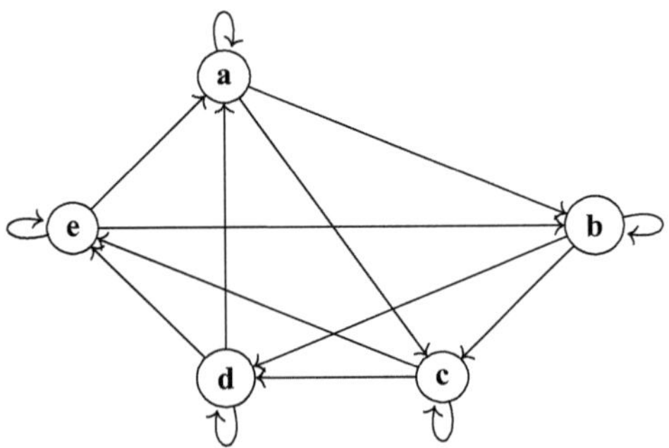


**Fig. 3.5** Digraph of the poset $(A, R)$ as in Example 3.9

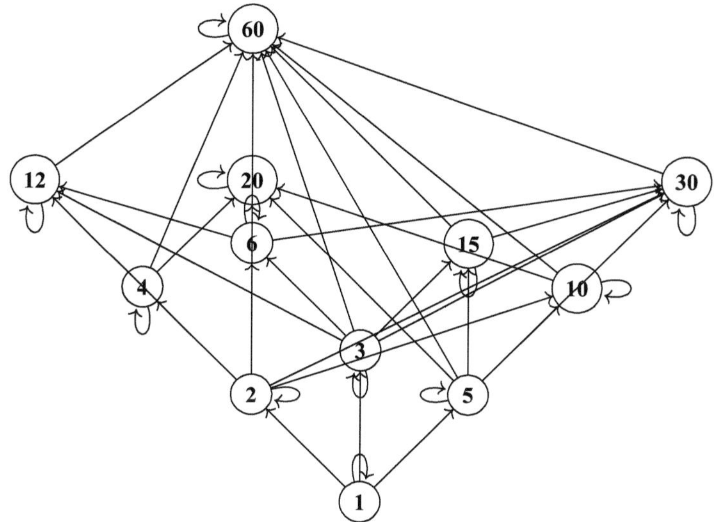


**Fig. 3.6** Digraph of the poset $(D_{60}, \mid)$

**Fig. 3.7** Digraph of the poset $(A, R)$

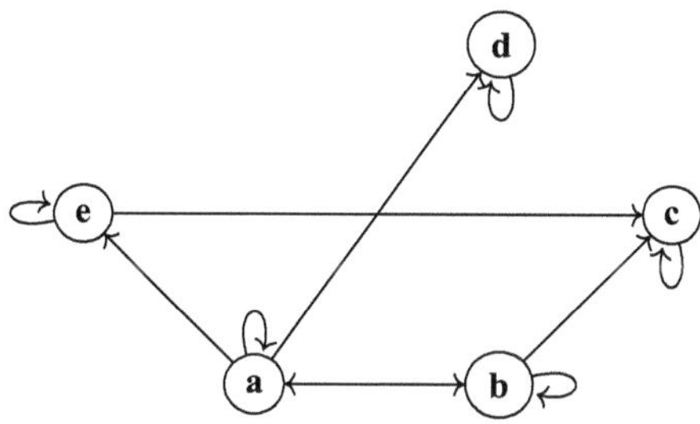


**Example 3.10** Let $A = \{a, b, c, d, e\}$ and $R$ be the relation given by

$$R = \Delta_A \cup \{(a, b), (a, d), (a, e), (b, a), (b, c), (e, c)\},$$

where $\Delta_A$ is the diagonal relation on $A$. The digraph of this relation is as shown in Fig. 3.7.

On the other hand, an algebraic representation of a binary relation $R$ on a finite set $A$ is a square matrix $M_R$ of order $|A|$, where both rows and columns of $M_R$ are indexed by the elements of the set $A$. It is called the *relation matrix* of the relation $R$. More generally, a formal definition of a matrix of a binary relation $R : A \to B$ is as given below.

**Definition 3.20** For $A = \{a_1, \ldots, a_n\}$ and $B = \{b_1, \ldots, b_m\}$, let $R : A \to B$ be a relation. The $n \times m$ matrix $M_R(A, B) := (m_{ij})$ given by

$$m_{ij} = \begin{cases} 1, & \text{if } (a_i, b_j) \in R \\ 0, & \text{otherwise} \end{cases}$$

is called the *relation matrix* of the relation $R$.

Clearly, the matrix $M_R(A, B)$ is unique subject to specific ordering assigned to the elements of the two sets $A$ and $B$. In particular, with respect to *list ordering* of the elements of the set $A = \{a_1, \ldots, a_n\}$, the *relation matrix* of a relation $R$ on $A$ is a unique square matrix of order $n$, which we usually write as $M_R$. In what follows, we may consider "*as listed*" as the *natural ordering* of a set $A$.

**Example 3.11** Let $A = \{a, b, c, d\}$, $B = \{0, 1, 2\}$, and a relation $R : A \to B$ be given by

$$R = \{(a, 0), (a, 1), (b, 1), (c, 0), (c, 2), (d, 1)\}.$$

We have the *relation matrix* $M_R(A, B)$ as

$$M_R(A, B) = \begin{pmatrix} 1\,1\,0 \\ 0\,1\,0 \\ 1\,0\,1 \\ 0\,1\,0 \end{pmatrix}$$

**Example 3.12** Let $A = \{1, 2, 3, 4\}$, and consider the relation $R$ given by

$$R = \{(x, y) \mid x, y \in A, \text{ with } x > y\}.$$

Notice that, in this case, we can write $R$ explicitly as

$$R = \{(2, 1), (3, 1), (3, 2), (4, 1), (4, 2), (4, 3)\}.$$

Therefore, the digraph $D_R$ is shown in Fig. 3.8, and *relation matrix* $M_R$ is given by

$$M_R = \begin{pmatrix} 0\,0\,0\,0 \\ 1\,0\,0\,0 \\ 1\,1\,0\,0 \\ 1\,1\,1\,0 \end{pmatrix} \qquad M_S = \begin{pmatrix} 0\,1\,0\,0 \\ 0\,0\,1\,0 \\ 0\,0\,0\,1 \\ 1\,0\,0\,0 \end{pmatrix}.$$

Also, the relation matrix $M_S$ of the relation

$$S = \{(1, 2), (2, 3), (3, 4, ), (4, 1)\}$$

is as given above.

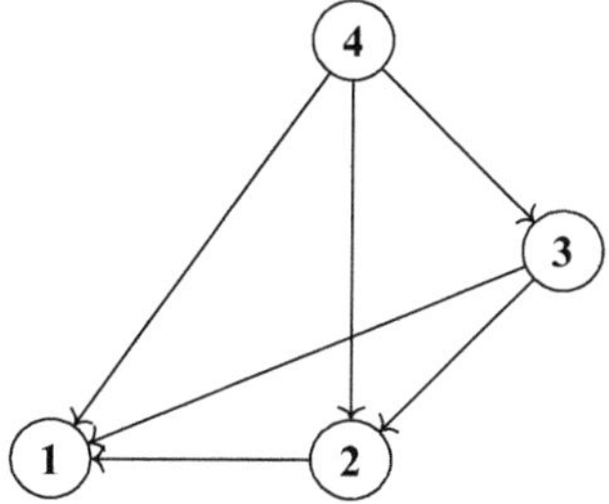


**Fig. 3.8** Digraph of the relation $R$ as in Example 3.12

Notice that the *relation matrix* $M_R = \left(m_{ij}\right)$ of a relation $R$ on a set $A = \{a_1, \ldots, a_n\}$ has the diagonal entry $m_{ii} = 1$ whenever we have $(a_i, a_i) \in R$, for $i = 1, \ldots, n$. Also, for $i \neq j \in \{1, \ldots, n\}$, $(a_i, a_j), (a_j, a_i) \in R$ implies that $m_{ij} = m_{ji} = 1$. The relation matrix $M_R$ of a relation $R$ in general helps read the following properties of the relation $R$:

1. $R$ is *reflexive* if and only if $m_{ii} = 1$, for all $i = 1, \ldots, n$.
2. $R$ is *irreflexive* if and only if $m_{ii} \neq 1$, for any $i = 1, \ldots, n$.
3. $R$ is *symmetric* if and only if $M_R$ is a symmetric matrix. That is, $m_{ij} = m_{ji}$, for all $i \neq j \in \{1, \ldots, n\}$. We may recall that a square matrix $M$ is said to be **symmetric** if $M = M^t$, where $M^t$ is the transpose of $M$.

As in the case of digraph of a relation, the relation matrix too is not very helpful in verifying the transitivity property of the relation. We may use the *composition* to find out the same.

**Theorem 3.6** *Let $R$ be a relation on a set $A$. Then $R$ is transitive if we have*

$$R^n \subseteq R, \qquad \text{for all} \quad n \geq 1, \tag{3.4.1}$$

*where $R^n$ is given by (3.2.2).*

***Proof*** Left for the reader as an exercise. □

Let $R$ and $S$ be two relations on a finite set $A$. In terms of the relation matrices $M_R$ and $M_S$, the *relation matrix* of the composition $R \circ S$ is the **Boolean product** of $M_R$ and $M_S$ (in that order). We write this *Boolean product* as $M_R * M_S$. Therefore, we have

$$M_{R \circ S} := M_R * M_S, \tag{3.4.2}$$

where the right side matrix is obtained by using the usual matrix multiplication, followed by applying the *logical addition* $+$ and the *logical multiplication* $\cdot$ as defined below:

$$0 + 0 = 0, \qquad 0 \cdot 0 = 0; \tag{3.4.3a}$$

$$1 + 0 = 1, \qquad 1 \cdot 0 = 0; \tag{3.4.3b}$$

$$1 + 1 = 1, \qquad 1 \cdot 1 = 1. \tag{3.4.3c}$$

We give simple examples to illustrate the concepts introduced above. For instance, let $A = \{1, 2, 3, 4, 5\}$, and consider the two relations $R$ and $S$ on $A$ given by

$$R = \big\{(1, 2), (2, 2), (3, 4)\big\} \quad \text{and} \quad S = \big\{(1, 3), (2, 5), (3, 1), (4, 2)\big\}.$$

The relation matrices of the relations $R$ and $S$ are, respectively, given by

$$M_R = \begin{pmatrix} 0\,1\,0\,0\,0 \\ 0\,1\,0\,0\,0 \\ 0\,0\,0\,1\,0 \\ 0\,0\,0\,0\,0 \\ 0\,0\,0\,0\,0 \end{pmatrix} \quad \text{and} \quad M_S = \begin{pmatrix} 0\,0\,1\,0\,0 \\ 0\,0\,0\,0\,1 \\ 1\,0\,0\,0\,0 \\ 0\,0\,0\,0\,0 \\ 0\,1\,0\,0\,0 \end{pmatrix}.$$

By Definition 3.14, we have that the composite relations $R \circ R$, $S \circ R$, $R \circ (S \circ R)$, $(R \circ S) \circ R$, $R \circ R$, $S \circ S$, and $R \circ R \circ R$ are given by

$$\begin{aligned} R \circ S &= \{(1, 5), (3, 2), (2, 5)\}, & S \circ R &= \{(1, 4), (3, 2), (4, 2)\}, \\ (R \circ S) \circ R &= \{(3, 2)\}, & R \circ (S \circ R) &= \{(3, 2)\}, \\ R \circ R &= \{(1, 2), (2, 2)\}, & S \circ S &= \{(1, 1), (3, 3), (4, 5)\}, \\ R \circ R \circ R &= \{(1, 2), (2, 2)\}. \end{aligned}$$

Now, using operations given in (3.4.3), the relation matrix of $R \circ S$ is given by

$$M_{R \circ S} = M_R * M_S = \begin{pmatrix} 0\,1\,0\,0\,0 \\ 0\,1\,0\,0\,0 \\ 0\,0\,0\,1\,0 \\ 0\,0\,0\,0\,0 \\ 0\,0\,0\,0\,0 \end{pmatrix} \begin{pmatrix} 0\,0\,1\,0\,0 \\ 0\,0\,0\,0\,1 \\ 1\,0\,0\,0\,0 \\ 0\,0\,0\,0\,0 \\ 0\,1\,0\,0\,0 \end{pmatrix} = \begin{pmatrix} 0\,0\,0\,0\,1 \\ 0\,0\,0\,0\,1 \\ 0\,1\,0\,0\,0 \\ 0\,0\,0\,0\,0 \\ 0\,0\,0\,0\,0 \end{pmatrix}.$$

Similarly, the relation matrix of $R \circ S$ is given by

$$M_{S \circ R} = M_S * M_R = \begin{pmatrix} 0\,0\,1\,0\,0 \\ 0\,0\,0\,0\,1 \\ 1\,0\,0\,0\,0 \\ 0\,0\,0\,0\,0 \\ 0\,1\,0\,0\,0 \end{pmatrix} \begin{pmatrix} 0\,1\,0\,0\,0 \\ 0\,1\,0\,0\,0 \\ 0\,0\,0\,1\,0 \\ 0\,0\,0\,0\,0 \\ 0\,0\,0\,0\,0 \end{pmatrix} = \begin{pmatrix} 0\,0\,0\,1\,0 \\ 0\,0\,0\,0\,0 \\ 0\,1\,0\,0\,0 \\ 0\,1\,0\,0\,0 \\ 0\,0\,0\,0\,0 \end{pmatrix}.$$

## 3.5 Closures of Relations

Let $A$ be a finite set. Suppose a relation $R$ on $A$ satisfies a property $\mathscr{P}$. By the $\mathscr{P}$-**closure** of $R$ we mean the smallest relation $S$ on $A$ such that $R \subseteq S$, and $S$ has the property $\mathscr{P}$. We are mainly interested to discuss the cases when the property $\mathscr{P}$ is *reflexivity, symmetric*, or *transitive*. The concepts we obtain are, respectively, that of the *reflexive closure*, *symmetric closure*, or the *transitive closure*, of the relation $R$.

**Definition 3.21** Let $R$ be a relation on a finite set $A$.

1. The **reflexive closure** of $R$ is the smallest *reflexive* relation $S$ on the set $A$ such that $S \supseteq R$. In this case, we may write $S := \overline{R}_{ref}$.
2. The **symmetric closure** of $R$ is the smallest *symmetric* relation $S$ on the set $A$ such that $S \supseteq R$. In this case, we may write $S := \overline{R}_{sym}$.
3. The **transitive closure** of $R$ is the smallest *transitive* relation $S$ on the set $A$ such that $S \supseteq R$. In this case, we may write $S := \overline{R}_{tran}$ (or simply $S := R^+$). Sometimes, we also write $R^+$ as $T$.

We can easily compute the reflexive and symmetric closures of a relation.

**Theorem 3.7** *Let $A$ be a finite set, and $R$ be a relation on $A$. Then we have*

$$\overline{R}_{ref} = R \cup \Delta_A \quad \text{and} \quad \overline{R}_{sym} = R \cup R^{-1}.$$

***Proof*** The proof of the first equality follows from the fact that a relation $S$ on $A$ is reflexive if and only if $\Delta_A \subseteq S$, and the proof of the second follows from the fact that a relation $S$ on $A$ is symmetric if and only if $S = S^{-1}$. □

Unfortunately, there is no convenient way to use $D_R$ or $M_R$ directly to compute the *transitive closure* $R^+$ of a relation $R$. In view of Theorem 3.6, one may use the $n$-fold compositions $R^n$ defined by (3.2.1). However, the procedure gets messy when $|A|$ is large or $|R| \geq 5$. Alternatively, one may apply *Warshall's algorithm*.

### *Warshall Algorithm*

Let $R$ be a relation on a nonempty set $A = \{a_1, \ldots, a_n\}$. The **Warshall algorithm** takes the *relation matrix* $M_R := \left(m_{ij}\right)$ of $R$ as the *initial input*. We may thus write $W^{[0]} := M_R$. We may recall that the entry $m_{ij}$ of $M_R$ is given by

$$m_{ij} := \begin{cases} 1, & \text{if } \left(a_i, a_j\right) \in R \\ 0, & \text{otherwise.} \end{cases} \tag{3.5.1}$$

For example, the relation matrix $M_R$ of the relation

$$R = \{(1, 1), (1, 2), (1, 4), (2, 1), (2, 3), (3, 1), (3, 2), (4, 2), (4, 4)\} \tag{3.5.2}$$

on set $A = \{1, 2, 3, 4\}$ is given by

$$M_R = \begin{bmatrix} 1\,1\,0\,1 \\ 1\,0\,1\,0 \\ 1\,1\,0\,0 \\ 0\,1\,0\,1 \end{bmatrix}. \tag{3.5.3}$$

A formal stepwise procedure of *Warshall's algorithm* is given a bit later. We first use the above matrix to illustrate the steps involved.

**Definition 3.22 (Row-Column Numbers)** For the relation matrix $M_R$ of a relation $R$ on a nonempty set $A$, the *row numbers* $r_i$ of the column $C_j$ refer to the row indices $i$ that contain a 1-entry, and the *column numbers* $c_j$ of the row $R_i$ refer to the column indices $j$ that contain a 1-entry.

For example, the row numbers of the first column $C_1$ of the relation matrix $M_R$ in (3.5.3) are 1, 2, and 3, because only rows $R_1$, $R_2$, and $R_3$ contain a 1-entry. Also, the column numbers of the first row $R_1$ of the relation matrix $M_R$ in (3.5.3) are 1, 2, and 4, because only columns $C_1$, $C_2$, and $C_4$ contain a 1-entry.

In Step-1, Warshall's algorithm computes the matrix $W^1$ from $W^0$, where we may get to replace a 0-*entry* of $W^0$ by a 1-entry. It never changes a 1-entry to a 0-entry. We may thus write

$$W^{[0]} = \begin{bmatrix} 1 & 1 & & 1 \\ 1 & & 1 & \\ 1 & 1 & & \\ & 1 & & 1 \end{bmatrix}. \tag{3.5.4}$$

We next find the *row numbers* $r_i$ of the column $C_1$ and the *column numbers* $c_j$ of the row $R_1$ (in that order only), followed by listing ordered pairs[2] $(r_i, c_j)$. Finally, if there is a 0 entry in matrix $W^1$ at the position $(r_i, c_j)$, then we replace it by a 1 entry. All the 1-entries in the matrix $W^1$ will stay put. We have obtained $W^1$. In particular, since the *row numbers* of the column $C_1$ of the matrix $W^1$ in (3.5.4) are 1, 2, 3, and the *column numbers* of the row $R_1$ are 1, 2, 4, we replace a 0-entry in matrix $W^{[1]}$ by a 1-entry at positions $\mathbf{(2, 2)}$, $\mathbf{(2, 4)}$, and $\mathbf{(3, 4)}$. We also say that *new edges* are created in the digraph $D_R$ of the relation $R$ for the pairs $(2, 2)$, $(2, 4)$, and $(3, 4)$. Notice that the algorithm don't change entries of the matrix $W^1$ at positions given by

$$(1, 1), \quad (1, 2), \quad (1, 4), \quad (2, 1), \quad (3, 1), \quad (3, 2).$$

[2] It is important to remember that Warshall algorithm works only when row numbers are taken as the first coordinate and the column numbers as the second coordinate.

Therefore, the first application of the algorithm yields the matrix

$$W^{[1]} = \begin{bmatrix} 1 & 1 & 0 & 1 \\ 1 & \mathbf{1} & 1 & \mathbf{1} \\ 1 & 1 & 0 & \mathbf{1} \\ 0 & 1 & 0 & 1 \end{bmatrix}. \tag{3.5.5}$$

We repeat the above procedure with the matrix $W^1$ as the input matrix. Accordingly, to obtain $W^{[2]}$, we start with the matrix

$$W^{[1]} = \begin{bmatrix} 1 & 1 & & 1 \\ 1 & 1 & 1 & 1 \\ 1 & 1 & & 1 \\ & 1 & & 1 \end{bmatrix}. \tag{3.5.6}$$

We now have that the *row numbers* of the column $C_2$ of the matrix $W^2$ in (3.5.6) are 1, 2, 3, 4, and the *column numbers* of the row $R_2$ of the matrix $W^2$ in (3.5.6) are 1, 2, 3, 4. Therefore, we replace a 0-entry of the matrix $W^2$ by a 1 at the positions $(\mathbf{1}, \mathbf{3})$, $(\mathbf{3}, \mathbf{3})$ $(\mathbf{4}, \mathbf{1})$, and $(\mathbf{4}, \mathbf{3})$. At this stage of the algorithm, only the pairs $(1, 3)$, $(3, 3)$, $(4, 1)$, and $(4, 3)$ have created *new edges* in the digraph $D_R$. Notice that the algorithm don't change entries of the matrix $W^1$ at positions given by

$$(1, 1), \quad (1, 2), \quad (1, 4), \quad (2, 1), \quad (2, 2), \quad (2, 3),$$
$$(2, 4), \quad (3, 1), \quad (3, 2), \quad (3, 4), \quad (4, 2), \quad (4, 4).$$

Hence, the second application of the algorithm yields the matrix:

$$W^{[2]} = \begin{bmatrix} 1 & 1 & \mathbf{1} & 1 \\ 1 & 1 & 1 & 1 \\ 1 & 1 & \mathbf{1} & 1 \\ \mathbf{1} & 1 & \mathbf{1} & 1 \end{bmatrix}. \tag{3.5.7}$$

It thus follows that the transitive closure $T$ of the relation $R$ in (3.5.2) is the full relation $A \times A$. In general, the "Warshall's Algorithm" works as follows:

For $k \geq 1$, the algorithm computes the matrix $W^k$ by using the row numbers of the $k$th column $C_k$ and the column numbers of the $k$th row $R_k$ of the matrix $W^{(k-1)}$ (in that order only). The following is the stepwise procedure to obtain $W^k$ from $W^{(k-1)}$:

STEP - 1 All 1 entries in $W^{(k-1)}$ are retained at their respective positions while computing the matrix $W^k$.

STEP - 2 List the row numbers of the $k$th column $C_k$ of the matrix $W^{(k-1)}$, and its column numbers of the $k$th row $R_k$.

STEP - 3 Pair each of the row number with each of the column number, and put a 1 in the corresponding position in $W^k$, if it is not already a 1-entry there. Let the remaining 0-entries stay put.

**Example 3.13** Consider the relation $R$ on a set $A = \{1, 2, 3, 4\}$, with the relation matrix given by

$$M_R = \begin{bmatrix} 0\,1\,0\,1 \\ 1\,0\,1\,0 \\ 1\,0\,0\,0 \\ 0\,0\,0\,0 \end{bmatrix}$$

In what follows, we use Warshall algorithm to find the *transitive closure* $T$ of the relation $R$. We first write $M_R = W^0$, and apply the above three steps to compute $W^1$. According to Step-1, we write

$$W^1 = \begin{bmatrix} & 1 & & 1 \\ 1 & & 1 & \\ 1 & & & \\ & & & \end{bmatrix}$$

Next, according to Step-2, we find the row numbers of the first column $C_1$, which are given by 2, 3, and also the column numbers of the first row $R_1$, which are given by 2, 4. Finally, according to Step-3, we pair each row number with each column number (in that order only), and replace a 0-entry in $W^1$ by a 1-entry at positions

$$(\mathbf{2}, \mathbf{2}), \quad (\mathbf{2}, \mathbf{4}), \quad (\mathbf{3}, \mathbf{2}), \quad \text{and} \quad (\mathbf{3}, \mathbf{4}),$$

because each of these four positions have a 0-entry. We thus obtain

$$W^1 = \begin{bmatrix} 0\,1\,0\,1 \\ 1\,\mathbf{1}\,1\,\mathbf{1} \\ 1\,\mathbf{1}\,0\,\mathbf{1} \\ 0\,0\,0\,0 \end{bmatrix}.$$

At this stage of the algorithm, the above four pairs have created new edges in the digraph $D_R$. Again, to construct $W^2$ from the matrix $W^1$ as obtained above, we repeat the same three steps. According to Step-1, we write

$$W^2 = \begin{bmatrix} & 1 & & 1 \\ 1 & 1 & 1 & 1 \\ 1 & 1 & & 1 \\ & & & \end{bmatrix}.$$

According to Step-2, we find the row numbers of the second column $C_2$ of the matrix $W^1$, which are given by 1, 2, 3, and also the column numbers of the second row $R_2$ of the matrix $W^1$, which are given by 1, 2, 3, 4. Finally, according to Step-3, we pair each row number with each column number (in that order only), and replace a

0-entry in $W^2$ by a 1-entry at positions **(1, 1)**, **(1, 3)**, and **(3, 3)**, because these are the only places where $W^2$ has a 0-entry. Notice that the pairs

$$(1, 2),\ (1, 4),\ (2, 1),\ (2, 2),\ (2, 3),\ (2, 4),\ (3, 1),\ (3, 2),\ (3, 4)$$

don't contribute any new edge to the digraph $D_R$. Therefore, after the second application of the algorithm, we obtain

$$W^2 = \begin{bmatrix} \mathbf{1} & 1 & \mathbf{1} & 1 \\ 1 & 1 & 1 & 1 \\ 1 & 1 & \mathbf{1} & 1 \\ 0 & 0 & 0 & 0 \end{bmatrix}.$$

It can be shown that the third application of the algorithm gives $W^3 = W^2$, and so we also have $W^4 = W^3$. Hence, the *transitive closure* $T$ of the relation $R$ is given by

$$\{(1, 1), (1, 2), (1, 3), (1, 4), (2, 1), (2, 2), (2, 3), (2, 4), (3, 1), (3, 2), (3, 3), (3, 4)\}.$$

In the remaining sections, we discuss two fundamental concepts on a nonempty set, namely, *equivalence relation* and *partial order*. We also show how the former is related to the concept of a partition of a set.

## 3.6 Equivalence Relations and Partitions

The next definition introduces the concept of *equivalence relation* on a set. It is an important idea that finds applications in many different types of problems such as related to data structure and design of algorithms.

**Definition 3.23** An **equivalence relation** on a nonempty set $A$ is a relation $R$ that is reflexive, symmetric, and transitive. In this case, it is customary to write $R$ as $\sim_R$ (or simply $\sim$).

In terms of notation $\sim$, the *reflexive, symmetric*, and *transitive* properties can be written, respectively, as $(i)\ a \sim a$; $(ii)\ a \sim b \Rightarrow b \sim a$; $(iii)\ a \sim b$ and $b \sim c \Rightarrow a \sim c$.

**Definition 3.24** Let $\sim$ be an equivalence relation on a nonempty set $A$, and $a \in A$. The set of elements $b \in A$ such that $b \sim a$ is called the **equivalence class** of $a$. The *equivalence class* of an element $a \in A$ is usually denoted by $[a]$. That is, we have

$$[a] := \{b \in A \mid b \sim a\}. \tag{3.6.1}$$

We prove some useful properties of equivalence classes of a set.

**Theorem 3.8** *Let $\sim$ be an equivalence relation on a nonempty set $A$. Then we have*

1. $a \in [a]$, *for any* $a \in A$. *In particular,* $[a] \neq \emptyset$.
2. $a \in A$ *and* $c \in [a] \quad \Rightarrow \quad [a] = [c]$, *for any* $a \in A$.
3. *For any* $a,\ b \in A$, *either* $[a] = [b]$ *or* $[a] \cap [b] = \emptyset$.

***Proof*** For (1), since $\sim$ is reflexive, we have $a \sim a$, for every $a \in A$, and so $a \in [a]$. For (2), let $c \in [a]$ so that $a \sim c$. Then, we also have $c \sim a$, because $\sim$ is symmetry. We first show that $[c] \subseteq [a]$. For instance, let $x \in [c]$. We thus have $x \sim c$. Now, since we also have $c \sim a$, it follows by transitivity of $\sim$ that $x \sim a$. Therefore, $x \in [a]$. We have thus shown that $[c] \subseteq [a]$. Similarly, if $x \in [a]$, then we have $x \sim a$, which together with $a \sim c$ implies that $x \sim c$, and so $x \in [c]$. This proves $[c] \subseteq [a]$. Hence, we have $[a] = [b]$. Finally, to complete the proof, suppose $[a] \cap [b] \neq \emptyset$. Let $c \in [a] \cap [b]$. Then, by (2), $c \in [a]$ implies that $[c] = [a]$. Similarly, $c \in [b]$ implies $[c] = [b]$. Therefore, it follows that $[a] = [b]\,(= [c])$. This proves (3). □

**Definition 3.25** Let $\sim$ be an equivalence relation on a nonempty set $A$. The set of *equivalence classes* of $A$ is called the **quotient set** of the set $A$. It is usually denoted by $A/\sim$. That is, we have

$$A/\sim := \big\{[a] \mid a \in A\big\}. \tag{3.6.2}$$

We give below some important equivalence relations, and also the related *quotient sets*. We start with the standard *equality relation* on a set.

**Example 3.14** (*Equality Relation*) Let $A$ be a nonempty set. Consider the relation $\sim$ on $A$ given by

$$\sim = \big\{(a, b) \mid a, b \in A, \text{ and } a = b\big\}.$$

This relation is reflexive, symmetric, and transitive:

(a) For every $a \in A$, we have $a = a$, and so $a \sim a$.
(b) If $a \sim b$ so that $a = b$, we have $b = a$, which implies $b \sim a$.
(c) If $a \sim b$ and $b \sim c$ then $a = b$ and $b = c$ implies $a = c$, and so $a \sim c$.

Therefore, the *equality relation* on any nonempty set is an equivalence relation, with associated equivalence classes given by

$$\big[a\big] = \big\{b \in A \mid a \sim b\big\} = \big\{b \in A \mid a = b\big\} = \big\{a\big\}, \quad \text{for} \quad a \in A.$$

In this case, the quotient set $A/\sim$ can be identified with the set $A$ itself.

The construction of the set of *rational numbers* $\mathbb{Q}$ from the set of integers $\mathbb{Z}$ uses the equivalence relation as described in the next example.

**Example 3.15** Consider the set of nonzero integers $\mathbb{Z}^*$, and define a relation $\sim$ on the set $A = \mathbb{Z} \times \mathbb{Z}^*$ as follows:

$$(a, b) \sim (c, d) \quad \Leftrightarrow \quad ad = bc.$$

We show that this relation is reflexive, symmetric, and transitive:

(a) Since $ab - ba = 0$, for all $(a, b) \in A$, we have $(a, b) \sim (a, b)$. That is, $\sim$ is a reflexive relation.
(b) If $(a, b) \sim (c, d)$ then $ad = bc$, and so $cb = da$, which implies $(c, d) \sim (a, b)$. That is, $\sim$ is a symmetric relation.
(c) If $(a, b) \sim (c, d)$ and $(c, d) \sim (e, f)$ then $ad = bc$ and $cf = de$ implies that $adcf = bcde$ or $af = be$. Therefore, we have $(a, b) \sim (e, f)$. That is, $\sim$ is a transitive relation.

Therefore, $\sim$ is an equivalence relation. It can be seen that the equivalence class of an element $(a, b) \in A$ contains elements of the form $(ka, kb)$, for $k \neq 0$.

The next example uses the concept of **divisibility**. Recall that a nonzero integer $y$ is said to *divide* an integer $x$ if we can write $x = my$, for some integer $m$. In this case, we write $y \mid x$. We also say that $y$ is a *divisor* of $x$.

**Example 3.16** (*Congruence Relation*) Let $n > 1$ be a fixed integer. Define a relation $\sim$ on the set of integers $\mathbb{Z}$ as given below:

$$a \sim b \quad \Leftrightarrow \quad n \mid a - b. \tag{3.6.3}$$

We show that the relation $\sim$ is an equivalence relation, i.e., it is reflexive, symmetric, and transitive.

1. Since $n \mid a - a = 0$, for all $a \in \mathbb{Z}$, it follows that $a \sim a$. Therefore, $\sim$ is a reflexive relation.
2. Let $a \sim b$.Then we have $n \mid a - b$, by (3.6.3), and so $n \mid b - a = -(a - b)$. It thus follows from (3.6.3) that $b \sim a$. Therefore, $\sim$ is a symmetric relation.
3. Suppose $a \sim b$ and $b \sim c$. Then we have $n \mid a - b$ and $n \mid b - c$, by (3.6.3). Therefore, by (A.1.1), we have $n$ also divides their sum, i.e., $n \mid a - b + b - c = a - c$. That is, $a \sim c$. Therefore, $\sim$ is a transitive relation.

With respect to equivalence relation (3.6.3), the equivalence class of an element $a \in \mathbb{Z}$ is given by

$$\begin{aligned}
[a] &= \{b \in \mathbb{Z} \mid b \sim a\} \\
&= \{b \in \mathbb{Z} \mid n \mid b - a\} \qquad \text{(by (3.6.3))} \\
&= \{b \in \mathbb{Z} \mid b - a = kn, \text{ for some integer } k\} \\
&= \{a + kn \mid k \in \mathbb{Z}\} \\
&= a + n\mathbb{Z},
\end{aligned}$$

where $n\mathbb{Z} = \{nk \mid k \in \mathbb{Z}\}$. The associated *quotient set*, denoted by $\mathbb{Z}_n$, is called the set of *integers modulo n*. Therefore, we have

$$\mathbb{Z}_n := \{[0], [1], \ldots, [n-1]\}, \tag{3.6.4}$$

where the $n$ distinct elements of $\mathbb{Z}_n$ are given by

$$\begin{aligned}
[0] &= 0 + n\,\mathbb{Z} = \{\ldots, -3n, -2n, -n, 0, n, 2n, 3n, \ldots\}, \\
[1] &= 1 + n\,\mathbb{Z} = \{\ldots, 1-3n, 1-2n, 1-n, 1, 1+n, 1+2n, 1+3n, \ldots\} \\
&\vdots \\
[n-1] &= (n-1) + n\,\mathbb{Z} = \{\ldots, -1-2n, -1-n, -1, n-1, 2n-1, 3n-1, 4n-1, \ldots\}.
\end{aligned}$$

Notice that $[n] = n + n\,\mathbb{Z} = [0]$, $[n+1] = (n+1) + n\,\mathbb{Z} = [1]$, and so on. In particular, with $A = \{2, 3, 4, 7, 8, 9\}$ and $n = 3$, we have

$$\begin{aligned}
[2] &= \{2, 8\} = [8], \\
[3] &= \{3, 9\} = [9], \\
[4] &= \{4, 7\} = [7], \\
[8] &= \{8\}.
\end{aligned}$$

**Definition 3.26** The relation (3.6.3) is also called the **congruence modulo n**. We say $a$ is **congruent** to $b$ modulo $n$, denoted by $\mathbf{a \equiv_n b}$, if $n \mid (b-a)$.

Said differently, two integers $a$ and $b$ are *congruent modulo n* if the two leave the same "residue" when divided by $n$. We may also write $a \equiv_n b$ as $a \bmod n = b \bmod n$ (or simply as $a = b \bmod n$). By (3.6.3), we have

$$a \equiv_n r \quad \Leftrightarrow \quad a = q\,n + r, \quad \text{where} \quad 0 \le r < n. \tag{3.6.5}$$

The element $[a] \in \mathbb{Z}_n$ is also called the **residue class modulo** $n$. The integer $n > 1$ is called the **modulus**. We also write the residue $r$ in (3.6.5) as $a \bmod n$. In particular, we have $a \bmod n = 0$ if and only if $n \mid a$. For example,

$$21 \bmod 7 = 0, \quad -20 \bmod 7 = 1, \quad \text{and} \quad 23 \bmod 7 = 2.$$

We may also write the three properties introduced in Example 3.16 as

(i) $a = a \bmod n$;
(ii) $a = b \bmod n \;\Rightarrow\; b = a \bmod n$;
(iii) $a = b \bmod n$ and $b = c \bmod n \;\Rightarrow\; a = c \bmod n$.

In view of definition (3.6.4) of the set $\mathbb{Z}_n$, it follows that the operator modn maps the set $\mathbb{Z}$ into the set of integers

$$\{0, 1, \ldots, n-1\}, \tag{3.6.6}$$

which we may again write as $\mathbb{Z}_n$. More precisely, each integer in the set $\mathbb{Z}_n$ represents a *residue class* mod$n$. Notice that it is customary that a particular residue class be represented by the smallest non-negative integer it contains. The next theorem proves some basic properties of integers in the set $\mathbb{Z}_n$ that we shall need in a later chapter to study certain important algebraic structures.

**Theorem 3.9** *For $a,\ b,\ c,\ d \in \mathbb{Z}_n$, we have*

1. $a = b \ \Rightarrow\ (a + c) = (b + c)$ *and* $ac \equiv bc$.
2. $a = b$ *and* $c = d \ \Rightarrow\ a \pm c = b \pm d$.
3. $a = b$ *and* $c = d \ \Rightarrow\ a\,c = b\,d$.
4. $a = b \ \Rightarrow\ a^k = b^k$, *for each integer* $k \geq 1$.

***Proof*** The assertions in (1) are trivial. For (2), since $a = b \bmod n$ and $c = d \bmod n$, we have

$$a = b + nk_1 \quad \text{and} \quad c = d + nk_2, \qquad \text{for some} \quad k_1, k_2 \in \mathbb{Z},$$

by (3.6.5). It thus follows that

$$\big(a \pm c\big) = \big(b \pm d\big) + n\big(k_1 \pm k_2\big),$$

which proves both cases of (1). A similar proof works for (2). For (3), let $k$ be a fixed integer, and $a = b \bmod n$. We then have

$$n \text{ divides } b - a \quad \Rightarrow \quad n \text{ divides } \big(b - a\big)\big(b^{k-1} + b^{k-1}a + \cdots + ba^{k-2} + a^{k-1}\big).$$

It thus follows that $a^k = b^k \bmod n$. This completes proof. □

The congruences in Theorem 3.9 have some interesting applications to the number theory, and also to cryptography as discussed in the last chapter.

**Example 3.17** We first show that every positive integer $a$ that is a *perfect square* is of the form $4k$ or $4k + 1$, for some integer $k$. For instance, we have $a^2$ is congruent modulo 4 equal to exactly one of the four values 0, 1, 2, or 3. That is,

$$a^2 \equiv_4 0, \quad a^2 \equiv_4 1, \quad a^2 \equiv_4 2^2 \equiv_4 0, \quad \text{or} \quad a^2 \equiv_4 3^2 \equiv_4 1,$$

by (3) of Theorem 3.9. Therefore, $a \neq 2 \bmod 4$ and $a \neq 3 \bmod 4$. Therefore, every perfect square is of the form $4k$ or $4k + 1$, for some integer $k$. Furthermore, if a positive integer $a$ is written in decimal form as

$$a = d_k\, 10^k + d_{k-1}\, 10^{k-1} + \cdots + d_1\, 10 + d_0, \qquad \text{where} \quad d_i \in \{0, 1, \ldots, 9\},$$

then we can use $10 = 1 \bmod 9 \ \Rightarrow\ 10^i = 1 \bmod 9$ to infer that

$$a = d_k + d_{k-1} + \cdots + d_1 + d_0 \bmod 9,$$

which proves $a$, and also the sum of digits in its decimal expansion are congruent to 9. In particular, it follows that $a$ is divisible by 9 if and only if the sum of its digits is divisible by 9.

We next introduce the concept of *partition* of a nonempty set.

**Definition 3.27** Let $A$ be a nonempty set. A **partition** of $A$ refers to a collection $\mathscr{P} := \{A_i \mid i \in I\}$ such that

(a) each $A_i$ is a nonempty subset $A$;
(b) $A_i \cap A_j = \emptyset$, for all $i \neq j \in I$;
(c) $A = \bigcup_{i \in I} A_i$.

We also say $A_i$'s are **partition sets** of $A$ with respect to the partition $\mathscr{P}$.

The concept of *partition* of a set $A$ finds many applications in theoretical computer science. For example, partition refinement techniques such as " lexicographic breadth-first search" (for graphs) is used to represent a partition of a set as a data structure. Also, several efficient algorithms on graphs and finite automata are based on partition refinement.

**Example 3.18** For $A = \{a, b, c, d, e, f, g, h, i\}$, the subsets $A_1, A_2, A_3, A_4$ given by

$$A_1 = \{a, c, d, e\}, \quad A_2 = \{b, f, g\}, \quad A_3 = \{h\}, \quad \text{and} \quad A_4 = \{i\}$$

form a partition of the set $A$. Clearly, there are many partitions of $A$. In general, for a set $A$ with at least three elements, we may have more than one partition.

It follows from (1) and (3) of Theorem 3.8 that a nonempty set $A$ with an equivalence relation $\sim$ is a disjoint union of equivalence classes. Therefore, every equivalence relation on a nonempty set defines a partition, with partitions given by the sets in $A/\sim$. The next theorem proves that the converse holds.

**Theorem 3.10** *Every partition of a nonempty set A defines an equivalence relation on A such that the associated equivalence classes are precisely the* partition sets.

***Proof*** For brevity, we may assume that $A$ is a finite nonempty set. Suppose $A_1, \ldots, A_n$ are *partition sets* of the set $A$. Define the relation $\sim$ on $A$ as follows:

$$a \sim b \quad \Leftrightarrow \quad a, b \in A_i, \quad \text{for some } i = 1, 2, \ldots, n.$$

We claim that $\sim$ is an equivalence relation. For instance, let $a \in A$. Then $a \in A_i$, for some $i$, and so $a \sim a$. Therefore, $\sim$ is reflexive. Next, let $a \sim b$. Then we have $a, b \in A_i$, for some $i$, and so $b \sim a$. Therefore, $\sim$ is symmetric. Finally, let $a \sim b$ and $b \sim c$. It thus follows that $a, b \in A_i$ and $b, c \in A_j$, for some $i, j \in \{1, \ldots, n\}$. However, since $A_i$ and $A_j$ are disjoint unless $i = j$, and we have $b \in A_i \cap A_j$, we must have $A_i = A_j$. It thus follows that $a, c \in A_i$, which proves that $a \sim c$. Therefore, $\sim$ is transitive. Hence, $\sim$ is an equivalence relation on $A$. It follows directly from the definition of the relation $\sim$ that the equivalence classes determined by elements $a \in A$ are precisely the sets in the given partition. This completes the proof. $\square$

Hence, the set of equivalence relations of a set and the set of its partitions are in one-to-one correspondence. There are some other ways to define an equivalence relation on a set:

(a) Suppose $R$ is a symmetric and transitive relation on a set $A$. We assert that if for any given $a \in A$ there is an element $b \in A$ such that $(a, b) \in R$, then $R$ is necessarily an equivalence relation. First, notice that if $R$ is already reflexive then the existence of $b$ is clear. For instance, we can take $b = a$. Otherwise, to show that $R$ is reflexive, let $a \in A$. Then by the given condition there is a $b \in A$ such that $aRb$. We also have $bRa$, by symmetry. Therefore, it follows that $aRa$, by transitivity. Hence, $R$ is an equivalence relation.

(b) Let $R$ be a reflexive relation on a set $A$, and suppose

$$(a, b) \text{ and } (a, c) \in R \qquad \Rightarrow \qquad (b, c) \in R.$$

We prove that $R$ is an equivalence relation. That is, we need to show that $R$ is symmetric and transitive. For instance, let $(a, b) \in R$. We also have $(a, a) \in R$, by reflexivity. It thus follows from the given condition that $(b, a) \in R$. Therefore, $R$ is symmetric. Next, to show that $R$ is transitive, let $(a, b)$, $(b, c) \in R$. Then, by symmetry, we have $(b, a)$, $(b, c) \in R$. Therefore, $(a, c) \in R$. Hence, $R$ is transitive. Notice that if $R$ is an equivalence relation, and if $(a, b), (a, c) \in R$, then we have $(b, a), (a, c) \in R$, by symmetry, and therefore $(b, c) \in R$, by transitivity.

## 3.7 Partially Ordered Set

Consider a nonempty set $P$. Recall that the set

$$\Delta_P := \{(x, x) \mid x \in P\} \tag{3.7.1}$$

is the *diagonal relation* on $P$. We can use $\Delta_P$ to reformulate reflexive and irreflexive properties as given below:

(a) a binary relation $R$ on $P$ is reflexive if and only if we have $\Delta_P \subset R$;
(b) a binary relation $R$ on $P$ is irreflexive if and only if $R \cap \Delta_P = \emptyset$.

In terms of the *inverse relation*

$$R^{-1} := \{(b, a) \in P \times P \mid \text{for } (a, b) \in R\}, \tag{3.7.2}$$

we have $R$ is symmetric if and only if $R = R^{-1}$. Also, $R$ is *antisymmetric* if

$$(a, b) \in R \text{ and } (b, a) \in R \quad \Rightarrow \quad a = b. \tag{3.7.3}$$

Clearly, the diagonal relation $\Delta_P$ is both symmetric and antisymmetric.

**Example 3.19** Consider a collection of items such as books or food items, each with a finite number of attributes. We may treat two items related if each attribute of one is also an attribute of the other. Clearly, this relation is reflexive, antisymmetric, and transitive. Also, the relation $\sim$ on the set $P = \{a, b, c, d, e\}$ given by

$$\sim = \Delta_P \bigcup \{(a, b), (a, c), (b, c), (b, d), (c, d), (c, e), (d, e), (d, a), (e, a), (e, b)\}$$

is reflexive, antisymmetric, and transitive. The usual order $<$ on the set of integers $\mathbb{Z}$ is neither reflexive nor antisymmetric. However, it is transitive.

The concept of "ordering" introduced below finds applications in many situations.

**Definition 3.28** A reflexive and transitive relation $R$ on $P$ is called a *pre-order*. An irreflexive and transitive relation $R$ on $P$ is called a *quasiorder*. An antisymmetric pre-order is called a **partial order**, and a nonempty set with a partial order is called a **partial ordered set** (or simply a **poset**).

An irreflexive transitive relation on a set is called a **strict order**. We can obtain a *strict order* from a partial order by deleting all diagonal elements. These types of order find applications in a variety of situations. Every strict order is asymmetric and also antisymmetric. Also, *reflexive closure* of a strict order is a partial order. The next examples illustrate the concepts introduced above.

**Example 3.20** (*Inclusion Order*) Let $X$ be a nonempty set. Let

$$R = \{(A, B) \mid A, B \in \mathscr{P}(X), \text{ with } A \subseteq B\},$$

where $\mathscr{P}(X)$ is the power set of $X$. The relation $R$ is *reflexive* because $A \subseteq A$, for all $A \in \mathscr{P}(X)$. Also, $R$ is *antisymmetric* because $A \subseteq B$ and $B \subseteq A$ implies that $A = B$. Further, $R$ is *transitive* because $A \subseteq B$ and $B \subseteq C$ implies that $A \subseteq C$. Therefore, $R$ is a partial order. However, since $\emptyset \subseteq A$ and $A \nsubseteq \emptyset$, the relation $R$ is not a symmetric relation. So, it is not an equivalence relation. Notice that $R$ is not irreflexive because $A \subseteq A$. We usually call $R$ the *inclusion relation*, and write it simply as $\subseteq$. Therefore, $(\mathscr{P}(X), \subseteq)$ is a poset, for any nonempty set $X$.

**Example 3.21** (*Usual Order*) On the set of real numbers $\mathbb{R}$, consider the relation

$$R = \{(x, y) \mid x, y \in \mathbb{R}, x \leq y\}.$$

This relation is *reflexive* because $a \leq a$, for all $a \in \mathbb{R}$. Also, it is *antisymmetric* because $a \leq b$ and $b \leq a$ implies that $a = b$. Further, $R$ is *transitive* because $a \leq b$

and $b \leq c$ implies that $a \leq c$. Therefore, the relation $R$ is a partial order on $\mathbb{R}$. However, $R$ is not a symmetric relation because $1 \leq 2$, but $2 \not\leq 1$. Hence, $R$ is not an equivalence relation. Also, notice that it is not irreflexive because $1 \leq 1$. We call the relation $R$ as *less than equal to*, and write it simply as $\leq$. Hence, $(\mathbb{R}, \leq)$ is a *poset*.

**Example 3.22** On any set $P \subseteq \mathbb{N}$, consider the relation $R$ given by

$$(a, b) \in R \quad \Leftrightarrow \quad a \text{ divides } b. \tag{3.7.4}$$

Clearly, $R$ is reflexive. Next, let $a$ divide $b$ so that $b = ak_1$, for some positive integer $k_1$, and also $b$ divides $a$ so that $a = bk_2$, for some positive integer $k_2$. But then, $a = a(k_1k_2)$ implies that $k_1k_2 = 1$. However, as both $k_1$ and $k_2$ are positive integers, we must have $k_1 = k_2 = 1$. Therefore, $a = b$, which shows that $R$ is antisymmetric. Now, if $a$ divides $b$ so that $b = ak_1$, for some positive integer $k_1$, and also $b$ divides $c$ so that $c = bk_2$, for some positive integer $k_2$; then $c = a(k_1k_2)$ implies that $a$ divides $c$. Therefore, $R$ is transitive. Hence, the relation $R$ in (3.7.4) is a partial order. It thus follows that every subset $P$ of $\mathbb{N}$ is a poset with respect to the relation "divides".

**Definition 3.29** (*Set of Divisors*) For integer $n > 1$, we write the set of the **divisors of n** as $D_n$. The set $D_n$ contains 1 and $n$ when $n$ is prime.

We prove in Theorem 5.26 that, if

$$n = p_1^{n_1} p_2^{n_2} \cdots p_k^{n_k},$$

where each $p_i$ is a *prime divisor* of $n$, then we have

$$|D_n| = (n_1 + 1)(n_2 + 1) \cdots (n_k + 1). \tag{3.7.5}$$

For example, since $60 = 2^2 \times 3 \times 5$, the set $D_{60}$ has $3 \times 2 \times 2 = 12$ elements. More explicitly, we have

$$D_{60} := \{1, 2, 3, 4, 5, 6, 10, 12, 15, 20, 30, 60\}. \tag{3.7.6}$$

By our discussion in Example 3.22, it follows that every set $D_n$ is a poset with respect to the relation *divides*.

**Notations.** In general, a partial order $R$ on a set $P$ is denoted by $\preceq$. We thus also write a *poset* as a pair $(P, \preceq)$. We write $a \preceq b$ when $(a, b) \in R$. When $a \prec b$, we say $a$ is smaller than $b$ or $b$ is bigger than $a$. We write $a \preceq b$ if either $a = b$ or $a \prec b$. We also write $a \succ b$ when $b \prec a$ and $a \succeq b$ if either $a = b$ or $a \succ b$.

**Definition 3.30** Let $(P, \preceq)$ be a poset. We say elements $a,\ b \in P$ are **comparable** with respect to the partial order $\preceq$ if we have

$$a \preceq b \quad \text{or} \quad b \preceq a. \tag{3.7.7}$$

Otherwise, $a$ and $b$ are said to be **incomparable**.

The term "partial" used in Definition 3.28 refers to fact that the condition (3.7.7) may not hold for every pair of elements, in general. That is, they may not be *comparable*. Also, a binary relation $R$ on $P$ is called a *connected relation* if it satisfies the condition

$$a \neq b \ \text{ in } \ P \quad \Rightarrow \quad (a, b) \in R \quad \text{or} \quad (b, a) \in R.$$

**Definition 3.31** Consider a nonempty set $P$, with a partial order $\preceq$, such that every pair of elements in $P$ is comparable in the sense of (3.7.7). We then call $\preceq$ a **total order** (or a **linear order**), and the pair $(P, \preceq)$ is called a **totally ordered set** (or a **linear ordered set**).

Notice that a binary relation on a set is a *linear order* if and only if it is a connected partial order. Clearly, the digraph of a linear order cannot contain a closed loop.

**Example 3.23** The usual order $\leq$ on a number set such as $\mathbb{N},\ \mathbb{Z},\ \mathbb{Q}$, or $\mathbb{R}$ is a familiar linear order. Also, the partial order $\preceq$ on $P = \{3, 6, 12, 24, 36\}$ given by

$$x \preceq y \quad \Longleftrightarrow \quad x \ \text{divides} \ y$$

is a total order, because we have $x$ divides $y$ or $y$ divides $x$, for every pair of elements $x, y \in P$.

Clearly, not every pair of elements of the poset $(\mathscr{P}, \subseteq)$ is comparable. However, it does contain many "chains" in the sense of the next definition.

**Definition 3.32** Let $(P, \preceq)$ be a poset.

(a) A set $C \subset P$ is called a **chain** if every pair of elements in $C$ is comparable. A *chain* $C \subset P$ is said to be a **maximal chain** if no superset of $C$ is a chain. The size of longest *maximal chain* in a poset $(P, \preceq)$ is called the **height** of $P$.
(b) A set $A \subset P$ is called a **antichain** if no two elements in $A$ are comparable. An *antichain* $A \subset P$ is said to be a **maximal antichain** if no superset of $A$ is an antichain. The size of longest *maximal antichain* in a poset $(P, \preceq)$ is called the **width** of $P$.

The next theorem is due to the American mathematician *Robert P. Dilworth* (1914–1993) [5]. The proof given below is due to *Fred Galvin* [6].

**Theorem 3.11** (Dilworth, 1950) *A finite poset P of* width *w can be partitioned into w chains.*

***Proof*** We use induction on $|P|$. The claim is trivial when $w = 1$, and so for $|P| = 1$. Let $a \in P$ be a maximal element, and let the size of a largest antichain in $P' = P - \{a\}$ be $k$. We then have $k \leq m$. By our induction hypothesis, $P'$ is the union of $k$ disjoint chains, say $C_1, \ldots, C_k$. Now, every $k$-element antichain in $P'$ consists of one element from each chain $C_i$. Suppose $a_i$ is the maximal element in $C_i$ that belongs to some $k$-element antichain in $P'$. It is not difficult to see that $A = \{a_1, \ldots, a_k\}$ is an antichain in $P'$. For a contradiction, suppose $a_i < a_j$, for some $i$ and $j$. By definition, $a_i$ is in some $k$-antichain $A_i$ and $a_j$ is in some $k$-antichain $A_j$. Since $A_j$ is a maximal antichain, it must intersect the chain $C_i$. Therefore, there is an $x \in C_i \cap A_j$. By definition of $a_i$, we have $x \leq a_i$. Hence, by transitivity $x < a_j$. This is the desired contradiction, because $x$ and $a_j$ are in the same antichain $A_j$. If $A \cup \{a\}$ is an antichain in $P$, then $k \leq m - 1$, and we are done: The chains $C_1, \ldots, C_k$ and $\{a\}$ give us a desired decomposition of $P$ into $k + 1 \leq m$ chains. Otherwise, we have $a > a_i$, for some $i$. Then

$$K = \{a\} \cup \{x \in C_i \mid x \leq a_i\}$$

is a chain in $P$, and there are no $k$-element antichains in $P - K$ (since $a_i$ was the maximal element of $C_i$ participating in such an antichain). By our induction hypothesis, $P - K$ is the union of $k - 1$ disjoint chains. Together with the chain $K$, these chains cover entire $P$, as desired. □

The Israeli mathematician *Micha Perles* gave a nice proof of an equivalent version of Theorem 3.11 [7], which is left for the reader as an exercise.

**Example 3.24** (*Lexicographic Order*) We introduce here an interesting linear order on the set of strings $\Sigma^*$, called the **lexicographic order**. As defined earlier in Example 3.3, the set $\Sigma^*$ consists of strings over an alphabet $\Sigma$ of all possible finite lengths. That is, we have

$$\Sigma^* = \Sigma^0 \cup \Sigma^1 \cup \Sigma^2 \cup \cdots \cup \Sigma^n \cup \cdots, \tag{3.7.8}$$

where $\Sigma^m$ is the set of strings of length $m \geq 0$, with the singleton $\Sigma^0$ containing the *empty string* $\varepsilon$. For two strings $s$ and $t$ in $\Sigma^*$, respectively, of lengths $p$ and $q$, we define the order $\preceq$ on the set $\mathscr{S}$ as follows:

$$s \preceq t := \begin{cases} t = sw, & \text{for some } w \in \Sigma^* \\ s = ua_iv_1, t = ua_jv_2, & \text{with } a_i \prec a_j, \text{ for some } u, v_1, v_2 \in \Sigma^* \end{cases} \tag{3.7.9}$$

It is also known as the **dictionary order**. It takes good effort to verify that $\preceq$ is indeed a linear order on the set $\Sigma^*$. The details are left for the reader as an exercise.

The linearly ordered set $\Sigma^*$ plays a significant role in various important formulations such as in *sorting the character data*. Also, strings constructed from finite number of characters such as $a, b, c, \ldots$ are used to develop a programming language. A limit on the number of characters used helps the *compiler* to determine when two strings are treated same. For $C$, only the first eight characters of a string are checked by the compiler. It also finds applications in the study of human genomics. Each gene is seen as a finite string in four alphabets that are initial letters of certain amino acids.

**Remark 3.2** In algebraic terms, the set $\Sigma^*$ is a monoid with respect to the binary operation *concatenation* as defined in (3.1.20), with the empty string $\varepsilon$ as the identity element. It is called the *free monoid* generated by the alphabet $\Sigma$.

It follows easily that, for any linear order set $(P, \prec)$, we have

(a) there is no $x \in P$ such that $x \prec x$ and
(b) $x \neq y$ in $P \quad \Rightarrow \quad$ either $x \preceq y$ or $y \preceq x$.

For a transitive relation, $(a)$ and $(b)$ are equivalent to the *trichotomy condition.*

**Definition 3.33** (*Ordered Set*) A transitive relation $\prec$ on a nonempty set $P$ is an **order** if, for any $x, y \in P$, the *trichotomy condition* as given below holds:

$$x \prec y, \qquad x = y, \qquad \text{or} \qquad y \prec x. \tag{3.7.10}$$

We then call the pair $(P, \prec)$ an **ordered set.**

## *Hasse Diagram*

The *Hasse diagram* of a finite poset $(P, \preceq)$ is a graphical tool, due to the German mathematician *Helmut Hasse* (1868–1940). It is a refined version of the *digraph* $D_{\preceq}$ of the relation $\preceq$ on the set $P$. We may recall that the *digraph* $D_{\preceq}$ has the elements of $P$ as vertices, and there is a directed edge from a vertex $a \in P$ to a vertex $b \in P$ if $a \preceq b$ (Definition 3.19). The following simple steps are applied to obtain the **Hasse diagram** of $(P, \preceq)$ from the digraph $D_{\preceq}$:

(a) Omit *loops* at all vertices of $D_{\preceq}$. No information is lost, because $\preceq$ is reflexive.
(b) Omit all *transitive edges*, i.e., the edges between two vertices that must be part of $D_{\preceq}$ due to the transitivity of the relation $\preceq$.
(c) Finally, the Hasse diagram is obtained by removing arrows from all the remaining edges, subject to that vertices are *oriented upwards.* That, in the Hasse diagram, the vertex $b$ is above the vertex $a$ whenever $a \preceq b$. Also, two vertices are at the same level if the corresponding elements in $P$ are *incomparable.*

We show below the Hasse diagrams of some simple posets.

**Example 3.25** Consider the partial order $\preceq$ on the set $P = \{a, b, c\}$ given by

$$\preceq = \big\{(a, a), (a, b), (a, c), (b, b), (c, c)\big\}.$$

Then the Hasse diagram of the poset $(P, \preceq)$ is the left side picture in Fig, 3.9. Next, for the partial order $\preceq$ on the set $P = \{3, 5, 15\}$ given by the relation "divides", the Hasse diagram is the right side picture in Fig, 3.9.

**Fig. 3.9** The Hasse diagrams of the first two posets in Example 3.25

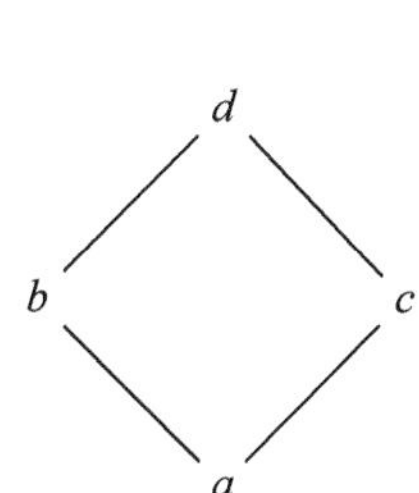


**Fig. 3.10** The Hasse diagrams of the first poset as in Example 3.26

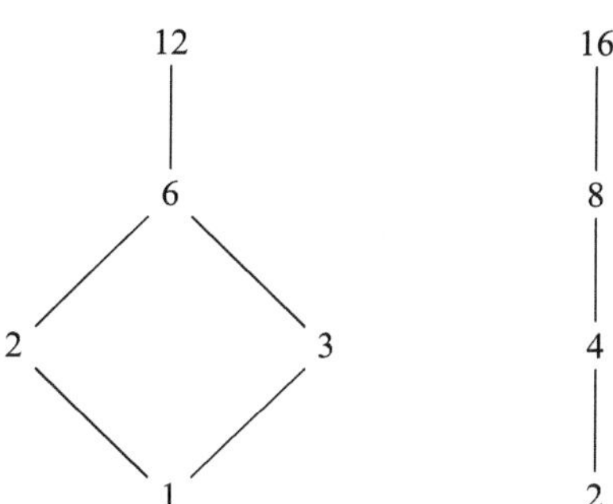


**Fig. 3.11** The Hasse diagram of the third poset in Example 3.26

**Example 3.26** Consider the relation $\preceq$ on the set $P = \{a, b, c, d\}$ given by

$$\preceq = \{(a,a), (a,b), (a,c), (b,b), (b,d), (c,c), (c,d), (d,d)\}.$$

Clearly, $(P, \preceq)$ is a poset, with the Hasse diagram as shown in Fig. 3.10. Next, the set $P = \{1, 2, 3, 6, 12\}$ is a poset with respect to relation $\preceq$ given by "divides". In this case, the Hasse diagram of the poset $(P, \preceq)$ is the left side picture in Fig. 3.11.

**Example 3.27** The set $P = \{2, 4, 8, 16\}$ and $\leq$ is a poset with respect to the relation "divides". Its Hasse diagram is the right side picture in Fig. 3.11. Notice that the poset $(\{3, 9, 27, 54\}, '' divides'')$ has the same Hasse diagram. Indeed, both posets are *totally ordered sets*. In general, the Hasse diagram of a *linearly ordered set* is a vertical line. Next, consider the poset $P = \{2, 3, 6, 12, 24, 36\}$ with respect to the relation "divides". The Hasse diagram of this poset is shown in Fig. 3.12.

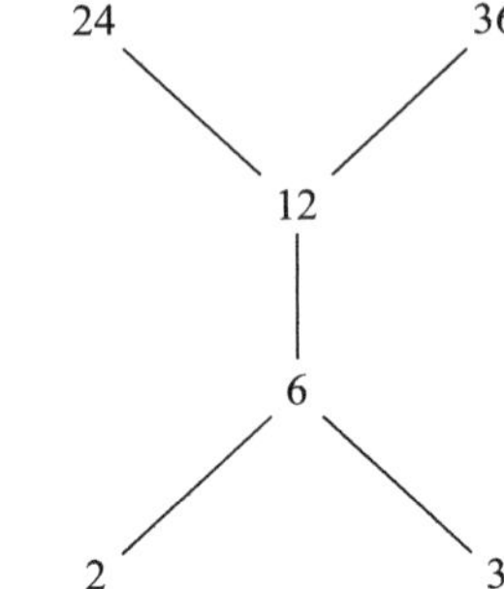


Fig. 3.12 The Hasse diagram of the poset in Example 3.27

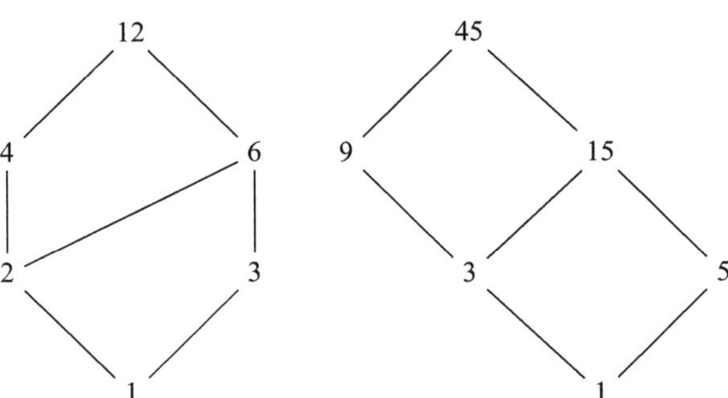


Fig. 3.13 The Hasse diagrams of the first two posets in Example 3.28

**Example 3.28** We know that $D_{12}$, $D_{36}$, $D_{45}$, and $D_{60}$ are posets with respect to the partial order "*divides*" (Example 3.22). Let $P$ denote $D_{12} = \{1, 2, 3, 4, 6, 12\}$, and $\leq$ be the relation "divides" on $P$. We thus have

$$\begin{aligned}\leq = \{&(1, 1), (1, 2), (1, 3), (1, 4), (1, 6), (1, 12),\\ &(2, 2), (2, 4), (2, 6), (2, 12), (3, 3), (3, 6), (3, 12),\\ &(4, 4), (4, 12), (6, 6), (6, 12), (12, 12)\}.\end{aligned}$$

The Hasse diagram of this poset is the left side picture in Fig. 3.13. Next, let $P = D_{45} = \{1, 3, 5, 9, 15, 45\}$, and $\leq$ be the relation "divides", so that we have

$$\begin{aligned}\leq = \{&(1, 1), (1, 3), (1, 5), (1, 9), (1, 15), (1, 45),\\ &(3, 3), (3, 9), (3, 15), (3, 45), (5, 5), (5, 15),\\ &(5, 45), (9, 9), (9, 45), (45, 45)\}.\end{aligned}$$

The Hasse diagram of the poset $(P, \leq)$ is the right side picture in Fig. 3.13.

**Example 3.29** Let $A = \{1\}$, and consider the poset $P = \mathscr{P}(A)$ with respect to relation $\subseteq$ so that we have

$$\subseteq = \{(\emptyset, \emptyset), (\emptyset, \{1\}), (\{1\}, \{1\})\}.$$

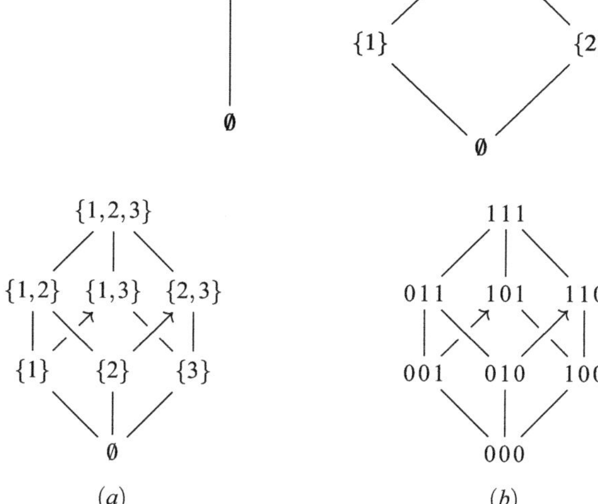


**Fig. 3.14** The Hasse diagram of the 1st and the 2nd posets in Example 3.28

**Fig. 3.15** Hasse diagrams of $(\mathscr{P}(\{1, 2, 3\}), \subseteq)$ and 3-bit strings

Then the Hasse diagram of this poset is the left side picture in Fig. 3.14. Next, let $A = \{1, 2\}$, and consider the poset $\big(\mathscr{P}(A), \subseteq\big)$ so that we have

$$\subseteq = \{(\emptyset, \emptyset), (\emptyset, \{1\}), (\emptyset, \{2\}), (\emptyset, A), (\{1\}, \{1\}), (\{1\}, A),$$

$$(\{2\}, \{2\}), (\{2\}, A), (\{A\}, \{A\})\}.$$

Then the Hasse diagram of this poset is the right side picture in Fig. 3.14. Also, if $A = \{1, 2, 3\}$, then $P = \mathscr{P}(A)$ is a poset with respect to relation $\subseteq$. The Hasse diagram is as shown in Fig. 3.15a. Notice that even though $\{3\} < \{1, 2, 3\}$ (since $\{3\} \subset \{1, 2, 3\}$), there is no edge directly between them because there are in-between elements: $\{2, 3\}$ and $\{1, 3\}$. However, there still remains an indirect path from $\{3\}$ to $\{1, 2, 3\}$. Also, the Hasse diagram of lexicographic ordering on 3-bit strings as described earlier in Example 3.24 shown in Fig. 3.15b.

## Exercises 3

3.1 Prove that

$$\Big(\bigcup_{i=1}^{n} A_i\Big)^c = \bigcap_{i=1}^{n} A_i^c \quad \text{and} \quad \Big(\bigcap_{i=1}^{n} A_i\Big)^c = \bigcup_{i=1}^{n} A_i^c.$$

3.2 Let $A = \{a, b, c, d\}$, $B = \{b, c, d, e\}$, and $C = \{a, d, f\}$. Verify the basic laws of set theory for these sets.
3.3 Show that $R \subseteq S$, $S \subset Q$ imply $R \subset Q$.
3.4 Let $A = \{a, b, c, d\}$, $B = \{b, c, d, e\}$. Find $P(A \cup B)$, $P(A \cap B)$, and $P(A \setminus b)$.
3.5 Let $A$ and $B$ be nonempty sets. Show that

a. $A \oplus A = \emptyset$ and $A \oplus \emptyset = A$.
b. $A \oplus B = B \oplus A$.
c. $A \oplus B = (A \cap B^c) \cup (A^c \cap B)$.
d. $A \oplus B = (A \cup B) \cap (A^c \cup B^c)$.
e. $A \oplus B = (A \cup B) - (A \cap B)$.
f. $A \oplus B = \emptyset$ if and only if $A = B$.

3.6 Let $A$, $B$, and $C$ be any three sets. Show that

a. $\emptyset \subseteq A$, $A \cap B \subseteq A$, $A \subseteq A \cup B$, and $A \setminus B \subseteq A$.
b. $A \subseteq A$ and $A \subseteq B$, $B \subseteq C \Rightarrow A \subseteq C$.
c. $A \subseteq B \quad \Leftrightarrow \quad A \cap B = A \quad \Leftrightarrow \quad A \cup B = B$.
d. $A \subseteq B \quad \Leftrightarrow \quad A \setminus B = \emptyset$.

3.7 Show that $(A \cap B) \cup C = A \cap (B \cup C)$ if and only if $C \subseteq A$.
3.8 Show that, for any two nonempty sets $A$ and $B$, we have

$$A \setminus B = A \cap B^c \quad \text{and} \quad A \setminus B = A \setminus (A \cap B).$$

3.9 Show that $(A \setminus B) \setminus C = A \setminus (B \cup C)$.
3.10 Show that $(A \setminus B) \setminus C = (A \setminus C) \setminus (B \setminus C)$.
3.11 (*involution law*) Let $A$ be a set with universal set $U$. Show that $\left(A^c\right)^c = A$.
3.12 Let $U$ be a universal set for the three sets $A$, $B$, and $C$. Show that

(i) $A \subseteq B \quad \Leftrightarrow \quad B^c \subseteq A^c$.
(ii) (*identity laws*) $A \cup \emptyset = A$, $\quad A \cap U = A$, $\quad A \cup U = U$, and $A \cap \emptyset = \emptyset$.
(iii) (*dominance*) $A \cup A^c = U$, and $A \cap A^c = \emptyset$.
(iv) $A \subseteq A^c \Leftrightarrow A = \emptyset$ and $A^c \subseteq A \Leftrightarrow A = U$.
(v) $A \cup A = A = A \cap A$.
(vi) (*commutative laws*) $A \cup B = B \cup A$ and $A \cap B = B \cap A$.
(vii) (*associative laws*) $(A \cup B) \cup C = A \cup (B \cup C)$ and $(A \cap B) \cap C = A \cap (B \cap C)$.
(viii) (*distributive laws*)

$$A \cup (B \cap C) = (A \cup B) \cap (A \cup C), \quad \text{and} \quad A \cap (B \cup C) = (A \cap B) \cup (A \cap C).$$

(ix) (*absorption laws*) $A \cup (A \cap B) = A$ and $A \cap (A \cup B) = A$.
(x) (*De Morgan's laws*) $(A \cup B)^c = A^c \cap B^c$ and $(A \cap B)^c = A^c \cup B^c$.

3.13 Use Venn diagram to verify De Morgan's laws.
3.14 Use Venn diagram to verify the distributive laws.

3.15 Consider the sets $A = \{a, b, c, d\}$, $B = \{b, c, d, e\}$, and $C = \{a, d, f\}$. Compute $A \times B$, $A \times C$, and $B \times C$. Also, find $(A \times B) \cap (B \times A)$ and $(A \times C) \cup (C \times A)$.

3.16 Let $A$, $X$ and $Y$ be three nonempty sets. Interpret the product $A \times X \times Y$, where $A$ is the set of airlines and $X$, $Y$ are sets of cities connected by air.

3.17 Let $X$, $Y$, and $Z$ be three nonempty sets.

(a) What can we say about the sets $X$ and $Y$ if $X \times Y = \emptyset$?
(b) Show that $X \times Y \neq Y \times X$ unless $X = Y$.
(c) Why $X \times Y \times Z$ and $(X \times Y) \times Z$ are not the same sets?

3.18 List all possible relations on the set $A = \{a, b, c, d\}$, and discuss their properties.

3.19 Give an example of a relation on the set $A = \{0, 1, 2, 3\}$ such that it is $(a)$ reflexive, $(b)$ irreflexive, $(c)$ neither reflexive nor irreflexive, $(d)$ symmetric, $(e)$ antisymmetric, $(f)$ symmetric and antisymmetric, $(g)$ symmetric but not antisymmetric, $(h)$ not symmetric but antisymmetric, and $(i)$ neither symmetric nor antisymmetric.

3.20 Consider the relations $R$ and $S$ on the set $X = \{1, 2, 3, 4\}$ given by $R = \{(x, y) \mid x - y \text{ is a multiple of } 2\}$ and $S = \{(x, y) \mid x - y \text{ is a multiple of } 3\}$. Find the relations $R \cap S$ and $R \cup S$.

3.21 On the set of integers $\mathbb{Z}$, consider the relations $R$ and $S$ given by

$$\begin{aligned} R &= \{(x, y) \mid x, y \in \mathbb{Z}, \text{ with } x \neq y\}; \\ S &= \{(x, y) \mid x, y \in \mathbb{Z}, \text{ with } x = y + 1 \text{ or } x = y - 1\}; \\ T &= \{(x, y) \mid x, y \in \mathbb{Z}, \text{ with } xy \geq 1\}; \\ U &= \{(x, y) \mid x, y \in \mathbb{Z}, \text{ and } x = y^2\}. \end{aligned}$$

Examine the above four relations for properties such as reflexive, irreflexive, symmetric, antisymmetric, and transitivity.

3.22 Let $R$ be a relation on a set $A$. Show that $R$ is reflexive and transitive, then $R^n = R$, for any $n \geq 1$.

3.23 Let $R$ and $S$ be relations on a set $A$. Show that

a. If $R$ and $S$ are reflexive then so are $R \cap S$ and $R \cup S$.
b. If $R$ and $S$ are symmetric then so are $R \cap S$ and $R \cup S$.
c. If $R$ and $S$ are transitive then so is $R \cap S$. Give an example to show that $R \cup S$ need not be transitive.

In particular, the intersection of two equivalence relations is an equivalence relation.

3.24 Let $R$ be a relation on a set $A$. Show that

a. $R$ is reflexive if and only if $R^{-1}$ is reflexive;
b. $R$ is symmetric if and only if $R = R^{-1}$;
c. $R$ is transitive if and only if $R^{-1}$ is transitive.

3.25 Consider the two relations $R$ and $S$ given by

$$R = \{(1,2), (2,2), (3,4)\} \quad \text{and} \quad S = \{(1,3), (2,5), (3,1), (4,2)\}.$$

Compute the matrices $M_{R \circ S}$ and $M_{S \circ R}$. Also, write the matrix of the inverse relation $R^{-1}$.

3.26 Prove that every quasiorder on a nonempty set is antisymmetric.

3.27 Let $X$ be a set with $n$ elements. How many reflexive relations could possibly be defined on $X$ ? How many symmetric relations could possibly be defined on $X$ ?

3.28 Let $X$ be a nonempty set. Show that a relation $R$ on $X$ is symmetric and transitive if and only if $R = R^{-1} \circ R$.

3.29 Prove each one of the statements as given below:

a. On the set $\mathbb{Z}$, define $a \sim b$ if and only if $a = b$ or $a = -b$. Show that the relation $\sim$ is an equivalence relation.
b. On the set $\mathbb{R}$, define $a \sim b$ if and only if $b - a \in \mathbb{Z}$. Show that the relation $\sim$ is an equivalence relation.
c. On the set $\mathbb{R}$, define $r \sim s$ if and only if $|r - s| < 1$. Show that the relation $\sim$ is not an equivalence relation.

3.30 Prove that the minimal number $m$ of disjoint chains covering a finite poset $P$ equals the maximal size $M$ of an antichain in $P$.

3.31 Draw all possible Hasse diagrams of a 3-element poset.

3.32 Let $A = \{a, b, c, d\}$, and consider the poset $P = \mathscr{P}(A)$ with respect to the relation $\subseteq$. Draw the Hasse diagram of the poset $(P, \subseteq)$.

## References

1. Devlin, K.: Sets, Functions, and Logic: An Introduction to Abstract Mathematics (3/e). Chapman & Hall/CRC (2003)
2. Nissanke, N.: Introductory Logic and Sets for Computer Scientists. Addison-Wesley (1999)
3. Stanat, D.F., McAllister, D.F.: Discrete Mathematics in Computer Science. Prentice-Hall (1977)
4. Enderton, H.B.: Elements of Set Theory. Academic Press Inc., NY (1977)
5. Dilworth, R.P.: A decomposition theorem for partially ordered sets. Ann. Math. **51**(1), 161–166 (1950)
6. Galvin, F.: A proof of Dilworth's chain decomposition theorem. Am. Math. Mon. **101**(4), 352–353 (1994)
7. Perles, M.A.: A proof of Dilworth's decomposition theorem for partially ordered sets. Isr. J. Math. **1**, 105–107 (1963)

# Chapter 4
# Functions

*For hundreds of pages the closely-reasoned arguments unroll, axioms and theorems interlock. And what remains with us in the end? A general sense that the world can be expressed in closely-reasoned arguments, in interlocking axioms and theorems.*

***Micheal Frayan*** *(1933– )*

The concept of a function is fundamental to everything we do in mathematics, and also in theoretical computer science. We need some special class of functions to study the *size* of a set; the *growth rate* of certain integer-argument functions that help analyze the computational complexity of an algorithm; design of optimal controls; and so on. Our emphasis in this chapter is to introduce certain important class of functions, discuss their properties, and prove related theorems. Some applications to problems in theoretical computer science are also included.

## 4.1 Preliminaries and Important Functions

In 1837, *Gustav Dirichlet* (1805–1859) defined a *function* as a "formula or an expression" that yields a unique *output value* for each assigned *input value*. This definition is found to be inadequate for various applications, including in computer science. We also need functions that are defined as various types of sequences, summations, and also as integrals. In 1888, *Richard Dedekind* defined a function as a *rule* $f$ that assigns to each element $x$ a nonempty set $X$ a unique element $f(x)$ of a nonempty set $Y$. For example, the function $\sigma : \mathbb{Z}^+ \to \mathbb{Z}^+$ given by $\sigma(a) = a + 1$, for $a \in \mathbb{Z}^+$,

 
V. Ravichandran and A. K. Razdan, *Fundamental Discrete Structures*,
University Texts in the Mathematical Sciences,
https://doi.org/10.1007/978-981-96-0069-4_4

is well-known Peano's *successor function*. The next definition gives a more abstract version of a function.

**Definition 4.1** Let $X$ and $Y$ be nonempty sets. A **function** $f : X \to Y$ is a relation $f \subseteq X \times Y$ such that every $x \in X$ appears as the ordinate of some ordered pair in $f$, and for $x \in X$ and $y_1, y_2 \in Y$,

$$\big(x, y_1\big),\ \big(x, y_2\big) \in f \quad \Rightarrow \quad y_1 = y_2. \tag{4.1.1}$$

In this case, we write

$$f = \big\{(x, y) \in X \times Y \mid y = f(x),\ \text{for } x \in X\big\}, \tag{4.1.2}$$

where $X := \mathrm{Dom}(f)$ is called the **domain** of $f$, $Y$ is called the *codomain* of $f$, and

$$\mathrm{Ran}(f) := \big\{y = f(x) \in Y \mid x \in X\big\} \tag{4.1.3}$$

is called the **range** of $f$. The set

$$\Gamma_f := \big\{(x, f(x)) : x \in X\big\} \subset X \times Y \tag{4.1.4}$$

is called the **graph** of the function $f$. Sometimes, we also write $f : X \to Y$ simply as $X \xrightarrow{f} Y$.

Since a relation $X \to Y$ in general is more likely to violate the first or the second condition of Definition 4.1, only a select few relations $f \subseteq X \times Y$ constitute a function. However, there can be at least as many functions $f : X \to Y$ as there are elements in the set $Y$: For any fixed $y \in Y$, the function $C_y : X \to Y$ defined by

$$C_y\big(x\big) := y, \quad \text{for all} \quad x \in X \tag{4.1.5}$$

is indeed a function, which is called the *constant function* on the set $X$ given at $y \in Y$. Whenever possible, one may choose two or more elements in the set $Y$, and define various other types of functions $f : X \to Y$. Let $X = \{1, 2, 3, 4\},\ Y = \{1, 4, 9, 16\}$, and $f : X \to Y$ be given by $f(x) = x^2$. We have

$$f(1) = 1, \quad f(2) = 4, \quad f(3) = 9, \quad \text{and} \quad f(4) = 16.$$

The graph of this function may be seen as in Fig, 4.1. Also, the graph of the *constant function* $C_a : \mathbb{R} \to \mathbb{R}$ given by $C_a(x) = x$ ($\forall x \in \mathbb{R}$) is the line $y = a$. In particular, the graph of the constant function $C_0 : \mathbb{R} \to \mathbb{R}$ is the *x-axis*.

In general, for an element $x \in X$, the element $y = f(x) \in Y$ is called the *image* of $x$ under $f$. Also, for any fixed $y \in Y$, the set

$$f^{-1}\big(y\big) = \big\{x \in X \mid f(x) = y\big\} \tag{4.1.6}$$

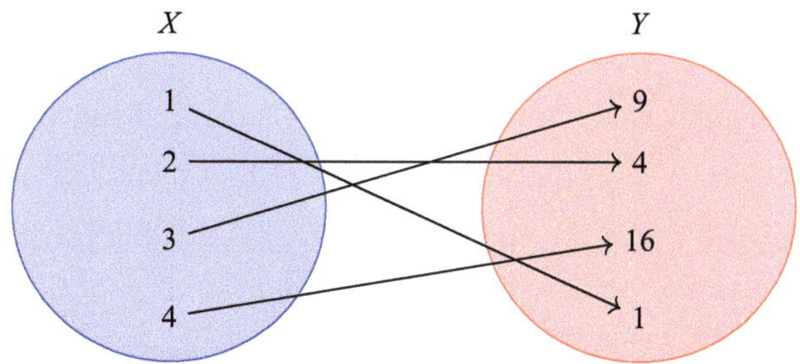


**Fig. 4.1** Graph of a simple function

consists of the *preimages* in $X$ of the element $y \in Y$. We may possibly have $f^{-1}(y) = \emptyset$, for some $y \in Y$. In some applications, we also call $x \in X$ as an *argument* or a *variable*, and $y \in Y$ as the *dependent variable* (or *value element*).

**Definition 4.2** Two functions $f : X \to Y$ and $g : U \to V$ are said to be **equal** if $X = U$; $Y = V$, and we have $f(x) = g(x)$, for all $x \in X(= U)$.

Moreover, when the two sets $X$ and $Y$ are equipped with analytical, algebraic, or a geometric structure, and $f : X \to Y$ is a *structure-preserving* function, we also use the terminology *mapping* or *transformation* for $f$.

**Definition 4.3** Let $X$ and $Y$ be two sets, and $f : X \to Y$ be a function. For any $A \subseteq X$, we call $f \cap (A \times Y)$ the **restriction function** of $f$, which is usually written as $f\big|_A$. In this case, we also say that $f$ is an *extension* of the function $f\big|_A$.

For example, let $X = \{1, 2, 3\}$ and consider the function $f : X \to X$ be given by

$$f(1) = 1, \qquad f(2) = 3, \qquad f(3) = 2$$

so that we can also write $f = \{(1, 1), (2, 3), (3, 2)\}$. Then, for $A = \{1, 3\}$, the *restriction function* $f\big|_A$ is given by

$$f\big|_A(1) = 1, \quad f\big|_A(3) = 2 \quad \text{or} \quad f\big|_A = \{(1, 1), (3, 2)\}.$$

**Definition 4.4** Let $A$ be a nonempty set. The function $I_A : A \to A$ given by

$$I_A := \{(a, a) \mid a \in A\} \tag{4.1.7}$$

is called the **identity function** of the set $A$. We may also write $I_A$ simply as $I$.

For example, the *identity function* $I_A$ of the set $A = \{1, 2, 3\}$ is given by

$$I_A(1) = 1, \quad I_A(2) = 2, \quad \text{and} \quad I_A(3) = 3.$$

Notice that the *graph* of the identity function $I : \mathbb{R} \to \mathbb{R}$ is the line $y = x$.

## Elementary Functions

An *elementary function* is roughly a class of functions that follow Dirichlet's definition. The most basic types of elementary functions are *polynomial functions*. A polynomial function of the form $y = f(x) = mx + c$ represents a *line*, and a polynomial function of the form $y = f(x) = ax^2 + bx + c$ represents a *parabola*. More generally, we have the next definition.

**Definition 4.5** (*Polynomial Functions*) A **polynomial function** is an expression of the form

$$p(x) = a_0x^n + a_1x^{n-1} + \cdots + a_n,$$

where $a_0, a_1, \ldots, a_n$ are called *coefficients* of the polynomial $p(x)$. We call $p(x)$ an *integer polynomial* if all $a_k \in \mathbb{Z}$, a *rational polynomial* if all $a_k \in \mathbb{Q}$, and a *real polynomial* if all $a_k \in \mathbb{R}$. The set of integer polynomials is denoted by $\mathbb{Z}[x]$, the set of rational polynomials is denoted by $\mathbb{Q}[x]$, and the set of real polynomials is denoted by $\mathbb{R}[x]$. When $a_0 \neq 0$, we say $p(x)$ is a polynomial of *degree* $n$.

Other types of elementary functions are *exponential, logarithm* (Figs. 4.2, 4.3), and *trigonometric functions*. Some of these are defined as *infinite series* of real numbers.

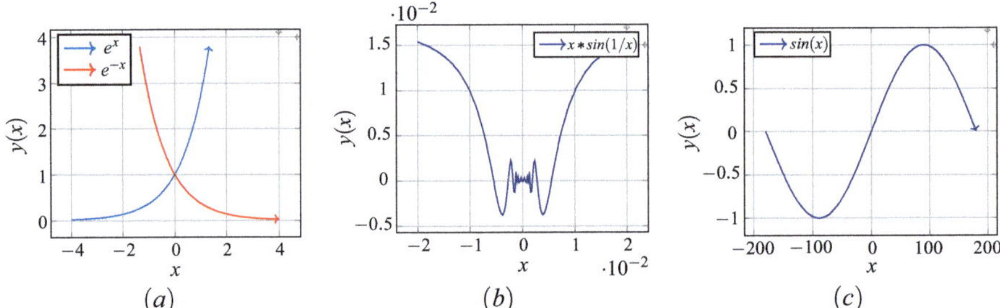


**Fig. 4.2** Graphs of some simple functions

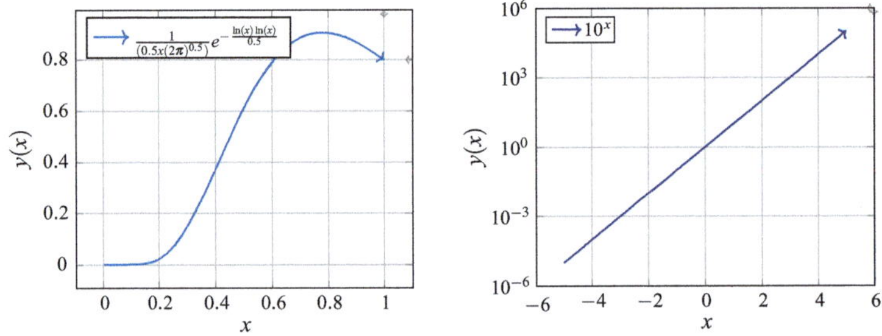


**Fig. 4.3** Graphs of some other simple functions

## *Floor and Ceiling Functions*

The two intimately related functions given in the next definition are based on the *Archimedean property* of the real numbers.

**Definition 4.6** The integer-valued functions given by

$$\lfloor x \rfloor = n, \quad \text{if} \quad x \geq n, \quad x \in \mathbb{R}; \tag{4.1.8a}$$

$$\lceil x \rceil = n, \quad \text{if} \quad x \leq n, \quad x \in \mathbb{R} \tag{4.1.8b}$$

are, respectively, called the **floor function** and the **ceiling function**. The floor function is also known as the *greatest integer function* (Fig. 4.4).

The two functions in Definition 4.6 are also known as the *stair functions*. In the next theorem, we state some interesting properties of the *floor function* and the *ceiling function* that prove very useful in computer science.

**Theorem 4.1** *For any $x \in \mathbb{R}$,*

1. $\lfloor x \rfloor = n \Leftrightarrow n \leq x < n + 1$; $\lceil x \rceil = n \Leftrightarrow n - 1 < x \leq n$.
2. $\lfloor x \rfloor = n \Leftrightarrow x - 1 < n \leq x$; $\lceil x \rceil = n \Leftrightarrow x \leq n < x + 1$.
3. $x - 1 < \lfloor x \rfloor \leq x \leq \lceil x \rceil < x + 1$.
4. $\lfloor -x \rfloor = -\lceil x \rceil$ *and* $\lceil -x \rceil = -\lfloor x \rfloor$.
5. $\lfloor x + n \rfloor = \lfloor x \rfloor + n$ *and* $\lceil x + n \rceil = \lceil x \rceil + n$.

In general, most assertions concerning the *floor functions* $\lfloor\ \rfloor$ can be proved by using the following relation:

$$x = n + \varepsilon, \quad \text{where} \quad n = \lfloor x \rfloor \in \mathbb{Z} \text{ and } 0 \leq \varepsilon < 1. \tag{4.1.9}$$

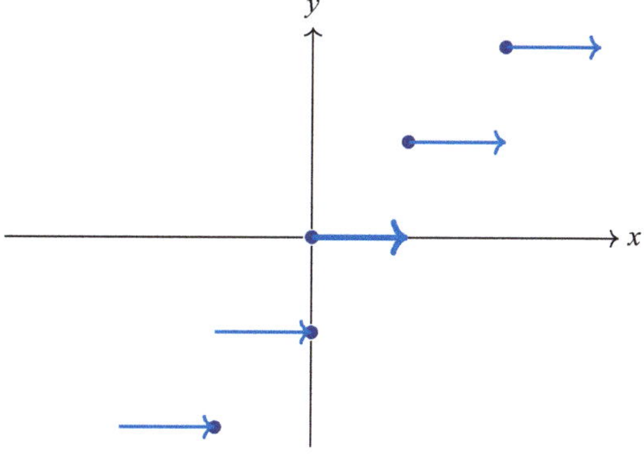


**Fig. 4.4** The greatest integer function $f(x) = \lfloor x \rfloor$

Similarly, to prove an assertion concerning the *ceiling functions*, we may use the relation as given below:

$$x = n - \varepsilon, \quad \text{where} \quad n = \lceil x \rceil \in \mathbb{Z} \text{ and } 0 \le \varepsilon < 1. \tag{4.1.10}$$

## *Sequences and Summation-Type Functions*

An important function that we shall use frequently is defined as an ordered list of elements of a nonempty set $X$, where the ordering is provided by the set of natural numbers $\mathbb{N}$. We call such type of function a *sequence* in the set $X$. We may view a *string of symbols*, introduced earlier in Example 3.3, as a *finite sequence*. In general, a sequence in a set $X$ is an infinite ordered list of elements of the set $X$. A formal definition is as given below.

**Definition 4.7** (*Sequence*) A **sequence** in a nonempty set $X$ is a function $x : \mathbb{N} \to X$. We usually write the *sequence* $x$ as $\langle x_n \rangle$, where

$$x_1 = x(1), \quad x_2 = x(2), \quad x_3 = x(3), \ \ldots$$

In this case, $x_n = x(n)$ is called the **n-term** of the sequence $x$. Two sequences $x = \langle x_n \rangle$ and $y = \langle y_n \rangle$ are said to be **equal** if $x$ and $y$ are equal as functions, i.e., we have $x(n) = y(n)$, for all $n \in \mathbb{N}$.

In particular, a *real sequence* is a sequence in the set of real numbers $\mathbb{R}$. Some simple illustrations of real sequences are as given below:

(a) $x(n) = 2n + 1$ is the sequence $\langle 3, 5, 7, \ldots \rangle$;
(b) $x(n) = (-1)^n \dfrac{n^2 - 1}{n}$ is the sequence $\langle 0, -3/2, 8/3, \ldots \rangle$;
(c) $x(n) = \begin{cases} \dfrac{1-n}{2}, & \text{if } n \text{ is odd} \\ \dfrac{n}{2}, & \text{if } n \text{ is even} \end{cases}$ is the sequence $\langle 0, 1, -1, 2, -2, \ldots \rangle$. Notice that this sequence lists the set of integers $\mathbb{Z}$ as $0, \ 1, \ -1, \ 2, \ -2, \ \ldots$;
(d) $x(n) = \cos n\pi = (-1)^n$ is the sequence $\langle -1, 1, -1, \ldots \rangle$. Notice that this sequence is a collection of the *discrete values* of the function $f(x) = \cos(\pi x)$, $x \in \mathbb{R}$, *sampled at points* $n = 1, 2, 3, \ldots$;
(e) $x(n) = \ln n$ is also an interesting sequence.

A large part of mathematics uses sequences of complex numbers, vectors, matrices, functions, and even sets. In some situation, it is more common to consider a sequence of the form $\langle x_0, x_1, x_2, \ldots \rangle$, which is called a *casual sequence*. The central theme of classical analysis is to use the concept of *convergence of a sequence* as a tool to model reality. Next, let $X$ be an *additive set*, i.e., the sum $x + y$ is a well-defined element of the set $X$, for any $x, y \in X$. A *sum function* over an additive set refers to an *infinite series* of the form:

$$\sum_{n=1}^{\infty} := x_1 + x_2 + x_3 + \cdots, \quad \text{with} \quad x_i \in X. \tag{4.1.11}$$

We call $x_n$ the *nth term* of the infinite series. In general, to make sense out of an "infinite sum" of elements of a set, we consider the sequence $\langle s_n \rangle$ given by

$$s_n = x_1 + x_2 + \cdots + x_n,$$

which is called the sequence of *partial sums* of the infinite series (4.1.11). Notice that we can write the $n$th term of the infinite series (4.1.11) as

$$x_n = s_{n+1} - s_n, \quad \text{for} \quad n \geq 1.$$

We say the infinite series (4.1.11) has the sum $S < \infty$ if the sequence of partial sums $\langle s_n \rangle$ converges to $S$ in some appropriate sense. Therefore, to deal with the complexities involved with the *summing problems* associated with an infinite series, we need to analyze its sequence of partial sums. For example,

$$1 + 1 + \frac{1}{1 \cdot 2} + \frac{1}{1 \cdot 2 \cdot 3} + \frac{1}{1 \cdot 2 \cdot 3 \cdot 4} + \cdots$$

is an important infinite series of real numbers that represents *Euler constant*. The infinite series expression for the function $f(x) = (1 + x)^q$, $q \in \mathbb{Q}$ is given by

$$(1 + x)^q = 1 + \frac{q}{1} x + \frac{q(q-1)}{1 \cdot 2} x^2 + \frac{q(q-1)(q-2)}{1 \cdot 2 \cdot 3} x^3 + \ldots \tag{4.1.12}$$

with $|x| < 1$. The above expansion was first obtained by Sir Issac Newton, while extending the Binomial Theorem for fractional exponents. In general, the *summation-type functions* help study the limiting cases of different type of infinite process occurring in various dynamical systems. We also use some *integral-type functions* such as *beta, gamma*, and *error functions*, in various important applications in engineering. In this text, we shall use infinite series only as *formal sums* to solve some combinatorial problems.

**Remark 4.1** The classical analysis is largely the study of functions defined in terms of infinite series of numbers, vectors, matrices, functions, sets, etc. For example, if

$$X = \mathbb{R}^I := \{f : f : I \to \mathbb{R}\}, \quad \text{where} \quad I \subseteq \mathbb{R},$$

then a sequence $\sigma = \langle f_n \rangle$ in the set $X$ is a function $\sigma : \mathbb{N} \to X$. In this case, we also need to define the function given by sum $\sum_{x \in I} f_n(x)$, if it exists.

## *Characteristic Function of a Set*

The function defined below is not a formula-based function.

**Definition 4.8** (*Characteristic Function*) Let $E$ be a nonempty set, and $A \subseteq E$. The **characteristic function**[1] of the set $A$, denoted by $\lambda_A$ (or $\chi_A$), is a function $\lambda_A : E \to \{0, 1\}$ given by

$$\lambda_A(x) = \begin{cases} 1, & \text{if } x \in A \\ 0, & \text{if } x \in E \setminus A \end{cases}.$$

It is also called the *indicator function* of the set $A$ and, in this case, we write $1_A$.

We can also use the *characteristic function* $\lambda_A$ to describe the set $A \subseteq E$ as

$$A = \{x \in E \mid \lambda_A(x) = 1\}.$$

In particular, it follows that we have

(a) $\lambda_A = 0 \quad \Leftrightarrow \quad A = \emptyset$;
(b) $\lambda_A = 1 \quad \Leftrightarrow \quad A = E$.

For $(a)$, we have

$$\lambda_A(x) = 0 \quad \Leftrightarrow \quad x \notin A \ (\forall\, x \in E) \quad \Leftrightarrow \quad A = \emptyset.$$

For $(b)$, we have

$$\lambda_A(x) = 1 \quad \Leftrightarrow \quad x \in A \ (\forall\, x \in E) \quad \Leftrightarrow \quad A = E.$$

Next, we show that

$$\lambda_{A^c} = 1 - \lambda_A.$$

For instance, we have

$$\begin{aligned} \lambda_{A^c}(x) &= \begin{cases} 1 & (x \in A^c) \\ 0 & (x \notin A^c) \end{cases} \\ &= \begin{cases} 1 & (x \notin A) \\ 0 & (x \in A) \end{cases} \\ &= 1 - \begin{cases} 0 & (x \notin A) \\ 1 & (x \in A) \end{cases} \\ &= 1 - \lambda_A(x). \end{aligned}$$

[1] The characteristic function of a set plays an important role in Lebesgue's theory of integration.

Further, since we have

$$\begin{aligned}\lambda_{(A^c)^c}(x) &= 1 - \lambda_{A^c}(x)\\ &= 1 - (1 - \lambda_A(x))\\ &= \lambda_A(x),\end{aligned}$$

it follows that $(A^c)^c = A$. The next theorem proves some basic set theoretic properties in terms of associated characteristic functions.

**Theorem 4.2** *Let $A$ and $B$ be two subsets of the set $E$. Then we have*

1. $\lambda_A(x) \leq \lambda_B(x)$, *for all $x \in E$, if and only if $A \subseteq B$.*
2. $\lambda_A = \lambda_B$ *if and only if $A = B$.*
3. $\lambda_{A\cap B} = \lambda_A \lambda_B$.
4. $\lambda_{A\cup B} = \lambda_A + \lambda_B - \lambda_A \lambda_B$.
5. $\lambda_{A\oplus B} = \lambda_A + \lambda_B - 2\lambda_A \cdot \lambda_B$.
6. $a\lambda_A + b\lambda_B = a\lambda_{A\setminus B} + (a+b)\lambda_{A\cup B} + b\lambda_{B\setminus A}$, *for any real numbers $a$ and $b$.*

***Proof*** For (1), we first assume that $A \subseteq B$, and let $x \in A$. Then, we have $\lambda_A(x) = 1$ and $\lambda_B(x) = 1$ so that $\lambda_A(x) = \lambda_B(x)$. Next, for $x \notin A$, we have $\lambda_A(x) = 0$, whereas $\lambda_B(x) \geq 0$. Therefore, we have $\lambda_A(x) \leq \lambda_B(x)$. Conversely, suppose $\lambda_A(x) \leq \lambda_B(x)$, for all $x \in E$. Then, for $x \in A$, $\lambda_A(x) = 1$, which shows that

$$1 \leq \lambda_B(x) \leq 1 \quad \text{or} \quad \lambda_B(x) = 1, \quad \text{or} \quad x \in B.$$

It thus follows that $A \subseteq B$. For (2), notice that $a = b$ is equivalent to $a \leq b$ and $b \leq a$, and similarly $A = B$ is equivalent to $A \subseteq B$ and $B \subseteq A$. Therefore, the assertion follows from (1). For (3), we consider the two cases as given below:

(a) Let $x \in A \cap B$. Then $x \in A$ and $x \in B$. Therefore, we have $\lambda_{A\cap B}(x) = 1$ and $\lambda_A(x) = 1 = \lambda_B(x)$. Thus we have

$$\lambda_{A\cap B}(x) = \lambda_A(x)\lambda_B(x).$$

(b) Let $x \notin A \cap B$. Then $x \notin A$ or $x \notin B$. Therefore, we have $\lambda_{A\cap B}(x) = 0$ and $\lambda_A(x) = 0$ or $\lambda_B(x) = 0$. We thus have

$$\lambda_A(x)\lambda_B(x) = 0 = \lambda_{A\cap B}(x).$$

Next, to prove (4), we have

$$\begin{aligned}\lambda_{A\cup B}(x) &= 1-\lambda_{(A\cup B)^c}\\ &= 1-\lambda_{A^c\cap B^c}(x)\\ &= 1-\lambda_{A^c}(x)\lambda_{B^c}(x)\\ &= 1-(1-\lambda_A(x))(1-\lambda_B(x))\\ &= \lambda_A(x)+\lambda_B(x)-\lambda_A(x)\lambda_B(x).\end{aligned}$$

The proofs of (5) and (6) are left to the reader as exercises. □

We can use property (2) in above theorem to prove *set identities* such as

$$A\cap(B\cap C)=(A\cap B)\cup(A\cap C).$$

For instance, we have

$$\begin{aligned}&\lambda_{A\cap(B\cap C)}(x)\\ &= \lambda_A(x)\lambda_{B\cup C}(x)\\ &= \lambda_A(x)(\lambda_B(x)+\lambda_C(x)-\lambda_{B\cap C}(x))\\ &= \lambda_A(x)\lambda_B(x)+\lambda_A(x)\lambda_C(x)-\lambda_A(x)\lambda_{B\cap C}(x)\\ &= \lambda_{A\cap B}(x)+\lambda_{A\cap C}(x)-\lambda_{A\cap(B\cap C)}(x)\\ &= \lambda_{A\cap B}(x)+\lambda_{A\cap C}(x)-\lambda_{(A\cap B)\cap(A\cap C)}(x)\\ &= \lambda_{(A\cap B)\cup(A\cap C)}(x).\end{aligned}$$

## *Choice Function*

We introduce here the concept of a *choice function*, and use the same to define the Cartesian product of an infinite family of nonempty sets. We start with giving an alternative description of the product of finite family $\mathscr{A}$ of nonempty sets $A_1, \ldots, A_n$. Taking $I = \{1, \ldots, n\}$ as the indexed set, we can write $\mathscr{A}$ as

$$\mathscr{A}=\{A_i \mid i\in I\}.$$

As introduced earlier in (3.12), the product $\prod_{i=1}^n A_i$ may also be defined as

$$\prod_{i=1}^n A_i := \{f : I\to \cup_{i\in I}A_i \mid f(i)=a_i\in A_i,\ \text{for } 1\le i\le n\}, \tag{4.1.13}$$

where each element $f : I \to \cup_{i\in I}A_i$ of the above set is called a **choice function**. More generally, when $I$ is an infinite set, an *indexed family* of nonempty sets $A_\alpha$ is the collection $\mathscr{A}$ written as

$$\mathscr{A} := \{A_\alpha \mid \alpha\in I\}$$

. In this case, we define the *Cartesian product* $\prod_{\alpha\in I} A_\alpha$ of the family $\mathscr{A}$ as

$$\prod_{\alpha\in I} A_\alpha := \{f : I \to \cup_{\alpha\in I} A_\alpha \mid f(\alpha) \in A_\alpha, \text{ for all } \alpha \in I\}. \tag{4.1.14}$$

The statement that $\prod_{\alpha\in I} A_\alpha \neq \emptyset$ is known as the *axiom of choice* (or simply AC):

**Axiom of Choice:** *For any indexed family of nonempty sets* $\{A_\alpha : \alpha \in I\}$, *there exists a choice function* $f : I \to \bigcup_{\alpha\in I} A_\alpha$ *such that* $f(\alpha) \in A_\alpha$, *for each* $\alpha \in I$.

That is, we can always *choose* one element from each set $A_\alpha$, no matter how many nonempty sets $A_\alpha$ the indexed family of sets $\mathscr{A}$ contains. In particular, if each $A_\alpha$ is a subset of some fixed set $U$, then there exists a *choice function* $c : \mathscr{P}(U) \setminus \{\emptyset\} \to U$ such that $c(A) \in A$. It should be clear that the converse holds. In 1904, Ernest Zermelo formulated the *axiom of choice* to prove *well-ordering theorem*, which is an important part of the ZF axiomatic theory of sets.

**Theorem 4.3** (Well-Ordering Theorem) *Every set can be* well ordered.[2]

We conclude our discussion here with a brief mention of some important theorems related to the axiom of choice, which are fundamental to axiomatic set theory. In 1938, *Kurt Gödel* (1906–1978) proved that AC is consistent with the ZF axioms of the set theory. In 1963, *Paul Cohen* (1934–2007) proved that AC is independent of the ZF axioms. The development of ZFC system is a culmination of works of many mathematicians and logicians of twentieth century such as Cantor, Dedekind, Frege, Peano, Russel, Whitehead, Hilbert, Gödel, Bernays, and Cohen. The letter C stands for the axiom of choice.

## 4.2 Properties of Functions

We start with the next definition.

**Definition 4.9** Let $f : X \to Y$ be a function.

1. We say $f$ is an *injective function* (or a *one-one function*) if every pair of distinct elements in $X$ gets mapped to distinct elements in $Y$. That is, for any $x_1, x_2 \in X$,

$$x_1 \neq x_2 \quad \Rightarrow \quad f(x_1) \neq f(x_2).$$

2. We say $f$ is a *surjective function* (or an *onto function*) if every element of $Y$ has a preimage in $X$. Or, equivalently, if $\text{Ran}(f) = Y$. A function that is not an onto function is called an *into function*.

[2] A partially ordered set is **well ordered** if every nonempty subset has a least element.

3. We say $f$ is a *bijective function* (or simply a *bijection*) if $f$ is both injective and surjective. In this case, we also say that $f$ defines an one-to-one correspondence between the sets $X$ and $Y$.

Notice that the condition in (1) is equivalent to the condition

$$f(x_1) = f(x_2), \quad \text{for} \quad x_1, x_2 \in X, \quad \Rightarrow \quad x_1 = x_2.$$

Further, not every injective function is surjective. For example, if $A$ and $B$ are two finite sets, with the set $B$ having at least one element more than the set $A$, then an injective function $f : A \to B$ cannot be an surjective function (Why?).

**Definition 4.10** Let $X$ and $Y$ be two sets, and $f : X \to Y$ be a function.

1. We say that $f$ is a *one-one function* or *injective function* (or simply an *injection*) if no two distinct elements of $X$ are mapped to the same element of $Y$. That is, we have
$$f(x_1) = f(x_2) \qquad \Rightarrow \qquad x_1 = x_2.$$
In this case, we also say $f$ gives a *one-one correspondence* from $X$ to $Y$.
2. We say that $f$ is an *onto function* or a *surjective function* (or simply a *surjection*) if every element $y \in Y$ has a preimage in $X$ with respect to $f$. That is, we have $\mathrm{Ran}(f) = Y$. We call $f$ an *into function* if it is not a surjection.
3. We say $f$ is a *bijective function* (or simply a *bijection*) if $f$ is both injective and surjective. In such case, we also say that the function $f$ is a *one-to-one correspondence* between the sets $X$ and $Y$.

**Example 4.1** The function $f : \mathbb{N} \to \mathbb{Z}$ given by

$$f(n) = \begin{cases} n/2, & \text{if } n \text{ is even} \\ (1-n)/2, & \text{if } n \text{ is odd} \end{cases}$$

is a bijective function. We first show that $f$ is one-one. For, notice that

$$f(n) > 0 \quad \text{if } n \text{ is even; and} \quad f(n) < 0 \quad \text{if } n \text{ is odd,}$$

for each $1 \neq n \in \mathbb{N}$. It thus follows that, if $f(r) = f(s) = a$, then $s$ and $r$ both must be 1 or both are even or odd, subjected to $a$ is zero, positive, or negative, respectively. In case both $s$ and $r$ are even, we have

$$s = 2\left(\frac{s}{2}\right) = 2f(s) = 2f(r) = 2\left(\frac{r}{2}\right) = r.$$

And, if both $s$ and $r$ are odd, we have

$$\begin{aligned} s &= (-2)\left[-\left(\frac{s-1}{2}\right)\right]+1 \\ &= -2[f(s)]+1 \\ &= -2[f(r)]+1 = r. \end{aligned}$$

Therefore, in each case, $f(r) = f(s) \Rightarrow r = s$. Next, we prove $f$ is surjective. For, notice that we have $2m \in \mathbb{N}$ when $m \in \mathbb{Z}$ is positive, and so $f(2m) = m$. Also, when $m \in \mathbb{Z}$ is zero or negative, we have $1 - 2m \in \mathbb{N}$, and so $f(1 - 2m) = m$, because $1 - 2m$ is odd. Hence, $f$ is a bijective function.

**Example 4.2** Let $X = \mathbb{N}$ and $Y = \{2n \mid n \in \mathbb{N}\}$. Consider the function $f : \mathbb{N} \to Y$ given by $f(n) = 2n$. We show that $f$ is a bijection. For, to show that $f$ is injective, let $f(n) = f(m)$, for $n, m \in \mathbb{N}$. Then, by definition, we have $2n = 2m$, which implies $n = m$. Therefore, $f$ is an injective function. Next, we show that $f$ is surjective. For, let $y = 2n_0 \in Y$, for some $n_0 \in \mathbb{N}$. Clearly, we have $f(n_0) = 2n_0 = y$. That is, $\beta$ is surjective. Hence, $f$ is a bijective function.

Consider a set $X$ with $m$ elements, and a set $Y$ with $n$ elements. For any fixed element $x_i \in X$, we can define a function $f_i : X \to Y$ by assigning it a unique element $y \in Y$ such that $f_i(x) = y$. Since this can be done in $n$ ways, there are exactly

$$n \times n \times \cdots \ (m \text{ terms}) = n^m$$

functions from the set $X$ to the set $Y$. In particular, for $X = \{a, b\}$ and $Y = \{0, 1\}$, there are exactly $2^2 = 4$ functions $f : X \to Y$, say $f_0, f_1, f_2, f_3$, which are given by

$$\begin{aligned} f_0 &= \{(a, 0), (b, 0)\}, & f_1 &= \{(a, 0), (b, 1)\}, \\ f_2 &= \{(a, 1), (b, 0)\}, & f_3 &= \{(a, 1), (b, 1)\}. \end{aligned}$$

Or, equivalently, we may write

$$\begin{aligned} f_0(a) &= 0, \quad f_0(b) = 0; & f_1(a) &= 0, \quad f_1(b) = 1, \\ f_2(a) &= 1, \quad f_2(b) = 0, & f_3(a) &= 1, \quad f_4(b) = 1. \end{aligned}$$

Clearly, we have

$$\begin{aligned} \text{Ran}\,(f_0) &= \{0\}, & \text{Ran}\,(f_1) &= \{0, 1\}, \\ \text{Ran}\,(f_2) &= \{0, 1\}, & \text{Ran}\,(f_3) &= \{1\}. \end{aligned}$$

Notice that both the functions $f_0$ and $f_3$ are injective and surjective, and so they are bijections. On the other hand, the functions $f_1$ and $f_2$ are neither one-one nor onto.

**Example 4.3** We may recall that the set of *counting numbers* is denoted by $\mathbb{N}_0$. That is, we have $\mathbb{N}_0$ is the set $\mathbb{N} \cup \{0\}$. Define the function $f : \mathbb{N}_0 \to \mathbb{N}_0$ as $f(j) = j^2 + 2$, for $j \in \mathbb{N}_0$. We check whether $f$ is one-one, onto, or bijection. First, since $f(j) \geq 2$, so there is no $j \in \mathbb{N}_0$ such that $f(j) = 0$. It thus follows that the function $f$ is not onto. Next, we show that the function is one-one. For instance, let $f(j) = f(k)$. Then, we have

$$j^2 + 2 = k^2 + 2 \quad \Rightarrow \quad j^2 = k^2 \quad \Rightarrow \quad j = k, \text{ because } \; j, k \geq 0.$$

Clearly, $f$ cannot be a bijective function.

For a nonempty set $X$, we write

$$2^X := \{f \mid f : X \to \{0, 1\}\}.$$

The next lemma proves that the *power set* $\mathscr{P}(X)$ of $X$ is in one-to-one correspondence with the set $2^X$.

**Lemma 4.1** *For any nonempty set* $X$, $|\mathscr{P}(X)| = 2^{|X|}$.

***Proof*** For any $A \in \mathscr{P}(X)$, we have the *characteristic function* $\lambda_A \in 2^X$. Conversely, given a $0 \neq f \in 2^X$, we have the set

$$A = \{x \in X : \; f(x) = 1\} \in \mathscr{P}(X).$$

Further, as $A^c = \{x \in X : \; f(x) = 0\}$, we have $\lambda_\emptyset = 0$, the *zero function*. Also, $\text{Ran}(\lambda_A) = \{0, 1\}$ if and only if $A$ is a proper subset of $X$. This completes the proof. □

## *Permutations*

We start with the next definition.

**Definition 4.11** Let $A$ be a nonempty set. A **permutation** on $A$ is a bijective function $\sigma : A \to A$.

For example, the function $\sigma : \mathbb{R} \to \mathbb{R}$ given by

$$\sigma(x) = 2x + 7, \qquad \text{for } \; x \in \mathbb{R}$$

is a permutation on the set $\mathbb{R}$. We usually write the set of permutations on a set $A$ as $\text{Perm}(A)$. Notice that $\text{Perm}(A) \neq \emptyset$, because the identity function $I_A : A \to A$ is a *permutation* on $A$. We also write $\text{Perm}(A)$ as $S_n$ when

$$A = [n] := \{1, 2, \ldots, n\}.$$

We may write a permutation $\sigma \in S_n$ in **two-row notation** as

$$\sigma = \begin{pmatrix} 1 & 2 & 3 & \cdots & n \\ \sigma(1) & \sigma(2) & \sigma(3) & \cdots & \sigma(n) \end{pmatrix}, \tag{4.2.1}$$

where the second row consists of elements $\sigma(1), \ldots, \sigma(n) \in [n]$ that are, respectively, the images of the elements $1, 2, \ldots, n$. In particular, the permutation $\sigma \in S_n$ given by single-step right shift can be written in *two-row notation* as

$$\sigma = \begin{pmatrix} 1 & 2 & 3 & \cdots & n-1 & n \\ 2 & 3 & 4 & \cdots & n & 1 \end{pmatrix}.$$

We can also express the above permutation using the **cycle notation**, i.e., by way of expressing it as a product of *disjoint cycles*. As we shall discover later, in comparison to the *array notation*, the *cycle notation* as introduced in the next definition is a far more convenient way to deal with permutations.

**Definition 4.12** A permutation $\sigma \in S_n$ is called a **cycle** of length $r$ if

$$\sigma(i_k) = i_{k+1}, \quad \text{for } 1 \le k < r, \ \sigma(i_r) = i_1, \quad \text{and } \sigma(i_m) = i_m, \tag{4.2.2}$$

for all $i_m \notin \text{Supp}(\sigma)$, where

$$\text{Supp}(\sigma) = \{i \in [n] \mid \sigma(i) \neq i\} \tag{4.2.3}$$

is called the **support** of the permutation $\sigma$. Also, the *cardinality* of the set Supp$(\sigma)$ is called the **degree** of $\sigma$.

Suppose a permutation $\sigma \in S_n$ is such that $i_1, i_2, \ldots, i_r \in \text{Supp}(\sigma)$, and we have

$$\sigma(i_1) = i_2, \quad \sigma(i_2) = i_3, \quad \ldots, \quad \sigma(i_{r-1}) = i_r, \quad \text{and } \sigma(i_r) = i_1,$$

for some $r \le n$. In this case, the permutation $\sigma$ in cycle notation is written as

$$(i_1 \, i_2 \quad i_r) \quad \text{or} \quad (\sigma(i_1)\sigma^2(i_1) \quad \sigma^{r-1}(i_1)), \tag{4.2.4}$$

where $i_1 \in \{1, 2, \ldots, n\}$ such that $\sigma(i_1) \neq i_1$, and $\sigma^k = \sigma \cdots \sigma$ ($k$-times). This permutation maps $i_1$ to $i_2$, $i_2$ to $i_3, \ldots, i_{r-1}$ to $i_r$, $i_r$ back to $i_1$, and *leaves fixed* all other elements of the set $[n]$.

**Definition 4.13** Two cycles $(i_1 \, i_2 \quad i_r)$ and $(j_1 \, j_2 \quad j_t)$ are said to be **disjoint** if

$$\{i_1, i_2, \ldots, i_r\} \cap \{j_1, j_2, \ldots, j_t\} = \emptyset. \tag{4.2.5}$$

The next lemma proves some simple facts related to above discussion.

**Lemma 4.2** *With notations as above, we have*

(a) *Cycles can be cycled in the following sense:*

$$\big(i_1\, i_2 \cdots i_r\big) = \big(i_r\, i_1 \cdots i_{r-1}\big) = \big(i_{r-1}\, i_r \cdots i_{r-2}\big) = \cdots. \tag{4.2.6}$$

(b) *Disjoint cycles commute. That is, if $\sigma$ and $\tau$ are disjoint cycles then $\sigma\tau = \tau\sigma$.*

***Proof*** The assertion $(a)$ is trivial. For $(b)$, the two permutations $\sigma$ and $\tau$ affect disjoint subsets of $\big[n\big]$ in view of the condition (4.2.5), and so the two may be performed in either order. This completes the proof. □

The next theorem proves that every $\sigma \in S_n$ can be written as the product of disjoint cycles. Such a representation is unique subject to the variations allowed by Lemma 4.2.

**Theorem 4.4** *Let $n > 1$ and $\sigma \in S_n$. There is a partition $\big\{S_1, \ldots, S_k\big\}$ of the set $\big[n\big]$, where we have*

$$\begin{aligned} S_1 &= \{a_{11}, \ldots, a_{1,r_1}\} \subset \big[n\big]; \\ S_2 &= \{a_{21}, \ldots, a_{2,r_2}\} \subset \big[n\big]; \\ &\vdots \\ S_k &= \{a_{k1}, \ldots, a_{k,r_k}\} \subset \big[n\big]. \end{aligned} \tag{4.2.7}$$

*That is, $\big[n\big] = S_1 \cup \cdots \cup S_k$ and $n = r_1 + \cdots + r_k$, and*

$$\sigma = \big(a_{11}\ \cdots\ a_{1,r_1}\big) \cdots \big(a_{k1}\ \cdots\ a_{k,r_k}\big). \tag{4.2.8}$$

***Proof*** We list all the *distinct* elements in $\big[n\big]$ by repeatedly applying $\sigma$ to 1, i.e., the "orbit" of 1 with respect to the permutation $\sigma$. We may write

$$O_1 = \big\{a_{10} = 1, a_{11} = \sigma(1), \ldots, a_{1,r_1} = \sigma^{r_1}(1)\big\}.$$

Next, we list the elements in the orbit of 2, and so on until we get to the *orbit of* $n$:

$$O_n = \big\{a_{n0} = n, a_{n1} = \sigma(n), \ldots, a_{n,r_n} = \sigma^{r_n}(1)\}.$$

In general, if we pick any two of these $n$ orbits $O_1, O_2, \ldots, O_n$, they will either be the same or disjoint. Denote all the *distinct orbits* by $O_1', O_2', \ldots, O_k'$. Notice that, in view of Lemma 4.2, it is immaterial in what order we write $O_i'$'s or the elements of an orbit. Suppose that

$$O'_1 = \{b_{11}, \ldots, b_{1,s_1}\}, \quad \text{with} \quad |O'_1| = s_1;$$
$$O'_2 = \{b_{21}, \ldots, b_{2,s_2}\}, \quad \text{with} \quad |O'_2| = s_2;$$
$$\vdots$$
$$O'_k = \{b_{k1}, \ldots, b_{k,s_k}\}, \quad \text{with} \quad |O'_k| = s_k.$$

To simplify the notation, we may relabel $a_{ij}$'s as $b_{ij}$'s so that we have

$$[n] = \cup_{I=1}^{k} O'_I = O'_1 \cup \cdots \cup O'_k,$$

and $s_1 + s_2 + \cdots + s_k = n$. The permutation $\sigma$ restricted to $O'_j$ gives the $s_j$-cycle $\{b_{j1}, \ldots, b_{j,s_j}\}$ $(1 \le j \le k)$. Since the subsets $O'_j$ constitute a partition of the set $[n]$, the above construction implies that

$$\sigma = (b_{11}\, b_{12} \quad b_{1,s_1}) \cdots (b_{k1}\, b_{k2} \quad b_{k,s_k}).$$

This proves the theorem. □

If any $O_j$ is a singleton, then we omit it from the notation. In such cases, we implicitly assumed that the permutation $\sigma$ leaves fixed these elements of $[n]$.

**Definition 4.14** The product in (4.2.8) is called a **cycle decomposition** of $\sigma$. If we rearrange the cardinalities $r_i$ of these sets $S_i$ in decreasing order, say $r'_1 \ge r'_2 \ge \cdots \ge r'_k$, then the *k-tuple* $(r'_1, r'_2, \ldots, r'_k)$ is called the *cycle structure* of $\sigma$, and the permutation $\sigma$ is called a $(r'_1, r'_2, \ldots, r'_k)$*-cycle*.

The next theorem proves that $|S_n| = n!$.

**Theorem 4.5** *Let a set A contains n elements. Then the set* Perm$(A)$ *of permutations of A contains n*! *elements.*

***Proof*** WLOG, we take $A = [n]$. An arbitrary permutation $\sigma \in S_n$ can be obtained as follows: First choose any one of the $n$ elements of $A$ for $\sigma(1)$; next, choose any one of the $n-1$ elements of $A \setminus \{1\}$ for $\sigma(2)$, and so on. It thus follows that the permutation $\sigma$ can be defined in

$$n \times (n-1) \times \cdots \times 2 \times 1 = n!$$

ways. This completes the proof. □

## Composition of Functions

We introduce here an important operation, namely, the *composition* of two function. In some specific case, it helps define an interesting algebraic structure on the set of functions with the same set as domain and codomain. We start with the next definition.

**Definition 4.15** Let $X$, $U$, $V$, and $Z$ be nonempty sets, and consider functions $f : X \to U$ and $g : V \to Z$ such that $\mathrm{Ran}(f) \subseteq \mathrm{Dom}(g)$. The **composition** of $f$ and $g$ (in that order), denoted by $g \circ f$, is a function from $X$ to $Z$ given by

$$g \circ f = \{(x, z) \mid (x \in X)(z \in Z)[\exists u \in U)(u = f(x), z = g(u)]\}.$$

In particular, when $U = V = Y$, say, we have

$$g \circ f = \{(x, z) \mid (x \in X)(z \in Z)(\exists y \in Y)[y = f(x), z = g(y)]\}. \tag{4.2.9}$$

Similarly, we can get the composition $f \circ g$ when $g : X \to Y$ and $f : Y \to Z$.

Consider the function $f : \mathbb{N} \to \mathbb{Z}$ given by $f(n) = n - 2$, and the function $g : \mathbb{Z} \to \mathbb{Z}$ given by $g(n) = n^2$. The composition $g \circ f$ is given by

$$(g \circ f)(n) = g(f(n)) = g(n - 2) = (n - 2)^2 = n^2 - 4n + 4.$$

Notice that, though the codomain of $g$ is the set $\mathbb{Z}$, but we have

$$\mathrm{Ran}\,(g) = \{n^2 \mid n \in \mathbb{Z}\} \subset \mathbb{N}.$$

It thus makes sense to consider the composition $f \circ g$ given by

$$(f \circ g)(n) = f(g(n)) = f(n^2) = n^2 - 2.$$

In particular, it follows that $f \circ g \neq g \circ f$, because of the following two reasons: $(a)$ the values of the two functions for the same argument are different and $(b)$ the domains of functions $f \circ g$ and $g \circ f$ are distinct sets. In general, both $(a)$ and $(b)$ should not hold for the compositions of two functions to be the same.

**Theorem 4.6** *The relation $g \circ f$ given by (4.2.9) is a function.*

***Proof*** For $x \in X$, there is a $y \in Y$ satisfying $y = f(x)$. Next, since $y \in Y$, there is an $z \in Z$ such that $z = g(y)$. Therefore, for each $x \in X$, there is an element $z \in Z$ such that $(x, z) \in g \circ f$. We also need to show that $z$ is unique. For, suppose $(x, z_1), (x, z_2) \in g \circ f$. Then, by (4.2.9), there are some $y_1, y_2 \in Y$ such that

$$y_1 = f(x), \ \ z_1 = g(x_1); \quad \text{and} \quad y_2 = f(x), \ \ z_2 = g(x_2).$$

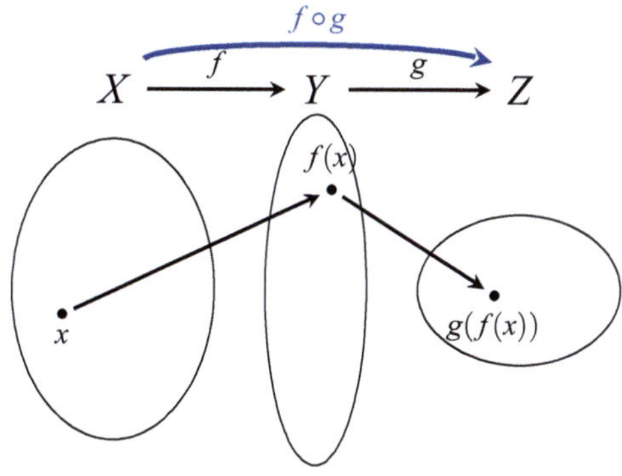


**Fig. 4.5** Pictorial representation of composition of two functions

However, since we have $y_1 = f(x) = y_2$, it follows that

$$z_1 = g(x_1) = g(x_2) = z_2.$$

This completes the proof. □

Notice that we have $z = g(y) = g(f(x))$. Conversely, if the above equation holds, then $(x, z) \in g \circ f$. We thus reformulate Definition 4.15 as given below.

**Definition 4.16** Given two functions $f : X \to Y$ and $g : Y \to Z$, the function $g \circ f : X \to Z$ given by

$$\big(g \circ f\big)(x) = g\big(f(x)\big), \qquad \text{for} \quad x \in X,$$

is called the *composition* of $f$ and $g$ (Fig. 4.5).

**Example 4.4** Let $X = \{1, 2, 3\}$, $Y = \{p, q\}$, and $Z = \{a, b\}$. Consider the two functions $f \subset X \times Y$ and $g \subset Y \times Z$, respectively, given by

$$f = \big\{(1, p), (2, p), (3, q)\big\} \qquad \text{and} \qquad g = \big\{(p, b), (q, b)\big\}.$$

Then, we have $g \circ f = \big\{(1, b), (2, b), (3, b)\big\}$. Next, let $A = \{1, 2, 3\}$, and consider the four functions $f,\ g,\ h,\ s \subset A \times A$ given by

$$f = \big\{(1, 2), (2, 3), (3, 1)\big\}, \qquad g = \big\{(1, 2), (2, 1), (3, 3)\big\},$$
$$h = \big\{(1, 1), (2, 2), (3, 1)\big\}, \qquad s = \big\{(1, 1), (2, 2), (3, 3)\big\}.$$

Then, we have

$$\begin{aligned}
f \circ f &= \{(1,3),(2,1),(3,2)\}, & f \circ g &= \{(1,1),(2,3),(3,2)\},\\
f \circ h &= \{(1,2),(2,1),(3,1)\}, & f \circ s &= \{(1,2),(2,3),(3,1)\},\\
g \circ f &= \{(1,3),(2,2),(3,1)\}, & g \circ g &= \{(1,1),(2,2),(3,3)\},\\
g \circ h &= \{(1,2),(2,1),(3,1)\}, & g \circ s &= \{(1,2),(2,1),(3,3)\},\\
h \circ f &= \{(1,2),(2,3),(3,2)\}, & h \circ g &= \{(1,2),(2,1),(3,2)\},\\
h \circ h &= \{(1,1),(2,2),(3,1)\}, & h \circ s &= \{(1,1),(2,2),(3,1)\},\\
s \circ f &= \{(1,2),(2,3),(3,1)\}, & s \circ g &= \{(1,2),(2,1),(3,3)\},\\
s \circ h &= \{(1,1),(2,2),(3,1)\}, & s \circ s &= \{(1,1),(2,2),(3,3)\}.
\end{aligned}$$

We also have

$$\begin{aligned}
f \circ g \circ h &= (f \circ g) \circ h = \{(1,3),(2,2),(3,3)\};\\
f \circ h \circ g &= (f \circ h) \circ g = \{(1,3),(2,2),(3,2)\}.
\end{aligned}$$

The next two theorems prove useful in our subsequent discussion.

**Theorem 4.7** *Let $X$, $Y$, $Z$ be three nonempty sets, and consider two functions $f : X \to Y$ and $g : Y \to Z$. Then, we have*

*(a) $f$ and $g$ are one-one implies $g \circ f$ is one-one;*
*(b) $f$ and $g$ are onto implies $g \circ f$ is onto;*
*(c) $f$ and $g$ are bijections implies $g \circ f$ is bijective.*

***Proof*** For $(a)$, let $x_1, x_2 \in X$ be such that

$$(g \circ f)(x_1) = (g \circ f)(x_2) \quad \text{or} \quad g(f(x_1)) = g(f(x_2)).$$

Now, since $g$ is one-one, it follows that $f(x_1) = f(x_2)$. Since $f$ is also a one-one function, we obtain $x_1 = x_2$. Therefore, $g \circ f$ is one-one. For $(b)$, let $z \in Z$. Then, as $g$ is onto, there is some element $y \in Y$ such that $g(y) = z$. Since $f$ is also an onto function, there is an element $x \in X$ such that $f(x) = y$. It thus follows that

$$(g \circ f)(x) = g(f(x)) = g(y) = z.$$

Therefore, $g \circ f$ is onto. Notice that $(c)$ follows directly from $(a)$ and $(b)$. □

**Theorem 4.8** *Let $X$, $Y$, $Z$ be three nonempty sets, and consider two functions $f : X \to Y$ and $g : Y \to Z$. Then, we have*

*(a) $g \circ f$ is one-one implies $f$ is one-one;*
*(b) $g \circ f$ is onto implies $f$ is onto.*

***Proof*** For $(a)$, let $x_1, x_2 \in X$ be such that $f(x_1) = f(x_2)$. Then, we have

$$g\big(f(x_1)\big) = g\big(f(x_2)\big) \quad \text{i. e.,} \quad \big(g \circ f\big)(x_1) = \big(g \circ f\big)(x_2).$$

Now, since $(g \circ f)$ is one-one, it follows that $x_1 = x_2$. Therefore, $f$ is one-one. For $(b)$, let $z \in Z$. Now, since $g \circ f$ is onto, there is some $x \in X$ such that $\big(g \circ f\big)(x) = z$. It thus follows that

$$g\big(f(x)\big) = z \quad \text{or} \quad g(y) = z,$$

where $y = f(x) \in Y$. Therefore, $g$ is onto. □

## *Composition of Permutations*

By Definition 4.11, a permutation $\sigma$ of a nonempty set $A$ is a bijective function $\sigma : A \to A$. Recall that the set of permutations of a set $A$ is denoted by $\text{Perm}(A)$. In particular, when

$$A = \big[n\big] := \big\{1, 2, \ldots, n\big\},$$

we write the set of permutations of a set $A$ as $S_n$. For $\sigma,\ \rho \in \text{Perm}(A)$, the composite $\rho * \sigma : A \to A$ is given by

$$(a)\big(\rho * \sigma\big) = \big(\rho(a)\big)\sigma, \qquad \text{with} \quad a \in A, \tag{4.2.10}$$

is again a permutation of the set $A$. Notice the *flip in order* of applications of the functions $\sigma$ and $\rho$.

**Example 4.5** Consider the two permutations $\pi_1, \pi_2 \in S_4$ given in *array notation* as

$$\pi_1 = \begin{pmatrix} 1 & 2 & 3 & 4 \\ 3 & 2 & 4 & 1 \end{pmatrix} \qquad \text{and} \qquad \pi_2 = \begin{pmatrix} 1 & 2 & 3 & 4 \\ 2 & 4 & 1 & 3 \end{pmatrix}$$

It follows from (8.1.6) that we have

$$\begin{aligned} \pi_1 * \pi_2 &= \begin{pmatrix} 1 & 2 & 3 & 4 \\ 3 & 2 & 4 & 1 \end{pmatrix} * \begin{pmatrix} 1 & 2 & 3 & 4 \\ 2 & 4 & 1 & 3 \end{pmatrix} \\ &= \begin{pmatrix} 1 & 2 & 3 & 4 \\ 3 & 2 & 4 & 1 \end{pmatrix} * \begin{pmatrix} 3 & 2 & 4 & 1 \\ 1 & 4 & 3 & 2 \end{pmatrix} \\ &= \begin{pmatrix} 1 & 2 & 3 & 4 \\ 1 & 4 & 3 & 2 \end{pmatrix}. \end{aligned}$$

Notice that, to obtain the composition $\pi_1 * \pi_2$, we have rewritten the first row of the permutation $\pi_2$ same as the second row of the permutation $\pi_1$, so that the permutation $\pi_1 * \pi_2$ can be obtained directly by writing the first row same as that of $\pi_1$ and the second row same as that of modified $\pi_2$. This is precisely the reason why the composition of permutations is also known as the *multiplication of permutation*.[3] Similarly, we have

$$\begin{aligned}\pi_2 * \pi_1 &= \begin{pmatrix}1\,2\,3\,4\\2\,4\,1\,3\end{pmatrix} * \begin{pmatrix}1\,2\,3\,4\\3\,2\,4\,1\end{pmatrix}\\ &= \begin{pmatrix}1\,2\,3\,4\\2\,4\,1\,3\end{pmatrix} * \begin{pmatrix}2\,4\,1\,3\\2\,1\,3\,4\end{pmatrix}\\ &= \begin{pmatrix}1\,2\,3\,4\\2\,1\,3\,4\end{pmatrix}.\end{aligned}$$

In particular, it follows that the composition of permutations is a "non-commutative" binary operation.

## *Inverse Function*

Let $X$ and $Y$ be two nonempty sets, and consider a function $f : X \to Y$. As a relation, $f$ has the inverse from $Y$ to $X$, which we may write as $\tilde{f}$. However, if $f$ is not onto, i.e., $\mathrm{Ran}(f) \neq Y$, then the domain of the relation $\tilde{f}$ is not whole of the set $Y$, and hence $\tilde{f}$ cannot be a function. Similarly, if $f$ is not one-one, then we can find elements $x_1, x_2 \in X$ and $y \in Y$ such that $f(x_1) = f(x_2) = y$. It thus follows that we have $(y, x_1), (y, x_2) \in \tilde{f}$. Therefore, in this case as well, $\tilde{f}$ is not a function. Hence, the relation $\tilde{f}$ is a function if and only if $f$ is a bijection. We are thus led to introduce the fundamental concept of the *inverse* of a function, provided it is a bijection (Fig. 4.6).

**Definition 4.17** Let $X$ and $Y$ be two nonempty sets, and $f : X \to Y$ be a bijection. A function $g : Y \to X$ given by the equivalence

$$g(y) = x \quad \Leftrightarrow \quad f(x) = y \tag{4.2.11}$$

is called the *inverse* of the function $f$. We usually write $g := f^{-1}$.

**Example 4.6** Let $X = \{1, 2, 3\}$, $Y = \{p, q, r\}$, and consider the function $f : X \to Y$ given by

$$f = \{(1, p), (2, q), (3, r)\}.$$

The inverse relation of $f$ is given by

$$\tilde{f} = \{(p, 1), (q, 2), (r, 3)\},$$

[3] We also say that the composition of permutations is obtained by the "*left action*" of the functions.

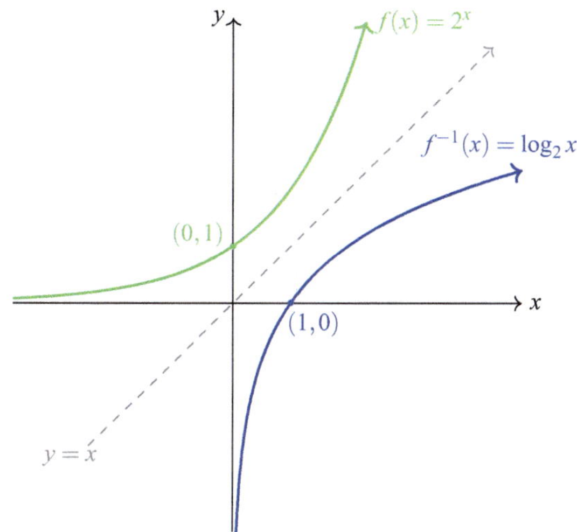


**Fig. 4.6** Inverse of a function

which is clearly a function, and so the inverse of the function $f$. Notice that the function $f$ is a bijection. Also, since the function $g$ given by

$$g = \{(1, p), (2, r), (3, q)\}$$

is a bijection, the inverse of $g$ is given by

$$\tilde{g} = \{(p, 1), (q, 3), (r, 2)\}.$$

**Example 4.7** Let $X = \{1, 2, 3\}$, $Y = \{p, q\}$, and consider the function $f : X \to Y$ given by

$$f = \{(1, p), (2, q), (3, p)\}.$$

In this case, the inverse relation of $f$ given by

$$\tilde{f} = \{(p, 1), (q, 2), (p, 3)\}$$

is not a function. Therefore, the inverse of the function $f$ does not exist. Notice that $f$ is onto, but not one-one. On the other hand, taking $X = \{1, 2\}$ and $Y = \{p, q, r\}$, the function $g : X \to Y$ given by $f = \{(1, p), (2, q)\}$ has the inverse relation given by $\tilde{f} = \{(p, 1), (q, 2)\}$, which is not a function. Therefore, the inverse of the function $f$ does not exist. Notice that, in this case, the function $f$ is one-one, but not onto.

**Example 4.8** Consider the function $f : \mathbb{N} \to \mathbb{Z}$ given by

$$f(n) = \begin{cases} n/2, & \text{if } n \text{ is even} \\ (1 - n)/2, & \text{if } n \text{ is odd.} \end{cases}$$

By reasoning as given earlier in Example 4.1, the inverse of $f$ is given by

$$f^{-1}(m) = \begin{cases} 2m, & \text{for } m \geq 1 \\ 1 - 2m, & \text{for } m \leq 0 \end{cases}.$$

**Example 4.9** Consider two permutations $p_1, p_2 \in S_4$ given by

$$p_1 = \begin{pmatrix} 1 & 2 & 3 & 4 \\ 3 & 2 & 4 & 1 \end{pmatrix} \quad \text{and} \quad p_2 = \begin{pmatrix} 1 & 2 & 3 & 4 \\ 2 & 4 & 1 & 3 \end{pmatrix}.$$

We shall find $(p_1 \circ p_2)^{-1}$ and $p_2^{-1} \circ p_1^{-1}$, and also verify that $(p_1 \circ p_2)^{-1} = p_2^{-1} \circ p_1^{-1}$. The inverse of $p_1$ is given by

$$p_1^{-1} = \begin{pmatrix} 3 & 2 & 4 & 1 \\ 1 & 2 & 3 & 4 \end{pmatrix} = \begin{pmatrix} 1 & 2 & 3 & 4 \\ 4 & 2 & 1 & 3 \end{pmatrix},$$

and the inverse of $p_2$ is given by

$$p_2^{-1} = \begin{pmatrix} 2 & 4 & 1 & 3 \\ 1 & 2 & 3 & 4 \end{pmatrix} = \begin{pmatrix} 1 & 2 & 3 & 4 \\ 3 & 1 & 4 & 2 \end{pmatrix}.$$

Now the composition is given by

$$\begin{aligned} p_2^{-1} \circ p_1^{-1} &= \begin{pmatrix} 1 & 2 & 3 & 4 \\ 3 & 1 & 4 & 2 \end{pmatrix} \circ \begin{pmatrix} 1 & 2 & 3 & 4 \\ 4 & 2 & 1 & 3 \end{pmatrix} \\ &= \begin{pmatrix} 1 & 2 & 3 & 4 \\ 3 & 1 & 4 & 2 \end{pmatrix} \circ \begin{pmatrix} 3 & 1 & 4 & 2 \\ 1 & 4 & 3 & 2 \end{pmatrix} \\ &= \begin{pmatrix} 1 & 2 & 3 & 4 \\ 1 & 4 & 3 & 2 \end{pmatrix} = (p_1 \circ p_2)^{-1}. \end{aligned}$$

Also, we have

$$p_1 \circ p_2 = \begin{pmatrix} 1 & 2 & 3 & 4 \\ 1 & 4 & 3 & 2 \end{pmatrix} \quad \Rightarrow \quad (p_1 \circ p_2)^{-1} = \begin{pmatrix} 1 & 2 & 3 & 4 \\ 1 & 4 & 3 & 2 \end{pmatrix}.$$

Therefore, $(p_1 \circ p_2)^{-1} = p_2^{-1} \circ p_1^{-1}$.

**Theorem 4.9** *Let $X$, $Y$ be two nonempty sets, and $f : X \to Y$ be a bijection. Then,*

1. *there is function $g : Y \to X$ such that the following two relations hold:*

$$g \circ f = I_X \quad \text{and} \quad f \circ g = I_Y. \tag{4.2.12}$$

   *In this case, we write $g := f^{-1}$, and call it the* ***inverse*** *of the function $f$.*
2. *$f^{-1} : Y \to X$ is a bijection.*

***Proof*** We first notice that both $g \circ f$ and the identity function $I_X$ are functions from $X$ to $X$. We also have

$$\big(g \circ f\big)(x) = g\big(f(x)\big) = g\big(y\big),$$

for each $x \in X$, where $y = f(x)$. Therefore, by (4.2.11), we have $y = f(x)$, and so $g(y) = x$. It thus follows that

$$\big(g \circ f\big)(x) = x = I_X\big(x\big), \qquad \text{for all} \quad x \in X.$$

Hence, $g \circ f = I_X$. Similarly, we can show that $f \circ g = I_Y$. This proves (1). For (2), to show that $f^{-1}$ is a bijection, we first show that $f^{-1}$ is one-one. For, let $y_1, y_2 \in Y$ be such that we have $f^{-1}(y_1) = f^{-1}(y_2)$. Then, by (4.2.11), we can find $x_1, x_2 \in X$ such that we have

$$f^{-1}\big(y_1\big) = x_1 \qquad \text{and} \qquad f^{-1}\big(y_2\big) = x_2.$$

It thus follows that

$$y_1 = f(x_1) \qquad \text{and} \qquad y_2 = f(x_2).$$

Therefore, we obtain

$$f^{-1}\big(y_1\big) = f^{-1}\big(f(x_1)\big) = x_1;$$
$$f^{-1}\big(y_2\big) = f^{-1}\big(f(x_2)\big) = x_2.$$

Now, assuming $y_1, y_2 \in Y$ are such that $f^{-1}(y_1) = f^{-1}(y_2)$, we have $x_1 = x_2$, by (4.2.11), and hence $f(x_1) = f(x_2)$ or, equivalently, $y_1 = y_2$. Hence, $f^{-1}$ is one-one. Next, we shall show that $f^{-1}$ is onto. For instance, let $x \in X$ be an arbitrary element. Since $f$ is a function, there exists $y \in Y$ such that $f(x) = y$ or, equivalently, $x = f^{-1}(y)$, by (4.2.11). Therefore, for $x \in X$, there exists $y \in Y$ such that $f^{-1}(y) = x$. Therefore, $f^{-1}$ is onto. Hence, $f^{-1}$ is bijective. □

**Theorem 4.10** *Let $X$, $Y$ be two nonempty sets, and $f : X \to Y$, $g : Y \to X$ be two functions. Then we have $g = f^{-1}$ if and only if*

$$g \circ f = I_X \qquad \textit{and} \qquad f \circ g = I_Y. \tag{4.2.13}$$

***Proof*** We first assume that, for some function $g : Y \to X$, we have

$$g \circ f = I_X \text{ and } f \circ g = I_Y.$$

Then, by using $g \circ f = I_X$ and $f^{-1} \circ f = I_X$, we obtain

$$(f^{-1} \circ f) \circ g = I_X \circ g = g,$$

and hence it follows that

$$g = (f^{-1} \circ f) \circ g = f^{-1} \circ (f \circ g) = f^{-1} \circ I_Y = f^{-1}.$$

Conversely, suppose $g = f^{-1}$, for some function $g : Y \to X$. Now, since the inverse satisfies

$$f^{-1} \circ f = I_X \quad \text{and} \quad f \circ f^{-1} = I_Y,$$

by replacing $g$ by $f^{-1}$, we obtain

$$g \circ f = I_X \quad \text{and} \quad f \circ g = I_Y.$$

□

**Theorem 4.11** *Let $X$, $Y$ be two nonempty sets, and $f : X \to Y$ and $g : Y \to Z$ be two bijective functions. Then, we have*

$$(g \circ f)^{-1} = f^{-1} \circ g^{-1}. \tag{4.2.14}$$

***Proof*** Since $g \circ f : X \to Z$ is a bijection, $(g \circ f)^{-1} : Z \to X$ is well defined. We also have $f^{-1} : Y \to X$ and $g^{-1} : Z \to Y$. Therefore, $f^{-1} \circ g^{-1} : Z \to X$ is well defined. It thus follows that both $(g \circ f)^{-1}$ and $f^{-1} \circ g^{-1}$ are functions from the set $Z$ to the set $X$. We know that

$$f \circ f^{-1} = I_Y, \quad f^{-1} \circ f = I_X, \quad g \circ g^{-1} = I_X, \quad g^{-1} \circ g = I_Y,$$

and also that the composite of any function with the identity function is function itself. Hence, we have

$$\begin{aligned}
(f^{-1} \circ g^{-1}) \circ (g \circ f) &= ((f^{-1} \circ g^{-1}) \circ g) \circ f \\
&= (f^{-1} \circ (g^{-1} \circ g)) \circ f \\
&= (f^{-1} \circ I_Y) \circ f \\
&= f^{-1} \circ f = I_X; \\
(g \circ f) \circ (f^{-1} \circ g^{-1}) &= ((g \circ f) \circ f^{-1}) \circ g^{-1} \\
&= (g \circ (f \circ f^{-1})) \circ g^{-1} \\
&= (g \circ I_Y) \circ g^{-1} \\
&= g \circ g^{-1} \\
&= I_X.
\end{aligned}$$

It thus follows from Theorem 4.10 that $(g \circ f)^{-1} = f^{-1} \circ g^{-1}$. □

In the next theorem, we state some simple facts that explain why set operations are more well behaved with respect to inverse functions.

**Theorem 4.12** *Let $f : X \to Y$ be an invertible function. Then*

1. $f^{-1}(A^c) = [f^{-1}(A)]^c$*; for all* $A \subseteq X$*;*
2. $\bigcup_\alpha f^{-1}(A_\alpha) = f^{-1}\Big[\bigcup_\alpha A_\alpha\Big]$*, for every class* $\{A_\alpha \subseteq X : \alpha \in I\}$*;*
3. $\bigcap_\alpha f^{-1}(A_\alpha) = f^{-1}\Big[\bigcap_\alpha A_\alpha\Big]$*, for every class* $\{A_\alpha \subseteq X : \alpha \in I\}$*.*

## 4.3 Cardinality of a Set

As said earlier in Definition 3.8, the cardinality of a set $X$ is the number of elements it contains. It is denoted by $|X|$ (or $\# X$). We also call the number $|X|$ as the *size* of the set $X$. For example, since there are 26 English letters, the size of the set of "English alphabets" is 26. In general, for a *finite set* $X$, the cardinality is determined by a *counting process*. That is, we count its elements as *one, two, three*, ..., and so on. Therefore, in layman terms, the size of a finite set can be specified as *five, fifty, hundred*, etc. We need plenty of time to do so literally when the size of the finite set $X$ is large. However, the following mathematical approach makes the task more precise and convenient.

**Definition 4.18** We say a set $X$ is a **finite set** if $X = \emptyset$ or there is an *injective* function $f : X \to \mathbb{N}$ given by $x_i \mapsto a_i$ such that

$$\mathrm{Ran}(f) = \big\{a_1, a_2, \ldots, a_n\big\}, \quad \text{where} \quad a_i \in \mathbb{N}.$$

In this case, the elements of $X$ are listed as $x_1, x_2, \ldots, x_n$ such that $x_i \mapsto a_i$, and we say the **cardinality** (or the **size**) of $X$ is the integer $n \geq 1$.

In all that follows, we use the notation $\big[0\big] := \emptyset$, and as earlier we have

$$\big[n\big] := \big\{1, 2, \ldots, n\big\}, \qquad \text{for} \quad n \in \mathbb{N}.$$

It follows that a set $X$ has size $n$ if there is a *bijective function* $f : X \to \mathbb{N}_n$. Therefore, if $|X| = |Y| = n$, then both the sets are in bijective correspondence with a subset of $\mathbb{N}$ containing $n$ positive integers. Hence, for two finite sets $X$ and $Y$, we have

$$|X| = |Y| \quad \Leftrightarrow \quad \text{there is a bijective function} \quad f : X \to Y.$$

More generally, if $X$ and $Y$ are two finite sets, and $f : X \to Y$ is an injective function that is *not surjective*, then $Y$ has *more elements* than the set $X$. In this case, we write $|X| < |Y|$. The discussion so far answers both the following fundamental questions for finite sets:

1. What is the size of a set?
2. How to compare sizes of two sets?

The case of *infinite sets* is entirely a different story. In this section, we apply concepts introduced in the previous section to discuss the *cardinality* of an infinite set $X$. The main point is to *compare* two infinite sets on the basis of their sizes. Therefore, our focus here is to find the answers to above two questions for *infinite sets*. It was George Cantor's insight to compare the sizes of two infinite sets [1], by using bijective functions as in Definition 4.18.

To start with, we first consider the infinite sets for which *counting process* makes some sense. For example, we can count the elements of the set of natural numbers $\mathbb{N}$ as $1, 2, 3, \ldots$ such that every $n \in \mathbb{N}$ is eventually counted. In such cases, we may define the *size* of an infinite set $X$ by comparing its cardinality with the set $\mathbb{N}$.

**Definition 4.19** We say $X$ is a *denumerable set* (or a *countably infinite* set) if there exists a bijective function : $X \to \mathbb{N}$. Further, we say $X$ is a *countable set* if it is finite or a countably infinite set.

The above definition provides a general scheme to *count* the elements of an infinite set that has the cardinality of the set $\mathbb{N}$. For instance, if $A$ is a *denumerable* set, and $f : \mathbb{N} \to A$ is a bijective function, then we can *enumerate* the elements of $A$ as $a_1, a_2, a_3, \ldots$. That is, we may write

$$A = \{a_n : a_n = f(n), \; n \in \mathbb{N}\},$$

where $f$ is called an *enumeration function* of the set $A$. This justifies occasional use of terminology *enumerable* for a countable set.

**Example 4.10** Some simple denumerable sets are as given below:

1. The set $\mathbb{N}$ is countably infinite as the identity function $I : \mathbb{N} \to \mathbb{N}$ is a bijection.
2. The set $2\mathbb{N} = \{2, 4, 6, \ldots\}$ is countably infinite, because the function $f : \mathbb{N} \to 2\mathbb{N}$ defined by $f(n) = 2n$ is bijective.
3. Since the function $f : \mathbb{N} \to \mathbb{Z}$ of Example 4.1 is bijective, the set of integers $\mathbb{Z}$ is countably infinite. Notice that $f$ maps odd positive integers $1, 3, 5, 7, \ldots$ respectively to the non-positive integers $0, -1, -2, -3, \ldots$, whereas it maps even positive integers $2, 4, 6, 8, \ldots$ respectively to the positive integers $1, 2, 3, 4, \ldots$.Therefore, in this case, the enumeration $f : \mathbb{N} \to \mathbb{Z}$ lists the countably infinite set of integers $\mathbb{Z}$ as

$$0, \; 1, \; -1, \; 2, \; -2, \; 3, \; -3, \cdots.$$

The statements of some important theorems about countably infinite sets are as given below. For their proofs, the reader may refer any standard book on real analysis such as [2].

**Theorem 4.13** *Every infinite subset of a countably infinite set is itself a countably infinite set.*

We can use any condition of the next theorem to test the countability of a set.

**Theorem 4.14** *Let A be a nonempty set. The following are equivalent:*

1. *A is countable.*
2. *There exists an injective function* $f : A \to \mathbb{N}$*.*
3. *There exists a surjective function* $f : \mathbb{N} \to A$*.*

*In particular, every infinite set contains a denumerable set.*

**Corollary 4.1** *Every subset of a countable set is countable.*

**Corollary 4.2** *The union of two countably infinite sets is countably infinite.*

**Theorem 4.15** *If* $A_1$ *and* $A_2$ *are countably infinite, then* $A_1 \times A_2$ *is countably infinite. In particular, the set* $\mathbb{Z} \times \mathbb{Z}$ *is countably infinite.*

In turn, the above theorem follows easily from the next lemma.

**Lemma 4.3** *The set* $\mathbb{N} \times \mathbb{N}$ *is countably infinite.*

***Proof*** Arrange the elements of $\mathbb{N} \times \mathbb{N}$ as shown in Fig. 4.7, and enumerate all pairs by using Cantor's *diagonal process* as indicated by the arrows. Therefore, we can *list* elements of $\mathbb{N} \times \mathbb{N}$ as

$$(1, 1),\ (1, 2),\ (2, 1),\ (3, 1),\ (2, 2),\ (1, 3),\ (1, 4),\ (2, 3),\ \ldots .$$

Hence, $\mathbb{N} \times \mathbb{N}$ is countably infinite. □

Theorem 4.14 also holds for countable family of sets.

**Theorem 4.16** *If, for each* $n \in \mathbb{N}$*, the set* $A_n$ *is a countably infinite subset of a set* $X$*, then* $\bigcup_{n=1}^{\infty} A_n$ *is a countable infinite subset of* $X$*.*

Finally, we consider how to compare sizes of infinite sets. In this case, we need to know when one infinite set can be treated as bigger than some other infinite set. A simple glance through the following three simple questions helps us understand why situation in the present case is very different from the finite sets:

**Fig. 4.7** $\mathbb{N} \times \mathbb{N}$ is countable

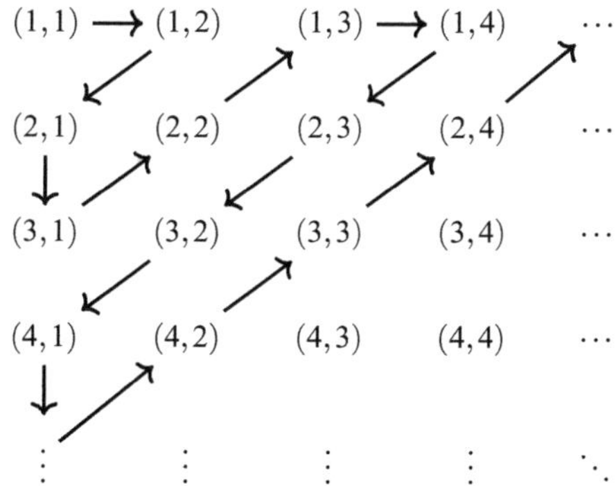

(a) Are there more even integers than odd integers?
(b) Are there more even integers than the natural numbers?
(c) Are there more rational numbers than negative integers?

For example, since $n \mapsto 2n : \mathbb{N} \to 2\mathbb{N}$ is a bijective function, the answer to question in $(b)$ is NO. Notice that the set $2\mathbb{N}$ is of the *half size* of the set $\mathbb{N}$; of course, the other half is the set of positive odd integers $\{1, 3, 5, \dots\}$. Therefore, our intuitive understanding fails while dealing with the infinite sets.

**Definition 4.20** Let $S$ and $T$ be two sets. We say $S$ is **equinumerous** to a subset of the set $T$ if there is an injective function $f : S \to T$. In this case, we write $|S| \leq |T|$. Also, two sets $S$ and $T$ are said to be *equinumerous* (or *equipotent sets*) if there exists an bijective function from $S$ onto $T$. In this case, we write $|S| = |T|$.

In particular, for finite sets, two sets $S$ and $T$ are *equinumerous* if and only if they have the same *number of elements*. However, the same is not true for infinite sets. For example, as seen in Example 4.10, we have $|\mathbb{N}| = |\mathbb{Z}|$. Also, since $n \mapsto n + 1$ is a bijective function from the set of *odd integers* to the set of *even integers*, we have that there are as many even integers as there are odd integers. The next theorem helps introduce the concept of the *size of a set*, finite or infinite, in general. However, the *relation* as defined below is not mathematically rigorous, because $\mathscr{F}$ is not a set in usual sense.

**Theorem 4.17** *The relation $\sim$ on the family $\mathscr{F}$ of sets given by*

$$S \sim T \quad \Leftrightarrow \quad |S| = |T|, \ \textit{for } S, T \in \mathscr{F}$$

*is an equivalence relation.*

***Proof*** For $S, T, U \in \mathscr{F}$, we prove that

(i) $\sim$ is *reflexive*, i.e., $S \sim S$.
(ii) $\sim$ is *symmetric*, i.e., $S \sim T \Rightarrow T \sim S$.
(iii) $\sim$ is *transitive*, i.e., $S \sim T$ and $T \sim U \Rightarrow S \sim U$.

First, notice that $(i)$ is *vacuously true* when $S = \emptyset$. Otherwise, the identity function $I : S \to S$ proves the assertion. We also know that, if $f : S \to T$ is bijective function, then so is the inverse $f^{-1} : T \to S$. This proves $(ii)$. Finally, if $f : S \to T$ and $g : T \to U$ are two bijective functions, then so is their composition $f \circ g : S \to T$. This proves $(iii)$. □

In particular, for any set $S \in \mathscr{F}$, we have $S \in [S]$. Furthermore, two equivalence classes $[S]$ and $[T]$ are equal or $[S] \cap [T] = \emptyset$. A proof of latter assertion may be given as follows: Suppose there is a pair of sets $S, T \in \mathscr{F}$ such that $[S] \cap [T] \neq \emptyset$. We then have, for $U \in [S] \cap [T]$,

$$U \in [S] \text{ and } U \in [T] \quad \Rightarrow \quad S \sim U \text{ and } U \sim T.$$

It thus follows that $S \sim T$, by $(iii)$. We may thus write

$$\mathscr{F} = \bigcup_{S \in \mathscr{F}} [S], \quad \text{where} \quad [S] \cap [T] = \emptyset, \quad \text{if} \quad S \nsim T.$$

All sets in any particular equivalence class are equinumerous, i.e., they have the same *size*. Therefore, we may assign a unique **cardinal number** to the equivalence class that is the **cardinality** of each set in the class.

**Example 4.11** Since $[\emptyset] = \{\emptyset\}$, the empty set $\emptyset$ is the only set having the *cardinal number* 0. Also, for the equivalence class consisting of singletons $\{x\}$, the *cardinal number* of is 1, and so on. In general, we can now understand the mathematically precise meaning of the *cardinal number* of a finite set. Next, recall that $X$ is a *denumerable set* if $X \in [\mathbb{N}]$, by Definition 4.19. We thus assign the cardinal number[4] $\aleph_0$ to each set in the equivalence class $\left[\mathbb{N}\right]$. Therefore, $\aleph_0$ is the *cardinality* of any countably infinite set such as the set of integers $\mathbb{Z}$, the set of even integers, and the set of odd integers.

**Definition 4.21** By a *transfinite number* we mean a cardinal $\alpha > \aleph_0$.

The two important facts about the *cardinal arithmetic* given below follow from Theorems 4.16 and 4.15:

$$k\,\aleph_0 = \aleph_0 \quad \text{and} \quad \aleph_0^k = \aleph_0, \quad \text{for all} \quad k \in \mathbb{N}.$$

We next assign a *cardinal number* to a set $X \notin \left[\mathbb{N}\right]$, which is called an **uncountable set**. Thanks to George Cantor's work on problems concerning assigning a cardinality to an uncountable set, several renowned mathematicians produced excellent work.

**Example 4.12** A *countable product* of countably infinite sets such as $\mathbb{N}$ is an *uncountable set*. In particular, the construction given earlier in Example 3.3 can be extended to define the set $\Sigma^\infty$ of strings of *infinite length* over a set of *k-alphabets*, $k \geq 2$. Cantor's *diagonal process* then implies that the set $\Sigma^\infty$ is uncountable.[5] The same argument applies to the set of real numbers with infinite *decimals*, provided we compute with base 10 in place of 2. Analogous to the set $\Sigma^\infty$ is the set of infinite strings over the alphabet $X = \{0, 1, \ldots, 9\}$.

We may start our discussion here with a formal definition of an infinite set, first proposed by the German mathematician *Richard Dedekind*.

**Definition 4.22** A set **infinite** if it is equinumerous to one of its proper subset.

---

[4] The symbol $\aleph_0$ is the first Hebrew letter, which is read as *aleph nought*.

[5] If it is countable, then there is bijection $f : \mathbb{N} \to \Sigma^\infty$. Let $s_n$ denote the $n$th symbol of the string $f(n)$. Define a new string $\sigma \in \Sigma^\infty$ such that the $n$th symbol of $\sigma$ is $s_n + 1$ $(modulo\ 2)$. Clearly, $\sigma \neq f(n)$ for any $n \in \mathbb{N}$, which proves that $f$ is not surjective, a contradiction.

The next two theorems prove that the condition stated in above definition is a characteristic feature of every infinite set. The reader is once again referred to [2] for a proof.

**Theorem 4.18** (Dedekind Theorem) *A set $S$ is infinite if and only if there exists a proper subset $A$ of $S$ that is equinumerous to $S$.*

**Theorem 4.19** (Cantor-Bernstein) *For any two sets $X$ and $Y$,*

$$|X| \leq |Y| \text{ and } |Y| \leq |X| \quad \Rightarrow \quad |X| = |Y|.$$

In general, $|S| \leq |T|$ implies that the set $T$ has at least as many elements as the set $S$. Theorem 4.19 provides a far more useful tool to compare the *sizes* of infinite sets. It was proved independently by George Cantor (1887), Richard Dedekind (1888), Sergei Bernstein (1897), and some others. Cantor's argument used the following *trichotomy of cardinal numbers*: Given any two sets $S, T$, exactly one of the following three possibilities hold:

$$|S| < |T|, \quad |S| = |T|, \quad |S| > |T|.$$

It is known that the above is equivalent to the well-ordering principle, and hence also to the Axiom of Choice. Peano's proof appeared in 1906, and that of Zermelo in 1908, who called it *Äquivalenzsatz.*

Now, let $X$ be an *uncountable set.* It is known, in general, $k^{\aleph_0}$ is a transfinite cardinal number, for all $k \geq 2$. That is, by Definition 4.21, we have

$$k^{\aleph_0} > \aleph_0, \qquad \text{for} \quad k \geq 2.$$

The next theorem is the statement for $k = 2$. It also tells us how big a set can be.

**Theorem 4.20** (Cantor 1891) *For every set $A$, $A \leq \mathscr{P}(A)$ and $A \nsim \mathscr{P}(A)$, where $\mathscr{P}(A)$ is the the power set of the set $A$.*

***Proof*** If $A = \emptyset$, the result is trivial because $\mathscr{P}(\emptyset) = \{\emptyset\}$. So, we assume $A \neq \emptyset$ and define a function $f : A \to \mathscr{P}(A)$, by $f(x) = \{x\}$, for $x \in A$. Then, $f$ is injective because

$$\{x\} = \{y\} \quad \Rightarrow \quad x = y.$$

Hence, $A \leq \mathscr{P}(A)$. Next, we use the method of contradiction to show that $A \nsim \mathscr{P}(A)$. For, if possible, suppose $A \sim \mathscr{P}(A)$, and let $g : A \to \mathscr{P}(A)$ be a bijective function. Now, we consider the set $B \subset A$ given by

$$B = \{x \in A \mid x \notin g(x)\},$$

and pick $a \in A$ such that $g(a) = B$. It thus follows that $a \in B \Leftrightarrow a \notin B$, which is impossible. This contradiction shows that our assumption is wrong, and hence the proof is complete. □

It follows from Theorem 4.20 that we have $\mathbb{N} \notin [\mathbb{R}]$. For instance, taking $A = \mathbb{N}$, there exists a set with cardinal number greater than $\aleph_0$. Of course, it is the cardinal number of the set $\mathscr{P}(\mathbb{N})$. Therefore, $\aleph_0 < \mathrm{c}$. Further, by Lemma 4.1, $\mathscr{P}(A)$ is in bijective correspondence with the set

$$2^A = \{f : f : A \to \{0, 1\} \text{ is a function}\},$$

and so, by a theorem of Bernstein, we have

$$|2^A| = 2^{|A|} \quad \Rightarrow \quad |\mathscr{P}(A)| = 2^{|A|},$$

for any set $A$. In particular, when $A = \mathbb{N}$, we obtain $|\mathscr{P}(\mathbb{N})| = 2^{\aleph_0}$. Therefore, $\aleph_0 < 2^{\aleph_0}$, by Theorem 4.20. On the other hand, by a deep result of Bernstein, $2^{\aleph_0} = \mathrm{c}$, which stands for the *continuum.*

To conclude our discussion, we denote the size of the uncountable set $\mathbb{R}$ by c. It thus makes sense to take c as the *cardinal number* of the equivalence class $\left[\mathbb{R}\right]$. In general, there are many outstanding problems related to transfinite cardinals. One such problem asks: *Does there exist a transfinite cardinal* $\alpha$ *such that* $\aleph_0 < \alpha < 2^{\aleph_0} = \mathrm{c}$? The following is a *conjecture* of Cantor:

**Continuum Hypothesis.** *There is no cardinal* $\alpha$, *with* $\aleph_0 < \alpha < \mathrm{c}$.

**Remark 4.2** The *continuum hypothesis* was first on Hilbert's list of 23 problems presented in 1900. Notice that if the superset of a set is countable, all its subsets are countable and therefore by Corollary 4.1 we have that every superset of an uncountable set is uncountable. So, there are many sets of cardinality $> \mathrm{c}$.

## 4.4 Hash Functions

In the context of *hashed data structure*, **hashing** is used to store data so that it can be accessed and processed efficiently by an algorithm. In general, it refers to a process that produces a fixed size output from an arbitrary size variable input by using a formula, known as a **hash function**. It is well known that the traditional data types such as *array* are not good enough to deal with ever-increasing size of a data. However, a *hash function* helps store data in an *array* or *linked list* such that each entry has its own unique index, in some suitable sense. Therefore, it searches an index or location of an item in the storage *in constant time*. Also, *hashing* provides a better synchronization than other non-linear data structures such as *search trees*.

In terms of time and space complexity, hashing is a disk-based data structure. To continue with our discussion, we first need to introduce some basic terminology such as given below:

1. A **key** is a string or integer that a mathematical formula takes as an input to determine an index or assigns a location to an item in a data. Said differently,

hashing maps numeric or alphanumeric keys to a *hash value* (or simply a *value*) that represents the original string. This value is used as a *hash index* (or simply an index), which is smaller than the original. In the case of an *array*, a key is an index $i$ and the value is the value at that index, which is usually denoted by $A[i]$.
2. The hash function returns the index of an element in $A[i]$, which is also known as a *hash table*. Said differently, a **hash table** is a hashed data structure that maps keys to values.

In general, depending on the problem in hand, we may apply many special types of hash functions that use numeric or alphanumeric *keys*. In what follows, we use notations introduced earlier in Example 3.16.

## *Division Method*

A simple way to generate a hash value is by using the *division method*, where hash function $h(k)$ uses the remainder obtained when the value $k$ is divided by an integer $m > 1$. That is, we have

$$h(k) = k \pmod{m}, \tag{4.4.1}$$

where $m$ denotes available number of *memory locations*. It is an interesting fact that the keys are more uniformly distributed when $m$ is a prime number. More specifically, the steps we follow to store the value corresponding to a key are as follows:

**Step-1.** The hash function (4.4.1) is used to calculate the hash value which acts as the index of the data structure such as a set of strings $\{ab, cd, efg\}$ where the value will be stored.
**Step-2.** Assign "values" to all alphabetical characters, say $a = 1,\ b = 2, \dots,$ etc.
**Step-3.** The numerical value of all characters of a string is obtained by usual sum:

$$ab = 1 + 2 = 3, \quad cd = 3 + 4 = 7, \quad efg = 5 + 6 + 7 = 18, \ \cdots$$

**Step-4.** Taking a table of size $m = 7$ to store these strings, the hash function used above is the sum of the characters in key *modulo* 7. That is, we apply the formula

$$Sum\big(string\ location\big) = string\ location \pmod{7}.$$

**Step-5.** Store string "ab" in $3 \pmod{7} = 3$, string "cd" in $7 \pmod{7} = 0$, and the string "efg" in $18 \pmod{7} = 4$.

The above procedure helps us to find the location of a given string by using a simple hash function, and also quickly find the value that is stored in that location. Therefore, hashing is a way to store pairs $(key, value)$ of the data in a table, which is known as *hash table*.

**Example 4.13** We consider here the problem of storing student's records on a Collage Server in a manner so as to ensure that each record is retrievable as efficiently as possible. In this case, each record is represented by an enrollment number as a *key* that helps identify a student. As said above, to solve this problem, we may use the hash function $h$ that assigns memory location $h(k)$ to the record with key $k$, where $h$ is given by (4.4.1). As said earlier, the function in (4.4.1) is one of the most commonly used hashing functions. In general, it is desirable to apply hashing that is easy to evaluate so that a data entry can be quickly located. Notice that above function meets this requirement. More precisely, to find $h(k)$, we only need to compute the remainder when $k$ is divided by $m$. Further, we must have that the hash function is onto so that all memory locations are accessible. The function in (4.4.1) satisfies this property as well. In particular, taking $m = 111$, since we have

$$h(064212848) = 064272848 \pmod{111} = 14,$$

the record of the student with enrollment number 064212848 is assigned to memory location 14, Similarly, since we have

$$h(031149212) = 037149212 \pmod{111} = 65,$$

the record of the student with enrollment number 037149212 is assigned to memory location 65.

## *Mid-Square Method*

The *division method* works for any value of $m$, and it is very fast. However, in general, it performs poorly because consecutive keys get mapped to consecutive values in the associated hash table. Moreover, it matters sometime what value we choose for $m$. A better hashing procedure is provided by *mid-square method* that uses middle $r$ digits of $k^2$ as the hash value. That is,

$$h(K) = h(k^2), \tag{4.4.2}$$

where the value of $r$ depends on the size of the hash table. In particular, if the table has 100 memory locations, we take $r = 2$ simply because two digits are needed to map the key $k = 60$ to the memory location:

$$k^2 = 60 \times 60 = 3600 \quad \Rightarrow \quad h(60) = 60.$$

In general, as all digits of the key contribute in generating the middle digits, this method performs well. However, as expected, key size is one serious limitation.

## Multiplication Method

The hashing provided by *multiplication method* chooses a constant $0 < a < 1$, extract the fractional part of $ka$, which is then multiplied by the size of the hash table $m$, followed by taking $\lfloor \text{resulting value} \rfloor$ as the hash value $h(k)$. That is, we have

$$h(k) = \lfloor m(ka \pmod 1) \rfloor. \tag{4.4.3}$$

In particular, with $k = 12345$, $a = 0.357840$, and $m = 100$, we obtain

$$\begin{aligned} h(12345) &= \lfloor 100(4417.5348 \pmod 1) \rfloor \\ &= \lfloor 100 \times 0.5348 \rfloor \\ &= 53. \end{aligned}$$

The benefit of working with this hashing technique is that it works for any value $0 < a < 1$, where some values may yield better results than others. Also, when the table size $m$ is a power of 2, the multiplication method computes the index associated with a key by using multiplication hashing, which is very fast.

## Collision

In most important cases, there are far more keys than the available memory locations. Therefore, a hash function cannot be injective. As such, more than one data entry may have to be assigned to the same memory location. In this case, we say a *collision* has occurred. A simple way to resolve a collision is to assign the first free location following the occupied memory location assigned by the hashing function. To see this, first notice that the hash function $h = h(k)$ in (4.4.1) maps the above enrollment number to the location 14, because

$$h(107405723) = 107405723 \pmod{111} = 14.$$

Since this location is already occupied by the data related to enrollment number 064212848, and the memory location 15 is the first location following memory location 14 that is free, the above hash function assigns location 15 to the record of the student with the enrollment number 107405723. We will discuss more about the *collision* at the end of the section.

Occurrences of collisions are also a shortcoming of *mid-square method*, though it is possible to reduce their numbers. Next, the hashing provided by *digit folding method* divides the key value $k$ into parts $k_1, \ldots, k_n$ such that the first $n - 1$ parts have the same number of digits, but $k_n$ can have lesser digits. Subsequently, we take the sum $s$ of these parts as the hash value $h(k)$ ignoring the last carry, if any. In particular, with $k = 12345$, we may take $k_1 = 12$, $k_2 = 34$, and $k_3 = 5$ so that

$$s = k_1 + k_2 + k_3 = 51 \quad \Rightarrow \quad h(k) = 51. \tag{4.4.4}$$

Notice the number of digits in any of the $k_1, \ldots, k_{n-1}$ varies depending upon the size of the hash table, subject to that $k_n$ can have a lesser number of digits.

At this point, it is natural to ask: *What properties a good hash function must have*? It is natural to call a hash function *perfect* if it provides a unique memory slot to each data item. However, knowing that the items and the collection will always change, we cannot construct a *perfect hash function* for any given data even if the size of the hash table is increased sufficiently to accommodate every possible data value. More generally, some of the properties that a good hash function must have are as follows:

1. The involved computability has to be *efficient*.
2. Keys have *uniform distribution*, i.e., table values are equally likely for each key.
3. There are *minimum collisions*.
4. The *load factor*[6] of a hash table determines the efficiency of the associated hash function in terms of distribution of the keys over the table. Therefore, the load factor of the related hash table must be low. Otherwise, the complexity increases, and we need to *rehash* using the same hash function, while doubling the size of the Array used earlier.
5. The *time complexity* and the *space complexity* of computing hash value should, respectively, be given by $O(n)$ and $O(1)$.

We now continue our discussion about the *collision*, i.e., when a hash function assigns different strings the same value in the hash table. Clearly, it creates problem in *searching, insertion, deletion*, and also while *updating a value*. As said earlier, this problem arises because hashing generates a small number for a big key so that two keys could possibly produce the same value. This situation is dealt with by using collision resolution technique such as *separate chaining* (open hashing) and *open addressing* (closed hashing). In *separate chaining*, the central idea is to use linked list for the records with the same hash value, which requires more memory outside the hash table. In the case of *open addressing*, all records are stored in the hash table itself, and the table slots are examined one by one until we find the desired record or it is clear that the table doesn't contain the one. When the hash table is searched sequentially starting from the original location of the hash, the process is known as the *linear probing*. On the other hand, *quadratic probing* such as

$$h(k) + 1^1, \quad h(k) + 2^2, \quad h(k) + 3^2, \; \ldots, \quad h(k) + k^2$$

takes the original hash index and adds successive values of an arbitrary quadratic polynomial until an open slot is found. This is also known as the *mid-square method*

[6] In the context of hashing, **load factor** of the hash table refers to the ratio of the number of items table contains to the size of the table. The usual default value of the load factor is 0.75.

simply because it looks for $i^2$th slot at $i$th stage of iteration, for $i = 0, 1, \ldots, m-1$, where $m$ is the size of the hash table. In this case, we always start from the original hash location unless it is already occupied. Next, the collision-resolving technique known as the *double hashing* uses two hash functions, say $h_1(k)$ and $h_2(k)$, such that if the location assigned by $h_1(k)$ is occupied then $h_2(k)$ is used in combination with $h_1(k)$ in the form:

$$h(k, i) := \big(h_1(k) + i\, h_2(k)\big) \pmod{m},$$

where the non-negative integer $i$ denotes a collision number. Its time complexity is $O(n)$. In particular, taking $m = 7$, and suppose we have to insert keys $27, 43, 92, 72$ by using the two hash functions given by

$$h_1(k) := k \pmod 7 \quad \text{and} \quad h_2(k) := 1 + \big(k \pmod 5\big).$$

Now, since we have

$$h(27) = 6 \pmod 7 \quad \text{and} \quad h(43) = 1 \pmod 7,$$

the key 27 takes slot 6 and the key 43 takes slot 1. However, since $h_1(92) := 6 \pmod 7$, collision occurs because slot 6 is not free. To resolve this collision, we use double hashing given by

$$\begin{aligned} h(k, i) &= \big(h_1(k) + i\, h_2(k)\big) \pmod 7 \\ &= \big(h_1(92) + i\, h_2(92)\big) \pmod 7 \\ &= \big(6 + \big(1 + 92 \pmod 5\big)\big) \pmod 7 \\ &= 9 \pmod 7 = 2, \end{aligned}$$

and so 92 is assigned the free slot 2. Similarly, since $h(72) = 2 \pmod 7$ gives a collision, it can be seen that double hashing resolves this collision by assigning the key 72 free slot 5.

**Remark 4.3** In some programming languages such as Python, hashing is also used to implement objects. Some other real-time applications of *hashed data structure* are such as that define *cache mapping* for fast access to the data, password verification, and also in cryptography as a message digest. In the last chapter, we will discuss more about application of hashing technique in cryptography. A major disadvantage of working with *hashed data structure* is that it is difficult to handle collisions efficiently when keys are large in number.

## 4.5 Recursive Functions

In this section, we discuss a class of *recursive functions* that were first used to model *computation* in the sense of the contemporary mathematical logic. The centrality of the concept of *recursion* to computation, and also its relationship to self-reference, is provided by *Kleene's recursion theorem* (1938), which also has applications to other areas of computer science. We may write the set of non-negative integers as $W := \mathbb{N} \cup \{0\}$, which is assumed to be a *recursively defined set*.[7]

**Definition 4.23** An *n-ary* function $f(x_1, \ldots, x_n)$ is said to be *number theoretic* if all the $n$ arguments $x_1, \ldots, x_n$, and also the resulting single output, take values in the set $W$.

The simplest case of *recursion* takes as input a single-variable number theoretic function $f$, and a number $n \in W$, to return the function $g$ given by

$$g(0) = n \quad \text{and} \quad g(m+1) = f\big(g(m)\big), \tag{4.5.1}$$

where the first part is called the *base case*, and the second determines how the value of recursive function $g$ is obtained at $m + 1$, given its value at the argument $m$. Notice that the function $g$ is a typical instance of an *iteration formula*. In particular, with $n = 3$ and $f : m \mapsto m^2$, we have $g$ is the $(m + 1)$th iterate of the function $m \mapsto 3m$, i.e., $g(m) = 3^{m+1}$. Next, to obtain a full primitive recursion scheme, we allow the value of $g$ in (4.5.1) at $m + 1$ depend not just on its own value at $m$, but also on the value $h(m)$, which gives

$$g(0) = n \quad \text{and} \quad g(m+1) = f\big(m, g(m)\big). \tag{4.5.2}$$

A typical example of above is the *factorial function*. Notice that a second possibility is that we allow the value of $g$ in (4.5.1) to depend on finitely many *parameters*, which may also be arguments to the base case. In this case, the recursion scheme is given by

$$g(m, 0) = \alpha(m) \quad \text{and} \quad g(m, p+1) = f\big(m, g(m, p)\big). \tag{4.5.3}$$

For example, by taking $\alpha(m_0) = m_0$ and $f(m_0, m_1) = S(m_1)$, the *addition function* $(m, p) \mapsto m + p$ can be defined in terms of the above scheme. In general, recursion operation takes as argument an $n$-ary function $f$, an $(n + 2)$-ary function $g$, and returns an $(n + 1)$-ary function $h$ such that we have

$$h(x_1, \ldots, x_n, 0) = f(x_1, \ldots, x_n); \tag{4.5.4a}$$

$$h(x_1, \ldots, x_n, y+1) = g\big(x_1, \ldots, x_n, y, h(x_1, \ldots, x_n)\big). \tag{4.5.4b}$$

---

[7] A set is said to **recursively defined** if it contains a *basis element*, all other elements can be produced by combining previously known elements in certain specific ways, and an object that cannot be produced in this manner is not an element of the set.

In this case, we say that $h$ is defined by a *recursive scheme*. The first $n$ variables $x_1, \ldots, x_n$ of $g$ are called *parameters*, the $(n+1)$th argument $y$ is called the *recursion variable*, and the $(n+2)$nd argument $h(x_1, \ldots, x_n)$ represents the previous value of the *recursive function h*. In particular, for a unary function $f$, we only need to find a function $h$ of two variables such that

$$h(0) = n \quad \text{and} \quad h(x+1) = g\big(x, h(x)\big),$$

where $n \in \mathbb{N}$. Notice that the above relation is same as (4.5.2).

## *Primitive Recursive Functions*

Our main focus is to introduce number theoretic *primitive recursive function* and *partial recursive function*. As we discover in a later chapter, these types of functions are important to the *computability theory*. An informal description about *primitive recursive functions* first appeared in the research work of Norwegian mathematician *Thoraff Skolem* (1887–1963), published in 1923. Though his work remained unnoticed for a long time, he is seen as an unwitting pioneer of theoretical computer science. We start here with the next definition.

**Definition 4.24** An $n$-ary number theoretic function is said to be a *partial function* if it is defined over a set $D \subset W^n$. When $D = W^n$, we say $f$ a *total function*.

For a simple illustration, consider the set $D \subset W^2$ given by

$$D = \big\{(x, y) \in W^2 \mid x \geq y\big\}, \tag{4.5.5}$$

and take function $g : D \to W$ defined by $g(x, y) = x - y$. Now, since $x - y \geq 0$ only when $x \geq y$, $g$ is a *partial function*. On the other hand, the function $f : W^2 \to W$ defined by $f(x, y) = x + y$ is clearly a *total function*. In general, fundamental to the main theory are some preliminary "primitive recursive functions" that are known as the *initial functions*.

**Definition 4.25** An *initial function* is a total function of the form as given below:

$$\begin{aligned} &\textbf{(Zero Function)} && Z(x) = 0, \ \text{for all} \ x \in W; \\ &\textbf{(Successor Function)} && S(0) = 1, \ \text{and} \ S(x+1) = S(x) + 1, \ \text{for all} \ x \in W; \\ &\textbf{(Projection Functions)} && \pi_k^n(x_1, \ldots, x_n) = x_k, \ \text{for} \ k = 1, \ldots, n. \end{aligned}$$

We usually write the above class of (infinite) functions as $I_{PR} = \big\{Z, S, \pi_k^n\big\}$.

The class of number theoretic functions, known as *primitive recursive functions*, is obtained from above types of functions by applying basic functional operations such as *composition* and simple *recursion*.

**Definition 4.26** (*Composition*) For $i = 1, 2, \ldots, m$, let

$$f_i : \mathrm{Dom}(f_i) \to \mathrm{Ran}(f_i)$$

be $n$-ary number theoretic functions, and

$$g : \mathrm{Dom}(g) \to \mathrm{Ran}(g)$$

be a similar $m$-ary function. The function $h : \mathrm{Dom}(h) \to \mathrm{Ran}(h)$ given by

$$h(x_1, \ldots, x_n) = g\big(f_1(x_1, \ldots, x_n), \ldots, f_m(x_1, \ldots, x_n)\big) \tag{4.5.6}$$

is called the *composition* of $g$ with $m$ functions $f_1, \ldots, f_m$, where we take

$$\mathrm{Dom}(h) := \mathrm{Ran}(f_1) \times \cdots \times \mathrm{Ran}(f_m) \subseteq \mathrm{Dom}(g).$$

Notice that we shall treat the composition (4.5.6) undefined if any $f_i$ or the function $h$ itself fails to return a value. For a simple illustration, consider the three total functions $f_1$, $f_2$, and $f_3$, respectively, given by

$$f_1(x, y) = x + y^2, \quad f_2(x, y) = x^2 + y, \quad \text{and} \quad f_3(x, y) = x + y.$$

Let $g$ be the total function given by $g(x, y, z) = xy + z^2$. Then the composition of $g$ with $f_1, f_2, f_3$ is the function $h : W^2 \to W$ given by

$$\begin{aligned} h(x, y) &= g\big(f_1(x, y), f_2(x, y), f_3(x, y)\big) \\ &= f_1(x, y) f_2(x, y) + \big[f_3(x, y)\big]^2 \\ &= \big(x + y^2\big)\big(x^2 + y\big) + \big(x + y\big)^2. \end{aligned}$$

Further, for the total function $f_1(x, y) = x^3 + y^3$, and the two partial functions $f_1, f_2 : D \to W$ given by

$$f_2(x, y) = x^3 - y^3 \quad \text{and} \quad g(x, y) = x^2 - y^2,$$

the composition of $g$ with $f_1, f_2$ is the partial function $h : D \to W$ given by

$$\begin{aligned} h(x, y) &= g\big(f_1(x, y), f_2(x, y)\big) \\ &= \big[f_1(x, y)\big]^2 - \big[f_2(x, y)\big]^2 \\ &= \big(x^3 + y^3\big)^2 - \big(x^3 - y^3\big)^2 \\ &= 4x^3 y^3. \end{aligned}$$

Notice that the set $D$ in the above description is same as defined in (4.5.5). As said earlier, simple recursion is also a functional operation. The next example illustrates how primitive recursion can be defined in terms of the initial functions, together with applying finitely many times the composition and primitive recursion operations.

**Example 4.14** We show here that the *addition function* $f(x, y) = x + y$ is primitive recursion by expressing it in terms of initial functions. Clearly, we have

$$f(x, 0) = x = \pi_1^1(x).$$

Since we also have

$$\begin{aligned} f(x, y + 1) &= x + (y + 1) = (x + y) + 1 \\ &= f(x + y) + 1 \\ &= S(f(x, y)) \\ &= S(\pi_3^3(x, y, f(x, y)), \end{aligned}$$

it follows that we may define the addition function $f$ as follows:

$$\begin{aligned} f(x, 0) &= \pi_1^1(x); \\ f(x, y + 1) &= S(\pi_3^3(x, y, f(x, y)). \end{aligned}$$

As it is possible to express $f$ in terms of initial functions $\pi_1^1, \pi_3^3, S$, and the composition of $S$ with $\pi_3^3$, we conclude that *addition function is a primitive recursion.* Similarly, it can see that the *multiplication function* $f(x, y) = xy$ is a primitive recursion. Notice that, in this case, we have

$$f(x, 0) = 0 = Z(x).$$

We also have

$$\begin{aligned} f(x, y + 1) &= x(y + 1) = xy + x \\ &= f(x, y) + x \\ &= g(\pi_3^3(x, y, g(x, y)), \pi_1^3(x, y, g(x, y))), \end{aligned}$$

where $g$ is the *addition function* that is shown to be a primitive recursion. Once again, since the multiplication function $f$ can be expressed in terms of composition of initial and recursions, it follows that *multiplication function is a primitive recursion.*

**Example 4.15** Let $f(x, y) = xy,\ g(x) = 0$, and $h(x, y, z) = x + z$. Then we have

$$f(x, 0) = 0 = g(x).$$

We also have

$$f(x, y+1) = x(y+1) = xy + x = f(x, y) + x = h(x, y, f(x, y)).$$

Therefore, $f$ is a primitive recursion defined in terms of $g$ and $h$.

**Definition 4.27** (*Primitive Recursion*) Any function in the class of functions PrimREC, containing the initial functions, that is closed under two functional operations *composition* and *recursion*, is called a *primitive recursion function.*

The above notion extends easily for $n$-ary relations defined over the set $W$.

**Definition 4.28** (*Primitive Recursive Relation*) A relation $R \subset W^n$ is called a *primitive recursive relation* if the *characteristic function* $\chi_R$ of $R$ given by

$$\chi_R(x_1, \ldots, x_n) = \begin{cases} 1 & \text{if } (x_1, \ldots, x_n) \in R, \\ 0 & \text{if } (x_1, \ldots, x_n) \notin R \end{cases}$$

is primitive recursive function.[8]

For example, it is easy to show that most sets and relations on the natural numbers such as the set of primes and the relation "*divides*" are primitive recursive. Also, for any fixed $k$, the relation $R$ given by

$$R := \{(k, y) \mid y > k\}$$

is primitive recursive. For instance, we apply the nonzero *test function* (or *sign function*) $\text{sg} : W \to W$ given by

$$\text{sg}(x) = \begin{cases} 0 & (x = 0), \\ 1 & (x \geq 1). \end{cases} \tag{4.5.7}$$

Also, the zero *test function* $\overline{\text{sg}} : W \to W$ is given by

$$\overline{\text{sg}}(x) = \begin{cases} 1 & (x = 0), \\ 0 & (x \geq 1). \end{cases} \tag{4.5.8}$$

Now, since we have

$$\text{sg}(y \dot{-} k) = \begin{cases} 1 & (y > k) \\ 0 & (y \leq k) \end{cases}$$

it follows that the primitive recursive function $\text{sg}(y \dot{-} k)$ is a characteristics function of the relation $R$. Therefore, $R$ is a primitive recursive relation. Also, the identity

[8] It is also possible to define an $n$-ary relation by an $n$-ary predicate, which is then called a primitive recursive relation.

relation $I := \{(x, x) \mid x \in W\}$ is primitive recursive. In this case, we may consider the primitive recursive relation given by

$$f(x, y) = \overline{\text{sg}}(|x - y|).$$

Now, since we have

$$f(x, y) = \begin{cases} 1, & (x = y) \\ 0, & (x \neq y) \end{cases}$$

it follows that $f$ is the characteristic function of $I$. Hence, the identity relation is primitive recursive. Further details are left as exercises for the reader.

## *Partial Recursive Functions*

It follows from the discussion above that there are only countably many primitive recursive functions. It is thus natural to seek *computable functions*[9] of the type $W^n \to W$ $(n > 0)$ that are not primitive recursive. Intuitively, we call a number theoretic function $f$ **computable** (or *effectively calculable*) if there exists an algorithm so that $f$ returns a value for every suitable input. We start with the next definition.

**Definition 4.29** An $(n+1)$-ary number theoretic function $g$ is called a *regular function* if there exists at least one non-negative integer $y$ such that we have

$$g(x_1, \ldots, x_n, y) = 0, \quad \text{for all } n\text{-tuples} \quad (x_1, \ldots, x_n) \in W^n.$$

Notice that not all number theoretic functions are regular. For example, we may consider the function $g$ given by

$$g(x, y) = |y^2 - x|, \qquad \text{for} \quad (x, y) \in W^2.$$

In this case, though $g$ is a total function, $|y^2 - x| \neq 0$ unless $x \in W$ is a *perfect square*. Therefore, we cannot find $y \in W$ such that $|y^2 - x| = 0$, for all $x \in W$. Hence, $g$ is not a regular function. On the other hand, the number theoretic function $y \dot{-} x$ is a regular function simply because

$$y = 0 \quad \Rightarrow \quad y \dot{-} x = 0, \quad \text{for all } x \in W.$$

The concept introduced in the next definition is central to the construction of the class of *partial recursive functions* from the class PrimREC.

[9] In specific cases, we also have computable functions such as *Turing-computable functions* and *$\mu$-recursive functions*. In fact, Turing machine may be seen as mathematical model of a partial recursion.

**Definition 4.30** ($\mu$ - *Operation*) We say an $n$-ary number theoretic function $f(x_1, \ldots, x_n)$ is obtained by *minimization* (or $\mu$ - *operation*) from an $(n+1)$-ary number theoretic function $g(x_1, \ldots, x_n, y)$ if

$$f(x_1, \ldots, x_n) = \begin{cases} \mu_y\big(g(x_1, \ldots, x_n, y) = 0\big), & \text{if there exists such a number } y; \\ \text{undefined}, & \text{otherwise} \end{cases}$$

where $\mu_y$ is the least $y \geq 0$.

Notice that $f$ is a total function when $g$ is a regular function. In general, $f$ is only a partial function. We may write PartREC for the smallest class of functions such that it contains all initial functions, and it is closed with respect to composition, primitive recursion, and $\mu$ -operation (or minimization).

**Definition 4.31** A function in the class PartREC is called a *partial recursive function*.

Notice that, for a total function as input, composition and primitive recursion produce a total function. However, when $\mu$ -operation is applied, we may end up with a partial function. Accordingly, a total function in the class PartREC is called a *total recursive functions* (or simply a *recursive function*). Clearly, the class of primitive recursive functions PrimREC is a proper subset of the class of recursive functions, which we may write simply as REC.

**Example 4.16** The function $f(x) = \frac{x}{2}$ is a partial recursive function. For instance, we may take $g(x, y) = |2y - x|$. Now, since $|2y - x| = 0$ only for even values of $x$, it follows that $g$ is not regular. We may define $f$ as given below:

$$f(x) = \mu_y\big(|2y - x| = 0\big).$$

It then follows that

$$f(x) = \frac{x}{2}, \qquad \text{for } x \text{ even.}$$

Therefore, $f(x) = \frac{x}{2}$ is partial recursive. Further, we claim that $\lfloor\sqrt{x}\rfloor$ is recursive function, where $\lfloor\sqrt{x}\rfloor$ denotes the greatest integer $\leq \sqrt{x}$. For instance, notice that $(y+1)^2 \dot{-} x$ is zero for $(y+1)^2 \leq x$, and it is nonzero for $(y+1)^2 > x$. Therefore, $\overline{sg}(y+1)^2 \dot{-} x)$ is 1 if $(y+1)^2 \leq x$, and it cannot equal 0. The smallest value of $y$ for which $(y+1)^2 > x$ is the required number $\lfloor\sqrt{x}\rfloor$. Hence,

$$\lfloor\sqrt{x}\rfloor = \mu_y\big(\overline{sg}((y+1)^2 \dot{-} x\big) = 0).$$

As the function $\lfloor\sqrt{x}\rfloor$ is defined for all $x$, it is a recursive function.

**Definition 4.32** (*Ackerman's Function*) *Ackerman's function* $A(x, y)$ is the total function given by

$$\begin{aligned} A\big(0, y\big) &= y + 1, \\ A\big(x + 1, 0\big) &= A\big(x, 1\big) \\ A\big(x + 1, y + 1\big) &= A\big(x, A(x + 1, y)\big). \end{aligned}$$

As a simple illustration, we compute $A(2, 2)$ in the next example.

**Example 4.17** Notice that, by using the relation $A(0, y) = y + 1$, we have

$$A(0, 0) = 1, \quad A(0, 1) = 2, \quad A(0, 2) = 3,$$

$$A(0, 3) = 4, \quad A(0, 4) = 5, \quad A(0, 5) = 6.$$

Next, by using the two relations given by

$$A(x + 1, 0) = A(x, 1) \quad \text{and} \quad A(x + 1, y + 1) = A(x, A(x + 1, y)),$$

it follows that

$$\begin{aligned} &A(1, 0) = A(0, 1) = 2; \\ &A(1, 1) = A(0, A(1, 0)) = A(0, 2) = 3; \\ &A(1, 2) = A(0, A(1, 1)) = A(0, 3) = 4; \\ &A(1, 3) = A(0, A(1, 2)) = A(0, 4) = 5; \\ &A(1, 4) = A(0, A(1, 3)) = A(0, 5) = 6; \\ &A(1, 5) = A(0, A(1, 4)) = A(0, 6) = 7. \end{aligned}$$

Finally, by using the relation

$$A(x + 1, y + 1) = A(x, A(x + 1, y)),$$

we obtain

$$\begin{aligned} &A(2, 0) = A(1, 1) = 3; \\ &A(2, 1) = A(1, A(2, 0)) = A(1, 3) = 5; \\ &A(2, 2) = A(1, A(2, 1)) = A(1, 5) = 7. \end{aligned}$$

Hence, we have $A(2, 2) = 7$.

Ackerman's function is an example of a recursive function that is not primitive recursive. Further, the concept of partial recursion extends easily to sets.

**Definition 4.33** We say $A \subseteq W$ is a *recursive set* (*partial recursive set*) if $\chi_A$ is recursive function (partial recursive function, respectively).

We consider only those sets whose elements can be mapped into $\mathbb{N}$. Recall that

$$\chi_{A^c} = 1 - \chi_A; \quad \chi_{A \cap B} = \chi_A * \chi_B; \quad \text{and} \quad \chi_{AUB} = \chi_A + \chi_B \dot{-} \chi_{A \cap B}.$$

Therefore, if sets $A$ and $B$ are recursive sets, then so are $A^c$, $A \cup B$, and $A \cap B$. Hence, the class of recursive sets is closed with respect to complementation, union, and intersection operations.

**Example 4.18** The sets of even and odd natural numbers are both recursive. For instance, the characteristic function of the set of odd natural numbers $\mathbb{O}$ is the parity function which is primitive recursive. Therefore, the set of odd numbers is primitive recursive. The set of even numbers $\sim \mathbb{O}$ or $\mathbb{E}$ is also primitive recursive since $\chi_{\mathbb{E}} = 1 - \chi_{\mathbb{O}}$ is primitive recursive. Further, the sets of divisors of a integer $n$ are recursive. For instance, let $A$ be the set of divisors of $n$. Recall that a number $x \leq n$ is a divisor of $n$ if and only if $|x * i - n|$ is equal to zero for one fixed value of $i$, $(1 \leq i \leq n)$. That is, $|x * i - n|$ is nonzero for all $1 \leq i \leq n$, if $x$ is not a divisor. Therefore, the characteristic function of the set $A$ is

$$\chi_A(x) = \sum_{i=1}^{n} \overline{\text{sg}}|x * i - n|.$$

Hence the set of divisors of $n$ is recursive.

## 4.6 Growth Rate of Functions

Very roughly, an *algorithm*[10] is a finite set of "well-defined instructions" that a machine needs to produce one or more outputs from a set of admissible inputs. The number of steps the related process takes to complete the computation is called the run time of the algorithm, which depends on the *input size*, say $n$, that is usually very large. The concept of *growth rate* refers to the *run-time function* $f(n)$ of an algorithm.

**Definition 4.34** (*Growth Rate*) For a fixed $n$, the **growth rate** (or *order of growth*) of $f(n)$ is a measure of efficiency of an algorithm determined in terms of the number of primitive operations or the number of steps machine executed.

The three types of algorithms we usually come across are classified as the *best case, worst case*, and the *average-case* algorithms. For some algorithms such as *merge sort*, all the three cases are asymptotically the same. On the other hand, for example,

[10] The term "algorithm" is derived from the name *al-Khowārizmī* of the Persian mathematician.

the worst case for *insertion sort* occurs when array is reverse sorted, and the best case when array is sorted in the same order.

(a) In the case of *worst-case analysis* of an algorithm, we estimate the upper bound on $f(n)$. For *linear search algorithm*, the worst case occurs when the element to be searched is not part of the array. In general, we must know case that effects a maximum number of operations to be executed. As expected, we generally use this to undertake complexity analysis of an algorithm.
(b) In the case of *best-case analysis* of an algorithm, we estimate the lower bound on the running time. For *linear search algorithm*, the best case happens when the element to be searched is the first of the array. Clearly, the number of operations to be executed in this case is constant, i.e., it is independent of the input size. This type of complexity analysis of an algorithm is of little practical use.
(c) In the case of *average-case analysis* of an algorithm, we first compute time for all types of possible inputs, and then find the average. Accordingly, we need estimated distribution of cases. For *linear search algorithm*, if all cases are uniformly distributed (even when element to be searched is not part of array), then we take sum over all cases and divide it by $n + 1$, where $n$ is the input size. As expected, we rarely use this type of complexity analysis of an algorithm.

We study growth rate of a class of functions defined on the set of non-negative integers $\mathbb{Z}^+ = \{0, 1, 2, \dots\}$ that help understand what it means to say a function grows at the same rate or slower than a function in the collection. In general, we choose a simple function that is as small as possible. These functions are also known as the **test functions**. Some commonly used *test functions* in the present context are such as shown in Fig. 4.8.

Notice that the *growth rate* of an algorithm is not about examining the total execution time of the algorithm. It is actually needed to compare the performances of different algorithms that do the same task or solve a computational problem. For example, *merge sort* can sort large arrays relatively quickly because it has a time complexity of $O(n \log n)$. Also, it is *stable sort* in the sense that the order of elements with equal values is preserved during the sort. In Appendix A.2, we have provided a brief introduction to some undefined technical terms. The interested reader may also refer to the three volumes of "*The Art of Computer Programming*" by *Donald Knuth* [3–5].

While analyzing an algorithm, we ignore constants and lower order terms of the function $f(n)$ so that it suffices to consider only the input size $n$. Also, since the behavior of an algorithm may be different for each possible input, we need to summarize the related behavior in simple, easily understood formulas. We use *asymptotic notations* such as defined below to describe the asymptotic run time of an algorithm. These are usually defined in terms of the function $f(n)$, with $n \in \mathbb{Z}^+ = \{0, 1, 2, \dots\}$. More generally, we consider functions $f(x)$ with the real argument $x$. We start with *Big-O notation* that describes the worst-case run-time function. It also helps explain the maximum time an algorithm takes to process all input values.

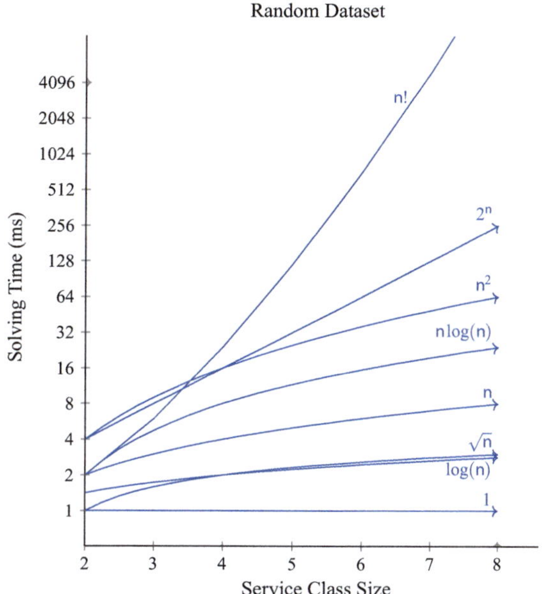


**Fig. 4.8** Growth rate of some test functions

**Definition 4.35** (*Big-O Notation*) Let $f(x)$ and $g(x)$ be two real-valued functions, with $x$ taking a non-negative integer or a real value. We say $f(x) = O\big(g(x)\big)$ if there are positive constants $C$ and $a$ such that we have

$$\big|f(x)\big| \leq C\,\big|g(x)\big|, \qquad \text{for all} \quad x > a.$$

The two constants $C$ and $a$ are called *witnesses*.

Clearly, when there is one pair of witnesses, there are infinitely many such pairs. It is straight forward to see that a polynomial

$$f(x) = a_n\,x^n + a_{n-1}\,x^{n-1} + \cdots + a_1\,x + a_0, \quad with\ \ a_n \neq 0,$$

is of Big–O order $x^n$, with witnesses given by

$$C = |a_n| + |a_{n-1}| + \cdots + |a_1| + |a_0| \quad \text{and} \quad k = 1.$$

Further, if $h$ is a similar function such that $g(x) < h(x)$, for large $x$, then we have $f(x) = O\big(h(x)\big)$, with the same witnesses. In the next example, we discuss growth of some of these functions.

**Example 4.19** We first consider $g_1(n) = n!$. In this case, we have

$$n! = n(n-1)\cdots 2\cdot 1 \leq n\cdot n\cdots n\cdot n = n^n \quad \Rightarrow \quad g_1(n) = O\big(n^n\big),$$

with witnesses given by $C = a = 1$. Next, for $g_2(n) = \log n!$, we have

$$\log n! \le \log n^n = n \log n \quad \Rightarrow \quad g_2(n) = O\big(n \log n\big),$$

with witnesses given by $C = a = 1$. Also, for $g_3(n) = n$, we have

$$n < 2^n \quad \Rightarrow \quad g_3(n) = O\big(2^n\big),$$

with witnesses given by $C = a = 1$. Similarly, for $g_4(n) = \log n$, we have

$$n < 2^n \quad \Rightarrow \quad \log n < n \log 2 \quad \Rightarrow \quad g_4(n) = O\big(n\big),$$

with witnesses $C = \log 2$ and $a = 1$. More generally, for $g_5(n) = \log_b n$, we have

$$\log_b n = \frac{\log n}{\log b} < \frac{n}{\log b} \quad \Rightarrow \quad \log n < n \log 2 \quad \Rightarrow \quad g_5(n) = O\big(n\big),$$

with witnesses $C = 1/\log b$ and $a = 1$. Notice that, however, $g(n) = n^2$ is not $O(n)$. For instance, if we can find witnesses $(C, a)$ so that

$$n > a \quad \Rightarrow \quad n^2 \le C\,n,$$

then it follows that $n \le C$, which cannot hold for all $n$ with $n > a$.

When algorithms consist of several procedures, the number of steps used by the algorithm of specified input size is the sum of the number of steps used by all such procedures. In the next two theorems, we determine the growth of the *sum* and the *product* of two functions.

**Theorem 4.21** *Let $f_1$, $f_2$ and $g_1$, $g_2$ be two real-valued functions, with argument $x$ taking an integer or a real value. Suppose $f_1(x) = O\big(g_1(x)\big)$ and $f_2(x) = O\big(g_2(x)\big)$, with respective witnesses $(C_1, a_1)$ and $(C_2, a_2)$. Then, there exist witnesses $(C, a)$ such that we have*

$$\big|(f_1 + f_2)(x)\big| \le C\,\big|h(x)\big|, \qquad \text{for all} \quad x > a,$$

*where $h(x) = \max\big(\big|g_1(x)\big|, \big|g_2(x)\big|\big)$.*

***Proof*** By the given conditions, we have

$$\begin{aligned} \big|f_1(x)\big| &\le C_1\,\big|g_1(x)\big|, \qquad \text{for all} \quad x > a_1; \\ \big|f_2(x)\big| &\le C_2\,\big|g_2(x)\big|, \qquad \text{for all} \quad x > a_2, \end{aligned}$$

which implies that

$$\begin{aligned}\left|(f_1 + f_2)(x)\right| &= \left|f_1(x) + f_2(x)\right| \\ &\le \left|f_1(x)\right| + \left|f_2(x)\right| \\ &\le C_1\left|g_1(x)\right| + C_2\left|g_2(x)\right|,\end{aligned}$$

for $x > a = \max(a_1, a_2)$. Therefore, for $C = C_1 + C_2$, we obtain

$$\left|(f_1 + f_2)(x)\right| \le C\left|h(x)\right|, \qquad \text{for all} \quad x > a.$$

Hence, $(f_1 + f_2)(x) = O(h(x))$. □

**Theorem 4.22** *Let $f_1$, $f_2$ and $g_1$, $g_2$ be two real-valued functions, with argument $x$ taking an integer or a real value. Suppose $f_1(x) = O(g_1(x))$ and $f_2(x) = O(g_2(x))$, with respective witnesses $(C_1, a_1)$ and $(C_2, a_2)$. Then, there exist witnesses $(C, a)$ such that we have*

$$\left|(f_1 \cdot f_2)(x)\right| \le C\left|h(x)\right|, \qquad \textit{for all} \quad x > a,$$

*where $h(x) = (g_1 \cdot g_2)(x)$.*

***Proof*** By the given conditions, we have

$$\begin{aligned}\left|f_1(x)\right| &\le C_1\left|g_1(x)\right|, \qquad \text{for all} \quad x > a_1; \\ \left|f_2(x)\right| &\le C_2\left|g_2(x)\right|, \qquad \text{for all} \quad x > a_2,\end{aligned}$$

which implies that

$$\begin{aligned}\left|(f_1 \cdot f_2)(x)\right| &= \left|f_1(x) \cdot f_2(x)\right| \\ &= \left|f_1(x)\right|\left|f_2(x)\right| \\ &\le (C_1C_2)\left|g_1(x)\right|\left|g_2(x)\right|,\end{aligned}$$

for $x > a = \max(a_1, a_2)$. Therefore, for $C = C_1C_2$, we obtain

$$\left|(f_1 \cdot f_2)(x)\right| \le C\left|h(x)\right|, \qquad \text{for all} \quad x > a.$$

Hence, $(f_1 \cdot f_2)(x) = O((g_1 \cdot g_2)(x))$. □

These two theorems help determine growth rate of combinations of functions.

**Example 4.20** We find here Big-O estimate of the function $f : \mathbb{Z}^+ \to \mathbb{R}$ given by

$$f(n) := 3n \log n! + n^2 + 3.$$

We already know that $\log n! = O(n \log n)$ (see Example 4.19). Also, by Theorem 4.22,

$$3n = O(n) \quad \Rightarrow \quad 3n \log n! = O(n \cdot n \log n) = O(n^2 \log n).$$

Further, for $C = 2$ and $a = 2$, we have

$$n^2 + 3 \leq C\, n^2, \qquad \text{for all} \quad n > a.$$

That is, $n^2 + 3 = O(n^2)$. It thus follows from Theorem 4.21 that Big-O estimate of the function $f$ is given by

$$f(n) = O(\max(n^2 \log n, n^2)) = O(n^2 \log n).$$

Similarly, it can be shown that the Big-O estimate of the function $f : \mathbb{Z}^+ \to \mathbb{R}$ given by

$$f(n) := n^2 \log(n^3 + 1) + n^3 + 4n^2 + 5$$

is $n^3$. The details are left as an exercise for the reader.

**Example 4.21** (*Asymptotic Bounds of Harmonic Number*) We consider here the $n$th *harmonic number* $H_n$ that we come across in many situations such as in the *coupon collector's problem*, as discussed in a later chapter. Recall that the $n$th *harmonic number* $H_n$ is given by

$$H_n := 1 + \frac{1}{2} + \cdots + \frac{1}{n}, \quad \text{for} \quad n \geq 1. \tag{4.6.1}$$

We first show that $H_n \sim \log n$. For, by the Stolz-Cesaro theorem, we have

$$\begin{aligned}
\lim_{n\to\infty} \frac{H_n}{\log n} &= \lim_{n\to\infty} \frac{H_{n+1} - H_n}{\log(n+1) - \log n} \\
&= \lim_{n\to\infty} \frac{1/(n+1)}{\log(1 + 1/n)} \\
&= \lim_{n\to\infty} \frac{1}{\log(1 + 1/n)^{n+1}} = [\log e]^{-1} = 1.
\end{aligned}$$

Therefore, we have $H_n = \Theta(\log n)$. Next, let $k = \lceil \log_2(n+1) \rceil$. We have

$$\begin{aligned}
H_n &:= 1 + \frac{1}{2} + \cdots + \frac{1}{n} \\
&\leq 1 + \frac{1}{2} + \cdots + \frac{1}{2^k - 1} \\
&\leq 1 + \left(\frac{1}{2} + \frac{1}{3}\right) + \left(\frac{1}{4} + \cdots + \frac{1}{7}\right) + \cdots + \left(\frac{1}{2^{k-1}} + \cdots + \frac{1}{2^k - 1}\right) \\
&\leq 1 + \left(\frac{1}{2} + \frac{1}{2}\right) + \left(\frac{1}{4} + \cdots + \frac{1}{4}\right) + \cdots + \left(\frac{1}{2^{k-1}} + \cdots + \frac{1}{2^{k-1}}\right) \\
&\leq 1 + 1 + \cdots + 1\,(k \text{ terms}) = k
\end{aligned}$$

Therefore, $H_n \leq \lceil \log_2(n+1) \rceil \sim \log_2 n$. Similarly, for $k = \lfloor \log_2 n) \rfloor$, we have

$$\begin{aligned}
H_n &:= 1 + \frac{1}{2} + \cdots + \frac{1}{n} \\
&\geq 1 + \frac{1}{2} + \cdots + \frac{1}{2^k} \\
&\geq 1 + \frac{1}{2} + \left(\frac{1}{3} + \frac{1}{4}\right) + \left(\frac{1}{5} + \cdots + \frac{1}{8}\right) + \cdots + \left(\frac{1}{2^{k-1}+1} + \cdots + \frac{1}{2^k}\right) \\
&\geq 1 + \frac{1}{2} + \left(\frac{1}{4} + \frac{1}{4}\right) + \left(\frac{1}{8} + \cdots + \frac{1}{8}\right) + \cdots + \left(\frac{1}{2^k} + \cdots + \frac{1}{2^k}\right) \\
&\geq 1 + \frac{1}{2} + \frac{1}{2} + \cdots + \frac{1}{2}\,(k \text{ terms}) = 1 + \frac{1}{2}k.
\end{aligned}$$

Therefore, $H_n \geq 1 + (1/2)\lfloor \log_2 n \rfloor \sim (1/2)\log_2 n$.

We next introduce *Big-Omega notation*, which is used to determine an algorithm's best-case time complexity. It helps explain the minimum time an algorithm requires to consider all input values.

**Definition 4.36** (*Big-Omega Estimate*) Let $f$ and $g$ be two real-valued functions, with argument $x$ taking a non-negative integer or a real value. We say $f(x) = \Omega\big(g(x)\big)$ if there are positive constants $C$ and $a$ such that we have

$$\left|f(x)\right| \geq C\left|g(x)\right|, \qquad \text{for all} \quad x > a.$$

The two constants $C$ and $a$ are called *witnesses*.

For example, we have

$$\log\left(n^4\right) + 2^n = 4\,\log\left(n\right) + 2^n \geq \log\left(n\right).$$

Therefore, with $C = a = 1$, it follows that

$$\log\left(n^4\right) + 2^n = \Omega\big(\log(n)\big).$$

For a function of interest to us in this section, Big-O estimate provides *upper bound* and $\Omega$ estimate provides *lower bound*. The next theorem gives an interesting connection between the two concepts.

**Theorem 4.23** *Let $f$ and $g$ be two real-valued functions, with argument $x$ taking an integer or a real value. If $f(x) = O\big(g(x)\big)$, with witnesses given by $(C, a)$, then $g(x) = \Omega\big(f(x)\big)$, with witnesses $(1/C, a)$. The converse holds too.*

***Proof*** Since $f(x) = O\big(g(x)\big)$, with witnesses $(C, a)$, we have

$$\big|f(x)\big| \le C\,\big|g(x)\big|, \qquad \text{for all} \quad x > a.$$

Or, equivalently, we have

$$\big(1/C\big)\big|f(x)\big| \le \big|g(x)\big|, \qquad \text{for all} \quad x > a,$$

which proves the assertion. Similarly, we can prove the converse. □

Finally, we introduce *Big-$\Theta$ notation*, which is used to determine an algorithm's average-case time complexity, i.e., when the set of functions lies in both $O(f)$ and $\Omega(f)$. Accordingly, it help explains the average time an algorithm requires to consider all input values.

**Definition 4.37** (*Big-Theta Estimate*) Let $f$ and $g$ be two real-valued functions, with argument $x$ taking a non-negative integer or a real value. We say $f(x) = \Theta\big(g(x)\big)$ if both $f(x) = O\big(g(x)\big)$ and $f(x) = \Omega\big(g(x)\big)$ hold. That is, there are pairs of positive constants $(C, a)$ and $(C', a')$ such that we have

$$\big|f(x)\big| \le C\,\big|g(x)\big|, \qquad \text{for all} \quad x > a; \tag{4.6.2a}$$

$$\big|f(x)\big| \ge C'\,\big|g(x)\big|, \qquad \text{for all} \quad x > a'. \tag{4.6.2b}$$

Clearly, both the conditions in (4.6.2) hold when $f$ is a polynomial. Therefore, we have $f(x) = \Theta\big(x^n\big)$. In general, $\Theta$ provides both lower and upper bounds of the function $f(x)$ in terms of the function $g(x)$, and vice versa. Also,

$$f(x) = \Theta\big(g(x)\big) \quad \Leftrightarrow \quad g(x) = \Theta\big(f(x)\big).$$

Therefore, we also say that $f(x)$ is of order $g(x)$.

**Example 4.22** What is shown in the last part of Example 4.21 is sufficient to conclude that $H_n = \Theta(\log n)$, where $H_n$ is the $n$th *harmonic number* defined by (4.6.1).

Notice that the exact base of the log is immaterial in big-$\Theta$ notation. We compute here some other asymptotic bounds of $H_n$. For instance, consider the two functions

$$f(n) := H_n - \ln n; \tag{4.6.3a}$$

$$g(n) := H_n - \ln(n+1). \tag{4.6.3b}$$

Notice that $f$ is strictly decreasing with $f(1) = 1$, and $g$ is strictly increasing with $g(1) = 1 - \ln 2$. We also have

$$\left(f(n) - g(n)\right) \to 0, \quad \text{as} \quad n \to \infty, \tag{4.6.4}$$

so that they share a common limit

$$\gamma := \int_1^\infty \left(-\frac{1}{x} + \frac{1}{\lfloor x \rfloor}\right) dx = 0.5772156649\ldots, \tag{4.6.5}$$

known as the **Euler-Mascheroni constant**. It thus follows that

$$\gamma + \ln n < H_n < \gamma + \ln(n+1). \tag{4.6.6}$$

Now, since we have

$$\begin{aligned} \ln(n+1) &= \ln\left(n \cdot \left(1 + \frac{1}{n}\right)\right) \\ &= \ln n + \ln\left(1 + \frac{1}{n}\right) \\ &= \ln n + \Theta\left(\frac{1}{n}\right), \end{aligned}$$

where in the last step we used the following series:

$$\ln(1+x) = x - \frac{x^2}{2} + \frac{x^3}{3} - \cdots, \quad \text{for} \quad -1 < x \le 1.$$

Notice that in our case $x = \frac{1}{n} \to 0$ as $n \to \infty$, so we can use the approximation $\ln(1 + 1/n) \approx 1/n$. The error bound in Taylor's theorem gives

$$\ln\left(1 + \frac{1}{n}\right) = 1 + \frac{1}{n} - \frac{\eta^2}{2}, \quad \text{for some} \quad 0 \le \eta \le \frac{1}{n}.$$

We thus have

$$\ln\left(1 + \frac{1}{n}\right) = \ln n + \frac{1}{n} - O\left(1 + \frac{1}{n^2}\right). \tag{4.6.7}$$

It follows from (4.6.6) and (4.6.7) that

$$H_n = \ln n + \gamma + O\left(\frac{1}{n}\right). \tag{4.6.8}$$

Although pseudocode for *merge sort* works correctly when the number of elements is not even, the related recurrence-based analysis is simplified provided the original problem has the input size a power of two. In this situation, each *divide step* yields two subsequences of size exactly $n/2$. Such an assumption does not affect the order of growth of the solution of the recurrence. The recurrence for *worst-case running time* of merge sort on $n$ numbers, say given by $t(n)$, may be set up as follows: Clearly, merge sort on one element takes constant time. When there are $n > 1$ elements, the break up of running time at the *divide, conquer*, and *combine* steps is, respectively, given by

(a) We compute the middle of the sub-array that takes constant time. Therefore, we have $d(n) = O(1)$.
(b) Since two sub-problems are recursively solved, each with input size $n/2$, it contributes $2t(n/2)$ to the running time.
(c) As already noted that the *merge procedure* on an $n$-element sub-array takes time $O(n)$, it follows that $c(n) = O(n)$.

Notice that, for the merge-sort analysis, when the functions $d(n)$ and $c(n)$ are added, we are actually adding a function that is $O(n)$ and a function that is $O(1)$. As this sum is a linear function of $n$, i.e., $O(n)$, adding it to $2t(n/2)$ term from the "*conquer step*" gives the following recurrence for the worst-case running time $t(n)$ of the merge sort:

$$t(n) = \begin{cases} O(1), & \text{if } n = 1 \\ 2\,t(n/2) + O(n), & \text{if } n > 1. \end{cases}$$

As shown later, $t(n)$ is $O(n \log_2 n)$. Therefore, for large enough inputs, merge sort outperforms *insertion sort*, which has the running time of order $O(n^2)$, in the worst case.

**Remark 4.4** The analysis of an algorithm in terms of its time efficiency and space requirements is dealt with in *complexity theory*, which is an integral part of the *theory of computation*. The *time complexity* studies the amount of time taken by an algorithm to run through the entire computation. It also measures the time taken to execute each statement of code in an algorithm. In general, analysis of even simple algorithms could be challenging. The mathematical tools needed include discrete combinatorics, elementary probability theory, algebraic dexterity, and the ability to identify the most significant terms in a particular formula.

## Exercises 4

4.1 Let $A = \{a, b, c, d\}$, $B = \{0, 1, 2\}$. List out all the relations from $A$ to $B$. Indicate which ones are functions. Also indicate if they are one-to-one, onto, one-to-one onto.

4.2 List all the relations from the set $A = \{0, 1, 2\}$ to $B = \{a, b, c\}$. Indicate which relation is a function. Further, for those that are functions, classify them as one-one, onto, and one-to-one correspondences.

4.3 Let $A = \{-1, 0, 1\}$. List out all the relations on $A$. Indicate which relation is a function. Further, for those that are functions, classify them as one-one, onto, and one-to-one correspondences.

4.4 Let $X = \{a, b\}$ and $Y = \{0, 1, 2\}$. Find all the functions $f : X \to Y$. Also find the range of each function so defined.

4.5 Let $X = \{a, b, c\}$ and $Y = \{0, 1\}$. Find all functions $f : X \to Y$. Also, determine which of the functions so defined are a surjection, an injection, or a bijection.

4.6 Recall that $\mathbb{N}_0$ denotes the set of natural numbers, including 0. Determine whether the following functions are one-one, onto, or a bijection:

a. $f : \mathbb{N}_0 \to \mathbb{N}_0$ given by $f(j) = j \pmod 3$.

b. $g : \mathbb{N}_0 \to \mathbb{N}_0$ given by

$$g(j) = \begin{cases} 1 & (j \text{ is even}) \\ 0 & (j \text{ is odd}) \end{cases}.$$

4.7 Let $\mathbb{Z}$ denote the set of all integers. Determine whether the function $f : \mathbb{Z} \to \mathbb{Z}$ given by

$$f(j) = \begin{cases} \dfrac{j}{2} & (j \text{ is even}) \\ \dfrac{j-1}{2} & (j \text{ is odd}) \end{cases}$$

is one-one, onto, or a bijective function.

4.8 Let $f : A \to B$ be a surjective function, and $b \in B$. The set $f^{-1}(b) \subseteq A$ given by

$$f^{-1}(b) := \{a \in A \mid f(a) = b\}$$

is known as the *fiber* of $b$ with respect to the function $f$. Prove these fibers form a partition of $A$.

4.9 Consider the set $\mathbb{Z}_4 = \{0, 1, 2, 3\}$. Determine whether the function $f : \mathbb{Z}_4 \to \mathbb{Z}_4$ given by $f(j) = 3j \pmod 4$ is one-one, onto, or a bijection.

4.10 Determine whether the two functions $f, g : \mathbb{N} \times \mathbb{N} \to \mathbb{N}$ given by

$$f(x, y) = x + y \quad \text{and} \quad g(x, y) = xy$$

are one-one, onto, or a bijection.

4.11 Give an example of a function $g : \mathbb{Z} \to \mathbb{Z}$ such that (*a*) $g$ is injective; (*b*) $g$ is surjective; (*c*) $g$ is bijective; (*d*) $g$ is injective, but not surjective; (*e*) $g$ is surjective, but not injective; and, (*f*) $g$ is neither injective nor surjective.

4.12 Let $f : X \to Y$ be a function, $A \subseteq X$, and $B \subseteq Y$. Show that

(a) $f^{-1}(Y \setminus B) = X \setminus f^{-1}(B)$;
(b) $f(X \setminus A) \supset Y \setminus f(A) \quad \Leftrightarrow \quad f$ is surjective;
(c) $f^{-1}(f(A)) = A$, if $f$ is injective;
(d) $f(f^{-1}(B)) = B$, if $f$ is surjective.

4.13 Let $f : X \to Y$ be an injective function. Find a surjective function $g : Y \to X$.

4.14 Show that the polynomial function $f : \mathbb{Z}^+ \times \mathbb{Z}^+ \to \mathbb{Z}^+$ given by

$$f(a, b) = \frac{(a + b - 2)(a + b - 1)}{b + 2}, \quad \text{for} \quad a, b \in \mathbb{Z}^+,$$

is a bijection.

4.15 Show that the following are equivalent:

(a) $A$ is a finite set, with $n$ elements.
(b) There is a surjective function $f : [n] \to A$.
(c) There is an injective function $f : A \to [n]$.

4.16 Let $A$ and $B$ be finite sets, and $f : A \to B$ be a function. Find necessary condition for $f$ to be an injective function.

4.17 Let $A$ be a finite set, and $f : A \to A$ be a function. Show that (*a*) if $f$ is a one-one function then it is necessarily onto and (*b*) if $f$ is an onto function then it is necessarily one-one.

4.18 Let $Z$ be the set of all integers. Let $f(x) = x + 2$, $g(x) = x^2$, $h(x) = x^3 + 1$. Compute $f \circ g$, $g \circ h$, $f \circ (g \circ h)$, and $(f \circ g) \circ h$. Compute $g \circ f$, $h \circ g$. Is the *commutative law* valid for the composition of functions?

4.19 Show that the composition of functions is an associative operation.

4.20 Let $f, g, h : \mathbb{R} \to \mathbb{R}$ be given by

$$f(x) = x + 2, \quad g(x) = x - 2, \quad h(x) = 3x.$$

Find composition of these functions. Also find $f \circ h \circ g$.

4.21 Let $\mathbb{R}$ denote the set of all real numbers, and $\mathbb{R}^+$ denote the set of all non-negative real numbers. Let $f : \mathbb{R} \to \mathbb{R}$ and $g : \mathbb{R}^+ \to \mathbb{R}^+$ be defined by $f(x) = -x^2$ and $g(x) = \sqrt{x}$. Compute $f \circ g$ and check if $g \circ f$ is defined.

4.22 Consider the following two permutations in the set $S_6$:

$$\pi_1 = \begin{pmatrix} 1\,2\,3\,4\,5\,6 \\ 3\,1\,5\,4\,6\,2 \end{pmatrix} \quad \text{and} \quad \pi_2 = \begin{pmatrix} 1\,2\,3\,4\,5\,6 \\ 5\,3\,1\,6\,2\,4 \end{pmatrix}.$$

Compute the *compositions* $\alpha^{-1} \circ \beta \circ \alpha$ and $\beta^{-1} \circ \alpha \circ \beta$.

4.23 Let $A = \{1, 2, 3, 4\}$, and consider the two permutations

$$\pi_1 = \begin{pmatrix} 1\,2\,3\,4 \\ 3\,2\,4\,1 \end{pmatrix} \quad \text{and} \quad \pi_2 = \begin{pmatrix} 1\,2\,3\,4 \\ 2\,4\,1\,3 \end{pmatrix}.$$

Find $\left(p_1 \circ p_2\right)^{-1}$ and $p_2^{-1} \circ p_1^{-1}$. Do the relation $\left(p_1 \circ p_2\right)^{-1} = p_2^{-1} \circ p_1^{-1}$ holds?

4.24 Find a bijective function $g : \mathbb{Z} \times \mathbb{Z} \to \mathbb{Z}$, and hence the inverse $g^{-1}$. Also, find a bijective function $h : \mathbb{Z} \to \mathbb{Z} \times \mathbb{Z}$ ($h \neq g^{-1}$), and hence the inverse $h^{-1}$.

4.25 Consider two functions $f, g : \mathbb{Z} \to \mathbb{Z}$ such that both are one-one. Is the composite function $f \circ g$ a one-one function? How about the converse statement?

4.26 Prove that $(a)$ the product of two even permutations is even; $(b)$ the product of two odd permutations is even; and, $(c)$ the product of an even and an odd permutation is odd.

4.27 Write the permutation $\begin{pmatrix} 1\,2\,3\,4 \\ 2\,4\,1\,3 \end{pmatrix}$ as product of cycles.

4.28 Let $A$ and $B$ be two subsets of a set $E$. Show that

a. $\lambda_{A \setminus B} = \lambda_A - \lambda_{A \cap B}$.
b. $\lambda_{A \oplus B} = \lambda_A + \lambda_B - 2\lambda_A\lambda_B$.
c. $(A \cup B)^c = A^c \cap B^c$.

4.29 Let $A$, $B$, and $C$ be subsets of a set $E$. Show that

a. $A \cap (B \oplus C) = (A \cap B) \oplus (A \cap C)$.
b. $A \oplus (B \oplus C) = (A \oplus B) \oplus C$.
c. $A \oplus B = A \oplus C \Rightarrow B = C$.

4.30 Find a function $f(x)$ such that $f(2) = 3$, $f(4) = 5$, $f(7) = 2$, and $f(x)$ assumes any arbitrary value for other arguments. Show that such $f$ is a primitive recursive function.

4.31 Compute $A(2, 1)$ and $A(2, 3)$, where $A(x, y)$ is *Ackermann's function.*

4.32 Recall that $W := \mathbb{N} \cup \{0\}$, where $\mathbb{N}$ is the set of natural numbers. Show that each of the following is a primitive recursive function:

a. The constant function.
b. The square function $f(y) = y^2$.
c. The nonzero *test function* $\mathrm{sg} : W \to W$ as defined in (4.5.7).
d. The function $f(x, y) = x^y$.
e. The zero *test function* $\overline{\mathrm{sg}} : W \to W$ as defined in (4.5.8).

4.33 Show that the *predecessor function* $\mathrm{P} : W \to W$ given by

$$\mathrm{P}(x) = \begin{cases} 0 & (x = 0), \\ x - 1 & (x \geq 1) \end{cases}$$

is primitive recursive.

4.34 Show that the odd and even *parity function* $\mathrm{Pr} : W \to W$ given by

$$\mathrm{Pr}(x) = \begin{cases} 0 & (x \text{ even}), \\ 1 & (x \text{ odd}) \end{cases}$$

is primitive recursive.

4.35 Show that the *proper subtraction function* $\dot{-}: W^2 \to W$ given by

$$x \dot{-} y = \begin{cases} 0 & (x < y), \\ x - y & (x \geq y) \end{cases}$$

is primitive recursive.

4.36 Show that the *absolute value function* $|.| : W^2 \to W$ is primitive recursive.

4.37 Show that the *min function* and *max function* are primitive recursive.

4.38 Show that the *greatest integer function* $\lfloor x/2 \rfloor$ is primitive recursive.

4.39 Show that the function $f(x, y)$ that equals the remainder when $y$ is divided by $x$ is a primitive recursive function.

4.40 Show that the function $f(x_1, x_2, y)$ defined by

$$f(x_1, x_2, y) = \begin{cases} x_2, & (x_1 > y), \\ (x_1 * y) + x_2, & (x_1 \leq y) \end{cases}$$

is primitive recursive.

4.41 Show that the function

$$f(x) = \begin{cases} \frac{x}{2} & (x \text{ is even}) \\ \frac{x-1}{2} & (x \text{ is odd}) \end{cases}$$

is primitive recursive.

4.42 Show that "the number of divisors of $x$" is primitive recursive.

4.43 Prove that the predicate "$x$ is prime" is primitive recursive.

4.44 Consider the function $f : \mathbb{Z}^+ \to \mathbb{R}$ given by

$$f(n) := n^2 \log(n^3 + 1) + n^3 + 4n^2 + 5.$$

Show that $f(n) = O(n^3)$.

4.45 Show that $1 + 2 + \cdots + n = \Theta(n^2)$.

4.46 Show that $n^2 + 5n \log n = \Theta(n^2)$.

## References

1. Enderton, H.B.: Elements of Set Theory. Academic, NY (1977)
2. Razdan, A.K., Ravichandran, V.: Fundamentals of Analysis. Springer (2022)
3. Knuth, D.E.: The Art of Computer Programming: Fundamental Algorithms (3/e). Addison-Wesley, Reading, Massachusetts (1997)
4. Knuth, D.E.: The Art of Computer Programming: Seminumerical Algorithms (3/e). Addison-Wesley (1997)
5. Knuth, D.E.: The Art of Computer Programming: Sorting and Searching (2/e). Addison-Wesley (1998)

# Chapter 5
# Elementary Combinatorics

*Mathematicians have tried in vain to this day to discover some order in the sequence of prime numbers, and we have reason to believe that it is a mystery into which the human mind will never penetrate.*

***Leonhard Euler*** *(1707–1783)*

*To understand recursion, one must first understand recursion.*

***Stephen Hawking*** *(1942–2018)*

In this chapter, we introduce some basic tools of combinatorial reasoning and ingenuity that help deal with intricacies of varied types of *enumeration problems*. We first introduce two fundamental "rules of counting". The fundamental *inclusion-exclusion principle* is about comparing sizes of the union and intersection of more than two nonempty sets. We also discuss some important identities involving multinomial coefficients. Combinatorics also offer tools to study existence of an *optimal arrangement* that satisfies some specified constraints. In large number of cases, we have an unknown sequence $(a_n)$ that satisfies a *recurrence relation*, and we need to find an explicit formula for $a_n$, subject to given *initial values*. In the last section, we discuss some important methods of solving recurrence relations.

The methods presented here help solve some important combinatorial problems concerning social sciences, biological sciences, and information theory [1–3]. In most applications, a useful approach is to gain insight from computations carried out at a much smaller scale. The reader may refer [4–7] to further their study about some advanced computing techniques.

 
V. Ravichandran and A. K. Razdan, *Fundamental Discrete Structures*,
University Texts in the Mathematical Sciences,
https://doi.org/10.1007/978-981-96-0069-4_5

## 5.1 Basic Rules of Counting

The ancient mathematicians from India, Arab, and Greek were familiar with some important combinatorial problems [8–10]. However, it is mainly due to the fundamental work of *Blaise Pascal* (1623–1662), *Jacob Bernoulli* (1654–1705), and *Leonhard Euler* (1707–1783) that the combinatorial techniques gained prominence among mathematicians. The relevant papers published during nineteenth and twentieth centuries introduced the world to important concepts such as discrete probability, error-correction codes, cryptography, and graph theory [11–15]. In this section, the focus is on certain finite discrete structures involving arrangements and selections of objects such as permutations and combinations, with or without repetitions.

### *Sum and Product Rules*

The two *counting rules* given below are fundamental to most *enumeration problems*:

**Sum Rule** Suppose $\{A_1, \ldots, A_k$ are mutually disjoint set, and there are $m_i$ different object in the set $A_i$, for $i = 1, 2, \ldots, k$. Then the number of ways of selecting an object from the $k$ sets $A_i$ is $m_1 + m_2 + \cdots + m_k$.

**Product Rule** Suppose an object $x_1$ can be selected in $m_1$ ways. Having selected $x_1$, a second object $x_2$ (different from $x_1$) can be selected in $m_2$ ways. Having selected $x_1$ and $x_2$, a third object $x_3$ (different from $x_1$ and $x_2$) can be selected in $m_3$ ways, and so on up to $k$th object $x_k$. Then the total number of ways to select the $k$ objects $x_1, x_2 \ldots, x_k$ is given by $m_1 m_2 \ldots m_k$.

We now give some applications of these two rules. To start with, we find the number of *bitstrings* of length $k$, with some specified constraints. Recall that a **bitstring** (or a **binary number**) $b$ is a finite string of bits 0 and 1. The *length* of a *bitstring b* is the number of bits appearing in the string $b$. In this case, we also call $b$ as a **k-binary digit** (or a base 2 **k-digit**).

**Example 5.1** In this illustration, we first find the number of bitstrings of length $k$ such that the first digit is 1. Due to the restriction that the first digit has to be 1, there is exactly one way to choose the first digit. Now, since the remaining $k - 1$ digits can be a 0 or a 1, there are 2 ways to choose the two. Therefore, by the *product rule*, the number of $k$-digit binary numbers is $2^{k-1}$.

More generally, a $k$-digit $b_1 b_2 \cdots b_k$ represents one of the following $2^k$ possibilities:

$$0,\ 1,\ 2,\ 3,\ \ldots,\ 2^k - 1,$$

with the associated *decimal equivalent* given by

$$b_1 b_2 \cdots b_k \equiv_2 b_1 \times 2^{k-1} + b_2 \times 2^{k-2} + \cdots + b_1 \times 2^1 + b_k \times 2^0.$$

On the other hand, for any natural number, we can find the corresponding *binary representation* by repeatedly dividing by 2. For example, a 7-digit binary representation of the numbers $1, 2, 3, 4, \ldots$ is, respectively, given by

$$1 = 0000001, \quad 10 = 0001010, \quad 11 = 0001011, \quad 100 = 1100100, \quad \ldots.$$

Similarly, a *ternary number* is a number represented in base 3. For example, a 7-digit ternary representation of the numbers $1, 2, \ldots, 15$ is, respectively, given by

$$\begin{aligned} 1 =, \quad 2 =, \quad 10 =, \quad 11 =, \quad 12 =, \quad 20 =, \\ 21 =, \quad 22 =, \quad 100 =, \quad 101 =, \quad 102 =, \\ 110 =, \quad 111 =, \quad 112 =, \quad \text{and} \quad 120 = . \end{aligned}$$

Of course, in the case of a ternary number of the form $c_1c_2 \cdots c_k$, we have $c_i \in \{0, 1, 2\}$. For example, to find the number of 4-digit bitstrings such as 0000, 0005, 0021 that have any of the digits 0 to 9 excepting 8, notice that each digit in any such 4-digit bitstring can be chosen in 9 ways. Therefore, by the product rule, this number is $9^4$. Further, if both the digits 8 and 9 are not allowed, the number is given by $8^4$.

**Example 5.2** Suppose MSS301, MAA111E, MAT101, and MGM503 are some subjects offered by an Institution, where the first letter is fixed as it codes the name of the Institution. The second and third characters may be any one of the 26 uppercase alphabets, and the next three positions can be filled by digits. The character E is appended at the end if English is the medium of instruction for a subject. Subject to all the above conditions, we ask: How many subject codes are possible? Notice that the each subject code has 7 characters in total, with the last one being a blank. Also, the first character is fixed and there are 26 ways of choosing each of the second and third characters. Finally, the other three characters can be chosen in 10 different ways each. Clearly, the last character can be chosen in two ways (either a blank or E). Therefore, the total number of possible subject codes is given by

$$26^2 \times 10^3 \times 2 = 1352000.$$

However, if we do not allow subject codes such as MSS000 wherein all the digits are zero, and put a restriction that all the subjects will be taught in English or Malay language, but not in both, then the number of possible subject codes is given by

$$26^2 \times (10^3 - 1) = 25974.$$

The modern-day decimal-based place value system has its origin in "*Aryabhatiya*" of *Aryabhatta* (476–540 AD).

## *Permutations and Combinations*

Consider the set of the English alphabets $a, b, \ldots, y, z$. A number of alphabets chosen with no regard for the order of selections is called a *combination* of alphabets. For example, the following are three combinations of the alphabets $c, d, f$:

$$c, f, d \quad d, c, f \quad \text{and} \quad f, d, c$$

However, when these arrangements are treated different, we say the three are *permutations* of the alphabets $c, d, f$.

We may recall that in general a **permutation** of a finite set $A$ is an ordered sequence of the elements of $A$ such that each element appears exactly once. However, we first discuss here permutations of objects *with repetitions*. For example, the following arrangements of two alphabets taken from the four alphabets $a$, $b$, $c$, $d$ are called the **2-permutations** with repetition:

$$\begin{array}{cccc} aa & ba & ca & da \\ ab & bb & cb & db \\ ac & bc & cc & dc \\ ad & bd & cd & dd \end{array}$$

Notice that we can have $4^2 = 16$ such types of permutations. Similarly, the **3-permutations** with repetition are given by

$$\begin{array}{cccccccc} aaa & baa & caa & daa & aba & bba & cba & dba \\ aca & bca & cca & dca & ada & bda & cda & dda \\ aab & bab & cab & dab & abb & bbb & cbb & dbb \\ acb & bcb & ccb & dcb & adb & bdb & cdb & ddb \\ aac & bac & cac & dac & abc & bbc & cbc & dbc \\ acc & bcc & ccc & dcc & adc & bdc & cdc & ddc \\ aad & bad & cad & dad & abd & bbd & cbd & dbd \\ acd & bcd & ccd & dcd & add & bdd & cdd & ddd \end{array}$$

Notice that we can have $4^3 = 64$ such types of permutations.

**Definition 5.1** (*Permutation with Repetition*) By a *k-permutation* of a set of $n$ objects, with repetition, we mean an arrangement of $k \leq n$ objects that are possibly the same. Two such arrangements are treated distinct when the objects are ordered differently. The number of all such $k$-permutations of $n$ objects is usually denoted by $\overline{A}_n^k$.

**Theorem 5.1** *The number of k-permutations with repetition of objects of n types is* $\overline{A}_n^k = n^k$.

***Proof*** Each object in a typical $k$-permutation with repetition of objects of $n$ types can be chosen in $n$ ways. Thus, by the product rule, the number of $k$-permutations with repetition of objects of $n$ types is $n^k$. □

**Example 5.3** We find here the number of ternary numbers with exactly $k$ digits. Notice that the first digit of the $k$-digit ternary number should be either 1 or 2 and thus there are exactly two ways of choosing the first digit. All the remaining $k - 1$ digits can be either 0 or 1 or 2 and thus there are 3 ways of choosing each of them. Thus by the product rule, the number of $k$-digit ternary numbers is $2 \times 3^{k-1}$. Alternately, the same conclusion may be obtained by using the formula for $\overline{A}_n^k$. A $k$-digit ternary number (with 0 as one of the possible initial digits) can be chosen in $3^k$ ways. The number of $k$ digit ternary numbers with initial digit 0 is $3^{k-1}$. Thus, the number of $k$-digit ternary numbers is $3^k - 3^{k-1} = 2 \times 3^{k-1}$.

**Theorem 5.2** *Suppose A and B are two sets, with $|A| = n$ and $|B| = m$. Then we have $|B^A| = m^n$, where $B^A$ denote the set of functions from A to B.*

***Proof*** Taking $f \in B^A$ and $A = \{a_1, \ldots, a_n\}$, we can write the images of these $n$ elements in the set $B$ as $f(a_1), \ldots, f(a_n)$. As the set $B$ contains $m$ elements, there are exactly $m$ ways to assign to $f(a_i)$ an element of $B$. Therefore, by the product rule, there are $m^n$ ways to choose the $n$-tuple $(f(a_1), \ldots, f(a_n))$. Hence, the number of functions from $A$ to $B$ is $m^n$. □

In particular, we may write $2^n$ for the *power set* of a set of $n$ elements. Next, we discuss *permutations without repetitions*. To start with, consider the case of 2-permutations without repetition from four alphabets, say $a$, $b$, $c$, and $d$, given by

$$\begin{array}{cccc} ab & ba & ca & da \\ ac & bc & cb & db \\ ad & bd & cd & dc \end{array}$$

Clearly, there are $4 \times 3 = 12$ such types of 2-permutations. Similarly, the following are the 3-permutations without repetition of the given 4 alphabets:

$$\begin{array}{cccccccccccc} abc & acb & adb & bac & bca & bda & cab & cba & cda & dab & dba & dca \\ abd & acd & adc & bad & bcd & bdc & cad & cbd & cdb & dac & dbc & dcb \end{array}$$

Clearly, there are $4 \times 3 \times 2 = 24$ such types of 3-permutations.

**Mathematica Example.** *The Mathematica command*

```
Permutations
```

*is used to generate permutation of a list of elements.*

1. *The command* `Permutations[list]` *generates a list of all possible permutations of the elements in list.*

2. *The command* `Permutations[list, n]` *gives all permutations containing at most n elements.*
3. *The command* `Permutations[list, {n}]` *gives all permutations containing exactly n elements.*

*For example, the command*

```
Permutations[{1, 2, 3}]
```

*produces the following permutations of 1, 2, 3:*

```
{{1, 2, 3}, {1, 3, 2}, {2, 1, 3},
 {2, 3, 1}, {3, 1, 2}, {3, 2, 1}}
```

*while the command*

```
Permutations[{a, b, c, d}, {3}]
```

*produces the output*

```
{{a, b, c}, {a, b, d}, {a, c, b}, {a, c, d}, {a, d, b},
 {a, d, c}, {b, a, c}, {b, a, d}, {b, c, a}, {b, c, d},
 {b, d, a}, {b, d, c}, {c, a, b}, {c, a, d}, {c, b, a},
 {c, b, d}, {c, d, a}, {c, d, b}, {d, a, b}, {d, a, c},
 {d, b, a}, {d, b, c}, {d, c, a}, {d, c, b}}
```

**Definition 5.2** (*Permutation without Repetition*) An arrangement of $k$ objects from a set of $n$ different objects is called a **k-permutation without repetition** whenever two such arrangements are treated distinct if the objects are different or they are ordered differently. In particular, an $n$-permutation without repetition of $n$ objects is called a **permutation of $n$ objects**.

**Theorem 5.3** *The number $A_n^k$ of k-permutations without repetition of n different objects is given by*

$$A_n^k = \frac{n!}{(n-k)!}.$$

***Proof*** The first object in a typical $k$-permutation without repetition of $n$ objects can be chosen in $n$ ways. The second object can be chosen in $n-1$ ways, the third object can be chosen in $n-2$ ways, ..., $k$th object can be chosen in $n-k+1$ ways. Thus by the product rule, the number of such $k$-permutations without repetition of $n$ objects is

$$A_n^k = n(n-1)(n-2)\cdots(n-k+1) = \frac{n!}{(n-k)!}.$$

□

Notice that two $k$-permutations without repetition from a set of $n$ different objects can differ in composition and order of the objects. However if $k = n$, then the two $n$-permutations can differ only in the order of the objects. Thus, a permutation of $n$ elements is just a rearrangement of the $n$ objects. Sometimes a permutation of $n$ objects is called a $n$-permutation or a permutation of order $n$. The number of $n$-permutations is $P_n = A_n^n = n!$.

**Example 5.4** An airline decides to give free tickets to three different destinations to three different participants of a conference attended by 70 participants. We ask: In how many ways can the tickets be given to the participants? For instance, let $D_1$, $D_2$, and $D_3$ be the three different destinations to which the airlines plans to give free tickets. The ticket to the destination $D_1$ can be given to any one of the 70 participants and the second ticket to the destination $D_2$ can be given to any one of the remaining 69 participants. Similarly the third ticket can be given to any one of the 68 participants. Thus there are

$$70 \times 69 \times 68 = 328440$$

ways of giving the tickets to the participants.

**Example 5.5** We find here in how many ways can 8 rooks be placed on a conventional chessboard so that no rook can attack each other. We know that the rook moves in a straight line, horizontally or vertically. Thus if the rooks are to be placed in the chessboard so that no rook can attack another, then there should be exactly one rook in every row and in every column. Let $a_i$ denote the square in the $i$th column in which the rook is placed. Then $(a_1, a_2, \ldots, a_8)$ is a permutation of the numbers $1, 2, \ldots, 8$. Since there are $P_8 = 8! = 40320$ permutation of the integers $1, 2, \ldots, 8$, we have 40320 ways of placing the rooks so that no rook can attack each other.

**Theorem 5.4** *Given $n$ objects of $k$ different types such that $n_1$ are of the first type, $n_2$ are of the second type, and so on. Then the number of $n$-permutations of these $n$ objects is given by*

$$P(n_1, n_2, \ldots, n_k) := \frac{n!}{n_1!n_2!\cdots n_k!}.$$

***Proof*** Notice that $n = n_1 + n_2 + \cdots + n_k$. Now, if all $n$ objects are distinct, say

$$a_1, a_2, \ldots, a_{n_1}; \quad b_1, b_2, \ldots, b_{n_2}; \quad \ldots \quad ; x_1, x_2, \ldots, x_{n_k},$$

then we have $n!$ permutations of the above $n$ elements. In the general case, since some objects could be identical, the number of $n$-permutations is less than $n!$. Indeed, when

$$\begin{aligned} a_1 &= a_2 = \cdots = a_{n_1} = a; \\ b_1 &= b_2 = \cdots = b_{n_2} = b; \\ &\vdots \\ x_1 &= x_2 = \cdots = x_{n_k} = x, \end{aligned}$$

there are $n_1!n_2!\cdots n_k!$ permutations of $a_1, \ldots, a_{n_1}; b_1, \ldots, b_{n_2}; \ldots; x_1, \ldots, x_{n_k}$ that gives rise to the same permutation of

$$a, a, \ldots a; \quad b, b, \ldots, b; \ \ldots; \ x, x, \ldots, x.$$

Therefore, the number of permutations of $n$ objects $a, \ldots, a; b, \ldots, b; \ldots; x, \ldots, x$ is

$$P(n_1, n_2, \cdots n_k) = \frac{n!}{n_1!n_2!\cdots n_k!}.$$

□

The next examples illustrate the above theorem.

**Example 5.6** Consider the symbols $a, A, b, c$. Then the 4-permutations of these 4 elements are

$$\begin{array}{cccccc} aAbc & abAc & acAb & Aabc & Abac & Acab \\ baAc & bAac & bcaA & caAb & cAab & cbaA \\ aAca & abcA & acbA & Aacb & Abca & Acba \\ bacA & bAca & bcAa & cabA & cAba & cbAa. \end{array}$$

If we replace $A$ by $a$, then we get the following 4-permutations of $a, a, b, c$:

$$\begin{array}{cccccc} aabc & abac & acab & baac & bcaa & caab \\ cbaa & aacb & abca & acba & baca & caba. \end{array}$$

The number of 4-permutations of $a, a, b, c$ is 12 and this number is given by

$$\frac{4!}{2!1!1!}.$$

We now consider problems involving combinations with and without repetitions.

**Definition 5.3** (*Combination without Repetition*) By a **k-combination** of $n$ objects we mean an unordered selection of $k$ objects from these $n$ objects, and the number of such types of combinations is denoted by $\binom{n}{k}$.

**Theorem 5.5** *The number of k-combination of n objects is given by*

$$\binom{n}{k} = \frac{n!}{k!(n-k)!}.$$

***Proof*** Since each $k$-combination of $n$ objects gives $k!$ permutations, we see that

$$k!\binom{n}{k} = A_n^k \quad \text{or} \quad \binom{n}{k} = \frac{A_n^k}{k!} = \frac{n!}{k!(n-k)!}.$$

□

**Example 5.7** Given the four letters $a, b, c, d$, the following are 3-combinations of 4 letters:

$$abc, abd, acd, bcd.$$

The number of 3-combinations of 4 elements is clearly $\binom{4}{3} = 4$. The 2-combinations of 4 letters are

$$ab, ac, ad, bc, bd, cd$$

and the number of 2-combinations of 4 letters is $\binom{4}{2} = 6$. Similarly, what is the number of ways of seating 24 students in a room that can accommodate 30 students is

$$\binom{30}{24} = \frac{30!}{24!6!} = \frac{25.26.27.28.29.30}{1.2.3.4.5.6} = 593775.$$

**Mathematica Example.** *The command* `Binomial[n,k]` *gives the binomial coefficient* $\binom{n}{k}$. *For example,* `Binomial[7,4]` *gives* 35. *A triangular array of the binomial coefficients, called Pascal's triangle, can be obtained by the following command:*

```
Column[Table[Binomial[n, k], {n, 0, 10},
            {k, 0, n}], Center]
```

*It produces the following output:*

```
                    {1},
                  {1, 1},
                {1, 2, 1},
              {1, 3, 3, 1},
            {1, 4, 6, 4, 1},
          {1, 5, 10, 10, 5, 1},
        {1, 6, 15, 20, 15, 6, 1},
      {1, 7, 21, 35, 35, 21, 7, 1},
    {1, 8, 28, 56, 70, 56, 28, 8, 1},
  {1, 9, 36, 84, 126, 126, 84, 36, 9, 1},
{1, 10, 45, 120, 210, 252, 210, 120, 45, 10, 1}
```

**Mathematica Example.** *The* `Subsets` *command is used to produce combinations. The command*

```
Subsets[{a, b, c, d}, {3}]
```

*produces*

```
{{a, b, c}, {a, b, d}, {a, c, d}, {b, c, d}}
```

*while the command*

```
Subsets[{a, b, c, d}, {2}]
```

*produces*

```
{{a, b}, {a, c}, {a, d}, {b, c}, {b, d}, {c, d}}
```

**Theorem 5.6** *For any integers n and k with $n \geq k$,*

$$\binom{n}{k} = P(k, n-k).$$

***Proof*** For a simple computational proof, notice that by the formulas for $P(n_1, \ldots, n_k)$ and $\binom{n}{k}$, we have

$$P(k, n-k) = \frac{n!}{k!(n-k)!} = \binom{n}{k}.$$

Alternatively, for a combinatorial proof, notice that the number $\binom{n}{k}$ is the number of ways of choosing $k$ objects from a set of $n$ objects. Let the given objects be ordered in some way and for each choice of $k$ objects, associate a sequence of zeros and ones as follows: The $i$th term of the sequence is 1 if the $i$th object is selected and it is zero otherwise. Therefore, the number of ways of choosing $k$ objects from the set of $n$ objects is the same as the number of sequences of length $n$ with $k$ ones and $n-k$ zeros and this number is $P(k, n-k)$. □

**Example 5.8** A pastry shop sells four kinds of pastries: napoleons, eclairs, shortcakes, and cream puffs. We find here number of different sets of seven pastries that one can buy. For instance, associate with each set of seven pastries purchased, a sequence of seven 1s and three 0s. We begin the sequence by writing as many ones as there are napoleons in the pastries purchased, followed by a zero. Then add as many ones as there are eclairs, followed by zero. Then write as many ones as there are shortcakes, followed by a zero. Finally, the sequence is completed by writing as many ones as there are cream puffs. If any one of the pastries is missing, we do not write any 1s. For example, if we buy two napoleons, two shortcakes, three cream puffs, the associated sequence is 1100110111. Thus every purchase corresponds to a sequence of seven 1s and three 0s. Therefore the number of purchase is

$$P(7, 3) = 10!/(7!3!) = 120.$$

**Theorem 5.7** *Suppose two sets A and B, respectively, contain n and m elements, with $n \leq m$. Then the number of different one-one functions from A to B is $m!/(m-n)!$. In particular, the number of different one-one functions from a set A with n elements to itself is $n!$.*

***Proof*** The function is one-to-one if every element in $B$ is assigned to at most one element in $A$. Consider the $n$ elements of $A$ in some order. The first element in $A$ can be mapped to any one of the $m$ elements in $B$. Since the function must be one-to-one, the next element in $A$ can only be mapped to one of the remaining $m-1$ elements. Thus the second element in $A$ has $m-1$ choices. Similarly, the $i$th element in $A$ has $m-i+1$ choices. Using the rule of product, we find that the number of different one-to-one functions from $A$ to $B$ is

$$m(m-1)(m-2)\cdots(m-n+1) = \frac{m!}{(m-n)!}.$$

The second result follows by taking $m = n$. □

Later, we use inclusion-exclusion principle to obtain a general formula to determine the number of surjective functions from a set $A$ of $n$ elements to a set $B$ of $m$ elements, with $n \geq m$ (Theorem 5.9).

**Definition 5.4** (*Combination with Repetition*) Given $n$ different type of objects, an arrangement of $k$ objects is called a *k-combination with repetition* of objects of $n$ types if the order of the objects in the arrangement does not matter. Notice that repetitions of objects are allowed. The number of $k$-combinations with repetitions of $n$ types of objects is denoted by $\overline{{}^nC_k}$.

**Theorem 5.8** *The number of k-combinations with repetition of objects of n types is*

$$\overline{{}^nC_k} = \binom{n+k-1}{k}.$$

***Proof*** For any choice of $k$ objects, associate a sequence of $k$ ones and $n-1$ zeros by writing as many ones as there are objects of a particular type and different types separated by a zero. Therefore, the number of $k$-arrangements of the given $n$ objects without regard to order is

$$\overline{{}^nC_k} = \frac{(n+k-1)!}{k!(n-1)!} = \binom{n+k-1}{k}.$$

□

**Example 5.9** Given four different types elements (denoted by $a, b, c, d$), the following are 3-combinations with repetition:

*aaa aab aac aad abb abc abd acc acd add*
*bbb bbc bbd bcc bcd bdd ccc ccd cdd ddd*

Since

$$\overline{{}^4C_3} = \binom{6}{3} = 20,$$

the number of 3-combinations with repetition of elements of 4 different types is 20.

## 5.2 Inclusion-Exclusion Principle

Suppose that there are $N$ objects and two properties $\alpha_1$ and $\alpha_2$ these objects can have. Let $N(\alpha_1)$, $N(\alpha_2)$ and $N(\alpha_1\alpha_2)$ denote the number of objects having the properties $\alpha_1$, $\alpha_2$ and both $\alpha_1$ and $\alpha_2$, respectively. Let $N(\alpha_1'\alpha_2')$ denote the number of objects without having any of the two properties $\alpha_1$ and $\alpha_2$. Let us find a relation among these numbers. The number of objects having the properties $\alpha_1$ and $\alpha_2$ is $N(\alpha_1\alpha_2)$. The number of objects having the property $\alpha_1$ but not $\alpha_2$ is

$$N(\alpha_1\alpha_2') = N(\alpha_1) - N(\alpha_1\alpha_2).$$

The number of objects having the property $\alpha_2$ but not $\alpha_1$ is

$$N(\alpha_1'\alpha_2) = N(\alpha_2) - N(\alpha_1\alpha_2).$$

Thus the number of objects with any one of the two properties is

$$\begin{aligned}
&N(\alpha_1'\alpha_2) + N(\alpha_1\alpha_2') + N(\alpha_1\alpha_2)\\
&= N(\alpha_1) - N(\alpha_1\alpha_2) + N(\alpha_2) - N(\alpha_1\alpha_2) + N(\alpha_1\alpha_2)\\
&= N(\alpha_1) + N(\alpha_2) - N(\alpha_1\alpha_2).
\end{aligned}$$

Thus the number of objects with none of these two properties is

$$N(\alpha_1'\alpha_2') = N - [N(\alpha_1) + N(\alpha_2)] + N(\alpha_1\alpha_2).$$

Alternatively, this can be proved as follows. It is clear that

$$N(\alpha_1'\alpha_2') = N - N(\alpha_1) - N(\alpha_1'\alpha_2).$$

However,

$$N(\alpha_1'\alpha_2) = N(\alpha_2) - N(\alpha_1\alpha_2)$$

and therefore

$$\begin{aligned} N(\alpha_1'\alpha_2') &= N - N(\alpha_1) - [N(\alpha_2) - N(\alpha_1\alpha_2)] \\ &= N - [N(\alpha_1) + N(\alpha_2)] + N(\alpha_1\alpha_2). \end{aligned}$$

To begin with, we give two simple illustrations. Suppose 60 students in a class can speak in French or English. Among these, 48 can speak only in French and 3 only in English. Now, since some know both languages whom we counted twice, there are 9 students who can speak both French and English. Next, to count the number of integers between 1 and 22 divisible by 3 or 5, we have seven integers 3, 6, 9, 12, 15, 18, 21 are divisible by 3 and four integers 5, 10, 15, 20 are divisible by 5. The total number of these is $7 + 4 = 13$. However, since 15 is counted twice, it follows that the number of integers between 1 and 22 divisible by 3 or 5 is $13 - 1 = 12$.

**Proposition 5.1** (Inclusion-Exclusion Principle) *Suppose that there are N objects and k properties $\alpha_1, \alpha_2, \ldots, \alpha_k$ these objects can have. Let $N(\alpha_i, \alpha_j, \ldots, \alpha_k)$ denote the number of objects having the properties $\alpha_i, \alpha_j, \ldots, \alpha_k$ and let $N(\alpha_i', \alpha_j', \ldots, \alpha_k')$ denote the number of objects without none of these properties $\alpha_i, \alpha_j, \ldots, \alpha_k$. Then*

$$\begin{aligned} N(\alpha_1'\alpha_2'\cdots\alpha_k') =& N - [N(\alpha_1) + N(\alpha_2) + \cdots + N(\alpha_k)] \\ &+ [N(\alpha_1\alpha_2) + N(\alpha_1\alpha_3) + \cdots + N(\alpha_{k-1}\alpha_k)] \\ &- [N(\alpha_1\alpha_2\alpha_3) + N(\alpha_1\alpha_2\alpha_4) + \cdots + N(\alpha_{k-2}\alpha_{k-1}\alpha_k) \\ &+ \cdots + (-1)^n N(\alpha_1\alpha_2\cdots\alpha_k). \end{aligned}$$

***Proof*** The proof is by induction on $k$. If $k = 1$, the result reduces to $N(\alpha_1') = N - N(\alpha_1)$ and this is trivial. When $k = 2$, the result is the one in the previous problem. Assume that the result is true for $k = m$ so that

$$\begin{aligned} &N(\alpha_1'\alpha_2'\cdots\alpha_m') \\ &= N - [N(\alpha_1) + N(\alpha_2) + \cdots + N(\alpha_m)] \\ &\quad + [N(\alpha_1\alpha_2) + N(\alpha_1\alpha_3) + \cdots + N(\alpha_{m-1}\alpha_m)] \\ &\quad - [N(\alpha_1\alpha_2\alpha_3) + N(\alpha_1\alpha_2\alpha_4) + \cdots + N(\alpha_{m-2}\alpha_{m-1}\alpha_m) \\ &\quad + \cdots + (-1)^m N(\alpha_1\alpha_2\cdots\alpha_m). \end{aligned}$$

Consider the set with the property $\alpha_{m+1}$. This set may have any of the properties $\alpha_1, \ldots, \alpha_m$. Using the induction hypothesis for this set, we have

$$\begin{aligned}
&N(\alpha_1'\alpha_2'\cdots\alpha_m'\alpha_{m+1}) \\
&= N(\alpha_{m+1}) - [N(\alpha_1\alpha_{m+1}) + N(\alpha_2\alpha_{m+1}) + \cdots + N(\alpha_m\alpha_{m+1})] \\
&\quad + [N(\alpha_1\alpha_2\alpha_{m+1}) + \cdots + N(\alpha_{m-1}\alpha_m\alpha_{m+1})] \\
&\quad - [N(\alpha_1\alpha_2\alpha_3\alpha_{m+1}) + \cdots + N(\alpha_{m-2}\alpha_{m-1}\alpha_m\alpha_{m+1}) \\
&\quad + \cdots + (-1)^m N(\alpha_1\alpha_2\cdots\alpha_m\alpha_{m+1}).
\end{aligned}$$

We will now prove the result for $k = m + 1$. Clearly

$$\begin{aligned}
&N(\alpha_1'\alpha_2'\cdots\alpha_m'\alpha_{m+1}') \\
&= N(\alpha_1'\alpha_2'\cdots\alpha_m') - N(\alpha_1'\alpha_2'\cdots\alpha_m'\alpha_{m+1}) \\
&= N - [N(\alpha_1) + N(\alpha_2) + \cdots + N(\alpha_{m+1})] \\
&\quad + [N(\alpha_1\alpha_2) + N(\alpha_1\alpha_3) + \cdots + N(\alpha_m\alpha_{m+1})] \\
&\quad - [N(\alpha_1\alpha_2\alpha_3) + N(\alpha_1\alpha_2\alpha_4) + \cdots + N(\alpha_{m-1}\alpha_m\alpha_{m+1}) \\
&\quad + \cdots + (-1)^{m+1} N(\alpha_1\alpha_2\cdots\alpha_{m+1}).
\end{aligned}$$

This shows that the result is true for $k = m + 1$ and therefore, by the principle of mathematical induction, it follows that the result is true for all $k$. □

In general, the *inclusion-exclusion principle* is a simple identity expressing the size of the union of a finite collection of finite sets in terms of sizes of unions and intersection of its sub-collections. First, consider the case of two sets, say $A$ and $B$, so that we have

$$|A \cup B| = |A| + |B| - |A \cap B|.$$

For three sets $A$, $B$, and $C$, we have

$$|A \cup B \cup C| = |A| + |B| + |C| - |A \cap B| - |A \cap C| - |B \cap C| + |A \cap B \cap C|.$$

More generally, if $A_1, A_2, \ldots, A_n$ are sets, then

$$\begin{aligned}
&|A_1 \cup A_2 \cdots \cup A_n| \\
&= |A_1| + |A_2| + \cdots + |A_n| \\
&\quad - (|A_1 \cap A_2| + |A_1 \cap A_3| + \cdots + |A_{n-1} \cap A_n|) \\
&\quad + (|A_1 \cap A_2 \cap A_3| + |A_1 \cap A_2 \cap A_4| + \cdots + |A_{n-2} \cap A_{n-1} \cap A_n|) \\
&\quad - \cdots + (-1)^n |A_1 \cap A_2 \cdots \cap A_n|.
\end{aligned}$$

**Example 5.10** In a class of 120 students, 25 students can speak Tamil, 40 students can speak Chinese, and 80 students can speak Malay. Also 10 students can speak both Tamil and Chinese, 20 students can speak both Tamil and Malay, and 15 students can speak both Chinese and Malay while 5 students can speak all three languages. We determine here the number of students who do not know any of these three languages. For instance, let $\alpha_1$ be the property of speaking in Tamil, $\alpha_2$ be the property of speaking in Chinese, and $\alpha_3$ be the property of speaking in Malay. Note that $\alpha_1\alpha_2$ denotes the property of speaking in Tamil and Chinese, $\alpha_1\alpha_3$ denotes the property of speaking in Tamil and Malay, and so on. Therefore

$$N = 120, \quad N(\alpha_1) = 25, \quad N(\alpha_2) = 40, \quad N(\alpha_3) = 80,$$

$$N(\alpha_1\alpha_2) = 10, \quad N(\alpha_1\alpha_3) = 20, \quad N(\alpha_2\alpha_3) = 15, \quad N(\alpha_1\alpha_2\alpha_3) = 5.$$

By using the above form of inclusion-exclusion principle, we have

$$\begin{aligned} N(\alpha_1'\alpha_2'\alpha_3') &= N - [N(\alpha_1) + N(\alpha_2) + N(\alpha_3)] \\ &\quad + [N(\alpha_1\alpha_2) + N(\alpha_1\alpha_3) + N(\alpha_2\alpha_3)] - N(\alpha_1\alpha_2\alpha_3) \\ &= 120 - [25 + 40 + 80] + [10 + 20 + 15] - 5 = 15. \end{aligned}$$

Therefore 15 students do not know any of the three languages.

**Example 5.11** We compute here the number of integers from 1 to 1000 that are not divisible by 2, 3, or 5. Let $\alpha_1$ be the property of being divisible by 2, $\alpha_2$ be the property of being divisible by 3, and $\alpha_3$ be the property of being divisible by 5. Notice that $\alpha_1\alpha_2$ denotes the property of being divisible by 6, $\alpha_1\alpha_3$ denotes the property of being divisible by 10, and so on. The number of integers from 1 to N that are divisible by another integer $k$ is the integral part of the number $N/k$. Therefore,

$$N = 1000, \quad N(\alpha_1) = 500, \quad N(\alpha_2) = 333, \quad N(\alpha_3) = 200,$$

$$N(\alpha_1\alpha_2) = 166, \quad N(\alpha_1\alpha_3) = 100, \quad N(\alpha_2\alpha_3) = 66,$$

$$N(\alpha_1\alpha_2\alpha_3) = 33.$$

By using the above form of *principle of inclusion and exclusion*, we have

$$\begin{aligned} N(\alpha_1'\alpha_2'\alpha_3') &= N - [N(\alpha_1) + N(\alpha_2) + N(\alpha_3)] \\ &\quad + [N(\alpha_1\alpha_2) + N(\alpha_1\alpha_3) + N(\alpha_2\alpha_3)] - N(\alpha_1\alpha_2\alpha_3) \\ &= 1000 - (500 + 333 + 200) + (166 + 100 + 66) - 33 \\ &= 266. \end{aligned}$$

Thus the number of integers from 1 to 1000 that is not divisible by 2, 3, or 5 is 266.

In Theorem 5.7, we obtained a general formula to determine the number of injective functions from a set of $n$ elements to a set of $m$ elements, by using the *product rule*. In the next theorem, we use inclusion-exclusion principle to obtain a general formula to determine the number of surjective functions from a set $A$ of $n$ elements to a set $B$ of $m$ elements, with $n \geq m$.

**Theorem 5.9** *The number of onto functions from A to B is given by*

$$\sum_{i=0}^{m}(-1)^i\binom{m}{i}(m-i)^n.$$

*In particular, the number of onto functions from A to itself is n!.*

***Proof*** We may write $B = \{b_1, b_2, \ldots, b_m\}$. At first, we note that there are $m^n$ functions from $A$ to $B$. Of these $m^n$ functions, there are $(m-1)^n$ functions for which $b_1$ is not an image of any element in $A$ (and these are the functions from $A$ to $\{b_2, b_3, \ldots, b_n\}$). Since there are $n$ elements in $B$, it follows that there are $\binom{m}{1}(m-1)^n$ functions for which one of the elements in $B$ is not an image of any element in $A$. Similarly, there are $(m-2)^n$ functions for which $b_1$ and $b_2$ are not images of any element in $A$ (and these are the functions from $A$ to $\{b_3, b_4, \ldots, b_n\}$). Since there are $m$ elements in $B$, it follows that there are $\binom{m}{2}(m-2)^n$ functions for which two of the elements in $B$ are not images of any element in $A$. In general, there are $\binom{m}{i}(m-i)^n$ functions for which $i$ elements in $B$ are not images of any element in $A$. Therefore, by inclusion-exclusion principle, it follows that the number of different onto functions from $A$ to $B$ is

$$\begin{aligned} m^n - \binom{m}{1}(m-1)^n + \cdots + (-1)^m\binom{m}{m}(m-m)^n \\ = \sum_{i=0}^{m}(-1)^i\binom{m}{i}(m-i)^n. \end{aligned}$$

□

**Definition 5.5** Euler's **totient function** $\varphi(n)$ counts the number of integers $1 \le a < n$ such that $\gcd(a, n) = 1$. By convention, $\varphi(1) = 1$.

In particular, we have

$$\varphi(p) = p - 1, \qquad \text{for every prime} \quad p, \tag{5.2.1}$$

because $1, 2, \ldots, p-1$ are relatively prime to $p$ when $p$ is a prime. In general, an explicit formula for the *totient function* $\varphi(n)$ is given by the next theorem.

**Theorem 5.10** *For a positive integer n, the function $\varphi(n)$ is given by*

$$\varphi(n) = n \cdot \left(1 - \frac{1}{p_1}\right) \cdots \left(1 - \frac{1}{p_k}\right), \tag{5.2.2}$$

*where $p_1, \ldots, p_k$ are the prime divisors of n.*

***Proof*** We use the inclusion-exclusion principle to show that (5.2.2) holds. Notice that a number is coprime to $n$ if and only if it is not divisible by any of the $k$ primes

$p_i$. Also, recall that the number of integers less than $n$ that are divisible by prime $p_i$ is exactly $n/p_i$. More generally, since the number of integers less than $n$ that is divisible by the prime numbers $p_1, \ldots, p_k$ equals $n/(p_1 \cdots p_k)$, the formula

$$\left|X \setminus \left(X_1 \cup \cdots \cup X_k\right)\right| = |X| - \sum_{\emptyset \neq I \subseteq \{1,\ldots,k\}} (-1)^{|I|+1} \left| \cap_{i \in I} X_i \right|, \tag{5.2.3}$$

implies that

$$\begin{aligned} \varphi(n) &= n - \sum_{\ell=1}^{k} \sum_{I \subseteq \{1,\ldots,k\}, |I|=\ell} (-1)^{\ell-1} \frac{n}{p_1 \cdots p_\ell} \\ &= n \cdot \sum_{\ell=0}^{k} \sum_{I \subseteq \{1,\ldots,k\}, |I|=\ell} (-1)^{\ell} \left(p_1^{-1} \cdots p_\ell^{-1}\right) \\ &= n \cdot \left(1 - \frac{1}{p_1}\right) \cdots \left(1 - \frac{1}{p_k}\right). \end{aligned}$$

The last equality follows from the fact that the term $p_1^{-1} \cdots p_\ell^{-1}$ occurs in the expansion of the product

$$\left(1 - \frac{1}{p_1}\right)\left(1 - \frac{1}{p_2}\right) \cdots \left(1 - \frac{1}{p_k}\right),$$

with a coefficient of $(-1)^{|I|}$. □

A more direct approach to obtain the formula (5.2.2) is as given below. To start with, we first compute $\varphi(p^2)$, where $p$ is a prime. In this case, the desired number is obtained by eliminating from the $p^2 - 1$ integers

$$1,\ 2,\ \ldots,\ p^2 - 1$$

those that are not relatively prime to $p^2$, namely,

$$p,\ 2p,\ 3p,\ \ldots,\ (p-1)p,$$

which are exactly $p - 1$ in number. We thus obtain the following number:

$$p^2 - 1 - (p - 1) = p^2 - p = p(p - 1).$$

We thus have $\varphi(p^2) = p(p - 1)$. The same argument gives

$$\varphi(p^k) = p^{k-1}(p - 1), \quad \text{for all} \quad k \geq 2. \tag{5.2.4}$$

We next consider the case when $n = pq$, where $p$ and $q$ are two distinct primes. In this case, $\varphi(n)$ is obtained by eliminating from the $pq - 1$ integers

$$1,\ 2,\ \ldots,\ pq - 1$$

those that have a factor in common with $pq$, namely,

$$p,\ 2p,\ \ldots,\ (q-1)p \quad \text{and} \quad q,\ 2q,\ \ldots,\ (p-1)q$$

Since in the above list there are no multiples of $p$ that are also multiples of $q$ such as 0, $pq$, etc. We thus obtain the following number:

$$pq - 1 - (p-1) - (q-1) = pq - p - q + 1 = (p-1)(q-1).$$

Therefore, with $n = pq$, we obtain

$$\varphi(n) = (p-1)(q-1) = \varphi(p)\varphi(q).$$

The above formula holds in general. That is, we have

$$\varphi(m\,n) = \varphi(m)\varphi(n), \quad \text{provided} \quad \gcd(m, n) = 1. \tag{5.2.5}$$

***Proof*** Since $\varphi(mn)$ counts the number of integers $1 \le x < n$ such that $\gcd(x, mn) = 1$, and the right side of (5.2.5) counts the number of ordered pairs of integers $(r, s)$ such that $1 \le r,\ s < n$, with $\gcd(r, m) = 1$ and $\gcd(s, n) = 1$, it suffices to show that the function

$$\psi : \mathbb{Z}^*_{mn} \to \mathbb{Z}^*_m \times \mathbb{Z}^*_n \tag{5.2.6}$$

given by $\psi(a) = (a, a)$ is bijective, where $a \in \mathbb{Z}^*_{mn}$. We may recall that the equivalence as given below holds in general:

$$\gcd(ab, n) = 1 \quad \Leftrightarrow \quad \gcd(a, n) = \gcd(b, n) = 1.$$

We first show that $\psi$ is surjective. For instance, let $r \in \mathbb{Z}_m$ and $s \in \mathbb{Z}_n$ be such that $a = r \bmod m$ and $a = s \bmod n$. Since $\gcd(m, n) = 1$, we can write $1 = x\,m + y\,n$, for some $x,\ y \in \mathbb{Z}$, by Lemma A.1. Now, taking $a = ryn + sxm$, we have

$$a = ryn + sxm \equiv ryn \equiv r \bmod m,$$

because $yn \equiv 1 \bmod m$, by Lemma A.1. Similarly, we have

$$a = ryn + sxm \equiv sxm \equiv s \bmod n.$$

We next show that $\psi$ is injective. For instance, suppose there are residue classes $a, b \in \mathbb{Z}_{mn}^*$ such that $a = b \bmod m$ and $a = b \bmod n$. That is, $m|(b-a)$ and $n|(b-a)$. Now, since $\gcd(m,n) = 1$, it follows that $a = b \bmod mn$. Therefore, $\psi$ is injective. Hence, the formula (5.2.5) holds. □

**Remark 5.1** (*Chinese Remainder Theorem*) The argument used above to prove that $\psi$ is surjective is known as the *Chinese remainder theorem*. For instance, consider the relatively prime residue classes $r \in \mathbb{Z}_m$ and $s \in \mathbb{Z}_n$. We show that there is an integer $a$ such that $a = r \bmod m$ and $a = s \bmod n$. The main idea is to write $a$ as a linear combination $x\,r + y\,s$, for some $x, y \in \mathbb{Z}$ so that reduction modulo $m$ we obtain

$$r \equiv a = x\,r + y\,s \bmod m. \tag{5.2.7}$$

For example, we may take $x = 1$ and $y = 0$. However, since we also need to have

$$s \equiv a = x\,r + y\,s \bmod n. \tag{5.2.8}$$

Therefore, a right way to go about the same is to use the fact that (5.2.7) is satisfied when $x = 1 \bmod m$ and $y = 0 \bmod m$. Similarly, (5.2.8) is satisfied when $x = 0 \bmod n$ and $y = 1 \bmod n$. Let's see if these four congruences are satisfied simultaneously for some values of $x$ and $y$. For instance, we have

$$x = 0 \bmod n \quad \text{and} \quad y = 0 \bmod m \quad \Leftrightarrow \quad x = an \quad \text{and} \quad t = bm;$$

for some integers $a, b$, which implies that

$$x = an \equiv 1 \bmod m \quad \text{and} \quad y = bm \equiv 1 \bmod n.$$

Now, since the above are solvable using the assumption $\gcd(m,n) = 1$, we have $n$ has an inverse $a$ modulo $m$, and $m$ has an inverse $b$ modulo $n$. We can use Lemma A.1 to compute these inverses.

Finally, with $n = p_1^{e_1} p_2^{e_2} \cdots p_k^{e_k}$, we obtain from (5.2.5) and (5.2.4) that

$$\begin{aligned}
\varphi(n) &= \varphi\big(p_1^{e_1}\big)\varphi\big(p_2^{e_2}\big)\cdots\varphi\big(p_k^{e_k}\big) \\
&= p_1^{e_1-1}\big(p_1-1\big)p_2^{e_2-1}\big(p_1-1\big)\cdots p_k^{e_k-1}\big(p_k-1\big) \\
&= p_1^{e_1} p_2^{e_2} \cdots p_k^{e_k} \frac{p_1-1}{p_1}\frac{p_2-1}{p_2}\cdots\frac{p_k-1}{p_k} \\
&= n \cdot \Big(1-\frac{1}{p_1}\Big)\cdots\Big(1-\frac{1}{p_k}\Big),
\end{aligned}$$

which proves (5.2.2).

**Definition 5.6** A **derangement** is a permutation of $n$ objects $a_1, a_2, \ldots, a_n$ such that $a_i$ does not appear in the $i$th position, for any $i$.

Said differently, a *derangements* of $n$ objects is a "fixed-point free" permutations of a set with $n$ elements. That is, all the elements in a derangement are displaced from their original positions. For example, $a_2a_3a_1$ and $a_3a_1a_2$ are derangements of the objects $a_1, a_2, a_3$.

**Example 5.12** The situation involving *derangements* given in this illustration is about a typical scientist who wrote letters to $n$ people, and sealed the envelopes without writing the addresses. Suppose letters are numbered from 1 to $n$. Assuming that the $n$ addresses are written randomly, we ask: *What is the probability that none of the addresses on the n letters is correct.* Said differently, we need to find the number of possible address permutations such that none of the letters is assigned the correct address.

The next theorem gives a formula for the number of derangements of $n$ objects.

**Theorem 5.11** *The number of derangements of n objects is given by*

$$n!\sum_{\ell=0}^{n}\frac{(-1)^{\ell}}{\ell!}. \tag{5.2.9}$$

***Proof*** We use the inclusion-exclusion principle to show that (5.2.9) holds. Recall that there are $n!$ permutations of $n$ objects. We proceed to determine the number of permutations for which a subset $I \subseteq \{1, 2, \dots, n\}$ is assigned the correct address. If $m = |I|$ addresses are correct, then the remaining $n - m$ letters can be assigned addresses in $(n - m)!$ ways. Therefore, the number of permutations for which a set $I$ receives the correct addresses is given by $(n - |I|)!$. Hence, the desired number obtained by formula (5.2.3) is given by

$$n! - \sum_{\ell=1}^{n}\sum_{I\subseteq\{1,\dots,k\},|I|=\ell}(-1)^{\ell-1}(n-\ell)!\,.$$

Since there are $\binom{n}{\ell}$ subsets of cardinality $\ell$, the above formula simplifies to

$$n! - \sum_{\ell=1}^{n}(-1)^{\ell-1}\frac{n!}{\ell!} = n!\sum_{\ell=0}^{n}\frac{(-1)^{\ell}}{\ell!}.$$

□

In particular, the probability that none of $n$ letters has correct address is given by

$$\sum_{\ell=0}^{n}\frac{(-1)^{\ell}}{\ell!} = 1 - 1 + \frac{1}{2} - \frac{1}{6} + \cdots \pm \frac{1}{n!}$$
$$\sim \frac{1}{e} = 0.367879.$$

Also, see Eq. 5.4.1 and Exercise 5.21, for some other application of the inclusion-exclusion principle.

## 5.3 Simple Combinatorial Identities

In this section, we introduce some simple techniques how to obtain certain interesting identities, and also expressions for the related *combinatorial numbers*. To begin with, the next theorem is about *binomial coefficients*.

**Theorem 5.12** *For any positive integers n and k, with $n \geq k$, we have*

$$\binom{n}{k} = \binom{n}{n-k}.$$

***Proof*** We first apply *computational method*. For instance, since we have

$$\binom{n}{k} = \frac{n!}{k!(n-k)!},$$

it follows that

$$\binom{n}{n-k} = \frac{n!}{(n-k)!(n-(n-k))!} = \frac{n!}{k!(n-k)!} = \binom{n}{k}.$$

Alternatively, to derive the same by applying *combinatorial argument*, notice that with every $k$-combination of $n$ objects there associate an $(n-k)$-combination of the $n-k$ remaining elements. Clearly, different $k$-combinations lead to different $n-k$-combinations, and conversely. Therefore, the number of $k$-combinations of $n$ objects is the same as the number of $n-k$-combinations. □

**Theorem 5.13** (Pascal Formula) *For any integers n and k with $n > k$,*

$$\binom{n}{k} = \binom{n-1}{k-1} + \binom{n-1}{k}.$$

***Proof*** We first apply computational method. For instance, since we have

$$\binom{n}{k} = \frac{n!}{k!(n-k)!},$$

it follows that

$$\begin{aligned}\binom{n-1}{k-1}+\binom{n-1}{k} &= \frac{(n-1)!}{(k-1)!(n-k)!}+\frac{(n-1)!}{k!(n-k-1)!}\\ &= \frac{(n-1)!}{k!(n-k)!}[k+n-k]\\ &= \frac{n!}{k!(n-k)!}\\ &= \binom{n}{k}.\end{aligned}$$

Alternatively, we may use a simple combinatorial argument. In this case, we group together the set of all $k$-combinations of $n$ objects into two groups, one containing $\binom{n-1}{k-1}$ elements and the other containing $\binom{n-1}{k}$. To do this, let us call one of the objects as a special object. Then consider the set of all $k$-combinations with the special object and the set of all $k$-combinations without the special object. Clearly the two sets have no elements in common and the sum of the number of elements in each of the sets gives the number of $k$-combinations of $n$ objects. The $k$-combinations (with the special element) in the first set can be obtained by including the special object to the $k-1$-combinations obtained from $n-1$ remaining objects. Therefore, there are $\binom{n-1}{k-1}$ combinations containing the special object.

Similarly, the $k$-combinations (without the special object) in the second set are the same as $k$-combinations of $n-1$ remaining objects. Thus there are $\binom{n-1}{k}$ combinations without the special object. □

**Theorem 5.14** *For any integer $n \geq 0$, we have*

$$\binom{n}{0}+\binom{n}{1}+\binom{n}{2}+\cdots+\binom{n}{n}=2^n.$$

***Proof*** We first apply principle of mathematical induction. Notice that the assertion holds trivially for $n = 0, 1$. Now, we may assume that the assertion holds for $n = k$. That is, suppose

$$\binom{k}{0}+\binom{k}{1}+\binom{k}{2}+\cdots+\binom{k}{k}=2^k.$$

Next, for $n = k + 1$, we have

$$\begin{aligned}\binom{k+1}{r} &= \binom{k}{r-1}+\binom{k}{r},\\ \binom{k+1}{0} &= \binom{k}{0}\end{aligned}$$

and

$$\binom{k+1}{k+1} = \binom{k}{k},$$

we have

$$\begin{aligned}
&\binom{k+1}{0} + \binom{k+1}{1} + \binom{k+1}{2} + \cdots + \binom{k+1}{k} + \binom{k+1}{k+1} \\
&= \binom{k}{0} + \left[\binom{k}{0} + \binom{k}{1}\right] + \left[\binom{k}{1} + \binom{k}{2}\right] + \cdots \\
&\quad + \left[\binom{k}{k-1} + \binom{k}{k}\right] + \binom{k}{k} \\
&= 2\left[\binom{k}{0} + \binom{k}{1} + \binom{k}{2} + \cdots + \binom{k}{k}\right] \\
&= 2^{k+1}.
\end{aligned}$$

This proves the result for $n = k + 1$. That is,

$$\binom{k+1}{0} + \binom{k+1}{1} + \binom{k+1}{2} + \cdots + \binom{k+1}{k+1} = 2^{k+1}.$$

Hence, by the principle of mathematical induction, the assertion holds for all positive integers $n$. Alternatively, we may use a simple combinatorial argument. Notice that the number of $n$-permutations with repetition of objects of two types is $2^n$. Group the $n$-permutations into classes with the samples in the $k$th class consisting of $k$ elements of the first type and $n - k$ elements of the second type. The elements in the $k$th class are the permutation of $k$ elements of the first type and $n - k$ elements of the second type. Thus the number of elements in the $k$th class is $P(k, n - k) = \binom{n}{k}$. The total number of $n$-permutations is therefore given by

$$\binom{n}{0} + \binom{n}{1} + \binom{n}{2} + \cdots + \binom{n}{n}.$$

Thus

$$\binom{n}{0} + \binom{n}{1} + \binom{n}{2} + \cdots + \binom{n}{n} = 2^n.$$

A yet another combinatorial argument is as follows: Consider a set of $n$ objects. The number of ways these $n$ elements can be grouped into two groups is clearly $2^n$. (This follows since any element can be assigned to group 1 or 2.) The $n$ objects can be grouped into two groups by taking $k$ objects to the first group and $n - k$ remaining objects to the second. This can be done in $\binom{n}{k}$ ways. Thus the total number of possible group is

$$\binom{n}{0}+\binom{n}{1}+\binom{n}{2}+\cdots+\binom{n}{n}.$$

It thus follows that

$$\binom{n}{0}+\binom{n}{1}+\binom{n}{2}+\cdots+\binom{n}{n}=2^n.$$

□

For example, let us consider the two objects $a$ and $b$. The 3-permutations with repetitions of these two objects are

$$aaa, aab, aba, bba, abb, bab, bba, bbb.$$

There are $2^3 = 8$ 3-permutations with repetitions of which $\binom{3}{0}$ has no $a$'s, $\binom{3}{1}$ with one $a$, $\binom{3}{2}$ with two $a$'s, and $\binom{3}{3}$ with 3 $a$'s. It thus follows that

$$\binom{3}{0}+\binom{3}{1}+\binom{3}{2}+\binom{3}{3}=2^3.$$

**Mathematica Example.** *Try the following Mathematica command*

```
Sum[Binomial[n, k] , {k, 0, n}]
```

*and see that you are getting the output* $2^n$.

**Theorem 5.15** *For any integers* $n, m \geq 1$,

$$\binom{n}{n}+\binom{n+1}{n}+\binom{n+2}{n}+\cdots+\binom{n+m-1}{n}=\binom{n+m}{n+1}.$$

***Proof*** We first use principle of mathematical induction. For instance, notice that the result is clearly true for $m = 1$. Assume that the result is true for $m = k$:

$$\binom{n}{n}+\binom{n+1}{n}+\binom{n+2}{n}+\cdots+\binom{n+k-1}{n}=\binom{n+k}{n+1}.$$

For $m = k + 1$, by making use of

$$\binom{n}{k}=\binom{n-1}{k-1}+\binom{n-1}{k},$$

we have

$$\begin{aligned}
&\binom{n}{n}+\binom{n+1}{n}+\binom{n+2}{n}+\cdots+\binom{n+k-1}{n}+\binom{n+k}{n}\\
&=\binom{n+k}{n+1}+\binom{n+k}{n}\\
&=\binom{n+k+1}{n+1}.
\end{aligned}$$

This proves assertion for $m = k + 1$, and hence it holds for all integers $m \geq 1$. Alternatively, we may apply combinatorial argument. In this case, the number of $m$-combinations with repetitions of objects of $n + 1$ types is

$$\overline{C}_{n+1}^{m} = \binom{n+m}{m}.$$

These $m$-combinations with repetitions can be put into classes. The $m$-combinations in the $k$th class have $k$-copies of the first object and the remaining $m - k$ places are filled by the remaining $n$ objects. Thus there are

$$\overline{C}_{n}^{m-k} = \binom{n+m-k-1}{m-k}$$

$m$-combinations with $k$ copies of the first object. Thus

$$\binom{n+m-1}{m}+\binom{n+m-2}{m-1}+\binom{n+m-3}{m-2}+\cdots+\binom{n-1}{0}=\binom{n+m}{m}.$$

Changing $n$ to $n + 1$ and $m$ to $m - 1$, we have

$$\binom{n}{0}+\binom{n+1}{1}+\binom{n+2}{2}+\cdots+\binom{n+m-1}{m-1}=\binom{n+m}{m-1}.$$

Since

$$\binom{n}{r}=\binom{n}{n-r},$$

the above identity becomes

$$\binom{n}{n}+\binom{n+1}{n}+\binom{n+2}{n}+\cdots+\binom{n+m-1}{n}=\binom{n+m}{n+1}.$$

□

**Mathematica Example.** *Try the following Mathematica command*

```
Sum[Binomial[n + k, n] , {k, 0, m - 1}]
```

*and see that you are getting the output*

```
(m Binomial[m+n,n])/(1+n).
```

*Is this number same as* $\binom{n+m}{n+1}$*?*

## *Identities using Binomial Theorem*

The binomial theorem can be used to prove identities about combinatorial numbers. We first prove

**Theorem 5.16** Binomial Theorem) *For any positive integer n,*

$$(a+x)^n = \binom{n}{0}a^n x^0 + \binom{n}{1}a^{n-1}x + \binom{n}{2}a^{n-2}x^2 + \cdots + \binom{n}{n}a^0 x^n.$$

***Proof*** This can be proved by induction. For $n = 1$, the result is obvious. Assume that it is true for $n = m$. That is,

$$(a+x)^m = \binom{m}{0}a^m + \binom{m}{1}a^{m-1}x + \cdots + \binom{m}{m}x^n.$$

Now, for $n = m + 1$, we have

$$\begin{aligned}
&(a+x)^{m+1}\\
&= (a+x)^m(a+x)\\
&= \left[a^m + \binom{m}{1}a^{m-1}x + \cdots + \binom{m}{m}x^m\right](a+x)\\
&= a^{m+1} + \binom{m}{1}a^m x + \binom{m}{2}a^{m-1}x^2 + \cdots + \binom{m}{m}ax^m\\
&\quad + a^m x + \binom{m}{1}a^{m-1}x^2 + \cdots + \binom{m}{m-1}ax^m + \binom{m}{m}x^{m+1}\\
&= a^{m+1} + \left[\binom{m}{1} + \binom{m}{0}\right]a^m x + \left[\binom{m}{2} + \binom{m}{1}\right]a^{m-1}x^2 + \cdots\\
&\quad + \left[\binom{m}{m} + \binom{m}{m-1}\right]ax^m + \binom{m}{m}x^{m+1}\\
&= a^{m+1} + \binom{m+1}{1}a^m x + \binom{m+1}{2}a^{m-1}x^2 + \cdots\\
&\quad + \binom{m+1}{m}ax^m + \binom{m+1}{m+1}x^{m+1}.
\end{aligned}$$

Thus the result is true for $n = m + 1$. By the principle of mathematical induction, the result is true for all positive integers $n \geq 1$. □

**Theorem 5.17** *For any positive integer n, we have*

$$\binom{n}{k} = \binom{n}{n-k}.$$

***Proof*** Since

$$(x + a)^n = (a + x)^n,$$

we see that

$$x^n + \binom{n}{1}x^{n-1}a + \cdots + \binom{n}{k}x^{n-k}a^k + \cdots$$
$$= a^n + \binom{n}{1}a^{n-1}x + \cdots + \binom{n}{k}a^{n-k}x^k + \cdots$$

By comparing the coefficient of $x^{n-k}$, we get

$$\binom{n}{k} = \binom{n}{n-k}.$$

□

**Theorem 5.18** *For any integers $n, k$ with $n \geq k \geq 1$,*

$$\binom{n+1}{k} = \binom{n}{k} + \binom{n}{k-1}.$$

***Proof*** Note that

$$(1 + x)^{n+1} = 1 + \binom{n+1}{1}x + \binom{n+1}{2}x^2 + \cdots + \binom{n+1}{k}x^k + \cdots$$
$$(1 + x)^n = 1 + \binom{n}{1}x + \binom{n}{2}x^2 + \cdots + \binom{n}{k}x^k + \cdots$$

and therefore we have

$$
\begin{aligned}
(1+x)^{n+1} &= \left[1+\binom{n}{1}x+\binom{n}{2}x^2+\cdots+\binom{n}{k}x^k+\cdots\right](1+x)\\
&= 1+\binom{n}{1}x+\cdots+\binom{n}{k}x^k+\cdots\\
&\quad +x+\cdots+\binom{n}{k}x^{k+1}+\cdots\\
&= 1+\left[\binom{n}{1}+\binom{n}{0}\right]x+\cdots+\left[\binom{n}{k}+\binom{n}{k-1}\right]x^k+\cdots.
\end{aligned}
$$

By comparing coefficient of $x^k$ in the two expressions for $(1+x)^{n+1}$, we get

$$\binom{n+1}{k}=\binom{n}{k}+\binom{n}{k-1}.$$

□

**Theorem 5.19** *For any positive integer n, we have*

$$\binom{n}{0}+\binom{n}{1}+\cdots+\binom{n}{n}=2^n.$$

***Proof*** The result follows by taking $x=1$ in the following identity:

$$(1+x)^n=\binom{n}{0}+\binom{n}{1}x+\binom{n}{2}x^2+\cdots+\binom{n}{n}x^n.$$

□

**Theorem 5.20** *For any positive integer n, we have*

$$\binom{n}{0}+\binom{n}{2}+\binom{n}{4}+\cdots=\binom{n}{1}+\binom{n}{3}+\cdots$$

***Proof*** Since

$$(1+x)^n=\binom{n}{0}+\binom{n}{1}x+\cdots,$$

we set, by taking $x=-1$,

$$0=\binom{n}{0}-\binom{n}{1}+\binom{n}{2}-\binom{n}{3}+\cdots$$

or

$$\binom{n}{0} + \binom{n}{2} + \binom{n}{4} + \cdots = \binom{n}{1} + \binom{n}{3} + \cdots .$$

□

Consider

$$\begin{aligned}(1+x)^{4m} = &\binom{4m}{0} + \binom{4m}{1}x + \binom{4m}{2}x^2 + \binom{4m}{3}x^3 + \binom{4m}{4}x^4 + \\ &+ \binom{4m}{5}x^5 + \binom{4m}{6}x^6 + \binom{4m}{7}x^7 + \binom{4m}{8}x^8 + \cdots .\end{aligned} \quad (5.3.1)$$

Let $x = i$. Note that $(1+i)^2 = 1 + 2i + i^2 = 2i$. We have

$$\begin{aligned}(1+x)^{4m} &= \left[(1+i)^4\right] \\ &= \left[(1+i)^2(1+i)^2\right] \\ &= [(2i)(2i)]^m \\ &= (-4)^m .\end{aligned}$$

Notice that

$$i^2 = -1, \quad i^3 = -1, \quad i^4 = 1$$

and in general

$$i^{2k} = (-1)^k, \quad i^{3k} = (-i)^k, \quad i^{4k} = 1.$$

Now, by taking $x = i$ in (5.3.1), we get

$$\begin{aligned}(-4)^m = &\binom{4m}{0} + \binom{4m}{1}i + \binom{4m}{2}(-1) + \binom{4m}{3}(-i) + \binom{4m}{4}(1) \\ &+ \binom{4m}{5}i + \binom{4m}{6}(-1) + \binom{4m}{7}(-i) + \binom{4m}{8}(1) + \cdots\end{aligned}$$

By comparing real part, we get

$$\binom{4m}{0} - \binom{4m}{2} + \binom{4m}{4} - \binom{4m}{6} + \binom{4m}{8} \cdots = (-4)^m .$$

By comparing imaginary part, we get

$$\binom{4m}{1} - \binom{4m}{3} + \binom{4m}{5} - \binom{4m}{7} \cdots = 0.$$

**Theorem 5.21** *For any positive integers n, m, and r with* $r \leq \min(m, n)$, *we have*

$$\binom{n+m}{r} = \binom{n}{0}\binom{m}{r} + \binom{n}{1}\binom{m}{r-1} + \cdots + \binom{n}{k}\binom{m}{r-k} + \cdots + \binom{n}{n}\binom{m}{r-n}.$$

***Proof*** Consider

$$(1+x)^{m+n} = (1+x)^m(1+x)^n.$$

By using binomial formula, we get

$$\begin{aligned}
&1 + \binom{n+m}{1}x + \binom{n+m}{2}x^2 + \cdots \\
&\quad + \binom{n+m}{k}x^k + \cdots + \binom{n+m}{n+m}x^{n+m} \\
=&\left[1 + \binom{m}{1}x + \binom{m}{2}x^2 + \cdots + \binom{m}{k}x^k + \cdots + \binom{m}{m}x^m\right] \\
&\times\left[1 + \binom{n}{1}x + \binom{n}{2}x^2 + \cdots + \binom{n}{k}x^k + \cdots + \binom{n}{n}x^n\right]. \\
=&1 + \left[\binom{m}{1} + \binom{n}{1}\right]x + \left[\binom{n}{2} + \binom{n}{1}\binom{m}{1}x + \binom{m}{2}\right]x^2 + \cdots \\
&+\left[\binom{n}{k} + \binom{m}{1}\binom{n}{k-1} + \binom{m}{2}\binom{n}{k-2} + \cdots \right. \\
&\left. + \binom{m}{0}\binom{m}{k}\right]x^k + \cdots
\end{aligned}$$

By comparing the coefficient of $x^r$, we get the required identity. □

As a special case of the above identity, we get, with $\zeta = n = m$,

$$\binom{2n}{n} = \binom{n}{0}\binom{n}{n} + \binom{n}{1}\binom{n}{n-1} + \cdots + \binom{n}{k}\binom{n}{n-k} + \cdots + \binom{n}{n}\binom{n}{n-n}$$

or

$$\binom{2n}{n} = \binom{n}{0}^2 + \binom{n}{1}^2 + \cdots + \binom{n}{k}^2 + \cdots + \binom{n}{n}^2.$$

**Mathematica Example.** *The above identity can be verified by using the command* `Sum[Binomial[n, k]^2, {k, 0, n}]`.

**Theorem 5.22** *If $\alpha$ is any real number and $|x| < a$, then*

$$(x+a)^{\alpha} = a^{\alpha} + \alpha a^{\alpha-1}x + \frac{\alpha(\alpha-1)}{2!}a^{\alpha-2}x^2 + \cdots + \frac{\alpha(\alpha-1)(\alpha-2)\cdots(\alpha-k+1)}{k!}a^{\alpha-k}x^k + \cdots$$

For $\alpha = n \geq 1$, the above reduces to the binomial formula discussed earlier. For $\alpha = -n < 0$, we get the following result.

$$\begin{aligned}\frac{1}{(x+a)^n} =& a^{-n} + (-n)a^{-n-1}x + \frac{(-n)(-n-1)}{2!}a^{-n-2}x^2 + \cdots \\ &+ \frac{(-n)(-n-1)(-n-2)\cdots(-n-k+1)}{k!}a^{-n-k}x^k \\ =& a^{-n} - na^{-n-1}x + (-1)^2\frac{(n)(n+1)}{2!}a^{-n-2}x^2 + \cdots \\ &+ \frac{(-1)^k n(n+1)\cdots(n+k-1)}{k!}a^{-n-k}x^k + \cdots\end{aligned}$$

Since

$$\binom{n+k-1}{k} = \frac{(n+k-1)!}{k!(n-1)!} = \frac{n(n+1)\cdots(n+k-1)}{k!},$$

we have

$$\left(1+\frac{x}{a}\right)^{-n} = 1 - \binom{n}{1}\left(\frac{x}{a}\right) + \binom{n+1}{2}\left(\frac{x}{a}\right)^2 - \cdots + (-1)^k\binom{n+k-1}{k}\left(\frac{x}{a}\right)^k + \cdots.$$

In particular, for $n > 0$,

$$\begin{aligned}\frac{1}{(1+x)^n} &= 1 - \binom{n}{1}x + \binom{n+1}{2}x^2 + \cdots + (-1)^k\binom{n+k-1}{k}x^k + \cdots, \\ \frac{1}{(1-x)^n} &= 1 + \binom{n}{1}x + \binom{n+1}{2}x^2 + \cdots + \binom{n+k-1}{k}x^k + \cdots.\end{aligned}$$

We now compute square roots of a number by making use of the binomial theorem for real power. Note that

$$\begin{aligned}(1+x)^{\frac{1}{2}} &= 1 + \frac{1}{2}x + \frac{\frac{1}{2}\left(\frac{1}{2}-1\right)}{2!}x^2 + \frac{\frac{1}{2}\left(\frac{1}{2}-1\right)\left(\frac{1}{2}-2\right)x^3}{3!} + \cdots \\ &= 1 + \frac{1}{2}x - \frac{1}{8}x^2 + \frac{1}{8}x^3 + \cdots\end{aligned}$$

Using this, we get

$$\begin{aligned}\sqrt{50} &= \sqrt{49+1}\\ &= 7\sqrt{1+\frac{1}{49}}\\ &= 7\left[1+\frac{1}{2}\times\frac{1}{49}-\frac{1}{8}\times\left(\frac{1}{49}\right)^2+\frac{1}{8}\left(\frac{1}{49}\right)^2+\cdots\right]\\ &= 7[1+0.010204]\\ &= 7.071428571.\end{aligned}$$

## 5.4 Constrained Enumeration Problems

Let $k$ and $n$ be positive integers, with $k \le n+1$. In the next example, we find the number of sequences involving $k$ different digits and $n$ different alphabets subject to constraint that no two digits appear consecutively in a sequence. Let us first consider the particular case when two digits are 0, 1 and the three alphabets are $a, b, c$. As given below, the three alphabets can be arranged in $3! = 6$ ways:

$$abc, \quad acb, \quad bac, \quad bca, \quad cab, \quad cba.$$

For each one of the above six arrangements, we have four places to insert the two digits 0 and 1. More specifically, the digit 0 can be inserted in four ways and the digit 1 can be inserted in three ways. Therefore, to form sequences such that no two digits appear consecutively, there are 12 ways to insert the two digits. Notice that we can insert the digits 0 (as the first digit) and 1 (as the second digit) in places as given below:

$$12, \quad 13, \quad 14, \quad 21, \quad 23, \quad 24, \quad 31, \quad 32, \quad 34, \quad 41, \quad 42, \quad 43.$$

For example, corresponding to arrangement $abc$, we obtain sequences given by

$$\begin{array}{cccccc} 0a1bc, & 0ab1c, & 0abc1, & 1a0bc1, & 1a0bc, & a0b1c,\\ a0bc1, & 1ab0c, & a1b0c, & ab0c1, & 1abc0, & a1bc0.\end{array}$$

Similarly, other arrangements generate 12 sequences with said property. Hence, there are in total $6 \times 12 = 72$ sequences involving two different digits and three different alphabets such that two digits don't appear consecutively.

**Example 5.13** Let $k$ and $n$ be positive integers, with $k \le n+1$. In the general case, when there are $k$ different digits and $n$ different alphabets, we have $n!$ ways of permuting the $n$ alphabets and, for each one of these permutations, there are $n+1$

ways to place the $k$ digits. Since the ordering in which digits are placed matters, there are

$$A_{n+1}^{k} = \frac{(n+1)!}{(n+1-k)!}$$

ways of doing so. Therefore, the number of sequences formed with $k$ digits and $n$ alphabets such that two digits don't appear consecutively is given by

$$\frac{n!(n+1)!}{(n+1-k)!}.$$

With $k$ and $n$ as above, in the next example, we find the number of sequences having $k$ ones and $n$ zeros such that none contains 11 as a part. Let us first consider the particular case when $k = 3$ and $n = 4$. Notice that, as there are four zeros, we have five places in the string $X0X0X0X0X$ to place the ones, where $X$'s are places for 0. So, there are $\binom{5}{3} = 10$ ways of placing the ones, there are 10 sequences with 3 ones and 4 zeros with no two consecutive ones. In fact, the possible placement of ones is

$$123, \quad 124, \quad 125, \quad 134, \quad 135, \quad 145, \quad 234, \quad 235, \quad 245, \quad 345$$

and the corresponding sequences are

$$\begin{array}{ccccc} 1010100, & 1010010, & 1010001, & 1001010, & 1001001, \\ 1000101, & 0101010, & 0101001, & 0100101, & 0010101. \end{array}$$

**Example 5.14** Let $k \leq n+1$. In the general case, when there are $k$ ones and $n$ zeros, there are $n+1$ places to insert the $k$ ones and there are $\binom{n+1}{k}$ ways of doing it. Thus the number of sequences of $k$ ones and $n$ zeros with no two consecutive ones is $\binom{n+1}{k}$.

Next, suppose $n$ people are seated at a round table. We find here the number of ways of selecting $k$ of them without including immediate neighbors. Let us first consider the case when four people are seated in a row, say at seats by $1, 2, 3, 4$. We can select 2 people without including immediate neighbors in 3 ways: 13, 14, 24 (here 13 means we select person sitting in seat numbers 1 and 3). If they were seated at a round table, then there are only 2 ways of doing it: 13, 24. Similarly when five people are seated in a line, then there are 6 ways of selecting people without including immediate neighbors: 13, 14, 15, 24, 25, 35. If they were at a round table, then there are only 5 ways: 13, 14, 24, 25.

In the general case, if $n$ people were seated in a row, then selection of $k$ of them without including immediate neighbors is $\binom{n-k+1}{k}$, provided $n \geq 2k-1$. This can be seen as follows. Each selection can be described as a sequence of $k$ ones and $n-k$ zeros: the $r$th term is one if $r$th person is selected and zero otherwise. Since immediate neighbors are not selected, the sequence cannot have consecutive ones. Thus, we need to find the number of sequences of $k$ ones and $n-k$ zeros without consecutive ones and this number is clearly $\binom{n-k+1}{k}$. We now solve the problem when the people are seated at a round table by making use of this formula.

**Example 5.15** First, we group our selection into two groups: selections involving the person seated on seat number 1 and the selections without the person in seat number 1. In the first case, the number of ways of choosing $k$ of them without including immediate neighbors is the same as the number of ways of choosing $k-1$ of the $n-3$ people (other than the person seated in seat 1 and his two neighbors). Since the $n-3$ people can be considered as seated in a row, there are

$$\binom{n-3-(k-1)+1}{k-1} = \binom{n-k-1}{k-1}$$

ways of doing this. Similarly in the second case, the number of ways of choosing $k$ of them without including immediate neighbors is the same as the number of ways of choosing $k$ of the $n-1$ people (other than the person seated in seat 1). Since the $n-1$ people can be considered as seated in a row, there are

$$\binom{n-1-k+1}{k} = \binom{n-k}{k}$$

ways of doing this. Thus there are

$$\binom{n-k-1}{k-1} + \binom{n-k}{k}$$

ways of selecting $k$ people among $n$ people seated at a round table without including immediate neighbors. Since

$$\begin{aligned}\binom{n}{r} &= \frac{n!}{r!(n-r)!}\\ &= \frac{n}{r}\frac{(n-1)!}{(r-1)!(n-r)!}\\ &= \frac{n}{r}\binom{n-1}{r-1},\end{aligned}$$

or

$$\binom{n-1}{r-1} = \frac{r}{n}\binom{n}{r},$$

we have

$$\binom{n-k-1}{k-1}+\binom{n-k}{k}=\frac{k}{n-k}\binom{n-k}{k}+\binom{n-k}{k}$$
$$=\frac{n}{n-k}\binom{n-k}{k}.$$

Thus, the number of ways of selecting $k$ people among $n$ people seated at a round table without including immediate neighbors is

$$\frac{n}{n-k}\binom{n-k}{k}.$$

**Theorem 5.23** *The number of derangements $D_n$ is given by*

$$D_n = n!\left[\frac{1}{0!}-\frac{1}{1!}+\frac{1}{2!}-\cdots+\frac{(-1)^n}{n!}\right]. \tag{5.4.1}$$

***Proof*** We use the inclusion-exclusion principle to show that (5.4.1) holds. Notice that the number of permutations of degree $n$ in which (at least) one element is fixed is $(n-1)!$; the number of permutations of degree $n$ in which (at least) two elements are fixed is $(n-2)!$, etc.; and the number of ways of choosing one, two, ... elements are $\binom{n}{1}, \binom{n}{2}, \ldots$, respectively. Since there are $n!$ permutations, by the inclusion-exclusion principle, the number of permutations in which all the elements are displaced from their original positions is

$$D_n = n! - \binom{n}{1}(n-1)! + \binom{n}{2}(n-2)! - \cdots + (-1)^n\binom{n}{n}(0!)$$
$$= \frac{n!}{0!}-\frac{n!}{1!}+\frac{n!}{2!}-\cdots+(-1)^n\frac{n!}{n!}$$
$$= n!\left[\frac{1}{0!}-\frac{1}{1!}+\frac{1}{2!}-\cdots+\frac{(-1)^n}{n!}\right].$$

The second equality follows from the identity:

$$\binom{n}{r}(n-r)! = \frac{n!}{(n-r)!r!}(n-r)! = \frac{n!}{r!}.$$

□

Notice that we have $D_n \approx n!/e$ when $n$ is sufficiently large.

**Theorem 5.24** *For $n \geq 3$, show that*

$$D_n = (n-1)\left[D_{n-1}+D_{n-2}\right].$$

***Proof*** A simple computational proof is as given below. Since

$$D_n = nD_{n-1} + (-1)^n,$$

we have

$$D_{n-1} = (n-1)D_{n-2} + (-1)^{n-1}$$

and hence

$$\begin{aligned} D_n &= (n-1)D_{n-1} + D_{n-1} + (-1)^n \\ &= (n-1)D_{n-1} + (n-1)D_{n-2} + (-1)^{n-1} + (-1)^n \\ &= (n-1)[D_{n-1} + D_{n-2}]. \end{aligned}$$

We may also have an alternative computational argument as given below. For instance, since

$$D_n = n!\left[\frac{1}{0!} - \frac{1}{1!} + \frac{1}{2!} - \cdots + \frac{(-1)^n}{n!}\right],$$

we have

$$\begin{aligned} &(n-1)[D_{n-1} + D_{n-2}] \\ &= (n-1)\left[(n-1)!\left[\frac{1}{0!} - \frac{1}{1!} + \frac{1}{2!} - \cdots + \frac{(-1)^{n-1}}{(n-1)!}\right]\right. \\ &\qquad \left. + (n-2)!\left[\frac{1}{0!} - \frac{1}{1!} + \frac{1}{2!} - \cdots + \frac{(-1)^{n-2}}{(n-2)!}\right]\right] \\ &= (n-1)[(n-1)! + (n-2)!]\left[\frac{1}{0!} - \frac{1}{1!} + \frac{1}{2!} - \cdots + \frac{(-1)^{n-2}}{(n-2)!}\right] \\ &\qquad + (-1)^{n-1}(n-1). \end{aligned}$$

Since

$$(n-1)\big[(n-1)! + (n-2)!\big] = (n-1)(n-2)!\big[n-1+1\big] = n!$$

and

$$(-1)^{n-1}(n-1) = n!\left[\frac{(-1)^{n-1}}{(n-1)!} + \frac{(-1)^n}{n!}\right],$$

we have

$$(n-1)\big[D_{n-1} + D_{n-2}\big] = n!\left[\frac{1}{0!} - \frac{1}{1!} + \frac{1}{2!} - \ldots + \frac{(-1)^n}{n!}\right] = D_n.$$

A combinatorial argument to prove the above equality goes as follows: Consider the derangements of the integers $1, 2, \ldots, n$ such that the first position is occupied by an integer $k$, with $k \neq 1$. Notice that there are $n - 1$ such classes of derangements, and each has the same number of elements. Next, we count the number of elements in the class of derangements such that the first position is occupied by 2. Such class is separated into two subclasses: The first consists of derangements in which the first position is occupied by 2, and the other with the second position occupied by 1. Clearly, this subclass has $D_{n-2}$ derangements. The second subclass consists of derangements in which the first position is occupied by 2 and the second position is not occupied 1. The number of permutation in this subclass is $D_{n-1}$. For instance, if a typical derangement is denoted by $2a_1a_2 \ldots a_{n-1}$, then $a_1a_2 \ldots a_{n-1}$ is a derangement of the objects $1, 3, 4, 5, \ldots, n$. Therefore, there are $D_{n-1} + D_{n-2}$ derangements such that the first position is occupied by 2. Hence, the number of derangements $D_n$ of $1, 2, \ldots, n$ is given by

$$(n-1)\big[D_{n-1} + D_{n-2}\big].$$

□

Considering the fact that the factorial of $n$ may also be written as

$$n! = (n-1)\big[(n-1)! + (n-2)!\big],$$

and since the identity

$$D_n = nD_{n-1} + (-1)^n$$

is similar to $n! = n(n-1)!$, $D_n$ is also known as the **subfactorial**. One may use the recurrence relation

$$D_n = (n-1)[D_{n-1} + D_{n-2}], \qquad \text{with} \quad D_1 = 0 \text{ and } D_2 = 1,$$

to compute the values $D_n$, for all $n \geq 3$. For example, we have

$$D_3 = 2, \quad D_4 = 9, \quad D_5 = 44, \quad D_6 = 265, \quad D_7 = 1854, \quad \text{and} \quad D_8 = 14833.$$

Next, let $D_{n,r}$ denote the number of derangements in which $r$ elements remain in their original positions and the remaining $n - r$ elements are displaced. Since the $r$ elements can be chosen in $\binom{n}{r}$ ways and the remaining $n - r$ elements can be deranged in $D_{n-r}$ ways, we have

$$D_{n,r} = \binom{n}{r} D_{n-r}.$$

We find next the number of permutations of the integers $1, 2, \ldots, n$ that do not contain any of the pairs $(1, 2), (2, 3), \ldots, (n-1, n)$. For, let $E_n$ denote the number of permutations of the integers $1, 2, \ldots, n$ which do not contain any of the pairs

$(1,2), (2,3), \ldots, (n-1,n)$. There are in total $P_n = n!$ permutations of the integers $1, 2, \ldots, n$. The permutations containing the pair $(1, 2)$ can be obtained by permuting $(1, 2)$ and the integers $3, 4, 5, \ldots, n$. Since there are $n-1$ objects to permute, we see that the pair $(1, 2)$ is contained in $P_{n-1}$ permutations. Since there are $\binom{n-1}{1}$ pairs, there are $\binom{n-1}{1}P_{n-1}$ permutations containing one of the $n-1$ pairs. Consider the pairs $(1, 2)$ and $(2, 3)$ or the pairs $(1, 2)$ and $(3, 4)$. In the first case, the permutations containing the pairs $(1, 2)$ and $(2, 3)$ are obtained by permuting 123 and the integers $4, 5, \ldots, n$. In the latter case, the permutations containing the pairs $(1, 2)$ and $(3, 4)$ are obtained by permuting 12, 34 and the integers $5, 6, \ldots, n$. In these two cases, the number of permutations is $P_{n-2}$. Since two pairs (out of the $n-1$ available pairs) can be chosen in $\binom{n-1}{2}$ ways, there are $\binom{n-1}{2}P_{n-2}$ permutations containing two of the $n-1$ pairs.

Similarly, if we consider $k$ pairs out of the $n-1$ pairs, the number of permutations containing all of these $k$ pairs is $\binom{n-1}{k}P_{n-k}$. Thus we have, by inclusion-exclusion principle,

$$E_n = P_n - \binom{n-1}{1}P_{n-1} + \binom{n-1}{2}P_{n-2} - \cdots + (-1)^{n-1}\binom{n-1}{n-1}P_1.$$

Since

$$\binom{n-1}{k}P_{n-k} = \frac{(n-1)!}{k!(n-k-1)!}(n-k)! = \frac{(n-1)!(n-k)}{k!}$$

we have

$$E_n = (n-1)!\left[n - \frac{n-1}{1!} + \frac{n-2}{2!} - \cdots + \frac{(-1)^{n-1}}{(n-1)!}\right].$$

Also since

$$\frac{(-1)^k(n-k)}{k!} = \frac{(-1)^k n}{k!} + \frac{(-1)^{k-1}}{(k-1)!}$$

and

$$D_n = n!\left[\frac{1}{0!} - \frac{1}{1!} + \frac{1}{2!} - \ldots + \frac{(-1)^{n-1}}{(n-1)!}\right] + (-1)^n,$$

we have

$$\begin{aligned} E_n &= n!\left[\frac{1}{0!} - \frac{1}{1!} + \frac{1}{2!} - \cdots + \frac{(-1)^{n-1}}{(n-1)!}\right] \\ &\quad + (n-1)!\left[\frac{1}{0!} - \frac{1}{1!} + \frac{1}{2!} - \cdots + \frac{(-1)^{n-2}}{(n-2)!}\right] \\ &= D_n - (-1)^n + D_{n-1} - (-1)^{n-1} \\ &= D_n + D_{n-1}. \end{aligned}$$

By using the above recurrence relation, one can compute the values for $E_n$. For example, $E_2 = 1$, $E_3 = 3$, $E_4 = 11$, $E_5 = 53$, $E_6 = 309$, $E_7 = 2119$, $E_8 = 16687$, $E_9 = 148329$, and $E_{10} = 1468457$.

**Theorem 5.25** *In a box office, $m + k$ people line up for tickets, $m$ of them have dollar notes, and $k$ of them have half-dollar coins. A ticket costs 1/2 dollar. At the beginning, there is no money in the box office. In the queue $hdhdhhd$ (where $h$ denotes half-dollar coin and $d$ denotes dollar note), every one presenting a dollar note will get change while in the queue $hddhdhhh$ the second person presenting a dollar note cannot get change. A queue in which every one presenting a dollar note gets change is a good queue; otherwise, the queue is a bad queue. Find the number of good queues.*

***Proof*** We first note that there are good queues only if $m \leq k$. Hence assume that $0 \leq m \leq k$. We need to find the number of sequences of $m$ $d$'s and $k$ $h$'s such that for any $r$, $1 \leq r \leq m + k$, the number of $d$'s in the first $r$ terms of the sequence is less than or equal to the number of $h$'s in the first $r$ terms. Note that there are $P(m, k) = \binom{m+k}{m}$ sequences of $m$ $d$'s and $k$ $h$'s. We solve this problem by finding the number of bad queues. We will show that the number of bad queues is

$$P(m-1, k+1) = \binom{m+k}{m-1}$$

so that the number of good queues becomes

$$\binom{m+k}{m} - \binom{m+k}{m-1} = \binom{m+k}{m} - \frac{m}{k+1}\binom{m+k}{m}$$
$$= \frac{k-m+1}{k+1}\binom{m+k}{m}.$$

We now count the number of bad queues by showing that every bad sequence (or queue) of $m$ $d$'s and $k$ $h$'s can be associated with a definite sequence of $m$ $d$'s and $k + 1$ $h$'s starting with a $d$ and conversely. Consider any bad sequence of $m$ $d$'s and $k$ $h$'s. Since the sequence is bad, there is a position $s$ at which the number of $d$'s in the first $s$ terms of the sequence is greater than the number of $h$'s in the first $s$ terms. This position must have a $d$ and the preceding positions must be occupied by an equal number of $d$'s and $h$'s, say $s$ $d$'s and $s$ $h$'s. Now consider appending a $h$ to the beginning of the sequence and consider the first $2s + 2$ positions of the new sequence of $m$ $d$'s and $k + 1$ $h$'s. By interchanging $d$ and $h$ in the first $2s + 2$ positions of the sequence, we get a sequence of $m$ $d$'s and $k + 1$ $h$'s starting with a $d$.

We now show that every sequence of $m$ $d$'s and $k + 1$ $h$'s starting with a $d$ corresponds to a bad sequence. Consider a sequence of $m$ $d$'s and $k + 1$ $h$'s starting with a $d$. Since $m \leq k$, there must be a segment consisting first $2s$ positions of the sequence with an equal number of $d$'s and $h$'s. By interchanging $d$ and $h$ in this segment and removing $h$ in the first position, we get a sequence of $m$ $d$'s and $k$ $h$'s. In this sequence, the number of $d$'s is one more than the number of $h$'s in the first $2s - 1$ terms. Thus the sequence is a bad sequence. Thus, we have shown that the number of bad sequences is the same as the number of sequences of $m$ $d$'s and $k + 1$ $h$'s starting with a $d$. Clearly this number is $\binom{m+k}{k-1}$. □

In the next example, we illustrate the above procedure by taking $m = 2$ and $k = 3$.

**Example 5.16** There are $\binom{5}{2} = 10$ queues in total of which the queues

$$ddhhh, dhdhh, dhhdh, dhhhd, hddhh$$

are bad queues and the queues

$$hdhdh, hdhhd, hhddh, hhdhd, hhhdd$$

are good queues. Consider the queues

$$ddhhh, dhdhh, dhhdh, dhhhd.$$

In the first position itself, there is a $d$ and therefore there are more $d$ than $h$ in the first 1 position. Thus by adding $h$ to these queues, we get the queues

$$hddhhh, hdhdhh, hdhhdh, hdhhhd$$

and by interchanging $d$ by $h$ in the first two positions, we get the queues

$$dhdhhh, dhhdhh, dhhhdh, dhhhhd.$$

These queues with 2 $d$'s and 4 $h$'s begin with a $d$. Consider the queue $hddhh$. At the third positions, the number of $d$ is more than the number of $h$. By adding $h$ to this queue and interchanging $d$ and $h$ in the first four positions in the new queue, we get $ddhhhh$. This queue with 2 $d$'s and 4 $h$'s begins with a $d$. Conversely, consider a sequence of 2 $d$'s and 4 $h$'s beginning with a $d$. For example, let us consider $dhhhdh$. In the segment consisting of first two positions, we have an equal number of $d$'s and $h$'s. By interchanging $d$ and $h$ in the first two positions and removing $h$ in the first position, we get the sequence $dhhdh$. Clearly, this is a bad sequence.

## 5.5 Distributions and Partitions

In general, a problem related to distributions of different objects is about arrangement, and distributions of identical of objects are about selection, *with repetition* in both cases. The idea is to break such a problem into various cases that are dealt with separately by applying known permutation- and combination-based formulas.

To start with, consider the problem of distributing $n$ different objects into $k$ boxes. In what follows, we describe two methods of finding the number of ways of placing $n_1$ of these objects in the first box, $n_2$ in the second box, ..., and $n_k$ in the $k$th box. Of course, we have $n = n_1 + n_2 + \cdots + n_k$. We assume there is no regard for the order how elements are placed in $k$ boxes.

**Method – I** For the first box, $n_1$ objects can be chosen from the $n$ objects in $\binom{n}{n_1}$ ways. For the second box, $n_2$ objects can be chosen from the remaining $n - n_1$ objects in $\binom{n-n_1}{n_2}$ ways. In general, for the $r$th box, $n_r$ objects can be chosen in

$$\binom{n - n_1 - n_2 - \cdots - n_{r-1}}{n_r}$$

ways. Therefore, by the product rule, the answer to our problem is

$$\begin{aligned}
&\binom{n}{n_1}\binom{n-n_1}{n_2}\cdots\binom{n-n_1-\cdots-n_k}{n_k} \\
&= \frac{n!}{n_1!(n-n_1)!}\frac{(n-n_1)!}{n_2!(n-n_1-n_2)!}\cdots\frac{(n-n_1-\cdots-n_{k-1})!}{n_k!(n-n_1-\cdots-n_{k-1}-n_k)!} \\
&= \frac{n!}{n_1!n_2!\cdots n_k!} = P(n_1, n_2, \ldots, n_k).
\end{aligned}$$

**Method – II** For every placement of $n$ different objects into $k$ boxes, we associate a permutation of $n_1$ 1's, $n_2$ 2's, ..., $n_k$ $k$'s. Notice that, in general, the $r$th term in the permutation is $j$ if the $r$th object is put into the $j$th box. For example, if $a, b, c, d, e, f$ are 6 different objects, we may place 3 objects in Box-1, 1 object in Box-2, and 2 objects in Box-3. More specifically, if $b, d, f$ are put in Box-1, $a$ in Box-2, and $c, e$ in Box-3, then the associated permutation of 3 ones, 1 two, and 2 threes is given by 213131. It thus follows that the answer to our problem is precisely the number of permutations of $n_1$ objects of first type, ..., $n_k$ objects of $k$th type. Therefore, once again, this number is given by $P(n_1, n_2, \ldots, n_k)$.

**Example 5.17** Suppose 20 different books have to be distributed among 4 students in a way so that the first student gets 8 books, the second and the third students get 3 books each, and the fourth student gets the remaining 6 books. We may take books as objects, and students for boxes. Then, by what we said above, there are

$$\frac{20!}{8!4!4!6!} = 145495350$$

ways to distribute 20 books among 4 students.

Next, suppose we need to distribute $n_1$ objects of the first type, $n_2$ objects of the second type, ..., $n_k$ objects of the $k$th type. In this case, we consider the following two variants of distributions problem:

1. In how many ways can these objects be distributed among two persons?
2. In how many ways can these objects be distributed among two persons if each of the two persons gets at least $s_1$ object of the first type, $s_2$ objects of the second type, ..., $s_k$ objects of the $k$th type?

The number of different ways of distributing the $n_1$ objects of the first type are

$$(0, n_1),\quad (1, n_1 - 1),\quad (2, n_1 - 2),\ \ldots,\quad (n_1, 0),$$

where the first value is the number of objects given to the first person, and the second is the number of objects given to the second person. Therefore, $n_1$ objects of the first type can be distributed among two persons in $n_1 + 1$ ways. In general, $n_i$ objects of type $i$ can be distributed among two persons in $n_i + 1$ ways. It thus follows that the number of ways these $n = n_1 + \cdots + n_k$ objects are distributed among two persons is given by

$$(n_1 + 1)(n_2 + 1)\cdots(n_k + 1).$$

Now, if each of the two persons gets at least $s_1$ objects of the first type, then the number of ways to distribute $n_1$ objects of the first type is given by

$$(s_1, n_1 - s_1),\quad (s_1 + 1, n_1 - s_1 - 1),\quad (s_2 + 2, n_1 - s_2 - 2),\ \ldots,\quad (n_1 - s_1, s_1).$$

Thus the $n_1$ objects of the first type can be divided among the two persons in exactly $n_1 - 2s_1 + 1$ ways. Similarly the $n_i$ objects of type $i$ can be divided among two persons in $n_i - 2s_i + 1$ ways. Thus the number of ways these objects be divided among two person is

$$(n_1 - 2s_2 + 1)(n_2 - 2s_2 + 1)\cdots(n_k - 2s_k + 1).$$

For the next application, we need the **fundamental theorem of arithmetic**: *Every integer $n > 1$ can be expressed uniquely as*

$$n = p_1^{n_1} p_2^{n_2} \cdots p_k^{n_k}, \tag{5.5.1}$$

*where each $p_i$ is a prime divisor of $n$*. Also, recall that $D_n$ denotes the set of divisors of $n$. The next theorem gives a formula for $|D_n|$.

**Theorem 5.26** *With notations as above, we have*

$$|D_n| = (n_1 + 1)(n_2 + 1)\cdots(n_k + 1). \tag{5.5.2}$$

***Proof*** Clearly, each divisor of $n$ is of the form

$$p_1^{m_1} p_2^{m_2} \cdots p_k^{m_k}.$$

Also, we have

$$n = (p_1^{m_1} p_2^{m_2} \cdots p_k^{m_k})(p_1^{n_1-m_1} p_2^{n_2-m_2} \cdots p_k^{n_k-m_k}), \quad (0 \le m_i \le n_i).$$

Since we also have

$$p_i^{n_i} = p_i p_i p_i \cdots p_i \quad (n_i \text{ times}),$$

the problem is about dividing $n_1$ objects of the first type (namely, $p_1$), $n_2$ objects of the second type (namely, $p_2$), ..., $n_k$ objects of type $k$ (namely, $p_k$) among two persons. It thus follows that the number of divisors of $n$ is given by (5.5.2). □

About selection type of distribution problems, we first consider the question: *In how many ways can n identical objects be distributed among k people*? Let us first discuss a particular case. Consider 6 identical objects, say $a, a, a, a, a, a$. To distribute these among 3 people, a scheme may be obtained by permuting the 6 objects using thrice a *separator*, denoted by |. For example, some instances of distributing these 6 objects are as given below:

$$a|aa|aaa, \quad aaa|aa|a, \quad a|aaaaa|, \quad ||aaaaaa, \quad a||aaaaa, \quad aaaaaa||.$$

Now, let $(x, y, z)$ be a triplet, where $x, y, z$, respectively, denote the number of objects received by the first, second, and the third persons. Then the triplets corresponding to above form of distributions are given by

$$(1, 2, 3), \quad (3, 2, 1), \quad (1, 5, 0), \quad (0, 0, 6), (1, 0, 5), \quad (6, 0, 0).$$

In general, if we have $n$ identical objects to be distributed among $k$ people, such distributions can be obtained by using $k-1$ separators. We essentially need to count the number of ways of permuting $n$ objects of the first type (namely, the given objects) and $k-1$ objects of the second type (the separators). This can be done in

$$P(n, k-1) = \binom{n+k-1}{n} = \binom{n+k-1}{k-1}$$

ways. Thus $n$ identical objects can be distributed among $k$ people in $\binom{n+k-1}{n}$ or $\binom{n+k-1}{k-1}$ ways.

We next consider the question: *In how many ways can n identical objects be distributed among k people so that each gets at least one object*? In this case, a distribution can be obtained by placing $k-1$ separators among the $n$ identical objects. The separators can be placed in $n-1$ places. Out of these $n-1$ places, we need to choose only $k-1$ places and this can be done in $\binom{n-1}{k-1}$ ways. It thus follows that the $n$ identical objects can be distributed among $k$ people in $\binom{n-1}{k-1}$ ways so that every people gets at least one object.

**Example 5.18** Assume that we have $n_1$ objects of the first type, $n_2$ objects of the second type, ..., $n_k$ objects of the $k$th type. We ask: *In how many ways can these objects be divided among r persons*? The $n_1$ objects of the first type can be divided among $r$ persons in $\binom{n_1+r-1}{r-1}$ ways. Similarly the $n_i$ objects of type $i$ can be divided among $r$ persons in

$$\binom{n_i + r - 1}{r - 1}$$

ways. Thus the number of ways these objects be divided among $r$ persons is

$$\binom{n_1 + r - 1}{r - 1}\binom{n_2 + r - 1}{r - 1}\cdots\binom{n_k + r - 1}{r - 1}.$$

In particular, when $n_1 = n_2 = \cdots = n_k = 1$, the previous problem gives: The number of ways of dividing $k$ different objects among $r$ persons is $(\binom{r}{r-1})^k = r^k$.

**Example 5.19** Suppose a mother gave 10 apples, 5 mangoes, 9 oranges, and 4 bananas to her two children. If there is no restriction on the ways the fruits are divided, then the children can divide the fruits in $(10 + 1)(5 + 1)(9 + 1)(4 + 1) = 3300$ ways. Also, if each of them should get at least one of every fruit, then the children can divide the fruits in $(10 - 2 + 1)(5 - 2 + 1)(9 - 2 + 1)(4 - 2 + 1) = 864$ ways. Next, suppose the number of children is three. In this case, the children can divide the fruits in

$$\binom{10 + 3 - 1}{3 - 1}\binom{5 + 3 - 1}{3 - 1}\binom{9 + 3 - 1}{3 - 1}\binom{4 + 3 - 1}{3 - 1}$$
$$= \binom{12}{2}\binom{7}{2}\binom{11}{2}\binom{6}{2} = 1143450$$

ways. Further, assume that every one gets at least one fruit of every type? Now, the children can divide the fruits in

$$\binom{10 - 1}{3 - 1}\binom{5 - 1}{3 - 1}\binom{9 - 1}{3 - 1}\binom{4 - 1}{3 - 1} = \binom{9}{2}\binom{4}{2}\binom{8}{2}\binom{3}{2} = 18144$$

ways.

Assume that we have $n_1$ objects of the first type, $n_2$ objects of the second type, $\ldots, n_k$ objects of the $k$th type. In how many ways can these objects be divided among $r$ persons such that every person gets at least one object of every type? The $n_1$ objects of the first type can be divided among $r$ persons in $\binom{n_1-1}{r-1}$ ways. Similarly the $n_i$ objects of type $i$ can be divided among $r$ persons in

$$\binom{n_i - 1}{r - 1}$$

ways. Thus the number of ways these objects be divided among $r$ persons is

$$\binom{n_1-1}{r-1}\binom{n_2-1}{r-1}\cdots\binom{n_k-1}{r-1}.$$

**Example 5.20** We find here the number of ways of sending $n$ photographs in $k$ different envelopes. Empty envelopes are not allowed. If the envelopes can be empty, then there are $k$ ways of placing each of the photographs in the envelopes and therefore the number of ways of distributing the $n$ photographs among the $k$ envelopes is $k^n$. Similarly, if $r$ envelopes are empty and other envelopes may be empty or not, then the number of ways of placing the $n$ photographs into the envelopes is $(k-r)^n$. Note that the $r$ empty envelopes can be chosen in $\binom{k}{r}$ ways. By using inclusion-exclusion principle, it follows that the number of ways of sending $n$ photographs in $k$ different envelopes (without having any empty envelope) is

$$k^n - \binom{k}{1}(k-1)^n + \binom{k}{2}(k-2)^2 - \cdots + (-1)^k\binom{k}{k-1}1^n.$$

Assume that we have $n_1$ objects of the first type, $n_2$ objects of the second type, $\ldots, n_k$ objects of the $k$th type. In how many ways can these objects be divided among $r$ persons if each person should get at least one object? If we require that every person gets at least one object of every type, this problem was solved already. The $n_1$ object of the first type can be divided among $r$ persons in $\binom{n_1+r-1}{r-1}$ ways. Similarly the $n_i$ objects of type $i$ can be divided among $r$ persons in

$$\binom{n_i+r-1}{r-1}$$

ways. Thus the number of ways these objects be divided among $r$ persons is

$$\binom{n_1+r-1}{r-1}\binom{n_2+r-1}{r-1}\cdots\binom{n_k+r-1}{r-1}.$$

If $k$ chosen persons do not get any object (and the remaining $r-k$ persons may or may not get an object), then the number of ways of doing this is the same as dividing the objects among $r-k$ persons and therefore this number is

$$\binom{n_1+r-k-1}{r-k-1}\binom{n_2+r-k-1}{r-k-1}\cdots\binom{n_k+r-k-1}{r-k-1}.$$

Notice that the $k$ persons who do not get any object can be chosen in $\binom{r}{k}$ ways. By using the principle of inclusion and exclusion, it follows that the number of ways these objects be divided among $r$ persons with each person getting at least one object is

$$
\begin{aligned}
&\binom{n_1+r-1}{r-1}\binom{n_2+r-1}{r-1}\cdots\binom{n_k+r-1}{r-1} \\
&\quad -\binom{r}{1}\binom{n_1+r-2}{r-2}\binom{n_2+r-2}{r-2}\cdots\binom{n_k+r-2}{r-2} \\
&\quad +\binom{r}{2}\binom{n_1+r-3}{r-3}\binom{n_2+r-3}{r-3}\cdots\binom{n_k+r-3}{r-3}-\cdots \\
&\quad +(-1)^{r-1}\binom{r}{r-1}\binom{n_1}{0}\binom{n_2}{0}\cdots\binom{n_k}{0}
\end{aligned}
$$

or

$$
\begin{aligned}
&\binom{n_1+r-1}{r-1}\binom{n_2+r-1}{r-1}\cdots\binom{n_k+r-1}{r-1} \\
&\quad -\binom{r}{1}\binom{n_1+r-2}{r-2}\binom{n_2+r-2}{r-2}\cdots\binom{n_k+r-2}{r-2} \\
&\quad +\binom{r}{2}\binom{n_1+r-3}{r-3}\binom{n_2+r-3}{r-3}\cdots\binom{n_k+r-3}{r-3}-\cdots \\
&\quad +(-1)^{r-1}\binom{r}{r-1}.
\end{aligned}
$$

Let us now find the number of ways of placing $n$ different objects in $k$ different boxes, where objects in each box are arranged in a definite order. Notice that a box may be empty. In this case, we permute the $n$ objects, using $k-1$ separators (denoted by $|$). For example, if the alphabets $a, b, c, d, e, f, g, h$ are to be placed in 3 different boxes, we permute the alphabets with 2 separators such as the permutations given by

$$gf\,|h|\,debca \qquad \text{and} \qquad |adefhg\,|\,bc.$$

Each one of such permutations gives a way of placing the alphabets into the 3 different boxes. Therefore, the number of ways placing the $n$ different objects in $k$ different boxes in the required manner is given by

$$P(\underbrace{1,1,1,\ldots,1}_{n\text{times}},k-1)=\frac{(n+k-1)!}{1!1!\cdots 1!(k-1)!}=\frac{(n+k-1)!}{(k-1)!}=A^n_{n+k-1}.$$

Finally, the case of placing $n$ identical objects in $k$ identical boxes is equivalent to expressing $n$ as the sum of at most $k$ positive integers in non-descending (or non-ascending) order. In general, positive integers $a_1, \ldots, a_i$ are said to form a *partition* of a positive integer $n$ if we have

$$n = a_1 + \cdots + a_i, \qquad \text{with } a_1 \le a_2 \le \cdots \le a_i.$$

For example, the following are some partitions of 5:

$$1+1+1+1+1, \quad 1+1+1+2, \quad 1+1+3, \quad 3+1+1, \quad 1+3+1, \quad 2+3,$$

Two such partitions are treated different when they differ in their order of their arrangements. For example, the two partitions $4+4+8+4$ and $8+4+4+4$ of 20 are different. Some other partitions of 20, with $2, 4, 12$ as the only summands, are given by

$$\begin{array}{lll} 2+2+4+12, & 2+2+12+4, & \\ 2+12+2+4, & 4+4+12, & \\ 4+2+2+12, & 12+2+2+4, & 12+4+4. \end{array}$$

In general, we may count the number of such partitions with 2, 4, 12 as the only summands. For instance, let $f(N)$ denote the number of partitions of $N$ with 2, 4, 12 as the only summands. A partition of $N$ with 2, 4, 12 as the only summands can begin in any of the summands $2, 4, 12$. The number of such partitions of $N$ that begin in the summand 2 is the same as the number of partitions of $N-2$ with $2, 4, 12$ as the only summands. This number is clearly $f(N-2)$. Similarly, the number of partitions of $N$ with $2, 4, 12$ as the only summands that begin in 4 (or 12) is $f(N-4)$ (or $f(N-12)$). It thus follows that $f(N)$ the recurrence relation

$$f(N) = f(N-2) + f(N-4) + f(N-12).$$

As empty sum is zero, we have $f(0) = 1$. Also, it can be seen directly that $f(1) = 0$, $f(2) = 1$, $f(3) = 0$, $f(4) = 2$, $f(5) = 0$, $f(6) = 3$, $f(7) = 0$, $f(8) = 5$, $f(9) = 0$, $f(10) = 8$, and $f(11) = 0$. One may use the fact $f(-k) = 0$ for any positive integer $k$ to obtain these equalities. In particular, for example, we have

$$\begin{aligned} f(6) &= f(4) + f(0) + f(-6) = 2+1+0 = 3, \\ f(8) &= f(6) + f(4) + f(-4) = 3+2 = 5 \end{aligned}$$

and

$$f(10) = f(8) + f(6) + f(-2) = 5+3+0 = 8.$$

It thus follows that

$$\begin{aligned} f(12) &= f(10) + f(8) + f(0) = 8+5+1 = 14 \\ f(13) &= f(11) + f(9) + f(1) = 0 \\ f(14) &= f(12) + f(10) + f(2) = 14+8+1 = 23 \\ f(15) &= f(13) + f(11) + f(3) = 0 \\ f(16) &= f(14) + f(12) + f(4) = 23+14+2 = 93 \end{aligned}$$

$$f(17) = f(15) + f(13) + f(5) = 0$$
$$f(18) = f(16) + f(14) + f(6) = 93 + 23 + 3 = 119$$
$$f(19) = f(17) + f(15) + f(7) = 0$$
$$f(20) = f(18) + f(16) + f(8) = 119 + 93 + 5 = 217.$$

More generally, if the summands are $n_1, n_2, \ldots, n_k$, then the formula to compute $f(N)$ is given by

$$f(N) = f(N - n_1) + f(N - n_2) + \cdots + f(N - n_k).$$

Once again, we have $f(-k) = 0$, for any $k > 0$, and $f(0) = 1$.

We next find the number of partitions of $N$, subject to the condition that two partitions that differ in the order of arrangement are treated different. In such partitions of $N$, the summands can be $1, 2, \ldots, N$ so that the number of summands varies from 1 to $N$. Clearly, the number of partitions with $k$ summands is the number of ways of placing $k - 1$ |'s in $1 + 1 + 1 + \cdots + 1$ ($N$ times), which is given by $C_{N-1}^{k-1}$. It thus follows that the number of partition of $N$ with said restriction is given by

$$C_{N-1}^0 + C_{N-1}^1 + C_{N-1}^2 + \cdots + C_{N-1}^{N-1} = 2^{N-1}.$$

**Example 5.21** A university applicant is required to pass 4 examinations. The numerical passing grades are 3, 4, and 5. The minimal cumulative score for admission to the university is 17. We find how many ways can the applicant accumulate the necessary 17 or more points. For instance, let $F(k, N)$ denote the number of ways in which it is possible to accumulate $N$ points after passing $k$ examinations. Then

$$f(k, N) = f(k - 1, N - 3) + f(k - 1, N - 4) + f(k - 1, N - 5).$$

Notice that $f(4, 20) = 1$. Since $f(3, 16) = 0$ (maximum point one can get in 3 examinations is 15), $f(3, 15) = 1$, $f(2, 11) = 0$, $f(2, 10) = 1$, we have

$$\begin{aligned} f(4, 19) &= f(3, 16) + f(3, 15) + f(3, 14) \\ &= 1 + f(3, 14) \\ &= 1 + f(2, 11) + f(2, 10) + f(2, 9) \\ &= 1 + 1 + 2 = 4. \end{aligned}$$

Similarly, we have

$$\begin{aligned} f(4, 18) &= f(3, 15) + f(3, 14) + f(3, 13) \\ &= 1 + [f(2, 11) + f(2, 10) + f(2, 9)] \\ &\quad + [f(2, 10) + f(2, 9) + f(2, 8)] \\ &= 1 + 2f(2, 10) + 2f(2, 9) + f(2, 8). \end{aligned}$$

Since

$$\begin{aligned} f(2,10) &= 1, \\ f(2,9) &= f(1,6) + f(1,5) + f(1,4) = 0 + 1 + 1 = 2 \end{aligned}$$

and

$$f(2,8) = f(1,5) + f(1,4) + f(1,3) = 1 + 1 + 1 = 3$$

we have

$$f(4,18) = 1 + 2 + 6 + 1 = 10.$$

Similarly

$$\begin{aligned} f(4,17) &= f(3,14) + f(3,13) + f(3,12) \\ &= [f(2,11) + f(2,10) + f(2,9)] \\ &\quad + f(2,10) + f(2,9) + f(2,8)] \\ &\quad + [f(2,9) + f(2,8) + f(2,7)] \\ &= 2f(2,10) + 3f(2,9) + 2f(2,8) + f(2,7) \\ &= 2 + 3[f(1,6) + f(1,5) + f(1,4)] + 2[f(1,5) \\ &\quad + f(1,4) + f(1,3)] + [f(1,4) + f(1,3) + f(1,2)] \\ &= 2 + 6 + 6 + 2 = 16. \end{aligned}$$

Thus, the number of ways the applicant can accumulate the necessary 17 or more points is $f(4,20) + f(4,19) + f(4,18) + f(4,17) = 1 + 4 + 10 + 16 = 31$.

**Example 5.22** A purse contains single coins of the following denominations: 1, 2, 3, 5, 10, 15, 20, and 50 cents. We find number of ways to pay for a 73 cent purchase with these coins. For instance, let $F(n_1, n_2, \ldots, n_m; N)$ denote the number of ways of paying $N$ cents using the coins of different denominations $n_1, n_2, \ldots, n_m$. The various ways of paying can be divided into two groups: one in which $n_m$ cent coin is used and the second in which $n_m$ cent coin is not used. In the former case, the payment of $N - n_m$ cents is to be made by the coins of denominations $n_1, \ldots, n_{m-1}$ while in the latter case the payment of $N$ cents should be made by using the coins of denominations $n_1, n_2, \ldots, n_{m-1}$. Thus we have

$$\begin{aligned} F(n_1, n_2, \ldots, n_m; N) = &\; F(n_1, n_2, \ldots, n_{m-1}; N - n_m) \\ &+ F(n_1, n_2, \ldots, n_{m-1}; N). \end{aligned}$$

Notice that

$$n_1 + n_2 + \cdots + n_m < N \Rightarrow F(n_1, n_2, \ldots, n_m; N) = 0$$

and

$$n_m > N \Rightarrow F(n_1, n_2, \ldots, n_m; N) = F(n_1, n_2, \ldots, n_{m-1}; N).$$

Now

$$\begin{aligned}
&F(1, 2, 3, 5, 10, 15, 20, 50; 73)\\
&= F(1, 2, 3, 5, 10, 15, 20; 23) + F(1, 2, 3, 5, 10, 15, 20; 73)\\
&= F(1, 2, 3, 5, 10, 15, 20; 23)\\
&= F(1, 2, 3, 5, 10, 15; 3) + F(1, 2, 3, 5, 10, 15; 23)\\
&= F(1, 2, 3; 3) + [F(1, 2, 3, 5, 10; 8) + F(1, 2, 3, 5, 10; 23)]\\
&= [F(1, 2; 0) + F(1, 2; 3)] + F(1, 2, 3, 5; 8)\\
&= 1 + 1 + [F(1, 2, 3; 3) + F(1, 2, 3; 8)]\\
&= 2 + 2\\
&= 4.
\end{aligned}$$

Note that, since

$$\begin{gathered}
1 + 2 + 3 + 5 + 10 + 15 + 20 = 56 < 73,\\
1 + 2 + 3 + 5 + 10 = 21 < 23,\\
1 + 2 + 3 = 6 < 8,
\end{gathered}$$

we have

$$\begin{gathered}
F(1, 2, 3, 5, 10, 15, 20; 73) = 0,\\
F(1, 2, 3, 5, 10; 23) = 0,\\
F(1, 2, 3; 8) = 0.
\end{gathered}$$

**Example 5.23** Assuming that the order of summands is immaterial, we find the number of partitions of a natural number $N$ into summands selected from among the numbers 1, 2, 3 ..., $n$. First notice that some of the summands can be equal. Let $\Pi_N^n$ denote the number of partitions of the number $N$ into summands selected from among the numbers 1, 2, 3 ..., $n$ and let $\Pi_0^n = 1$. Since the summands in a partition of $N$ cannot be more than $N$, it is clear that the number of partitions of $N$ is clearly $\Pi_N^N$. Therefore, if $n > N$, then $\Pi_N^n = \Pi_N^N$. Also, $\Pi_N^1 = 1$. Hence assume that $1 < n \leq N$. In this case, the class of partitions of $N$ into summands selected from among the numbers 1, 2, 3 ..., $n$ can be grouped into two classes: The class of partitions of $N$ in which the summand $n$ is used, and the other class consisting of partitions of $N$ in which the summand $n$ is not used. In the first case, the number of partitions is the same as the number of partitions of $N - n$ into summands selected from among the numbers 1, 2, 3 ..., $n$. Clearly, this number is $\Pi^n_{N-n}$. In the latter case, the number of partitions is the same as the number of partitions of $N$ into summands selected from among the numbers 1, 2, 3 ..., $n - 1$ and there are $\Pi^{n-1}_N$ such partitions. It thus follows that we have

$$\Pi^n_N = \Pi^n_{N-n} + \Pi^{n-1}_N.$$

We can also represent a partition of an integer by a graph consisting of a set of rows of points listed in order of increasing size, which are known as a *Ferrers diagram*. For example, consider the following partition of 19:

$$19 = 1 + 1 + 2 + 3 + 3 + 4 + 5.$$

This partition can be represented by the graph shown in Fig. 5.1.

Notice that the $i$th row from the top corresponds to the $i$th summand and the number of dots in a row corresponding to summand $k$ is exactly $k$. All the dots are aligned at the first column. We call such a graph *normal*. By changing $i$th row from the bottom of the original graph as the $i$th column of a new graph, we obtain what is called the *dual* (or the *conjugate*) of the original graph, which gives another partition of the integer. For example, see Fig. 5.2. Notice that two Ferrers diagrams have the same dual if and only if they are *equal*.

**Example 5.24** We use graphical method to show that the number of partitions of an integer $N$ into at most $m$ summands equals the number of partitions of $N + m$ into $m$ summands. For instance, notice that the graph of a partition of the integer $N$ into at most $m$ summands has at most $m$ rows. If we add a column of $m$ dots to the graph (as in Fig. 5.3), then we get a graph of a partition of $N + m$ with exactly $m$ rows.

Similarly, by deleting the first column (with $m$ dots) in the graph of a partition of $N + m$ with exactly $m$ rows, we get the graph of a partition of the integer $N$ into at most $m$ summands. Thus, the two classes of graphs have the same number of

**Fig. 5.1** Graphical representation of $19 = 1 + 1 + 2 + 3 + 3 + 4 + 5$

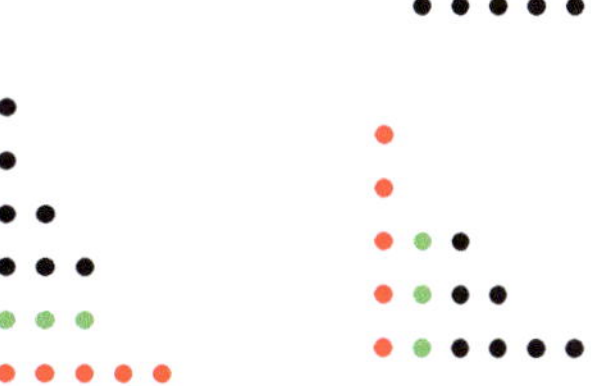

**Fig. 5.2** Graph of $15 = 1 + 1 + 2 + 3 + 3 + 5$ and its dual

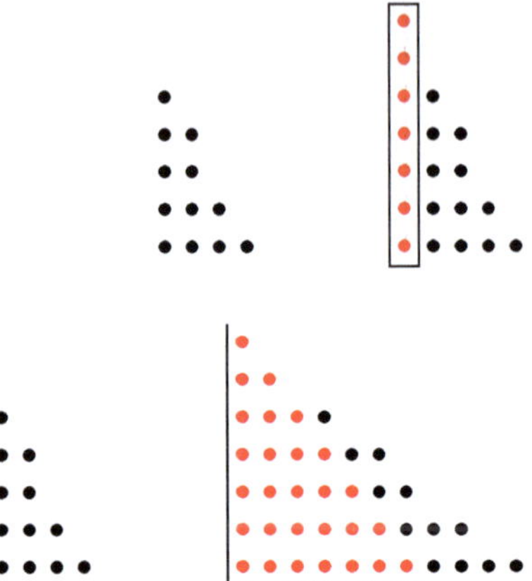

**Fig. 5.3** A partition of 12 with at most 7 summands and the corresponding partition of $19 = 12 + 7$ with exactly 7 summands

**Fig. 5.4** A partition of 12 with at most 7 summands and the corresponding partition of $19 = 12 + 7$ with exactly 7 summands

elements. This proves that the number of partitions of an integer $N$ into at most $m$ summands equals the number of partitions of $N + m$ into $m$ summands.

**Example 5.25** We use graphical method to show that the number of partitions of an integer $N$ into at most $m$ summands equals the number of partitions of $N + m(m + 1)/2$ into $m$ unequal summands. For instance, notice that the graph of a partition of the integer $N$ into at most $m$ summands has at most $m$ rows. If we add an isosceles right-angled triangle of dots consisting of $m$ rows and $m$ columns to the graph (as in Fig. 5.4), then we get a graph of a partition of $N + m(m + 1)/2$ with $m$ unequal summands (upon normalizing the new graph.)
Similarly, by deleting an isosceles right-angled triangle of dots consisting of $m$ rows and $m$ columns in the graph of a partition of $N + m(m + 1)/2$ with $m$ unequal summands, we get the graph of a partition of the integer $N$ into at most $m$ summands. Thus the two classes of graphs have the same number of elements. This proves that the number of partitions of an integer $N$ into at most $m$ summands equals the number of partitions of $N + m(m + 1)/2$ into $m$ unequal summands.

## 5.6 Methods of Solving Recurrence Relations

We may recall that a *real sequence* is a function $a : \mathbb{N} \to \mathbb{R}$. We write the image $a(n)$ of an element $n \in \mathbb{N}$ as $a_n$, i.e., $a_n = a(n)$, for $n \in \mathbb{N}$. For example, $a_n = n^2$ is the real sequence given by

$$(1, 4, 9, 16, 25, \dots).$$

Similarly, we can define a sequence of complex numbers. In some cases, we also consider *causal sequences*, which have the set $\mathbb{N}_0 = \{0, 1, 2, \dots\}$ as the domain.

Notice that the *range* of a sequence $a$ is the set given by

$$a(\mathbb{N}) := \{a(n) \mid n \in \mathbb{N}\}.$$

For example, the range of the sequence $a_n = (-1)^n$ is the set $\{-1, 1\}$, while the range of the sequence $(n^2)$ is the set $\{1, 4, 9, \dots\}$. In this section, we solve some interesting *recurrence relation* of an unknown sequence $(a_n)$. In what follows, by a *solution* of a recurrence relation we mean an explicit formula for the $n$th term $a_n$. For example, $a_n$ may represent

(a) Time required to solve a problem of size $n$ such as related to the time complexity of an algorithm.
(b) Ways to select $n$ objects in a certain situation, say ways to distribute $n$ identical objects into seven different boxes, subject to that an even number of objects don't exceed five in the first three boxes and eight in the remaining four boxes.
(c) Integers between 0 and 9999 with digit sum $n$.
(d) $n$-bit strings that don't contain the substring 111.
(e) *Good routes* to move from the lower left corner of an $n \times n$ square grid to the upper right corner subject to the restriction that movement is allowed only to the right or upward, without moving above a diagonal line from the lower left corner to the upper right corner. In this case, we call $a_n$ the $n$th *Catalan number*.
(f) Moves needed to solve the $n$-disk problem related to *Tower of Hanoi* with three pegs.
(g) Regions into which the plane is divided by $n$ lines, subject to the restriction that a pair of lines meet in a point, but no three meet in a point.
(h) The minimum number of links required by $n$ nodes to make known all available files.

In general, counting problems such as involving special types of constraints in selection and arrangement belong to this class of problems. We cannot solve them using the techniques introduced in the previous sections.

In this section, we describe two important methods of solving a linear recurrence relation with *constant coefficients*: The *method of characteristic equation* and the *generating function method*. In general, a *generating function* (ordinary or exponential) for a real or complex sequence $(a_n)$ is a formal power series with $a_n$ being the $n$th coefficient, which is a powerful combinatorial tool to solve broad spectrum of enumerative problems. We will also discuss certain *divide-and-conquer algorithms* modeled by recurrence relations. The motivated reader may read the excellent texts such as [16–19], for a much deeper understanding about the topics discussed in this chapter.

## *Recurrence Relations*

Informally, a recurrence relation is a recursive formula that helps count the number of ways to do a task of size $n$, given how the same can be done for tasks of smaller sizes. Of course, it is not possible to recursively compute the next number in a sequence unless we have some *initial values* to start with the process. Therefore, to solve a counting problem by recursive computing, the solution of the involved recurrence relation provides a way to find the desired number for successive larger values of $n$. As in the case of mathematical induction, it involves base case, which are usually given as *initial conditions* of the problem. We may recall that a recursive function is a function that invokes itself.

**Definition 5.7** A *recurrence relation* for $(a_n)$ is an equation that expresses $a_n$ in terms of some of its previous terms, and also a known function of $n$, say $g(n)$. That is, we have

$$a_n = f\big(a_{n-1}, a_{n-2}, \ldots, a_{n-k}; g\big), \qquad \text{with} \;\; k \geq 1, \tag{5.6.1}$$

where $f$ is a linear or non-linear polynomial. The positive integer $k$ is called the *order* of the recurrence relation (5.6.1).

We give below some classical problems that involve a recurrence relation.

**Example 5.26** (*Geometric Progression*) We consider the problem related to the computation of return on an investment with interest compounding annually. We may recall that a sequence $(a_n)$ is said to be in *geometric progression* if $a_n/a_{n-1}$ is a constant, say $r$, so that we can write $a_n = r\,a_{n-1}$. This equation is the recurrence relation for geometric sequence. Taking the first term of the sequence as $a_0 = a$, we have

$$\begin{aligned} a_1 &= ra_0 = ar, \\ a_2 &= ra_1 = ar^2, \\ a_3 &= ra_2 = ar^3, \\ a_4 &= ra_3 = ar^4. \end{aligned}$$

Proceeding this way, it follows that $a_n = ar^n$. The sequence $a_n = ar^n$ is the solution of the recurrence relation $a_n = r\,a_{n-1}$ satisfying the condition $a_0 = a$.

**Example 5.27** (*Towers of Hanoi*) Suppose $n$ disks are placed on one of the three pegs (or rods), say $A$, $B$, and $C$, which are arranged in ascending order of their sizes. That is, the smallest one is at the top of the Peg-A. The *Towers of Hanoi* problem is about finding the minimum number of moves required to stack these $n$ disks on to another peg subject to the conditions that only one disk can be moved at a time and, at no stage of the process, a larger size disk can be placed over a disk of smaller size. For instance, let $H_n$ denote the minimum number of moves required. Obviously, we have $H_1 = 1$. Next, suppose there are two disks, say Disk-1 and Disk-2, with Disk-2 bigger than the Disk-1. In this case, we first move Disk-1 to Peg-B, and then Disk-2 to Peg-C. Finally, move Disk-1 to Peg-C. It thus follows that we need 3 moves, i.e., $H_2 = 3$. The seven moves are required for the problem with three disks. In general, starting with moving the top $n-1$ disks from Peg-A to Peg-B in $H_{n-1}$ moves, move the last disk to Peg-C. Subsequently, move all the $n-1$ disks on Peg-B to Peg-C, which requires $H_{n-1}$ moves. Therefore, the total number of moves required is given by the recurrence relation

$$H_n = H_{n-1} + 1 + H_{n-1} = 2H_{n-1} + 1.$$

**Example 5.28** (*Fibonacci Sequence*) The sequence introduced here is named after an Italian mathematician *Leonardo Fibonacci* (1170–1250), which first arose in a puzzle about the ideal growth model of a pair of rabbits over the duration of 1 year. It is assumed that $(a)$ every month each mature pair of rabbits gives birth to a new pair of rabbits of opposite sex; $(b)$ each new pair of rabbits become mature in a month; and $(c)$ taking gestation period of a rabbit to be 1 month, they start giving birth to additional pairs of rabbits from third month of their birth. We may also assume that no rabbit dies. Then, beginning with a single pair of new born rabbits of opposite sex in the first month, we need to find how many pairs of rabbits will be born in $n$ months? Now, considering our assumption that a new born and 1-month old rabbits do not give birth to a new pair of rabbits, while they give birth to a new pair of rabbits from the third month, the number of new born, 1-month old, and mature rabbits, for a period of 10 months, are as listed in the table below:

| Month | New born | 1-month old | Mature | Total |
|---|---|---|---|---|
| 0 | 0 | 0 | 0 | 0 |
| 1 | 1 | 0 | 0 | 1 |
| 2 | 0 | 1 | 0 | 1 |
| 3 | 1 | 0 | 1 | 2 |
| 4 | 1 | 1 | 1 | 3 |
| 5 | 2 | 1 | 2 | 5 |
| 6 | 3 | 2 | 3 | 8 |
| 7 | 5 | 3 | 5 | 13 |
| 8 | 8 | 5 | 8 | 21 |
| 9 | 13 | 8 | 13 | 34 |
| 10 | 21 | 13 | 21 | 55 |

Taking $F_n$ to denote the number of pairs in the $n$th month, it follows from the above table that we have

$$F_0 = 0, \quad F_1 = 1, \quad F_2 = 1, \quad \text{and} \quad F_3 = 2.$$

In general, we have that the number of pairs in the $n$th month is the sum of the number of rabbits in the previous month, i.e., $F_{n-1}$, and the number of *newborn pairs* in the

$n$th month. However, since the number of newborn rabbits in the $n$th month is the same as the number of pairs in the $n-2$nd month, we obtain the recurrence relation:

$$F_n = F_{n-1} + F_{n-2}. \tag{5.6.2}$$

For centuries, the Fibonacci sequence has captivated mathematicians, artists, and scientists, because of its appearance in many unexpected situations. For example, the number of petals in a flower such as the *lily* consistently follows the Fibonacci sequence. Also, a sunflower provides a good example of related *spiraling patterns* observed in the head of a flower. Similarly, the seed pods on a *pine cone* are arranged in a spiral pattern. The way tree branches form or split such as in the *sneeze wort* follows Fibonacci sequence. This similar pattern is exhibited in the case of root systems and even algae.

**Definition 5.8** Let $a = \big(a_n\big)$ be a real or complex sequence. A general *linear recurrence relation* for the sequence $a$ of *order* $k$ is an equation of the form:

$$c_0 a_n + c_1 a_{n-1} + \cdots + c_k a_{n-k} = g(n), \tag{5.6.3}$$

where $c_0, c_1, \ldots, c_k$ are independent of $a_k$'s, with both $c_0$ and $c_k$ being nonzero.

An equivalent form of (5.6.3) is the recurrence relation given by

$$c_0 a_{n+k} + c_1 a_{n+k-1} + \cdots + c_k a_n = g(n). \tag{5.6.4}$$

In the case when the coefficients $c_i$'s are all constants, the recurrence relation (5.6.3) is called a linear recurrence relation of order $k$ with *constant coefficients*; otherwise, we say the recurrence relation is with variable coefficients. Further, when $g \equiv 0$, we say (5.6.3) is *homogeneous recurrence relation*. In the general case, we also say the recurrence relation is *non-homogeneous*.

**Example 5.29** The recurrence relation $a_n = a_{n-1} + a_{n-2}$ is of order 2, while $a_n = na_{n-1}$ is of order 1. However, the recurrence relation given by

$$a_n = a_{n-1} + a_{n-2} + \cdots + a_2 + a_1$$

has no order. The recurrence relation given by

$$a_n - 2a_{n-1} - 3a_{n-2} - 4a_{n-3} = 0$$

is a homogeneous linear recurrence relation of order 3 with constant coefficients, while the recurrence relation

$$a_n - na_{n-2} = n^2$$

is a non-homogeneous linear recurrence relation of order 2 with variable coefficients. On the other hand, each of the following recurrence relations is non-linear:

$$a_{2^k} = a_{2^{k-1}} + 1, \quad \text{with} \quad a_1 = 0;$$
$$a_n = a_0 a_{n-1} + a_1 a_{n-2} + \cdots + a_{n-1} a_0, \quad \text{with} \quad a_0 = 0 \text{ and } a_1 = 1.$$

When $n$ is a power of 2, a practical aspect of the first of above two recurrence relations is that tiling an $n \times n$ deficient board with trominoes can be done recursively because we can find four $(n/2) \times (n/2)$ sub-boards within the original $n \times n$ board. Notice the similarity of the tiling algorithm to the inductive step of the proof that every $n \times n$ deficient board can be tiled with trominoes when $n$ is a power of 2.

**Example 5.30** We derive here a second-order linear recurrence relation for the sequence given by $a_n = A2^n + B3^n$. We need to eliminate the constants $A$ and $B$. For instance, as we have

$$a_{n+1} = A2^{n+1} + B3^{n+1} = 2A2^n + 3B3^n$$
$$a_{n+2} = A2^{n+2} + B3^{n+2} = 4A2^n + 9B3^n,$$

it follows that

$$a_{n+2} - 5a_{n+1} + 6a_n$$
$$= 4A2^n + 9B3^n - 5(2A2^n + 3B3^n) + 6(A2^n + B3^n) = 0.$$

Thus, the recurrence relation satisfied by $a_n = A2^n + B3^n$ is

$$a_{n+2} - 5a_{n+1} + 6a_n = 0.$$

**Example 5.31** Consider the sequence $b_n = n!$. Then

$$b_n = n! = n(n-1)! = nb_{n-1}.$$

Also $b_0 = 0! = 1$. The recurrence relation satisfied by $b_n = n!$ is

$$b_n = nb_{n-1};$$

it is a homogeneous first-order linear recurrence with variable coefficient.

To solve a counting problem by recursive computing, the solution of the involved recurrence relation provides a way to find the desired number for successive larger values of $n$. As in the case of mathematical induction, it involves base case, which is usually given in terms of some *initial conditions*. Of course, it is not possible to recursively compute the next number unless we have some *initial values* to start with the process.

## *Method of Characteristic Equation*

In what follows, we discuss methods of solving linear recurrence relations of small orders. In this section, we start with the method of *characteristic equation*. A more powerful method uses the fundamental concept of *generating function* of a sequence, which we discuss in the next section. Here, we first consider the second-order homogeneous linear recurrence relation with constant coefficients of the form:

$$a_n = c_1 a_{n-1} + c_2 a_{n-2}. \tag{5.6.5}$$

A sequence $a_n$ satisfying (5.6.5) is called a **solution** of the relation (5.6.5). For example, the sequence

$$a_n = A(-1)^n + B(-3)^n$$

satisfies the recurrence relation

$$a_n + 4a_{n-1} + 3a_{n-2} = 0$$

and therefore

$$a_n = A(-1)^n + B(-3)^n$$

is a solution of the recurrence relation:

$$a_n + 4a_{n-1} + 3a_{n-2} = 0.$$

**Theorem 5.27** *If $a_n^{(1)}$, $a_n^{(2)}$ are two solutions of the second-order homogeneous linear recurrence relation with constant coefficients given by (5.6.5), then, for every pair of constants A and B,*

$$a_n = Aa_n^{(1)} + Ba_n^{(2)}$$

*is also a solution of (5.6.5). In particular, if $a_n^{(1)}$ and $a_n^{(2)}$ are solutions of (5.6.5), and $\alpha$ is a constant, then $a_n^{(1)} + a_n^{(2)}$ and $\alpha a_n^{(1)}$ are also solutions of (5.6.5).*

***Proof*** Since $a_n^{(1)}$, $a_n^{(2)}$ are solutions of (5.6.5), we have

$$a_n^{(1)} = c_1 a_{n-1}^{(1)} + c_2 a_{n-2}^{(1)}$$

$$a_n^{(2)} = c_1 a_{n-1}^{(2)} + c_2 a_{n-2}^{(2)}$$

and hence using

$$a_n = Aa_n^{(1)} + Ba_n^{(2)}$$

and the above two equations, we have

$$\begin{aligned} a_n &= Aa_n^{(1)} + Ba_n^{(2)} \\ &= A(c_1 a_{n-1}^{(1)} + c_2 a_{n-2}^{(1)}) + B(c_1 a_{n-1}^{(2)} + c_2 a_{n-2}^{(2)}) \\ &= c_1(Aa_{n-1}^{(1)} + Ba_{n-1}^{(2)}) + c_2(Aa_{n-2}^{(2)} + Ba_{n-2}^{(2)}) \\ &= c_1 a_{n-1} + c_2 a_{n-2}. \end{aligned}$$

This proves that

$$a_n = Aa_n^{(1)} + Ba_n^{(2)}$$

is a solution of (5.6.5). □

**Definition 5.9** A solution that involves two arbitrary constants is called the **general solution** of the recurrence relation (5.6.5).

Each *particular solution* of (5.6.5) comes from the general solution for some special values of the arbitrary constants.

**Theorem 5.28** *The sequence $a_n = r_1^n$ is a solution of the second-order homogeneous linear recurrence relation with constant coefficients given by (5.6.5) if the number $r = r_1$ satisfies the equation $r^2 - c_1 r - c_2 = 0$. In addition, if $r = r_1$ is a repeated root of the equation $r^2 - c_1 r - c_2 = 0$, then $a_n = nr_1^n$ is also a solution of (5.6.5).*

***Proof*** Let $a_n = r_1^n$. It then follows that

$$\begin{aligned} a_n - c_1 a_{n-1} - c_2 a_{n-2} &= r_1^n - c_1 r_1^{n-1} - c_2 r_1^{n-2} \\ &= r_1^{n-2}(r_1^2 - c_1 r_1 - c_2) \\ &= 0. \end{aligned}$$

Let $r = r_1$ be a repeated root of $r^2 - c_1 r - c_2 = 0$. Then, we have $r_1^2 - c_1 r_1 - c_2 = 0$ and $2r_1 - c_1 = 0$. The last equation follows because the repeated roots of $r^2 - c_1 r - c_2 = 0$ are given by $r_1 = c_1/2$. Next, let $a_n = nr_1^n$. Then we have

$$\begin{aligned} a_n - c_1 a_{n-1} - c_2 a_{n-2} &= nr_1^n - c_1(n-1)r_1^{n-1} - c_2(n-2)r_1^{n-2} \\ &= r_1^{n-2}\left[nr_1^2 - (n-1)r_1 c_1 - (n-2)c_2\right] \\ &= r_1^{n-2}\left[(n-2)(r_1^2 - c_1 r_1 - c_2) + (2r_1 - c_1)r_1\right] = 0. \end{aligned}$$

This completes the proof. □

In view of the above theorem, to solve the recurrence relation

$$a_n = c_1 a_{n-1} + c_2 a_{n-2}, \tag{5.6.6}$$

we first write the following equation, called the **characteristic equation**,

$$r^2 - c_1 r - c_2 = 0.$$

As this is a quadratic equation, let the two roots be given by $r_1$ and $r_2$. Now, if $r_1 \neq r_2$, then we have

$$r_1^n \quad \text{and} \quad r_2^n$$

are two solutions of (5.6.6). Therefore, the general solution of (5.6.6) is given by

$$a_n = Ar_1^n + Br_2^n,$$

where $A$ and $B$ are constants. Next, if $r_1 = r_2$, then $r_1^n$ and $nr_1^n$ are two solutions. Therefore, the general solution of (5.6.6) is given by

$$a_n = Ar_1^n + Bnr_1^n = (A + Bn)r_1^n,$$

where $A$ and $B$ are constants.

**Example 5.32** (*Case of Unequal Roots*) We find here the general solution of the recurrence relation

$$a_n = 3a_{n-1} - 2a_{n-2}.$$

For, in this case, the characteristic equation is

$$r^2 = 3r - 2 \quad \text{or} \quad r^2 - 3r + 2 = 0.$$

It has two roots given by $r = 1, 2$. Therefore, the general solution is given by

$$a_n = A1^n + B2^n = A + B2^n, \quad \text{for} \quad n \geq 0,$$

where $A$ and $B$ are constants.

**Example 5.33** (*Case of Equal Roots*) We solve here the recurrence relation

$$a_n = 6a_{n-1} - 9a_{n-2}, \quad \text{with} \quad a_0 = 0 \text{ and } a_1 = 1.$$

In this case, the characteristic equation is given by

$$r^2 - 6r + 9 = 0 \quad \text{or} \quad (r - 3)^2 = 0.$$

It has two equal root given by $r = 3, 3$. It thus follows that the general solution is given by

$$a_n = (A + Bn)3^n.$$

Now, since $a_0 = 0$ and $a_1 = 1$, we have

$$A = 0 \text{ and } (A + B)3 = 1 \quad \Rightarrow \quad B = \frac{1}{3}.$$

Therefore, a particular solution is given by

$$a_n = \frac{1}{3}n3^n = n3^{n-1}, \quad \text{for} \quad n \geq 0.$$

Notice that, since $a_n = n3^{n-1}$ implies that $a_0 = 0,\ a_1 = 1$, and

$$\begin{aligned} a_n - 6a_{n-1} + 9a_{n-2} &= n3^{n-1} - 6(n-1)3^{n-2} + 9(n-2)3^{n-3} \\ &= 3^{n-1}[n - 2n + 2 + n - 2] = 0, \end{aligned}$$

we have verification of our solution.

Before proceeding further, we recall some terminologies about complex numbers that are needed in our subsequent discussion. A complex number is of the form $z = a + ib$, where $a, b$ are real numbers, and $i = \sqrt{-1}$. The *modulus* of a complex number $z$, denoted by $|z|$, is the positive square root of the real number $a^2 + b^2$. That is, $|z| = +\sqrt{a^2 + b^2}$. The *argument* of a nonzero complex number $z$ is the value of $\theta$ satisfying the two equations:

$$a = |z| \cos \theta \quad \text{and} \quad b = |z| \sin \theta.$$

In particular, the arguments of the complex numbers $1$, $1 + i$, and $i$ are, respectively, given by $0$, $\pi/4$, and $\pi/2$. Notice that a nonzero complex number $z$ with argument $\theta$ can be written as

$$z = |z|(\cos \theta + i \sin \theta) = |z|e^{i\theta}.$$

Therefore, by de Moivre's theorem, we have

$$(e^{i\theta})^n = (\cos \theta + \sin \theta)^n = \cos(n\theta) + i \sin(n\theta) = e^{in\theta}.$$

**Example 5.34** (*Case of Complex Roots*) We solve here the recurrence relation

$$a_n + 4a_{n-1} + 8a_{n-2} = 0, \quad \text{with} \quad a_0 = 1 \text{ and } a_1 = 0.$$

In this case, the characteristic equation is given by

$$r^2 + 4r + 8 = 0,$$

which has the two complex numbers as roots given by

$$r = 2 \pm 2i = 2\sqrt{2}e^{\pm\frac{\pi i}{4}}.$$

It thus follows that the general solution is given by

$$\begin{aligned}
a_n &= A\left(2\sqrt{2}e^{\frac{\pi i}{4}}\right)^n + B\left(2\sqrt{2}e^{-\frac{\pi i}{4}}\right)^n \\
&= (2\sqrt{2})^n\left(Ae^{\frac{n\pi i}{4}} + Be^{-\frac{n\pi i}{4}}\right) \\
&= (2\sqrt{2})^n\left(A\left(\cos\left(\frac{n\pi}{4}\right) + i\sin\left(\frac{n\pi}{4}\right)\right) + B\left(\cos\left(\frac{n\pi}{4}\right) - i\sin\left(\frac{n\pi}{4}\right)\right)\right) \\
&= (2\sqrt{2})^n\left((A+B)\cos\left(\frac{n\pi}{4}\right) + i(A-B)\sin\left(\frac{n\pi}{4}\right)\right) \\
&= (2\sqrt{2})^n\left(C\cos\left(\frac{n\pi}{4}\right) + D\sin\left(\frac{n\pi}{4}\right)\right).
\end{aligned}$$

Now, since $a_0 = 1$, we have $C = 1$, and $a_1 = 0$ yields $2(C + D) = 0$ or $D = -1$. Therefore, we have

$$a_n = (2\sqrt{2})^n\left(\cos\left(\frac{n\pi}{4}\right) - \sin\left(\frac{n\pi}{4}\right)\right)$$

is a particular solution of the given recurrence relation.

**Example 5.35** We find here the complete solution of the recurrence relation

$$F_n = F_{n-1} + F_{n-2}, \quad \text{with} \quad F_0 = 1 \text{ and } F_1 = 2.$$

In this case, the characteristic equation is given by

$$r^2 = r + 1 \quad \text{or} \quad \text{or} \quad r^2 - r - 1 = 0.$$

The two roots $\alpha$ and $\beta$ of this equation are, respectively, given by

$$\frac{1 \pm \sqrt{1+4}}{2} = \frac{1 \pm \sqrt{5}}{2}.$$

We may write

$$\alpha = \frac{1+\sqrt{5}}{2} \quad \text{and} \quad \beta = \frac{1-\sqrt{5}}{2}.$$

Then the general solution is given by

$$F_n = A\alpha^n + B\beta^n,$$

where $A$ and $B$ are constants. Now, since $F_0 = 1$ and $F_1 = 2$, we have

$$A + B = 1, \tag{5.6.7}$$

$$A\alpha + B\beta = 2. \tag{5.6.8}$$

By multiplying the first of these two equations by $\alpha$, we get

$$A\alpha + B\alpha = \alpha. \tag{5.6.9}$$

Next, by subtracting (5.6.9) from (5.6.8), we obtain $B(\alpha - \beta) = \alpha - 2$. That is, by using $1 - \beta = \alpha$,

$$B = \frac{\alpha - 2}{\alpha - \beta} = \frac{1 - \beta - 2}{\alpha - \beta} = -\frac{(1 + \beta)}{\alpha - \beta}.$$

Using this value of $B$ in (5.6.7), we obtain

$$A = 1 - B = 1 + \frac{(1 + \beta)}{\alpha - \beta} = \frac{1 + \alpha}{\alpha - \beta}.$$

Finally, since $\alpha - \beta = \sqrt{5}$, we have

$$\begin{aligned} F_n &= \frac{1 + \alpha}{\sqrt{5}}\alpha^n - \frac{1 + \beta}{\sqrt{5}}\beta^n \\ &= \frac{1}{\sqrt{5}}\left[\alpha^n - \beta^n\right] + \frac{1}{\sqrt{5}}\left[\alpha^{n+1} - \beta^{n+1}\right]. \end{aligned}$$

Notice that

$$\begin{aligned} F_0 &= \frac{1}{\sqrt{5}}(1 - 1) + \frac{1}{\sqrt{5}}(\alpha - \beta) = 1 \\ F_1 &= \frac{1}{\sqrt{5}}(\alpha - \beta) + \frac{1}{\sqrt{5}}(\alpha + \beta)(\alpha - \beta) = 1 + 1 = 2. \end{aligned}$$

We illustrate use of *diagonalization of matrices* in the next example.

**Example 5.36** We solve here the recurrence equation

$$x_{n+2} = 2x_{n+1} + 3x_n \quad \text{with} \quad x_0 = x_1 = 1.$$

For each $n$, let

$$u_n = \begin{pmatrix} x_{n+1} \\ x_n \end{pmatrix}.$$

Using the given equation, we get

$$u_{n+1} = \begin{pmatrix} x_{n+2} \\ x_{n+1} \end{pmatrix} = \begin{pmatrix} 2x_{n+1} + 3x_n \\ x_{n+1} \end{pmatrix} = \begin{pmatrix} 2 & 3 \\ 1 & 0 \end{pmatrix} \begin{pmatrix} x_{n+1} \\ x_n \end{pmatrix} = Au_n$$

where

$$A = \begin{pmatrix} 2 & 3 \\ 1 & 0 \end{pmatrix}.$$

Hence $u_n = Au_{n-1} = A^2u_{n-2} = \cdots = A^n u_0$. Since $x_0 = x_1 = 1$, we have $u_0 = [1, 1]^t$. We now compute $A^n$. The eigenvalues of $A$ are 3, $-1$ and the corresponding eigenvectors are $[3, 1]^t$ and $[-1, 1]^t$. Let

$$P = \begin{pmatrix} 3 & -1 \\ 1 & 1 \end{pmatrix}$$

be the matrix formed by the eigenvectors. Then $A = PDP^{-1}$ and hence

$$\begin{aligned} A^n &= PD^nP^{-1} \\ &= \begin{pmatrix} 3 & -1 \\ 1 & 1 \end{pmatrix} \begin{pmatrix} 3^n & 0 \\ 0 & (-1)^n \end{pmatrix} \begin{pmatrix} 1/4 & 1/4 \\ -1/4 & 3/4 \end{pmatrix} \\ &= \frac{1}{4} \begin{pmatrix} 3^{n+1} - (-1)^{n+1} & 3^{n+1} + 3(-1)^{n+1} \\ 3^n - (-1)^n & 3^n + 3(-1)^n \end{pmatrix}. \end{aligned}$$

Therefore

$$\begin{aligned} u_n = A^n u_0 &= \frac{1}{4} \begin{pmatrix} 3^{n+1} - (-1)^{n+1} & 3^{n+1} + 3(-1)^{n+1} \\ 3^n - (-1)^n & 3^n + 3(-1)^n \end{pmatrix} \begin{pmatrix} 1 \\ 1 \end{pmatrix} \\ &= \frac{1}{4} \begin{pmatrix} 2.\ 3^{n+1} + 2(-1)^{n+1} \\ 2.\ \ 3^n + 2(-1)^n \end{pmatrix}. \end{aligned}$$

Therefore, we obtain a particular solution as

$$x_n = \frac{3^n + (-1)^n}{2}, \quad \text{for} \quad n \geq 0.$$

## *Solving Non-homogeneous Recurrence Relations*

We start our discussion here with the second-order non-homogeneous linear recurrence relation with constant coefficients of the form:

$$a_n = c_1 a_{n-1} + c_2 a_{n-2} + g(n). \tag{5.6.10}$$

As in the case of homogeneous equation, $a_n$ satisfying (5.6.10) is called a *solution* of the recurrence relation (5.6.10). We start with the associated homogeneous recurrence relation given by

$$a_n = c_1 a_{n-1} + c_2 a_{n-2}. \tag{5.6.11}$$

As in previous section, a solution of (5.6.10) is called a **general solution** if it has two arbitrary constants, and every **particular solution** of (5.6.10) comes from the general solution by assigning two special values to these two constants.

In the next theorem, we prove if a solution of the non-homogeneous recurrence relation (5.6.10) is known, then every solution of the equations is a sum of a particular solution and the general solution of the associated homogenous recurrence relation (5.6.11).

**Theorem 5.29** *The general solution of the second-order non-homogeneous linear recurrence relation with constant coefficients of the form (5.6.10) is given by*

$$a_n = a_n^{(h)} + a_n^{(p)},$$

*where $a_n^{(h)}$ is the general solution of the associated homogeneous recurrence relation (5.6.11), and $a_n^{(p)}$ is any particular solution of (5.6.10).*

***Proof*** Since $a_n^{(h)}$ is the general solution of the associated homogeneous recurrence relation

$$a_n = c_1 a_{n-1} + c_2 a_{n-2},$$

we have

$$a_n^{(h)} = c_1 a_{n-1}^{(h)} + c_2 a_{n-2}^{(h)}.$$

Our assumption that $a_n^{(p)}$ is a particular solution of (5.6.10) gives

$$a_n^{(p)} = c_1 a_{n-1}^{(p)} + c_2 a_{n-2}^{(p)} + g(n).$$

Therefore, using the last two equations, we see that $a_n = a_n^{(h)} + a_n^{(p)}$ satisfies

$$\begin{aligned} a_n &= a_n^{(h)} + a_n^{(p)} \\ &= (c_1 a_{n-1}^{(h)} + c_2 a_{n-2}^{(h)}) + (c_1 a_{n-1}^{(p)} + c_2 a_{n-2}^{(p)} + g(n)) \\ &= c_1 (a_{n-1}^{(h)} + a_{n-1}^{(p)}) + c_2 (a_{n-2}^{(h)} + a_{n-2}^{(p)}) + g(n) \\ &= c_1 a_{n-1} + c_2 a_{n-2} + g(n). \end{aligned}$$

This proves that $a_n = a_n^{(h)} + a_n^{(p)}$ is the general solution of (5.6.10). □

We list in Table 5.1 choice of trial particular solutions that one may take for a recurrence relation $a_n = c_1 a_{n-1} + c_2 a_{n-2} + g(n)$.

**Table 5.1** The trial particular solutions for $a_n = c_1a_{n-1} + c_2a_{n-2} + g(n)$

| $g(n)$ | $a_n^{(p)}$ |
|---|---|
| $c$, a constant | A |
| $n^k$, $k$ an integer | $A_kn^k + A_{k-1}n^{k-1} + \cdots + A_1n + A_0$ |
| $r^n$ | $Ar^n$ |
| $n^kr^n$ | $(A_kn^k + A_{k-1}n^{k-1} + \cdots + A_1n + A_0)r^n$ |
| $\sin \alpha n$ or $\cos \alpha n$ | $A \sin \alpha n + B \cos \alpha n$ |
| $r^n \sin \alpha n$ or $r^n \cos \alpha n$ | $r^n(A \sin \alpha n + B \cos \alpha n)$ |

**Example 5.37** Consider the non-homogeneous recurrence relation given by

$$a_n = 5a_{n-1} - 6a_{n-2} + 5 \cdot 4^n.$$

We first find the general solution of the associated homogeneous recurrence relation. For instance, the characteristic equation $r^2 - 5r + 6 = 0$ has two roots given by $r = 2, 3$. It thus follows that we have

$$a_n^{(h)} = A2^n + B3^n,$$

where $A$ and $B$ are constants. Next, to find a particular solution of the given relation, we have $g(n) = 5 \cdot 4^n$. In this case, $r = 4$ is not a root of the characteristic equation. Therefore, by the scheme given in Table 5.1, we take $a_n^{(p)} = D\,4^n$ as a trial particular solution. Substituting this into the given relation, we obtain

$$D4^n = 5D4^{n-1} - 6D4^{n-2} + 5 \cdot 4^n \quad \text{or} \quad 16D = 20D - 6D + 90.$$

Therefore, we obtain $D = 45$, and hence $a_n^{(p)} = 45 \cdot 4^n$ is a particular solution. Hence, the general solution of the recurrence relation is obtained as

$$a_n = a_n^{(h)} + a_n^{(p)} = A2^n + B3^n + 45 \cdot 4^n.$$

**Example 5.38** Consider the non-homogeneous recurrence relation given by

$$a_n - 6a_{n-1} + 8a_{n-2} = n^2 2^n.$$

We first find the general solution of the associated homogeneous recurrence relation. The characteristic equation $r^2 - 6r + 8 = 0$ has two roots given by $r = 2, 4$. It thus follows that we have

$$a_n^{(h)} = A2^n + B4^n,$$

where $A$ and $B$ are constants. Next, to find a particular solution of the given relation, we have $g(n) = n^2 2^n$. In this case, $r = 2$ is a simple root of the characteristic equation. Therefore, by the scheme given in Table 5.1, we take

$$a_n^{(p)} = n2^n(Dn^2 + En + F) = 2^n(Dn^3 + En^n + Fn),$$

as a trial particular solution. Substituting this into the given relation, we obtain

$$\begin{aligned} &2^n(Dn^3 + En^2 + Fn) - 6 \cdot 2^{n-1}(D(n-1)^3 + E(n-1)^2 + F(n-1)) \\ &+ 8 \cdot 2^{n-2}(D(n-2)^3 + E(n-2)^2 + F(n-2)) = n^2 2^n. \end{aligned}$$

Cancelling out the factor $2^n$, we obtain

$$\begin{aligned} &Dn^3 + En^2 + Fn - 3(D(n-1)^3 + E(n-1)^2 + F(n-1)) \\ &+ 2(D(n-2)^3 + E(n-2)^2 + F(n-2)) = n^2. \end{aligned}$$

After simplification, the above gives

$$-13D + 5E - F + (15D - 2E)n - 3Dn^2 = n^2.$$

Now, comparing the coefficients of $n^2$, $n$, and 1, we have

$$-3D = 1, \quad 15D - 2E = 0, \quad -13D + 5E - F = 0.$$

Solving the above equations, we get

$$D = -\frac{1}{3}, \quad E = -\frac{15}{6}, \quad F = -\frac{49}{6}.$$

It thus follows that a particular solution of the given recurrence relation is given by

$$a_n^{(p)} = -\frac{1}{6}n2^n(2n^2 + 15n + 49).$$

Hence, the general solution of the recurrence relation is obtained as

$$a_n = A2^n + B4^n - \frac{1}{3}n2^{n-1}(2n^2 + 15n + 49).$$

**Example 5.39** Consider the non-homogeneous recurrence relation given by

$$a_n = 4a_{n-1} - 4a_{n-2} + 2^n.$$

We first find the general solution of the associated homogeneous recurrence relation. The characteristic equation $r^2 - 4r + 4 = 0$ has a double root given by $r = 2$. It thus follows that we have

$$a_n^{(h)} = A2^n + Bn2^n = (A + Bn)2^n,$$

where $A$ and $B$ are constants. Next, to find a particular solution of the given relation, we have $g(n) = 2^n$. In this case, $r = 2$ is a double root of the characteristic equation. Therefore, by the scheme given in Table 5.1, we take $a_n^{(p)} = Dn^2 2^n$ as a trial particular solution. Substituting this into the given recurrence relation, we have

$$Dn^2 2^n = 4D(n-1)^2 2^{n-1} - 4D(n-2)^2 2^{n-2} + 2^n.$$

Cancelling the factor $2^n$, and collecting the terms with $D$ on one side, we obtain

$$D(n^2 - 2(n-1)^2 + (n-2)^2) = 1 \quad \text{or} \quad D(n^2 - 2n^2 + 4n - 2 + n^2 - 4n + 4) = 1,$$

which gives $D = 1/2$. It thus follows that a particular solution of the given recurrence relation is given by

$$a_n^{(p)} = \frac{1}{2} n^2 2^n = n^2 2^{n-1}.$$

Hence, the general solution of the recurrence relation is obtained as

$$a_n = a_n^{(h)} + a_n^{(p)} = (A + Bn)2^n + n^2 2^{n-1}.$$

**Example 5.40** Consider the non-homogeneous recurrence relation given by

$$a_n + a_{n-1} - 2a_{n-2} = 9.$$

We first find the general solution of the associated homogeneous recurrence relation. The characteristic equation $r^2 + r - 2 = 0$ has two roots given by $r = 1, -2$. It thus follows that we have

$$a_n^{(h)} = A + B(-2)^n,$$

where $A$ and $B$ are constants. Next, to find a particular solution of the given relation, we have $g(n) = 9 \cdot 1^n$. In this case, $r = 1$ is a simple root of the characteristic equation. Therefore, by the scheme given in Table 5.1, we take $a_n^{(p)} = Dn$ as trial particular solution. Substituting this into the given recurrence relation, we have

$$Dn + D(n-1) - 2D(n-2) = 9 \qquad \text{or} \qquad 3D = 9.$$

Therefore, we obtain $D = 3$. It thus follows that a particular solution of the given recurrence relation is given by $a_n^{(p)} = 3n$. Hence, the general solution of the recurrence relation is obtained as

$$a_n = a_n^{(h)} + a_n^{(p)} = A + B(-2)^n + 3n.$$

**Mathematica Example.** *The recurrence relation can be solved using Mathematica. Try the following command to get the solution for the previous two examples:*

```
RSolve[{a[n]+a[n-1]-2a[n-2]==9}, a[n], n]
RSolve[{a[n]-6a[n-1]+8a[n-2]==n^2 2^n}, a[n], n]
```

## 5.7 Generating Function Method

The concept of *generating function* introduced below was first given in the early eighteenth century by *Abrahan de Moivre* (1667–1754). However, it got prominence due to the work of *Leonhard Euler* (1707–1783). In general, such type of "functions" help analyze discrete problems that involve sequences of numbers, say of objects satisfying a specific property. As such, they find applications in many engineering branches, including computer science. Our main focus is to describe how generating functions help solve linear recurrence relations, and also discuss some important properties inasmuch as are needed in the sequel.

### *Ordinary Generating Function*

Consider a sequence $(a_n)$ of real or complex numbers. The **ordinary generating function** (or simply the **generating function**) of a sequence $(a_n)$ is a *formal power series* $G(x)$ of the form:

$$G(x) := \sum_{n=0}^{\infty} a_n x^n = a_0 + a_1 x + a_2 x^2 + \cdots \tag{5.7.1}$$

We give below the generating functions of some simple sequences:

1. The generating function for the sequence $1, 1, \ldots,$ is given by

$$1 + x + x^2 + \cdots = \frac{1}{1-x}. \tag{5.7.2}$$

More generally, the generating function for the sequence $1, a, a^2, a^3, \ldots$ is given by

$$1 + ax + a^2x^2 + \cdots = \frac{1}{1 - ax}. \tag{5.7.3}$$

Taking $a = -1$, it follows that the generating function for the sequence

$$1,\ -1,\ 1,\ -1,\ 1,\ -1,\ \ldots$$

is given by

$$1 - x + x^2 - x^3 + x^4 - \cdots = \frac{1}{1 + x}. \tag{5.7.4}$$

2. The generating function of the sequence defined by $a_n = n + 1$ is given by

$$1 + 2x + 3x^2 + 4x^3 + \cdots = \frac{1}{(1 - x)^2}. \tag{5.7.5}$$

3. The generating function for the sequence defined by $a_n = 1/n!$ is given by

$$1 + \frac{x}{1!} + \frac{x^2}{2!} + \cdots = e^x. \tag{5.7.6}$$

We next discuss some algebraic manipulation with generating functions that help solve some interesting linear recurrence relations. Let $G(x)$ and $H(x)$ be, respectively, the generating functions of sequences $(a_n)$ and $(b_n)$. That is,

$$\begin{aligned} G(x) &= a_0 + a_1x + a_2x^2 + \cdots = \sum_{n=0}^{\infty} a_nx^n \\ H(x) &= b_0 + b_1x + b_2x^2 + \cdots = \sum_{n=0}^{\infty} b_nx^n. \end{aligned} \tag{5.7.7}$$

The **sum** of two generating functions in (5.7.7) is the formal power series given by

$$\sum_{n=0}^{\infty} (a_n + b_n)x^n = (a_0 + b_0) + (a_1 + b_1)x + \cdots, \tag{5.7.8}$$

which may be written as $G(x) + H(x)$. That is, $G(x) + H(x)$ is the generating function for the sequence $(a_n + b_n)$. More generally, if $\alpha, \beta$ are constants, then the function $\alpha G(x) + \beta H(x)$ is the generating function for the sequence $(\alpha a_n + \beta b_n)$. The **product** of two generating functions in (5.7.7) is the formal power series given by

$$\sum_{n=0}^{\infty} c_n x^n = a_0 + a_1 x + a_2 x^2 + \cdots, \quad \text{where}$$
$$c_n = \sum_{k=0}^{n} a_k b_{n-k}, \quad \text{for} \quad n \geq 0, \tag{5.7.9}$$

which may be written as $G(x)H(x)$. In particular, we have

$$\sum_{n=1}^{\infty} a_n x^n = a_1 x + a_2 x^2 + \cdots = G(x) - a_0; \tag{5.7.10a}$$

$$\sum_{n=2}^{\infty} a_n x^n = a_2 x^2 + a_3 x^3 + \cdots = G(x) - a_0 - a_1 x; \tag{5.7.10b}$$

$$\sum_{n=1}^{\infty} a_{n-1} x^n = a_0 x + a_1 x^2 + \cdots = x\,G(x); \tag{5.7.10c}$$

$$\sum_{n=2}^{\infty} a_{n-1} x^n = a_1 x^2 + a_2 x^3 + \cdots = x\left[G(x) - a_0\right]; \tag{5.7.10d}$$

$$\sum_{n=2}^{\infty} a_{n-2} x^n = a_0 x^2 + a_1 x^3 + \cdots = x^2\,G(x). \tag{5.7.10e}$$

We shall use the above equations in the next part of the section. Also, the generating function $G(x)^2$ is the formal power series given by

$$\sum_{n=0}^{\infty} d_n x^n = a_0 + a_1 x + a_2 x^2 + \cdots, \quad \text{where}$$
$$d_n = a_0 a_n + a_1 a_{n-1} + \cdots + a_n a_0. \tag{5.7.11}$$

Therefore, we have

$$\begin{aligned}
d_0 &= a_0{}^2, \\
d_1 &= a_0 a_1 + a_1 a_0 = 2a_0 a_1, \\
d_2 &= a_0 a_2 + a_1 a_1 + a_2 a_0 = 2a_0 a_2 + a_1{}^2, \\
d_3 &= a_0 a_3 + a_1 a_2 + a_2 a_1 + a_3 a_0 = 2(a_0 a_3 + a_1 a_2). \\
&\vdots
\end{aligned}$$

That is, we have

$$G(x)^2 = a_0^2 + 2a_0 a_1 x + \left(2a_0 a_2 + a_1^2\right)x^2 + 2\left(a_0 a_3 + a_1 a_2\right)x^3 + \cdots.$$

**Theorem 5.30** *Let $G(x)$ and $H(x)$ be as in (5.7.7), with $b_0 \neq 0$. Then the quotient $G(x)/H(x)$ exists.*

***Proof*** Suppose $F(x) = G(x)/H(x)$ is the generating function of a sequence $(c_n)$ so that we can write

$$F(x) = c_0 + c_1 x + c_2 x^2 + \cdots$$

Rewriting $F(x) = G(x)/H(x)$ as $G(x) = H(x)F(x)$, we have

$$\begin{aligned} & a_0 + a_1 x + a_2 x^2 \cdots \\ & = (b_0 + b_1 x + b_2 x^2 + \cdots)(c_0 + c_1 x + c_2 x^2 + \cdots) \\ & = b_0 c_0 + (b_0 c_1 + b_1 c_0)x + \cdots + (b_n c_0 + b_{n-1} c_1 + \cdots + b_0 c_n)x^n + \cdots \end{aligned}$$

By comparing the coefficients of $x^k$ for $k \geq 0$, we obtain

$$\begin{aligned} a_0 &= b_0 c_0 \\ a_1 &= b_0 c_1 + b_1 c_0 \\ a_2 &= b_0 c_2 + b_1 c_1 + b_2 c_0 \\ &\vdots \\ a_n &= b_0 c_n + \cdots + b_n c_0. \end{aligned} \tag{5.7.12}$$

Since $b_0 \neq 0$, it follows from the first equation in (5.7.12) that $c_0 = a_0/b_0$. Using this in the second equation in (5.7.12), we get

$$b_0 c_1 \;=\; a_1 - b_1 c_0 \;=\; a_1 - b_1 \frac{a_0}{b_0} \;=\; \frac{a_1 b_0 - b_1 a_0}{b_0},$$

which gives

$$c_1 = \frac{a_1 b_0 - b_1 a_0}{{b_0}^2}.$$

Using $c_0$ and $c_1$ in the third equation in (5.7.12), we get

$$\begin{aligned} b_0 c_2 &= a_2 - b_1 c_1 - b_2 c_0 \\ &= a_2 - \frac{b_1(a_1 b_0 - b_1 a_0)}{{b_0}^2} - b_2 \left(\frac{a_0}{b_0}\right), \end{aligned}$$

which gives

$$c_2 = \frac{1}{b_0^3}\left[a_2 b_0^2 - b_1 a_1 b_0 + b_1^2 a_0 - b_2 b_0 a_0\right].$$

Continuing this way, we can solve for all $c_n$. Hence, the quotient $F(x)$ exists. □

We now describe the *generating function method*.

**Example 5.41** We solve the first-order recurrence relation given by

$$H_n = 2H_{n-1} + 1, \qquad \text{with} \quad H_1 = 1. \tag{5.7.13}$$

Notice that, since $H_n = 2H_{n-1} + 1$ implies $1 = H_1 = 2H_0 + 1$, we have $H_0 = 0$. We may write the generating function of the sequence $\big(H_n\big)$ as $H(x)$. That is,

$$H(x) = \sum_{n=0}^{\infty} H_n x^n = H_0 + H_1 x + \cdots .$$

As the given recurrence relation is of first order, we multiply (5.7.13) by $x, x^2, x^3, \ldots,$ and sum all equations. We thus obtain

$$\sum_{n=1}^{\infty} H_n\, x^n = 2\sum_{n=1}^{\infty} H_{n-1} x^n + \sum_{n=1}^{\infty} x^n.$$

By using the first relation in (5.7.10) given by

$$\sum_{n=1}^{\infty} H_n x^n = H(x) - H_0$$

and also the relation

$$\sum_{n=1}^{\infty} H_{n-1} x^n = x\, H(x),$$

it follows that

$$H(x) - H_0 = 2x\, H(x) + \frac{x}{1-x} \quad \text{i.e.,} \quad (1-2x)G(x) = \frac{x}{1-x},$$

because $H_0 = 0$. Therefore, we obtain

$$\begin{aligned} G(x) &= \frac{x}{(1-x)(1-2x)} \\ &= \frac{1}{1-2x} - \frac{1}{1-x} \\ &= \sum_{n=0}^{\infty} (2^n - 1)x^n. \end{aligned}$$

Hence, the solution of the recurrence relation (5.7.13) is $H_n = 2^n - 1$.

**Example 5.42** Consider the second-order recurrence relation given by

$$a_n - a_{n-1} - 6a_{n-2} = -2^n, \qquad a_0 = 1, a_1 = -3.$$

Let $G(x)$ denote the generating function of the sequence $\left(a_n\right)$. As the given recurrence relation is of second order, we multiply (5.7.14) by $x^2, x^3, x^4, \ldots,$ and sum all equations. We thus obtain

$$\sum_{n=2}^{\infty} a_n x^n - \sum_{n=2}^{\infty} a_{n-1} x^n - 6\sum_{n=2}^{\infty} a_{n-2} x^n = -\sum_{n=2}^{\infty} 2_n x^n.$$

Rewriting the above in terms of $G(x)$, we have

$$G(x) - a_0 - a_1 x - x[G(x) - a_0] - 6x^2 G(x) = -\frac{4x^2}{1-2x},$$

which gives

$$\begin{aligned} G(x)[1 - x - 6x^2] &= 1 - 3x - x - \frac{4x^2}{1-2x} \\ &= \frac{(1-4x)(1-2x) - 4x^2}{1-2x} \\ &= \frac{1 - 6x + 4x^2}{1-2x}. \end{aligned}$$

That is, we have

$$G(x) = \frac{1 - 6x + 4x^2}{(1-2x)(1-x-6x^2)} = \frac{1 - 6x + 4x^2}{(1-2x)(1+2x)(1-3x)}.$$

To resolve this fraction into partial fractions, we write

$$\frac{1 - 6x + 4x^2}{(1-2x)(1+2x)(1-3x)} = \frac{A}{1-2x} + \frac{B}{1+2x} + \frac{C}{1-3x}.$$

It thus follows that

$$A = \left.\frac{1 - 6x + 4x^2}{(1+2x)(1-3x)}\right|_{x=\frac{1}{2}} = \frac{1 - 3 + 1}{(2)\left(1 - \frac{3}{2}\right)} = \frac{-1}{2\left(\frac{-1}{2}\right)} = 1$$

$$B = \left.\frac{1 - 6x + 4x^2}{(1-2x)(1-3x)}\right|_{x=-\frac{1}{2}} = \frac{1 + 3 + 1}{(2)\left(1 + \frac{3}{2}\right)} = \frac{5}{2\left(\frac{5}{2}\right)} = 1$$

$$C = \left.\frac{1 - 6x + 4x^2}{(1-2x)(1+2x)}\right|_{x=\frac{1}{3}} = \frac{1 - 2 + \left(\frac{4}{9}\right)}{1 - (4)\left(\frac{1}{9}\right)} = \frac{-1 + \frac{4}{9}}{1 - \left(\frac{4}{9}\right)} = -1.$$

Therefore, we obtain

$$G(x) = \frac{1}{1-2x} + \frac{1}{1+2x} - \frac{1}{1-3x}.$$

By comparing the coefficients of the same power of $x$, we have

$$G(x) = \sum_{n=0}^{\infty} \left[2^n + (-2)^n - 3^n\right] x^n.$$

Hence, the solution of the recurrence relation is given by

$$a_n = 2^n + (-2)^n - 3^n.$$

**Example 5.43** Consider the second-order recurrence relation of the Fibonacci sequence $\left(F_n\right)$ given by

$$F_n = F_{n-1} + F_{n-2}, \qquad \text{with} \quad F_0 = 0 \text{ and } F_1 = 1. \tag{5.7.14}$$

If $F(x)$ is the generating function of the sequence $\left(F_n\right)$, then we have

$$F(x) = F_0 + F_1 x + F_2 x^2 + \cdots .$$

As the given recurrence relation is of second order, we multiply (5.7.14) by $x^2, x^3, x^4, \ldots$, and sum all equations. We thus obtain

$$\sum_{n=2}^{\infty} F_n\, x^n = \sum_{n=2}^{\infty} F_{n-1} x^n + \sum_{n=2}^{\infty} F_{n-2} x^n.$$

Now, using the three relations given by

$$\begin{aligned}
\sum_{n=2}^{\infty} F_n x^n &= F(x) - F_0 - x F_1; \\
\sum_{n=2}^{\infty} F_{n-1} x^n &= x(F(x) - F_0); \\
\sum_{n=2}^{\infty} F_{n-2} x^n &= x^2 F(x),
\end{aligned}$$

it follows that we have

$$F(x) - F_0 - F_1 x = x\left(F(x) - F_0\right) + x^2 F(x).$$

However, since $F_0 = 0$ and $F_1 = 1$, we obtain

$$F(x)[1 - x - x^2] = x \quad \text{or} \quad F(x) = \frac{x}{1 - x - x^2}.$$

If we write the two roots of the quadratic equation $x^2 + x - 1 = 0$ as

$$\alpha = \frac{-1 + \sqrt{5}}{2} \quad \text{and} \quad \beta = \frac{-1 - \sqrt{5}}{2},$$

then we have the two relations given by

$$\alpha\beta = -1 \quad \text{and} \quad \alpha + \beta = 1.$$

Therefore, we can also write the two roots of $x^2 + x - 1 = 0$ as

$$-\beta,\ -\alpha \quad \text{or} \quad \frac{1}{\alpha}, \frac{1}{\beta}.$$

That is, we have

$$x^2 + x - 1 = \left(x - \frac{1}{\alpha}\right)\left(x - \frac{1}{\beta}\right) = \frac{(1 - \alpha x)(1 - \beta x)}{\alpha\beta}.$$

Also, since $\alpha\beta = -1$, we have

$$1 - x - x^2 = (1 - \alpha x)(1 - \beta x).$$

It thus follows that

$$\begin{aligned} F(x) &= \frac{x}{(1 - \alpha x)(1 - \beta x)} \\ &= \frac{1}{\sqrt{5}}\left[\frac{1}{1 - \alpha x} - \frac{1}{1 - \beta x}\right] \\ &= \frac{1}{\sqrt{5}}\sum_{0}^{\infty}(\alpha^n - \beta^n)x^n. \end{aligned}$$

Hence, the solution of the recurrence relation (5.7.14) is obtained as

$$\begin{aligned} F_n &= \frac{1}{\sqrt{5}}\left[\alpha^n - \beta^n\right] \\ &= \frac{1}{\sqrt{5}}\left[\left(\frac{1 - \sqrt{5}}{2}\right)^n - \left(\frac{1 + \sqrt{5}}{2}\right)^n\right]. \end{aligned}$$

## Exercises 4

5.1 How many passwords are there if a password should be of six to eight characters long made up of upper case letters and digits with the condition that there must be at least one digit?

5.2 How many four digit even numbers have no repeated digits?

5.3 In how many ways can $n-2$ men and two women be seated in a row so that the two women are not seated next to each other?

5.4 In how many ways can $m$ adults and $n$ children seated in a round table so that no two children are seated next to each other?

5.5 Show that the number of permutations of the integers $1, 2, \ldots, n$ which do not contain $r$ preassigned pairs from $(1, 2), (2, 3), \ldots, (n-1, n)$ is

$$P_n - \binom{r}{1} P_{n-1} + \binom{r}{2} P_{n-2} - \cdots + (-1)^r \binom{r}{r} P_{n-r}.$$

For $r = n-1$, this number is $E_n$.

5.6 Show that the number of permutations of the integers $1, 2, \ldots, n$ which do not contain any of the pairs $(1, 2), (2, 3), \ldots, (n-1, n), (n, 1)$ is

$$P_n - \binom{n}{1} P_{n-1} + \binom{n}{2} P_{n-2} - \cdots$$

$$+ (-1)^{n-1} \binom{n}{n-1} P_1 = n D_{n-1}.$$

5.7 Give a computational proof of the above two identities using the formula for $D_n$.

5.8 Show that the number of permutations of $n$ elements in which $r$ prescribed elements are displaced and the remaining $n-r$ elements are either displaced or remain fixed is

$$n! - \binom{r}{1}(n-1)! + \binom{r}{2}(n-2)! - \cdots$$

$$+ (-1)^r \binom{r}{r}(n-r)!.$$

For $r = n$, this number is $D_n$.

5.9 In a box office, $m + k$ people line up for tickets, $m$ of them have dollar notes, and $k$ of them have half-dollar coins. A ticket costs 1/2 dollar. At the beginning, there are $q$ half-dollar coins in the box office. A queue in which every one presenting a dollar note gets change is a good queue; otherwise, the queue is a bad queue. Find the number of good queues.

5.10 A mother got 10 apples, 5 mangoes, 9 oranges, and 4 bananas to her three children. In how many ways can the children divide these fruits among them such that every one gets at least one fruit?

5.11 Find the number of ways of placing $n$ different objects in $k$ different boxes with the objects in each box arranged in a definite order if no box can be empty.

5.12 Find the number of ways of placing some or all of the $n$ different objects in $k$ different boxes with the objects in each box arranged in a definite order. (A box can be empty.)

5.13 Show that the number of partitions of $N$ into $n$ summands equals the number of partitions of $N$ into summands not larger than $n$ of which least one is equal to $n$.

5.14 Show that the number of partitions of $N$ into even summands equals the number of partitions of $N$ in which each summand enters an even number of times.

5.15 In how many ways can 3 d within a week be selected if a selection must not include 2 consecutive days.

5.16 Let $D_n$ denote the number of derangements of $n$ objects. Show that

$$D_n = nD_{n-1} + (-1)^n.$$

5.17 The squares of a chessboard are painted with 8 colors. The squares of each row are painted with all the 8 colors and no two consecutive squares of one column are painted the same color. In how many ways can this be done?

5.18 Let $D_{n,r}$ denote the number of derangements in which $r$ elements remain in their original positions and the remaining $n-r$ elements are displaced. Show that

$$n! = \sum_{r=0}^{n} \binom{n}{r} D_{n-r}.$$

5.19 Show that

$$n! = \sum_{r=0}^{n} r\binom{n}{r} D_{n-r}.$$

5.20 Let $E_n$ denote the number of permutations of the integers $1, 2, \ldots, n$ which do not contain any of the pairs $(1, 2), (2, 3), \ldots, (n-1, n)$. Show that $E_n$ satisfies the recurrence relation given by

$$E_n = (n-1)E_{n-1} + (n-2)E_{n-2}, \qquad E_2 = 1, \quad E_3 = 3, \quad \text{for} \quad n \geq 4.$$

5.21 Suppose $n$ children go on a merry-go-round. They decide to switch seats so that each has a new companion in front of him. In how many ways can this be done?

5.22 With meanings of the symbols as defined in the text, show that

$$Q_n = D_{n-1} - D_{n-2} + \cdots + (-1)^{n-3} D_2.$$

5.23 What is the number of permutations of the letters in the longest word "Pneumonoultramicroscopicsilicovolcanoconiosis"?

5.24 If $x_1, x_2$ are real numbers, show that $\max\{x_1, x_2\} = x_1 + x_2 - \min\{x_1, x_2\}$. Extend it to $n$ real numbers.

5.25 State and prove the inclusion-exclusion principle for $n$ sets.

5.26 Among 50 students in a class, 26 got an A in the first examination and 21 got A in the second examination. If 17 students did not get A in either examination, how many students got an A in both examinations?

5.27 Among 100 students, 32 study mathematics, 20 study statistics, 45 study accountancy, 15 study mathematics and accountancy, 7 study mathematics and statistics, 10 study statistics and accountancy, and 30 do not study any of these three subjects. Find the number of students studying all the three subjects and the number of students studying exactly one of the three subjects.

5.28 In a survey of 100 students, it was found that 40 studied Mathematics, 64 studied Physics, 35 studied Chemistry, 1 studied all the three subjects, 25 studied Mathematics and Physics, 3 studied Mathematics and Chemistry, and 20 studied Physics and Chemistry. Find the number of students who studied Chemistry only and the number of students who studied none of these subjects.

5.29 Out of 35 people who can either drink coffee or tea, 25 can drink coffee and 15 can drink tea. How many of them can drink both coffee and tea? How many of them can drink coffee but not tea? How many of them can drink tea but not coffee?

5.30 In a class of 50 students, 20 students know Malay, 16 know Chinese, and 15 know Tamil. Also 7 of them know both Malay and Chinese, 8 of them know both Malay and Tamil, and 5 of them know both Chinese and Tamil. If 3 of them know these three languages, how many of the students do not know any of these three languages?

5.31 Out of 30 students, 15 take a course in arts, 8 take a course in biology, and 6 take a course in chemistry. It is known that three students take a course in all the three subjects. Show that seven or more students take a course in none of these three subjects.

5.32 A bag contains balls colored with one or more of the 3 colors. Of them, 10 balls are colored with red, 12 balls are colored with blue, and 13 balls are colored with green. Also there 5 balls colored with both blue and red, 6 balls colored with red and green, and 8 balls colored with green and blue. If there are 3 balls colored with all the 3 colors, then how many balls are there in the bag?

5.33 Compute the number of integers from 1 to 250 that are divisible by any one of the integers 2, 3, 5, and 7.

5.34 Compute the number of integers from 1 to 1000 that are not divisible by 5, 6, and 8.

5.35 Twenty people went to a fruit stall which sells three different kinds of fruits. It was found that 5 of them bought apples, 8 bought oranges, and 6 bought bananas. If 3 of them bought all the 3 fruits, how many of them did not buy any fruits?

5.36 How many binary strings of length 6 either start with the digit 1 or end with 00?

5.37 Assuming that the order of summands is immaterial, find the number of partitions of a natural number $N$ into unequal summands selected from among the numbers 1, 2, 3 …, $n$.

5.38 Show that the sum of partitions of an integer $N$ in which the summands do not exceed $n$ equals the number of partitions of $N$ in which the number of summands does not exceed $n$.

5.39 Obtain a linear recurrence relation satisfied by the sequence $a_n = An3^n$.

5.40 Obtain a second-order linear recurrence relation satisfied by sequence $a_n = (A + Bn)2^n$.

5.41 By eliminating the constants $A$ and $B$, find a second-order recurrence relation satisfied by the following sequences:

a. $a_n = A2^n + B(-2)^n$.
b. $a_n = A3^n + Bn3^n$.
c. $a_n = A(-1)^n + B$.
d. $a_n = A(-2)^n + B3^n + n^2$.
e. $a_n = A2^n + B3^n + n2^n$.

5.42 Find a recurrence relation satisfied by $a_k = \binom{n}{k}$.

5.43 Solve the recurrence relation (5.6.2) satisfied by Fibonacci number sequence $(F_n)$.

5.44 Solve the following second-order homogeneous linear recurrence relations.

a. $a_n + 4a_{n-1} + 3a_{n-2} = 0$.
b. $a_n - 2a_{n-1} - 3a_{n-2} = 0$.
c. $a_n + 2a_{n-1} - 3a_{n-2} = 0$.
d. $a_n + 4a_{n-1} + 4a_{n-2} = 0$.

5.45 Solve the following second-order homogeneous linear recurrence relations satisfying the given conditions. Use the method of characteristic equations and also the method of generating functions.

a. $a_n - a_{n-1} - 6a_{n-2} = 0$, with $a_0 = 1, a_1 = -2$.
b. $a_n - a_{n-1} - 6a_{n-2} = 0$, with $a_0 = 1, a_1 = 3$.
c. $a_n - a_{n-1} - 6a_{n-2} = 0$, with $a_0 = 5, a_1 = 0$.
d. $a_n - 8a_{n-1} + 16a_{n-2} = 0$, with $a_0 = 0, a_1 = 4$.
e. $a_n - 8a_{n-1} + 16a_{n-2} = 0$, with $a_0 = 3, a_1 = 20$.
f. $a_n + 4a_{n-1} + 4a_{n-2} = 0$, with $a_0 = -2, a_1 = 2$.

5.46 Solve the following first-order non-homogeneous linear recurrence relations. Use the method of characteristic equations and also the method of generating functions.

a. $a_n - a_{n-1} = n, a_0 = 0.$
b. $a_n - 3a_{n-1} = 3^n, \quad \text{with} \quad a_0 = 1.$
c. $a_n + a_{n-1} = n^2, \quad \text{with} \quad a_0 = 0.$
d. $a_n + a_{n-1} = n + 1, \quad \text{with} \quad a_0 = 1.$

5.47 Solve the following second-order non-homogeneous linear recurrence relations.

a. $a_n - 5a_{n-1} + 6a_{n-2} = 7^n.$
b. $a_n - 7a_{n-1} + 10a_{n-2} = 3^n.$
c. $a_n - 6a_{n-1} + 9a_{n-2} = 3^n.$
d. $a_n - 5a_{n-1} + 6a_{n-2} = 2^n + n.$
e. $a_n - a_{n-1} - 2a_{n-2} = n^2 + 2n.$

5.48 Solve the following second-order non-homogeneous linear recurrence relations satisfying the given conditions. Use the method of characteristic equations and also the method of generating functions.

a. $a_n - 2a_{n-1} + a_{n-2} = 2n, \quad \text{with} \quad a_0 = 3, a_1 = 0$
b. $a_n - 4a_{n-1} + 4a_{n-2} = 2^n, \quad \text{with} \quad a_0 = 1, a_1 = 2.$

5.49 Use generating function method to solve the recurrence relation

$$a_n - 9a_{n-1} = 9^n, \qquad \text{with} \ a_0 = 1.$$

5.50 Use *generating function method* to solve the recurrence relation

$$a_n - 5a_{n-1} + 6a_{n-2} = 0, \qquad \text{with} \ a_0 = 0, a_1 = 1.$$

5.51 Use generating function method to solve the recurrence relation

$$a_n - 3a_{n-1} = -4n, \qquad \text{with} \ a_0 = -2.$$

## References

1. Byer, O.D., Smeltzer, D.L., Kenneth, L.W.: Journey into Discrete Mathematics. MAA Press, AMS, Providence, Rhode Island (2018)
2. Howson, A.G., Kahane, J.P., Lauginie, P., de Turckheim, E.: Mathematics as a Service Subject. Cambridge University Press (1988)
3. Lewis, H., Zax, R.: Essential Discrete Mathematics for Computer Science. Princeton University Press (2019)
4. Stanley, R.P.: Enumerative Combinatorics (2/e), vol. 1. Cambridge University Press (2011)
5. Stanley, R.P.: Enumerative Combinatorics, vol. 2. Cambridge Studies in Advanced Mathematics 62. Cambridge University Press (1999)
6. van Lint, J.H., Wilson, R.M.: A Course in Combinatorics. Cambridge University Press, Cambridge (1992)

7. Wallis, W. D. and George, J. C., *Introduction to Combinatorics* (2/e), Chapman and Hall/CRC, 2017
8. Brualdi, R.: Introductory Combinatorics (3/e). Prentice-Hall (1999)
9. Cameron, P.J.: Combinatorics: Topics Techniques Algorithms. Cambridge University Press, Cambridge (1994)
10. Kheyfits, A.: A Primer in Combinatorics. de Gruyter (2010)
11. Aigner, M.: A Course in Enumeration. GTM-238, Springer, Berlin (2007)
12. Feller, W.: An Introduction to Probability Theory (3/e), vol. 1. Wiley (1968)
13. Johnsonbaugh, R.: Discrete Mathematics (8/e). Pearson (2017)
14. Pless, V.S.: Introduction to the Theory of Error-Correcting Codes (3/e). Wiley, New York (1998)
15. Tokareva, N.: Bent Functions: Results and Applications to Cryptography. Academic Press (2015)
16. Alongi, J.M., Nelson, G.S.: Recurrence and Topology. Graduate Studies in Mathematics, vol. 85. AMS (2007)
17. Graham, R.L., Knuth, D.E., Patashnik, O.: Concrete Mathematics: A Foundation for Computer Science (2/e). Addison-Wesley (1994)
18. Kauers, M., Paule P.: The Concrete Tetrahedron. Springer, Berlin (2011)
19. Wilf, H.S.: Generatingfunctionology. Academic Press, Inc. (1994)

# Chapter 6
# Discrete Probability

*The most important questions of life are indeed, for the most part, really only problems of probability.*

***Pierre-Simon Laplace*** *(1749–1827)*

*That even after the observation of the frequent or constant conjunction of objects, we have no reason to draw any inference concerning any object beyond those of which we have had experience.*

***David Hume*** *(1711–1776), A Treatise of Human Nature.*

The probability theory is a mathematical study of unpredictable events occurring in the interface of the real world with various subject areas. Informally, probability is a measure of the "likelihood" of occurrence of such an event. Our emphasis in this chapter is the *discrete probability theory*. We introduce some basic tools inasmuch as needed to discuss some significant probabilistic problems related to computer science. We start with basic aspects of Kolmogorov's probability model and prove certain core results involving the *expectation* and *variance* of a random variable. The core idea is to use *average behavior* of problems to analyze, even when the associated distributions are not known. We conclude the chapter with some useful concentration inequalities and limit theorems.

The topics presented here find applications to problems concerning randomized algorithms [1, 2], machine learning [3], computer vision [4], error-correction coding [5], cryptography [6], optimization [7], robotics [8], and also computational linguistics [9]. Also, in the case of performance modeling, these probabilistic ideas help predict delays, schedule jobs and resources, and also to manage provision capacity [10]. The interested reader may refer [11, 12] or [13], for any missing details and further study.

 
V. Ravichandran and A. K. Razdan, *Fundamental Discrete Structures*,
University Texts in the Mathematical Sciences,
https://doi.org/10.1007/978-981-96-0069-4_6

## 6.1 Probability Models

In this part, our emphasis is to discuss the fundamental work of mathematicians who contributed significantly to the development of *discrete probability theory*. We begin by introducing some basic terminology. A **random experiment** is an experiment whose outcomes cannot be predicted with certainty. An **event** is a collection of unpredictable outcomes of a *random experiment*. Events of some familiar random experiments are given below:

(a) "getting at least two heads"; in flipping a coin three times.
(b) "the sum of face numbers is 5"; in rolling a pair of six-sided dice.
(c) "getting three aces"; in picking three cards from a deck of 52 cards.
(d) "getting a blue ball"; in picking from a box containing some blue and red balls.
(e) "getting defective items"; while sampling from a production lot.
(f) "getting a winning number"; in a lottery draw.
(g) "the recovery rate" of patients; in a clinical trial of a new drug.

It is common to assume that, even though the outcomes of a random experiment are unpredictable, all possible outcomes are known (however large). For example, the number of 5-letter words that can be formed using 26 English alphabets written in any order (without replacements) is

$$26 \times 25 \times 24 \times 23 \times 22 = 7{,}893{,}600.$$

Similarly, since each player in the game of bridge plays with 13 cards, the number of possible hands is

$$\binom{52}{13} = \frac{52!}{13!\,39!} = 635{,}013{,}559{,}600.$$

Also, if a fair coin is tossed any finite number of times until a head comes up, then the outcomes are infinite set of strings over the alphabet $\{H, T\}$ of the form

$$H,\ TH,\ TTH,\ TTTH,\ TTTTH, \ldots, \tag{6.1.1}$$

where H stands for a Head and T stands for a Tail. In general, to quantify the "unpredictability" of outcomes of a random experiment, the mathematicians use the language of set theory.

**Definition 6.1** The **sample space** of a random experiment, denoted by $\Omega$, is the set of all possible outcomes. Each element of $\Omega$ is called a **sample point**, and a set $E \subseteq \Omega$ is an **event**. We say $E$ is an *elementary event* if it consists of a single *sample point*; otherwise, it is called a *compound event*.

It is assumed that the *sample space* $\Omega$ represents the universe of study appropriately. The set $\emptyset \subset \Omega$ corresponds to the *impossible event*, and the whole set $\Omega$ corresponds to the *sure event*.

**Example 6.1** In case of *roll of an unbiased six-sided die, toss of a pair of distinguishable coins*, and *drawing a red or a blue ball subject to throw of a six-sided die*, the three sample spaces are respectively given by

$$\begin{aligned}\Omega_1 &= \{1, 2, 3, 4, 5, 6\};\\ \Omega_2 &= \{HH,\ HT,\ TH,\ TT\};\\ \Omega_3 &= \{(x, B),\ (x, R) \mid 1 \le x \le 6\},\end{aligned}$$

where B and R denote a blue ball and a red ball, respectively. Also, in the case of the random experiment wherein a ball is drawn from a bag consisting of 5 blue and 7 white balls, if a coin throw shows H; and, we throw a die if it shows T, the sample space $\Omega$ is given by

$$\Omega = \{HB1,\ HB2,\ \dots,\ HB5,\ HW1,\ HW2, \dots,\ HW7,\ T1,\ T2,\ \dots,\ T6\},$$

where W denotes a white ball. In general, all we need to know is whether a particular outcome belongs to an *event*. For example, in the case of the sample space $\Omega_2$, if the focus is to study the sample points that correspond to the occurrence of exactly one head then the event of interest is $E = \{\, HT,\ TH\,\}$.

The *probability* of an event is a number in the interval [0, 1], which estimates "likelihood" of occurrence of the event. A "very likely" event has a probability near 1; and, an event with a probability near 0 is "very unlikely" to occur. The abstract model of probability theory is due to the Soviet mathematician *Andrey N. Kolmogorov* (1903–1987), published in 1933. This model of Kolmogorov uses mathematical formulation based on sets and functions. Definition 6.5 takes into consideration all crucial aspects of measurements related to "uncertain outcomes" of a random experiment, and all our practical needs. However, before the publication of Kolmogorov's fundamental work, most probability models in use lacked precision.

(A) The traditional *empirical probability* specifies the "relative frequency" of occurrence of an event. More precisely, if there are $n$ *trials*[1] of a random experiment, and $m$ of these are in favor of a particular outcome ($m \le n$), then we say the **empirical probability** of the event is $p := m/n$. That is, we have

$$p := \frac{\text{number of favorable outcomes}}{\text{total number of possible outcomes}}. \tag{6.1.2}$$

For example, in a single roll of an unbiased six-sided die, the total number of outcomes is $n = 6$. In this case, there are $m = 3$ favorable outcomes to obtain an odd face number. So, the probability of the associated event is $3/6 = 1/2 = 0.5$, by (6.1.2). Also, the probability "to pick a blue ball" from a box containing 7 blue and 3 red balls is $7/10$, because out of 10 possible outcomes, the number

[1] Each performance of a random experiment is called a **trial**.

of favorable outcomes is once again 3. In general, the two numbers $n$ and $m$ may be very large.

(B) In the case of **classical probability** (or **theoretical probability**), all outcomes are assumed to be *equally likely*, i.e., no outcome occurs more or less frequently than the rest. In certain situations, such an assumption makes sense. For example, in a coin toss experiment, we may take the *classical probability* of getting a H (or a T) to be $1/2$, provided the number of trials is sufficiently large. This is known as the *law of large numbers*, which states that the empirical probability of success in a series of Bernoulli trials will converge to the theoretical probability. For similar reasons, the classical probability of getting a 6 (or any other number face) in a throw of an unbiased die may be taken to be $1/6$.

**Remark 6.1** The probability model (6.1.2) has two serious limitations, namely, $(i)$ the numerator changes with every trial even when the physical conditions of the experiment are same; and, $(ii)$ the denominator must be finite. Thus empirical probability is too restrictive to meet all practical needs. Also, classical probability is of little significance to develop a general theory of probability. Of course, it is unrealistic to assume that all outcomes are "equally likely".

We conclude this part with some other important definitions.

**Definition 6.2** Let $\Omega$ be a sample space, and $E$, $F$ be two events.

1. the event $E' = \Omega \setminus E$ is called the **complementary event** (or event not E);
2. the event $E \cup F$ is called a **composite event** $E$ or $F$;
3. the event $E \cap F$ is called a **compound event** $E$ and $F$; and,
4. the event $E \setminus F$ corresponds to the occurrence of event $E$, but not $F$.

In general, for any two *complementary events* $E$ and $F$, the event $E \cup F = \Omega$ is a *sure event*, and $E \cap F = \emptyset$ is an *impossible event.*

**Definition 6.3** Two events of a random experiment are said to be **mutually exclusive** if the occurrence of one excludes the possibility of the occurrence of the other. That is, two events are called *mutually exclusive* if they are *disjoint sets.*

Notice that mutually exclusive events do not share favorable outcomes, and so such events are also known as the *incompatible events.* For example, simple events of a random experiment are always incompatible; they also have another interesting property, namely, they are *exhaustive* and *mutually exclusive.*

**Definition 6.4** Let $\Omega$ be the sample space of a random experiment. We say events $E_1, \ldots, E_n$ are **exhaustive** if at least one of these $n$ events necessarily occurs whenever the experiment is performed. That is, we have

$$\Omega = E_1 \cup E_2 \cup \cdots \cup E_n.$$

Further, these $n$ events are said to be *mutually exclusive* and *exhaustive* if they are exhaustive and pairwise disjoint.

### Kolmogorov's Model

Kolmogorov's model of probability is based on sets and functions, which caters to a wide variety of applications. More precisely, the foundation of Kolmogorov's probability theory is based on "abstractions" related to the following three basic notions: $(i)$ the *sample space* $\Omega$ of a random experiment consisting of all possible outcomes or observations; $(ii)$ a collection of subsets of $\Omega$, denoted by $\mathscr{E}$, where a set $E \in \mathscr{E}$ is termed as an *event* of the experiment; and, $(iii)$ a function $\mathbb{P} : \Omega \to \mathbb{R}$ such that the real number $\mathbb{P}(w)$ satisfies the condition

$$0 \leq \mathbb{P}(w) \leq 1, \quad \text{for all} \quad w \in \Omega. \tag{6.1.3}$$

Intuitively, $\mathbb{P}(w)$ represents the "likelihood" of occurrence of the outcome $w$. When $\Omega$ is finite, we also assume that

$$\sum_{w \in \Omega} \mathbb{P}(w) = 1. \tag{6.1.4}$$

In abstract terms, the function $\mathbb{P}$ is called a *probability measure* defined on the sample space $\Omega$. We also call it a *probability distribution*[2] on $\Omega$, because it distributes the probabilities of 1 among all possible outcomes $w \in \Omega$. In many situations, it is convenient to assume that the probability distribution is "uniform", i.e., every outcome $w \in \Omega$ is "equally likely". Therefore, for a *uniform probability distribution* $\mathbb{P}$, we have

$$\mathbb{P}(w) := \frac{1}{|\Omega|}, \quad \text{for all} \quad w \in \Omega. \tag{6.1.5}$$

More generally, for an event $E \in \mathscr{E}$, the real number

$$\mathbb{P}(E) := \sum_{w \in E} \mathbb{P}(w) \tag{6.1.6}$$

measures the relative size of $E$ with respect to the number $|\Omega|$. It is called the *probability* of the event $E$. In particular, we obtain

$$\mathbb{P}(\emptyset) = 0 \quad \text{and} \quad \mathbb{P}(\Omega) = 1. \tag{6.1.7}$$

Notice that there may be other events $E$ such that $\mathbb{P}(E) = 1$.

**Example 6.2** The sample space of "flipping a coin three times" is given by

$$\Omega = \{ HHH, \ HHT, \ HTH, \ THH, \ HTT, \ THT, \ TTH, \ TTT \}$$

---

[2] As discussed later, the term "probability distribution" in general refers to the cumulative distribution of a random variable.

In this case, the event "at least one T appears" is the subset

$$E = \{\, HHT,\ HTH,\ THH,\ HTT,\ THT,\ TTH,\ TTT \,\}$$

so that $\mathbb{P}(E) = 7/8$. So, the event "no tails appear" is the set $F = \{\, HHH \,\}$ so that $\mathbb{P}(F) = 1/8$. We also have $F = \Omega \backslash E$. Said differently, the event $F$ is actually the event "not E". Similarly, the event "at least two tails" is complementary to the event "at most one tail". For the random experiment of "flipping a coin five times", the event "heads turns up exactly once" given by

$$F = \{\, HTTTT,\ THTTT,\ TTHTT,\ TTTHT,\ TTTTH \,\}$$

has the probability $5/32$.

**Example 6.3** For a student scoring low in two subjects, say S-1 and S-2, it is found that the probabilities of failing in the two subjects are, respectively, given by 0.4 and 0.3. It is also known that the probability of failing in S-1 and S-2 is 0.2. We find the probabilities of the events $A, B, C, D$ given by

(a) $A =$ The student may fail in S-1 or S-2.
(b) $B =$ The student does not fail in any of the two subjects.
(c) $C =$ The student may fail in S-1, but passes S-2.
(d) $D =$ The student may fail only in one of the two subjects.

Let $E_1,\ E_2$ be events that the student may fail in subjects S-1, S-2. Then, we have

$$\mathbb{P}\big(E_1\big) = 0.4,\ \ \mathbb{P}\big(E_2\big) = 0.3, \quad \text{and} \ \ \mathbb{P}\big(E_1 \cap E_2\big) = 0.2.$$

It thus follows from the respective definitions given above that

$$\begin{aligned}
\mathbb{P}(A) &= \mathbb{P}(E_1 \cup E_2) \\
&= \mathbb{P}(E_1) + \mathbb{P}(E_2) - \mathbb{P}(E_1 \cap E_2) \\
&= 0.4 + 0.3 - 0.2 = 0.5; \\
\mathbb{P}(B) &= \mathbb{P}((E_1 \cup E_2)') \\
&= 1 - \mathbb{P}(E_1 \cup E_2) \\
&= 1 - 0.5 = 0.5; \\
\mathbb{P}(C) &= \mathbb{P}(E_1 \cap E_2') \\
&= \mathbb{P}(E_1) - \mathbb{P}(E_1 \cap E_2) \\
&= 0.4 - 0.2 = 0.2; \\
\mathbb{P}(D) &= \mathbb{P}(B) - \mathbb{P}(E_1 \cap E_2) = 0.5 - 0.2 = 0.3.
\end{aligned}$$

The event $B$ in the last equation is the union of the disjoint events $D$ and $E_1 \cap E_2$.

Kolmogorov's *probability model* is free from the kind of limitations mentioned in Remark 6.1. In particular, there is no restriction on the cardinality of the *sample space* of a random experiment, and outcomes need not be *equally likely*. Also, the probability of an event is not merely an estimate. In general, since $\mathbb{P}(E)$ is the "relative size" of the set $E$ with respect to the size of the set $\Omega$, it makes sense to take $\Omega$ as the "universe" to answer all questions concerning probabilities of various events of the random experiment for which $\Omega$ is the sample space.

**Definition 6.5** (*Kolmogorov's Probability Model*) Let $\mathscr{E}$ be a collection of subsets of a nonempty set $\Omega$. A set function $\mathbb{P} : \mathscr{E} \to \mathbb{R}$ is called a **probability measure** if it has the following three properties:

1. (**Positivity**) $\mathbb{P}(E) \geq 0$, for each $E \in \mathscr{E}$.
2. (**Normalization**) $\mathbb{P}(\Omega) = 1$.
3. (**Additivity**) The $\mathbb{P}$-value of the union of two mutually disjoint sets $E, F \in \mathscr{E}$ is the sum of $\mathbb{P}$-values of the sets $E$ and $F$, $i.e.,$ : $\mathbb{P}(E \cup F) = \mathbb{P}(E) + \mathbb{P}(F)$.

We call the triplet $(\Omega, \mathscr{E}, \mathbb{P})$ as a **probability space**, $\mathbb{P}(E)$ is the **probability** of the set $E$, and the above three properties are called the **axioms of probability**.

Notice that the above abstract definition is motivated by the "relative frequency model" of the probability theory. In this broader sense, all sets $E \subset \Omega$ for which $\mathbb{P}(E)$ is defined are called **events**. In this chapter, we are mainly dealing with the case when the set $\Omega$ is at most countably infinite. It is thus safe to assume that $\mathscr{E} = 2^{\Omega}$ (see also Remark 6.2).

**Definition 6.6** (*Discrete Probability Space*) A *probability space* $\big(\Omega, \mathscr{E}, \mathbb{P}\big)$ is called a **discrete probability space** if the set $\Omega$ is at most countably infinite; $\mathscr{E} = 2^{\Omega}$; and, the *probability measure* $\mathbb{P}$ is defined for every set $E \subseteq \Omega$.

In particular, we have $\emptyset \in \mathscr{E}$ so that $\mathbb{P}(\emptyset)$ is defined. In general, for a *discrete probability space*, the *probability function* $\mathbb{P} : \mathscr{E} \to \mathbb{R}$ can be defined in terms of elementary events such that (6.1.6) holds, for every set $E \subseteq \Omega$. It also satisfies the familiar properties such as given below:

(a) (**Complementary probability**) For any event $E$, we have

$$\begin{aligned} 1 = \mathbb{P}(\Omega) &= \mathbb{P}(E) + \mathbb{P}(E'), \quad \text{by (2) and (3) of Definition 6.5} \\ \Rightarrow \quad \mathbb{P}(E') &= 1 - \mathbb{P}(E). \end{aligned}$$

So, the probability of "not E" is $1 - \mathbb{P}(E)$;

(b) We have that $\mathbb{P}(E) \in [0, 1]$, for all events $E \in \mathscr{E}$. For,

$$1 - \mathbb{P}(E) = \mathbb{P}(E') \quad \text{and} \quad \mathbb{P}(E') \geq 0, \quad \text{by (1) of Definition 6.5.}$$

so $1 - \mathbb{P}(E) \geq 0$, which implies $\mathbb{P}(E) \leq 1$. Also, it follows from $(a)$ that $\mathbb{P}(\emptyset) = 0$, by taking $E = \Omega$, and using (2) of Definition 6.5. Recall that, when $\Omega$ is finite, an impossible event is same as an event with probability 0. However, the two notions are practically different whenever $\Omega$ is an infinite set. A similar dichotomy exists for a sure event and an event with probability 1;

(c) (**Monotonicity**) To prove that the function $\mathbb{P}$ is *monotone*, let $E \subset F$ be two events in $\mathscr{E}$. Then

$$\mathbb{P}(E \cup (F \backslash E)) = \mathbb{P}(E) + \mathbb{P}(F \backslash E), \quad \text{by (3) of Definition 6.5,}$$

and so $\mathbb{P}(F) = \mathbb{P}(E \cup (F \backslash E)) \geq \mathbb{P}(E)$, by (1) of Definition 6.5. That is, for any events $E, F \in \mathscr{E}$, we have

$$E \subset F \quad \Rightarrow \quad \mathbb{P}(E) \leq \mathbb{P}(F). \tag{6.1.8}$$

(d) (**Inclusion-Exclusion Principle**) For any two events $E, F \in \mathscr{E}$,

$$\mathbb{P}(E \cup F) = \mathbb{P}(E) + \mathbb{P}(F) - \mathbb{P}(E \cap F). \tag{6.1.9}$$

To prove this, we may recall that

$$E \cup F = E \cup (F \backslash E),$$

where $E, F \in \mathscr{E}$ are *mutually exclusive*. So, we have

$$\mathbb{P}(E \cup F) = \mathbb{P}(E \cup (F \backslash E)) = \mathbb{P}(E) + \mathbb{P}(F \backslash E), \quad \text{by (3) of Definition 6.5.}$$

Also, since the two sets $F = (E \cap F) \cup (F \backslash E)$ and $E \cap F$ and $F \backslash E$ in the collection $\mathscr{E}$ are mutually exclusive, it follows that

$$\mathbb{P}(F) = \mathbb{P}((E \cap F) \cup (F \backslash E)) = \mathbb{P}(E \cap F) + \mathbb{P}(F \backslash E).$$

Therefore, (6.1.9) holds. It is possible to extend the *inclusion-exclusion principle* for three or more events.

Indeed, by putting the elusive probability function $\mathbb{P}$ on an axiomatic footing, Kolmogorov did provide real assurance that one may study probability as sensibly as one could study any other subject in mathematics.

**Remark 6.2** In general, $\mathscr{E} \neq 2^{\Omega}$ (**No-Go Theorem**). More specifically, the collection $\mathscr{E}$ is a *$\sigma$-algebra* consisting of "measurable events". When the space $\Omega$ is uncountably infinite, we assume $\emptyset \in \mathscr{E}$, and that it is closed with respect to complementation and countable unions. Also, measure $\mathbb{P}$ is assumed to be *countable additivity*: For every sequence of mutually disjoint events $E_n \in \mathscr{E}$, we have

$$\mathbb{P}\Big(\bigcup_{n=1}^{\infty} E_n\Big) = \sum_{n=1}^{\infty} \mathbb{P}(E_n). \tag{6.1.10}$$

This is also known as the **continuity axiom**. In this case, we simply use the terminology *probability space*.

**Example 6.4** We first consider coin toss experiment. Suppose 0 represents the outcome $H$ and 1 represents the outcome $T$, with probabilities given by

$$\mathbb{P}(\{H\}) = \mathbb{P}(\{0\}) := p;$$
$$\mathbb{P}(\{T\}) = \mathbb{P}(\{1\}) := 1 - p.$$

Then $(\{0, 1\}, 2^{\{0,1\}}, \mathbb{P})$ is a discrete probability space, where $\mathbb{P}$ is called the *Bernoulli measure*. Next, let $\Omega = \mathbb{N}$ and $\mathscr{E} = 2^{\Omega}$. For $n \in \mathbb{N}$, we define

$$\mathbb{P}(\{n\}) = a_n, \quad \text{where} \quad a_n \in \mathbb{R}^+, \tag{6.1.11}$$

subject to that

$$\sum_{n\in\mathbb{N}} \mathbb{P}(\{n\}) = 1.$$

For example, one may take $a_n = 1/2^n$; or $a_n = (1-a)^{n-1}a$, with $0 < a < 1$. The probability measure in the later case is known as the *geometric measure* with parameter $a$. Also, for $\Omega = \mathbb{N} \cup \{0\}$ and $\mathscr{E} = 2^{\Omega}$, the probability measure defined by

$$\mathbb{P}(\{k\}) = \frac{e^{-\lambda}\lambda^k}{k!}, \quad \text{with} \quad k \in \mathbb{N} \cup \{0\}, \tag{6.1.12}$$

is called the **Poisson measure** with parameter $\lambda > 0$. Finally, for $\Omega = \{0, 1, 2, \ldots, n\}$ with $n \in \mathbb{N} \cup \{0\}$, and $\mathscr{E} = 2^{\Omega}$, the probability measure defined by

$$\mathbb{P}(\{k\}) = \binom{n}{k} p^k (1-p)^{n-k}, \quad \text{where} \quad 0 < p < 1, \tag{6.1.13}$$

is called the **binomial measure** with parameters $n$ and $p$.

**Example 6.5** Three candidates $C_1$, $C_2$, $C_3$ are completing for one post. Suppose the first two candidates have the same chance of success, whereas $C_3$ has only half the chance of $C_1$ or $C_2$. It thus follows by normalization property of $\mathbb{P}$ that

$$\begin{aligned} 1 &= \mathbb{P}(C_1) + \mathbb{P}(C_2) + \mathbb{P}(C_3) \\ &= 2\mathbb{P}(C_3) + 2\mathbb{P}(C_3) + \mathbb{P}(C_3) \\ \Rightarrow \quad \mathbb{P}(C_3) &= \frac{1}{5}. \end{aligned}$$

We also have $\mathbb{P}(C_1) = \mathbb{P}(C_2) = 2/5$. It thus follows that

$$\mathbb{P}(C_1 \text{ or } C_3 \text{ gets posted}) = \mathbb{P}(C_1) + \mathbb{P}(C_3) = \frac{3}{5}.$$

To a practical mind, Kolmogorov's theory may appear only to defer the problem of the construction of probability models that serve to inform us about the physical and social world. However, as a matter of fact, we can derive classical probability from this axiomatic approach alone (see Exercise 6.1). Also, in the context of the axiomatic approach, it is important to remember that we regard equally likely a matter outside the axiomatic formulation. In fact, it is out of the scope of the book to have a discussion about the same.

## *Independent Events*

In general, there is no easy way to express the probability $\mathbb{P}(A \cap B)$ in terms of probabilities $\mathbb{P}(A)$ and $\mathbb{P}(B)$.

**Definition 6.7** We say two events $A$ and $B$ are **independent** if

$$\mathbb{P}(A \cap B) = \mathbb{P}(A)\mathbb{P}(B). \tag{6.1.14}$$

More generally, two sub-collections $A$ and $B$ of the collection $\mathscr{E}$ are independent if every pair of events $(A, B)$ satisfy (6.1.14), for all $A \in \mathscr{A}$ and all $B \in \mathscr{B}$. Two events are called **dependent** if they are not independent.

**Example 6.6** Suppose a study of mortality data of a hospital predicts that the probability of dying from a heart attack is 0.2, and the probability of the patient being a smoker is 0.6. If smoking and dying of heart attack were *independent events*, then the probability of being a smoker and dying of heart attack would be 0.12. If the probability of this intersection is greater than 0.12, then the two are *dependent events*.

**Example 6.7** With reference to the sample space of random experiment "flipping a coin five times", let

$$x \in \{HHH,\ HHT,\ HTH,\ THH,\ HTT,\ THT,\ TTH,\ TTT\}.$$

Then the event $A$ in which the first flip shows H is given by

$$A = \{HHx,\ HTx\},$$

and the event $B$ in which the second flip shows H is given by

$$B = \{HHx,\ THx\}.$$

Since each of these two events contains 16 outcomes, we have

$$\mathbb{P}(A) = \mathbb{P}(B) = \frac{16}{32} = \frac{1}{2}.$$

The event in which the first two flips show H is the intersection $A \cap B$. That is,

$$A \cap B = \{HHx \mid x \in \{H, T\}^3\}$$

Since it contains 8 outcomes, we obtain

$$\mathbb{P}(A \cap B) = \frac{8}{32} = \frac{1}{4} = \frac{1}{2} \cdot \frac{1}{2} = \mathbb{P}(A) \cdot \mathbb{P}(B).$$

Therefore, $A$ and $B$ are *independent events*. However, the two events

$$E = \{TTTTT,\ HHTTT\} \quad \text{and} \quad F = \{TTTTT,\ HTTTT\}.$$

are dependent events, as one may verify easily.

**Example 6.8** Consider the experiment of flipping a fair coin three times. Each outcome has a probability of $1/8$. Let $E$, $F$ be respectively events such that there are more Hs than Ts and the results of the first two tosses are the same. Clearly, we have

$$\mathbb{P}(E) = \frac{1}{2}, \quad \mathbb{P}(F) = \frac{1}{2}, \quad \text{and} \quad \mathbb{P}(E \cap F) = \frac{1}{4}.$$

It thus follows that the events $E$ and $F$ are independent. However, both events $E$ and $F$ involve the results of the first two tosses and it is not possible to make a convincing argument that one of these events has no effect on the other. Observe that if $A$ is the event Hs on the last toss so that

$$\mathbb{P}(A) = \frac{1}{2} \quad \text{and} \quad \mathbb{P}(E \cap A) = \frac{3}{8},$$

which implies $E$ and $A$ are not independent, i.e., they are **dependent**.

### *Birthday Paradox*

We conclude our discussion in this part with a brief description of the **birthday paradox**, which is also known as the *birthday problem*. Suppose $\mathbb{P}_n$ denotes the probability that no two of $n < 365$ people have the same birthday. We may assume that none of them was born on February 29. In this case, the outcomes are $365^n$ possible choices of birthdays for $n$ individuals, which we may take *equally likely*. Therefore, the allocation of birthdays to these $n$ people is independent. We are not taking "twins" into consideration. We ask: *What is the probability that two or more of these n people have the same birthday?* To answer this question, it is more convenient to first compute $\mathbb{P}_n$. Notice that $n$ distinct birthdays may be chosen in $\binom{365}{n}$ ways, and the same can be allocated to $n$ people in $n!$ ways. It thus follows that there are

$$\binom{365}{n} \times n! = 365 \times 364 \times \cdots \times (365 - n + 1)$$

possible combinations such that no two of $n$ people share their birthday. Therefore,

$$\begin{aligned}\mathbb{P}_n &= \frac{365 \times 364 \times \cdots \times (365 - n + 1)}{365^n} \\ &= \Big(1 - \frac{1}{365}\Big)\Big(1 - \frac{2}{365}\Big)\cdots\Big(1 - \frac{n-1}{365}\Big)\end{aligned}$$

Notice that the first factor on the right side of the last equation is the probability that no two people share their birthday; the second factor is the probability that a third person doesn't share a birthday with any of the first two, and so on. Now, using the standard inequality $1 - x \le e^{-x}$, for $x \in \mathbb{R}$, we find that

$$\begin{aligned}\mathbb{P}_n &= \Big(1 - \frac{1}{365}\Big)\Big(1 - \frac{2}{365}\Big)\cdots\Big(1 - \frac{n-1}{365}\Big) \\ &\le e^{-1/365}e^{-2/365}\cdots e^{-(n-1)/365} \\ &= e^{-\frac{n(n-1)}{2\times 365}}\end{aligned}$$

Hence, for $\mathbb{P}_n \le 0.5$, we must have

$$n(n-1) \ge 2 \times 365 \ln 2.$$

The two roots of the equation $n^2 - n - 2 \times 365 \ln 2 = 0$ are given by

$$n_1 = \frac{1 + \sqrt{1 + 8 \times 365 \ln 2}}{2} \quad \text{and} \quad n_2 = \frac{1 - \sqrt{1 + 8 \times 365 \ln 2}}{2},$$

with $n_1 \approx 23$. That is, there must be at least 23 people for the odds to favor at least two of them to have the same birthday. More precisely, we obtain $\mathbb{P}_{23} = 50.7\%$. Also, $\mathbb{P}_{30} = 70.6\%$, though most find it difficult to believe that among 30 people two are likely to share their birthday. However, it is difficult to prove that we need $n = 88$ to ensure that at least three people share their birthday. There are several different versions of the *birthday paradox* [7].

## 6.2 Conditional Probability

In this part, we discuss the *conditional probability* of an event $A$, given that event $B$ with positive probability has already occurred. It is denoted by $\mathbb{P}(A \mid B)$ (read as *probability of $A$ given $B$*). It is assumed that $A$ and $B$ are neither independent nor mutually exclusive. The main result is Bayes' theorem, which provides a useful formula to compute $\mathbb{P}(A \mid B)$, in general. We start our discussion with a simple illustration.

**Example 6.9** Aman and Aniket are planning a weekend dinner. They agreed, the one getting more heads in "three flips of a fair coin" will pay the bill. In this case, the set of outcomes is given by

$$\{HHH,\ HHT,\ HTH,\ THH,\ HTT,\ THT,\ TTH,\ TTT\}.$$

Each outcome is equally likely. The two events

$$A := \text{Aman pays} \quad \text{and} \quad N := \text{Aniket pays}$$

are respectively given by

$$A = \{HHH,\ HHT,\ HTH,\ THH\} \quad \text{and} \quad N = \{HTT,\ THT,\ TTH,\ TTT\}.$$

Clearly, each has a fifty-fifty chance to pay. Suppose the first toss shows H. We now re-assess their chances to pay. The main idea is that, in this "conditional situation", the revised sample space is given by

$$E = \{HHH,\ HHT,\ HTH,\ HTT\}$$

The sample points not in $E$ are insignificant to the "revised events" $E \cap A$ and $E \cap N$ given by

$$E \cap A = \{HHH,\ HHT,\ HTH\}, \qquad E \cap N = \{HTT\},$$

with the respective "conditional probabilities" given as $3/4$ and $1/4$. So, Aman will have to pay for the dinner.

This illustration suggests, in order to calculate the probability $\mathbb{P}(A \mid B)$, we only need to consider the modified sample space containing those outcomes of $B$ that are favorable to occurrence of the event $A$. Notice that, for an outcome to be in $A$ given that it is in $B$, the same outcome has to be in the intersection $A \cap B$. We give another illustration.

**Example 6.10** In a flip of three fair coins, we find the probability of the event $A :=$ at least two Hs appear, subject to that in the event $B$ first coin shows T. In this case, the sample space is given by

$$\Omega = \{HHH, HHT, HTH, THH, HTT, THT, TTH, TTT\}.$$

Each outcome is equally likely so that has the probability $1/8$. We thus have

$$\mathbb{P}(A) = \mathbb{P}(\{HHH\}) + \mathbb{P}(\{HHT\}) + \mathbb{P}(\{HTH\}) + \mathbb{P}\{THH\}) = \frac{1}{2};$$
$$\mathbb{P}(B) = \mathbb{P}(\{HTT\}) + \mathbb{P}(\{THT\}) + \mathbb{P}(\{TTH\}) + \mathbb{P}(\{TTT\}) = \frac{1}{2}.$$

We also have $\mathbb{P}(A \cap B) = \mathbb{P}(\{THH\}) = 1/8$. Recall that, while computing the probability of $A$, we cannot consider the outcome such that the first coin doesn't show T. This restriction leads to consider the set $B$ as the sample space for computing the "conditional probabilities". So, taking $B$ as the sample space, the probability of $A$ is $1/4$, because the only sample point of $B$ favorable to the event $A$ is $THH$. That is, $\mathbb{P}(A|B) = 1/4$.

In general, suppose $E$ and $F$ are two events of the same random experiment. The **conditional probability** of the event $E$ given that $F$ has occurred is given by

$$\mathbb{P}(E \mid F) = \frac{\text{number of sample points favorable to } E \cap F}{\text{number of sample points favorable to } F}.$$

Dividing the numerator and the denominator by $|\Omega|$, it follows that

$$\mathbb{P}(E \mid F) = \frac{|E \cap F|/|\Omega|}{|F|/|\Omega|} = \frac{\mathbb{P}(E \cap F)}{\mathbb{P}(F)}. \tag{6.2.1}$$

Notice that (6.2.1) holds only when $\mathbb{P}(F) \neq 0$, *i.e.*, $F \neq \emptyset$. This ratio don't modify the relative likelihood of outcomes in the event $F$. If $\mathbb{P}(F) = 0$, then we take $\mathbb{P}(E \mid F) = 0$. It though appears somewhat arbitrary, but this is needed to make the related formulas work in all cases.

**Example 6.11** Suppose a ball is drawn from a box containing ten identical balls that are numbered $1, 2, \ldots, 10$. In this case, the sample space is given by

$$\Omega = \{1, 2, 3, 4, 5, 6, 7, 8, 9, 10\}$$

Now, given that the ball drawn shows a number $> 4$, we ask: *What is the probability that it is an odd number*? To answer this question, we may take

$$E = \text{ the number is odd} \quad \text{and} \quad F = \text{ the number is } > 4.$$

We need to find $\mathbb{P}(E \mid F)$. For, since

$$\begin{aligned} E &= \{1, 3, 5, 7, 9\}; \\ F &= \{5, 6, 7, 8, 9, 10\}; \\ E \cap F &= \{5, 7, 9\}, \end{aligned}$$

it follows that

$$\mathbb{P}(E \mid F) = \frac{\mathbb{P}(E \cap F)}{\mathbb{P}(F)} = \frac{3/10}{6/10} = \frac{1}{2}.$$

**Example 6.12** Consider the experiment "roll of two unbiased six-sided dice". We find the conditional probability of the event $A :=$ the sum of face numbers exceeds 6, given that the first roll shows 3. In this case, we have

$$\begin{aligned} A &= \{(i, j) \mid 1 \leq i, j \leq 6; \text{ with } i + j \geq 7\}; \\ B &= \{(3, j) \mid 1 \leq j \leq 6\}. \end{aligned}$$

We also have $A \cap B = \{(3, 4), (3, 5), (3, 6)\}$. Therefore, we obtain

$$\mathbb{P}(A|B) = \frac{A \cap B}{\mathbb{P}(B)} = \frac{3/36}{6/36} = \frac{1}{2}.$$

**Example 6.13** A family has two children. We ask: *What is the probability that both are boys, given at least one is a boy?* In this case, there are four possible combinations of sexes, so the sample space is given by

$$\Omega = \{GG, GB, BG, BB\}$$

We may assume that each outcome is equally likely. We have $E = \{BB\}$ and $F = \{GB, BG, BB\}$ so that $E \cap F = \{BB\}$. Therefore, the probability that both are boys, given at least one is a boy, is

$$\mathbb{P}(E|F) = \frac{E \cap F}{\mathbb{P}(F)} = \frac{1/4}{3/4} = \frac{1}{3}.$$

Hence, contrary to the common perception that $\mathbb{P}(E|F) = 1/2$, we have the probability is $1/3$. Consider the question: *What is the probability that both are boys, given that the first child is a boy?* It is a little surprising to find that the answer in this case is $1/2$.

**Remark 6.3** The definition (6.2.1) is based on the concept of "equally likely". More generally, the same definition applies in cases wherein the sample points are not necessarily equally likely, subject to which we know how to compute the probabilities of the involved events.

The next lemma proves some interesting properties of conditional probability.

**Lemma 6.1** *Let $\Omega$ be the sample space of a random experiment, and $E$ and $F$ are two events of the experiment. Then, we have*

*(a)* $\mathbb{P}(\Omega \mid F) = \mathbb{P}(F \mid F) = 1$;
*(b)* $\mathbb{P}((E \cup F) \mid A) = \mathbb{P}(E \mid A) + \mathbb{P}(F \mid A) - \mathbb{P}((E \cap F) \mid A)$, *where $A$ is an event with $\mathbb{P}(A) \neq 0$; in particular,* $\mathbb{P}((E \cup F) \mid A) = \mathbb{P}(E \mid A) + \mathbb{P}(F \mid A)$, *when $E \cap F = \emptyset$;*
*(c)* $\mathbb{P}(E' \mid F) = 1 - \mathbb{P}(E \mid F)$.

***Proof*** Both $(a)$ and $(b)$ follow directly from the definition (6.2.1), using the following distributive law of sets in the latter case:

$$(E \cup F) \cap A = (E \cap A) \cup (F \cap A).$$

For (c), use $(a)$ together with $E \cup E' = \Omega$, followed by using the second part of $(b)$. The following is an interesting corollary of this proposition: Let $A$ of an event of a sample space $\Omega$, with $\mathbb{P}(A) > 0$, and

$$E_A = \{\, E \cap A \mid E \subset \Omega\}.$$

By the definition (6.2.1), we have $\mathbb{P}(E \mid A) \geq 0$, for any event $E$ of $\Omega$. It thus follows from $(a)$ and the second part of $(b)$ of Proposition 6.1 that the trinity $(A, E_A, \mathbb{P}(-- \mid A))$ satisfy all the axioms of probability. □

The independence of two events introduced earlier in Definition 6.7 is closely related to the concept of conditional probability. In Example 6.6, the conditional probability of dying of a heart attack given that the patient is a smoker is higher than the unconditional probability of dying of a heart attack. Whenever intuitively reasonable, we may apply independence to facilitate probability computations.

**Example 6.14** Suppose a card is drawn from a well-shuffled deck of cards, and $E$, $F$ be *equally likely* events the *card drawn is a spade* and *an ace*, respectively. Clearly,

$$\mathbb{P}(E) = \frac{13}{52} = \frac{1}{4} \quad \text{and} \quad \mathbb{P}(F) = \frac{4}{52} = \frac{1}{13}.$$

Further, we have $\mathbb{P}(E \cap F) = 1/52$. It thus follows that

$$\mathbb{P}(E \mid F) = \frac{\mathbb{P}(E \cap F)}{\mathbb{P}(F)} = \frac{1/52}{1/13} = \frac{1}{4}.$$

Since $\mathbb{P}(E \mid F) = \mathbb{P}(E) = 1/4$, it follows that the event $F$ has not affected the probability of the occurrence of the event $E$. We also have

$$\mathbb{P}(F \mid E) = \frac{\mathbb{P}(E \cap F)}{\mathbb{P}(E)} = \frac{1/52}{1/4} = \frac{1}{13} = \mathbb{P}(F).$$

That is, the event $E$ has not affected the probability of occurrence of the event $F$, and the probability of occurrence of $E$ is not affected by the occurrence of the event $F$. Therefore, the two events $E$ and $F$ are independent. The converse statement may not be true, in general.

The next lemma relates conditional probability with the independence of events.

**Lemma 6.2** *Suppose events $E$ and $F$ are such that $\mathbb{P}(F) \neq 0$. These two events are independent if and only if $\mathbb{P}(E \mid F) = \mathbb{P}(E)$.*

***Proof*** Suppose that events are independent, i. e., $\mathbb{P}(E \cap F) = \mathbb{P}(E) \cdot \mathbb{P}(F)$. Then,

$$\mathbb{P}(E \mid F) = \frac{\mathbb{P}(E \cap F)}{\mathbb{P}(F)} = \frac{\mathbb{P}(E) \cdot \mathbb{P}(F)}{\mathbb{P}(F)} = \mathbb{P}(E).$$

Conversely, if the condition holds, then we have

$$\mathbb{P}(E \mid F) = \mathbb{P}(E) \quad \Rightarrow \quad \mathbb{P}(E \cap F) = \mathbb{P}(E) \cdot \mathbb{P}(F),$$

using the definition (6.2.1). Therefore, $E$ and $F$ are independent events. For events $E$ and $F$ with positive probabilities, we thus have

$$\mathbb{P}(E \mid F) = \mathbb{P}(E) \iff \mathbb{P}(F \mid E) = \mathbb{P}(F).$$

Therefore, for events $E$ and $F$ with positive probabilities, either of the two equalities implies *independence* of these events. □

**Lemma 6.3** *If $E$ and $F$ are independent events then*

*(a) $E$ and $F'$ are independent events;*
*(b) $E'$ and $F$ are independent events; and,*
*(c) $E'$ and $F'$ are independent events.*

***Proof*** For $(a)$, suppose $E$ and $F$ are independent events. Then,

$$\begin{aligned}
&\mathbb{P}(E \cap (F \cup F') = \mathbb{P}(E) \\
\Rightarrow\ &\mathbb{P}((E \cap F) \cup (E \cap F')) = \mathbb{P}(E) \quad \text{(distributivity)} \\
\Rightarrow\ &\mathbb{P}(E \cap F) + \mathbb{P}(E \cap F') = \mathbb{P}(E) \quad \text{(Axiom A3)} \\
&\quad \Rightarrow\ \mathbb{P}(E \cap F') = \mathbb{P}(E) - \mathbb{P}(E \cap F) \\
&\quad \Rightarrow\ \mathbb{P}(E \cap F') = \mathbb{P}(E) - \mathbb{P}(E) \cdot \mathbb{P}(F) \quad \text{(independence of } E \text{ and } F) \\
&\qquad\qquad = \mathbb{P}(E)(1 - \mathbb{P}(F)) = \mathbb{P}(E) \cdot \mathbb{P}(F').
\end{aligned}$$

Therefore, $E$ and $F'$ are independent events. The other two cases follow by interchanging the roles of $E$ and $F$. □

It is often, but not always, intuitively clear when two events are independent.

**Example 6.15** Suppose a biased coin has the probability $p$ to show-up a head $H$, and the probability $q$ for a tail $T$. If such a coin is tossed twice, by using the *relative frequency* model, it is reasonable to assign to the outcome $HH$ the probability $p^2$, to the outcome $HT$ the probability $pq$, and so on. Let $E$ be the event that heads turns up on the first toss, and $F$ the event that tails turns up on the second toss, then

$$\mathbb{P}(E) = p^2 + pq = p \quad \text{and} \quad \mathbb{P}(F) = pq + q^2 = q$$

implies that $\mathbb{P}(E \cap F) = pq = \mathbb{P}(E) \cdot \mathbb{P}(F)$. It thus follows that the two events are independent, as expected. However, if $E$ is the event wherein the first toss is a head and $F$ is the event such that the two outcomes are same, then

$$\mathbb{P}(F \mid E) = \frac{\mathbb{P}(F \cap E)}{\mathbb{P}(E)} = \frac{1/4}{1/2} = \frac{1}{2} = \mathbb{P}(F).$$

Therefore, once again, events $E$ and $F$ are independent. But, in this case, the fact is not so obvious.

The next lemma is known as the *multiplication rule* of probability.

**Lemma 6.4** (Multiplication Rule) *Suppose events $E$, $F$ of a sample space $\Omega$ are such that $\mathbb{P}(E) \neq 0$ and $\mathbb{P}(F) \neq 0$. Then,*

$$\mathbb{P}(E \cap F) = \mathbb{P}(E)\mathbb{P}(E \mid F) = \mathbb{P}(F)\mathbb{P}(F \mid E).$$

*In particular, if $E$ and $F$ are independent, then*

$$\mathbb{P}(E \mid F) = \mathbb{P}(F) \quad \textit{and} \quad \mathbb{P}(F \mid E) = \mathbb{P}(E).$$

***Proof*** The first equality follows directly from the definition (6.2.1). On the other hand, the second equality follows by using the assumption $\mathbb{P}(F) \neq 0$ and the symmetry with regard to $E$ and $F$ in the definition (6.2.1). The next example illustrates the above rule. □

**Example 6.16** Suppose two balls are drawn one after the other without replacement from an urn containing 8 *black* and 6 *white* balls, and we wish to find the probability that both drawn balls are black. That is, if $E$ and $F$ denote the events that the first and second ball drawn are black, respectively, we then have to find $\mathbb{P}(E \cap F)$. Now $\mathbb{P}(E) = 8/14$, and the drawn ball is not replaced before the second draw, so $\mathbb{P}(F \mid E) = 7/13$. It thus follows from the multiplication rule that

$$\mathbb{P}(E \cap F) = \mathbb{P}(E)\mathbb{P}(F \mid E) = \frac{8}{14} \times \frac{7}{13} = \frac{4}{13}.$$

The illustration in the next example is well-known *Monty Hall Problem*.

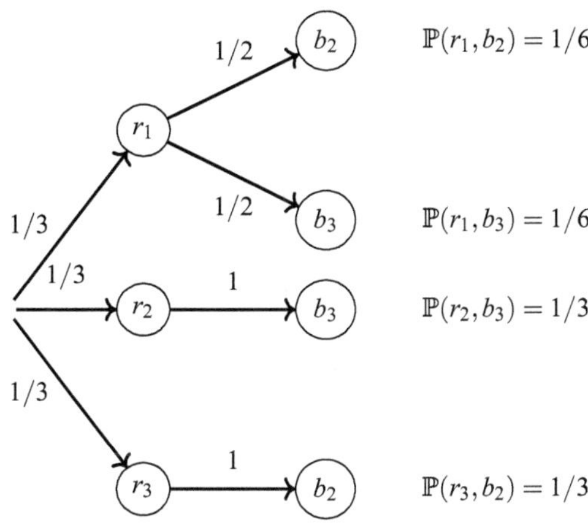


**Fig. 6.1** Tree representation of Monty hall problem

**Example 6.17** (*Monty Hall Problem*) It is common to find the host of a game show on television to give the participants a choice to flip their selection of a box. Exactly one of the three boxes, say $B_1$, $B_2$, and $B_3$, contains the real reward. More specifically, once a participant chooses a box, the host opens one of the other two boxes that contain a worthless reward. At this moment, there are two closed boxes in front of the participant, and she is offered to reconsider her choice. *Should the participant stick to the original choice or go ahead with a flip?* To help us analyze the situation, we use conditional probability. Suppose $B_1$ is the original choice. If it contains the real reward, then there is an equal probability that the host opens box $B_2$ or $B_3$. However, if the box $B_2$ contains the real reward, then the host opens the box $B_3$, with the probability 1. The same situation holds for $B_3$ and $B_2$. For $i = 1, 2, 3$ and $j = 2, 3$, suppose the value of the reward in box $B_i$ is $r_i$, and the chance that the host opens box $j$ be $b_j$. Of course, in no case, the host would open the box $B_1$. We may represent the sequences of choices by a *rooted tree* as shown in Fig. 6.1. This is known as the *probability tree*.[3] The leaves of this tree correspond to the paths associated with the possible outcomes. Therefore, the sample space is given by

$$\Omega = \big\{(r_1, b_2),\ (r_1, b_3),\ (r_2, b_3),\ (r_3, b_2)\big\},$$

and the probability of an outcome is obtained by multiplying the probabilities assigned along the corresponding path. We thus have

$$\mathbb{P}\big((r_1, b_2)\big) = \frac{1}{6}; \qquad \mathbb{P}\big((r_1, b_3)\big) = \frac{1}{6};$$
$$\mathbb{P}\big((r_2, b_3)\big) = \frac{1}{3}; \qquad \mathbb{P}\big((r_3, b_2)\big) = \frac{1}{3}$$

[3] The **probability trees** prove useful while analyzing problems wherein we have to make sequences of choices involving various probabilities. See Sect. 10.4 for any missing details.

If the host reveals the contents of the box $B_2$, then the following events are of interest:

$$A = \{(r_1, b_2),\ (r_1, b_3)\},\quad B = \{(r_3, b_2)\},\quad \text{and}\quad C = \{(r_1, b_2),\ (r_3, b_2)\}.$$

Therefore, the answer to the question of whether the participant should switch boxes or stick to $B_1$ is given by the conditional probabilities as given below:

$$\mathbb{P}(A|C) = \frac{A \cap C}{C} = \frac{1/6}{1/2} = \frac{1}{3};$$
$$\mathbb{P}(B|C) = \frac{B \cap C}{C} = \frac{1/3}{1/2} = \frac{2}{3}.$$

The first one gives the probability that the reward is in $B_1$, given that the host opens box $B_2$; and, the second is the probability that the reward is in $B_3$, given that the host opens box $B_2$. Hence, the participant has a better chance of winning a big reward by switching the boxes. Notice that the conclusion remains the same had the host opened the box $B_3$. It should be clear that the above probabilities may also be computed by constructing the labels carried out in reverse order. Indeed, there is a one-to-one correspondence between paths in the reverse and forward trees. So, each arc is assigned the same probability.

The multiplication rule extends easily to more than two events.

**Example 6.18** We compute the probability of drawing successively (without replacement) from a pack of 52 well-shuffled cards first two jacks followed by a king. For, let $J_1$ denote the event that the card drawn is a *jack* and $K$ be the event that the card drawn is a *king*, then we need to find

$$\mathbb{P}(J_1 \cap J_2 \cap K).$$

Now the probability that a jack is obtained in the first draw is $\mathbb{P}(J_1) = 4/52 = 1/13$, and the probability that again a jack $J_2$ is obtained in the second draw from $52 - 1 = 51$ deck of cards is $\mathbb{P}(J_2 \mid J_1) = 3/51$. Further, the conditional probability of the third drawn card to be a king is $\mathbb{P}(K \mid J_1 \cap J_2) = 4/50 = 2/25$. Thus, by multiplication law of probability for three events, the desired probability is

$$\mathbb{P}(J_1 J_2 K) = \mathbb{P}(J)\mathbb{P}(J_2 \mid J_1)\mathbb{P}(K \mid J_1 \cap J_2) = \frac{4}{52} \times \frac{3}{51} \times \frac{4}{50} = \frac{2}{13 \times 17 \times 25},$$

where $J_1 J_2 K$ is a shorthand notation for the intersection $J_1 \cap J_2 \cap K$.

**Example 6.19** Alina is appearing for a *typing proficiency test* that has to be qualified in three attempts. It is reasonable to assume that she may very likely pass on the first attempt so that the sample space is given by

$$\Omega = \{p,\ fp,\ ffp,\ fff\},$$

where $fp$ denotes the outcome that she fails in the first and passes in the second attempt, etc. Clearly, the chance of her eventually passing the test would be $3/4$, provided all outcomes are taken to be equally likely. Instead, suppose there is a study that suggests the probability of passing the test is 0.8. It is assumed that no matter how many times one may fail the chance of passing the test at the next attempt is still 0.8. Therefore, we have

$$\mathbb{P}(\{p\}) = 0.8, \ \ \mathbb{P}(\{fp\}) = 0.2 \times 0.8 = 0.16;$$
$$\mathbb{P}(\{ffp\}) = (0.2)^2 \times 0.8 = 0.032, \ \ \mathbb{P}(\{fff\}) = (0.2)^3 = 0.008.$$

Hence, the probability that she may eventually get the certificate is

$$\mathbb{P}(\{p,\ fp,\ ffp\}) = 0.8 + 0.16 + 0.032 = 0.992.$$

Of course, Alina may eventually get the certificate unless she fails *three times*, and so the probability is $1 - 0.008 = 0.992$.

We may once again represent the sample space of the *two-step experiment* given in the next example as a *rooted tree*. Recall that all possible outcomes are given by the leaves of this tree, and the arcs labels are the associated probabilities. These probabilities may also be computed by traversing the arcs in reverse order. So, in both directions, arcs are labeled with the same probability.

**Example 6.20** Suppose Urn-I contains 3 white and 5 red balls, and Urn-II contains 2 white and 1 red balls. We can draw a ball randomly from any of the two urns. For each urn, Fig. 6.2 shows the probabilities of drawing a ball of a particular color. Let

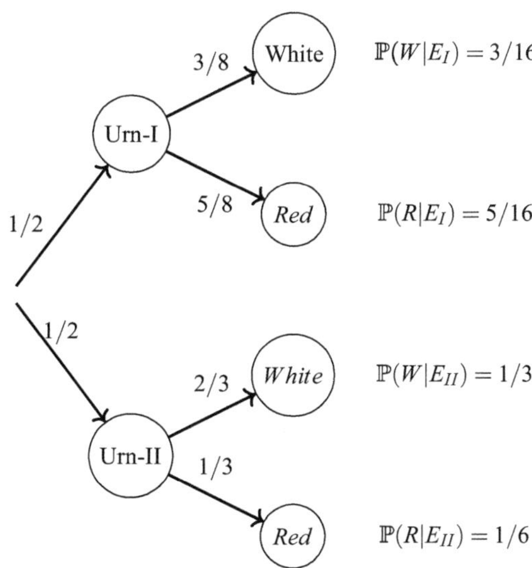


**Fig. 6.2** An illustration of Bayesian statistical inference

$W$ is the event that a white ball is drawn, and $E_I$ is the event that it is drawn from Urn-I. The conditional probability $\mathbb{P}(W|E_I) = 3/8$ is the label of an arc as shown in Fig. 6.2. On the other hand, given the color of a ball, we wish to compute the *inverse probability* that it was picked from a particular urn, which is also known as the *Bayes' probability*. It is a simple illustration of *Bayesian inference*. In this case, we proceed as follows:

$$\begin{aligned}\mathbb{P}(E_I \mid W) &= \frac{\mathbb{P}(E_I \cap W)}{\mathbb{P}(W)} \\ &= \frac{\mathbb{P}(E_I \cap W)}{\mathbb{P}(E_I \cap W) + \mathbb{P}(E_{II} \cap W)} \\ &= \frac{3/8}{(3/8) + (2/3)} = \frac{9}{25}.\end{aligned}$$

We have thus computed $\mathbb{P}(E_I \mid W)$. From the forward tree, we find that the probability of a white ball is

$$\frac{1}{2} \times \frac{3}{8} + \frac{1}{2} \times \frac{2}{3} = \frac{25}{48}.$$

The probabilities labels at the second level are found by simple division. For, if $x$ is the probability label of the top arc at the second level, we must have

$$\frac{25}{48} \times x = \frac{3}{16} \quad \Rightarrow \quad x = \frac{9}{25}.$$

It thus follows that $\mathbb{P}(E_I \mid W) = 9/25$, which is in agreement with our previous computation. The reverse tree thus displays the inverse probabilities.

## 6.3 Bayes' Theorem

In Example 6.20, we illustrated how to answer the question: *Given the probabilities of the outcome of the second stage of a two-stage random experiment, how to find the probability for an outcome at the first stage*. As said there, such inverse probabilities are also called the Bayes probabilities. We now discuss what are Bayes probabilities, in general. Some preparation is needed. We start with the next definition.

**Definition 6.8** A complete set of **alternatives** (or a **hypotheses**) refers to a collection of events $E_1, \ldots, E_n$ that partition the sample space $\Omega$. That is,

1. $\Omega = E_1 \cup E_2 \cup \cdots \cup E_n$; and

2. $E_i \cap E_j = \emptyset$, for all $1 \leq i \neq j \leq n$.

To have the events $E_1, \ldots, E_n$ constitute a *complete set of alternatives* for a sample space $\Omega$, these must be mutually exclusive and exhaust the space $\Omega$. For certain results to hold, we may also assume that $\mathbb{P}(E_i) > 0$, for each $i = 1, \ldots, n$. Let $F \subset \Omega$ be an event with nonzero probability. It thus follows from the definition that

$$\mathbb{P}(F) = \mathbb{P}(E_1 \cap F) + \cdots + \mathbb{P}(E_n \cap F).$$

By (6.2.1), we also have

$$\mathbb{P}\big(E_i \cap F\big) = \mathbb{P}\big(F|E_i\big)\mathbb{P}\big(E_i\big).$$

Therefore, we obtain

$$\mathbb{P}(F) = \mathbb{P}(F \mid E_1)\mathbb{P}(E_1) + \cdots + \mathbb{P}(F \mid E_n)\mathbb{P}(E_n). \tag{6.3.1}$$

This is known as Bayes' *rule of exclusive and exhaustive clauses.*

**Lemma 6.5** (Law of Total Probability) *If $E_1, \ldots, E_n$ is a complete set of alternatives, and $F \subset \Omega$ is an event with nonzero probability, then* (6.3.1) *holds.*

In particular, if $E$ and $F$ are two events such that $\mathbb{P}(E) < 1$ and $\mathbb{P}(F) < 1$, then

$$\mathbb{P}(E) = \mathbb{P}(E \mid F)\mathbb{P}(F) + \mathbb{P}(E \mid F')\mathbb{P}(F'). \tag{6.3.2}$$

The next example illustrates the above equality.

**Example 6.21** The work at a construction site has to be completed by the stipulated time. It is given that there is 55% chance that workers may strike. There is 75% chance of completing work on time, if there is no worker's strike; and, only 45% chance, if there is a strike. To find the probability that the construction is completed on time, let $E$ be the event that the construction is completed on time, and $F$ be the event there is a strike. By (6.3.2), we have

$$\begin{aligned}\mathbb{P}(E) &= \mathbb{P}(E \mid F)\mathbb{P}(F) + \mathbb{P}(E \mid F')\mathbb{P}(F') \\ &= 0.45 \times 0.55 + 0.75 \times 0.45 \\ &= 0.2475 + 0.3375 \\ &= 0.585.\end{aligned}$$

There is thus 58.5% chance that the construction gets completed on time.

Suppose $E$ and $F$ be two events with nonzero probabilities. By (6.2.1), we have

$$\mathbb{P}(E \cap F) = \mathbb{P}(E \mid F)\mathbb{P}(F) = \mathbb{P}(F \mid E)\mathbb{P}(E),$$

which implies that

$$\mathbb{P}(E \mid F) = \frac{\mathbb{P}(F \mid E)\mathbb{P}(E)}{\mathbb{P}(F)}. \tag{6.3.3}$$

It is known as the Bayes' *rule of retrodiction*. Together with (6.3.1), it gives

$$\mathbb{P}(E \mid F) = \frac{\mathbb{P}(F \mid E)\mathbb{P}(E)}{\mathbb{P}(E)\mathbb{P}(F \mid E) + \mathbb{P}(E')\mathbb{P}(F \mid E')}, \tag{6.3.4}$$

which is known as the *Bayes' law* (Theorem 6.1), named after the English statistician *Thomas Bayes* (1701–1761).

**Example 6.22** Suppose one out of the two coins is fair, and for the second we have

$$\mathbb{P}(\{H\}) = 0.6 = 1 - \mathbb{P}(\{T\})$$

Suppose picking any one coin is an equally likely event. A coin is picked, and it is flipped four times. Out of the four outcomes, three are $H$. Let $E$ is the event that fair coin is picked, and $F$ be the event that $3/4$ outcomes are $H$. It follows from (6.3.4) that the probability $\mathbb{P}(E \mid F)$ is given by

$$\begin{aligned}\mathbb{P}(E \mid F) &= \frac{\mathbb{P}(F \mid E)\mathbb{P}(E)}{\mathbb{P}(F)\mathbb{P}(F \mid E) + \mathbb{P}(F)\mathbb{P}(F \mid E')}\\ &= \frac{\binom{4}{3}(1/2)^4}{\binom{4}{3}(1/2)^4 + \binom{4}{3}(0.6)^3(0.4)}\\ &= \frac{2^{-4}}{2^{-4} + (0.6)^3(0.4)}.\end{aligned}$$

The next theorem extends (6.3.4) to a set of alternatives with positive probabilities.

**Theorem 6.1** (Bayes' Theorem) *Let $E_1, \ldots, E_n$ be a complete set of alternatives for a sample space $\Omega$, with $\mathbb{P}(E_i) > 0$ for each $i$, and $F$ be any event of nonzero probability. Then, for each $i = 1, 2, \ldots, n$, we have*

$$\mathbb{P}\big(E_i \mid F\big) = \frac{\mathbb{P}\big(F \mid E_i\big)\mathbb{P}\big(E_i\big)}{\mathbb{P}\big(E_1\big)\mathbb{P}\big(F \mid E_1\big) + \cdots + \mathbb{P}\big(E_n\big)\mathbb{P}\big(F \mid E_n\big)}.$$

***Proof*** Use (6.3.3), with $E_i$ in place of $E$, followed by applying (6.3.1). □

The rule of retrodiction (6.3.3) is central to the Bayesian inference. We may take the event $E$ to describe a state such as "having a disease", and the event $F$ indicating results of a related pathology test. It is desired to infer the "a posteriori probability" $\mathbb{P}(E|F)$ of the state of the disease given the test results, in terms of the "a prior probability" $\mathbb{P}(E)$ and the **likelihood function** $\mathbb{P}(F|E)$. The *likelihood function* $\mathbb{P}(F|E)$

measures the likelihood of the test $F$, subject to that we know the state $E$, and $\mathbb{P}(E)$ measures our prior knowledge about having the particular disease. Since $\mathbb{P}(F|E')$ is known, we usually apply (6.3.2) to obtain the probability $\mathbb{P}(F)$.

**Example 6.23** Statistical study shows that one out of 1000 people may have a certain rare disease. It is also found that in 99% of the cases, the test result is positive if a patient is affected by the disease. However, it may so happen that the test may also be positive in 2% of the cases even when a patient is healthy. We need to find the probability that a patient with a positive test is actually affected by the disease. We may write $E$ for the event that a patient has the disease, and $A,\ B$ be events that the test is positive or negative. We have

$$\mathbb{P}(E) = 0.001, \quad \mathbb{P}(A|E) = 0.99, \quad \text{and} \quad \mathbb{P}(A|E') = 0.02,$$

and we need to find $\mathbb{P}(E|A)$. By the rule of retrodiction, we obtain

$$\begin{aligned}\mathbb{P}(E|A) &= \frac{\mathbb{P}(A|E)\mathbb{P}(E)}{\mathbb{P}(A)} \\ &= \frac{\mathbb{P}(A|E)\mathbb{P}(E)}{\mathbb{P}(A|E)\mathbb{P}(E) + \mathbb{P}(A|E')\mathbb{P}(E')} \qquad \text{(by (6.3.4))} \\ &= \frac{0.99 \times 0.001}{0.99 \times 0.001 + 0.02 \times 0.999} \\ &\approx \frac{1}{20} = 5\%.\end{aligned}$$

Since this probability is very low, we must question the reliability of the test. The idea is to let a small portion of the population to undergo a better test, but only those whose test shows positive. This is because

$$\mathbb{P}(A) = 0.99 \times 0.001 + 0.02 \times 0.999 \approx 0.003.$$

Also, if we redo the above calculations considering the probabilities:

$$\mathbb{P}(E) = 0.00001, \quad \mathbb{P}(A|E) = 0.99, \quad \text{and} \quad \mathbb{P}(A|E') = 0.01,$$

then it follows that $\mathbb{P}(E|A) \approx 0.000099$. So, it is less likely that the patient is sick, and it is more likely that the test result is not correct.

More generally, given the **prior probabilities** $\mathbb{P}(E_i)$ of possible causes (or hypotheses) $E_i$ and also the **posteriori probabilities** of that each leads to the effect $F$, Theorem 6.1 is a way to calculate the probability that a given cause led to a particular effect. In medical diagnosis, a doctor finds the *prior probabilities* of the various diseases before the tests, and the probabilities for specific test outcomes, given a particular disease. What the doctor is interested to know is the *posterior probability* for the particular disease, given the outcomes of the tests.

**Example 6.24** Suppose three Engineering Institutions $I_1$, $I_2$ and $I_3$ conducts entrance test of 300, 700 and 1000 students in an year, with 15%, 36% and 47% failures, respectively. The name of a student of one of these Institutions is picked at random, and it is found that he had failed the test. Now, in order to find the probability that he has appeared for the test of the Institution $I_3$, let $E_j$ be the event the student picked at random appeared for the test of Institution $I_j$, for $i = 1, 2, 3$, and let $F$ be the student picked at random has failed. By Theorem 6.1, the desired probability is given by

$$\begin{aligned}\mathbb{P}(E_3 \mid F) &= \frac{\mathbb{P}(F \mid E_3)\mathbb{P}(E_3)}{\sum_{i=1}^{3} \mathbb{P}(F \mid E_i)\mathbb{P}(E_i)} \\ &= \frac{0.47 \times \dfrac{1000}{2000}}{0.15 \times \dfrac{300}{2000} + 0.36 \times \dfrac{700}{2000} + 0.47 \times \dfrac{1000}{2000}} \\ &= \frac{0.47 \times 0.5}{0.15 \times 0.15 + 0.36 \times 0.35 + 0.47 \times 0.5} \\ &= 0.6128.\end{aligned}$$

## 6.4 Random Variables and Distributions

Most engineering processes involve randomness. The term "stochastic" is derived from the Greek word *stókhos* meaning "guess", which simply refers to the fact that the outcomes of a random experiment are uncertain. In general, while studying a *stochastic problem*, we consider all relevant qualitative and quantitative features that may influence its solution, if any. The former types of features are called **attributes**, and the latter types are called **variables**. A variable taking definite values is said to be a *deterministic variable*, while the one with *uncertain values* is called a *random variable*. The governing equations of most stochastic problems use a mix of both types of variables.[4] The concept of the random variable is needed to attach numerical values to the outcomes of a random experiment. A formal definition is given below.

**Definition 6.9** Let $(\Omega, \mathbb{P})$ be a discrete probability space. A **random variable** is a real-valued function defined on the set $\Omega$. It is usually denoted by $X$ so that we have $X : \Omega \to \mathbb{R}$. A random variable $X$ is said to be **discrete** if it takes at most countably many values. A random variable $X$ is said to be **continuous** if $X(w) \in (a, b) \subseteq \mathbb{R}$, for all $w \in \Omega$.

[4] More generally, a **stochastic process** refers to a collection of random variables indexed by $t \geq 0$. We need such a formalism to model various types of random phenomena such as those related to the *growth of a bacterial population, electrical current fluctuating* due to thermal noise, or the *movement of a gas molecule*. Moreover, seemingly random changes in financial markets have motivated the extensive use of stochastic processes.

Notice that the human heart is a nice blend of the two types of processes: The blood flow is a continuous phenomenon, whereas heart beats are discrete. In all that follows, we are mainly dealing with *discrete random variables*. Such types of variables find applications in many disciplines such as biology, chemistry, ecology, neuroscience, physics, image processing, signal processing, control theory, information theory, computer science, and telecommunications. For example, a random variable $X$ may represent the number of defective items produced in a production run; the number of vehicles passing through a crossing during peak hours; or, the number of customers arriving or leaving at a service facility such as in a hospital, a bank, or a ticket-booking counter.

**Example 6.25** (*Quicksort Algorithm*) In this illustration, we use **quicksort algorithm** to arrange a finite sequence $S = \big(a_1, a_2, \ldots, a_n\big) \subset \mathbb{R}$ in ascending order. We now describe how *quicksort algorithm* works:

(a) When $S = \emptyset$ or $n = 1$, return $S$;
(b) Pick a *pivot* $a = a_i \in S$;
(c) Reorder $S$ so that for every $a_j \neq a \in S$, if $a_j < a$, then $a_j$ is moved to a list $S_1$; else, if $a_j > a$, then $a_j$ is moved to a second list $S_2$;
(d) the same algorithm is applied recursively to $S_1$ and $S_2$;
(e) return the sorted list $S_1,\ a,\ S_2$.

We may use a rooted tree to represent pivots at each step of the algorithm applied to the sequence $S = \big(1, 5, 9, 2, 3, 8, 7, 14, 12, 10\big)$. Other than the root node, each node corresponds to our choice of a pivot. At each internal node, the list $S_1$ is shown on the left side, and the list $S_2$ on the right side, as two labels. The node corresponding to a pivot $a$ has a left child, if $|S_1| \geq 2$; and, it has a right child, if $|S_2| \geq 2$. A leaf node is such that $|S_1| \leq 1$ and $|S_2| \leq 1$. We call such a rooted tree a **computation tree** (see Fig. 6.3). Notice that it is largely like a *binary search tree*. We can retrieve

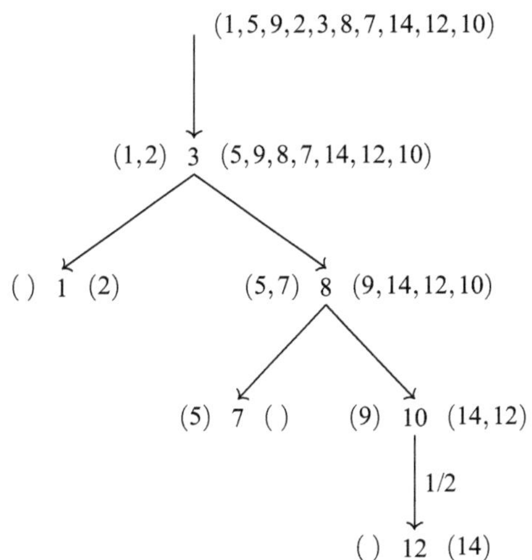


**Fig. 6.3** The comparison tree of a *quicksort run*

the sorted list using an *inorder tree traversal* of the *computation tree*. Some more runs of this algorithm applied to a few other lists are enough to conclude that the number of comparisons needed depends on our choice of a pivot. If at each step a pivot is chosen so that the sizes of lists $S_1$ and $S_2$ are roughly the same, then the number of comparisons is $O(n \ln n)$. However, in worst-case scenario, this number may increase to $n(n-1)/2$.

**Remark 6.4** We apply *randomized algorithms* to better the *average performance*. A pivot at each step is chosen by using an algorithm such as **rand** that returns a random element from some list $Y = \big(y_1, \ldots, y_j\big)$. In this context, the variable of interest is $X$ which counts the number of comparisons done by the algorithm. To analyze the related average running time, it is necessary to determine when the first (or the last) element of $Y$ is chosen as a pivot [2].

We use data for attributes and variables to analyze a stochastic problem. The data for random variables help identify an appropriate distribution. Subsequently, the probabilities of the data values of the involved random variables are obtained using the "shape" of the distribution curve. When $X$ is a random variable and $a \in \mathbb{R}$, the statement $X = a$ makes no sense unless we assign a "value" to it that represents the "likelihood of $X$ taking value $a$". This particular value is denoted by $\mathbb{P}(X = a)$, which is read as the probability that $X$ takes the value $a$. It is a measure of randomness (or *uncertainty*) that the random variable $X$ takes value $a$. Also, since the set

$$X^{-1}\big(a\big) = \big\{w \in \Omega \mid X(w) = a\big\} \tag{6.4.1}$$

is an event, we have

$$\begin{aligned}\mathbb{P}\big(X = a\big) &:= \mathbb{P}\big(X^{-1}(a)\big) = \mathbb{P}\big(\{w \in \Omega \mid X(w) = a\}\big);\\ \mathbb{P}\big(X \le a\big) &:= \mathbb{P}\big(X^{-1}((-\infty, a])\big) = \mathbb{P}\big(\{w \in \Omega \mid X(w) \le a\}\big)\end{aligned} \tag{6.4.2}$$

When $X$ is a *discrete variable*, if a large number of identical trials of a random experiment are conducted, then the probability $\mathbb{P}(X = a) = p_a$ means on an average $X = a$ occurs in a fraction $p_a$ of total number of trials. If $X$ is given by the data $(x_i, f_i)$, then we write

$$\mathbb{P}\big(X = x_i\big) = \frac{f_i}{N}, \quad \text{where} \quad N = \sum f_i. \tag{6.4.3}$$

In this case, we say $f(a) = \mathbb{P}(X = a)$ is the **probability mass function** (or simply pmf) of the discrete random variable $X$. For example, Fig. 6.4a depicts the *probability mass function* for the following data:

$$\begin{aligned}\mathbb{P}\big(X = -2\big) &= \frac{1}{16}, \quad \mathbb{P}\big(X = -1\big) = \frac{4}{16}, \quad \mathbb{P}\big(X = 0\big) = \frac{6}{16},\\ \mathbb{P}\big(X = 1\big) &= \frac{4}{16}, \quad \mathbb{P}\big(X = 2\big) = \frac{1}{16}.\end{aligned}$$

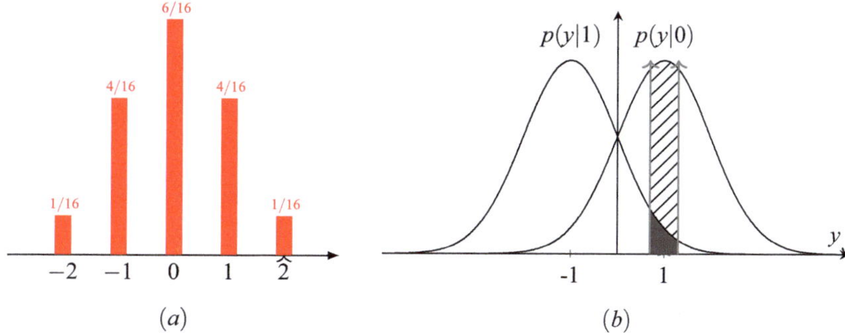


**Fig. 6.4** Probability mass and density functions

On the other hand, for a continuous variable $X$, the function

$$F(x) = \mathbb{P}(X \le x) \tag{6.4.4}$$

is called the **cumulative distribution function** (or simply **distribution function**) of variable $X$. We abbreviate *cumulative distribution function* as cdf. The function $F$ has four important properties as given below:

(a) $0 \le F(x) \le 1$. More generally, we have

$$\lim_{x\to-\infty} F(x) = 0 \quad \text{and} \quad \lim_{x\to\infty} F(x) = 1; \tag{6.4.5}$$

(b) It is piecewise constant with jumps, but it is right-continuous, i.e.,

$$F(a) = \lim_{h\to 0^+} F(a+h); \tag{6.4.6}$$

(c) It is *monotonic non-decreasing*, i.e., $x_1 < x_2 \Rightarrow F(x_1) \le F(x_2)$. In particular,

$$F(a-) = \lim_{h\downarrow 0} F(a-h) = \lim_{h\to 0^+} F(a-h); \tag{6.4.7}$$

(d) $\mathbb{P}(a \le X \le b) = F(b) - F(a)$.

When $X$ is discrete with *probability mass function* $f(x_i) = p_i$, we have

$$\begin{aligned} F(x) &= \sum_{x_i \le x} f(x_i) \\ \Rightarrow \quad p_i &= \mathbb{P}(X = x_i) = F(x_i) - F(x_{i-1}). \end{aligned} \tag{6.4.8}$$

Therefore, $F$ is the *distribution function* of the variable $X$. Notice that the right side in (6.4.8) is the limit of an infinite sum when there are countably infinite $x_i$'s. On the other hand, given a distribution function $F$ of a discrete variable $X$, we may define the probability mass function as

$$f(x) = F(x) - F(x-), \tag{6.4.9}$$

where $F(x-)$ is as defined in ($c$) above. More generally, when the cdf $F$ of a random variable $X$ has discontinuities, we treat $X$ as discrete in the sense that $X(w) \neq 0$, for at most countably many $w \in \Omega$. In this case, the pmf of $X$ is seen as a discrete version of a *probability density function* (or simply pdf).

**Definition 6.10** Let $X$ is a continuous variable taking values in an interval $(a, b)$. The **probability density function** of $X$ is a continuous function $f$ such that

$$\mathbb{P}\left(x - \frac{dx}{2} \leq X \leq x + \frac{dx}{2}\right) = f(x)\, dx. \tag{6.4.10}$$

In this case, $y = f(x)$ is called the *probability curve*, and $f(x)\, dx = d\, F(x)$ is known as the *probability differential*.

It follows from the above definition that

$$\mathbb{P}(a \leq X \leq b) = \int_a^b f(x)\, dx. \tag{6.4.11}$$

Clearly, we have $\mathbb{P}(X = x_0) = 0$ so that

$$\begin{aligned} \mathbb{P}(a \leq X \leq b) &= \mathbb{P}(a < X \leq b) \\ &= \mathbb{P}(a \leq X < b) \\ &= \mathbb{P}(a < X < b). \end{aligned} \tag{6.4.12}$$

Also, the distribution $F$ of a continuous variable $X$ is given by

$$F(x) = \mathbb{P}(X \leq x) = \int_{-\infty}^{x} f(x)\, dx. \tag{6.4.13}$$

In summary, the distribution of a random variable $X$ is given by

$$\begin{aligned} &(x_i, p_i), \text{ if } X \text{ is discrete given by the data } (x_i, f_i); \\ &(x, f(x)), \text{ if } X \text{ is continuous given by the data } f(x),\ x \in (a, b). \end{aligned} \tag{6.4.14}$$

**Example 6.26** In a screw manufacturing unit of an industry, an item is considered to be *defective* if it doesn't meet the diameter specification, within the assigned *tolerance limits*. Let the number of non-defective items be denoted by $X$. We first find a distribution for $X$ that *best fits* the observed values. Subsequently, we use probabilities as read from the distribution to compute the percentage of average defective items. The quality manager then decides what necessary measures are needed to improve the production processes. Notice that mass-production in general involves random variations due to *small* (but uncontrollable) differences in raw material or manufacturing processes involved.

With reference to Example 6.25, we use the computation tree to compute the probabilities: Simply add to every edge the probability that one of the elements of the corresponding list, say $Y$, was chosen uniformly, namely $1/|Y|$. So, given an input list $S$ of length $n$, the sample space $\Omega$ is the set of all computation trees $\mathscr{T}$ with root label $S$. We assign a probability to the trees $\mathscr{T}$ in $\Omega$ as follows: If $n = 0, 1$, then there is a single tree, and its probability is 1. If $n \geq 2$, then for every leaf multiply the probabilities along the path from the root to that leaf and then add up the probabilities assigned to these leaves. This is $P(\mathscr{T})$. In general, it is assumed that the underlying probability distribution is uniform and that any two consecutive choices are independent.

## *Binomial Distribution*

A random experiment with exactly two outcomes is called a **Bernoulli trial**, named after the Swiss mathematician *Jacob (Jacques) Bernoulli* (1654–1705). For example, whether it would rain today or not is a Bernoulli trial. More generally, in a concrete situation, outcomes of a Bernoulli trial may be given as a pair such as (Yes, No), (Head, Tail), (On, Off), (Defective, Non-defective), or (Occurrence, Non-occurrence). There is no loss of generality in assuming that the "success" refers to the term in the first coordinate of these pairs; and, the "failure" refers to the term in the first coordinate. The probability of success for a *Bernoulli trial* remains the same every time we conduct the random experiment, which we may denote by $p$. We also write $q := 1 - p$, which is the probability of failure for each Bernoulli trial. The corresponding discrete random variable is known as the *binomial variable*, which is a *two-parameter* random variable.

**Definition 6.11** (*Binomial Variable*) The **binomial variable** with parameters $n$ and $p$, usually denoted by $B(n, p)$, counts the number of successes in $n$ Bernoulli trials, where $p$ denotes the probability of success.

To derive the probability mass function of $X \sim B(n, p)$, we may once again consider the coin toss experiment with $n$ flips so that the associated sample space $\Omega$ consists of strings over the alphabet $\{H, T\}$ of length $n$. As said above, the probability of getting $H$ is taken as $p$, and the probability of getting $T$ is taken as $q = 1 - p$. Further, it is assumed that the coin flips are *independent* so that the probabilities of $w \in \Omega$ may be obtained by replacing $H$ with $p$ and $T$ with $q$ in the string $w$. Therefore, the random variable $X$ is such that $X(w)$ counts the number of H's in $w$. Now, since there are $\binom{n}{k}$ subsets with $k$ elements in a set of $n$ elements, the probability that $w$

has $k$ occurrences of H's is $p^k q^{n-k}$. It thus follows that the probability of getting $k$ successes in $n$ Bernoulli trials is given by

$$\mathbb{P}(X = k) = \binom{n}{k} p^k\, q^{n-k}, \quad \text{for} \quad k = 0, 1, \ldots, n, \tag{6.4.15}$$

which is same the *binomial measure*, as defined earlier. Hence, the *probability mass function* $f$ of the *binomial variable* $X \sim B(n, p)$ is given by

$$f(k) = \binom{n}{k} p^k\, q^{n-k}, \quad \text{for} \quad k = 0, 1, \ldots, n; \tag{6.4.16}$$

and, 0, otherwise. Applications include the problems related to finding the number of defective items in a production lot; estimation of reliability of systems; and also in radar detection (see Fig. 6.5a). The next example is a simple illustration.

**Example 6.27** With reference to Example 6.26, let $X$ denote the number of defective screws found in a sample of 10. We may assume that past observations suggest that 10% of the screws produced are "defective". That is, we have $p = 0.1$ so that $q = 1 - p = 0.9$. The probabilities of defective screws in various situations computed using (6.4.16) are given below:

$$\begin{aligned}
\mathbb{P}(X = 1) &= \binom{10}{1}(0.1)\,(0.9)^9 = 0.3874;\\
\mathbb{P}(X = 0) &= \binom{10}{0}\,(0.9)^{10} = 0.3486;\\
\mathbb{P}(X \le 2) &= \mathbb{P}(X = 0) + \mathbb{P}(X = 1) + \mathbb{P}(X = 2)\\
&= 0.3874 + 0.3486 + \binom{10}{2}(0.1)^2\,(0.9)^8\\
&= 0.3874 + 0.3486 + 0.1937 = 0.9297.
\end{aligned}$$

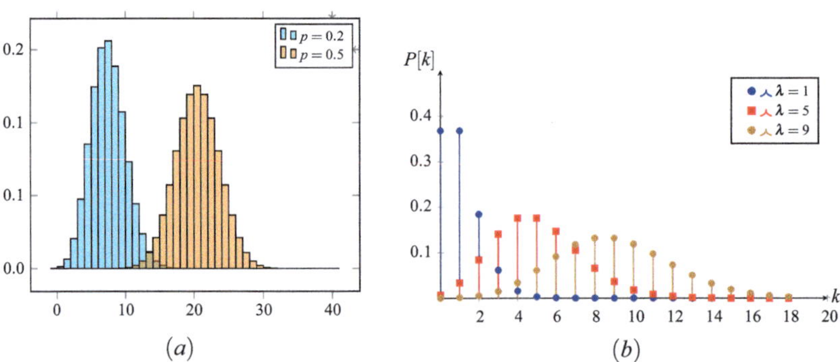


**Fig. 6.5** Binomial and Poisson distributions

The last equality shows that, in a sample of 10 screws, there is 92.97% chance of getting less than 3 defective screws.

Notice that if a random experiment consists of $n$ Bernoulli trials, and the same experiment is repeated $N$ times, then the frequency of $X = k$ is given by

$$f_k = N \times \binom{n}{k} p^k q^{n-k}, \; k = 0, 1, \ldots, n. \tag{6.4.17}$$

**Example 6.28** The probability of failure in a course is known to be 20%, Suppose there are 25 batches of 6 students each who appeared in the examination. We find below the number of batches such that 4 or more students passed the examination. Notice that, in this case, we have $p = 0.8$ and $q = 0.2$ so that the probability of having $\geq 4$ successes out of 6 is given by

$$\binom{6}{4}(0.8)^4 \cdot (0.2)^2 + \binom{6}{5}(0.8)^5 \cdot (0.2) + \binom{6}{6}(0.8)^6 = 0.90112.$$

Also, since there are 25 batches of 6 students each, the number of batches in which this can happen is given by (6.4.17). This number is

$$25 \times 0.90112 = 22.528 \approx 23.$$

We can thus expect to have $\geq 4$ successes in 23 batches.

Recall that if a biased coin is flipped a finite number of times until a head comes up then the sample space $\Omega$ is an infinite set consisting of strings $w$ over the alphabet $\{H, T\}$ of the form (6.1.1). As before, let $p$ be the probability of getting H so that the probability of getting a $T$ is $q = 1 - p$. Also, since coin flips are assumed to be *independent*, the probabilities of $w \in \Omega$ are obtained by replacing $H$ with $p$ and $T$ with $q$ in the string $w$. In this case, if the random variable $X$ is such that $X(w) = n$ if and only if $|w| = n$; 0, otherwise. Said differently, $X$ is the number of trials until a $H$ is obtained. Therefore, the *probability mass function* $f$ of $X$ is given by

$$f(n) = (1 - p)^n p, \quad \text{for} \quad n \geq 1; \tag{6.4.18}$$

and, 0, otherwise. This is a special case of the *geometric distributions*. These types of distributions find many applications in real-world problems such as when $X$ counts the number of calls needed to reach out to a customer or the number of attempts needed to win a game.

## *Poisson Distribution*

We now introduce an important class of *one-parameter* discrete random variables, known as the *Poisson variables*. We assume that all events of interest are independent and uniformly distributed over specified time intervals. Said differently, we don't allow clustering around events in one particular period of time. Recall that $B(n, p)$ is the binomial variable with parameters $n$ and $p$.

**Definition 6.12** (*Poisson Variable*) A **Poisson variable** is the limit $\lim_{n\to\infty} B(n, p)$ such that $\lambda := np$ is a constant of *moderate* size. It is denoted by $P(\lambda)$.

For example, if the events are the arrival or departure of customers at a service counter of a bank during a specific time period so that the Poisson variable $P(\lambda)$ denotes the number of customers, then the parameter $\lambda$ represents the "arrival rate" of the customers. We again consider the coin toss experiment such that $n$ is large and the probability of success $p$ is very small, so that $\lambda := np$ is constant. In this case, we can approximate the distribution $f$ of the variable $B(n, p)$ as follows:

$$f(k) = \frac{\lambda^k}{k!} e^{-\lambda}, \quad \text{for } k = 0, 1, 2, \ldots, \tag{6.4.19}$$

which is the probability of $k$ successes in the limiting case as $n \to \infty$, subject to $\lambda := np$ is of *moderate* size. The function $f$ is the probability mass function of the Poisson variable $P(\lambda)$, which is also called the *Poisson distribution* with the parameter $\lambda$. It is named after the French mathematician *Siméon D. Poisson* (1781–1840). The proof of the formula (6.4.19) is left for the reader as a simple problem (Exercise 6.5($a$)). The next example is a simple illustration.

**Example 6.29** Suppose 2% of blades manufactured by a Company are known to be defective so that, in order to find the probability that at most 5 defective blades will be found in a box of 200 blades, we have

$$\lambda = np = 200 \times \frac{2}{100} = 4.$$

We thus obtain

$$\begin{aligned}\mathbb{P}(X \le 5) &= 1 + \frac{4}{1!} e^{-4} + \frac{4^2}{2!} e^{-4} + \frac{4^3}{3!} e^{-4} + \frac{4^5}{5!} e^{-4} \\ &= 0.7851,\end{aligned}$$

which is a little on the higher side.

The types of random variables that follow a Poisson distribution play a crucial role when problems involve counting of the number of events occurring in a specified time interval or situation. Some simple illustrations are given below:

(i) The number of births in a hospital in one year;
(ii) The number of customers entering a post office each day;
(iii) The number of calls received in a telephone exchange during a period of time;
(iv) The number of misprints on a group of pages in a book;
(v) The number of cars passing through an intersection in a city;
(vi) The number of defective bolts per box of the sample.

The Poisson distribution has some nice mathematical properties. As shown latter, the way how the distributions of some processes are approximated by a Poisson distribution is quite a useful idea.

## *Independence of Random Variables*

The **joint distribution** $f_{X,Y}$ of two random variables $X$ and $Y$ defined on the same discrete probability space $(\Omega, \mathbb{P})$ is given by

$$\begin{aligned} f_{X,Y}(x, y) &= \mathbb{P}(X = a \text{ and } Y = b) \\ &= \mathbb{P}(\{w \in \Omega \mid X(w) = a \text{ and } Y(w) = b\}), \end{aligned} \tag{6.4.20}$$

for any $a, b \in \mathbb{R}$. We say $X$ and $Y$ are **independent** if

$$\mathbb{P}(X = a \text{ and } Y = b) = \mathbb{P}(X = a)\mathbb{P}(Y = b), \quad \text{for all } a, b \in \mathbb{R}. \tag{6.4.21}$$

For example, if $X_1$ and $X_2$ respectively represent the values obtained on rolling two dice, then they are independent. For, we have

$$\mathbb{P}(X_1 = a \text{ and } X_2 = b) = \frac{1}{36} = \frac{1}{6} \cdot \frac{1}{6} = \mathbb{P}(X_1 = a)\mathbb{P}(X_2 = b).$$

On the other hand, in the case of the flip of a biased coin, if $X$ counts the number of heads that occurred and $Y$ counts the number of tails that occurred, then we have $\mathbb{P}(X = 1 \text{ and } Y = 1) = 0$, but

$$\mathbb{P}(X = 1)\mathbb{P}(Y = 1) = p(1 - p).$$

Therefore, $X$ and $Y$ are not independent. Notice that, however, if the same coin is flipped $N$ times, where $N$ follows the Poisson distribution with parameter $\lambda$, then the variables $X$ and $Y$ are independent ([12], Sect. 3.2). On the other hand, two continuous random variables $X$ and $Y$ are said to be *independent* if we have

$$\mathbb{P}(X \leq a \text{ and } Y \leq b) = \mathbb{P}(X \leq a)\,\mathbb{P}(Y \leq b), \quad \text{for all } a, b \in \mathbb{R}, \tag{6.4.22}$$

where the *joint distribution* $F_{X,Y}$ of the two is given by

$$\begin{aligned} F_{X,Y}(a,b) &= \mathbb{P}(X \le a \text{ and } Y \le b) \\ &= \mathbb{P}(\{w \in \Omega \mid (X(w) \le a) \text{ and } (Y(w) \le b)\}), \end{aligned} \tag{6.4.23}$$

for all $a, b \in \mathbb{R}$. Further, we say $X$ and $Y$ are **jointly continuous** with **joint density function** $f_{X.Y}$ (a non-negative integrable function) if we have

$$F_{X,Y}(x,y) := \int_{-\infty}^{x} \int_{-\infty}^{y} f_{X.Y}(u,v)\, du\, dv, \qquad \text{for all} \quad x, y \in \mathbb{R}. \tag{6.4.24}$$

The proof of the next theorem is left for the reader as a simple exercise.

**Theorem 6.2** *Let* $(\Omega, \mathbb{P})$ *be a discrete probability space. Suppose the* marginal mass functions *and a* joint mass function *of two random variables X and Y on* $\Omega$ *are respectively given by* $f_X$, $f_Y$, *and* $f_{X,Y}$. *Then X and Y are independent if and only if*

$$f_{X,Y}(x,y) := f_X(x) f_Y(y), \qquad \textit{for all} \quad x, y \in \mathbb{R}. \tag{6.4.25}$$

In the discrete case, the **marginal mass functions** $f_X$ and $f_Y$ are obtained from $f_{X,Y}$ using the relation

$$f_X(x) = \sum_y f_{X,Y}(x,y) \quad \text{and} \quad f_Y(y) = \sum_x f_{X,Y}(x,y). \tag{6.4.26}$$

On the other hand, the **marginal density functions** $f_X$ and $f_Y$ are obtained from $f_{X,Y}$ using the relation

$$f_X(x) = \int_{-\infty}^{\infty} f_{X,Y}(x,y)\, dy \quad \text{and} \quad f_Y(y) = \int_{-\infty}^{\infty} f_{X,Y}(x,y)\, dx. \tag{6.4.27}$$

The two correspond to the *marginal distribution functions* $F_X$ of $X$ and $F_Y$ of $Y$ in the following ways:

$$\begin{aligned} F_X(x) &= \mathbb{P}(X \le x) = F_{X,Y}(x, \infty); \\ F_Y(y) &= \mathbb{P}(Y \le y) = F_{X,Y}(-\infty, y) \end{aligned} \tag{6.4.28}$$

For example, consider two random variables $X$ and $Y$ with *joint density function* $f_{X.Y}$ given by

$$f_{X,Y}(x,y) = \frac{1}{y} e^{-y-\frac{x}{y}}, \qquad \text{for} \quad 0 < x\, y < \infty$$

then the *marginal density function* $f_Y$ of $Y$ is given by

$$f_Y(y) = \int_{-\infty}^{\infty} f_{X,Y}(x, y)\, dx$$
$$= \int_0^{\infty} \frac{1}{y} e^{-y-\frac{x}{y}} = e^{-y}, \quad \text{for} \quad y > 0.$$

In general, it follows easily that $X$ and $Y$ are *independent* if

$$F_{X,Y}(x, y) = F_X(x)\, F_Y(y), \qquad \text{for all} \quad x, y \in \mathbb{R}, \tag{6.4.29}$$

which in the case of continuous variables is equivalent to

$$f_{X,Y}(x, y) = f_X(x)\, f_Y(y), \qquad \text{for all} \quad x, y \in \mathbb{R}, \tag{6.4.30}$$

We next discuss some of the most important concepts about random variables.

## 6.5 Expectation, Variance, and Moments

A statistical measure of data such as the *mean, median*, or a *mode*, refers to a central value that describes the data in some appropriate sense. The *mean* is the arithmetic average; the *median* is the middle value, when the data is arranged in ascending or descending order; and, a data value with the highest frequency is called a *mode*. For example, the *mean, median*, and the *mode* of the data $1, 4, 3, 5, 1$ are respectively $2.8$, $3$, and $1$. The mean of data is the most frequently used measure, simply because it is the only one to represent the data in its entirety. In the case of symmetrically distributed data such as for a *normal variate*, all the three central values are the same. In general, the position of the other two relative to the mean depends on the shape of the distribution. So, their interpretations are different in various situations. The median for a *skewed data* lies between the mean and the mode, where the mean is pulled by the *data outliers*. By **outliers** we mean extreme data entries that are notably different from the expected data values. Data results are very sensitive to outliers because the mean is affected by the skewness of the distribution. However, the median and a modal value are less affected by skewness.

### *Expectation of Random Variables*

Let $(\Omega, \mathbb{P})$ be a finite discrete probability space, and $X : \Omega \to \mathbb{R}$ be a random variable. However, it is much improved than the usual averages, as mentioned above. If $X(w_i) = x_i$, then we have

$$\frac{X(w_1) + X(w_2) + \cdots + X(w_n)}{n} = \frac{a_1 + a_2 + \cdots + a_n}{n},$$

is the usual average. However, since this value has not taken into consideration the probabilities $p_i := \mathbb{P}(X = x_i)$, it doesn't capture the true essence of the "average" of the random variable $X$. An appropriate approach is to take the average of a random variable $X$ as the '*weighted average* of the values $x_i$ that it takes in $\mathbb{R}$, with the probabilities $p_i$ as the **weights**. The **expectation** of a random variable $X$ is thus defined as the number

$$\sum_{x \in X(\Omega)} x\,\mathbb{P}(X = x) = \sum_{w \in \Omega} X(w)\,\mathbb{P}(w), \tag{6.5.1}$$

where $X(\Omega) = \{x_1, x_2, x_3, \ldots\} \subset \mathbb{R}$. It is denoted by $E(X)$, which we also call the **mean value** of the random variable $X$. It follows immediately from (6.5.1) that the *mean value* of a constant random variable $X(w) = c$ ($\forall\, w \in \Omega$) is $c$ itself, as it should be. More generally, if $f$ is the *probability mass function* of $X$, then we take the *expectation* of $X$ as

$$E\big(X\big) := \sum_{x} x\, f(x), \qquad \text{for all } x \text{ with } \ f(x) > 0. \tag{6.5.2}$$

Notice that, when $\Omega$ is countably infinite, we also need to consider the "absolute convergence" of the right side sum in (6.5.2). In this setting, we define the *median* of $X$ as the set of elements $x \in X(\Omega)$ such that

$$\mathbb{P}\big(X \le x\big) \ge \frac{1}{2} \ \text{ and } \ \mathbb{P}\big(X \ge x\big) \ge \frac{1}{2}. \tag{6.5.3}$$

Furthermore, when $\Omega$ is uncountable and $X$ is absolutely continuous with a *probability density function* $f$, the expectation of $X$ is given by integral

$$E\big(X\big) := \int_{-\infty}^{\infty} x\, f(x)\, dx. \tag{6.5.4}$$

Suppose a biased coin is flipped once, with $p$ as the probability of landing H. Consider the random variable $X$ given by $X(H) = 1$ and $X(T) = 0$. Then the expectation of $X$ is given by

$$E\big(X\big) := 1 \cdot \mathbb{P}(X = 1) + 0 \cdot \mathbb{P}(X = 0) = 1 \cdot p + 0 \cdot (1 - p) = p. \tag{6.5.5}$$

In general, let $X$ be a random variable counting the number of Hs in a sequence of $n$ flips. We know that the probability mass function $f$ of $X$ is given by

$$f\big(k\big) := \binom{n}{k} p^k q^{n-k}, \qquad \text{for } \ k = 0, 1, \ldots, n,$$

where $q = 1 - p$. We thus have

$$
\begin{aligned}
E(X) &:= \sum_{k=0}^{n} k\, f(k) = \sum_{k=0}^{n} k \binom{n}{k} p^k q^{n-k} \sum_{k=0}^{n} k \binom{n}{k} \frac{p^k}{q^k} q^n \\
\Rightarrow \quad \frac{np}{q^n} &= \sum_{k=0}^{n} k \binom{n}{k} \frac{p^k}{q^k} \\
\Rightarrow \quad E(X) &= np. \qquad (6.5.6)
\end{aligned}
$$

Notice that the second relation in the above derivation is obtained by taking $x = p/q$ in the right side identity of the following implication:

$$
n(1+x)^{n-1} = \sum_{k=0}^{n} k \binom{n}{k} x^{k-1} \quad \Rightarrow \quad nx(1+x)^{n-1} = \sum_{k=0}^{n} k \binom{n}{k} x^k
$$

The computation of the expectation may not be easy in all cases, especially when the probability mass function $f$ of a random variable $X$ is complicated. For example, we have $E(X) = \infty$ for the mass function $f$ is given by

$$
f(k) := \frac{1}{k(k+1)}, \qquad \text{for } k = 1, 2, \ldots
$$

Linearity of $E$ is a crucial property that helps in numerous cases to simplify its computation.

**Theorem 6.3** (Linearity of Expectation) *Let $(\Omega, \mathbb{P})$ be a discrete probability space, $X, Y$ be two random variables defined on $\Omega$, and $\alpha \in \mathbb{R}$. We then have*

$$
\begin{aligned}
E(X+Y) &= E(X) + E(Y); \\
E(\alpha X) &= \alpha E(X),
\end{aligned}
\qquad (6.5.7)
$$

*where $X + Y$ and $\alpha X$ are defined in natural way.*

***Proof*** For the first relation in (6.5.7), by definition, we have

$$
\begin{aligned}
E(X+Y) &= \sum_{z} z\, \mathbb{P}(X+Y=z) \\
&= \sum_{x} \sum_{y} (x+y)\, \mathbb{P}(X=x \text{ and } Y=y) \\
&= \sum_{x} \sum_{y} x\, \mathbb{P}(X=x \text{ and } Y=y) + \sum_{x} \sum_{y} y\, \mathbb{P}(X=x \text{ and } Y=y) \\
&= \sum_{x} \sum_{y} x\, \mathbb{P}(X=x \text{ and } Y=y) + \sum_{y} \sum_{x} y\, \mathbb{P}(X=x \text{ and } Y=y) \\
&= \sum_{x} x \sum_{y} \mathbb{P}(X=x \text{ and } Y=y) + \sum_{y} y \sum_{x} \mathbb{P}(X=x \text{ and } Y=y)
\end{aligned}
$$

Considering the partition of $\Omega$ given by the events $A_x = \{x \mid X = x\}$, we obtain

$$\sum_y \mathbb{P}(X = x \text{ and } Y = y) = \mathbb{P}(X = x).$$

Similarly, the partition of $\Omega$ given by the events $B_y = \{y \mid Y = y\}$, we have

$$\sum_x \mathbb{P}(X = x \text{ and } Y = y) = \mathbb{P}(Y = y).$$

Substituting these two relations above, we obtain

$$\begin{aligned} E(X+Y) &= \sum_x x\,\mathbb{P}(X = x) + \sum_y y\,\mathbb{P}(Y = y) \\ &= E(X) + E(Y). \end{aligned}$$

The second assertion is trivial for $\alpha = 0$. So, we may take $\alpha \neq 0$. Now,

$$\begin{aligned} E(\alpha X) &= \sum_x x\,\mathbb{P}(\alpha X = x) \\ &= \sum_x x\,\mathbb{P}(X = x/\alpha) \\ &= \alpha \sum_x (x/\alpha)\mathbb{P}(X = x/\alpha) \\ &= \alpha \sum_{x'} x'\,\mathbb{P}(X = x') \\ &= \alpha\, E(X). \end{aligned}$$

This completes the proof. □

We give a simple illustration to demonstrate how Theorem 6.3 facilitates the computation of $E$. For, let $n$ random variables $X_1, \ldots, X_n$ be defined by

$$X_i(w) = \begin{cases} 1, & \text{if the } i\text{th flip shows a Head} \\ 0, & \text{otherwise} \end{cases} \tag{6.5.8}$$

Now, since the random variable $X := X_1 + \cdots + X_n$ counts the number of Heads appearing in $n$ trials of the coin flip experiment, the linearity of $E$ implies that

$$E(X) = E(X_1) + \cdots + E(X_n),$$

and so we obtain $E(X) = np$, because $E(X_i) = p$, for all $i$, by (6.5.5). In the next definition, we generalize the notion of *indicator variable* as in (6.5.8).

**Definition 6.13** (*Indicator Variable*) Let $(\Omega, \mathbb{P})$ be a discrete probability space. For any event $A \subseteq \Omega$, the **indicate variable** (or **indicator function**) of $A$ is the random variable $I_A$ given by

$$I_A(w) = \begin{cases} 1, & \text{if } w \in A \\ 0, & \text{if } w \notin A \end{cases} \tag{6.5.9}$$

Notice that $E(I_A) = \mathbb{P}(A)$. For, we have

$$E(I_A) = \sum_{w \in \Omega} I_A(w)\mathbb{P}(w) = \sum_{w \in A} \mathbb{P}(w) = \mathbb{P}(A).$$

We can use this simple fact, together with the linearity of $E$, to compute the expectation of a random variable by expressing it as a sum of *indicator functions*. For example, let $\Omega = S_n$, the set of *permutations* of the set $[n] = \{1, 2, \dots, n\}$, for $n \geq 1$. It is a probability space with the *uniform probability measure* $\mathbb{P}$ given by

$$\mathbb{P}(w) := \frac{1}{n!}, \qquad \text{for every} \quad w \in \Omega. \tag{6.5.10}$$

We call the points in $\Omega$ as **random permutations**. Recall that $k \in [n]$ is a *fixed point* of a permutation $w \in \Omega$ if we have $w(k) = k$. Let $X$ be a random variable such that $X(w)$ counts the number of fixed points of the permutation $w \in \Omega = S_n$. To compute $E(X)$, we introduce random variables $X_k$ given by

$$X_k(w) = \begin{cases} 1, & \text{if } w(k) = k \\ 0, & \text{if } w(k) \neq k \end{cases}, \quad \text{for} \quad k = 1, 2, \dots, n \tag{6.5.11}$$

We thus have $X = X_1 + \cdots + X_n$ so that by the linearity of $E$ we obtain

$$E(X) = E(X_1) + \cdots + E(X_n) = n \cdot \frac{1}{n} = 1,$$

because $E(X_k) = \mathbb{P}(X_k = 1)$, and there are $(n-1)!$ permutations that leave $k$ fixed so that $\mathbb{P}(X = 1) = 1/n$. Said differently, on average, a *random permutation* has one fixed point.

**Theorem 6.4** *Let $(\Omega, \mathbb{P})$ be a discrete probability space, and $X : \Omega \to \mathbb{Z}^+$ be a random variable. Then the expectation $E(X)$ (if it exists) is given by*

$$E(X) = \sum_{i=1}^{\infty} \mathbb{P}(X \geq i). \tag{6.5.12}$$

***Proof*** By definition, for any integer $n \geq 1$, we have

$$\begin{aligned}\sum_{j=1}^{n} j\,\mathbb{P}(X=j) &= \sum_{j=1}^{n}\sum_{i=1}^{j} \mathbb{P}(X=j)\\ &= \sum_{i=1}^{n}\sum_{j=i}^{n} \mathbb{P}(X=j)\\ &= \sum_{i=1}^{n} \mathbb{P}(n \geq X \geq i)\end{aligned}$$

So, letting $n \to \infty$, we get

$$\begin{aligned}\sum_{i=1}^{\infty} \mathbb{P}(X=j) &= \sum_{i=1}^{\infty}\sum_{j=i}^{\infty} \mathbb{P}(X=j)\\ &= \sum_{j=1}^{\infty}\sum_{i=1}^{j} \mathbb{P}(X=j)\\ &= \sum_{j=1}^{\infty} j\,\mathbb{P}(X=j) = E(X)\end{aligned}$$

This completes the proof. □

Intuitively, the above theorem says that the expectation $E(X)$ of $X$ is precisely the area above the graph of the distribution function $F(k) = \mathbb{P}(X \leq k)$ and below the line $F = 1$. For a simple illustration, let a random variable $X$ stands for the first occurrence of a Head (success) in a finite sequence of flips of a biased coin. In this case, the mass function of $X$ is given by

$$f(n) = (1-p)^{n-1}p, \qquad \text{for} \quad n \geq 1,$$

which is the same as given in (6.4.18). Now, we have

$$\begin{aligned}\mathbb{P}(X \geq i) &= \sum_{j=i}^{\infty} (1-p)^{j-1}p\\ &= p(1-p)^{i-1}\sum_{j=0}^{\infty}(1-p)^{j}\\ &= p(1-p)^{i-1}\cdot\frac{1}{1-(1-p)}\\ &= (1-p)^{i-1}\end{aligned}$$

It thus follows from Theorem 6.4 that the expectation $E(X)$ is given by

$$E(X) = \sum_{i=1}^{\infty} \mathbb{P}(X \geq i) = \sum_{i=1}^{\infty} (1-p)^{i-1}$$
$$= \frac{1}{1-(1-p)} = \frac{1}{p}$$

Therefore, on average, it takes $1/p$ flips for the first occurrence of a Head.

**Definition 6.14** Let $(\Omega, \mathbb{P})$ be a discrete probability space, and $X$ be a random variable on $\Omega$. For any function $g : \mathbb{R} \to \mathbb{R}$, the composite function $g \circ X : \Omega \to \mathbb{R}$ is called the **random variable given by the function** $g$. It is denoted by $g(X)$.

In general, computation of the probability mass function of the random variable $g(X)$ in terms of a mass function $f$ of $X$ is a difficult problem. However, it is easy to compute the expectation of $g(X)$.

**Theorem 6.5** *Let $(\Omega, \mathbb{P})$ be a discrete probability space, and $X$ be a random variable on $\Omega$. For any function $g : \mathbb{R} \to \mathbb{R}$, the expectation $E(g(X))$ of the random variable $g(X)$ (if it exists) is given by*

$$E\big(g(X)\big) = \sum_{x} g(x) f(x), \tag{6.5.13}$$

*where $f$ is a probability mass function of $X$.*

***Proof*** By definition, we have

$$\begin{aligned} E\big(g(X)\big) &= \sum_{y} y \cdot \mathbb{P}\big(g \circ X = y\big) \\ &= \sum_{y} y \cdot \mathbb{P}\big(\{w \in \Omega \mid g(x) = y,\ X(w) = x\}\big) \\ &= \sum_{y} \sum_{x, y = g(x)} y \cdot \mathbb{P}\big(\{w \in \Omega \mid X(w) = x\}\big) \\ &= \sum_{y} \sum_{x, y = g(x)} g(x) \cdot \mathbb{P}\big(X = x\big) \\ &= \sum_{x} g(x) \cdot \mathbb{P}\big(X = x\big) \\ &= \sum_{x} g(x)\, f(x) \end{aligned}$$

This completes the proof. □

Continuing with our previous illustration involving the random variable $X$ with the mass function given by (6.4.18), it follows from Theorem 6.5 that

$$\begin{aligned}
E(X^2) &= \sum_{i=1}^{\infty} i^2 (1-p)^{i-1} p \\
&= \sum_{i=1}^{\infty} (i-1+1)^2 (1-p)^{i-1} p \\
&= \sum_{i=1}^{\infty} (i-1)^2 (1-p)^{i-1} p + \sum_{i=1}^{\infty} 2(i-1)(1-p)^{i-1} p + \sum_{i=1}^{\infty} (1-p)^{i-1} p \\
&= \sum_{i=1}^{\infty} j^2 (1-p)^j p + 2\sum_{i=1}^{\infty} j(1-p)^j p + 1 \qquad \text{(letting } j = i-1\text{)} \\
&= (1-p)E(X^2) + 2(1-p)E(X) + 1,
\end{aligned}$$

which implies $E(X^2) = (2-p)/p^2$, because we know $E(X) = 1/p$.

**Example 6.30** (*Randomized Quicksort Algorithm*) In this illustration, let us continue our discussion with what we said earlier about the **randomized quicksort algorithm** (see Remark 6.4). More specifically, we shall compute the expectation of the number $X$ of comparisons needed while running the *randomized quicksort algorithm*, as described earlier in Example 6.25. We may recall that the input to the algorithm is a finite sequence $S = (a_1, \dots, a_n)$ of distinct real numbers, and the sequence $(b_1, \dots, b_n)$ is the list of $a_i$'s sorted in increasing order. To compute the expectation $E(X)$, we decompose the variable $X$ as a sum of indicator variables $X_{i,j}$ such that $X_{i,j} = 1$ if and only if $b_i$ and $b_j$ are ever compared; and, we may take $X_{i,j} = 0$, otherwise. So, we have

$$X = \sum_{j=2}^{n} \sum_{i=1}^{j-1} X_{i,j} \quad \Rightarrow \quad E(X) = \sum_{j=2}^{n} \sum_{i=1}^{j-1} E(X_{i,j})$$

Notice that we have

$$E(X_{i,j}) = \mathbb{P}(b_i \text{ and } b_j \text{ are ever compared}),$$

and $b_i$ and $b_j$ are ever compared if and only if either $b_i$ or $b_j$ is chosen as the pivot when $B_{i,j} := \{b_i, b_{i+1}, \dots, b_j\}$ is a subset of the set of elements of the (left or right) sublist considered for the choice of a pivot. This is because if the next pivot $b$ is larger than $b_j$, then all the elements in $B_{i,j}$ are placed in the list to the left of $b$, and if $b$ is smaller than $b_i$, then all the elements in $B_{i,j}$ are placed in the list to the right of $b$. So, if $b_i$ and $b_j$ are ever compared, some pivot $b$ must belong to the set $B_{i,j}$, and every $b_k \neq b$ in the list will be compared with $b$. But, if the pivot $b$ is distinct from $b_i$ and $b_j$, then $b_i$ is placed in the left sublist and $b_j$ in the right sublist, so $b_i$ and $b_j$ will never be compared. It thus remains to compute the probability that the next pivot chosen in the sublist $B_{i,j}$ is $b_i$ (or that the next pivot chosen is $b_j$, but the two probabilities are equal). Since the pivot is one of the values in $B_{i,j}$ and since each of these is equally likely to be chosen (by hypothesis), we have

$$\mathbb{P}(b_i \text{ is chosen as the next piot in } B_{i,j}) = \frac{1}{j-i+1},$$

Since $b_i$ and $b_j$ are ever compared if and only if either $b_i$ is chosen as a pivot or $b_j$ is chosen as a pivot, and since these two events are mutually exclusive, we have

$$E(X_{i,j}) = \mathbb{P}(b_i \text{ and } b_j \text{ are ever compared}) = \frac{2}{j-i+1}.$$

It thus follows that

$$\begin{aligned} E(X) &= \sum_{j=2}^{n}\sum_{i=1}^{j-1} E(X_{i,j}) \\ &= 2\sum_{j=2}^{n}\sum_{k=2}^{j} \frac{1}{k} \qquad (\text{take } k = j-i+1) \\ &= 2\sum_{j=2}^{n}\sum_{j=k}^{n} \frac{1}{k} = 2\sum_{k=2}^{n} \frac{n-k+1}{k} \\ &= 2(n+1)\sum_{k=1}^{n} \frac{1}{k} - 4n = 2(n+1)H_n - 4n, \end{aligned}$$

where $H_n$ is the $n$th *harmonic number*. Using the inequality

$$\ln n + \frac{1}{n} \le H_n \le \ln n + 1,$$

we have $H_n = \ln n + \Theta(1)$. Therefore,

$$E(X) = 2n \ln n + \Theta(n).$$

Hence, the expected number of comparisons done by the randomized quicksort is $2n \ln n + \Theta(n)$.

Next, let $(\Omega, \mathbb{P})$ be a discrete probability space, and $X$, $Y$ be two random variables on $\Omega$. For any function $g : \mathbb{R} \times \mathbb{R} \to \mathbb{R}$, we may extend the concept introduced earlier in Definition 6.14 to define the random variable $g(X, Y)$ on $\Omega$ as the composite function $g \circ (X \times Y) : \Omega \times \Omega \to \mathbb{R}$. In this case, the expectation $E(g(X, Y))$ of the random variable $g(X, Y)$ (if it exists) is given by

$$E\big(g(X,Y)\big) = \sum_{x,y} g\big(x, y\big)\, f_{X,Y}\big(x, y\big), \tag{6.5.14}$$

where $f_{X,Y}$ is a joint mass function of $X$ and $Y$.

### *Variance of Random Variable*

Let $(\Omega, \mathbb{P})$ be a discrete probability space, and $X$ be a random variable on $\Omega$, with probability mass function given by $f$. In what follows, we may write $E(X) = \mu$. The **variance** of $X$, denoted by $\mathrm{Var}(X)$, is used to study the *spread* of the distribution of $X$. It is defined as

$$\mathrm{Var}\big(X\big) = E\big((X-\mu)^2\big), \tag{6.5.15}$$

if the right side exists. The **standard deviation** of $X$, denoted by $\sigma_X$ (or simply $\sigma$), is the *positive* square root of the *variance* $\mathrm{Var}(X)$. That is, we have

$$\sigma := +\sqrt{\mathrm{Var}\big(X\big)} = +\sqrt{E\big((X-\mu)^2\big)}. \tag{6.5.16}$$

Using the linearity of $E$ and the fact that $E(c) = c$, we obtain

$$\begin{aligned}\mathrm{Var}\big(X\big) &= E\big((X-\mu)^2\big)\\ &= E\big(X^2 - 2X\mu + \mu^2\big)\\ &= E\big(X^2\big) - 2E\big(X\big)\mu + \mu^2\\ &= E\big(X^2\big) - \mu^2\end{aligned} \tag{6.5.17}$$

In particular, it follow that $\mathrm{Var}(X) \leq E(X^2)$. For example, if $X$ represents the value on a dice, then $\mu = E(X) = 7/2$ and

$$E\big(X^2\big) = \frac{1}{6}\Big(1^2 + 2^2 + 3^2 + 4^2 + 5^2 + 6^2\Big) = \frac{91}{6},$$

implies that the variance of $X$ is given by

$$\mathrm{Var}\big(X\big) = E\big(X^2\big) - \mu^2 = \frac{91}{6} - \Big(\frac{7}{2}\Big)^2 = \frac{35}{12}.$$

As shown earlier, for a random variable $X$ with *binomial distribution*, we have

$$E(X^2) = np((n-1)p+1).$$

We also know that $\mu = E(X) = np$. Therefore, we obtain

$$\begin{aligned}\text{Var}(X) &= E(X^2) - \mu^2 \\ &= np((n-1)p+1) - (np)^2 \\ &= np(1-p).\end{aligned}$$

That is, $\text{Var}(X) = npq$, where $q = 1 - p$. Also, if $X$ is a random variable with *geometric distribution* given by (6.4.18), then

$$\mu = E(X) = \frac{1}{p} \quad \text{and} \quad E(X^2) = \frac{2-p}{p^2} \tag{6.5.18}$$

implies that we have

$$\begin{aligned}\text{Var}(X) &= E(X^2) - \mu^2 \\ &= \frac{2-p}{p^2} - \frac{1}{p^2} = \frac{1-p}{p^2}.\end{aligned} \tag{6.5.19}$$

Therefore, for a variable $X$ with *geometric distribution*, $\text{Var}(X) = (1-p)/p^2$. Further, if $X$ is a random variable with *Poisson distribution* given by (6.4.19), then

$$\mu = E(X) = \lambda \quad \text{and} \quad E(X^2) = \lambda(\lambda+1)$$

implies that, in this case, we have

$$\text{Var}(X) = E(X^2) - \mu^2 = \lambda(\lambda+1) - \lambda^2 = \lambda.$$

Therefore, for a variable $X$ with *Poisson distribution*, $\text{Var}(X) = \lambda$. The quantity $E(X^2)$ in general is called the *second moment* of $X$. More generally, the particular case of Theorem 6.5 when $g(X) = X^k$ holds special significance.

**Definition 6.15** Let $X$ be a random variable defined on a discrete probability space $(\Omega, \mathbb{P})$. For $k \geq 1$, the **kth moment** $\nu_k$ (about the origin) and the **kth central moment** $\mu_k$ (about the mean) of $X$ are respectively given by

$$\nu_k := E(X^k) \quad \text{and} \quad \mu_k = E((X-\mu_1)^k). \tag{6.5.20}$$

In most cases, our main concern is $\mu = \mu_1$ and $\mu_2$. Notice that $\sigma = +\sqrt{\mu_2}$. Also, the two quantities $\mu_3$ and $\mu_4$ help study the *skewness* $(\mu_3/\mu^3)$ and the *kurtosis* $((\mu_4/\mu^4) - 3)$ of the distribution of the random variable $X$.

**Theorem 6.6** *Let* $(\Omega, \mathbb{P})$ *be a discrete probability space, and* $X$, $Y$ *be two* independent *random variables on* $\Omega$. *Then we have*

$$\mathrm{Var}\left(X+Y\right) = \mathrm{Var}\left(X\right) + \mathrm{Var}\left(Y\right). \tag{6.5.21}$$

***Proof*** It follows easily that $E(XY) = E(X)E(Y)$. So, we have

$$\begin{aligned} E\big((X+Y)^2\big) &= E\big(X^2 + 2XY + Y^2\big) \\ &= E\big(X^2\big) + 2E\big(XY\big) + E\big(Y^2\big) \\ &= E\big(X^2\big) + 2E\big(X\big)E\big(Y\big) + E\big(Y^2\big) \end{aligned}$$

It thus follows that

$$\begin{aligned} \mathrm{Var}\left(X+Y\right) &= E\big((X+Y)^2\big) - \big(E(X+Y)\big)^2 \\ &= E\big(X^2\big) + 2E\big(X\big)E\big(Y\big) + E\big(Y^2\big) - \big[\big(E(X)\big)^2 + 2E(X)E(Y) + \big(E(Y)\big)^2\big] \\ &= E\big(X^2\big) - \big(E(X)\big)^2 + E\big(Y^2\big) - \big(E(Y)\big)^2 \\ &= \mathrm{Var}\left(X\right) + \mathrm{Var}\left(Y\right), \end{aligned}$$

as claimed. □

Recall that the two random variables $X$ and $Y$ representing the values on two dice are independent so that we have $E(XY) = E(X)E(Y) = 49/2$, because all probabilities are equal to $1/6$. We also have that $E(X+Y) = 7$. It thus follows that, in this case, we have

$$\mathrm{Var}\left(X+Y\right) = \mathrm{Var}\left(X\right) + \mathrm{Var}\left(Y\right) = \frac{35}{12} \cdot \frac{35}{12} = \frac{35}{6}.$$

The next theorem summarizes some useful results.

**Theorem 6.7** *Let* $(\Omega, \mathbb{P})$ *be a discrete probability space, and* $X$ *and* $Y$ *be two random variables on* $\Omega$. *Then we have*

*(1) If* $X \geq 0$ *over* $\Omega$, *then* $E(X) \geq 0$*;*
*(2) If* $X$ *has a constant value* $c$, *then* $E(X) = c$*;*
*(3) If* $X \leq Y$ *over* $\Omega$, *then* $E(X) \leq E(Y)$. *That is,* $E$ *is a* monotonic function*;*
*(4) For any scalar* $\alpha \in \mathbb{R}$, *we have*

$$\mathrm{Var}\left(\alpha X\right) = \alpha^2 \ \mathrm{Var}\left(X\right). \tag{6.5.22}$$

*That is,* Var() *is a* quadratic function.

***Proof*** The assertions (1)–(3) are straightforward. For a proof of (4), we have

$$\begin{aligned}\text{Var}\big(\alpha X\big) &= E\big((\alpha X - E(\alpha X))^2\big)\\ &= E\big(\alpha^2(X - E(X))^2\big)\\ &= \alpha^2 E\big((X - E(X))^2\big)\\ &= \alpha^2 \text{ Var}\big(X\big).\end{aligned}$$

This completes the proof. □

The next illustration is an interesting application of expectation values of geometrically distributed random variables. We may recall that a random variable $X$ is said to follow a *geometric distribution* with parameter $p$ if its probability mass function is given by (6.4.18). In this case, we say $X$ is a **geometric random variable**, and write $X \sim \text{Geom}(p)$.

**Example 6.31** (*Coupon Collector's Problem*) The reference to our discussion here is a contest involving collecting $m$ different types of coupons, each packed in some household product. It is assumed that one can any number of products and, each time a coupon is obtained, it is equally likely to be one among these $m$ coupons. Let the random variable $X$ counts the number of products one needs to buy to ensure that at least one of each type of coupon is obtained. The *coupon collector's problem* asks: *What is the expected value of $X$?* As our analysis of the problem shows, it is increasingly difficult to estimate how many products one would need to buy on average to collect all $m$ different coupons. The basic idea is to use the *divide and conquer* strategy, i.e., we consider the sum

$$X := X_0 + X_1 + \cdots + X_{m-1},$$

where the random variable $X_i$ counts the number of additional coupons needed, after $i$ distinct types have been collected, until another new type is obtained, for $i = 0, 1, \ldots, m-1$. Notice that the probability of finding a new coupon decays as we collect more and more coupons. Each variable $X_i$ is given by a *geometric random variable* $\text{Geom}(p_i)$, where the probability of success $p_i$ for each trial is given by $p_i = (m-i)/m$. By (6.5.18), we already know that

$$E\big(X_i\big) = \frac{1}{p_i} = \frac{m}{m-i}.$$

It thus follows that

$$E\big(X\big) = \sum_{i=0}^{m-1} E\big(X_i\big) = \sum_{i=0}^{m-1} \frac{m}{m-i} = m \sum_{i=1}^{m} \frac{1}{i} = m\, H_m,$$

where $H_m$ is the $m$th *harmonic number*. Now, by (4.6.8), we have

$$H_m \approx \ln m + \gamma + \frac{1}{2m} + o\left(\frac{1}{m^2}\right),$$

and so we obtain

$$E(X) = m\, H_m \approx m \ln m + m\,\gamma + \frac{1}{2} + o\left(\frac{1}{m}\right),$$

where $\gamma \approx 0.577215665$ is the *Euler-Mascheroni constant*. In particular, when $m = 50$, we have $\ln 50 = 3.912$ so that $m \ln m \approx 196$. Also, when $m = 100$ so that $\in 100 = 4.6052$ implies $m \ln m \approx 461$. This explains why a product manufacturing company makes money when the product is expensive. Since $X_i$'s are independent, we can also compute the variance of $X$, with little extra effort. For, by (6.5.19), we have

$$\text{Var}\left(X_i\right) = \frac{1 - p_i}{p_i^2} = \left(1 - \frac{m - i}{m}\right) / \frac{m^2}{(m - i)^2} = \frac{m\, i}{(m - i)^2},$$

it thus follows that

$$\text{Var}\left(X\right) = \sum_{i=0}^{m-1} \text{Var}\left(X_i\right) = m \sum_{i=0}^{m-1} \frac{i}{(m - i)^2}.$$

To compute the right side sum, we may proceed as follows:

$$\begin{aligned} \sum_{i=0}^{m-1} \frac{i}{(m - i)^2} &= \sum_{i=0}^{m-1} \frac{m}{(m - i)^2} - \sum_{i=0}^{m-1} \frac{m - i}{(m - i)^2} \\ &= \sum_{i=0}^{m-1} \frac{m}{(m - i)^2} - \sum_{i=0}^{m-1} \frac{1}{m - i} \\ &= m \sum_{j=1}^{m} \frac{1}{j^2} - \sum_{j=1}^{m} \frac{1}{j} \end{aligned}$$

Now, using Euler's formula

$$\lim_{m \to \infty} \sum_{j=1}^{m} \frac{1}{j^2} = \sum_{j=1}^{\infty} \frac{1}{j^2} = \frac{\pi^2}{6},$$

we conclude that

$$\mathrm{Var}\,(X) = \frac{m^2\pi^2}{6} + \Theta(m \ln m).$$

This relation helps to understand how the number of products one needs to buy varies from trial to trial. If $\mathrm{Var}(X)$ is small, then it can expected that one needs to buy the same number of products in each trial. On the other hand, if $\mathrm{Var}(X)$ is large, then the number of products that one needs to buy could vary widely from trial to trial. In general, since $\mathrm{Var}(X)$ is proportional to $m^2$, it grows faster for larger values of $m$. So, as $m$ gets larger, the number of products that one needs to buy to collect all required $m$ coupons becomes more and more unpredictable. Nevertheless, this seemingly simple problem finds applications in many fields, including computer science, economics, and finance.

**Example 6.32** (*Random Permutations*) Recall the concept of *random permutations* $w \in S_n$, which is a discrete probability space with respect to the probability defined by (6.5.10). The random variable of interest is $X$ such that $X(w)$ counts the number of fixed points of the permutation $w$. We have shown that $E(X) = 1$. The computation of the standard deviation is left to the reader. The real difficulty, in this case, is that the random variables $X_k$ given by

$$X(w) = \begin{cases} 1, & \text{if } w(k) = k \\ 0, & \text{otherwise} \end{cases}$$

are not independent. It follows easily that $\sigma = 1$ ([14], Chap. 8).

In general, when two random variables $X$ and $Y$ defined on the same discrete probability space $(\Omega, \mathbb{P})$ are *not independent*, it can be shown that

$$\mathrm{Var}\,(X + Y) = \mathrm{Var}\,(X) + \mathrm{Var}\,(Y) + 2E\big((X - E(X))(Y - E(Y))\big). \quad (6.5.23)$$

The last term on the right side is well-known in probability theory.

**Definition 6.16** (*Covariance*) Given two random variables $X$ and $Y$ on the same discrete probability space $(\Omega, \mathbb{P})$, their **covariance** $\mathrm{Cov}(X, Y)$ is defined by

$$\mathrm{Cov}\,(X, Y) = E\big((X - E(X))(Y - E(Y))\big) = E(XY) - E(X)E(Y). \quad (6.5.24)$$

In particular, $X$ and $Y$ are said to be **uncorrelated** if $\mathrm{Cov}(X, Y) = 0$.

Notice that $\mathrm{Var}(X) = \mathrm{Cov}(X, X)$, and $X$ and $Y$ are *uncorrelated* if the two variables are independent. Also, (6.5.23) implies that

$$\mathrm{Var}\,(X + Y) = \mathrm{Var}\,(X) + \mathrm{Var}\,(Y), \quad (6.5.25)$$

if $X$ and $Y$ are uncorrelated. However, $\mathrm{Cov}(X, Y) = 0$ need not imply that $X$ and $Y$ are independent. For example, we may consider the random variables $X$ and $Y$ defined on $\Omega = \{-1, 0, 1\}$ by

$$\mathbb{P}(X = -1) = \mathbb{P}(X = 0) = \mathbb{P}(X = 1) = \frac{1}{3};$$

$$Y = \begin{cases} 0, & \text{if } X \neq 0 \\ 1, & \text{if } X = 0 \end{cases}$$

Now, since $XY = 0$, we have $E(XY) = 0$. Also, since $E(X) = 0$, we have

$$\mathrm{Cov}\left(X, Y\right) = E\left(XY\right) - E\left(X\right)E\left(Y\right) = 0.$$

Clearly, $X$ and $Y$ are not independent. In general, a better measure of *independence* of two random variables $X$ and $Y$ defined on the same discrete probability space $(\Omega, \mathbb{P})$ is given by the **correlation coefficient** between $X$ and $Y$, denoted by $\rho(X, Y)$. It is defined by

$$\rho\left(X, Y\right) = \frac{\mathrm{Cov}\left(X, Y\right)}{\sqrt{\mathrm{Var}\left(X\right)}\sqrt{\mathrm{Var}\left(Y\right)}}, \tag{6.5.26}$$

provided $\mathrm{Var}(X) \neq 0$ and $\mathrm{Var}(Y) \neq 0$. The next theorem proves that $\left|\rho(X, Y)\right| \leq 1$.

**Theorem 6.8** (Cauchy-Schwarz Inequality) *Let $X$ and $Y$ be two random variables defined on the same discrete probability space $(\Omega, \mathbb{P})$. Then we have*

$$\left|E\left(XY\right)\right| \leq \sqrt{E(X^2)}\sqrt{E(Y^2)}, \tag{6.5.27}$$

*where the equality holds if and only if $E\left((\alpha X + \beta Y)^2\right) = 0$, for some $\alpha, \beta \in \mathbb{R}$ (not both zero).*

***Proof*** We consider the function $f : \mathbb{R} \to \mathbb{R}$ given by

$$f\left(\lambda\right) = E\left((X + \lambda\, Y)^2\right), \qquad \text{for } \ \lambda \in \mathbb{R}. \tag{6.5.28}$$

Notice that $f \geq 0$, because $(X + \lambda\, Y)^2 \geq 0$. Since

$$\begin{aligned} E\left((X + \lambda\, Y)^2\right) &= E\left(X^2 + 2\lambda\, XY + \lambda^2\, Y^2\right) \\ &= E\left(X^2\right) + 2\lambda\, E\left(XY\right) + \lambda^2\, E\left(Y^2\right), \end{aligned}$$

if $E(Y^2) = 0$, then we must also have $E(XY) = 0$; otherwise, we can choose $\lambda$ so that $E(X^2) + 2\lambda\, E(XY) < 0$. In this case, the inequality is trivial. On the other hand, if $E(Y^2) > 0$, then for $f(\lambda)$ to be non-negative the quadratic equation

$$\lambda^2 E(Y^2) + 2\lambda E(XY) + E(X^2) = 0 \tag{6.5.29}$$

must have at most one real root. Or, equivalently,

$$4\big(E(XY)\big)^2 - 4E(X^2)E(Y^2) \leq 0,$$

which in turn is equivalently to (6.5.27). Next, if the equality holds, then either $E(Y^2) = 0$ so that $E\big((\alpha X + \beta Y)^2\big) = 0$, for $\alpha = 0$ and $\beta = 1$; or, $E(Y^2) > 0$ so that the quadratic equation in (6.5.29) has a unique real root, say $\lambda_0$, which gives $E\big((X + \lambda_0 Y)^2\big) = 0$. Conversely, if $E\big((\alpha X + \beta Y)^2\big) = 0$, for some $\alpha, \beta \in \mathbb{R}$ (not both zero), then either $E(Y^2) = 0$ so that we have $E(XY) = 0$, as shown above; or, the quadratic equation (6.5.29) has a real root, which implies we must have

$$\big(E(XY)\big)^2 - E(X^2)E(Y^2) = 0.$$

In each case, it follows that the equality holds in (6.5.27). □

Applying Theorem 6.8 to the random variables $X - E(X)$ and $Y - E(Y)$, we obtain

$$\big|\rho(X, Y)\big| \leq 1, \tag{6.5.30}$$

where the equality holds if and only if $\mathbb{P}\big(\alpha X + \beta Y = \gamma\big) = 1$, for some $\alpha, \beta, \gamma \in \mathbb{R}$ such that $\alpha^2 + \beta^2 \neq 0$.

## *Moments Generating Functions*

As discussed in this part, the particular case of Theorem 6.5 when $g(X) = e^{tX}$ hold special significance. We obtain the concept of *moments generating function* (or simply MGF) of the random variable $X$, denoted by $M_X(t)$, for $t \in \mathbb{R}$.

**Definition 6.17** The **moments generating function** $M_X(t)$ of a random variable $X$ is the expectation of the random variable $Y := e^{tX}$. That is, we have

$$M_X(t) = E\big(e^{tX}\big), \qquad \text{for } t \in \mathbb{R}. \tag{6.5.31}$$

When $X$ is discrete variable given by a mass function $f$, we have

$$M_X(t) = \sum_x e^{tx} f(x); \tag{6.5.32a}$$

when $X$ is a continuous variable given by a density function $f$, we have

$$M_X(t) = \int_{-\infty}^{\infty} e^{tx} f(x)\, dx. \tag{6.5.32b}$$

The sum or integral on the right side of (6.5.32a) and (6.5.32b), respectively, may not exits for all $t \in \mathbb{R}$. However, if $M_X(t)$ converges absolutely on some subinterval $(-t, t) \subseteq \mathbb{R}\,(t > 0)$, then its $n$th derivative at $t = 0$ is given by

$$\begin{aligned} M_X^{(n)}(0) &= \frac{d^n(M_X(t))}{d\,t^n}\bigg|_{t=0} \\ &= \sum_x x^n e^{tx} f(x)\bigg|_{t=0} \\ &= \sum_x x^n f(x) = E(X^n). \end{aligned} \tag{6.5.33}$$

Therefore, the $n$th moment $\nu_n$ of $X$ is given by $E(X^n) = M_X^{(n)}(0)$. Further, inside the interval of convergence, the Taylor series of $M_X(t)$ is given by

$$\begin{aligned} M_X(t) &= \sum_{k=0}^{\infty} \frac{E(X^k)}{k!}\,t^k \\ &= 1 + \frac{E(X)}{1\,!} + \frac{E(X^2)}{2\,!} + \frac{E(X^3)}{3\,!} + \cdots \end{aligned}$$

which shows that $M_X(t)$ is the *exponential generating function* of the sequence of moments $\big(E(X^n)\big)$ [14]. It thus follows from (??) that the $n$th moment $\nu_n$ is given by

$$\nu_n = \frac{d^n(M_X(t))}{d\,t^n}\bigg|_{t=0}, \tag{6.5.34}$$

and hence the name. Also, since for any real number $a$, we have

$$\begin{aligned} M_{X-a}(t) &= \sum e^{t(x_k - a)} f(x_k) \\ &= e^{-a\,t} \sum e^{t x_k} f(x_k) \\ &= e^{-a\,t}\, M_X(t), \end{aligned}$$

it follows that the $n$th moments (about the value $a$) is given by

$$\mu_{\mathbf{n}}{}' = e^{-a\,t} \left[\frac{d^n(M_X(t))}{d\,t^n}\right]_{t=0} \tag{6.5.35}$$

More generally, if $U$ is the random variable obtained from $X$ by *scaling* it by $b \neq 0$ and *shifting the origin* to $x = a$, i.e., $U = a + b\,X$, then we have

$$\begin{aligned} M_U(t) &= E\big(e^{tU}\big) \\ &= \sum_x e^{t(a+bx)}\, f(x) \\ &= e^{at} \sum_x e^{(bt)x}\, f(x) \\ &= e^{at}\, M_X\big(bt\big). \end{aligned}$$

Notice that, if $M_X(t)$ is defined on the interval $\big[-t, t\big] \subseteq \mathbb{R}$, then $M_U(t)$ is defined on the interval $\big[-t/b, t/b\big]$. On the other hand, if $X$ is a continuous variable with density function $f$, then $M_X(-t)$ is the Laplace transform of the function $f$. To illustrate the above discussion, we use two important discrete distributions.

**Example 6.33** Consider the binomial variable $X$ counting the number of Heads in a sequence of coin flips of length $n$. We know that the probability of getting $k$ successes in $n$ trials is given by

$$\mathbb{P}\big(X = k\big) = \binom{n}{k} p^k\, q^{n-k}, \quad \text{for} \quad k = 0, 1, \ldots, n,$$

where $p$ is the probability of success and $q = 1 - p$ the probability of *failure* in a single trial. In this case, the function $f$ is the mass function of the variable $X$. It thus follows from (??) that

$$\begin{aligned} M_X(t) &= \sum_{k=0}^{n} e^{t\,k}\, f(k) \\ &= \sum_{k=0}^{n} e^{t\,k} \binom{n}{k} p^k\, q^{n-k} \\ &= \sum_{k=0}^{n} \binom{n}{k} (p\, e^t)^k\, q^{n-k} \\ &= (p\, e^t + q)^n, \end{aligned}$$

which is the MGF of the random variable $X$. Next, it follows from (6.5.34) that

$$\begin{aligned} \nu_1 &= \frac{d(M_X(t))}{d\,t}\bigg|_{t=0} \\ &= \big[\, n\, p\, e^t (p\, e^t + q)^{n-1}\, \big]_{t=0} \\ &= np; \end{aligned}$$

$$\begin{aligned} \nu_2 &= \frac{d^2(M_X(t))}{d\,t^2}\Big|_{t=0} \\ &= np\left[\,(n-1)\,p\,e^{2t}(p\,e^t+q)^{n-2}+e^t(p\,e^t+q)^{n-1}\,\right]_{t=0} \\ &= (np)^2+npq; \\ \nu_3 &= \frac{d^2(M_X(t))}{d\,t^2}\Big|_{t=0} \\ &= np\left[n^2p^2-2p^2-3p+1\right]; \end{aligned}$$

We next consider the case when $g(X) = X^k$ in Theorem 6.5.

**Example 6.34** Consider the random variable $X$ as in Example 6.33. This time, we compute $E(X^k)$. As before, let $p$ the probability of success in a single trial, and $q = 1 - p$. By definition, we have

$$\begin{aligned} E\big(X^k\big) &= \sum_{i=0}^{n} i^k\, f(i) \\ &= \sum_{i=0}^{n} i^k \binom{n}{i} p^i\, q^{n-i} \\ &= np \sum_{i=0}^{n} i^{k-1}\binom{n-1}{i-1} p^{i-1}\, q^{n-i} \qquad \left(\text{using } i\binom{n}{i} = n\binom{n-1}{i-1}\right) \\ &= np \sum_{j=0}^{n-1} (j+1)^{k-1}\binom{n-1}{j} p^j\, q^{n-1-j} \qquad (\text{using } j := i-1) \\ &= np\, E\big((Z+1)^{k-1}\big), \end{aligned}$$

using Theorem 6.5, where $Z$ is a random variable with binomial distribution on sequences of length $n-1$, and the probability of success $p$. We have thus obtained an inductive method to compute $E\big(X^k\big)$, for $k \geq 1$. In particular, it follows that

$$E\big(X^2\big) = npE\big(Z+1\big) = np\big[(n-1)p+1\big].$$

**Example 6.35** (*Poisson Distribution*) We first obtain the MGF of the Poisson variable $X$ given by the probability mass function

$$f\big(k\big) = e^{-\lambda}\,\frac{\lambda^k}{k\,!}, \quad \text{with } \lambda > 0 \text{ and } \; k = 0, 1, 2, \; \ldots .$$

We know that the *MGF* of the Poisson variable $X = P(\lambda)$ is

$$M_X\big(t\big) = e^{\lambda\left[e^t-1\right]}, \quad \text{with } \; \lambda > 0. \tag{6.5.36}$$

Thus, its *mean* $= \nu_1 = \lambda$, and the second moment about the *mean* i.e., the *variance* $\sigma^2 = \lambda$. In fact, its third moment is about the *mean* $\mu_3 = \lambda$. Now, since the *Poisson variable* $X = P(\lambda)$ has a single parameter distribution viz., $\lambda$, it follows that its mean or *variance* or the third moment about the *mean* $\mu_3$ is sufficient to compute any probability. Moreover, we also find the *second moment* about the *mean* $\mu = \nu_1$.

## 6.6 Some Useful Inequalities and Limit Theorems

In this concluding part of the chapter, we show that the expected value of large numbers of random variables is more predictable than what it is for the individual variables. More specifically, it follows from the *laws of large numbers* that the average of independent (and identically distributed in basic version) random variables is asymptotically the expected value.[5] More generally, other than the average, for functions of dependent random variables with a fixed number of observations, we consider specific types of **concentration inequalities**. The description given in the next part presents two such basic *concentration inequalities*. Later on, we also introduce some advanced techniques related to Chernoff bounds. To illustrate the utility of these results, some applications from electrical engineering, computer science, and statistics, are also included.

### *Markov and Chebyshev Inequalities*

For a random variable $X$ and a given real number $\alpha$, we refer to the probability $\mathbb{P}(X \geq \alpha)$ or $\mathbb{P}(X \leq \alpha)$ as **tail probability**. More specifically, $\mathbb{P}(X \geq \alpha)$ is called an *upper tail* (or *right tail*); and, $\mathbb{P}(X \leq \alpha)$ is called a *lower tail* (or *left tail*). In the majority of cases, tail probabilities provide upper bounds on the probability of deviation of functions of independent random variables from their expectations. Such inequalities are critical to the analysis of various types of problems in machine learning, and also in the design of some new efficient algorithms.

We prove below two basic inequalities, which are widely used to bound the *tail probability* of a random variable. The first one is known as the *Markov's inequality*, which is named after the Russian mathematician *Andrey Markov* (1856–1922). Markov's inequality provides an upper bound on the tail probability of any random variable $X \geq 0$. In simple terms, it says the heavier the tail larger the expectation. As shown later, this simple inequality helps deduce more sophisticated tailbounds.

---

[5] The asymptotic rate at which "reduction in randomness" takes place follows from the central limit theorem and large deviations techniques, as discussed in the later part of the section.

**Theorem 6.9** (Markov's Inequality) *Let $X \geq 0$ be a random variable defined on a probability space $(\Omega, \mathbb{P})$, and $\alpha > 0$. Then we have*

$$\mathbb{P}\big(X \geq \alpha\big) \leq \frac{E(X)}{\alpha}, \tag{6.6.1}$$

*provided $E(X)$ is finite.*

***Proof*** Let $Y := \alpha\, I_A$, where $I_A$ is the *indicator variable* of the event $A := \{X \geq \alpha\}$, as defined earlier in (6.5.9). That is, we have

$$Y\big(w\big) = \begin{cases} \alpha, & \text{if } X(w) \geq \alpha \\ 0, & \text{if } X(w) < \alpha \end{cases}$$

so that $Y \leq X$. Using monotonicity and linearity of $E$, we thus obtain

$$E\big(X\big) \geq E\big(Y\big) = E\big(\alpha\, I_A\big) = \alpha\, E\big(I_A\big) = \alpha\, \mathbb{P}\big(X \geq \alpha\big)$$

which proves (6.6.1), upon dividing the last equation by $\alpha > 0$. □

The inequality (6.6.1) makes sense only when $\alpha > E(X)$. It is tight in the sense that, for each chosen $\alpha > 0$, we can find a random variable $X \geq 0$ such that (6.6.1) is indeed an equality. For, take $X$ such that it places the whole of its probability mass at either $\alpha$ or 0. We then have

$$E\big(X\big) = \alpha\, \mathbb{P}\big(X = \alpha\big) = \alpha\, \mathbb{P}\big(X \geq \alpha\big),$$

so equality in (6.6.1) holds. More generally, if $F$ is a strictly monotonically increasing non-negative function, then it can be shown that for any random variable $X \geq 0$ and $\alpha > 0$ we have

$$\mathbb{P}\big(X \geq \alpha\big) = \mathbb{P}\big(F(X) \geq F(\alpha)\big) \leq \frac{E(F(X))}{F(\alpha)}. \tag{6.6.2}$$

We next prove the second important concentration inequality, known as the *Chebyshev's inequality*. It is named after the Russian mathematician *Pafnuty Chebyshev* (1821–1894), though it was first formulated by the French statistician *Irénée-Jules Bienaymé* (1796–1878) in 1853 [15].

**Theorem 6.10** (Chebyshev's Inequality) *Let $X$ be a random variable defined on a probability space $(\Omega, \mathbb{P})$, and $\alpha > 0$. Then we have*

$$\mathbb{P}\big((X - E(X))^2 \geq \alpha\big) \leq \frac{\mathrm{Var}\,\big(X\big)}{\alpha}, \tag{6.6.3}$$

*provided both $E(X)$ and* $\mathrm{Var}(X)$ *are finite.*

***Proof*** We have

$$\begin{aligned}\text{Var}\,(X) &= \sum_{w\in\Omega} \big(X(w) - E(X)\big)^2\mathbb{P}(w)\\ &\geq \sum_{\left(X(w)-E(X)\right)^2\geq\alpha} \big(X(w) - E(X)\big)^2\mathbb{P}(w)\\ &\geq \sum_{\left(X(w)-E(X)\right)^2\geq\alpha} \alpha\,\mathbb{P}(w)\\ &= \alpha\,\mathbb{P}\big(\big(X(w) - E(X)\big)^2 \geq \alpha\big),\end{aligned}$$

which proves (6.6.3). □

Chebyshev's inequality (6.6.3) makes sense only when $\alpha > \sqrt{\text{Var}(X)}$. In simple terms, it says a random variable with a small variance is rarely far from its expected value. In this case also, the bound is tight. For, let $X$ be such that

$$\mathbb{P}\big(X = \alpha\big) = \mathbb{P}\big(X = -\alpha\big) = \frac{p}{2}, \quad \text{and} \quad \mathbb{P}\big(X = 0\big) = 1 - p.$$

We then have $E(X) = 0,\ \text{Var}(X) = p\,\alpha^2$, and

$$\mathbb{P}\big(\big|X - E(X)\big| \geq \alpha\big) = p = \frac{\text{Var}\,(X)}{\alpha^2}.$$

In this case, more tighter bounds can be obtained, provided the distribution of $X$ is known. Also, it follows from Theorem 6.10 that the interval $\big[\mu - 2\sigma, \mu + 2\sigma\big]$ contains 75% of data, and the interval $\big[\mu - 3\sigma, \mu + 3\sigma\big]$ contains 88.9% of the data. We may recall that 68-95-99.7-*Rule*[6] holds only for a normal distribution (Definition 6.18). For, we write the inequality (6.6.3) as

$$P\big(\,|X - E(X)| \geq \alpha\big) \leq \frac{\text{Var}\,(X)}{\alpha^2}, \tag{6.6.4}$$

which is a special case of the inequality (6.6.2): Take $F(x) = (x - E(X))^2$. Now, using $\alpha = c\sigma = c\sqrt{\text{Var}(X)}$ $(c > 0)$ and $\mu = E(X)$, we can write (6.6.4) as

$$P\big(\,|X - \mu| \geq c\sigma\big) \leq \frac{1}{c^2} \quad \text{or} \quad P\big(\,|X - \mu| < c\sigma\big) \geq 1 - \frac{1}{c^2}. \tag{6.6.5}$$

[6] The **68-95-99.7-Rule** says that 68.27% of data lies within $\sigma$-limits, i.e., within the interval $\big[\mu - \sigma, \mu + \sigma\big]$; 95.45% within $2\sigma$-limits, i.e., within the interval $\big[\mu - 2\sigma, \mu + 2\sigma\big]$; and, 99.73% within $3\sigma$-limits, i.e., within the interval $\big[\mu - 3\sigma, \mu + 3\sigma\big]$.

The first inequality in (6.6.5) shows that the probability of the random variable $X$ falling more than $c$ standard deviations from its mean $\mu$ (in both directions) is at most $1/c^2$. The second inequality says that the values of $X$ lie within $c\sigma$ of its expected value $\mu$ with probability at least $1 - (1/c^2)$. In particular, with $c = 10$, the values of $X$ lie in the interval $\big[\mu - 10\sigma, \mu + 10\sigma\big]$ at least 99% of the time. For a simple illustration, let $X$ and $Y$ represent the values on two fair dice in a single roll. Since $X$ and $Y$ are independent, for $Z = X + Y$, we have

$$E\big(Z\big) = 7 \quad \text{and} \quad \text{Var}\,\big(Z\big) = \frac{35}{6}.$$

Further, for $n$ independent trials of the same experiment, the total value of the $n$ rolls has the expected value $7n$ and the variance is given by $35n/6$. It thus follows that the sum lies between the values

$$7n - 10\sqrt{\frac{35n}{6}} \quad \text{and} \quad 7n + 10\sqrt{\frac{35n}{6}}$$

at least 99% of the time. In particular, for $n = 10^6$, the total value lies between 6.976 million and 7.024 million more than 99% of the time. In general, with $F(x) = x^q$ $(q > 0)$ in (6.6.2), Chebyshev's inequality applies to a wide range of distributions, including the normal distributions. In specific cases, we choose the value of $q$ to optimize the upper bound so obtained. A related concept provides the foundation of **Chernoff bounding method**, as discussed in the last part of the section.

**Example 6.36** (*Sample Average*) Let $X$ be a random variable defined on a discrete probability space $(\Omega, \mathbb{P})$. For $n \geq 1$, consider the *product space* $(\Omega^n, \mathbb{P}_n)$, where

$$\mathbb{P}_n\big(w_1, \ldots, w_n\big) = \mathbb{P}\big(w_1\big) \cdots \mathbb{P}\big(w_1\big), \quad \text{for} \quad w_i \in \Omega. \tag{6.6.6}$$

On this space, we define $n$ independent random variables given by

$$X_k\big(w_1, \ldots, w_n\big) = X\big(w_k\big), \quad \text{for} \quad k = 1, 2, \ldots, n, \tag{6.6.7}$$

which we may take as $n$ independent *samples* taken from the *population* $\Omega$. Now, let $Y = X_1 + \cdots + X_n$. We know that $E(Y) = n\,E(X)$ and $\sigma_Y = \sqrt{n}\,\sigma$. It thus follows from Theorem 6.10 that

$$\sigma - 10\,\frac{\sigma}{\sqrt{n}} \leq \frac{Y}{n} \leq \sigma + 10\,\frac{\sigma}{\sqrt{n}}, \tag{6.6.8}$$

at least 99% of the time, which implies that for large enough $n$ the average of $n$ samples will almost always be very close to the expected value $\mu = E(X)$.

More generally, the above type of "tail inequalities" find many applications such as in *Monte Carlo simulation* when a random variable $X$ is the **sample average** of $n$

*independent and identically distributed* random variables[7] $X_1, \ldots, X_n$, i.e., we have $X = Y/n$. Clearly, if the mean of each $X_i$ is $\mu$, then so does the variable $X$. We thus can use the *sample average* to estimate $\mu$, which help to show how unlikely it is for an estimate to be far away from the mean.

**Remark 6.5** Since Markov's inequality only involves the expected value $E[X]$, no improvement in its bound is expected by increasing $n$. On the other hand, assuming each $\mathrm{Var}(X_i)$ is finite, the standard deviation of $X$ is given by $\sqrt{\mathrm{Var}(X)}/n$. It then follows from Chebyshev's inequality that

$$\mathbb{P}\big(|X - \mu| \geq \alpha\big) \leq \frac{\mathrm{Var}(X)}{n\alpha^2}.$$

It thus follows that the likelihood of actual values of $X$ are far from its mean is inversely proportional to $n$.

## *Limit Theorems*

The behavior of the average sum of $n$ independent random variables, as illustrated in Example 6.36, is a special case of more general types of results known as the *laws of large numbers*. The most basic type of *limit theorem* is called the *weak law of large numbers*, which is due to *Jacob (Jacques) Bernoulli*.

**Theorem 6.11** (Weak Law of Large Numbers) *Let $X_1, X_2, \ldots$ be a sequence of IID random variables, with the common mean $\mu$ and variance* $\mathrm{Var}(X) = \sigma^2$, *and let $S_n := X_1 + \cdots + X_n$. Then, for every $\varepsilon > 0$, we have*

$$\lim_{n\to\infty} \mathbb{P}\left(\left|\frac{S_n}{n} - \mu\right| \geq \varepsilon\right) = 0. \tag{6.6.9}$$

***Proof*** As earlier, we have $E\big(S_n/n\big) = \mu$, and since $X_i$'s are independent, we also have $\mathrm{Var}(S_n/n) = \sigma^2/n$. Therefore, by applying Chebyshev's inequality,[8] we obtain

$$\mathbb{P}\left(\left|\frac{S_n}{n} - \mu\right| \geq \varepsilon\right) \leq \frac{\sigma^2}{n\varepsilon^2}$$

which proves (6.6.9). □

[7] The phrase "independent and identically distributed", abbreviated as IID in probability books, is often used for a collection of random variables to convey that these variables are independent, and they all have the same distribution.

[8] Since Chebyshev inequality was not available to Jacques Bernoulli, he had to develop very ingenious techniques to prove a weaker version of this theorem.

Another closely related remarkable limit theorem is due to the French mathematicians *Abraham de Moivre* (1667–1754), first published in 1738, and *Pierre-Simon Laplace* (1749–1827), who improved Moivre's in 1812. It is about the distributions of the random variable $X$ given by

$$X = \frac{S_n - n\,\mu}{\sigma\sqrt{n}}, \tag{6.6.10}$$

where $X_i$ are IID random variables, with the common mean $\mu$ and variance $\mathrm{Var}(X) = \sigma^2$. In this case, the mean of $S_n$ is $n\,\mu$ and its variance is $n\,\sigma^2$. We need some preparation to prove the same. For, we define a famous distribution for a continuous random variable $X$, known as the *normal distribution*.

**Definition 6.18** (*Normal Distribution*) We say a continuous random variable $X$ is **normally distributed** (or simply a **normal variate**), with the mean $E(X) = \mu$ and the variance $\mathrm{Var}(X) = \sigma^2$, if the probability density function of $X$ is a **Gaussian** given by

$$f(x) = \frac{1}{\sigma\sqrt{2\pi}}\, exp\left[-\frac{1}{2}\left(\frac{x-\mu}{\sigma}\right)^2\right], \quad \text{for} \quad -\infty < x < \infty. \tag{6.6.11}$$

In this case, we write $X$ as $N(\mu, \sigma)$.

The concept of normal distribution was first introduced by de Moivre in 1733. However, it was the German mathematician *Carl Friedrich Gauss* (1777–1855) who showed the importance of these types of distributions in 1809, and hence the name *Gaussian distributions*. These distributions[9] have bell-shaped probability curve, which is perfectly symmetric about the *mean line* $x = \mu$ (see Fig. 6.6a). For a *normal variate* $X \sim N(\mu, \sigma)$, the random variable $Z = (X - \mu)/\sigma$ has the probability density function given by

$$\phi(z) = \frac{1}{\sqrt{2\pi}}\, e^{-\frac{1}{2}z^2}, \quad \text{for} \quad -\infty < z < \infty. \tag{6.6.12}$$

It is called the **standard normal variate**. Notice that $Z \sim N(0, 1)$. The function $\phi(z)$ for the *standard normal variate* $Z$ decays to zero very quickly and it has a bell-shaped curve about the $y$-axis, with the inflection points at $z = \pm 1$ and the maximum $1/\sqrt{2\pi}$ (see Fig. 6.6b). The CDF of $\phi$ is the function

$$\Phi(z) = \frac{1}{\sqrt{2\pi}} \int_0^z e^{-\frac{1}{2}y^2}\, dy. \tag{6.6.13}$$

[9] Since the mean, median, and mode of a normal variate is same, it is the most preferred choice to approximate a random variable whose distributions are not known.

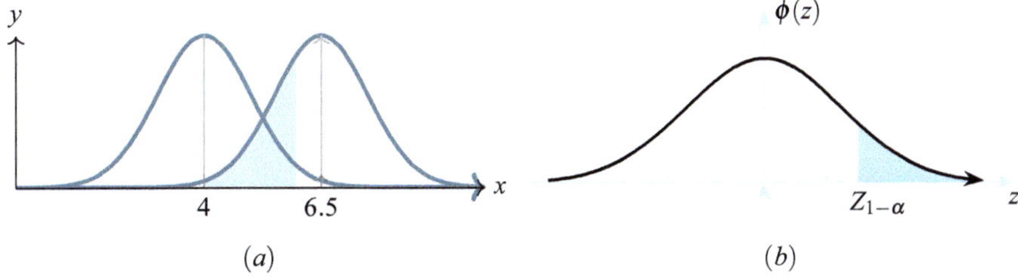


**Fig. 6.6** Normal curves for $\mu = 4,\ \mu = 6.5$, and standard normal variate

Since $\mathbb{P}(X > b) = 1 - \Phi(b)$, the above relation helps compute the tabular values of $Z$. For example, we have $\mathbb{P}(X > 1) = 0.1587$ and $\mathbb{P}(X > 3) = 0.0013$. Now, let $\langle X_i \rangle$ be an infinite sequence of independent Bernoulli random variables, each with success parameter $p \in (0, 1)$. Let $q = 1 - p$. We know that, for each $i$, $E(X_i) = p$ and $\text{Var}(X_i) = pq$. With $S_n = X_1 + \cdots + X_n$, we also know that

$$E\big(S_n\big) = np \quad \text{and} \quad \text{Var}\,\big(S_n\big) = npq. \tag{6.6.14}$$

For each $n \in \mathbb{N}$, define the random variable

$$Y_n = \frac{S_n - np}{\sqrt{npq}}. \tag{6.6.15}$$

It thus follows from the linearity of $E$ that

$$E\big(Y_n\big) = \frac{1}{\sqrt{npq}}\, E\big(S_n - np\big) = \frac{1}{\sqrt{npq}}\, \big(np - np\big) = 0. \tag{6.6.16}$$

Also, using (6.6.14), we have

$$\text{Var}\,\big(Y_n\big) = \frac{1}{\sqrt{npq}}\, \text{Var}\,\big(S_n - np\big) = \frac{1}{\sqrt{npq}}\, \text{Var}\,\big(S_n\big) = 1. \tag{6.6.17}$$

Notice that the random variables $Y_n$ depend on $n$ in a non-trivial way in the sense that the *maximum* of $Y_n$ is $(n - np)/\sqrt{npq} = \sqrt{nq/p}$, which certainly depends on $n$. However, if follows from (6.6.16) and (6.6.17) that $E(Y_n)$ and $\text{Var}(Y_n)$ are independent. To prove *de Moivre-Laplace theorem*, we shall use the following formula mainly due to de Moivre[10]:

$$n! \approx \sqrt{2\pi}\, n^{n+1/2} e^{-n}, \tag{6.6.18}$$

[10] It is known as the *Stirling's formula*, though the only contribution of Stirling was to identify the value of the approximation constant as $\sqrt{2\pi}$.

The random variable $Y_n$ as in (6.6.15) takes the values

$$y_{n,k} = \frac{k - np}{\sqrt{npq}}, \quad \text{for} \quad 0 \leq k \leq n. \tag{6.6.19}$$

These values are evenly spaced with the distance between them given by

$$\Delta_n \, y = \frac{1}{\sqrt{npq}}. \tag{6.6.20}$$

Also, since $\mathbb{P}\big(S_n = k\big) = \binom{n}{k} p^k q^{n-k}$, we have

$$\mathbb{P}\big(Z_n = y_{n,k}\big) = \binom{n}{k} p^k q^{n-k}. \tag{6.6.21}$$

The next theorem helps prove Theorem 6.13.

**Theorem 6.12** (Local de Moivre Theorem) *For any $\rho > 0$ and $\varepsilon > 0$, there exists an integer $n_0 := n_0(\rho, \varepsilon, p)$ such that, for all $n \geq n_0$ and $k$ with $y_{n,k} \leq \rho$, we have*

$$\big(1 - \varepsilon\big)\phi\big(y_{n,k}\big)\Delta_n \, y \leq \mathbb{P}\big(Y_n = y_{n,k}\big) \leq \big(1 + \varepsilon\big)\phi\big(y_{n,k}\big)\Delta_n \, y, \tag{6.6.22}$$

*where $\phi(x)$ is as in* (6.6.12)

**Corollary 6.1** *Let the notations be as in Theorem 6.12, and $g$ be a piecewise continuous function on $\mathbb{R}$ such that for some finite $c$ we have $|g(x)| \leq c$, for all $x \in \mathbb{R}$, then we have*

$$\lim_{n\to\infty} Eg\big(Y_n\big) = \frac{1}{\sqrt{2\pi}} \int_{\mathbb{R}} g\big(x\big)e^{-x^2/2}dx. \tag{6.6.23}$$

**Theorem 6.13** (de Moivre-Laplace Theorem) *Let the random variables $Y_n$ be as in* (6.6.15) *expressed in terms of an infinite sequence $\langle X_i \rangle$ of IID Bernoulli variables. Then for all $a < b \in \mathbb{R}$, we have*

$$\lim_{n\to\infty} \mathbb{P}\left(a < Y_n \leq b\right) = \Phi\big(b\big) - \Phi\big(a\big). \tag{6.6.24}$$

For proofs of the above two theorems, the reader is referred to ([16], Chap. 7). Since $\Phi$ is continuous, for any $a$ and any $\varepsilon > 0$, we can find a $\delta$ so that $\Phi(a + \delta) - \Phi(a - \delta) < \varepsilon$. It follows that for all sufficiently large $n$, $\mathbb{P}(Y_n = a) \leq P(a - \delta < Y_n \leq a + \delta) < \varepsilon$, and therefore, $\lim_{n\to\infty} \mathbb{P}(Y_n = a) = 0$, for all $a$. Hence, we can also write (6.6.24) as

$$\lim_{n\to\infty} \mathbb{P}\left(a \leq Y_n \leq b\right) = \Phi\big(b\big) - \Phi\big(a\big). \tag{6.6.25}$$

Theorem 6.13 a special case of next fundamental limit theorem, known as the (classical) *central limit theorem*, which is due to the Finnish mathematician *Jarl W. Lindeberg* (1876–1932) and the French mathematician *Paul Lévy* (1886–1971). The first rigorous proof of Theorem 6.14 is given by the Russian mathematician *Aleksandr Lyapunov* (1857–1918) in 1901. In simple terms, it says when the number of observations is sufficiently large, the averages of samples of observations of random variables independently drawn from independent distributions become normally distributed.

**Theorem 6.14** (Central Limit Theorem) *Consider a sequence of independent random variables $X_1, X_2, \ldots$. Suppose they all have the same distribution, with the common mean $\mu$ and the common variance $\sigma^2$. Then the distribution of the random variable $X$ in* (6.6.9) *tends to the standard normal variate $Z$ as $n \to \infty$. That is, for every real number $\alpha$, we have*

$$\lim_{n\to\infty} \mathbb{P}\left(\frac{S_n - n\,\mu}{\sigma\sqrt{n}} \le \alpha\right) = \frac{1}{\sqrt{2\pi}}\int_0^z e^{-\frac{1}{2}y^2}\,dy. \tag{6.6.26}$$

For a proof, the reader is referred to ([11], Chap. 8) or ([12], Chap. 5). In practical terms, Theorem 6.14 says the variables representing physical quantities that are "sum" of many different types of independent processes often have their "collective distributions" nearly *normally distributed*.

## *Chernoff Bounds*

The most widely used *Chernoff bounds technique* is about computing bounds on the *tail probabilities* of a random variable $X$. Notice that we can apply Markov's inequality to the random variable $e^X$ even when $X$ is negative. More generally, we can apply it to the random variable $e^{tX}$, for any $t > 0$, and then optimize the bound over all choices of $t$. For, let $\alpha > 0$. We have a bound on the right tail given by

$$\begin{aligned}\mathbb{P}\big(X \ge \alpha\big) &= \mathbb{P}\big(e^{tX} \ge e^{t\alpha}\big)\\ &\le \frac{E\big(e^{tX}\big)}{e^{t\alpha}} = e^{-t\,\alpha} M_X\big(t\big).\end{aligned}$$

Since this holds for all $t > 0$, we obtain

$$\mathbb{P}\big(X \ge \alpha\big) \le \min_{t>0} e^{-t\,\alpha} M_X\big(t\big) \tag{6.6.27}$$

Similarly, a bound on the left tail is given by

$$\mathbb{P}\big(X \le \alpha\big) \le \min_{t<0} e^{-t\,\alpha} M_X\big(t\big) \tag{6.6.28}$$

Choosing appropriate values for $t$ yields bounds for specific distributions. While a value of $t$ that minimizes $e^{-t\alpha}M_X(t)$ is the best possible bound, we often pick a suitable $t$ that gives a more convenient form. The bounds so derived are referred to as **Chernoff bounds**. So, when we talk about a *Chernoff bound* for a random variable, it may be one of many bounds derived using this technique.

We will pursue the inequality (6.6.27) to develop the most commonly applied versions of Chernoff bounds, namely, for the tail distribution of **Poisson trials**. These refer to the sum of independent 0–1 random variables $X_1, \ldots, X_n$, with $\mathbb{P}(X_i = 1) := p_i$. Notice that the *Bernoulli trials* are a special case of *Poisson trials*, where the independent 0–1 random variables have the same distribution. We also recall that the binomial distribution gives the number of successes in $n$ independent Bernoulli trials. Let $X = \sum_{i=1}^{n} X_i$ so that we have

$$\mu = E(X) = E\left(\sum_{i=1}^{n} X_i\right) = \sum_{i=1}^{n} E(X_i) = \sum_{i=1}^{n} p_i. \tag{6.6.29}$$

Also, by (6.5.36), the MGF of $X_i$ is given by

$$M_{X_i}(t) = p_i e^t + (1 - p_i) = 1 + p_i(e^t - 1) \le e^{p_i(e^t-1)}, \tag{6.6.30}$$

the last inequality follows using $e^x \ge 1 + x$, for all $x$. It thus follows using the independence of $X_i$ that

$$\begin{aligned} M_X(t) &= \prod_{i=1}^{n} M_{X_i}(t) \le \prod_{i=1}^{n} e^{p_i(e^t-1)} \\ &= \exp\left(\sum_{i=1}^{n} p_i(e^t - 1)\right) \\ &= e^{(e^t-1)\mu}, \end{aligned} \tag{6.6.31}$$

In actual practice, we need to bound the probability that $X$ deviates from its mean $\mu$, so we will apply the above observations to the random variable $X - \mu \ge \delta\mu$, for a given $\delta > 0$.

**Theorem 6.15** (Chernoff Bounds) *With notations as introduced in the above discussion, the following Chernoff bounds hold:*

*(a) For any* $\delta > 0$,

$$\mathbb{P}(X \ge (1+\delta)\mu) \le \left(\frac{e^{\delta}}{(1+\delta)^{1+\delta}}\right)^{\mu}. \tag{6.6.32}$$

*(b) For any* $0 < \delta \leq 1$,

$$\mathbb{P}\big(X \geq (1+\delta)\mu\big) \leq e^{-\mu\delta^2/3}. \tag{6.6.33}$$

*(c) For* $\rho \geq 6\mu$,

$$\mathbb{P}\big(X \geq \rho\big) \leq 2^{-\rho}. \tag{6.6.34}$$

***Proof*** To prove $(a)$, using Markov's inequality, we have

$$\begin{aligned}\mathbb{P}\big(X \geq (1+\delta)\mu\big) &= \mathbb{P}\big(e^{tX} \geq e^{t(1+\delta)\mu}\big)\\ &\leq \frac{E\big(e^{tX}\big)}{e^{t(1+\delta)\mu}}\\ &\leq \frac{e^{(e^t-1)\mu}}{e^{t(1+\delta)\mu}},\end{aligned}$$

by (6.6.31). Now, take $t = \ln(1+\delta)$ to obtain (6.6.32). The proof of $(b)$ and $(c)$ is left for the reader as an exercise. □

These bounds are most effective when the MGF $M_X(t)$ is finite for $t > 0$ and $t < 0$. As with the Markov and Chebyshev inequalities, the Chernoff bounds are tight. For (6.6.27) with $\alpha > 0$, let

$$\mathbb{P}\big(X = \alpha\big) = p_\alpha \quad \text{and} \quad \mathbb{P}\big(X = 0\big) = 1 - p_\alpha.$$

We then have

$$\begin{aligned}M_X\big(t\big)e^{-t\alpha} &= \big[p_\alpha\, e^{t\alpha} + \big(1 - p_\alpha\big)e^0\big]e^{-t\alpha}\\ &= p_\alpha + \big(1 - p_\alpha\big)e^{-t\alpha}\end{aligned}$$

Therefore, the upper bound in (6.6.27) is tight, because it is arbitrarily close to $p_\alpha$ as $t \to \infty$. Similarly, we can construct an example for the lower bound in (6.6.28).

## Exercises 6

6.1. Show that the *classical probability* is a special case of Kolmogorov's model.
6.2. Suppose a card from a well-shuffled deck of cards is picked at random, and it is found to be red. Find the probability that it is the jack of the diamond.
6.3. Let $E$, $F$ be two events of a sample space $\Omega$. If the occurrence of event $F$ makes $E$ more likely, then does the occurrence of the event $E$ make $F$ more likely? Justify your answer. (Hint: The given hypothesis is equivalent to $P(E \mid F) > 1/2$.)

6.4. If an event $E$ is independent of itself, show that $P(E)$ is 1 or 0. Conversely, if $P(E)$ is 1 or 0, show that $E$ is independent of all events $F$.

6.5. Given an example of two dependent events for the experiment involving the tossing of an unbiased coin twice.

6.6. State a problem analogous to the situation above and solve it. (Hint: For example, take *traffic jam* instead of a strike, and change the construction situation to the situation *reaching railway station* on time).

6.7. Consider the three identical bags $B_1$, $B_2$ and $B_3$, with the first bag containing 3 red ball, the second 3 blue balls, and the third 1 red & 2 blue balls. We may choose a bag at random and pick a ball out of it. Now, if it is a red ball, what is the probability that the remaining ball in the box is also red?

6.8. If the probability of hitting a target is 10%, and 10 shots are fired independently, find the probability that the target would be hit at least once.

6.9. Suppose half the population of a city consumes chocolates, and there are 100 investigators each interviewing 10 individuals out there to find the number of people who consume chocolate. How many investigators would report that $\leq 3$ people consume chocolates ?

6.10. If 20% of bolts produced by a machine are known to be *defective*, determine the probability that out of 10 bolts chosen at random only $(i)$ 1; $(ii)$ 0; $(iii)$ at most 2 bolts would be defective.

6.11. Let $X \sim B(n, p)$ be a binomial variable with probability mass function given by (6.4.16). Find the MGF of $X$, and hence show that the mean and the variance of $X$ are, respectively, given by $np$ and $npq$. Also, show that

$$E\left(\frac{1}{1+X}\right) = \frac{1-(1-p)^{n+1}}{(n+1)p}.$$

6.12. Show that the Poisson variable $P(\lambda)$ is the limiting case of a binomial variable $X \sim B(n, p)$, subject to $\lambda := pn$ is a constant as $n \to \infty$.

6.13. Let $X \sim P(\lambda)$ be the Poisson variable. Show that $E(X) = \lambda$ and $E(X^2) = \lambda(\lambda + 1)$.

6.14. Suppose a random variable $X$ has the mean 8 and the variance 4. Which discrete distribution would best fit to $X$? Also, find the binomial distribution of a random variable $X$ if $E(X) = 12$ and the standard deviation $\sigma = 2$.

6.15. Five hundred TV sets are inspected as they come off the production line, and the number of defects per set is recorded below

| No. of defects | 0 | 1 | 2 | 3 | 4 |
|---|---|---|---|---|---|
| No. of Sets | 368 | 72 | 52 | 7 | 1 |

Estimate the average number of defects per set, and expected frequencies of 0, 1, 2, 3, and 4 defects, using $e^{-0.402} = 0.6689$.

6.16. Suppose a data related to *number of road accidents* happening in a city follow the *Poisson distribution* with $\lambda = 4$. Out of total of 100 days observations, find the probability that there would be $(i)$ 0; $(ii)$ $\geq 2$; and $(iii)$ $\leq 3$ accidents. (Use $e^{-4} = 0.01832$)

6.17. For a skilled typist, the number of mistakes made per day in the past 300 days are as given below:

| No. of Mistakes | 0 | 1 | 2 | 3 | 4 | 5 | 6 |
|---|---|---|---|---|---|---|---|
| No. of Days | 143 | 90 | 42 | 12 | 9 | 3 | 1 |

Fit a Poisson distribution to this data, and compute the theoretical frequencies.

6.18. Describe the *exponential distribution* of a continuous random variable $X$, with parameter $\lambda > 0$, and hence compute its moment generating function. Also, find the second and third central moments of $X$.

6.19. For a factory manufacturing cricket balls, there is a small chance of 0.002 for any ball to be defective. The balls are supplied in a pack of 10. Use suitable distribution to compute the probability that the pack may contain $(i)$ 0; $(ii)$ 1; and $(iii)$ 2 defective balls, in a consignment of 10000 packets.

6.20. Find the MGF of the Poisson variable $X \sim P(\lambda)$ with mass function given by (6.4.19). Also, find the second central moment of $X$.

6.21. Suppose the *mean height* of soldiers is $68.22\,inches$, with variance of $10.8\,inches$. Estimate the number of soldiers over $6\,ft$ tall, when it given that $F(0.35) = 0.1368$ and $F(1.15) = .3746$.

6.22. In a sample test on 2000 bulbs, it was found that the life of a particular make was *normally distributed*, with an average life ($\mu$) of 2040 hours and S.D. ($\sigma$) of 60 hours. Estimate the number of bulbs likely to burn for $(i)$ more than 2150 hours; $(ii)$ less than 1950 hours; and, $(iii)$ more than 1920 hours and less than 2160 hours. You may use

$$F(1.83) = 0.4664, \quad F(1.5) = 0.4332, \quad \text{and} \quad F(2) = 0.4772.$$

6.23. For a normally distributed variable $X$, 31% of the values are under 45 and 8% are over 64. Find the mean and the variance of the variable $X \sim N(n, p)$. It is given that $F(0.5) = 0.19$ and $F(1.4) = 0.42$.

6.24. Let a random variable $X$ has the probability density function given by

$$f(x) = \begin{cases} x, & 0 < x < 1 \\ 2 - x, & 1 \leq x < 2 \\ 0, & \text{elsewhere} \end{cases}$$

Obtain the MGF of $X$. Also, determine $\mu = E(X)$ and $\sigma^2 = \mathrm{Var}(X)$.

6.25. Let $X,\ Y : \Omega \to \mathbb{R}$ be two independent random variables, and let $\phi,\ \psi : \mathbb{R} \to \mathbb{R}$ be any two functions. Show that the random variables $\phi(X)$ and $\psi(Y)$ are also independent.

6.26. Suppose two random variables $X$ and $Y$ defined on the same discrete probability space are independent. Show that $E(XY) = E(X)E(Y)$. Also, give an example to show that the same may not hold in general.

6.27. Suppose two random variables $X$ and $Y$ defined on the same discrete probability space are independent. Show that $M_{X+Y}(t) = M_X(t)M_Y(t)$, for $t \in \mathbb{R}$.

6.28. The marks of students in a class are normally distributed, with the mean $\mu = 75$ and the standard deviation $\sigma = 15$. Suppose the lowest passing mark is 60. Find the percentage of students who are likely to fail.

6.29. It is known that the monthly income of a group of 10000 people is normally distributed, with the mean $\mu = 750$ and the standard deviation $\sigma = 50$. Show that around 95% have a monthly income of more than 668 and only 5% have a monthly income of more than 832.

6.30. Prove that the binomial variate $X \sim B(n,\ p)$ approaches to the normal variate $N(\mu, \sigma)$, as $n \to \infty$, with $p \neq q$.

6.31. Use $X \sim N(\mu, \sigma^2)$ to show that Chebyshev's inequality produces a stronger bound.

6.32. Use Chebyshev's inequality to prove Theorem 6.11.

6.33. With the situation and notations as in Example 6.31, use Markov and Chebyshev inequalities

$$\mathbb{P}\big(X \geq cmH_m\big) \leq \frac{1}{c} \quad \text{and} \quad \mathbb{P}\big(|X - mH_m| \geq cm\big) \leq \frac{\pi^2}{6c^2},$$

where $c$ is the least number of products that one needs to buy.

6.34. Let $X_1,\ X_2,\ X_3,\ \ldots$ be IID random variables such that it is equally likely they take any value in $\lfloor n \rceil = \{1, 2, \ldots, n\}$. Let $Y_{n,i}$ denote the counts the first time that $i$ elements of $\lfloor n \rceil$ are picked. That is,

$$Y_{n,i} := \inf\big\{k \geq 1 \mid \big|\{X_1, \ldots, X_k\}\big| = i\big\},$$

with $Y_{n,0} := 0$. Show that, for any $\varepsilon > 0$, we have

$$\mathbb{P}\left(\left|Y_{n,n} - n\sum_{j=1}^{n} j^{-1}\right| \geq \varepsilon\, n \log n\right) \to 0, \quad \text{as} \quad n \to \infty.$$

6.35. Prove Chernoff bounds given in (6.6.33) and (6.6.34).

6.36. Let the conditions be as in Theorem 6.15. Prove the following, for $0 < \delta < 1$:

$$\mathbb{P}\big(X \leq (1-\delta)\mu\big) \leq \left(\frac{e^{-\delta}}{(1-\delta)^{1-\delta}}\right)^{\mu};$$
$$\mathbb{P}\big(X \leq (1-\delta)\mu\big) \leq e^{-\mu\delta^2/2}.$$

Hence deduce that

$$\mathbb{P}\big(|X-\mu| \geq \delta\mu\big) \leq 2\,e^{-\mu\delta^2/3}.$$

## References


1. Mitzenmacher, M., Upfal, E.: Probability and Computing: Randomized Algorithms and Probabilistic Analysis. Cambridge University Press (2005)
2. Motwani, R., Raghavan, P.: Randomized Algorithms. Cambridge University Press (2004)
3. Aggarwal, C.C.: Probability and Statistics for Machine Learning. Springer (2024)
4. Prince, S.J.D.: Computer Vision: Models, Learning, and Inference. Cambridge University Press (2012)
5. Shu, L.: Error Control Coding. Pearson India (2011)
6. Stinson, D.R., Paterson, M.B.: Cryptography: Theory and Practice (4/e). CRC Press (2019)
7. Ross, S.: Applied Probability Models with Optimization Applications. Dover Publications Inc, New York (1992)
8. Thrun, S., Burgard, W., Fox, D.: Probabilistic Robotics. MIT Press (2005)
9. Clark, A., Fox, C., Lappin, S. (Eds.): The Handbook of Computational Linguistics and Natural Language Processing. Wiley (2013)
10. Molloy, M.K.: Fundamentals of Performance Modeling. Macmillan Publishing Company (1989)
11. Ross, S.: A First Course in Probability (8/e). Pearson Prentice Hall (2010)
12. Grimmett, G., Stirzaker, D.: Probability and Random Processes (3/e). Oxford University Press (2001)
13. Vankatesh, S.S.: The Theory of Probability: Explorations and Applications. Cambridge University Press (2012)
14. Graham, R.L., Knuth, D.E., Patashnik, O.: Concrete Mathematics: A Foundation for Computer Science (2/e). Addison-Wesley (1994)
15. Knuth, D.E.: The Art of Computer Programming: Fundamental Algorithms (3/e). Addison-Wesley, Reading, Massachusetts (1997)
16. Chung, K.L.: Elementary Probability Theory with Stochastic Processes. Springer (1974)

# Chapter 7
# Lattices and Boolean Algebras

*Of the many forms of false culture, a premature converse with abstractions is perhaps the most likely to prove fatal to the growth of a masculine vigour of intellect.*

George Boole (1815–1864)

This chapter is about some interesting properties of lattices and Boolean algebras. The main emphasis is on techniques needed to model various types of *precedence relations* using the fundamental concept of partial ordering, as discussed earlier in Sect. 3.7. A *lattice* is a special type of a partially ordered set; and, a *Boolean algebra* is a special type of lattice. The lattices find applications in information flow and topological sorting, and Boolean algebras help solve "optimization problems" related to the design of digital circuits [1]. The reader may refer [2, 3], or [4], for further details on these topics.

## 7.1 Extremal Elements

Let $(P, \preceq)$ be a poset, where $\preceq$ is read as *precedes*. In fact, $a \preceq b$ is same as $a = b$ or $a \prec b$, i.e., *a strictly precedes b*. We may also write

$$a \prec b \quad \Leftrightarrow \quad a \preceq b \text{ and } a \neq b.$$

 
V. Ravichandran and A. K. Razdan, *Fundamental Discrete Structures*,
University Texts in the Mathematical Sciences,
https://doi.org/10.1007/978-981-96-0069-4_7

So, if two distinct elements $a, b \in P$ are comparable, then we have $a \prec b$ or $b \prec a$. We can also write $b \prec a$ as $a \succ b$, which is read as *a strictly succeeds b*. In general, $a \succeq b$ is same as $b \preceq a$.

**Definition 7.1** (*Cover*) Let $(P, \preceq)$ be a poset, and $a, b \in P$. We say $b$ **covers** $a$ if $a \prec b$, and there is no $c \in P$ such that $a \prec c$ and $c \prec b$. In this case, we say $a$ an *immediate predecessor* of $b$ or $b$ is an *immediate successor* of $a$.

Notice that the digraph of a strict order does not contain any *closed loop*. That is, there is no $a \in P$ such that $a \prec a$. In what follows, we use the *Hasse diagram* of a poset to study its *extremal elements*.

**Definition 7.2** (*Maximal and Minimal Elements*) Let $(P, \preceq)$ be a poset. An element $b \in P$ is called a **maximal element** if it has no immediate successor, i.e., there is no $c \in P$ such that $b \prec c$. Similarly, an element $a \in P$ is called a **minimal element** if it has no immediate predecessor, i.e., there is no $c \in P$ such that $a \succ c$.

In general, the minimal or maximal element of a poset may not be unique. For example, the poset with the Hasse diagram as shown in Fig. 7.1 has $a$ and $b$ as maximal elements, whereas $c$ and $d$ are the minimal elements. Clearly, distinct minimal (or maximal) elements of a poset are incomparable.

**Definition 7.3** (*Maximum and Minimum Elements*) Let $(P, \preceq)$ be a poset. A unique maximal element of $P$ is called the **greatest element** (or the **maximum element**), and a unique minimal element is called the **least element** (or the **maximum element**).

The *Minimum element* and the *Maximum element* of a poset $(P, \preceq)$ are usually written respectively as 0 and 1. The Hasse diagrams of two posets as shown in Fig. 7.2($i$)

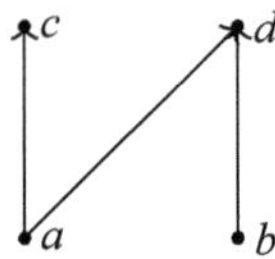


**Fig. 7.1** Hasse diagram of $\Delta_A \cup \{(a, c), (a, d), (b, d)\}$

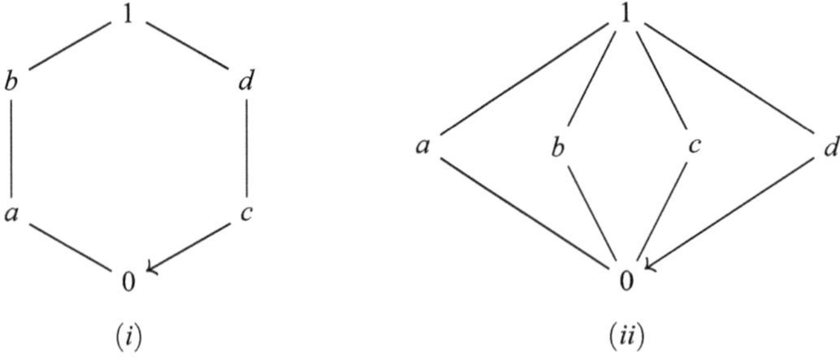


**Fig. 7.2** Maximum and minimum elements in hexagon and diamond

and $(ii)$ display 0 and 1 as the minimum and maximum elements. Since such types of elements appear at the lowest and highest level of the Hasse diagram of a poset, they may not exist always. The next definition generalizes the concept introduced above.

**Definition 7.4** Let $(P, \preceq)$ be a poset, and $A \subseteq P$. An element $a \in P$ is called a *Minimal element* of the set $A$ if $a \in A$, and there is no $x \in A$ such that $a \prec x$. Also, an element $\alpha \in P$ is called the **minimum element** (or the *least element*) of the set $A$ if $\alpha \in A$, and we have

$$\alpha \preceq x, \qquad \text{for all} \quad x \in A.$$

Similarly, an element $b \in P$ is called a *Maximal element* of the set $A$ if $b \in A$, and there is no $y \in A$ such that $b \prec y$. And, an element $\beta \in P$ is called the **maximum element** (or the *greatest element*) of the set $A$ if $\beta \in A$, and we have

$$y \preceq \beta, \qquad \text{for all} \quad y \in A.$$

Not every subset of a poset may have a *minimum element* or a *maximum element*.

**Theorem 7.1** *Let $(P, \preceq)$ be a poset, and $A \subseteq P$. Then we have*

1. *A has a unique maximum element, whenever it exists;*
2. *if A has a unique minimum element, whenever it exists.*

***Proof*** Suppose $a_1, a_2$ are two maximum elements of the set $A$ so that we have

$$a \preceq a_1 \quad \text{and} \quad a \preceq a_2, \qquad \text{for all} \quad a \in A.$$

In particular, taking $a = a_2$, it follows from $a \preceq a_1$ that $a_2 \preceq a_1$. Similarly, by taking $a = a_2$ in $a \preceq a_1$, we obtain $a_1 \preceq a_2$. Therefore, by antisymmetry property, we conclude from $a_1 \preceq a_2$ and $a_2 \preceq a_1$ that $a_1 = a_2$. Therefore, the maximum element is unique, if it exists. This proves (1). A similar proof may be given for (2). □

We may recall that, for $P \subseteq \mathbb{R}$, the symbol $\preceq$ denotes the usual "less than or equal to" $\leq$. Also, for $P = \mathscr{P}(A)$, $\preceq$ is the usual set inclusion relation. That is, $S \preceq T$ in $\mathscr{P}(A)$ means $S \subseteq T$. Notice that the ordering in $\mathscr{P}(A)$ is reversed when $\subseteq$ is replaced with $\supseteq$ so that we can write $A \succeq B$ when $A \supseteq B$. A similar remark applies to $\geq$ in the case when $P \subseteq \mathbb{R}$.

**Example 7.1** With respect to usual order $\leq$, the set $\{0, 1, 2, \ldots, 9\}$ has 0 as the minimum element and 9 as the maximum element. On the other hand, the set of integers $\mathbb{Z}$ has no maximum or minimum element. However, the set of natural numbers $\mathbb{N}$ has 1 as the minimum element,[1] but it has no maximum element. Also, with respect to *inclusion relation* $\subseteq$, the minimum and the maximum elements of a collection $P$ of subsets of a nonempty set $A$ are respectively given by their *intersection* and *union*.

[1] **Well Ordering Property** states that every nonempty subset of $\mathbb{N}$ has a minimum element.

Further, with respect to the ordering given by the relation "*divides*", each finite set $A = \{n_1, \dots, n_k\} \subset \mathbb{N}$ has the minimum and maximum elements respectively given by

$$\alpha = \text{GCD}\left\{n_1, \dots, n_k\right\} \quad \text{and} \quad \beta = \text{LCM}\left\{n_1, \dots, n_k\right\}.$$

The next definition introduces the concept of *greatest lower bound* (or *Infimum*) and *least upper bound* (or *supremum*) of a subset of a poset.

**Definition 7.5** Let $(P, \prec)$ be a poset, and $A \subset P$.

1. An element $u \in P$ is called an **upper bound** of the set $A$ if $a \preceq u$, for all $a \in A$. We say $A$ is **bounded above** if it has an *Upper bound*. The **supremum** of a set $A$, denoted by $\sup(A)$, is given by

$$\sup(A) := \min\left\{u \in X \mid u \text{ is an upper bound of } A\right\}. \tag{7.1.1}$$

   Clearly, for a *Bounded above* set $A$, $\beta = \sup A$ satisfies the property that $u \preceq \beta$, for each upper bound $u$ of the set $A$. We also call $\beta$ as the **least upper bound** (abbreviated as LUB) of the set $A$.
2. An element $\ell \in X$ is a **lower bound** of the subset $A$ if $\ell \leq a$, for all $a \in A$. We say $A$ is **bounded below** if it has a *Lower bound*. The **infimum** of a set $A$, denoted by $\inf(A)$, is given by

$$\inf(A) := \max\left\{\ell \in X \mid \ell \text{ is a lower bound of } A\right\}. \tag{7.1.2}$$

   Clearly, for a *Bounded below* set $A$, $\alpha = \inf A$ satisfies the property that for each lower bound $\ell$ of the set $A$ we have $\ell \preceq \alpha$. We also call $\alpha$ as the **greatest lower bound** (abbreviated as GLB) of the set $A$.

We always have $\inf A \leq \sup A$, if the two exist. Finally, a set $A \subseteq P$ is called a **bounded set** if it is both bounded above and bounded below.

**Example 7.2** Let $A = \{a, b, c\}$, and $P = \mathscr{P}(A)$. We know that $\left(P, \subseteq\right)$ is a poset, with the Hasse diagram shown in Fig. 7.3. For the set $B = \left\{\{b, c\}, \{b\}, \{c\}\right\} \subset P$, the upper bounds are given by $\{b, c\}$ and $\{a, b, c\}$. Clearly, $\emptyset$ is the lower bound. Further, for the set $C = \left\{\{a, c\}, \{c\}\right\}$, the upper bounds are given by $\{a, c\}$ and $\{a, b, c\}$, while $\{c\}$ and $\emptyset$ are the lower bounds.

**Example 7.3** Consider the poset $(P, \preceq]$, where $P = \{2, 3, 6, 12, 24, 36\}$ and $\preceq$ is the relation "divides". The Hasse diagram of this poset is shown in Fig. 7.4. Let $A = \{2, 3, 6\} \subset P$. Notice that $A$ has the elements $6, 12, 24, 36 \in P$ as upper bounds, but the lower bound of $A$ doesn't exist. We thus conclude that

$$\sup\left(A\right) = \min\left\{6, 12, 24, 36\right\} = 6;$$

while $\inf(A)$ doesn't exist. Also, for the set $B = \{12, 6\} \subset P$, we have

**Fig. 7.3** The Hasse diagram of the poset in Example 7.2

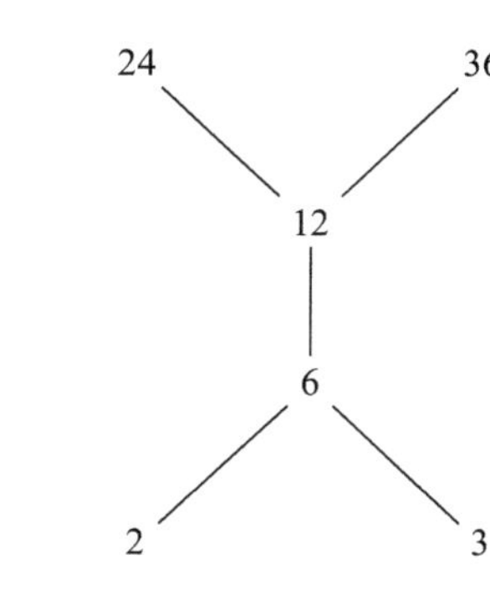


**Fig. 7.4** The Hasse diagram of the poset in Example 7.3

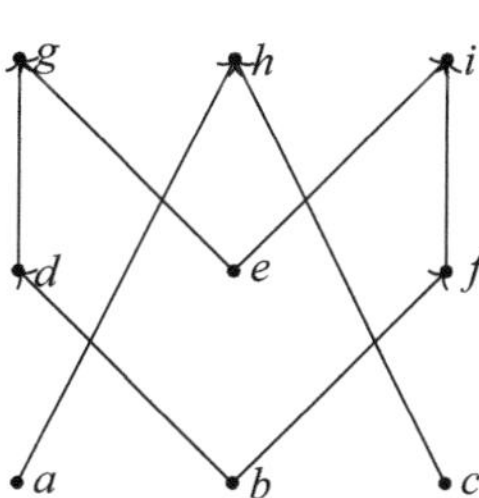


**Fig. 7.5** The Hasse diagram of the poset in Example 7.4

$$\sup\big(B\big) = 12 \quad \text{and} \quad \inf\big(C\big) = 6.$$

**Example 7.4** Consider the poset $P$ with the Hasse diagram as shown in Fig. 7.5. Since the set $A_1 = \{d, e, f\} \subset P$ has no lower bound, we have $\inf(A_1)$ does not exist. Also, the set $A_2 = \{a, c\} \subset P$ has no lower bound, and so $\inf(A_2)$ does not exist. However, since the element $h$ is the only upper bound of the set $A_2$, we have $\sup(A_2) = \{h\}$. Further, since the set $A_3 = \{b, d\} \subset P$ has the element $b$ as the lower bound, we have $\inf(A_3) = \{b\}$. Moreover, since the elements $d, g$ are upper bounds of the set $A_3$, it follows that $\sup(A_3) = \{d\}$.

**Theorem 7.2** *Let $(P, \preceq)$ be a poset, and $a_1, a_2 \in P$. We have*

1. $\inf\big\{a_1, a_2\big\}$ *is unique, whenever it exists;*

2. $\sup\{a_1, a_2\}$ *is unique, whenever it exists.*

***Proof*** If possible, suppose $\ell_1$ and $\ell_2$ are two greatest lower bounds of $a_1$ and $a_2$. We then have

$$\ell_1 \preceq a_1, \quad \ell_1 \preceq a_2 \quad \text{and} \quad \ell_2 \preceq a_1, \quad \ell_2 \preceq a_2.$$

It thus follows that $\ell_1$, $\ell_2$ are lower bounds of the set $\{a_1, a_2\}$. Further, since $\ell_1$ is the greatest lower bound, then $\ell_1 \preceq \ell_2$. But, $\ell_2$ is also a greatest lower bound, so $\ell_2 \preceq \ell_1$. It thus follows from the antisymmetry of $\preceq$ that $\ell_1 = \ell_2$. This proves (1). A similar proof can be written for (2). □

**Definition 7.6** (*Principle of Duality*) The **principle of duality** for posets states that, if $(P, \preceq)$ is a poset, then the set $P$ is also a poset with respect to the reverse of the partial order $\preceq$. That is, $(P, \succeq)$ is also a poset, which is called the **dual poset**.

Since the partial order $\preceq$ on $P$ and the partial order $\succeq$ on its *Dual* are related by

$$x \preceq y \quad \Leftrightarrow \quad y \succeq x,$$

the least elements of one is the greatest elements of the other, and vice-versa. Said differently, the concepts of the least element and the greatest element are *dual* of each other. Also, the concepts of maximal and minimal elements in a poset are dual. Further, for any set $A \subseteq P$, $\inf(A)$ with respect to the relation $\preceq$ is the $\sup(A)$ with respect to the relation $\succeq$, and vice-versa. Therefore, if a statement is valid over a poset $(P, \preceq)$, then the *Dual statement* obtained by reversing the ordering and making other relevant changes, is a valid statement over the *dual poset* $(P, \preceq)$.

## 7.2 Lattice Theory

Lattices play a significant role in solving computation problems such as related to the *structural configuration, hierarchy*, and *scheduling*. We first fix some notions. Let $(P, \preceq)$ be a poset, and $a, b \in P$. It follows from Theorem 7.2 that $\inf\{a, b\}$ is a unique element of $P$, if it exists. We may write

$$\inf\{a, b\} := a \wedge b, \tag{7.2.1}$$

where $a \wedge b$ is read as $a$ **meet** $b$. We may also write the *meet* $a \wedge b$ as the product $a \cdot b$. Again, by Theorem 7.2, we also have that $\sup\{a, b\}$ is unique, if it exists. We may write

$$\sup\{a, b\} := a \vee b, \tag{7.2.2}$$

where $a \vee b$ is read as $a$ **join** $b$. Sometimes, we also write the *join* $a \vee b$ as the sum $a + b$. We now give a formal definition of a *lattice*.

**Definition 7.7** (*Lattice*) A *lattice* is a poset $(L, \preceq)$ such that, for every pair of elements $a, b \in L$, both $\sup\{a, b\}$ and $\inf\{a, b\}$ exist.

Said differently, on a lattice $(L, \preceq)$, both $\vee$ and $\wedge$ are binary operations. For the sake of completeness, we may also write a lattice $L$ as the quadruple $(L, \preceq, \vee, \wedge)$. In what follows, we prove some important theorems concerning these two operations. Notice that a lattice cannot have isolated points, unless it is a point lattice. Therefore, the Hasse diagram of a lattice is always a connected graph.

**Example 7.5** Let $A$ be a nonempty set, and $L = \mathscr{P}(A)$. With respect to the relation $\subseteq$, the set $L$ is a lattice. In this case, the *join* and *meet* operations are respectively given by the union and intersection of two sets. Therefore, we may write this lattice $L$ as the quadruple $(\mathscr{P}(A), \subseteq, \cup, \cap)$. Also, it follows from Example 7.1 that every set $L \subseteq \mathbb{N}$ is a lattice with respect to the relation "divides". In this case, the *join* and *meet* operations are respectively given by the *lcm* and the *gcd* of two numbers. Therefore, we may write this lattice $L$ as the quadruple $(L, \text{divides}, \text{lcm}, \text{gcd})$. In particular, each poset $(D_n, \text{ divides})$ is a lattice, which is called the **divisor lattice** for $n > 1$. Figure 7.6 shows the Hasse diagrams of the *divisor lattice* for $n = 6, 10, 15$; Fig. 7.7 shows the Hasse diagrams of the divisor lattice for $n = 8, 12, 24$; and,

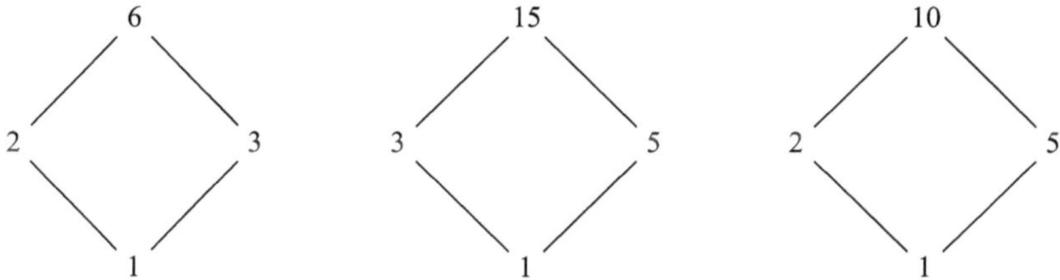


**Fig. 7.6** The Hasse diagrams of the lattices $D_6$, $D_{15}$, and $D_{10}$

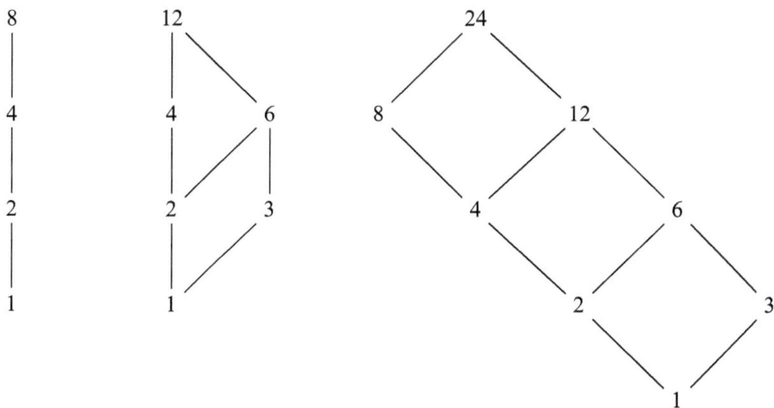


**Fig. 7.7** The Hasse diagrams of the lattices $D_8$, $D_{12}$, and $D_{24}$

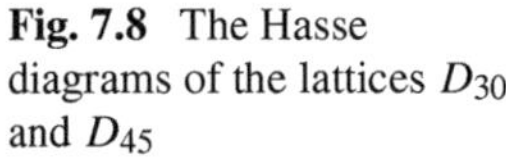

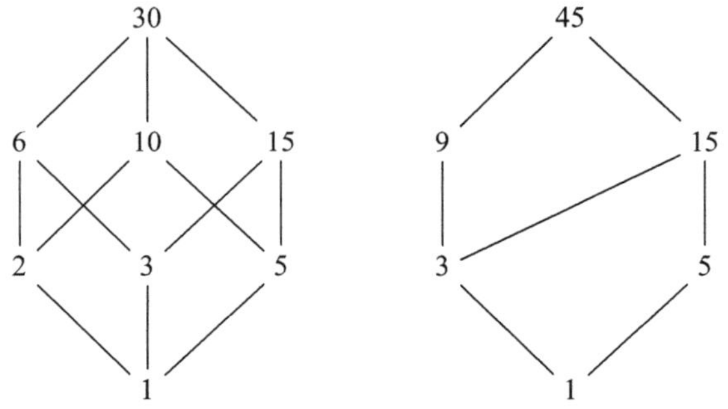

**Fig. 7.8** The Hasse diagrams of the lattices $D_{30}$ and $D_{45}$

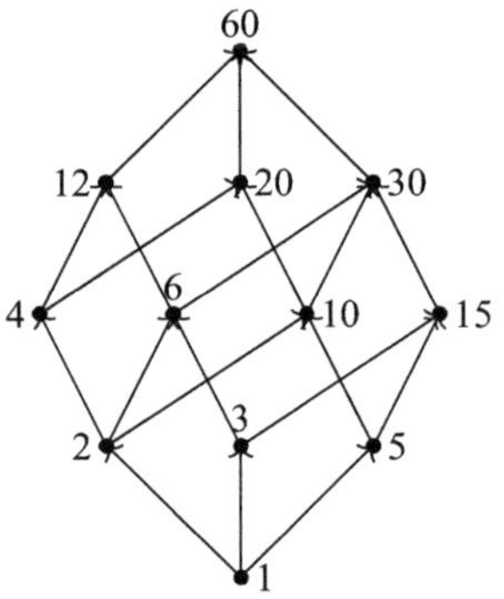

**Fig. 7.9** Hasse diagram of the lattice $D_{60}$

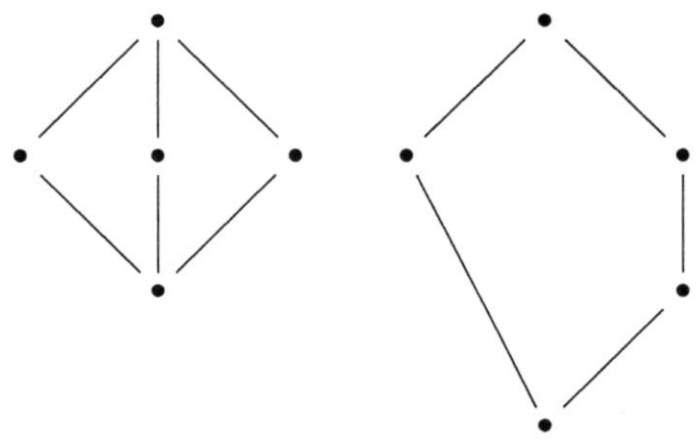

**Fig. 7.10** Hasse diagram of two important lattices

Fig. 7.8 shows the Hasse diagrams of the divisor lattice for $n = 30, 45$. It follows from Fig. 7.9 that the poset in (3.7.6) is also a lattice. The poset with the Hasse diagram shown in Fig. 7.10 are also lattices.

**Example 7.6** The posets with the Hasse diagrams as shown in Fig. 7.11 are *not* lattices, because the elements $x$ and $y$ have no upper bound, and hence $x \vee y$ does not exists. Also, the poset with the Hasse diagram as shown in Fig. 7.12 is not a lattice, because the elements $x$ and $y$ have three elements $u_1, u_2$, and $u_3$ as upper bounds of the set $\{x, y\}$, but the minimum of the set $\{u_1, u_2, u_3\}$ does not exist.

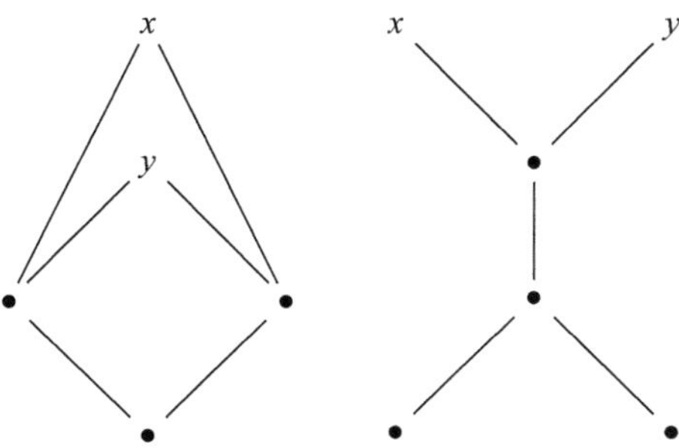


**Fig. 7.11** Two posets that are not lattices

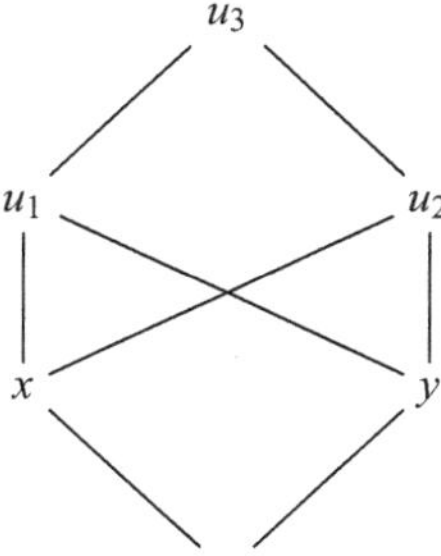


**Fig. 7.12** A poset that is not a lattice

**Theorem 7.3** *Let $(L, \preceq, \vee, \wedge)$ be a lattice. For every finite subset $S \subseteq L$, both* $\inf(S)$ *and* $\sup(S)$ *exist.*

***Proof*** Let $S = \{a_1, \ldots, a_n\}$. We may define

$$\wedge_{i=1}^{1} a_i = a_1, \quad \wedge_{i=1}^{n} a_i = \wedge_{i=1}^{n-1} a_i \wedge a_n, \quad \text{for } n > 1;$$
$$\vee_{i=1}^{1} a_i = a_1, \quad \vee_{i=1}^{n} a_i = \vee_{i=1}^{n-1} a_i \vee a_n, \quad \text{for } n > 1.$$

Then $\inf(S)$ and $\sup(S)$ are respectively given by

$$\inf(S) = \wedge_{i=1}^{n} a_i = a_1 \wedge \cdots \wedge a_n;$$
$$\sup(S) = \vee_{i=1}^{n} a_i = a_1 \vee \cdots \vee a_n.$$

This completes the theorem. □

**Definition 7.8** A lattice $L$ is said to be **complete** if every nonempty subset of $L$ has a greatest lower bound and a least upper bound.

It follows from Theorem 7.3 that every finite lattice is complete.

**Definition 7.9** Let $(L, \preceq, \vee, \wedge)$ be a lattice. A set $S \subseteq L$ is called a **sublattice** of $L$ if $S$ is closed with respect to operations $\vee$ and $\wedge$. That is, for all $x, y \in S$, both $x \vee y$ and $x \wedge y$ belongs to $S$.

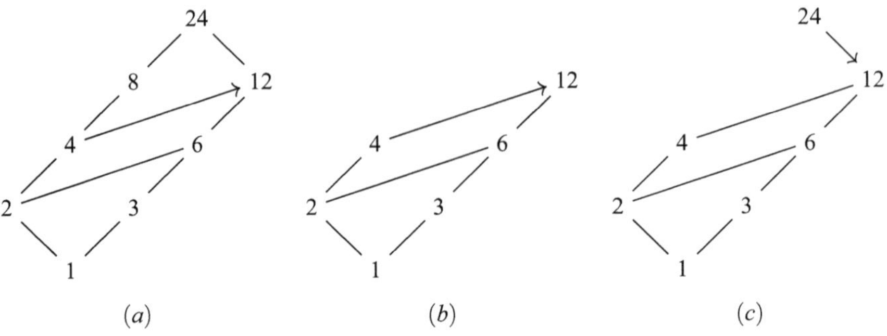


**Fig. 7.13** Some sublattices of the lattice ($D_{24}$, divides)

It follows from the above definition that a sublattice itself is a lattice. For example, every *divisor lattice* ($D_n$, divides) is a *sublattice* of the lattice ($\mathbb{N}$, divides). However, not every subset of a lattice $L$ may be a sublattice. Figure 7.13a shows the Hasse diagram of the divisor lattice ($D_{24}$, divides). Two sublattices of this lattice such as $\{2, 4, 8, 24\}$ and $\{1, 3, 6, 12\}$ are *chains of length* 4. Also, the two sublattices $\{1, 2, 4, 8, 24\}$ and $\{1, 3, 6, 12, 24\}$ are *chains of length* 5. Figure 7.13b and c show respectively two sublattices with 6 and 7 elements. This lattice also has sublattices with 3 elements. The reader is encouraged to specify some.

**Definition 7.10** (*Product Lattice*) Let $(L_1, \preceq_1, \vee_1, \wedge_1)$ and $(L_2, \preceq_2, \vee_2, \wedge_2)$ be two lattices. The **product lattice** is the poset $(L_1 \times L_2, \preceq)$ such that $\preceq$ is lexicographic order; the join $\vee$ and the meet $\wedge$ operations on $L_1 \times L_2$ are given by

$$\begin{aligned} (a_1, b_1) \vee (a_2, b_2) &= (a_1 \vee_1 a_2, b_1 \vee_2 b_2); \\ (a_1, b_1) \wedge (a_2, b_2) &= (a_1 \wedge_1 a_2, b_1 \wedge_2 b_2), \end{aligned} \tag{7.2.3}$$

where $(a_1, b_1), (a_2, b_2) \in L_1 \times L_2$.

In particular, we can talk about the *power lattices* $L^n$ of a lattice $L$. For example, with $L = \{0 < 1\}$ and $\preceq$ as lexicographic order, we have $L^2 = \{(0, 0), (0, 1), (1, 1)\}$, and $L^3$ is a cub. The corresponding Hasse diagrams are as shown in Fig. 7.14.

**Definition 7.11** (*Lattice Homomorphism*) Let $(L_1, \preceq_1, \vee_1, \wedge_1)$ and $(L_2, \preceq_2, \vee_2, \wedge_2)$ be two lattices. A function $f : L_1 \to L_2$ is called a **lattice homomorphism** if, for any $a, b \in L$, we have

$$\begin{aligned} f(a \wedge_1 b) &= f(a) \wedge_2 f(b); \\ f(a \vee_1 b) &= f(a) \vee_2 f(b). \end{aligned} \tag{7.2.4}$$

A **lattice isomorphism** $f : L_1 \to L_2$ is a bijective *lattice homomorphism*. Two lattices $L_1$ and $L_2$ are called **isomorphic lattices** if there is a *lattice isomorphism* $L_1 \to L_2$ or $L_2 \to L_1$.

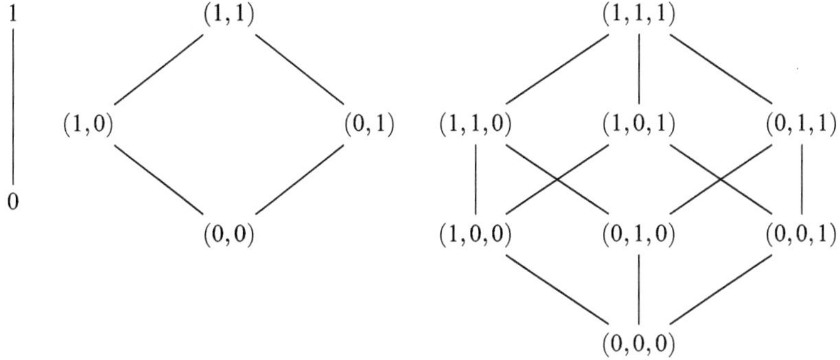


**Fig. 7.14** The Hasse diagrams of power lattices $L^1$, $L^2$, and $L^3$

As shown later, two isomorphic lattices have the same order properties, because a lattice homomorphism is an *order preserving* function. Also, a lattice homomorphism $f : L_1 \to L_2$ is an isomorphism if and only if there is a lattice homomorphism $g : L_2 \to L_1$ such that $f \circ g = \mathrm{Id}_{L_2}$ and $g \circ f = \mathrm{Id}_{L_1}$.

**Example 7.7** The three lattices with Hasse diagrams shown in Fig. 7.6 are isomorphic. In general, lattices with the same Hasse diagram are *isomorphic lattices*. However, no two of the three lattices with Hasse diagrams shown in Fig. 7.7 are isomorphic.

By Definition 7.6, and the subsequent discussion, it follows that every lattice $(L, \preceq)$ has a *dual lattice* given by $(L, \succeq)$. So, $\vee$ and $\wedge$ are *dual operations*, as are the two relations $\preceq$ and $\succeq$. We next state the *duality principle* for lattices.

**Theorem 7.4** (Duality Principle) *If a statement holds for a lattice* $(L, \preceq, \vee, \wedge)$*, then the* dual statement *obtained by interchanging the relations* $\preceq$ *and* $\succeq$*, and also the operations* $\vee$ *and* $\wedge$*, also holds.*

The lattices $(L, \preceq, \vee, \wedge)$ and $(L, \succeq, \wedge, \vee)$ are called the **dual lattices**.

## 7.3 Properties of Lattices

Let $L := (L, \preceq, \vee, \wedge)$ be a lattice. Considering the significance of what we do here to the topics discussed in subsequent sections, we write the binary operations join $\vee$ and meet $\wedge$ as $\oplus$ and $*$, respectively. That is, for $a, b \in L$, we take

$$a \oplus b = a \vee b \quad \text{and} \quad a * b := a \wedge b. \tag{7.3.1}$$

So, a lattice $L$ may also be written as $(L, \preceq, \oplus, *)$. Our emphasis here is to prove some fundamental properties of lattices. We start with the *consistency property*.

**Theorem 7.5** (Consistency Property) *Let $(L, \preceq, \oplus, *)$ be a lattice, and $a, b \in L$. We then have*

$$a \preceq b \iff a * b = a \iff a \oplus b = b. \tag{7.3.2}$$

***Proof*** We first show that

$$a \preceq b \iff a * b = a.$$

For, suppose $a \preceq b$. Now, as $a \preceq a$, we have $a \preceq a * b$. However, it follows from the definition of $a * b$ that $a * b \preceq a$. Therefore, we obtain

$$a \preceq b \quad \Rightarrow \quad a * b = a.$$

On the other hand, suppose $a * b = a$. Now, as $a = a * b \preceq b$, so we have $a \preceq b$. It thus follows that

$$a * b = a \quad \Rightarrow \quad a \preceq b.$$

Combining the above two results, we conclude that

$$a \preceq b \quad \Leftrightarrow \quad a * b = a.$$

Next, we show that

$$a \preceq b \iff a \oplus b = b.$$

For, suppose $a \preceq b$. Now, as $b \preceq b$, we have $b \preceq a \oplus b$. However, by definition of $a \oplus b$, we have $a \oplus b \preceq b$. Therefore, we obtain

$$a \preceq b \quad \Rightarrow \quad a \oplus b = b.$$

On the other hand, suppose $a \oplus b = b$. Now, as $a \preceq a \oplus b = b$, we have $a \preceq b$. It thus follows that

$$a \oplus b = b \quad \Rightarrow \quad a \preceq b.$$

Hence, we have

$$a \preceq b \iff a \oplus b = b.$$

□

We next prove that both the operations $\oplus$ and $*$ satisfy the *idempotent law*.

**Theorem 7.6** (Idempotent Law) *Let $L$ be a lattice, and $a \in L$. We then have*

$$a \oplus a = a \quad \text{and} \quad a * a = a. \tag{7.3.3}$$

***Proof*** As the two equalities in (7.3.3) are dual of each other, it suffices to prove one so that the other follows from it by applying the *duality principle* (Theorem 7.4).

However, for the sake of completeness, we give a proof for both. Now, to prove the first, we have that $a$ is an upper bound of the set $\{a\}$ because $a \preceq a$, by reflexive property. Further, if $b$ is any upper bound of the set $\{a\}$ then $a \preceq b$ implies that $a \preceq a$ and $a \preceq b$. That is, any other upper bound of the set $\{a\}$ is bigger than $a$, which shows that $a$ is the least upper bound of the set $\{a\}$. Hence, we have

$$a \oplus a = \sup\{a\} = a.$$

Next, to prove the second equality, we have $a$ is a lower bound of the set $\{a\}$ because $a \preceq a$, by reflexive property. Further, if $b$ is any other lower bound of the set $\{a\}$, then $b \preceq a$ implies that $a \preceq a$ and $b \preceq a$. That is, any other lower bound of the set $\{a\}$ is smaller than $a$, which shows that $a$ is the greatest lower bound of the set $\{a\}$. Hence, we have

$$a * a = \inf\{a\} = a.$$

□

The next theorem proves that the operations $\oplus$ and $*$ are *commutative*.

**Theorem 7.7** (Commutative Law) *Let $(L, \preceq)$ be a lattice, and $a, b \in L$. Then we have*

$$a * b = b * a \quad \text{and} \quad a \oplus b = b \oplus a. \tag{7.3.4}$$

***Proof*** Notice that, as $\{a, b\} = \{b, a\}$, we have

$$a * b = \inf\{a, b\} = \inf\{b, a\} = b * a.$$

Similarly, we have

$$a \oplus b = \sup\{a, b\} = \sup\{b, a\} = b \oplus a.$$

□

The next theorem proves that the operations $\oplus$ and $*$ are *associative*.

**Theorem 7.8** (Associative law) *Let $(L, \preceq)$ be a lattice. For $a, b, c \in L$, we have the following:*

$$\begin{aligned} a * \big(b * c\big) &= \big(a * b\big) * c; \\ a \oplus \big(b \oplus c\big) &= \big(a \oplus b\big) \oplus c. \end{aligned} \tag{7.3.5}$$

***Proof*** We first prove $a * (b * c) = (a * b) * c$. For, let

$$x = a * (b * c) \quad \text{and} \quad y = (a * b) * c.$$

It thus follows that

$$\begin{aligned}
& x = a * (b * c) \\
\Rightarrow\quad & x \preceq a \quad \text{and} \quad x \preceq (b * c) \\
\Rightarrow\quad & x \preceq a, \quad x \preceq b, \quad \text{and} \quad x \preceq c \\
\Rightarrow\quad & x \preceq (a * b) \quad \text{and} \quad x \preceq c. \\
\Rightarrow\quad & x \preceq (a * b) * c \\
\Rightarrow\quad & x \preceq y;
\end{aligned}$$

and we also have

$$\begin{aligned}
& y = (a * b) * c \\
\Rightarrow\quad & y \preceq (a * b) \quad \text{and} \quad y \preceq c \\
\Rightarrow\quad & y \preceq a, \quad y \preceq b, \quad \text{and} \quad y \preceq c \\
\Rightarrow\quad & y \preceq a \quad \text{and} \quad y \preceq (b * c). \\
\Rightarrow\quad & y \preceq a * (b * c) \\
\Rightarrow\quad & y \preceq x.
\end{aligned}$$

Therefore, we have $x \preceq y$ and $y \preceq x$ so that $x = y$, by antisymmetry property. Hence, we obtain

$$a * (b * c) = x = y = (a * b) * c.$$

Next, we show that $a \oplus (b \oplus c) = (a \oplus b) \oplus c$. For, let

$$x = a \oplus (b \oplus c) \quad \text{and} \quad y = (a \oplus b) \oplus c.$$

It thus follows that

$$\begin{aligned}
& x = a \oplus (b \oplus c) \\
\Rightarrow\quad & a \preceq x \quad \text{and} \quad (b \oplus c) \preceq x \\
\Rightarrow\quad & a \preceq x, \quad b \preceq x, \quad \text{and} \quad c \preceq x \\
\Rightarrow\quad & (a \oplus b) \preceq x \quad \text{and} \quad c \preceq x. \\
\Rightarrow\quad & (a \oplus b) \oplus c \preceq x \\
\Rightarrow\quad & y \preceq x;
\end{aligned}$$

and we also have

$$\begin{aligned} & y = (a \oplus b) \oplus c \\ \Rightarrow\ & (a \oplus b) \preceq x \quad \text{and} \quad c \preceq x \\ \Rightarrow\ & a \preceq x, \quad b \preceq x, \quad \text{and} \quad c \preceq x \\ \Rightarrow\ & a \preceq x \quad \text{and} \quad (b \oplus c) \preceq x. \\ \Rightarrow\ & a \oplus (b \oplus c) \preceq x \\ \Rightarrow\ & x \preceq y. \end{aligned}$$

Therefore, we have $x \preceq y$ and $y \preceq x$ so that $x = y$, by antisymmetry property. Hence, we obtain

$$a \oplus (b \oplus c) = x = y = (a \oplus b) \oplus c.$$

Notice that it follows from the above equality and the duality principle that

$$a * (b * c) = (a * b) * c.$$

□

We next prove that the operations $\oplus$ and $*$ satisfy *absorption laws*.

**Theorem 7.9** (Absorption Law) *Let* $(L, \preceq)$ *be a lattice, and* $a, b \in L$. *Then we have*

$$a * (a \oplus b) = a \quad \textit{and} \quad a \oplus (a * b) = a. \tag{7.3.6}$$

***Proof*** For any $a \in L$, we have $a \preceq a$. Since $a \oplus b$ is the least upper bound (and hence an upper bound) for $a$ and $b$, we have $a \preceq a \oplus b$. Also, from $a \preceq a$ and $a \preceq a \oplus b$, we obtain $a \preceq a * (a \oplus b)$. We know that, by the definition of $*$, the element $a * (a \oplus b)$ is the greatest lower bound (and hence a lower bound) for the set $\{a, a \oplus b\}$ so that we have $a * (a \oplus b) \preceq a$. Therefore, we have

$$a \preceq a * (a \oplus b) \quad \text{and} \quad a * (a \oplus b) \preceq a,$$

which by antisymmetry property implies $a * (a \oplus b) = a$. It follows from the duality principle that we also have $a \oplus (a * b) = a$. □

The next theorem proves *isotonicity property* for the operations $\oplus$ and $*$.

**Theorem 7.10** (Isotonicity) *Let* $(L, \preceq)$ *be a lattice, and* $a, b \in L$. *Then we have*

$$b \preceq c \quad \Rightarrow \quad a * b \preceq a * c \quad \textit{and} \quad a \oplus b \preceq a \oplus c. \tag{7.3.7}$$

***Proof*** We first prove that $b \preceq c \Rightarrow a * b \preceq a * c$. For, as $a \preceq b$ is equivalent to $a * b = a$ by (7.3.2), it is enough to show that

$$(a * b) * (a * c) = a * b.$$

Suppose $b \preceq c$ so that $b * c = b$, then by using Theorems 7.8 and 7.7, we obtain

$$\begin{aligned}(a * b) * (a * c) &= a * (b * (a * c)) \\ &= a * ((a * c) * b) \\ &= (a * a) * (c * b) \\ &= a * (b * c) \\ &= a * b.\end{aligned}$$

To prove the second assertion, suppose $b \preceq c$ so that $b \oplus c = c$. In this case, we have to show that $a \oplus b \leq a \oplus c$ or equivalently $(a \oplus b) \oplus (a \oplus c) = a \oplus c$. Using Theorems 7.8, 7.6, and 7.7, we obtain

$$\begin{aligned}(a \oplus b) \oplus (a \oplus c) &= ((a \oplus b) \oplus a) \oplus c \\ &= (a \oplus (a \oplus b)) \oplus c \\ &= ((a \oplus a) \oplus b) \oplus c \\ &= (a \oplus b) \oplus c \\ &= a \oplus (b \oplus c) \\ &= a \oplus c.\end{aligned}$$

□

We next give an *axiomatic definition* of a lattice. That is, we introduce a lattice as an algebraic system with two binary operations that satisfy certain *axioms*, namely, *commutativity,associativity*, and *absorption*.

**Definition 7.12** (*Axiomatic Definition*) A **lattice** $L$ is an algebraic system with two binary operations $\oplus$ and $*$ such that, for all $a, b, c \in L$, we have

1. $a \oplus b = b \oplus a$ and $a * b = b * a$;
2. $a \oplus (b \oplus c) = (a \oplus b) \oplus c$ and $a * (b * c) = (a * b) * c$;
3. $a \oplus (a * b) = a$ and $a * (a \oplus b) = a$.

In this case, we write a *lattice* as $(L, \oplus, *)$.

The next theorem provides the first step toward proving the equivalence of the above approach with the one introduced earlier in Definition 7.7. The main idea is to use the two operations to define a partial order relation such that it has all the desired properties (Exercise 7.9).

**Theorem 7.11** *Suppose* $(L, *, \oplus)$ *is a lattice in the sense of above definition. We may define a relation* $R$ *on* $L$ *as follows:*

$$a\,R\,b \quad \Leftrightarrow \quad a * b = a. \tag{7.3.8}$$

*Then* $R$ *is a partial order relation on* $L$.

***Proof*** To complete the proof, we need to show that

1. $R$ is reflexive, i.e., $aRa$, for all $a \in L$;
2. $R$ is symmetric, i.e., for $a, b \in L$, $aRb$ and $bRa$ implies that $a = b$;
3. $R$ is transitive, i.e., for $a, b, c \in L$, $aRb$ and $bRc$ implies that $aRc$.

For (1), by absorption law, we have

$$a * a = a * [a \oplus (a * a)] = a,$$

and hence $a * a = a$ or $aRa$. For (2), if $aRb$ and $bRa$, then we have

$$a * b = a \quad \text{and} \quad b * a = b.$$

Therefore, by commutative law, we get $a = a * b = b * a = b$. For (3), if $aRb$ and $bRc$, then we have

$$a * b = a \quad \text{and} \quad b * c = b.$$

Using the above, we obtain

$$\begin{aligned} a * c &= (a * b) * c \\ &= a * (b * c) \\ &= a * b = a. \end{aligned}$$

It thus follows that $aRc$. □

## 7.4 Distributive and Modular Lattices

We now prove some important inequalities that are valid in any lattice. The next theorem proves *distributive inequalities.*

**Theorem 7.12** (Distributive Inequalities) *Let* $(L, \preceq)$ *be a lattice, and* $a, b, c \in L$. *Then we have*

$$a \oplus (b * c) \leq (a \oplus b) * (a \oplus c), \tag{7.4.1}$$

$$a * (b \oplus c) \geq (a * b) \oplus (a * c). \tag{7.4.2}$$

***Proof*** We clearly have

$$a \preceq \sup\{a, b\} = a \oplus b \quad \text{and} \quad a \preceq \sup\{a, c\} = a \oplus c.$$

The above inequalities yield

$$a \preceq (a \oplus b) * (a \oplus c).$$

Further, since we have

$$b * c = \inf\{b, c\} \preceq b \preceq \sup\{a, b\} = a \oplus b;$$
$$b * c = \inf\{b, c\} \preceq c \preceq \sup\{a, c\} = a \oplus c,$$

it follows that

$$b * c \preceq (a \oplus b) * (a \oplus c).$$

Finally, using $a \preceq (a \oplus b) * (a \oplus c)$ and $b * c \preceq (a \oplus b) * (a \oplus c)$, we obtain

$$a \oplus (b * c) \preceq (a \oplus b) * (a \oplus c).$$

The other inequality follows from the first by applying the principle of duality. □

**Definition 7.13** A lattice $(L, *, \oplus]$ is called a **distributive lattice** if, for any $a, b, c \in L$, we have

$$a * (b \oplus c) = (a * b) \oplus (a * c); \tag{7.4.3}$$
$$a \oplus (b * c) = (a \oplus b) * (a \oplus c). \tag{7.4.4}$$

Said differently, $L$ is a *distributive lattice* if each one of the two operations $*$ and $\oplus$ distributive over the other.

**Example 7.8** The lattice $\big(\mathscr{P}(X), \subseteq, \cup, \cap\big)$ of the power set of a nonempty set $X$ is a distributive lattice. In general, the direct product of any two distributive lattices is a distributive lattice.

It follows easily that the conditions (7.4.3) and (7.4.4) are equivalent. Therefore, to check a given lattice for distributivity, it is enough to verify any one of the two conditions, for all possible combinations of the elements of the given lattice. The next example illustrates the procedure.

**Example 7.9** The lattice $L_1$ with the Hasse diagram shown in Fig. 7.15 is a *non-distributive lattice*. For, we have

$$x_1 * (x_2 \oplus x_3) = x_1 * 1 = x_1;$$

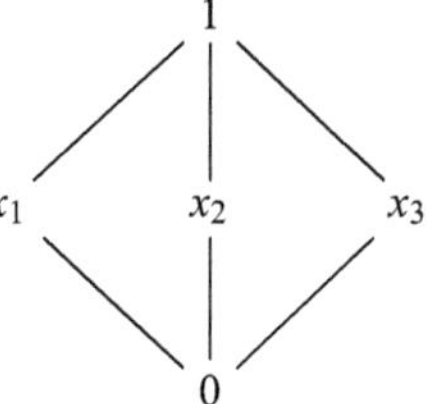


**Fig. 7.15** A non-distributive lattices $L_1$

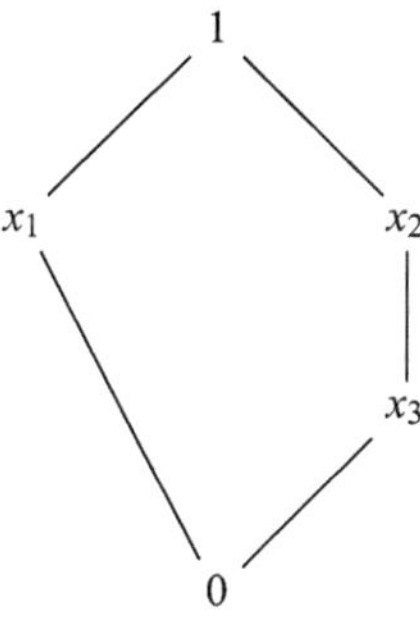


**Fig. 7.16** A non-distributive lattices $L_2$

and also that

$$(x_1 * x_2) \oplus (x_1 * x_3) = 0 \oplus 0 = 0.$$

It thus follows that

$$x_1 * (x_2 \oplus x_3) \neq (x_1 * x_2) \oplus (x_1 * x_3).$$

Therefore, $L_1$ is a non-distributive lattice. Also, the lattice $L_2$ with the Hasse diagram shown in Fig. 7.16 is not distributive. In this case, we have

$$x_3 * (x_1 \oplus x_2) = x_3 * 1 \ = x_3;$$

and also that

$$(x_3 * x_1) \oplus (x_3 * x_2) = 0 \oplus x_2 = x_2.$$

It thus follows that

$$x_3 * (x_1 \oplus x_2) \neq (x_3 * x_1) \oplus (x_3 * x_2).$$

Therefore, $L_2$ is not distributive.

**Theorem 7.13** (Cancellation Property) *Let* $(L, *, \oplus)$ *be distributive lattice, and* $a, b, c \in L$. *Then we have*

$$a \oplus b = a \oplus c \quad \text{and} \quad a * b = a * c \quad \Rightarrow \quad b = c.$$

***Proof*** Let

$$\begin{aligned}
b &= b \oplus (b * a) && \text{(Absorption law)}\\
&= b \oplus (a * b) && \text{(Commutative law)}\\
&= b \oplus (a * c) && (a * b = a * c)\\
&= (b \oplus a) * (b \oplus c) && \text{(Distributive law)}\\
&= (a \oplus b) * (c \oplus b) && \text{(Commutative law)}\\
&= (a \oplus c) * (c \oplus b) && (a \oplus b = a \oplus c)\\
&= (c \oplus a) * (c \oplus b) && \text{(Commutative law)}\\
&= c \oplus (a * b) && \text{(Distributive law)}\\
&= c \oplus (a * c) && (a * b = a * c)\\
&= c \oplus (c * a) && \text{(Commutative law)}\\
&= c && \text{(Absorption law)}
\end{aligned}$$

Therefore $a \oplus b = a \oplus c$ and $a * b = a * c \Rightarrow b = c$. □

## 7.5 Bounded and Complemented Lattices

Let $X$ be a nonempty set, and consider the lattice $\big(\mathscr{P}(X), \subseteq, \cup, \cap\big)$ of all subsets of the set $X$. We know that $\emptyset$ is the least element of this lattice and $X$ is the greatest element. In this case, we say $\mathscr{P}(X)$ is a bounded lattice, with its bounds given by $\emptyset$ and $X$. Also, it is complemented in the sense that every element $A \in \mathscr{P}(X)$ has the complement given by the set $A' = X \setminus A$. In all that follows in this part of the section, we generalize the above concepts for an arbitrary lattice.

**Definition 7.14** A lattice $L := (L, *, \oplus)$ is called a *bounded lattice* if it has both the least element and the greatest element, denoted by 0 and 1, respectively. The elements 0 and 1 are also known as the bounds of the lattice $L$.

Clearly, in any bounded lattice $(L, *, \oplus, 0, 1)$, we have

$$a \oplus 0 = a, \quad a * 1 = a \qquad \text{and} \qquad a \oplus 1 = 1, \quad a * 0 = 0.$$

**Definition 7.15** In a bounded lattice $(L, *, \oplus, 0, 1)$, an element $b \in L$ is said to be a *complement* of an element $a \in L$ if

$$a * b = 0 \qquad a \oplus b = 1.$$

We write a complement of the element $a$ as $a'$.

Clearly, in any bounded lattice $(L, *, \oplus, 0, 1)$, we have

$$1 \oplus 0 = 0 \oplus 1 = 1, \qquad 0 * 1 = 0 * 1 = 0.$$

Therefore, 0 and 1 are complements to each other.

**Definition 7.16** A bounded lattice $(L, *, \oplus, 0, 1)$ is called a *complemented lattice* if every element of $L$ has at least one complement.

**Example 7.10** Let $L = (\{0, 1\}, \leq)$, and consider the lattice $(L^n, \preceq_n)$ of all $n$-tuples of 0 and 1, where the partial order $\leq$ is the usual "less than" relation while $\preceq_n$ is the partial order obtained from $\leq$ for the product lattice $L^n$. It is a bounded lattice with $(0, \ldots, 0)$ and $(1, \ldots, 1)$ as its bounds. Further, $L^n$ is complemented because every element has a complement obtained by interchanging 0's and 1's.

**Example 7.11** All the three lattices with the Hasse diagrams shown in Fig. 7.6 are complemented. For example, with reference to the left side lattice, we have elements $2, 3 \in D_6$ that are complements of each other. Also, for the lattice $L_1$ in Example 7.9 with the Hasse diagram as shown in Fig. 7.15, any one of the three elements $x_1$, $x_2$, and $x_3$ has the other two elements as complements. Notice that, in particular, it follows that *complement of an element in an arbitrary lattice may not be unique*. Further, the lattice $L_2$ in Example 7.9 with the Hasse diagram as shown in Fig. 7.16 is complemented. In this case, $x_1$ has two complements $x_2$ and $x_3$, while $x_1$ is the only complement for each one of the two elements $x_2$ and $x_3$.

The next theorem gives a far more convenient way of showing that certain lattices are non-distributive. In particular, it simplifies arguments given in Example 7.9.

**Theorem 7.14** *In a distributive lattice, the complement of an element is unique if it exists.*

***Proof*** Let $(L, *, \oplus)$ be a distributive lattice, and $a \in L$. If possible, suppose $b$ and $c$ are complements of $a$. Then we have

$$a * b = 0 \quad a \oplus b = 1 \qquad \text{and} \qquad a * c = 0 \quad a \oplus c = 1.$$

Therefore, it follows that

$$a * b = a * c, \quad a \oplus b = a \oplus c.$$

Hence, by Theorem 7.13, we obtain $b = c$. The uniqueness of a complement is thus proved. □

For the next theorem, recall that every chain is a lattice (Exercise 7.9).

**Theorem 7.15** *Every chain is a distributive lattice.*

***Proof*** Let $(L, \preceq)$ be a chain, and for any $x, y, z \in L$. We consider the following two possible cases:

$$(i)\ x \preceq y \text{ or } x \preceq z \quad \text{and} \quad (ii)\ y \preceq x \text{ and } z \preceq x.$$

If $(i)$ holds, we have $x \preceq y$ or $x \preceq z$, and so $x \preceq (y \oplus z)$. Therefore, we obtain $x * (y \oplus z) = x$. Also, when $x \preceq y$, by absorption law, we have

$$(x * y) \oplus (x * z) = x \oplus (x * z) = x.$$

When $x \preceq z$, by absorption law, we have

$$(x * y) \oplus (x * z) = (x * y) \oplus x = x.$$

It thus follows that

$$x * (y \oplus z) = x = (x * y) \oplus (x * z).$$

In the present case, it proves that one form of the distributive law. The other distributive law follows by applying the duality principle. Next, in case $(ii)$ holds, we have $y \preceq x$ and $z \preceq x$ so that $(y \oplus z) \preceq x$. Therefore, we have $x * (y \oplus z) = (y \oplus z)$ and

$$(x * y) \oplus (x * z) = (y * z),$$

which proves that

$$x * (y \oplus z) = (x * z) = (y * z).$$

As said above, the other distributive law follows by applying the duality principle. □

The next two theorems prove *modular inequalities*.

**Theorem 7.16** (Modular Inequality–I) *Let $(L, \preceq)$ be a lattice, and $a, b, c \in L$. Then we have*

$$a \preceq c \Rightarrow a \oplus (b * c) \preceq (a \oplus b) * c. \tag{7.5.1}$$

***Proof*** It follows from the inequality (7.4.1) that

$$a \oplus (b * c) \preceq (a \oplus b) * (a \oplus c).$$

Therefore, as $a \preceq c$, we have $a \oplus c = c$, and hence

$$a \oplus (b * c) \preceq (a \oplus b) * c.$$

□

**Theorem 7.17** (Modular Inequalities–II) *Let $(L, \preceq)$ be a lattice, and $a, b, c \in L$. Then we have*

$$(a * b) \oplus (a * c) \preceq a * [b \oplus (a * c)], \qquad (7.5.2)$$
$$(a \oplus b) * (a \oplus c) \succeq a \oplus [b * (a \oplus c)], \qquad (7.5.3)$$

***Proof*** It follows from Theorem 7.9 that $a \oplus (a * c) = a$. Also, by Theorems 7.7 and 7.12, we have

$$\begin{aligned}(a * b) \oplus (a * c) &= (a * c) \oplus (a * b)\\ &\preceq [(a * c) \oplus a] * [(a * c) \oplus b]\\ &\preceq [a \oplus (a * c)] * [b \oplus (a * c)]\\ &= a * [b \oplus (a * c)].\end{aligned}$$

The second inequality follows from the first by applying the duality principle. □

**Definition 7.17** A lattice $(L, *, \oplus)$ is called a *modular lattice* if we have

$$a \preceq c \quad \Rightarrow \quad a \oplus (b * c) = (a \oplus b) * c, \qquad (7.5.4)$$

for any $a, b, c \in L$.

**Example 7.12** We already know that the lattice $L_1$ with the Hasse diagram as shown in Fig. 7.15 is a non-distributive modular lattice. We next show that it satisfies the condition (7.5.4). Notice that, if $a = c$, then we are done using the absorption law. Therefore, we may assume that $a \prec c$. In this case, the two possible cases involving $x_1$ are as follows: $0 \prec x_1$ and $x_1 \prec 1$. The other cases involving $x_2$ and $x_3$ can be dealt with similarly. Now, if $0 \prec x_1$, then we have

$$0 \oplus (x * x_1) = x * x_1 = (0 \oplus x) * x_1.$$

Similarly, if $x_1 \prec 1$, then we have

$$x_1 \oplus (x * 1) = x_1 \oplus x = (x_1 \oplus x) * 1.$$

Hence, $L_1$ is modular. On the other hand, the lattice $L_2$ with the Hasse diagram as shown in Fig. 7.16 is not a modular lattice. In this case, we have $x_3 \preceq x_2$, but

$$x_3 \oplus (x_1 * x_2) = x_3 \oplus 0 = x_3$$

and we also have

$$(x_3 \oplus x_1) * x_2 = 1 * x_2 = x_2.$$

Therefore, we see that $x_3 \preceq x_2$, but

$$x_3 \oplus (x_1 * x_2) \neq (x_3 \oplus x_1) * x_2.$$

Hence, $L_2$ is not a modular lattice.

**Theorem 7.18** *Every distributive lattice is modular.*

***Proof*** Let $(L, *, \oplus)$ be a distributive lattice, and $a, b, c \in L$ with $a \preceq c$. Then we have $a \oplus c = c$, and so it follows from the distributive law that

$$a \oplus (b * c) = (a \oplus b) * (a \oplus c) = (a \oplus b) * c.$$

Therefore, $L$ is modular. □

**Corollary 7.1** *Every chain is a modular Lattice.*

***Proof*** By Theorem 7.15, every chain is a distributive lattice, and hence it is modular by the above theorem. □

**Theorem 7.19** *In a complemented distributive lattice, each element has a unique complement.*

***Proof*** Suppose $(L, *, \oplus)$ is a complemented distributive lattice, and $a \in L$. By definition, we know that $a$ has at least one complement. If possible, let $a$ has two *distinct* complements, say $a_1$ and $a_2$. Then we have

$$\begin{aligned} a * a_1 &= 0 \quad \text{and} \quad a \oplus a_1 = 1; \\ a * a_2 &= 0 \quad \text{and} \quad a \oplus a_2 = 1. \end{aligned}$$

It thus follows that

$$\begin{aligned} a_1 &= a_1 * 1 \\ &= a_1 * (a \oplus a_2) \\ &= (a_1 * a) \oplus (a_1 * a_2) \\ &= 0 \oplus (a_1 * a_2) \\ &= (a_2 * a) \oplus (a_1 * a_2) \\ &= (a_2 * a) \oplus (a_2 * a_1) \\ &= a_2 * (a \oplus a_1) \\ &= a_2 * 1 = a_2, \end{aligned}$$

which contradicts our assumption that $a_1 \neq a_2$. Hence, $a$ has a unique complement. □

**Theorem 7.20** (Involution Law) *In a complemented distributive lattice, every element is an* involution *with respect to unary operation $a \mapsto a'$, for $a \in L$.*

***Proof*** Suppose $(L, *, \oplus)$ is a complemented distributive lattice, and $a \in L$. Since

$$a' \oplus a = 1 \qquad \text{and} \qquad a' * a = 0,$$

by definition of a complement, the assertion follows directly from Theorem 7.19. □

**Theorem 7.21** (De Morgan's Law) *Let $(L, *, \oplus)$ be a complemented distributive lattice. Then, for any $a, b \in L$, we have*

$$\big(a \oplus b\big)' = a' * b' \qquad \textit{and} \qquad \big(a * b\big)' = a' \oplus b'.$$

***Proof*** We have

$$\begin{aligned}
(a \oplus b) * (a' * b') &= (a * (a' * b')) \oplus (b * (a' * b')) \\
&= ((a * a') * b') \oplus ((b * (b' * a')) \\
&= (0 * b') \oplus ((b * b') * a') \\
&= 0 \oplus (0 * a') \\
&= 0 \oplus 0 = 0.
\end{aligned}$$

We also have that

$$\begin{aligned}
(a \oplus b) \oplus (a' * b') &= (a \oplus (a' * b')) \oplus (b \oplus (a' * b')) \\
&= ((a \oplus a') * (a \oplus b')) \oplus ((b \oplus a') * (b \oplus b')) \\
&= (1 * (a \oplus b')) \oplus ((b \oplus a') * 1) \\
&= (a \oplus b') \oplus (b \oplus a') \\
&= a \oplus (b' \oplus b) \oplus a' \\
&= (a \oplus 1) \oplus a' \\
&= a \oplus a' = 1.
\end{aligned}$$

It thus follows that

$$(a \oplus b) * (a' * b') = 0 \qquad \text{and} \qquad (a \oplus b) \oplus (a' * b') = 1.$$

That is, $a \oplus b$ and $a' * b'$ are complements to each another. Hence, $(a \oplus b)' = a' * b'$. The second result follows from the duality principle. □

**Theorem 7.22** *Let $(L, *, \oplus)$ be a complemented distributive lattice, and $a, b \in L$. Then the following are equivalent:*

1. $a \preceq b$.
2. $a * b' = 0$.
3. $a' \oplus b = 1$.
4. $b' \preceq a'$.

***Proof*** To prove (1) $\Rightarrow$ (2), suppose $a \preceq b$. Then we have $a \oplus b = b$. Therefore,

$$b * b' = 0 \quad \Rightarrow \quad \big(a \oplus b\big) * b' = b * b' = 0,$$

and hence, by distributive law, we have

$$\big(a * b'\big) \oplus \big(b * b'\big) = 0.$$

It thus follows from $b * b' = 0$ that

$$\big(a * b'\big) = \big(a * b'\big) \oplus 0 = \big(a * b'\big) \oplus \big(b * b'\big) = 0.$$

Next, to prove (2) $\Rightarrow$ (3), suppose $a * b' = 0$. Then we have $(a * b')' = 0' = 1$. Also, by De Morgan's law, we obtain $a' \oplus (b')' = 1$ or $a' \oplus b = 1$, by Theorem 7.20. Further, to prove (3) $\Rightarrow$ (4), suppose $a' \oplus b = 1$. Then we have

$$\begin{aligned} b' &= 1 * b' \\ &= \big(a' \oplus b\big) * b' \\ &= \big(a' * b'\big) \oplus \big(b * b'\big) \\ &= \big(a' * b'\big) \oplus 0 \\ &= a' * b' \preceq a'. \end{aligned}$$

Finally, to prove (4) $\Rightarrow$ (1), suppose $b' \preceq a'$. Then we have $a' * b' = b'$, and by taking complements and applying De Morgan's law, we obtain

$$\begin{aligned} b &= \big(b'\big)' \\ &= \big(a' * b'\big)' \\ &= \big(a'\big)' \oplus \big(b'\big)' \\ &= a \oplus b \preceq b. \end{aligned}$$

□

Lattices we introduced above find applications in various subject areas. In particular, some with certain nice properties have a richer algebraic structure that is of prime importance to the switching theory. This is the topic for the next section.

## 7.6 Boolean Algebras

A *Boolean algebra* $B$ may be defined as a distributive complemented lattice, which is called the *lattice Boolean algebra*. Recall that each element $a$ in a distributive complemented lattice has a unique complement $a'$ so that we have a well-defined unary operation $^- : B \to B$ given by $a \mapsto \bar{a}$, where $\bar{a} = a'$.

**Definition 7.18** A distributive complemented lattice $(L, \preceq, \vee, \wedge, ', 0, 1)$ is called a *Boolean algebra*, which we may write as $(B, \oplus, *, ^-, 0, 1)$ such that the sum $\oplus$ and the product $*$ correspond respectively to join $\vee$ and meet $\wedge$ operations. This is called **lattice Boolean algebra**.

A Boolean algebra may also be defined as an algebraic system.

**Definition 7.19** A set $B$ with at least two distinct elements, usually denoted by 0 and 1, is called a **Boolean algebra** if there are two binary operations $\oplus, \ * : B \times B \to B$ and a unary operation $^- : B \to B$ such that the following axioms hold:

(B1) Both $\oplus$ and $*$ are associative:

$$(a \oplus b) \oplus c = a \oplus (b \oplus c) \text{ and } \quad (a * b) * c = a * (b * c),$$

for all $a, b, c \in B$;

(B2) Both $\oplus$ and $*$ are commutative:

$$a \oplus b = b \oplus a \text{ and } \quad a * b = b * a,$$

for all $a, b \in B$;

(B3) Both $\oplus$ and $*$ distribute over each other:

$$a * (b \oplus c) = (a * b) \oplus (a * c)$$
$$a \oplus (b * c) = (a \oplus b) * (a \oplus c),$$

for all $a, b, c \in B$;

(B4) Both $\oplus$ and $*$ have identity elements:

$$a \oplus 0 = a \text{ and } \quad a * 1 = a.$$

(B5) Each element $a \in B$ has a complement $\bar{a}$:

$$a \oplus \bar{a} = 1 \text{ and } \quad a * \bar{a} = 0.$$

In this case, we also write Boolean algebra as $(B, \oplus, *, ^-, 0, 1)$.

By taking $\vee$ as the sum $+$ and $\wedge$ as the product $*$, it follows from our discussion in the previous section that all the five axioms $(B1)$–$(B5)$ hold, with the least element $0 \in L$

as the *zero element*, and the greatest element $1 \in L$ as the *identity element*. Therefore, every distributive complemented lattice is a Boolean algebra. For emphasis, we may also write $B$ as $B_L$. Conversely, we may write the poset associated with a Boolean algebra $(B, \oplus, *, ^-, 0, 1)$ as $(B, \preceq)$, where 0 is the least element and 1 is the greatest element (Theorem 7.11).

**Example 7.13** (*Boolean Algebra of Sets*) Let $B$ be the power set $\mathscr{P}(A)$ of a set $A = \{a, b, c\}$. That is, we have

$$\mathscr{P}(A) = \big\{\emptyset, \{a\}, \{b\}, \{c\}, \{a, b\}, \{b, c\}, \{a, c\}, A\big\}.$$

Since $\big(\mathscr{P}(A), \subseteq, \cap, \cup\big)$ is a distributive complemented lattice, with the least element $\emptyset$ and the greatest element $A$, it follows that $(B, \cup, \cap, ^c, \emptyset, A)$ is a Boolean algebra. In general, with respect to binary operations $\cup$ and $\cap$, and unary operation given by usual "complementation", the set $\mathscr{P}(A)$ is a Boolean algebra. As above, we take $0 \equiv \emptyset,\ 1 \equiv A$. This is called the Boolean algebra of subsets of a set $A$.

According to Theorem 7.24, up to an isomorphism of Boolean algebras, every finite Boolean algebra is an algebra of sets. In particular, a finite Boolean algebra has $2^n$ elements, for some $n \geq 2$ (Corolary 7.2). For $n = 2$, we have the simplest Boolean algebra. It is defined on the set $B = \{0, 1\}$, with operations $\oplus$, $*$, and $^-$, as given in the following tables. It is called the **2-point** Boolean algebra.

| $\oplus$ | 0 | 1 |
|---|---|---|
| 0 | 0 | 1 |
| 1 | 1 | 1 |

| $*$ | 0 | 1 |
|---|---|---|
| 0 | 0 | 0 |
| 1 | 0 | 1 |

| $x$ | $\bar{x}$ |
|---|---|
| 0 | 1 |
| 1 | 0 |

**Example 7.14** (*Boolean Algebra of Bit Strings*) Let $B^n$ denotes the set of **bit strings** of length $n$ over the set $B$. That is, we have

$$B^n := \big\{x_1x_2\cdots x_n \mid \ x_i \in B\big\}.$$

The above set of *bit strings* has a natural Boolean algebra structure, where the addition, product, and the complementation, are defined pointwise. That is, we use the Boolean algebra structure of the two-point set $B^1$ to define these three operations as follows:

$$\begin{aligned} x_1\cdots x_n + y_1\cdots y_n &= \big(x_1 + y_1\big)\cdots\big(x_n + y_n\big); \\ \big(x_1\cdots x_n\big)\cdot\big(y_1\cdots y_n\big) &= \big(x_1\cdot y_1\big)\cdots\big(x_n\cdot y_n\big); \\ \big(x_1\cdots x_n\big)' &= x_1'\cdots x_n'. \end{aligned}$$

The zero element **0** and the identity element **1** of $B^n$ are respectively given by

$$\mathbf{0} := 0\cdots 0 \qquad \text{and} \qquad \mathbf{1} := 1\cdots 1.$$

**Remark 7.1** The German mathematician *Gottfried Wilhelm Leibniz* (1646–1716) was among the first who realized the need to formalize logical *argument forms*. He wanted to create a universal symbolic language of science (*characteristica generalis* or *lingua generalis*) that would reduce all debates about the subject to a matter of mere calculation by automating the rules of reasoning [5]. The first breakthrough appeared when the self-taught English mathematician *George Boole* published opus *An Investigation of the Laws of Thought* in 1854, wherein he developed a *system of calculation* to discuss the fundamental laws of those operations of the mind by which reasoning is performed in terms of operations $+$, $\times$, and $-$, with meanings taken in a purely logical sense. These symbols were introduced to rewrite statements involving connectives "and", "or", and "except". However, Boole's use of "or" was mostly in the exclusive sense.

In what follows, the term algebra would always mean a Boolean algebra.

## 7.7 Laws of Boolean Algebras

Let $(B, \oplus, *, ^{-}, 0, 1)$ be a Boolean algebra. We now state some other important laws that hold in every Boolean algebras. For any $a, b, c \in B$, we have

1. (**Associative Laws**) $(a \oplus b) \oplus c = a \oplus (b \oplus c)$; $(a * b) * c = a * (b * c)$.
2. (**Commutative Laws**) $a \oplus b = b \oplus a$; $a * b = b * a$.
3. (**Distributive Laws**):

$$a * (b \oplus c) = (a * b) \oplus (a * c); \quad a \oplus (b * c) = (a \oplus b) * (a \oplus c).$$

4. (**Identity Laws**) $a \oplus 0 = a$ and $a * 1 = a$.
5. (**Complement Laws**) $a \oplus \bar{a} = 1$ and $a * \bar{a} = 0$.
6. (**Idempotent Laws**) $a \oplus a = a$ and $a * a = a$.
7. (**Dominance Laws**) $a \oplus 1 = 1$ and $a * 0 = 0$.
8. (**Absorption Laws**) $a \oplus (a * b) = a$ and $a * (a \oplus b) = a$.
9. (**DeMorgan's Laws**) $\overline{(a \oplus b)} = \bar{a} * \bar{b}$ and $\overline{(a * b)} = \bar{a} \oplus \bar{b}$.
10. (**Involution Law**) $\overline{(\bar{a})} = a$.
11. (**0-1 Law**) $\bar{0} = 1$ and $\bar{1} = 0$.

The distributive law for addition over multiplication may appear somewhat unusual since it has no counterpart in the case of usual addition and multiplication of a number set. More generally, in what follows, it is best to avoid any analogies with usual arithmetic. Instead, we may use the *propositional algebra* as our model to draw analogies, where the operations $\oplus$, $*$, and $^{-}$, are seen as the logical connectives "or", "and", and "negation", respectively. We have already given proof for most of these laws in the case of lattices (Sect. 7.3). The same works for Boolean algebras, with minor modifications, whenever needed.

**Definition 7.20** Let $(B, \oplus, *, ^{-}, 0, 1)$ be a Boolean algebra and $S \subseteq B$. If $S$ contains the elements 0 and 1 and is closed under the operations $\oplus$, $*$ and $^{-}$, then $(S, \oplus, *, ^{-}, 0, 1)$ is called sub-Boolean algebra.

**Definition 7.21** (*Direct product*) The direct product of the two Boolean algebras given by

$$(B_1, \oplus_1, *_1, ^{-}, 0_1, 1_1) \text{ and } (B_2, \oplus_2, *_2, ^{-}, 0_2, 1_2).$$

is the Boolean algebra given by $(B_1 \times B_2, \oplus, *, ^{-}, 0, 1)$ such that, for any $(a_1, b_1), (a_2, b_2) \in (B_1 \times B_2)$, we have

$$\begin{aligned}(a_1, b_1) \oplus (a_2, b_2) &= (a_1 \oplus_1 a_2, b_1 \oplus_2 b_2);\\ (a_1, b_1) * (a_2, b_2) &= (a_1 *_1 a_2, b_1 *_2 b_2);\\ \overline{(a_1, b_1)} &= (\overline{a}_1, \overline{b}_1).\end{aligned}$$

Furthermore, we have $0 = (0_1, 0_2)$ and $1 = (1_1, 1_2)$.

**Definition 7.22** Let $(B_1, \oplus_1, *_1, ^{-}, 0_1, 1_1)$ and $(B_2, \oplus_2, *_2, ^{-}, 0_2, 1_2)$ be two Boolean algebras. A mapping $f : B_1 \to B_2$ is called a **Boolean homomorphism** if, for all $a, b \in B$, we have

$$\begin{aligned}f\big(a \oplus_1 b\big) &= f\big(a\big) \oplus_2 f\big(b\big);\\ f\big(a *_1 b\big) &= f\big(a\big) *_2 f\big(b\big);\\ f\big(\bar{a}\big) &= \overline{f(a)}.\end{aligned}$$

Furthermore, $f(0_1) = 0_2$ and $f(1_1) = 1_2$. A one-to-one *Boolean homomorphism* $f : B_1 \to B_2$ is called a **Boolean isomorphism**. In this case, we say $B_1$ and $B_2$ are **isomorphic** Boolean algebras.

We next state a fundamental result for Boolean algebras, which is known as *Stone's representation theorem* (Theorem 7.24). However, we need some preparation to do so.

**Definition 7.23** Let $(B, \oplus, *, ^{-}, 0, 1)$ be a Boolean algebra. A nonzero element $a \in B$ is called an **atom** if, for every $x \in B$, we have $x * a = a$ or $x * a = 0$. We say $B$ is **atomic** if, for every $0 \neq x \in B$, it is possible to find an atom $b \in B$ such that $b \leq x$.

Clearly, the condition $x * a = 0$ is true only when $x$ and $a$ are not connected. This occurs when $x$ is another atom or if it is a successor of atoms different from $a$. The immediate successors of the zero element 0 in any Boolean algebra are atoms. In particular, for any nonempty set $A$, the singletons in the Boolean algebra $(\mathscr{P}(A), \subseteq, \cup, \cap)$ are atoms, which are also *algebraic generators* in the sense that each element of $\mathscr{P}(A)$ can be expressed uniquely as join of singleton sets. The next theorem proves an interesting fact for finite Boolean algebras.

**Theorem 7.23** *Every finite Boolean algebra is atomic.*

***Proof*** Let $(B, \oplus, *, ^{-}, 0, 1)$ be a Boolean algebra, with $B$ a finite set. If possible, suppose $B$ is not atomic. Then there is an element $0 \neq x_0 \in B$ such that $x_0$ contains no atom.hence there is some $x_1$ such that $0 < x_1 < x_0$. Similarly, there is some $x_2$ such that $0 < x_2 < x_1 < x_0$. Proceeding in this way, we obtain an infinite sequence $x_0, x_1, x_2, \ldots$ such that $x_0 > x_1 > x_2 > \cdots$, which contradicts the fact that $B$ is finite. Hence for every non- zero $x \in B$ there is some atom $b$ such that $b \leq x$. Hence $B$ is atomic. □

**Theorem 7.24** (Stone Representation Theorem) *Let* $(B, \oplus, *, ^{-}, 0, 1)$ *be a Boolean algebra, and let A be the set of all atoms in this Boolean algebra. Then B is isomorphic to* $(P(A), \cup, \cap, ^{c}, \emptyset, A)$.

**Corollary 7.2** *Every finite Boolean algebra* $(B, \oplus, *, ^{-}, 0, 1)$ *has* $2^n$ *elements for some positive integer n.*

***Proof*** Let $A$ be the set of all atoms of $B$ and let $O(A) = n$ Then there are exactly $2^n$ elements(subsets) in $P(A)$, and by the previous theorem $(B, \oplus, *, ^{-}, 0, 1)$ is isomorphic to

$$(P(A), \cup, \cap, ^{c}, \emptyset, A).$$

Therefore $B$ has exactly $2^n$ elements where $n$ is the number of elements in $A$. □

**Corollary 7.3** *Every Boolean algebra of order* $2^n$ *are isomorphic to each other.*

***Proof*** By the previous theorem, every Boolean algebra of order $2^n$ is isomorphic to $(P(A), \cup, \cap, ^{c}, \emptyset, A)$ where $A$ is set of all atoms and $O(A) = n$. Hence, all Boolean algebras of order $2^n$ is isomorphic to each other. □

## 7.8 Boolean Expressions and Boolean Functions

Suppose $x_1, \ldots, x_n$ are variables taking values in a Boolean algebra $(B, \oplus, *, ^{-}, 0, 1)$. A *literal* refers to variable $x_i$ or its complement $\overline{x}_i$. In particular, a variable $x_i$ that takes values in the 2-point Boolean algebra $B^1 = \{0, 1\}$ is called a **Boolean variable**. In the context of *switching algebra*, an **input condition** refers to an assignment of the value 1 or 0 to a *Boolean variable* $x_i$, where 1 indicates active *input condition* and 0 inactive input condition of a two-value device such as an *OR gate* or an *AND gate*. The total number of these types of strings of length $n$ is $2^n$.

**Definition 7.24** (*Boolean Expression*) A **Boolean expression** in $n$ variables, denoted by $E(x_1, \ldots, x_n)$, is a combination of literals, and elements of $B$, formed by using sum and product operations of $B$.

For example, $a \oplus \overline{(c \oplus a)} * b$ is a *Boolean expression* in variables $a, b$, and $c$, and so is $a * (b * a) \oplus c \oplus 0$. In practical terms, a Boolean expression is an algebraic representation of a **logic circuit**, where variables are taken as inputs to various types of component logic gates, and each sub-circuit (also known as stages or levels) contributes to the whole expression involving sum, product, or complementation of the variables.

**Definition 7.25** (*Boolean Function*) A **Boolean function** over a Boolean algebra $B$ of degree $n$ is an $n$-variable function $F : B^n \to B$. We also say that $F$ is a *Boolean function* of order $n$. A unique tabular representation of values of $F$ for all possible input conditions of the $n$ variables is called the **truth table** of the Boolean function $F$.

In particular, to each Boolean expression $E(x_1, \ldots, x_n)$, there corresponds a Boolean function $F_E : B^n \to B$ given by

$$F_E\big(a_1, \ldots, a_n\big) = E\big(a_1, \ldots, a_n\big), \tag{7.8.1}$$

where $a_1, \ldots, a_n \in B$. When each $x_i$ is a Boolean variable, $F$ is called the **switching function** for the *logic circuit* represented by the Boolean expression $E$. In this latter case, the variables $x_i$'s take value 1 or 0 depending on the input condition of the $n$ variables. Hence, there are $2^{2^n}$ Boolean functions involving $n$ variables.

**Example 7.15** For $n = 0$ variable, we take $F_0 \equiv 0$ and $F_1 \equiv 1$ as the switching functions. For $n = 1$, the four switching functions of a variable $x$ are given by

$$F_0\big(x\big) = 0, \quad F_1\big(x\big) = \bar{x}, \quad F_2\big(x\big) = x, \quad F_3\big(x\big) = 1.$$

For $n = 2$ variables $x$ and $y$, $F_0(x, y) = 0$ is the trivial *switching function*. The other fifteen switching functions are given by

$$\begin{aligned}
&F_1\big(x, y\big) = \bar{x}\,\bar{y}, \quad F_2\big(x, y\big) = \bar{x}\,y, \quad F_3\big(x, y\big) = \bar{x}\,y + \bar{x}\,\bar{y} = \bar{x};\\
&F_4\big(x, y\big) = x\bar{y}, \quad F_5\big(x, y\big) = x\bar{y} + \bar{x}\bar{y} = \bar{y}, \quad F_6\big(x, y\big) = x\bar{y} + \bar{x}y;\\
&F_7\big(x, y\big) = x\bar{y} + \bar{x}y + \bar{x}\bar{y} = \bar{x} + \bar{y}, \quad F_8\big(x, y\big) = xy, \quad F_9\big(x, y\big) = xy + \bar{x}\bar{y};\\
&F_{10}\big(x, y\big) = xy + \bar{x}y, \quad F_{11}\big(x, y\big) = xy + \bar{x}y + \bar{x}\bar{y} = \bar{x} + y;\\
&F_{12}\big(x, y\big) = xy + x\bar{y} = x, \quad F_{13}\big(x, y\big) = xy + x\bar{y} + \bar{x}\bar{y} = x + \bar{y};\\
&F_{14}\big(x, y\big) = xy + x\bar{y} + \bar{x}y = x + y, \quad F_{15}\big(x, y\big) = xy + x\bar{y} + \bar{x}y + \bar{x}\bar{y}.
\end{aligned}$$

The above functions may also be expressed as *truth tables*, by evaluating them for the four input conditions of variables $x$ and $y$ (see Fig. 7.1). An example of a switching function in three variables $x,\ y,\ u$ may be given by a Boolean expression of the form

$$f\big(x, y, u\big) = xy + \bar{x}u + x\bar{u},$$

**Table 7.1** Truth Tables of $F_3$, $F_6$, $F_{10}$, and $F_{15}$

| $x$ | $y$ | $F_3(x, y)$ |
|---|---|---|
| 1 | 1 | 1 |
| 1 | 0 | 1 |
| 0 | 1 | 0 |
| 0 | 0 | 0 |

| $x$ | $y$ | $F_6(x, y)$ |
|---|---|---|
| 1 | 1 | 0 |
| 1 | 0 | 1 |
| 0 | 1 | 1 |
| 0 | 0 | 0 |

| $x$ | $y$ | $F_{10}(x, y)$ |
|---|---|---|
| 1 | 1 | 1 |
| 1 | 0 | 0 |
| 0 | 1 | 1 |
| 0 | 0 | 0 |

| $x$ | $y$ | $F_{15}(x, y)$ |
|---|---|---|
| 1 | 1 | 1 |
| 1 | 0 | 1 |
| 0 | 1 | 1 |
| 0 | 0 | 1 |

and we compute its values for the eight input conditions that can be assigned to three variables $x$, $y$, and $u$.

There is a one-to-one correspondence between the switching algebra and the propositional algebra. We replace 1 by T and 0 by F to obtain the *truth table* of the proposition representing a logic circuit as a Boolean expression.

**Definition 7.26** (*Dual*) Let $E$ be a Boolean expression. The **dual** of $E$ is the expression obtained from $E$ by interchanging operations $\oplus$ and $*$, and also 0 and 1, whenever present. We usually write the *dual* of an expression $E$ as $E^d$.

For example, the following equations are duals of each other:

(a) $a \oplus (b * c) = a$ and $a * (b \oplus c) = a$;
(b) $a \oplus \overline{(b \oplus a) * b} = 1$ and $a * \overline{(b * a) \oplus b} = 0$;
(c) $\overline{a * b} * b = a \oplus b$ and $\overline{a \oplus b} \oplus b = a * b$.

By Theorem 7.4, we know that the lattices $(L, \preceq, \vee, \wedge)$ and $(L, \succeq, \wedge, \vee)$ are *dual lattices* so that $\vee$ and $\wedge$ are *dual operations*, as are the two relations $\preceq$ and $\succeq$. We next state the *duality principle* for Boolean algebras.

**Theorem 7.25** (Duality Principle) *If $S$ is a valid statement over a Boolean algebra $(B, \oplus, *, ^-, 0, 1)$, then so is the dual statement $S^d$.*

Said differently, if $F$ and $G$ are two Bolean functions such that $F = G$, then $F^d = G^d$.

**Definition 7.27** (*Equivalent Expressions*) Two Boolean expressions $E_1(x_1, \ldots, x_n)$ and $E_2(x_1, \ldots, x_n)$ are said to be **equivalent** if the corresponding Boolean functions are same, i.e., $F_{E_1} = F_{E_2}$.

Corresponding to the four Boolean functions given as *truth tables* in Fig. 7.1, all the Boolean expressions are **non-equivalent**. The significance of *equivalent expressions* would be clear from the applications of Boolean algebras to the switching theory, as discussed later in Sect. 7.9. We first develop tools needed to solve the related "simplification problems".

**Definition 7.28** (*Minterm & Maxterm*) A **minterm** in variables $x_1, \ldots, x_n$ is a Boolean expression of the form

$$y_1 \cdot y_2 \cdot \ldots \cdot y_k, \quad \text{where each } y_i \text{ is a literal.} \tag{7.8.2}$$

A **maxterm** in variables $x_1, \ldots, x_n$ is a Boolean expression of the form

$$y_1 + y_2 + \cdots + y_k, \quad \text{where each } y_i \text{ is a literal.} \tag{7.8.3}$$

Notice that, there are $2^n$ different *minterms* or *maxterms*.

**Definition 7.29** (*Minterm Normal Form*) A **minterm normal form** (or a *sum of products* form) of a Boolean expression $E(x_1, \ldots, x_k)$ is the sum of the type

$$a * m, \quad \text{where } a \in B \text{ and } m \text{ is a minterm.} \tag{7.8.4}$$

That is, it is of the form $\oplus_{i=1}^{p}\big(a_j * m_j\big)$, where $p = 2^k$, and $m_1, \ldots, m_p$ are the minterms generated by variables $x_1, \ldots, x_k$.

For example, we have

$$F\big(x, y, z\big) = (\bar{x} * \bar{y} * \bar{z}) \oplus (\bar{x} * \bar{y} * z) \oplus (\bar{x} * y * \bar{z}) \oplus \cdots$$

is a sum of minterms.

**Definition 7.30** (*Maxterm Normal Form*) A **maxterm normal form** (or a *product of sums* form) of a Boolean expression $E(x_1, \ldots, x_k)$ is the sum of the type

$$a \oplus M, \quad \text{where } a \in B \text{ and } m \text{ is a maxterm.} \tag{7.8.5}$$

That is, it is of the form $\prod_{i=1}^{p}\big(a_j \oplus M_j\big)$, where $p = 2^k$, and $M_1, \ldots, M_p$ are the maxterms generated by variables $x_1, \ldots, x_k$.

For example, we have

$$F(x, y, z) = (x \oplus y \oplus z) * (x \oplus y \oplus \bar{z}) * (x \oplus \bar{y} \oplus z) * \cdots$$

is a product of maxterms.

**Example 7.16** We simplify the Boolean expressions

$$x \oplus (\bar{x} * y) \quad \text{and} \quad (\bar{x} * \bar{y} * z) \oplus (\bar{x} * y * z) \oplus (x * \bar{y}),$$

subject to that at each step of the *reduction process* the resulting expression is equivalent to the previous one, and the final one involves the minimum number of literals. For, we have

$$\begin{aligned} x \oplus (\bar{x} * y) &= (x \oplus \bar{x}) * (x \oplus y) && \text{(distributive laws)} \\ &= 1 * (x \oplus y) && \text{(complement law)} \\ &= (x \oplus y) && \text{(identity law).} \end{aligned}$$

We also see that

$$\begin{aligned} &(\bar{x} * \bar{y} * z) \oplus (\bar{x} * y * z) \oplus (x * \bar{y}) \\ &= \left[(\bar{x} * z) * \bar{y}\right] \oplus \left[(\bar{x} * z) * y\right] \oplus \left[x * \bar{y}\right] \\ &= \left[(\bar{x} * z) * (\bar{y} \oplus y)\right] \oplus \left[x * \bar{y}\right] \quad \text{(distributive law)} \\ &= \left[(\bar{x} * z) * 1\right] \oplus \left[x * \bar{y}\right] \\ &= \left(\bar{x} * z\right) \oplus \left[x * \bar{y}\right] \end{aligned}$$

**Example 7.17** We simplify the Boolean function

$$f\left(x_1, x_2, x_3\right) = (\overline{x_3} * x_2) \oplus (\overline{x_1} * x_3) \oplus (x_2 * x_3).$$

to obtain the *minterm normal form*. For, we have

$$\begin{aligned} f(x_1, x_2, x_3) &= (\overline{x_3} * x_2) \oplus (\overline{x_1} * x_3) \oplus (x_2 * x_3) \\ &= (\overline{x_3} * x_2) \oplus (x_2 * x_3) \oplus (\overline{x_1} * x_3) && \text{(commutative law)} \\ &= (x_2 * \overline{x_3}) \oplus (x_2 * x_3) \oplus (\overline{x_1} * x_3) && \text{(commutative law)} \\ &= (x_2 * (\overline{x_3} \oplus x_3)) \oplus (\overline{x_1} * x_3) && \text{(distributive law)} \\ &= (x_2 * 1) \oplus (\overline{x_1} * x_3) && \text{(complement law)} \\ &= x_2 \oplus (\overline{x_1} * x_3). && \text{(identity law)} \end{aligned}$$

The minterm normal form is given by

$$\begin{aligned} f(x_1, x_2, x_3) &= (\overline{x_3} * x_2) \oplus (\overline{x_1} * x_3) \oplus (x_2 * x_3) \\ &= [(\overline{x_3} * x_2) * (x_1 * \overline{x_1})] \oplus [(\overline{x_1} * x_3) * (x_2 * \overline{x_2})] \\ &\quad \oplus [(x_2 * x_3) * (x_1 * \overline{x_1})] \\ &= (\overline{x_3} * x_2 * x_1) \oplus (\overline{x_3} * x_2 * \overline{x_1}) \oplus (\overline{x_1} * x_3 * x_2) \\ &\quad \oplus (\overline{x_1} * x_3 * \overline{x_2}) \oplus (x_2 * x_3 * x_1) \oplus (x_2 * x_3 * \overline{x_1}) \\ &= (x_1 * x_2 * \overline{x_3}) \oplus (\overline{x_1} * x_2 * \overline{x_3}) \oplus (\overline{x_1} * x_2 * x_3) \\ &\quad \oplus (\overline{x_1} * \overline{x_2} * x_3) \oplus (x_1 * x_2 * x_3) \oplus (\overline{x_1} * x_2 * x_3) \\ &= (x_1 * x_2 * \overline{x_3}) \oplus (\overline{x_1} * x_2 * \overline{x_3}) \oplus (\overline{x_1} * x_2 * x_3) \\ &\quad \oplus (\overline{x_1} * \overline{x_2} * x_3) \oplus (x_1 * x_2 * x_3). \end{aligned}$$

**Definition 7.31** (*Canonical Form*) A **canonical form** is a Boolean expression written as a sum of products of minterms (or simply SOP) or as a product of sums of maxterms (or simply POS).

Minimal SOPs or POSs of a Boolean function are used to design an optimal circuit.

**Example 7.18** We express the Boolean function $F(x, y, z) = (x * y) \oplus (x' * z)$ as a product of maxterm form. We first convert the function in to OR terms using the distributive law.

$$\begin{aligned} F &= (x * y) \oplus (x' * z) \\ &= (x * y) \oplus x') * (x * y) \oplus z) \\ &= (x \oplus x') * (y \oplus x') * (x \oplus z) * (y \oplus z) \\ &= (y \oplus x') * (x \oplus z) * (y \oplus z) \end{aligned}$$

Now the function has three variables $x$, $y$ and $z$. Each OR term is missing one variable, therefore

$$\begin{aligned} (y \oplus x') &= (x' \oplus y) = (x' \oplus y) \oplus (z * z') \\ &= (x' \oplus y \oplus z) * (x' \oplus y \oplus z') \\ (x \oplus z) &= (x \oplus z) \oplus (y * y') \\ &= (x \oplus z \oplus y) * (x \oplus z \oplus y') \\ &= (x \oplus y \oplus z) * (x \oplus y' \oplus z) \\ (y \oplus z) &= (y \oplus z) = (y \oplus z) \oplus (x * x') \\ &= (y \oplus z \oplus x) * (y \oplus z \oplus x') \\ &= (x \oplus y \oplus z) * (x' \oplus y \oplus z). \end{aligned}$$

Therefore, we have

$$\begin{aligned} F &= (y \oplus x') * (x \oplus z) * (y \oplus z) \\ &= (x' \oplus y \oplus z) * (x' \oplus y \oplus z') * (x \oplus y \oplus z) \\ &\quad * (x \oplus y' \oplus z) * (x \oplus y \oplus z) * (x' \oplus y \oplus z) \\ &= M_4 * M_5 * M_0 * M_2 \\ &= \prod(0, 2, 4, 5). \end{aligned}$$

It thus follows that

$$F(x, y, z) = \prod(0, 2, 4, 5).$$

Notice that the complement of a function expressed as the sum of minterms equals the sum of minterms missing from the original function.

Simplification of a circuit in terms of reduction in the number of variables is not always a good approach. So, the application of Boolean algebra techniques to the *synthesis problem* is not satisfactory in every situation. In fact, criteria related to the economy and simplicity of a circuit depend on the types of components involved. At best, we can hope to devise a simple and rapid way to construct a wide variety of *near-minimal* algebraic forms.

## 7.9 Applications to Switching Theory

In this section, our emphasis is to introduce an important graphical tool that is used to simplify the design of digital circuits. We may recall that a *digital circuit* consists of parallel and serial combinations of switching elements such as logic gates. The technique presented here is closely related to the algebraic modifications of a Boolean expression, with some significant differences. The main idea is to use a 2D or a 3D graphical representation of the *truth table* of a Boolean function. It is known as the *Karnaugh map* (or simply a *K-map*) of the Boolean function. A *K-map* is also termed as the Venn diagram of a Boolean function. As shown below, the K-map technique proves to be advantageous at least in terms of time efficiency.

### *Karnaugh Map*

Fig. 7.17a is a K-map that shows graphically the truth table of a Boolean function in two variables $x$ and $y$. The boxes represent the possible $2^2 = 4$ input conditions. Fig. 7.17b is a K-map that shows graphically the truth table of a Boolean function in three variables $u$, $v$, $w$. In this case, the boxes represent the $2^3 = 8$ input conditions. In each case, two adjacent squares, horizontal or vertical, differ only in a single bit. Further, in the 3-variable case, to avoid the geometric inconsistency, the K-map is visualized as a *cylinder*, obtained by identifying the left and right vertical edges. We may recall that a *synthesis problem* begins with the specification of outputs for each of the possible input conditions.
More generally, a K-map of a Boolean function is a square-shaped box-diagram (or two copies of the same), where each sub-square with 1 represents an input condition for which the output value of the Boolean function is 1. Said differently, the boxes with 1 are formed by the product of minterms. Therefore, a K-map has a box corresponding to a row of miniterm in the truth table. A similar correspondence holds between the truth table rows and associated K-map squares when maxterms are used. An unfilled box is assumed to contain a 0. A definition is given below.

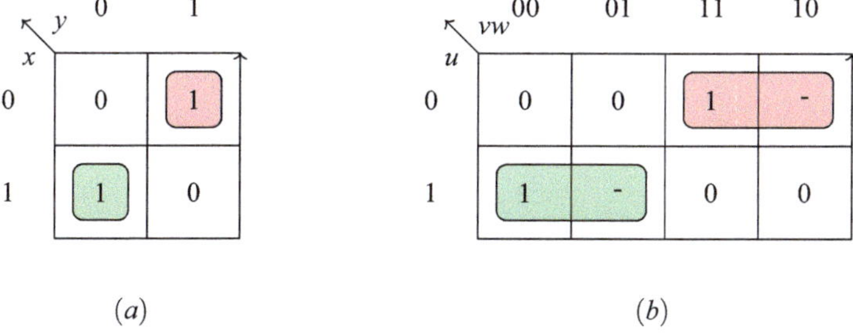


**Fig. 7.17** A 2-variable and a 3-variable K-maps

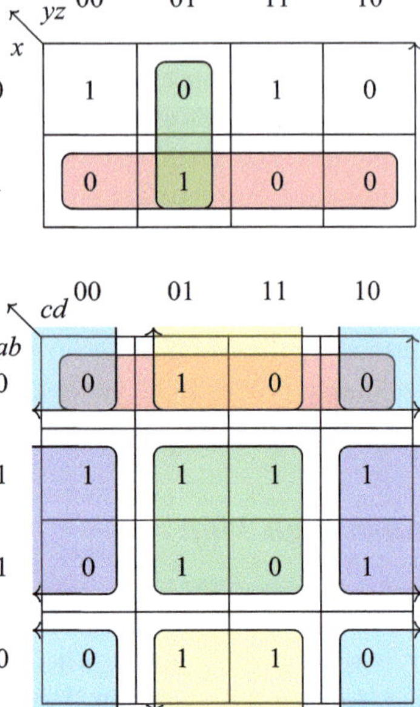


**Fig. 7.18** The K-maps for a Boolean functions of 3 variables

**Fig. 7.19** The K-map for a Boolean function of 4 variables

**Definition 7.32** (*Karnaugh Map*) The **Karnaugh map** (or simply a **K-map**) of a Boolean expression $E(x_1, \ldots, x_n)$ is a grid of $2^n$ squares representing the Boolean function $F_E : B^n \to B$ (see (7.8.1)). For each of the $2^n$ input values taken from the Boolean algebra $B^n$, the corresponding square has 1 or 0 as the value of $E$. A *minterm* is represented by a square with the value 1, and a *maxterm* is represented by a square with the value 0.

When $n = 2$, a K-map is a *square* such that each box has two adjacent boxes; when $n = 3$, a K-map is a *cylinder* obtained by identifying the shorter edges of the $2^3$-grid such that each box has three adjacent boxes (see Fig. 7.18). When $n = 4$, a K-map is a *torus* obtained by identifying opposite edges of the $2^4$-grid such that each box has four adjacent boxes (see Fig. 7.19), and so on. Notice that the 4-variable case is simply an extension of the 3-variable case. A 5-variable K-map of a Boolean function $f(x, y, u, v, w)$ has two parts, the left representing miniterms containing $\bar{x}$ and the right representing miniterms containing $x$. We may view the two halves stacked one on the top of the other. Similarly, a 6-variable K-map for a Boolean function $f(x, y, z, u, v, w)$ has four quadrants, each representing a combination of variables $x$ and $y$. Such quadrants are viewed as being stacked on top of each other, with vertically adjacent cells being glued together. For most problems, K-maps involving more than six variables are impractical.

**Remark 7.2** The simplest case of a K-map of a Boolean function arises when the corresponding expression is in a canonical form. Each minterm or maxterm corresponds to one cell of the K-map. For example, consider the Boolean function

$$f(x, y, u) = \sum (0, 3, 5) = \prod (1, 2, 4, 6, 7)$$

We know that the maxterm list may be derived from a given minterm list, and vice-versa. In the case of the SOP form of the function $f$, the associated K-map is Fig. 7.18, which is obtained such that each cell corresponds to one row of the truth table of the function $f$. Similarly, we obtain the K-map of $f$ when its POS form is taken into consideration.

The next definition introduces some important terminologies.

**Definition 7.33** An **implicant** is a product term (with one or more literals) that could be used to cover minterms of the function. A **prime implicant** is an *implicant* that is not a part of (covered by) any other implicant of the function. A prime implicant that covers at least one minterm, which is not covered by any other prime implicant, is known as an **essential prime implicant**. Finally, a **cover** of a Boolean function is a set of prime implicants such that each minterm of the function is contained in (or covered by) at least one prime implicant.

**Example 7.19** In Fig. 7.20, there are 11 implicants given by

$$\begin{aligned} \text{5 Minterms}: &\quad \bar{x}y\bar{u},\ xy\bar{u},\ x\bar{y}\bar{u},\ xyu,\ x\bar{y}u; \\ \text{5 Groups of 2 minterms}: &\quad y\bar{u},\ x\bar{u},\ xu,\ xy,\ x\bar{y}; \\ \text{1 Group of 4 minterms}: &\quad x \end{aligned}$$

As said earlier, variables get eliminated when miniterms (implicants) are combined (in powers of 2). So, a *prime implicant* is formed when implicants are combined to form a maximal group, which we see on the K-map as a set of squares that is not a subset of any set containing a larger number of squares. Therefore, prime implicants represent the largest groupings of minterms that can be derived for the function. For example, in the case of the K-map shown in Fig. 7.20, $x$ and $y\bar{u}$ are the only prime implicants, where the former covers the implicants given by

$$\bar{x}y\bar{u},\quad \bar{x}y\bar{u},\quad xy\bar{u},\quad x\bar{y}\bar{u},\quad xyu,\quad x\bar{y}u,\quad x\bar{u},\quad xu,\quad xy \text{ and } x\bar{y},$$

and the prime implicant $y\bar{u}$ covers implicant $\bar{x}y\bar{u}$ and $xy\bar{u}$.

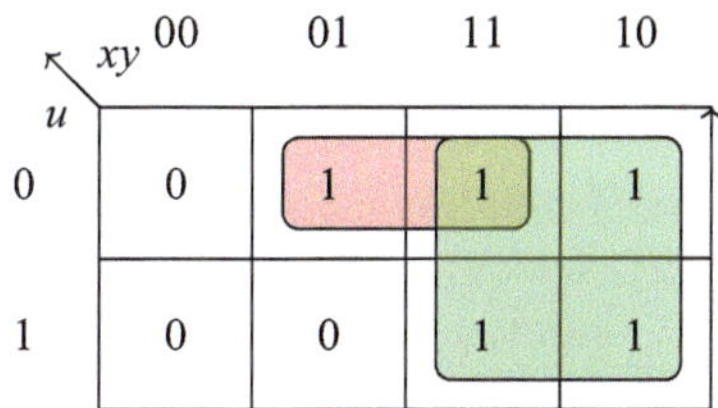


**Fig. 7.20** Implicant and prime implicant

Next, to identify an *essential prime implicant* on a K-map, we need to use the condition that it covers at least one minterm that is circled only once.

**Example 7.20** The prime implicant $y\bar{u}$ in the case of K-map shown in Fig. 7.20 is essential because it is the only implicant that covers minterm 1, and the prime implicant $x$ is essential because it is the only prime implicant that covers minterms 2, 6, and 7. Clearly, we must select all essential prime implicants to form a *cover* of a Boolean function. Also, by what is said above, the set of prime implicants $\{x, y\bar{u}\}$ is a cover of the Boolean function with the K-map as shown in Fig. 7.20. We shall use these terminologies later to describe an interesting *simplification technique*.

The following steps apply while simplifying a Boolean function using a K-map:

(a) Each box (miniterm) of a K-map for 2-variable case has two logically adjacent boxes (minterms). Each box (miniterm) of a K-map for a 3-variable case has three logically adjacent boxes (minterms), and so on.
(b) While combining boxes of a K-map, we always group in the power of 2. Grouping two boxes eliminates one variable; grouping four boxes eliminates two variables, and so on. In general, grouping $2^n$ boxes eliminates $n$ variables.
(c) Group together as many boxes as possible. The larger the group, the fewer would be the number of literals in the resulting product terms.
(d) To cover all boxes (minterms of the function), make as few groups as possible. A minterm is said to have been **covered** if it is included in at least one group. Less the number of groups, the fewer would be the number of product terms in the simplified function. In this and the previous step, we may use a minterm any number of times, subject to that it is being used at least once. Stop when all minterms are used once.
(e) While combining boxes of a K-map, begin with ones that have the least number of adjacent squares (so-called the *loneliest* square of the $K$-map). Minterms with multiple adjacent minterms (called *adjacencies*) offer more possible combinations and should therefore be combined later in the simplification process.

In particular, to obtain a minimal SOP expression for a Boolean function, we aim to find a minimum set of prime implicants that covers the function. The two algorithms stated below help to follow the above five guidelines to simplify the simplification process and to ensure a high likelihood of finding the minimum cover for any given Boolean function.

**Algorithm-1** It generally finds a minimal solution in an easy way. However, as random choices are allowed in Step-1 & and Step-2, it is not an optimal option to find the minimum cover of a Boolean function.

(i) Count the number of *adjacencies* for each minterm of the K-map.
(ii) Select an uncovered minterm with the lowest number of adjacencies. Choose randomly when more than one choice is available.

(iii) Generate a prime implicant for the chosen minterm and put it in the cover. In case it is covered by more than one prime implicant, select the one that covers the most uncovered minterms.
(iv) Keep repeating the previous two steps until all minterms are covered.

**Algorithm-2** It is an alternative for the above algorithm for the case when the number of prime implicants is not too large. As it generates more terms, we need to work more to identify all prime implicants that form a minimal cover. However, again, it is not optimal because Step-3 involves randomness.

(i) Circle all prime implicants on the K-map.
(ii) Identify and select all essential prime implicants for the cover.
(iii) Select a minimum subset of the remaining prime implicants to complete the cover. That is, to cover those minterms not already covered by the essential prime implicants.

The next example illustrates the application of the two algorithms.

**Example 7.21** Consider the Boolean function given by

$$f\big(x, y, u, v\big) = \sum \big(2, 3, 4, 5, 7, 8, 10, 13, 15\big).$$

We use the K-map shown as Fig. 7.21a, and the Algorithm-1, to simplify this function. We first count the adjacencies for each minterm. Clearly, each of the two minterms $m_4$ and $m_8$ have one adjacency; each of the five minterms $m_2$, $m_3$, $m_{10}$, $m_{13}$ and $m_{15}$ have two adjacencies; and, each of the two minterms $m_5$ and $m_7$ have three adjacencies. By Step-1 of Algorithm-1, we start with minterms $m_4$ and $m_8$. The prime implicant 4–5 as shown in Fig. 7.21b (in red) is the only one that covers $m_4$, and the prime implicant 8–10 (shown in green) is the only one that covers $m_8$. Therefore, at this stage, we add these two prime implicants to the cover by circling them on the K-map. Also, a simple glance through the remaining uncovered minterms suggests that the four of these have two adjacencies. So, at the next stage, we can make a random selection, say $m_2$, which is covered by two prime implicants 2–3 and 2–10. Now, as the prime implicant 2–3 covers more of the remaining uncovered miniterms, we select it for the cover by circling it. Next, as each of the two minterms $m_{13}$ and $m_{15}$ have two adjacencies, we can again make a random selection, say $m_{13}$, which is covered by the prime implicant 5–7–13–15. Hence, by adding this prime implicant to the cover, the simplification process is complete: The *minimum cover* is given by

$$\big\{4-5,\ 8-10,\ 2-3,\ 5-7-13-15\big\}.$$

Notice that it covers all the minterms. We next describe a procedure how to obtain product terms represented by the prime implicants. To start with, notice that the prime implicant 2–3 is located outside $x$ and $y$, but inside $u$, and so the product term is $\bar{x}\bar{y}u$. Also, the prime implicant 4–5 is located outside $x$ and $u$, but inside $y$, and so the product term is $\bar{x}y\bar{u}$. Further, the prime implicant 5–7–13–15 is inside $y$ and $v$,

**Fig. 7.21** K-maps illustrating Algorithm-1

| $uv \backslash xy$ | 00 | 01 | 11 | 10 |
|---|---|---|---|---|
| 00 | 0 | 0 | 1 | 1 |
| 01 | 1 | 1 | 1 | 0 |
| 11 | 0 | 1 | 1 | 0 |
| 10 | 1 | 0 | 0 | 1 |

(*a*)

| $uv \backslash xy$ | 00 | 01 | 11 | 10 |
|---|---|---|---|---|
| 00 | 0 | 0 | 1 | 1 |
| 01 | 1 | 1 | 1 | 0 |
| 11 | 0 | 1 | 1 | 0 |
| 10 | 1 | 0 | 0 | 1 |

(*b*)

and so the product term is $yv$. Finally, the prime implicant 8–10 is located outside $x$ and outside $y$ and $v$, and so the product term is $x\bar{y}\bar{v}$. Hence, the minimized function is

$$f(x, y, u, v) = \bar{x}\bar{y}u + \bar{x}y\bar{u} + yv + x\bar{y}\bar{v}.$$

**Example 7.22** Consider the Boolean function given by

$$f(x, y, u, v) = \sum\big(2, 3, 4, 5, 7, 8, 10, 13, 15\big).$$

We use the K-map shown as Fig. 7.22a, and the Algorithm-2, to simplify this function. We begin by circling all prime implicants. These prime implicants, listed according to the minterms covered, are given by

$$\{2-3,\ 3-7,\ 4-5,\ 5-7-13-15,\ 8-10,\ 2-10\}.$$

In Stage-2 of the process, we select the essential prime implicants. We can see from the K-map Fig. 7.22a that the prime implicant 4–5 is essential because it covers $m_4$, which is not covered by any other prime implicant. The same holds for the prime implicants 8–10 and 5–7–13–15. Accordingly, these essential prime implicants must be part of the cover for the given function (see the K-map Fig. 7.22b). Finally, in Stage-3, a minimum number of prime implicants are selected that cover the remaining minterms. We can see that $m_2$ and $m_3$ remain uncovered. Notice that, the prime implicants 2–3 and 2–10 cover $m_2$, while prime implicants 2–3 and 3–7 cover $m_3$. Clearly, prime implicant 2–3 covers both, as shown in the K-map Fig. 7.22c. Hence, once again, the minimum cover contains essential prime implicants 4–5, 5–7–13–15, and 8–10.

**Remark 7.3** An identical procedure applies while using K-maps to find a minimal expression for a Boolean function given in *product of sums* form. In this case, we use maxterms, in place of minterms, and so our focus is now on cells containing a 0. Now we use terms such as implicate, prime implicate, etc., which may be defined in an analogous manner. For example, we may define an *implicate* as the sum of one or more literals that could be used to cover maxterms of a Boolean function. Also, Algo-1 and Algo-2 can be easily modified to produce minimum covers from what one gets *product of sums*.

## Exercises 7

7.1 Identify all the minimal and maximal elements of the poset with the Hasse diagram as shown in Fig. 7.23.

7.2 Let $\mathbb{Q}^+$ denote the set of positive rational numbers. Show that, given any rational number $r > 1$, the set $A$ given by $A = \{x \in \mathbb{Q}^+ : x^2 < r\}$ has no greatest element, and the set $B$ given by $B = \{x \in \mathbb{Q}^+ : x^2 > r\}$ has no least element.

7.3 If $S$ and $T$ are two bounded nonempty subsets of $\mathbb{R}$ and $S \subset T$, show that $\inf T \preceq \inf S \preceq \sup S \preceq \sup T$.

7.4 Let $S$ be a subset of positive real numbers and $1/S = \{1/s \mid s \in S\}$. If $S$ is bounded above, show that $1/S$ is bounded below and $\inf(1/S) = 1/\sup S$.

7.5 If $S$ and $T$ are two nonempty subsets of $\mathbb{R}$ and $S \subset T$, show that $\inf T \leq \inf S$ if $T$ is bounded below, and $\sup S \leq \sup T$ if $T$ is bounded above.

7.6 If $S$ and $T$ are two bounded nonempty subsets of $\mathbb{R}$ and $S \subset T$, show that $\inf T \leq \inf S \leq \sup S \leq \sup T$.

7.7 If $S$ and $T$ are two subsets of $\mathbb{R}$ and $s \leq t$ for each $s \in S$ and $t \in T$, show that $S$ is bounded above and $T$ is bounded below. Further, show that $\sup S \leq \inf T$.

7.8 If $S$ and $T$ are two subsets of $\mathbb{R}$ which are bounded above, find $\sup(S \cup T)$.

**Fig. 7.22** K-maps illustrating Algorithm-2

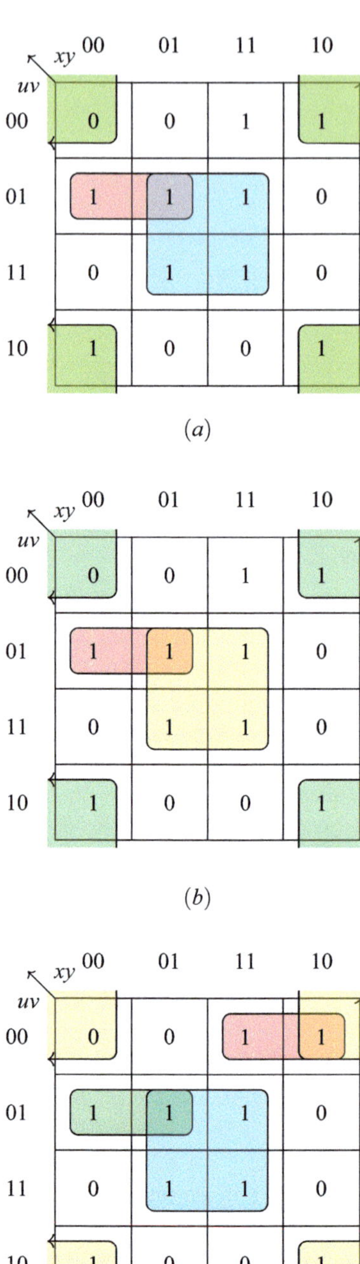

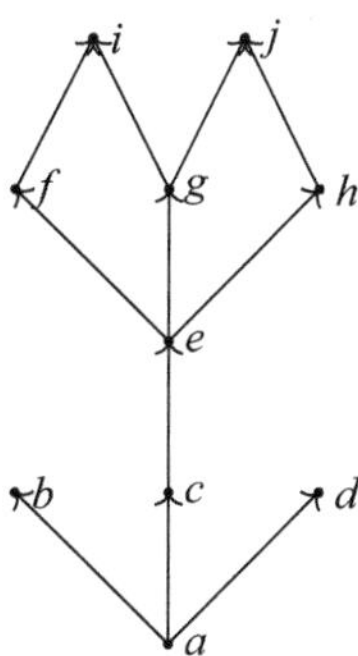


**Fig. 7.23** The Hasse diagram for Exercise 7.1

7.9 For two nonempty subsets $S$ and $T$ of $\mathbb{R}$ , define $S+T=\{s+t : s \in S, t \in T\}$. If $S$ and $T$ are bounded below, find $\inf(S+T)$ in terms of $\inf S$ and $\inf T$. What can you say about $\sup(S+T)$? If $S=\{a\}$, write $S+T=a+T$. Give a suitable definition of $S-T$ for bounded subsets $S$ and $T$ of $\mathbb{R}$. Find $\inf(a+T)$, $\sup(a+T)$, $\sup(S-T)$ and $\inf(S-T)$ in terms of supremum and infimum of $S$ and $T$.

7.10 If $S$ is bounded above subset of $\mathbb{R}$, show that $kS=\{ks : s \in S\}$ is bounded above and $\sup(kS)=k \sup S$ for each $k>0$. What happens if $k \leq 0$? Is $S+S=2S$?

7.11 For two nonempty subsets $S$ and $T$ of positive real numbers in $\mathbb{R}$ , define $ST=\{st : s \in S, t \in T\}$. Find $\inf(ST)$ in terms of $\inf S$ and $\inf T$. What can you say about $\sup(ST)$?

7.12 Let $S$ be a subset of positive real numbers and $1/S=\{1/s : s \in S\}$. If $S$ is bounded above, show that $1/S$ is bounded below and $\inf(1/S)=1/\sup S$.

7.13 Show that, in any poset, the greatest element (or the least element) is always unique if it exists.

7.14 Prove that, in any lattice, the LUB of two elements is always unique.

7.15 Show that every chain is a lattice.

7.16 Show that the lattices with Hasse diagrams as in Fig. 7.10 are non-isomorphic.

7.17 Show that, in any lattice $L(\leq, \vee, \wedge)$, if $a \leq b$ and $c \leq d$, then $a \wedge c \leq b \wedge d$.

7.18 Show that, in a lattice if $L(\leq, \vee, \wedge)$, we have

$$a \leq c \quad \Leftrightarrow \quad a \oplus (b * c) \leq (a \oplus b) * c,$$

for any $a, b, c \in L$.

7.19 Prove that the product of two distributive lattices is also a distributive lattice.

7.20 Let $L(\leq, \vee, \wedge, 0, 1)$ be a distributive complemented lattice. Show that, for any $a, b \in L$, we have

$$a \leq b \quad \Leftrightarrow \quad a * b' = 0 \quad \Leftrightarrow \quad a' \oplus b = 1 \quad \Leftrightarrow \quad b' \leq a'.$$

7.21 Let $L(\leq, \vee, \wedge)$ be a lattice, and $a, b, c \in L$. Show that the equations

$$a * (b \oplus c) = (a * b) \oplus (a * c) \quad \text{and} \quad a \oplus (b * c) = (a \oplus b) * (a \oplus c)$$

are equivalent.

7.22 Let $L(\leq, \vee, \wedge)$ be a lattice, and $a, b, c \in L$. Show that

$$a \preceq b \preceq c \quad \Rightarrow \quad a \oplus b = b * c.$$

Also, show that $(a * b) \oplus (b * c) = b = (a \oplus b) * (a \oplus c)$.

7.23 Let $L(\leq, \vee, \wedge)$ be a lattice. Show that

$$a \preceq b \text{ and } c \preceq d \quad \Rightarrow \quad a * c \preceq b * d.$$

7.24 Let $L(\leq, \vee, \wedge)$ be a lattice, and $a, b, c \in L$. Show that, $a \preceq b$ and $a \preceq c$ implies that $a \preceq b \oplus c$ and $a \preceq b * c$.

7.25 Let $L(\leq, \vee, \wedge)$ be a lattice, and $a, b, c \in L$. Show that, $b \preceq a$ and $c \preceq a$ implies that $b \oplus c \preceq a$ and $b * c \preceq a$.

7.26 Let $L(\leq, \vee, \wedge)$ be a lattice, and $a, b, c \in L$. Show that $(a * b) \oplus (c * d) \preceq (a \oplus c) * (b \oplus d)$.

7.27 Let $L(\leq, \vee, \wedge)$ be a lattice, and $a, b, c \in L$. Show that

$$(a * b) \oplus (b * c) \oplus (c * a) \preceq (a \oplus b) * (b \oplus c) * (c \oplus a).$$

7.28 Let $L(\leq, \vee, \wedge)$ be a lattice. Show that the absorption law implies the idempotent laws.

7.29 Let $L(\leq, \vee, \wedge)$ be a lattice, and $a, b \in L$. Show that $a * b = a$ if and only if $a \oplus b = b$.

7.30 Suppose $(L, \oplus, *)$ a lattice in the sense of Definition 7.12, with partial order $R$ given by (7.3.8). Show that $a * b$ and $a \oplus b$ are, respectively, the greatest lower bound and the least upper bound of the set $\{a, b\}$.

7.31 Explain why there cannot be a 3-point or a 5-point lattice that is also a Boolean algebra.

7.32 Show that $B = \{1, 2, 3, 6\}$ is a Boolean algebra with respect to following operations:

$$a \oplus b = \text{lcm}\left\{a, b\right\}, \quad a * b = \gcd\left\{a, b\right\}, \quad \bar{a} = \frac{6}{a}.$$

Also, explain why $B = \{1, 2, 4, 8\}$ is a not a Boolean algebra with respect to $\oplus$ and $*$ as above, but $\bar{x} = 8/x$.

7.33 Give a proof of the 0-1 law.

7.34 Show that a lattice $L(\leq, \vee, \wedge)$ is distributive if and only if

$$(a * b) \oplus (b * c) \oplus (c * a) = (a \oplus b) * (b \oplus c) * (c \oplus a)$$

for all $a, b, c \in L$.

7.35 Let $L(\leq, \vee, \wedge)$ be a distributive lattice, and $a, b, c \in L$. Show that

$$(a * b) \oplus (b * c) \oplus (c * a) = (a \oplus b) * (b \oplus c) * (c \oplus a).$$

7.36 Show that the collection of complemented elements in a bounded distributive lattice forms a sublattice.

7.37 Let $(B, \oplus, *, ^{-}, 0, 1)$ be a Boolean algebra, and $x \in B$. Show that the complement $\bar{x}$ is unique.

7.38 Let $(B, \oplus, *, ^{-}, 0, 1)$ be a Boolean algebra, and $x, y, z \in B$. Show that

a. $x \oplus y = x \oplus z$ and $\overline{x} \oplus y = \overline{x} \oplus z \quad \Rightarrow \quad y = z$
b. $x \oplus y = 0 \quad \Leftrightarrow \quad x = 0$ and $y = 0$
c. $x * y = 1 \quad \Leftrightarrow \quad x = 1$ and $y = 1$
d. $x \leq \overline{y} \quad \Leftrightarrow \quad x * y = 0$
e. $x \leq y \quad \Leftrightarrow \quad \overline{x} \oplus y = 1$.

7.39 Let $(B, \oplus, *, ^{-}, 0, 1)$ be a Boolean algebra, and $a, b, c \in B$. Write the dual of the following expressions:

a. $a * (b \oplus c) = (a * b) \oplus (a * c)$.
b. $a \oplus a' = 1$.
c. $a \oplus ((a \oplus b)') = a' * b'$.
d. $a * (a' \oplus b) = a * b$.
e. $(a \oplus 1) * (a \oplus 0) = a$
f. $(a \oplus b) * (b \oplus c) = (a * c) \oplus b$
g. $(a * b) \oplus (b * c) = (a \oplus c) * b$.

7.40 Show that the direct product of two Boolean algebras is also a Boolean algebra.

7.41 Express the following as a Boolean expression involving a minimal number of literals.

a. $x * (\bar{x} \oplus y)$
b. $x * y) \oplus (\bar{x} * z) \oplus (x * y)$
c. $((x * y) \oplus x) * (\bar{x} \oplus (z * x))$

7.42 Let $(B, \oplus, *, ^{-}, 0, 1)$ be a Boolean algebra, and $a, b \in B$. Show that

a. $a * \bar{b} = 0 \Leftrightarrow \bar{a} \oplus b = 1$.
b. $a = b \Leftrightarrow ab' + a'b = 0$.
c. $a \oplus b = b \Leftrightarrow a * b = a \Leftrightarrow \bar{a} \oplus b = 1$.

7.43 Let $(B, \oplus, *, ^{-}, 0, 1)$ be a Boolean algebra, and $a, b, c \in B$. Show that $(a + b)(a' + c) = ac + a'b + bc$.

7.44 Let $(B, \oplus, *, ^{-}, 0, 1)$ be a Boolean algebra, and $a, b, c \in B$. Show algebraically that $ab' + bc' + ca' = a'b + b'c + c'a$.

7.45 Let $(B, \oplus, *, ^{-}, 0, 1)$ be a Boolean algebra, and $x_1, x_2, x_3c \in B$. Simplify the Boolean function $x_1 * x_2) \oplus (x_1 \oplus (x_2' * x_3))$, algebraically.

7.46 For the following Boolean functions, find the complement and reduce each to a Boolean expression involving a minimum number of literals:

a. $((b * c') \oplus (a * d)) * ((a * b') \oplus (c * d'))$.
b. $(b' * d) \oplus (a' * b * c') \oplus (a * c * d) \oplus (a' * b * c)$.

7.47 Let $(B, \oplus, *, ^{-}, 0, 1)$ be a Boolean algebra, and $a, b, c \in B$. Express the following as a sum of minterms and a product of maxterms.

a. $F(a, b, c) = (\bar{a} \oplus b) * (\bar{b} \oplus c)$;
b. $F(a, b, c) = ((a * b) \oplus c) * (b \oplus (a * c))$;
c. $F(a, b, c, d) = (d * (\bar{a} \oplus b)) \oplus (\bar{b} * d)$.

7.48 Expand the Boolean function $f(x, y, z) = xy + yz$ as the canonical SOP from.

7.49 Find the complement of the function

$$F_1 = (x' * y * z') \oplus (x' * y' * z) \text{ and } \quad F_2 = x * ((y' * z') \oplus (y * z)).$$

7.50 Let $(B, \oplus, *, ^{-}, 0, 1)$ be a Boolean algebra, and $x, y, u, v \in B$. Find a minimal *sum of products* for the Boolean expression given by $f(x, y, u, v) = \sum(1, 2, 3, 6)$.

7.51 Let $(B, \oplus, *, ^{-}, 0, 1)$ be a Boolean algebra, and $x, y, u, v \in B$. Find minimal *sum of products* for the Boolean expression given by $f(x, y, u, v) = \sum(0, 1, 2, 7, 8, 9, 10, 15)$.

7.52 Let $(B, \oplus, *, ^{-}, 0, 1)$ be a Boolean algebra, and $x, y, u, v \in B$. Find minimal *sum of products* for the Boolean expression given by $f(x, y, u, v) = \sum(0, 4, 5, 7, 8, 10, 14, 15)$.

7.53 Use Algorithm-1 and Algorithm-2 to simplify the Boolean function

$$f(x, y, u, v) = \sum\left(0, 5, 7, 8, 10, 12, 14, 15\right).$$

7.54 Use Karnaugh map to obtain the minimal SOP expression for the function

$$F(x, y, z) = \sum(0, 2, 5, 6).$$

7.55 Use K-map to obtain the minimal SOP expression for the function

$$F(u, v, w) = u\bar{v}\bar{w} + uv\bar{w} + uvw + \bar{u}\bar{v}w.$$

7.56 Use K-map to obtain the minimal SOP expression for the function

$$F(a, b, c, d) = \sum(0, 1, 2, 3, 4, 5, 6, 7, 8, 9, 11).$$

# References

1. Mendelson, E.: Boolean Algebra and Switching Circuits. Schaum's Outline Series, McGrw-Hill Inc (1970)
2. Davey, B.A., Priestley, H.A.: Introduction to Lattices and Order (2/e). Cambridge University Press (2002)
3. Dilworth, R.P., Crawley, P.: Algebraic Theory of Lattices. Prentice-Hall (1973)
4. Whitesitt, J.E.: Boolean Algebra and Its Applications. Dover Publications Inc. (2010)
5. Russell, B.: A Critical Exposition of the Philosophy of Leibniz (2/e). Allen and Unwin, London (1937)

# Chapter 8
# Algebraic Structures

*As for everything else, so for a mathematical theory: Beauty can be perceived but not explained.*

***Arthur Cayley*** *(1821–1895)*

In this chapter, we introduce certain fundamental *Algebraic structures* such as groups, rings, and fields. We also prove some related important theorems. Our treatment of the topics covered here is very crisp. The reader is referred to standard textbooks such as [1–5], for any missing details and further study.

## 8.1 Semigroups, Monoids, and Groups

The abstract algebra at an introductory level such as in this text is a study of classical *Algebraic structures*. The related *Algebraic techniques* find numerous applications in most branches of science and engineering, including computer science. For example, "Boolean" is a fundamental structure that consists of the set $\{0, 1\}$, together with three logic operations AND, OR, and NOT. It finds applications in the analysis and design of digital circuits.

### *Binary Operations*

In this section, we introduce some basic terminology and fix notations for use in the sequel. Let $X$ be a nonempty set. We write

 
V. Ravichandran and A. K. Razdan, *Fundamental Discrete Structures*,
University Texts in the Mathematical Sciences,
https://doi.org/10.1007/978-981-96-0069-4_8

$$X^n := X \times \cdots \times X \ (n \text{ times}), \quad \text{for} \quad n \geq 1.$$

We say $*$ is an **n-ary operation** on $X$ if, for each $n$-tuple $(x_1, \ldots, x_n) \in X^n$, the product of the form

$$x_1 * x_2 * \cdots * x_n \tag{8.1.1}$$

is well defined in some appropriate sense. In particular, when $n = 2$, $*$ is called a **binary operation** on the set $X$. The usual *addition* and *multiplication* on a set $X$ consists of *numbers, vectors, square matrices*, or *polynomials*, are two simple *Binary operations* that are known to us from our school mathematics. We also know the binary operation "composition" on a set of *self-maps*, i.e., functions from a set to itself. We give below some other familiar binary operations on a set, as specified alongside:

(i) $\vee$ (disjunction) and $\wedge$ (conjunction) on a collection of propositions.
(ii) $\cup$ (union) and $\cap$ (intersection), on a collection of sets;
(iii) $\vee$ (join) and $\wedge$ (meet), on a lattice;
(iv) $+$ (sum) and $\cdot$ (product), on a Boolean algebra;

More generally, considering the type of elements a set contains or the nature of the problem in hand, we may also have to deal with other binary operations on some of these sets (see examples given below).

**Definition 8.1** Let $X$ be a nonempty set. A function $* : X \to X$ is called a **unary operation** on the set $X$. It is usually denoted by $x \mapsto x^*$, for $x \in X$.

For example, the "standard complementation" is a *Unary operation* on the power set of a nonempty set $X$ or on a bounded complemented lattice or on a Boolean algebra. Also, on the collection of propositions, the *negation* ($p \mapsto \neg\, p$) is a unary operation. Recall that most of the aforementioned binary operations satisfy *Algebraic properties* such as associativity, commutativity, and also distributivity.

**Definition 8.2** A nonempty set together with one or more operations that satisfy certain algebraic properties is called an **algebraic structure**.

Let $X$ be a nonempty set. In general, the product in (8.1.1) may not be an element of the set $X$, for every choice of $n$-tuple $(x_1, \ldots, x_n) \in X^n$. The binary operation *subtraction* $(-)$ on the set of natural numbers $\mathbb{N}$ is an example. Also, the binary operation *division* $(\div)$ on the set of *nonzero* integers $\mathbb{Z} - \{0\}$ is an example.

**Definition 8.3** We say a binary operation $*$ on $X$ has the **closure property** if

$$x * y \in X, \qquad \text{for all} \quad x,\ y \in X. \tag{8.1.2}$$

We also say that the $*$ is a **closed binary operation** on the set $X$.

Said differently, a binary operation $*$ on $X$ has the *Closure property* if $*$ is a function from the set $X \times X$ to $X$. In this case, it is convention to write the image $*(x, y) \in X$

of the element $(x, y) \in X \times X$ simply as the *product* $x * y$ (or simply as $x\,y$). A nonempty set $X$ together with a *Closed binary operation* $*$ is called a **groupoid**. We usually write a *Groupoid* as the pair $(X, *)$.

**Example 8.1**

(i) With respect to the binary operations given by usual *addition* (+), *subtraction* (−), or *multiplication* (×), each one of the number sets $\mathbb{Z}$, $\mathbb{Q}$, $\mathbb{R}$, or the $\mathbb{C}$, is a groupoid. However, the set of natural numbers $\mathbb{N}$ is not a groupoid with respect to the binary operation *subtraction* (−), because the difference between two natural numbers need not be a natural number. Also, with respect to the binary operation *division* (÷) given by $a * b = a \div b$, the set $\mathbb{Z} - \{0\}$ is not a groupoid.

(ii) With respect to the binary operations given by usual *division* (÷), the set of *nonzero* real numbers $\mathbb{R}^*$, and also the set of the set of *nonzero* complex numbers $\mathbb{C}^*$, are groupoids.

(iii) Consider the set of **nth roots of unity** $U_n$, i.e., the roots of the equation $x^n - 1 = 0$. Clearly, $U_2 = \{1, -1\}$, and $U_3 = \{1, \omega, \omega^2\}$, where

$$\omega = 1/2\big(-1 + i\sqrt{3}\big) \quad \text{and} \quad \omega^2 = 1/2\big(-1 - i\sqrt{3}\big),$$

are two complex roots of the equation $x^2 + x + 1 = 0$. In general, the set of $n$th *Roots of unity* $U_n$ consists of $n$ *equidistant points* on the unit circle in the Argand plane $\mathbb{C}$ with center at the origin. More generally, taking

$$\omega = e^{2\pi i/n} = \cos\big(2\pi/n\big) + i\,\sin\big(2\pi/n\big), \tag{8.1.3}$$

every next point in $U_n$ is obtained from the previous one by applying *anti-clockwise rotation* through an angle $(2\pi)/n$, starting from the point $\omega^n = 1$. We thus have

$$U_n = \big\{\omega,\, \omega^2,\, \ldots,\, \omega^{n-1},\, \omega^n = 1\big\}. \tag{8.1.4}$$

Notice that, for $z_1, z_2 \in U_n$, we have $z_1^n = 1$ and $z_2^n = 1$, which implies $\big(z_1 z_2\big)^n = 1$. Therefore, $z_1 z_2 \in U_n$. Hence, $U_n$ is a groupoid with respect to the binary operation given by the usual multiplication of complex numbers.

(iv) As said earlier in (3.6.6), we have

$$\mathbb{Z}_n = \big\{0, 1, \ldots, n-1\big\}$$

is the set of residue class modulo $n > 1$, where $a \in \mathbb{Z}_n$ is given by

$$a := a + n\mathbb{Z} = \big\{a + nk \mid k \in \mathbb{Z}\big\}.$$

That is, $a \in \mathbb{Z}_n$ is the set of integers that leave residue $a$ when divided by $n$. On the set $\mathbb{Z}_n$, the two binary operations $+_n$ and $\times_n$ are defined by

$$a +_n b = r \quad \text{and} \quad a \times_n b = s, \tag{8.1.5}$$

where $r$ is the residue obtained when $a+b$ is divided by $n$, and $s$ is the residue obtained when $ab$ is divided by $n$. These two binary operations on the set $\mathbb{Z}_n$ are respectively called the **addition modulo n** and **multiplication modulo n**. For example, we have

$$1 +_2 1 = 0, \quad 2 \times_3 2 = 1, \quad 2 +_6 5 = 1, \quad \text{and so on.}$$

In general, since the two operations are defined modulo $n$, we have both $a +_n b$ and $a \times_n b$ are elements of the set $\mathbb{Z}_n$. Therefore, $\big(\mathbb{Z}_n, a +_n b\big)$ and $\big(\mathbb{Z}_n, a \times_n b\big)$ are groupoids. The arithmetic on the set $\mathbb{Z}_n$ with respect to operations defined in (8.1.5) is called the **modular arithmetic**. The German mathematician *Carl Friedrich Gauss* (1777–1855) first developed the *Modular arithmetic* in his book "*Disquisitiones Arithmeticae*", which was published in 1801.

(v) We may recall that a *permutation* of a set $A$ with at least two elements is a bijective function from the set $A$ to itself. Also, $\mathrm{Perm}(A)$ denotes the set of permutations of the set $A$. As before, the composition of two permutations is defined using the "right action" notation. That is, if $\sigma,\ \rho \in \mathrm{Perm}(A)$, then the composite $\sigma\,\rho \in \mathrm{Perm}(A)$ is obtained by first applying the permutation $\sigma$ followed by the permutation $\rho$. Accordingly, we may write

$$(a)\big(\sigma\,\rho\big) = \big((a)\sigma\big)\rho, \qquad \text{for} \quad a \in A. \tag{8.1.6}$$

By Theorem 4.5, we also know that there are $n!$ permutations when

$$A = \big[n\big] := \big\{1, 2, \ldots, n\big\}.$$

In this case, we write $\mathrm{Perm}(A)$ simply as $S_n$. In particular, the six permutations of the set $A = \big[3\big]$ are given by

$$\begin{aligned} e &= \begin{pmatrix} 1 \mapsto 1 \\ 2 \mapsto 2 \\ 3 \mapsto 3 \end{pmatrix}, \quad \tau_1 = \begin{pmatrix} 1 \mapsto 2 \\ 2 \mapsto 1 \\ 3 \mapsto 3 \end{pmatrix}, \quad \tau_2 = \begin{pmatrix} 1 \mapsto 3 \\ 2 \mapsto 2 \\ 3 \mapsto 1 \end{pmatrix}, \\ \tau_3 &= \begin{pmatrix} 1 \mapsto 1 \\ 2 \mapsto 3 \\ 3 \mapsto 2 \end{pmatrix}, \quad \rho_1 = \begin{pmatrix} 1 \mapsto 3 \\ 2 \mapsto 3 \\ 3 \mapsto 2 \end{pmatrix}, \quad \rho_2 = \begin{pmatrix} 1 \mapsto 2 \\ 2 \mapsto 3 \\ 3 \mapsto 1 \end{pmatrix} \end{aligned} \tag{8.1.7}$$

As shown in the left side picture of Fig. 8.1, in geometric terms, the three permutations $\tau_1$, $\tau_2$, and $\tau_3$, are the three *transvections* of an equilateral triangle with respect to three lines of symmetries passing through the three vertices of the triangle; and, the three permutations $\rho_1$, $\rho_2$, and the identity function $e$, represent the three anticlockwise *rotations* of the triangle through angles

$120^o$, $240^o$, and $360^o$, respectively. We will give a generalizes of this construction later in Example 8.6 (v). In *two-row notation*, the above six permutations are written as

$$\begin{aligned} e &= \begin{pmatrix} 1\ 2\ 3 \\ 1\ 2\ 3 \end{pmatrix}, & \tau_1 &= \begin{pmatrix} 1\ 2\ 3 \\ 2\ 1\ 3 \end{pmatrix}, & \tau_2 &= \begin{pmatrix} 1\ 2\ 3 \\ 3\ 2\ 1 \end{pmatrix}, \\ \tau_3 &= \begin{pmatrix} 1\ 2\ 3 \\ 1\ 3\ 2 \end{pmatrix}, & \rho_1 &= \begin{pmatrix} 1\ 2\ 3 \\ 3\ 1\ 2 \end{pmatrix}, & \rho_2 &= \begin{pmatrix} 1\ 2\ 3 \\ 2\ 3\ 1 \end{pmatrix} \end{aligned} \tag{8.1.8}$$

We may also write these six permutations in $S_3$ in *cycle form* as

$$\begin{aligned} e &= (1)(2)(3), & \tau_1 &= (1\ 2), & \tau_2 &= (1\ 3), \\ \tau_3 &= (2\ 3), & \rho_1 &= (1\ 3\ 2), & \rho_2 &= (1\ 2\ 3) \end{aligned} \tag{8.1.9}$$

In general, since the composition of two bijections is also a bijection, $S_n$ is a groupoid with respect to the binary operation given by "composition". As said earlier, for each permutation, the second row in *two-row notation* consists of the images of the elements $1, 2, 3$, respectively; and so, the composition of two permutations $\sigma$ and $\tau$ is given by

$$\sigma\tau = \begin{pmatrix} 1 & 2 & 3 \\ \tau(\sigma(1)) & \tau(\sigma(2)) & \tau(\sigma(3)) \end{pmatrix}.$$

(vi) With respect to the binary operations given by usual addition and subtraction, the set of $m \times n$ real or complex matrices $M_{m\times n}$ is a groupoid. Also, with respect to the binary operations given by usual multiplication, the set $M_n$ ($n \geq 2$) of real or complex square matrices of order $n$ is a groupoid.

(vii) With respect to the binary operations given by usual addition and multiplication, the set of polynomials with real or complex *coefficients* is a groupoid.

(viii) Let $X$ be a nonempty set, and $A := \{f \mid f : X \to X\}$ be the set of *Self-map* of the set $X$. Clearly, $A$ is closed with respect to the binary operation given by the *composition*. That is, we have $f \circ g \in A$, for $f, g \in A$. Recall that

$$\big(f \circ g\big)(x) = f\big(g(x)\big), \qquad \text{for } x \in X.$$

Therefore, $(A, \circ)$ is a groupoid.

When a groupoid $(X, *)$ is finite, with a small number of elements, it is more convenient to express the *closure property* of the operation $*$ by way of writing the products $x * y$ $(x, y \in X)$ in a *tabular form*. Such a tabular representation of a finite groupoid $X$ with respect to the binary operation $*$ is called the **product table** (or the **Cayley table**) of the groupoid $X$.

**Example 8.2**

(i) With respect to the usual multiplication of complex numbers, we may write the product table of the groupoid $U_n$ as introduced in Example 8.1 (iii), for any value of $n \geq 2$. In particular, Table 8.1 is the product table of the groupoid $U_4 = \{1, -1, i, -i\}$, where $i = \sqrt{-1}$.

(ii) Let the two groupoids $(\mathbb{Z}_n, +_n)$ and $(\mathbb{Z}_n, \times_n)$ be as introduced earlier in Example 8.1 (iv). The product tables of these groupoids with respect to *Addition modulo* $+_n$ and *Multiplication modulo* $\times_n$ can be written for any value of $n \geq 2$. In particular, the product tables of $\mathbb{Z}_4$ with respect to the operations $+_4$ and $\times_4$ are given in Table 8.2.

(iii) With respect to the operation "composition", we may write the product table of the groupoid $S_n$ as introduced in Example 8.1 (v), for any value of $n \geq 2$. In particular, Table 8.3 is the product table of $S_3$ with respect to the operation given by (8.1.6).

The next definition states the property that is crucial to ensure that the product of three or more elements in a groupoid is well defined. More precisely, it follows from

**Table 8.1** Product table of $U_4$ with respect to complex multiplication

| $\cdot$ | $1$ | $-1$ | $i$ | $-i$ |
|---|---|---|---|---|
| $1$ | $1$ | $-1$ | $i$ | $-i$ |
| $-1$ | $-1$ | $1$ | $-i$ | $i$ |
| $i$ | $i$ | $-i$ | $-1$ | $1$ |
| $-i$ | $-i$ | $i$ | $1$ | $-1$ |

**Table 8.2** Product tables of $\mathbb{Z}_4$ for operations $+_4$ and $\times_4$

| $+_4$ | 0 | 1 | 2 | 3 |
|---|---|---|---|---|
| 0 | 0 | 1 | 2 | 3 |
| 1 | 1 | 2 | 3 | 0 |
| 2 | 2 | 3 | 0 | 1 |
| 3 | 3 | 0 | 1 | 2 |

| $\times_4$ | 0 | 1 | 2 | 3 |
|---|---|---|---|---|
| 0 | 0 | 0 | 0 | 0 |
| 1 | 0 | 1 | 2 | 3 |
| 2 | 0 | 2 | 0 | 2 |
| 3 | 0 | 3 | 2 | 1 |

**Table 8.3** Product table for $S_3$ with respect to "composition"

| $\circ$ | $e$ | $\tau_1$ | $\tau_2$ | $\tau_3$ | $\rho_1$ | $\rho_2$ |
|---|---|---|---|---|---|---|
| $e$ | $e$ | $\tau_1$ | $\tau_2$ | $\tau_3$ | $\rho_1$ | $\rho_2$ |
| $\tau_1$ | $\tau_1$ | $e$ | $\rho_2$ | $\rho_1$ | $\tau_3$ | $\tau_2$ |
| $\tau_2$ | $\tau_2$ | $\rho_1$ | $e$ | $\rho_2$ | $\tau_1$ | $\tau_3$ |
| $\tau_3$ | $\tau_3$ | $\rho_2$ | $\rho_1$ | $e$ | $\tau_2$ | $\tau_1$ |
| $\rho_1$ | $\rho_1$ | $\tau_2$ | $\tau_3$ | $\tau_1$ | $\rho_2$ | $e$ |
| $\rho_2$ | $\rho_2$ | $\tau_1$ | $\tau_2$ | $\rho_1$ | $e$ | $\rho_1$ |

the property (8.1.10) that each product $x * y * z$ in a groupoid $(X, *)$ is well defined, i.e., it doesn't matter how we put braces around the two pairs of elements $x * y$ and $y * z$.

**Definition 8.4** Let $(X, *)$ be a groupoid. We say $*$ is an **associative operation** if it satisfies the following property:

$$\big(x * y\big) * z = x * \big(y * z\big), \qquad \text{for all} \quad x, y, z \in X. \tag{8.1.10}$$

A groupoid $(S, *)$ with an *Associative operation* $*$ is called a **semigroup**.

For example, all groupoids mentioned in previous examples are *Semigroups*. Also, we can ignore the order of elements in a product in a groupoid $(X, *)$ if the binary operation $*$ satisfies the property (8.1.11).

**Definition 8.5** Let $(X, *)$ be a groupoid. We say $*$ is a **commutative operation** if it satisfies the following property:

$$x * y = y * x, \qquad \text{for all} \quad x, y \in X. \tag{8.1.11}$$

It is common practice to use *Additive notation* $(+)$ for a *Commutative operation* $*$. A semigroup $(S, *)$ is called an *Abelian semigroup* if $*$ is a commutative operation. Recall that the integer 0 has the property that

$$a + 0 = a = 0 + a, \qquad \text{for any number} \quad a.$$

In this case, for each one of the number sets $\mathbb{Z}$, $\mathbb{Q}$, $\mathbb{R}$, and $\mathbb{C}$, we say 0 is the *Additive identity* for the usual addition $(+)$. Similarly, for each groupoid in Example 8.1, 1 is the *Multiplicative identity* for the usual multiplication $(\times)$:

$$a \times 1 = a = 1 \times a, \qquad \text{for any number} \quad a.$$

In general, the *Identity element* with respect to additive notation $(+)$ is called the **additive identity**, and the identity element with respect to multiplicative notation $(\times \text{ or } \cdot)$ is called the **multiplicative identity**. A formal definition is given next.

**Definition 8.6** Let $(X, *)$ be a groupoid.

(a) An element $e_\ell \in X$ is called a *Left identity* of $X$ for the operation $*$ if we have

$$e_\ell * x = x, \qquad \text{for all} \quad x \in X.$$

(b) An element $e_r \in X$ is called a *Right identity* of $X$ for the operation $*$ if we have

$$x * e_r = x, \qquad \text{for all} \quad x \in X.$$

(c) An element $e \in X$ is called an *Identity element* for the operation $*$ if we have

$$e * x = x * e = x, \qquad \text{for all} \quad x \in X. \tag{8.1.12}$$

The next theorem proves that we necessarily have $e_\ell = e_r = e$.

**Theorem 8.1** *Let* $(X, *)$ *be a groupoid, with* $e_\ell$ *and* $e_r$ *respectively as left and right identity elements for the operation* $*$. *Then we have* $e_l = e_r$; *and, if* $e = e_r = e_l$, *then* $e$ *is an identity.*

***Proof.*** Since $e_l$ is the left identity, for all $x \in X$, we have $e_l * x = x$. In particular, taking $x = e_r$, we obtain

$$e_l * e_r = e_r.$$

Again, since $e_r$ is the right identity, we have $x * e_r = x$, for all $x \in X$. In particular, taking $x = e_l$, we obtain

$$e_l * e_r = e_l.$$

It thus follows that $e_l = e_l * e_r = e_r$. Now, if $e = e_r = e_l$, then

$$e * x = x = x * e, \qquad \text{for all} \quad x \in X,$$

implies that $e$ an identity for the operation $*$. □

**Definition 8.7** A semigroup $(M, *)$ with an identity element $e \in M$ is called a **monoid**. In this case, we usually write the monoid as $(M, *, e)$. A monoid $(M, *, e)$ with a commutative operation $*$ is called an *Abelian monoid*.

**Example 8.3**

(i) On the set of integers $\mathbb{Z}$, consider the binary operation $*$ given by $a * b = a + b - 2$, for $a, b \in \mathbb{Z}$. Clearly, $(\mathbb{Z}, *)$ is a groupoid. Also, for $a, b, c \in \mathbb{Z}$, we have

$$a * \big(b * c\big) = a * \big(b + c - 2\big) = a + \big(b + c - 2\big) - 2 = a + b + c - 4;$$
$$\big(a * b\big) * c = \big(a + b - 2\big) * c = \big(a + b - 2\big) + c - 2 = a + b + c - 4.$$

Therefore, $(\mathbb{Z}, *)$ is a semigroup. Now, suppose $e \in \mathbb{Z}$ is the identity element with respect to the operation $*$ so that we have $a * e = a$. That is, $a + e - 2 = a$ or $e = 2$. Therefore, $2 \in \mathbb{Z}$ is the identity element for the operation $*$. Hence, $(\mathbb{Z}, *, 2)$ is a monoid.

(ii) On the set $M = \mathbb{Q} \times \mathbb{Q}$, consider the binary operation $*$ given by

$$(a, b) * (x, y) = (ax, ay + b), \qquad \text{for} \quad a, b, x, y \in \mathbb{Q}. \tag{8.1.13}$$

Since $ax, ay + b \in \mathbb{Q}$, for all $a, b, x, y \in \mathbb{Q}$, it follows that $*$ satisfies the *closure property*. Next, let $a, b, x, y, u, v \in \mathbb{Q}$. Then, by definition of $*$, we have

$$\begin{aligned}\big((a, b) * (x, y)\big) * (u, v) &= (ax, ay + b) * (u, v)\\ &= \big(axu, axv + ay + b\big);\\ (a, b) * \big((x, y) * (u, v)\big) &= (a, b) * (xu, xv + y)\\ &= \big(axu, axv + ay + b\big),\end{aligned}$$

which shows that $*$ is an associativity operation. Now, suppose $(e_1, e_2) \in M$ is the identity for the operation $*$ so that we have

$$(a, b) * (e_1, e_2) = (e_1, e_2) * (a, b) = (a, b), \qquad \text{for all} \quad a, b \in \mathbb{Q}.$$

That is, for all $a, b \in \mathbb{Q}$, we have

$$a = ae_1 \;\text{ and }\; b = ae_2 + b \qquad \Rightarrow \qquad e_1 = 1 \;\text{ and }\; e_2 = 0.$$

Therefore, $(M, *, (1, 0))$ is a monoid.

(iii) Consider the groupoid $(A, \circ)$, as introduced earlier in Example 8.1$(e)$. Now, for $f, g, h \in A$ and $x \in X$, we have

$$\begin{aligned}\big((f \circ g) \circ h\big)(x) &= \big(f \circ g\big)\big(h(x)\big)\\ &= f\big(g\big(h(x)\big)\big) = f\big((g \circ h)(x)\big)\\ &= \big(f \circ \big(g \circ h\big)\big)(x).\end{aligned}$$

Therefore, $(f * g) * h = f * (g * h)$. It thus follows that $(A, \circ)$ is a semigroup. Also, the identity function $e := I_X \in A$ given by $e(x) = x$ for all $x \in X$ is such that we have

$$\begin{aligned}\big(f \circ e\big)(x) &= f\big(e(x)\big) = f(x)\\ &= e\big(f(x)\big) = \big(e \circ f\big)(x).\end{aligned}$$

Therefore, we also have $f * e = f = e * f$. That is, $e \in A$ is the identity element. Hence, $(A, \circ, e)$ is a monoid.

(iv) Let $X$ be a nonempty set, and $S = \mathscr{P}(X)$ be the power set of $X$. Now, since we have

$$\big(A \cap B\big) \cap C = A \cap \big(B \cap C\big), \qquad \text{for} \quad A, B, C \in S,$$

it thus follows that $(S, \cap)$ is a semigroup. We also have that

$$A \cap X = A = X \cup A, \qquad \text{for all} \quad A \in S.$$

Thus $X \in S$ is the identity element. Therefore, $(S, \cap)$ is a monoid.

(v) A natural example of a monoid is the set of finite sequences $\mathscr{A}^*$ defined over a set of symbols $\mathscr{A}$. These finite sequences are also known as strings defined over the "alphabet" $\mathscr{A}$. With respect to associative binary operation given by *concatenation*, $\mathscr{A}^*$ is a monoid, where the *empty string* is the identity element. Notice that the set of nonempty strings $\mathscr{A}^+$ is a semigroup that is not a monoid, which is also called the *Free semigroup* generated by the alphabet $\mathscr{A}$.

(vi) The semigroup $(\mathbb{Z}_n, +_n)$ is a monoid with $0 \in \mathbb{Z}_n$ as the additive identity. Also, the semigroup $(\mathbb{Z}_n, \times_n)$ is a monoid with $1 \in \mathbb{Z}_n$ as the multiplicative identity.

(vii) The semigroup $(U_n, \cdot)$ is a monoid with 0 as the additive identity.

(viii) It follows from Table 8.3 that the semigroup $(S_3, \circ)$ has the identity function $e$ as the multiplicative identity.

**Theorem 8.2** *Every* Monoid *has a unique identity element.*

***Proof.*** Let $M$ be a monoid with respect to a binary operation $*$. If possible, suppose both $e_1$ and $e_2$ are identity elements of $M$ for the operation $*$. Then, for all $x \in X$, we have

$$e_1 * x = x = x * e_1; \tag{8.1.14}$$

$$e_2 * x = x = x * e_2. \tag{8.1.15}$$

In particular, taking $x = e_2$ in (8.1.14), it follows that $e_1 * e_2 = e_2 = e_2 * e_1$. Similarly, taking $x = e_1$ in (8.1.15), it follows that $e_2 * e_1 = e_1 = e_1 * e_2$. Therefore, we obtain

$$e_1 = e_1 * e_2 = e_2.$$

Hence, the identity element of a monoid is unique. □

**Remark 8.1** The formal study of semigroups began in the early twentieth century with Cayley's theorem for semigroups. Finite semigroups are particularly of special significance in theoretical computer science. They find applications to the theory of finite automata via the *syntactic monoid*. Also, Markov processes link semigroups to the probability theory [6]. Similar to the Jordan-Hölder decomposition theorems, Krohn-Rhodes theory provides a classification of finite semigroups.

In many applications, it is important to know when an element of a monoid $(M, *, e)$ is *Invertible* with respect to the operation $*$, and the identity element $e$.

**Definition 8.8** Let $(M, *, e)$ be a monoid.

(a) An element $x \in M$ is said to be **left invertible** if there is an element $x_\ell \in M$ such that $x_\ell * x = e$. In this case, $x_\ell$ is called a *Left inverse* of the element $x$.
(b) An element $x \in M$ is said to be **right invertible** if there is an element $x_r \in M$ such that $x * x_r = e$. In this case, $x_r$ is called a *Right inverse* of the element $x$.
(c) An element $x \in M$ is said to be an **invertible element** if there is an element $y \in M$ such that we have

$$x * y = y * x = e. \tag{8.1.16}$$

We call $y$ an *Inverse* of the element $x$, and write $y := x^{-1}$.

The next theorem proves that an element of a monoid is invertible if it is both left and right invertible.

**Theorem 8.3** *Let $(M, *, e)$ be a monoid, and $x \in M$ be an* Invertible element. *Then the left and right inverses of $x$ are equal. Also, the inverse of $x$ is unique.*

***Proof.*** If $x_\ell$ and $x_r$ are respectively the left and right inverses of $x$, then we have

$$x_\ell * x = e \quad \text{and} \quad x * x_r = e.$$

It thus follows that

$$x_\ell = x_\ell * e = x_\ell * (x * x_r) = (x_\ell * a) * x_r = e * x_r = x_r.$$

We next prove the uniqueness of the inverse. For, suppose both $y_1$ and $y_2$ are inverses of $x$. Then we have

$$x * y_1 = y_1 * x = e \quad \text{and} \quad x * y_2 = y_2 * x = e.$$

We thus conclude that

$$y_1 = e * y_1 = (y_2 * x) * y_1 = y_2 * (x * y_1) = y_2 * e = y_2.$$

This completes the proof. □

The inverse of an element $x \in M$ with respect to additive notation $(+)$ is called an **additive inverse**, which we may write as $-x$. Also, the inverse of an element $x \in M$ with respect to multiplicative notation is called a **multiplicative inverse**, which we may write as $x^{-1}$. It thus follows from (8.1.16) that

$$x^{-1} * x = x * x^{-1} = e,$$

so that the inverse of $x^{-1}$ is $x$, i.e., $\left(x^{-1}\right)^{-1} = x$.

**Example 8.4**

(i) We find invertible elements of the monoid $M = \big(\mathbb{Q} \times \mathbb{Q}, *, (1, 0)\big)$ as in Example 8.3(ii). For, if $(a, b) \in M$ is invertible with respect to the operation $*$, then for some $(x, y) \in M$ we have

$$(a, b) * (x, y) = (x, y) * (a, b) = (1, 0).$$

That is, $ax = 1$ and $ay + b = 0$, which implies $x = 1/a$ and $y = -b/a$. Therefore, an element $(a, b) \in M$ is *invertible* with respect to the operation $*$, provided $a \neq 0$.

(ii) For each number set other than $\mathbb{N}$, every element is invertible with respect to usual addition, and the inverse is given by the "negative". Also, for each number set other than $\mathbb{N}$ and $\mathbb{Z}$, every nonzero element is invertible with respect to usual multiplication, and the inverse is given by the "reciprocal".

(iii) Every square matrix over a number set other than $\mathbb{N}$ and $\mathbb{Z}$ is invertible, provided its *determinant* is nonzero. The inverse of a square matrix may be computed by using the *adjoint relation* or the *Cayley-Hamilton theorem.*

(iv) Every element of the monoids $\big(\mathbb{Z}_n, +_n, 0\big)$ and $(U_n, \cdot, 1)$ is invertible, and we can read the additive inverse and the multiplicative inverse directly from the respective product tables. We may recall that Euler's *Totient function* $\phi(n)$ counts the integers $1 \leq a < n$ such that $\gcd(a, n) = 1$. In particular, we have $\phi(p) = p - 1$, for every prime $p$. Therefore, every nonzero element of the monoid $\big(\mathbb{Z}_p, \times_p, 1\big)$ is invertible, where $p$ is a prime. For, if $\gcd(a, b) = 1$ then it is possible to find integers $x$ and $y$ such that $a\,x + b\,y = 1$, by Lemma A.1. In particular, we obtain $a\,x \equiv 1 \bmod p$, where $x \in \mathbb{Z}_p^*$.

(v) It follows from Table 8.3 that every element of the monoid $\big(S_3, \circ, e\big)$ is invertible. For example, the inverse of $\tau_1$ is $\tau_1$ itself, and the inverse of $\rho_1$ is $\rho_2$. In general, using *two-row* notation, the inverse of a permutation is obtained by reading pre-images of elements $1, 2, \ldots, n$ from the *second row*. For example, consider two elements $\alpha$ and $\beta$ of the symmetric group $S_6$ given by

$$\alpha = \begin{pmatrix} 1\,2\,3\,4\,5\,6 \\ 3\,1\,5\,4\,6\,2 \end{pmatrix} \quad \text{and} \quad \beta = \begin{pmatrix} 1\,2\,3\,4\,5\,6 \\ 5\,3\,1\,6\,2\,4 \end{pmatrix}$$

Then we have

$$\alpha^{-1} = \begin{pmatrix} 1\,2\,3\,4\,5\,6 \\ 2\,6\,1\,4\,3\,5 \end{pmatrix} \quad \text{and} \quad \beta^{-1} = \begin{pmatrix} 1\,2\,3\,4\,5\,6 \\ 3\,5\,2\,6\,1\,4 \end{pmatrix}$$

In particular, we can compute elements $\alpha^{-1}\beta\alpha$ and $\beta^{-1}\alpha\beta$.

The concept introduced in the next definition is closely related to invertibility.

**Definition 8.9** Let $(M, *, e)$ be a monoid. An element $a \in M$ is said to be **cancellable** if it is both **left cancellable** and **right cancellable**, That is, for every $x, y \in S$, we have

$$a * x = a * y \;\Rightarrow\; x = y \quad \text{and} \quad x * a = y * a \;\Rightarrow\; x = y.$$

The next theorem proves in a monoid every invertible element is cancellable.

**Theorem 8.4** *Let $(M, *, e)$ be a monoid, and element $a \in M$ be invertible. Then $a$ is a* Cancellable element.

***Proof.*** We first prove $a \in M$ is *Left cancellable*. For, let $x, y \in X$, and $a * x = a * y$. Pre-multiplying by $a^{-1}$, we have

$$a^{-1} * (a * x) = a^{-1} * (a * y),$$

so that associativity of $*$ implies that

$$(a^{-1} * a) * x = (a^{-1} * a) * y,$$

which gives $e * x = e * y$, by using (8.1.16). It thus follows from (8.1.12) that $x = y$. Similarly, we can show that $a \in M$ is *Right cancellable*. That is,

$$x * a = y * a \quad \Rightarrow \quad x = y.$$

This completes the proof. □

The abstract definition of a (finite) group was first given by the prolific British mathematician *Arthur Cayley* (1821–1895) in a paper published in 1854.

**Definition 8.10** A nonempty set $G$ is called a **group** if there is a binary operation $*$ on $G$ such that it satisfies the following four conditions:

1. For all $a, b \in G$, $a * b \in G$. That is, $*$ has the closure property;
2. For all $a, b, c \in G$, $a * (b * c) = (a * b) * c$. That is, $*$ has the associativity property;
3. There is an element $e \in G$ such that $a * e = e * a = a$, for all $a \in G$. That is, $*$ has the identity element $e \in G$;
4. For each $a \in G$, there is an element $b \in G$ such that $a * b = b * a = e$. That is, $*$ is invertible with respect to the identity element $e \in G$.

We say $G$ is an **Abelian group** if in addition $*$ is commutative. That is,

5. $a * b = b * a$, for all $a, b \in G$.

Said differently, an *Abelian group* is a commutative monoid wherein every element is invertible. In particular, every group is a monoid, but not conversely. For example, the set of non-negative integers $\mathbb{Z}^+ := \mathbb{N} \cup \{0\}$ satisfies associative property with respect to the usual addition of integers; has 0 as the identity; but, no nonzero element in $\mathbb{Z}^+$ has an inverse. We may also write a group as $(G, *)$ when emphasis on the binary operation $*$ is required.

**Example 8.5**

(i) The usual addition $+$ of integers satisfies the properties as given below: For $a, b, c \in \mathbb{Z}$, we have

a. (*Closure Property*) $a + b \in \mathbb{Z}$.
b. (*Associativity Property*) $a + (b + c) = (a + b) + c$.
c. (*Additive Identity*) The element $0 \in \mathbb{Z}$ is such that it satisfies $a + 0 = 0 + a = a$. For this reason, 0 is also called the *Additive identity* of the semigroup $(\mathbb{Z}, +)$.
d. (*Additive Inverse*) For $a \in \mathbb{Z}, -a \in \mathbb{Z}$ is such that $a + (-a) = 0 = (-a) + a$. For this reason, $-a$ is called the *Additive inverse* of the element $a$.
e. (*Commutativity Property*) $a + b = b + a$.

Therefore, the set $\mathbb{Z}$ is a commutative group with respect to the usual addition $+$, which we may write as $(\mathbb{Z}, +)$.

(ii) The usual multiplication $\times$ of integers satisfies the properties as given below: For $a, b, c \in \mathbb{Z}$, we have

a. (*Closure Property*) $a \times b \in \mathbb{Z}$.
b. (*Associativity Property*) $a \times (b \times c) = (a \times b) \times c$.
c. (*Multiplicative Identity*) The element $1 \in \mathbb{Z}$ is such that it satisfies $a \times 1 = 1 \times a = a$. For this reason, $1 \in \mathbb{Z}$ is also called the *Multiplicative identity* of the semigroup $(\mathbb{Z}, \times)$.
d. (*Commutativity Property*) $a \times b = b \times a$.

Therefore, the set $\mathbb{Z}$ is a commutative group with respect to usual multiplication $\times$, which we may write as $(\mathbb{Z}, \times)$.

(iii) The set $\mathbb{R}$ is an Abelian group with respect to the usual addition of real numbers. Therefore, for $a, b \in \mathbb{R}$, we have

a. $a + b \in \mathbb{R}$, for all $a, b \in \mathbb{R}$;
b. $a + (b + c) = (a + b) + c$, for all $a, b, c \in \mathbb{R}$;
c. $0 \in \mathbb{R}$ satisfies the condition $a + 0 = 0 + a = a$, for all $a \in \mathbb{R}$, so that 0 is the *additive identity*;
d. each $a \in \mathbb{R}$ satisfies the condition $a + (-a) = 0 = (-a) + a$, so that $-a \in \mathbb{R}$ is the *additive inverse* of $a$;
e. $a + b = b + a$, for all $a, b \in \mathbb{R}$.

Similarly, $(\mathbb{Q}, +)$, $(\mathbb{Q}^+, \times)$ and $(\mathbb{Q}^*, \times)$ are Abelian groups. Also, the set $\mathbb{C}$ is an Abelian group with respect to the usual addition of complex numbers. It can also be shown that the complex numbers of the form $m + ni$, with $m, n \in \mathbb{Z}$, form an Abelian group with respect to the usual addition of complex numbers.

(iv) Notice that all properties for the usual addition $+$ in the previous example hold for the set of *nonzero* real numbers $\mathbb{R}^* := \mathbb{R} \setminus \{0\}$ with respect to usual multiplication, where 1 is the *multiplicative identity* and $a^{-1} \in \mathbb{R}$ is the *multiplicative inverse* of $0 \neq a \in \mathbb{R}^*$. Therefore $(\mathbb{R}^*, \times)$ is also an Abelian group. Similarly, the set of *nonzero* complex numbers $\mathbb{C}^* := \mathbb{C} \setminus \{0\}$ is an Abelian group with respect to usual multiplication, where 1 is the *multiplicative identity* and $z^{-1} = \overline{z}/|z|^2 \in \mathbb{C}$ is the *multiplicative inverse* of $0 \neq z \in \mathbb{C}^*$.

(v) Let $G = \mathscr{P}(X)$ be the power set of a nonempty set $X$. We use the *symmetric difference*, denoted by $\oplus$, as a binary operation on $G$ given. That is, for $A, B \in G$, we take

$$A \oplus B := \big(A \setminus B\big) \cup \big(B \setminus A\big) = \big(A \cup B\big) \setminus \big(A \cap B\big).$$

It is easy to verify that the above operation is associative and commutative and that $(G, \oplus)$ is a group with $\emptyset$ as the identity element because

$$A \oplus \emptyset = A = \emptyset \oplus A, \qquad \text{for all} \quad A \in G.$$

(vi) The set of $m \times n$ matrices with entries in $\mathbb{Z}$ is a commutative group under usual addition of matrices. It is known as the additive group of integer matrices of the type $m \times n$. The same holds when the entries of such types of matrices are taken from the set $\mathbb{Q}$, $\mathbb{R}$, or $\mathbb{C}$, which are respectively called the additive group of rational, real, or complex matrices. Also, the set $M_2$ of $2 \times 2$ matrices over any one of the above number sets of the form

$$\begin{pmatrix} a & b \\ c & d \end{pmatrix}, \qquad \text{with} \quad ad - bc = 1,$$

is a group with respect to usual matrix multiplication. For, as the determinant of the product of two matrices is the product of their determinant, we see that $A_1, A_2 \in M_2 \Rightarrow A_1 A_2 \in M_2$. It is in general true that the matrix multiplication is associative. Also $\begin{pmatrix} 1 & 0 \\ 0 & 1 \end{pmatrix}$ is the identity element, and the inverse of the matrix $\begin{pmatrix} a & b \\ c & d \end{pmatrix} \in M_2$ is the matrix $\begin{pmatrix} d & -b \\ -c & a \end{pmatrix} \in M_2$. Notice that, since

$$\begin{pmatrix} 0 & 1 \\ -1 & 0 \end{pmatrix} \begin{pmatrix} 1 & 1 \\ 0 & 1 \end{pmatrix} = \begin{pmatrix} 0 & 1 \\ -1 & -1 \end{pmatrix} \neq \begin{pmatrix} -1 & 1 \\ -1 & 0 \end{pmatrix} = \begin{pmatrix} 1 & 1 \\ 0 & 1 \end{pmatrix} \begin{pmatrix} 0 & 1 \\ -1 & 0 \end{pmatrix},$$

it follows that $M_2$ is a *Non-commutative group*.

(vii) Let $\mathbf{k}$ denote the field of real numbers $\mathbb{R}$ or the field of complex numbers $\mathbb{C}$. In general, the set of all invertible square matrices of order $n$ over the field $\mathbf{k}$, denoted by $GL(n, \mathbf{k})$, is a group with respect to usual matrix multiplication. It is known as the *General linear group* of degree $n$ over the field $\mathbf{k}$. A matrix $A \in GL(n, \mathbb{R})$ is called an *orthogonal matrix* if we have

$$A^t A = A A^t = I_n,$$

where $A^t$ as usual denotes the transpose of the matrix $A$. The set of orthogonal matrices, denoted by $\mathscr{O}(n, \mathbb{R})$, forms a group. It is known as the *orthogonal group* of degree $n$.

(viii) Consider replacing unknowns $x_i$ in an $n$-variable $x_1, \ldots, x_n$ linear combination given by

$$x_i \mapsto X_i' = a_{1i}x_1 + a_{2i}x_2 + \cdots + a_{ni}x_n, \quad \text{with} \quad a_{ij} \in \mathbb{R} \text{ or } \mathbb{C}. \tag{8.1.17}$$

With notions as given below

$$X = \big(x_1, \ldots, x_n\big), \quad X' = \big(x_1', \ldots, x_n'\big), \quad \text{and} \quad A = \big(a_{ij}\big),$$

we may write (8.1.17) in matrix notation as $X' = XA$. In linear algebra, we recognize $A$ as the matrix of "change of basis" over an $n$ dimensional vector space $V$ over the field $\mathbf{k}$. If $X'' = X'B$ is another *substitution* with matrix $B$, then the product of the two *substitutions* is given by $X'' = XAB$, with the associated matrix $AB$ [7]. The **substitution group** on a given set of $n$ symbols $x_1, \ldots, x_n$ is the group of $\mathbf{k}$-*automorphisms* of a $n$-dimensional vector space, which we can abstractly identify with the group $GL(n, \mathbf{k})$.

(ix) Let $X$ be a nonempty set and consider the set

$$A = \mathbb{R}^X := \big\{f \mid f : X \to \mathbb{R}\big\}$$

The **pointwise sum** and the **pointwise product** of two functions $f,\ g \in A$ are respectively defined by

$$\big(f + g\big)(x) = f\big(x\big) + g\big(x\big) \quad \text{and} \quad \big(f \cdot g\big)(x) = f\big(x\big) \cdot g\big(x\big),$$

for $x \in X$. Notice that $+$ and $\cdot$ on the right of the above two equations are usual addition and multiplication of real numbers. Clearly, the two are binary operations on $A$. The set $A$ is a group with respect to *pointwise sum* $+$ .

**Definition 8.11** The **order** of a group $G$, denoted by $o(G)$, is the cardinality $|G|$ of the set $G$. We say $G$ is a **finite group** if $o(G) < \infty$. Otherwise, it is called an **infinite group**.

Most illustrations of groups in Example 8.5 are infinite. However, in this book, our focus is to discuss applications of some special classes of *finite groups* such as given below. We may recall that, for finite groups of smaller order, it is more convenient to represent a group by its *product table*. When all entries of the table are elements of the group, the *closure property* holds. The operation is commutative if the product table is symmetric. Also, for the identity element, the row (and also the column) are identical. The inverse of a row element is the column element such that the corresponding entry in the product table is the identity element.

**Example 8.6**

(i) As introduced earlier in Example 8.1 (iii), the set of $n$th roots of unity $U_n$ is a group with respect to the usual multiplication of complex numbers, which is an associative operation. The number $1 \in U_n$ is the identity element, and the inverse of $\omega^i \in U_n$ is given by the element $\omega^j \in U_n$ such that $j \equiv i$ (modulo $n$). In particular, the set of fourth roots of unity $U_4$ given by

$$= \{1, -1, i, -i\}, \qquad \text{with } i = \sqrt{-1}.$$

is a finite Abelian group of order 4, with the identity element $1 \in U_4$. The inverse of an element can be read directly from the *product table* Table 8.1. For example, the inverse of $i$ in the first column is $-i$ because their product in the fifth column corresponds to the identity element 1; and so on.

(ii) Consider the set $C^1 := \{z \in \mathbb{C} \mid |z| = 1\}$. It follows easily that, with respect to usual multiplications of complex numbers, $C^1$ is grouped with $1 \equiv (1, 0)$ as the identity element. This is usually referred to as the **circle group**.

(iii) As discussed earlier in Example 8.1 (iv), the set $\mathbb{Z}_n$ is a finite Abelian group of order $n$ with respect to the *addition modulo* $+_n$ (Exercise 8.6).

(iv) As discussed earlier in Example 8.1 (v), the set $\text{Perm}(A)$ is a group with respect to the binary operation "composition". It is called the *permutation group* on the set $A$. Recall that on any collection functions, whenever defined, the "composition" is associative. The identity function $i_A : A \to A$ is the identity element and by Theorem 4.9 each element of the set $\text{Perm}(A)$ is invertible with respect to the operation (8.1.6). Therefore, the set $\text{Perm}(A)$ is a group with respect to the operation "composition". In particular, $S_n$ is a group, which is called the **symmetric group** of degree $n$. For $n \geq 3$, this is a non-Abelian group (see also Example 4.5). The French mathematician *Évariste Galois* (1811–1832) was the first to apply group properties of permutations in his fundamental work related to the roots of an algebraic equation. He proved that a *quintic, in general, may not be soluble by radicals* [4].

(v) The set of $2n$ symmetries of a regular polygon on $n$ vertices, denoted by $D_{2n}$, is a group with respect to the binary operation given by "composition". This is known as the **dihedral group** of degree $n$. We also say that $D_{2n}$ is the *group of symmetries* of the regular $n$-gon. In particular, the *dihedral group* $D_6$ contains the three *rotational symmetries* of an equilateral triangle with vertices $v_1$, $v_2$ and $v_3$, given by

$$v_1 \to v_2 \to v_3; \qquad v_2 \to v_3 \to v_1; \qquad v_3 \to v_1 \to v_2$$

The other three elements of $D_6$ are the *axial symmetries* given by

$$v_1 \text{ fixed}, v_2 \to v_3; \quad v_2 \text{ fixed}, v_1 \to v_3; \quad v_3 \text{ fixed}, v_1 \to v_2.$$

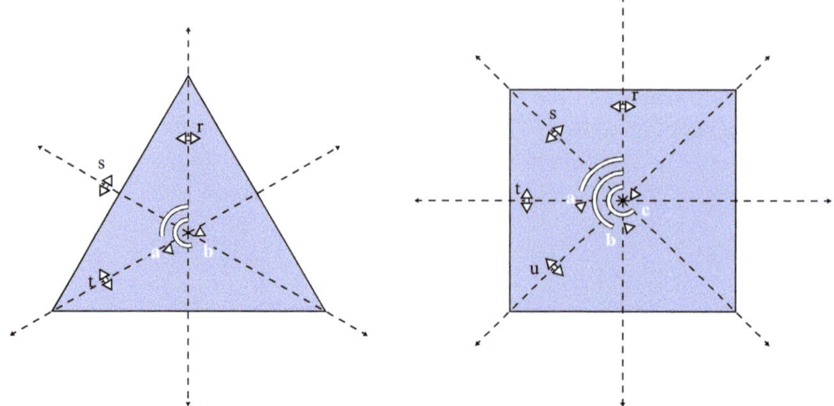


**Fig. 8.1** Symmetries of a regular polygon on 3 and 4 vertices

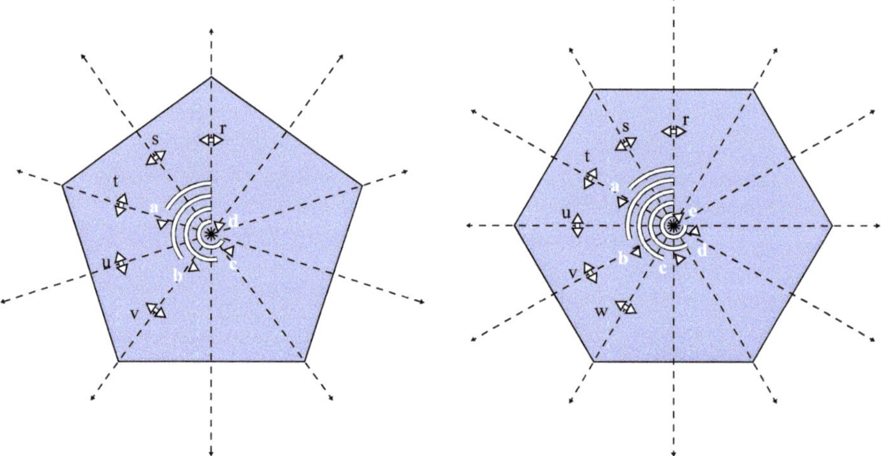


**Fig. 8.2** Symmetries of a regular polygon on 5 and 6 vertices

We have $D_6$ as an Abelian group with 6 elements. Figure 8.1 shows symmetries of a regular polygon on 3 and 4 vertices. Also, Fig. 8.2 shows symmetries of a regular polygon on 5 and 6 vertices.

(vi) Consider the three $2 \times 2$ matrices over the set of real numbers given by

$$\mathbf{i} = \begin{pmatrix} i & 0 \\ 0 & -i \end{pmatrix}, \; \mathbf{j} = \begin{pmatrix} 0 & 1 \\ -1 & 0 \end{pmatrix}, \; \mathbf{k} = \begin{pmatrix} 0 & i \\ i & 0 \end{pmatrix},$$

together with the identity matrix $I$, and their negatives $-\mathbf{i}$, $-\mathbf{j}$, $-\mathbf{k}$ and $-I$. The set

$$Q = \{I, -I, \mathbf{i}, -\mathbf{i}, \mathbf{j}, -\mathbf{j}, \mathbf{k}, -\mathbf{k}\} \subset GL(2, \mathbb{C})$$

forms a group $Q$ of order 8, with $\mathbf{i}^2 = \mathbf{j}^2 = \mathbf{k}^2 = -I$, and

$$\mathbf{i}\,\mathbf{j} = -\mathbf{j}\,\mathbf{i} = \mathbf{k}; \quad \mathbf{j}\,\mathbf{k} = -\mathbf{k}\,\mathbf{j} = \mathbf{i}; \quad \mathbf{k}\,\mathbf{i} = -\mathbf{i}\,\mathbf{k} = \mathbf{j}.$$

The group $Q$ is called the *group of quaternion*.

We conclude this section by proving some basic results related to groups. The next theorem proves that *cancellation laws* hold in any group.

**Theorem 8.5** *Let $(G, *)$ be a group, and $a, b, c \in G$. Then we have*

$$a * c = b * c \quad \Rightarrow \quad a = b \qquad \textit{and} \qquad c * a = c * b \quad \Rightarrow \quad a = b.$$

***Proof.*** We first prove *right cancellation law*. For, post-multiplying the relation $a * c = b * c$ by $c^{-1}$, we obtain

$$\big(a * c\big) * c^{-1} = \big(b * c\big) * c^{-1},$$

which gives

$$a * \big(c * c^{-1}\big) = b * \big(c * c^{-1}\big),$$

by associativity of $*$. Therefore, by definition of inverse, we obtain

$$a * e = b * e \qquad \text{or} \qquad a = b,$$

by using the definition of identity. A similar argument proves that the *left cancellation law* holds too. □

**Corollary 8.1** *Each row (and also each column) of the product table of a finite group $(G, *)$ is a* permutation *of the set $G$.*

***Proof.*** Let $G = \{a_1, a_2, \ldots\}$. Suppose two elements in some row of the product table of $G$ equal, say $a_i * a = a_j * a$. It then follows from Theorem 8.5 that $a_i = a_j$. Therefore, all elements in any row of the product table for $G$ are distinct. The same holds for all elements in any column. Since the injective self-map of a finite set is necessarily surjective, the proof is complete. □

We already know that the identity element and inverse of an element in a group $G$ are unique, and also that $\big(x^{-1}\big)^{-1} = x$, for all $x \in G$. The property stated in the next theorem is also a useful fact.

**Theorem 8.6** *Let $(G, *)$ be a group. Then, for any $a, b \in G$, we have $\big(a * b\big)^{-1} = b^{-1} * a^{-1}$. More generally, if $a_1, a_2, \ldots, a_n \in G$, then we have*

$$(a_1 * a_2 * a_3 * \cdots * a_n)^{-1} = a_n^{-1} * a_{n-1}^{-1} * \cdots * a_1^{-1}.$$

***Proof.*** Since $a^{-1} = b$ if and only if $a * b = e = b * a$, we need to show that

$$(a * b) * (b^{-1} * a^{-1}) = (b^{-1} * a^{-1}) * (a * b) = e.$$

For, we have

$$\begin{aligned}(a * b) * (b^{-1} * a^{-1}) &= a * (b * (b^{-1} * a^{-1}))\\ &= a * ((b * b^{-1}) * a^{-1})\\ &= a * (e * a^{-1})\\ &= a * a^{-1} = e.\end{aligned}$$

Similarly, it can be shown that $(b^{-1} * a^{-1}) * (a * b) = e$. The general case can be proven by a simple induction on $n$. □

**Theorem 8.7** *Let* $(G, *)$ *be a group, and* $a \in G$*. Then we have* $a^2 = e$ *if and only if* $a = a^{-1}$.

***Proof.*** First, let $a^2 = e$, Then we have

$$\begin{aligned}a &= a * e = a * (a * a^{-1})\\ &= (a * a) * a^{-1} = a^2 * a^{-1}\\ &= e * a^{-1} = a^{-1}.\end{aligned}$$

Conversely, $a = a^{-1}$ implies that $a^2 = a * a = a * a^{-1} = e$. □

An *involution* in a group $G$ is an element $a$ such that $a^2 = e$. It follows from Theorem 8.7 that a group $G$ consisting only of involutions is Abelian. For, since every element $a \in G$ is its own inverse, we have $a^{-1} = a$ and $b^{-1} = b$, for every $a, b \in G$. It thus follows that

$$(a * b) = (a * b)^{-1} = b^{-1} * a^{-1} = b * a.$$

Therefore, $G$ is an Abelian group.

**Theorem 8.8** *Let* $(G, *)$ *be a group, and* $a, b \in G$*. Then both equations*

$$a * x = b \quad \text{and} \quad y * a = b$$

*have a unique solution in* $G$*.*

***Proof.*** We prove the statement for the equation $a * x = b$. A similar argument applies to the other equation. Now, for $x = a^{-1} * b \in G$, we have

$$a * x = a * (a^{-1} * b) = (a * a^{-1}) * b = e * b = b.$$

It thus that $x = a^{-1} * b$ is a solution of the equation $a * x = b$. Next, to prove the uniqueness part of the assertion, let $x_1, x_2 \in G$ be two solutions of the equation $a * x = b$ so that we have

$$a * x_1 = b \quad \text{and} \quad a * x_2 = b.$$

Therefore, $a * x_1 = a * x_2$, which implies $x_1 = x_2$, by using the left cancellation law. □

We next specify a way to construct new groups from the known ones. For $i = 1, \ldots, n$, let $(G_i, *_i)$ be $n$ groups. Consider the Cartesian product

$$G = G_1 \times \cdots \times G_n, \tag{8.1.18}$$

and let the (pointwise) binary operation $*$ on $G$ be defined by

$$(a_1, \ldots, a_n) * (b_1, \ldots, b_n) = (a_1 * b_1, \cdots, a_n * b_n), \tag{8.1.19}$$

where $(a_1, \ldots, a_n), (b_1, \ldots, b_n) \in G$. The next theorem proves that $(G, *)$ is a group, which is known as the **direct product** of the $n$ groups $G_1, \ldots, G_n$.

**Theorem 8.9** *The product set G as in* (8.1.18) *is a group with respect to the* operation $*$ *given by* (8.1.19). *Also, the* direct product $G$ *is an Abelian group when each group* $G_i$ *is so.*

***Proof.*** For $a = (a_1, \ldots, a_n)$, $b = (b_1, \ldots, b_n)$, $c = (c_1, \ldots, c_n) \in G$, using the associativity of $n$ operations $*_i$, we have

$$\begin{aligned}
(a * b) * c &= (a_1 * b_1, \ldots, a_n * b_n) * (c_1, \ldots, c_n) \\
&= ((a_1 * b_1) * c_1, \ldots, (a_n * b_n) * c_n) \\
&= (a_1 * (b_1 * c_1), \ldots, a_n * (b_n * c_n)) \\
&= (a_1, \ldots, a_n) * (b_1 * c_1, \ldots, b_n * c_n) \\
&= a * (b * c),
\end{aligned}$$

and so $*$ is associative. Next, if $e_i$ is the identity element of the group $G_i$, then for $e = (e_1, \ldots, e_n) \in G$ we have

$$\begin{aligned}
e * a &= (e_1, \ldots, e_n) * (a_1, \ldots, a_n) \\
&= (e_1 * a_1, \ldots, e_n * a_n) \\
&= (a_1, \ldots, a_n) = a.
\end{aligned}$$

Similarly, we have $a * e = a$. Therefore, $e \in G$ is the identity element. Now, since

$$(a_1, \ldots, a_n) * (a_1^{-1}, \ldots, a_n^{-1}) = (a_1 * a_1^{-1}, \ldots, a_n * a_n^{-1}) = (e_1, e_2, \cdots, e_n) = e,$$

and also $\left(a_1^{-1}, \ldots, a_n^{-1}\right) * \left(a_1, \ldots, a_n\right) = e$, it follows that the inverse of the element $(a_1, \ldots, a_n) \in G$ is the element $\left(a_1^{-1}, \ldots, a_n^{-1}\right) \in G$. Therefore, $(G, *)$ is a group. Finally, if each group $G_i$ is Abelian, then we have

$$a * b = \left(a_1 * b_1, \ldots, a_n * b_n\right) = \left(b_1 * a_1, \ldots, b_n * a_n\right) = b * a.$$

Hence, $G$ is also an Abelian group. □

**Example 8.7 (Group of Bit Strings)** Let $\Sigma := \{0, 1\}$, and consider the finite set $S = \Sigma^n$ of $n$-bit strings. Let the operation $\oplus$ denote the "addition mod 2" so that, for $x_i, y_i \in \{0, 1\}$, we have

$$x_i \oplus y_i := \begin{cases} 0 & (x_i = y_i) \\ 1 & (x_i \neq y_i) \end{cases} \tag{8.1.20}$$

We can extend the operation $\oplus$ to the set $S$. For $x, y \in S$ given by

$$x = x_1 x_2 \cdots x_n \qquad \text{and} \qquad y = y_1 y_2 \cdots y_n,$$

we define

$$x \oplus y = \left(x_1 \oplus y_1\right)\left(x_2 \oplus y_2\right) \cdots \left(x_n \oplus y_n\right). \tag{8.1.21}$$

Clearly, $\oplus$ is associative. Therefore, $\left(S, \oplus\right)$ is a semigroup. In fact, the $n$-bit string of $\mathbf{0} \in \Sigma^n$ denoted by

$$\mathbf{0} := 00 \cdots 0$$

is the identity element of this semigroup. We may also write $\mathbf{0}$ as $\mathbf{0}_n$, whenever emphasis on the length is needed. Also, and

$$x \oplus x = \left(x_1 \oplus x_1\right)\left(x_2 \oplus x_2\right) \cdots \left(x_n \oplus x_n\right) = \mathbf{0}.$$

That is, each $x \in S$ is an *involution*. Hence, $\left(S, \oplus\right)$ is an Abelian group.

## 8.2 Substructures and Homomorphisms

A *substructure* of an algebraic structure is a nonempty subset that inherits all algebraic properties under the *induced operation*, and a *homomorphism* tells us how identical two algebraic structures are. As shown in a later section, these concepts help us understand the properties of an algebraic structure in terms of related "quotient structures". In this section, we study various types of sub-structures of algebraic structures, and also homomorphisms among related quotient structures.

## Substructures

We may recall that, for example, both $\mathbb{Z}$ and $\mathbb{Q}$ are groupoids with respect to the same operation $+$ that makes the set $\mathbb{R}$ a groupoid. This is so mainly because each is closed with respect to the operation $+$. A formal definition for semigroups and monoids is given below.

**Definition 8.12** Let $(S, *)$ be a semigroup. A nonempty subset $T \subseteq S$ is a *subsemigroup* of $S$ if it is closed with respect to the (induced) binary operation $*$. That is, for all $x, y \in T$, we have $x * y \in T$. Similarly, for $(M, *, e)$ is a monoid, a nonempty subset $T \subseteq M$ is called a *submonoid* if it is closed with respect to the (induced) binary operation $*$, and also $e \in T$.

**Example 8.8** Consider a commutative monoid $(M, *, e)$, and let $T$ be the set of *idempotent elements* of $M$. That is, we have

$$T = \{a \in M \mid a * a = a\}.$$

Since $e * e = e$, we have $e \in T$. Also, $a, b \in T$ implies $a^2 = a$ and $b^2 = b$, and so

$$\begin{aligned}(a * b)^2 &= (a * b) * (a * b) = a * (b * (a * b)) \\ &= a * ((a * b) * b) = a * (a * (b * b)) \\ &= (a * a) * (b * b) = a^2 * b^2 \\ &= a * b,\end{aligned}$$

which shows that $a * b \in T$. Therefore, $T$ is a submonoid of the monoid $M$.

In our pursuit to have an insight into a larger algebraic structure by looking at its constituent *smaller parts* having similar structural properties, we next introduce the concept of a *subgroup* of a group.

**Definition 8.13** Let $(G, *)$ be a group, and $H \subseteq G$ be a nonempty set. We say $H$ is a **subgroup** of the group $G$ if $(H, *)$ itself is a group. Sometimes, we may denote this by using the notation $H \leq G$.

In particular, both $H = \{e\}$ and $H = G$ are subgroups of the group $G$. These are called the **trivial subgroups**. In the next example, we give some other examples of *non-trivial* subgroups of the standard groups as introduced earlier.

**Example 8.9**

(i) Since every number set in the following ascending chain

$$\mathbb{Z} \subset \mathbb{Q} \subset \mathbb{R} \subset \mathbb{C}$$

is a groups with respect to usual addition $+$, each is a subgroup of the groups above in the chain. In particular, $\mathbb{Z}$ is a subgroup of the group $(\mathbb{Q}, +)$; $\mathbb{R}$ is a subgroup of the group $(\mathbb{C}, +)$; and so on. Also, for any natural number $n > 1$, the set $H = n\mathbb{Z}$ is a subgroup of the group $(\mathbb{Z}, +)$. In fact, it follows easily that every subgroup of the additive group $\mathbb{Z}$ is of the form $n\mathbb{Z}$, for some integer $n \geq 0$ (Theorem 8.12).

(ii) The set $U_2 = \{1, -1\}$ is a subgroup of the group $U_4 = \{1, -1, i, -i\}$ of the four roots of unity with respect to the usual multiplication of complex numbers, because it is clear from the product table of $U_2$, as read from Table 8.1, that all the four *group axioms* are satisfied. However, notice that the same is not true for the subset $H = \{-1, i\}$ of the group $U_4$, because $-i = (-1) \times i \notin H$.

(iii) It follows from the product tables read from Table 8.3 that the following subsets are subgroups of the group $S_3$:

$$H_1 = \{e, \tau_1\}, \quad H_2 = \{e, \tau_2\}, \quad H_3 = \{e, \tau_3\}, \tag{8.2.1a}$$

$$K = \{e, \rho_1, \rho_2\}. \tag{8.2.1b}$$

(iv) With notations as introduced earlier in Example 8.5 (vii), the subgroup of *orthogonal group* $\mathcal{O}(n, \mathbb{R})$ consisting of matrices with determinant $+1$ is called the *special orthogonal group*. It is denoted by $\mathcal{SO}(n, \mathbb{R})$. Also, the subset of the *general linear group* $GL(n, \mathbb{R})$ consisting of the matrices with determinant $+1$ forms a subgroup. It is denoted by $SL(n, \mathbb{R})$, and is known as the *special linear group*. These *matrix groups* play a significant role in solving various types of problems in many different applications related to physics and computer science.

**Theorem 8.10** *Let $(G, *)$ be a group, and $H \subseteq G$ be nonempty. A necessary and sufficient condition for $H$ to be a subgroup of $G$ is that, for all $a, b \in H$, we have $(i)$ $a * b \in H$; and, $(ii)$ $a^{-1} \in H$.*

***Proof.*** First, suppose $H$ is a subgroup of $G$. That is, the set $H$ itself is a group under the induced operation $*$. In particular, by the closure property, we have $a * b \in H$, for every $a, b \in H$. Next, since the relations $a * e = e * a = a$ hold in particular for all $a \in H$, it follows that $e$ is the identity element for the group $H$. Also, since $(H, *)$ is a group, the inverse of $a \in H$ necessarily belongs to the set $H$. Therefore, the conditions (i) and (ii) hold. Conversely, suppose $H \subseteq G$ is a nonempty set that satisfies the two given conditions. So, if $h \in H$ then $h^{-1} \in H$, by (ii), and so $e = h * h^{-1} \in H$, by (i). Of course, the associativity of the operation $*$ holds for the elements of $H$, because the same is true for all elements of the group $G$. Therefore, $(H, *)$ itself is a group. Hence, $H$ is a subgroup of the group $G$. □

We can combine the two conditions in Theorem 8.10 into a single condition.

**Theorem 8.11** *Let $(G, *)$ be a group, and $H \subseteq G$ be a nonempty set. Then $H$ is a subgroup of the group $G$ if and only if we have $a * b^{-1} \in H$, for every $a, b \in H$.*

***Proof.*** We first assume that $H$ be a subgroup of $G$. That is, $(H, *)$ is a group. Now, since for every $b \in H$ we have $b^{-1} \in H$, it follows that $a * b^{-1} \in H$, for every

$a, b \in H$. Conversely, suppose the condition holds. To show that $(H, *)$ is a subgroup of $(G, *)$, it suffices to prove that

1. for every $a, b \in H$, we have $a * b \in H$;
2. the identity element $e$ lies in $H$; and,
3. for every $a \in H$, we have $a^{-1} \in H$.

We first show that (2) holds. For, since $H$ is nonempty, there is some $a \in H$. It thus follows from the given condition that $e = a * a^{-1} \in H$. Next, let $a \in H$. Then $a^{-1} = e * a^{-1} \in H$. Therefore, the inverse of every element in $H$ lies in $H$ itself. Let $a, b \in H$ so that $b^{-1} \in H$. Notice that $(b^{-1})^{-1} = b$. Hence, $a * b = a * (b^{-1})^{-1} \in H$, which proves that $H$ is closed under the (induced) operation $*$. This completes the proof. □

**Corollary 8.2** *Let* $(G, *)$ *be a* finite group*, and* $H \subseteq G$ *be a nonempty set. Then* $H$ *is a subgroup of the group* $G$ *if and only if, for every* $a, b \in H$*, we have* $a * b \in H$.

***Proof.*** If $H$ is a subgroup of $G$, then $H$ is closed under the (induced) operation $*$. Conversely, suppose $H$ is a nonempty finite subset of a group $(G, *)$ such that the condition holds. In view of Theorem 8.11, it suffices to show that $a \in H$ implies that $a^{-1} \in H$. For, since $H$ is closed with respect to operation $*$, it follows that all the positive powers of $a$ lie in $H$. However, since $H$ is finite, we must have $a^i = a^j$, for some $i < j$. That is, $a^{j-i} = e$, with $j - i > 0$. It thus follows that $e \in H$, and all non-negative powers of $a$ lie in $H$. Therefore, we have

$$a^{-1} = a^{j-i} * a^{-1} = a^{j-i-1}.$$

Finally, since $j - i - 1 \geq 0$, $a^{-1}$ is a non-negative power of $a$, it belongs to $H$. This completes the proof. □

**Theorem 8.12** *Every nonzero subgroup of the group* $(\mathbb{Z}, +)$ *is of the form* $n\mathbb{Z}$*, for some positive integer* $n$.

***Proof.*** Let $H$ be a nonzero subgroup of the group $(\mathbb{Z}, +)$. Since $H$ contains a nonzero integer, it follows from the *well-ordering principle* that $H$ contains a least positive integer, say $n_0$. We claim that $H = n_0\mathbb{Z}$. Since $n_0 \in H$, and it is a subgroup, we have $n_0\mathbb{Z} \subseteq H$. To prove the reverse inclusion, let $a \in H$. By Euclid's division algorithm, we can find integers $q$, $r \in \mathbb{Z}$ such that

$$a = n_0 q + r, \qquad \text{where} \quad 0 \leq r < n_0.$$

But then, $a - n_0 q = r \in H$ implies that $r = 0$, because $n_0$ is the least positive integer in $H$, Therefore, $a = n_0 q \in n_0\mathbb{Z}$. That is, $H \subseteq n_0\mathbb{Z}$. This completes the solution. □

In particular, we have $H = 2\mathbb{Z}$ and $K = 3\mathbb{Z}$ are subgroups of the group $(\mathbb{Z}, +)$. However, their union $H \bigcup K$ given by

$$2\mathbb{Z} \cup 3\mathbb{Z} = \{0, 2, -2, 3, -3, 4, -4, 6, 6, \ldots\}$$

is not a subgroup, because $2 + 3 = 5 \notin H \bigcup K$. Therefore, the union of two subgroups of a group in general need not be a subgroup. However, notice that $2\mathbb{Z} \bigcup 4\mathbb{Z} = 2\mathbb{Z}$ is a subgroup of $\mathbb{Z}$ simply because $4\mathbb{Z} \subset 2\mathbb{Z}$. More generally, if $H$ and $K$ are subgroups of a group $G$, then $H \bigcup K$ is a subgroup of $G$ if and only if either $H \subseteq K$ or $K \subseteq H$. The next theorem proves that the arbitrary intersection of subgroups of a group $G$ is always a subgroup.

**Theorem 8.13** *Let $(G, *)$ be a group, and $H_i$ be subgroups of $G$, for $i \in I$. Then $\left(\bigcap H_i, *\right)$ is a subgroup of $G$.*

***Proof.*** Since each $H_i$ a subgroup of $G$, the identity element $e$ belongs to all $H_i$, and hence $\bigcap H_i$ is nonempty. It is thus enough to show that, for every $a, b \in \cap H_i$, we have $a * b^{-1} \in \bigcap H_i$. For, let $a, b \in \cap H_i$ so that $a, b \in H_i$, for every $i \in I$. Now, as each $H_i$ is a subgroup, we have $a * b^{-1} \in H_i$, for every $i \in I$, and hence $a * b^{-1} \in \bigcap H_i$. This completes the proof. □

Now, let $H$ and $K$ be two nonempty subsets of a group $(G, *)$. When $H = \{h\}$, we may define the product $H * K := h * K$ simply as the set $\{h * k \mid k \in K\} \subseteq G$. Similarly, we can define the product $H * k$, for $k \in K$. For example, with $G = S_3$ and $H = \{e, (1\,2)\}$, we have

$$\begin{aligned} eH &= H = (1\,2)H; \\ (1\,3)H &= \{(1\,3\,2), (1\,3)\} = (1\,3\,2)H; \\ (2\,3)H &= \{(1\,2\,3), (2\,3)\} = (1\,2\,3)H. \end{aligned} \tag{8.2.2}$$

More generally, we may define the product $H * K$ as follows:

$$H * K = \{h * k \mid h \in H \text{ and } k \in K\}. \tag{8.2.3}$$

In terms of the above notation, $H$ is a subgroup of the group $G$ if and only if we have

$$HH^{-1} \subseteq H, \quad \text{where} \quad H^{-1} = \{h^{-1} \mid h \in H\}.$$

However, the product $HK$ in general may not be a subgroup of $G$ even if both $H, K$ are subgroups. The next theorem gives a simple criterion to verify when the product of two subgroups of a group is a subgroup.

**Theorem 8.14** *Let $H$ and $K$ be two subgroups of a group $(G, *)$. Then the set $H * K$ defined by* (8.2.3) *is a subgroup of $G$ if and only if $H * K = K * H$.*

***Proof.*** We first assume that $H * K = K * H$. Now, as $e \in H \bigcap K$, we have $e = e * e \in H * K$. Therefore, to prove that $H * K$ is a subgroup of $G$, we only need to show that $x, y \in H * K$ implies $x * y^{-1} \in H * K$. For, let $x, y \in H * K$ so that we have

$$x = h_1 * k_1 \quad \text{and} \quad y = h_2 * k_2, \qquad \text{for some} \quad h_1, h_2 \in H, \; k_1, k_2 \in K.$$

Also, using the condition $H * K = K * H$, $h \in H$ and $k \in K$ implies that we can write $h * k = k_1 * h_1$, for some $k_1 \in K$ and $h_1 \in H$. It thus follows that

$$\begin{aligned} x * y^{-1} &= \big(h_1 * k_1\big) * \big(h_2 * k_2\big)^{-1} \\ &= \big(h_1 * k_1\big) * \big(k_2^{-1} * h_2^{-1}\big) \\ &= h_1 * \big(k_1 * k_2^{-1}\big) * h_2^{-1}\big) \\ &= h_1 * h * k = \big(h_1 * h\big) * k \in H * K. \end{aligned}$$

Therefore, we have shown that $H * K$ is a subgroup of $G$. Conversely, suppose $H * K$ is a subgroup of $G$, and let $x \in H * K$. We can write $x = h * k$, where $h \in H$ and $k \in K$. Then, as $H$, $K$ and $H * K$ are subgroups of $G$, we see that $h^{-1} * k^{-1} \in H * K$ and

$$x = h * k = \big(h^{-1} * k^{-1}\big)^{-1} \in K * H.$$

Hence, $H * K \subseteq K * H$. Similarly, it can be shown that $K * H \subseteq H * K$. Hence, we have $K * H = H * K$. This completes the proof. □

**Corollary 8.3** *If $H$ and $K$ are subgroups of an Abelian group $G$, then $H * K$ is a subgroup of $G$.*

***Proof.*** In this case, we have

$$H * K = \big\{h * k \mid h \in H, k \in K\big\} = \big\{k * h \mid h \in H, k \in K\big\} = K * H.$$

Therefore, $H * K$ is a subgroup of $G$. □

The next theorem follows directly from Theorems 8.9 and 8.11.

**Theorem 8.15** *For $i = 1, \ldots, n$, let $H_i$ be a subgroup of $\big(G_i, *_i\big)$. Then $H = H_1 \times \cdots \times H_n$ is a subgroup of the* direct product $G = G_1 \times \cdots \times G_n$.

***Proof.*** By Theorem 8.9, we know that the set $G$ is a group with identity element $e = \big(e_1, \ldots, e_n\big)$, where $e_i \in G_i$ be the identity element for $i = 1, 2, \ldots, n$. Notice that, since $H_i$ is a subgroup of $G_i$, we have $e \in H$. Therefore, $H \neq \emptyset$. Now, to prove that $H$ is a subgroup of the direct product $G$, it is enough to prove that $a * b^{-1} \in H$, for every $a = (a_1, \ldots, a_n)$, $b = (b_1, \ldots, b_n) \in H$. Of course, since $H_i$ is a subgroup of $G_i$, we have $a_i * b_i^{-1} \in H_i$, for $i = 1, 2, \ldots, n$. It thus follows that

$$a * b^{-1} = \big(a_1 * b_1^{-1}, \ldots, a_n * b_n^{-1}\big) \in H. \square$$

**Definition 8.14** Let $(G, *)$ be a group, and $a, b \in G$. We say $b$ is **conjugate** of $a$ if there is some $x \in G$ such that $b = x^{-1} * a * x$. We also write $b := a^x$.

The proof of the next theorem is left for the reader as an exercise.

**Theorem 8.16** *For $a,\ b \in G$, let*

$$a \sim b \quad \Leftrightarrow \quad b = a^x := x^{-1} * a * x, \ \textit{for some} \ \ x \in G. \tag{8.2.4}$$

*The relation $\sim$ is an equivalence relation.*

The equivalence classes associated with the relation (8.2.4) are called the **conjugacy classes** of the group. More precisely, for $a \in G$, the *conjugacy class* of $a$ is the set

$$a^G := \{a^x := x^{-1} * a * x \mid x \in G\}.$$

The concept introduced in Definition 8.14 gives an interesting partition of a finite group, leading to well-known *class equation* of a finite group. For its derivation, we also need the concept introduced in the next definition.

**Definition 8.15** Let $(G, *)$ be a group, and let

$$\zeta(G) := \{a \in G \mid a * x = x * a, \ \text{for all}\ x \in G\}.$$

This is known as the **center** of the group $G$.

Clearly, a group $G$ is Abelian if and only if $G = \zeta(G)$.

## *Homomorphisms*

The main point is to know when two algebraic structures $X$ and $Y$ may be treated *algebraically identical*. A simple answer is that the elements of these sets are in one-to-one correspondence in a manner so that the product tables with respect to all operations underlying the two algebraic structures are the same. More specifically, suppose $*$ and $\triangle$ are respectively the binary operations on $X$ and $Y$. A function $f : X \to Y$ is called a **homomorphism** if it satisfies the condition

$$f(x * y) = f(x) \triangle f(y), \qquad \text{for all} \ \ x,\ y \in X. \tag{8.2.5}$$

We also say that $f$ *preserve operations*. In particular, if $(S, *)$ and $(T, \triangle)$ are semigroups, then a function $f : S \to T$ satisfying the condition (8.2.5) is called a **semigroup homomorphism**. A surjective *semigroup homomorphism* is called a **semigroup epimorphism**; an injective semigroup homomorphism is called a **semigroup monomorphism**; and, a bijective semigroup homomorphism is called a **semigroup isomorphism**. We say two semigroups $(S, *)$ and $(T, \triangle)$ are **isomorphic**, and write $S \cong T$, if there is a *semigroup isomorphism* $f : S \to T$.

**Example 8.10** Consider the power set $\mathscr{P}(S)$ of a nonempty set $S$. Recall that both $(\mathscr{P}(S), \cup)$ and $(\mathscr{P}(S), \cap)$ are semigroups. Let $g : \mathscr{P}(S) \to \mathscr{P}(S)$ be given by $g(A) = A^c$. Then, by DeMorgan's law, we have

$$g(A \cup B) = (A \cup B)^c = A^c \cap B^c = g(A) \cap g(B),$$

which shows that $g$ is a semigroup homomorphism. Notice that, by involution property of complementation operation, $g$ is an epimorphism.

**Proposition 8.1** *Let $(S, *)$ be a semigroup, and $\triangle$ be a binary operation on a nonempty set $T$. Suppose $g : S \to T$ is a surjective function such that it satisfies condition* (8.2.5)*, for all $a, b \in S$. Then, we have $(T, \triangle)$ a semigroup.*

***Proof.*** By condition (8.2.5), we have

$$g(a * b) = g(a) \triangle g(b), \qquad \text{for all} \quad a, b \in S.$$

Notice that each element of $T$ is of the form $g(x)$, for some $x \in S$. Now, we have

$$\begin{aligned}(g(a) \triangle g(b)) \triangle g(c) &= g(a * b) \triangle g(c) \\ &= g((a * b) * c) = g(a * (b * c)) \\ &= g(a) \triangle g(b * c) \\ &= g(a) \triangle (g(b) \triangle g(c)).\end{aligned}$$

Therefore, $\triangle$ is associative, and hence $(T, \triangle)$ a semigroup. □

**Proposition 8.2** *Let $(S, *)$, $(T, \triangle)$ be two semigroups, and $g : S \to T$ be a semigroup homomorphism. If $e \in S$ is an identity element of $S$, then $g(e)$ satisfies*

$$g(a) \triangle g(e) = g(e) \triangle g(a).$$

*Moreover, if $g$ is onto, then $g(e)$ is an identity element of $T$.*

***Proof.*** Since $a$ is an identity element, we have

$$a * e = a = e * a, \qquad \text{for all} \quad a \in S.$$

It thus follows that

$$g(a) \triangle g(e) = g(a * e) = g(a) = g(e * a) = g(e) \triangle g(a).$$

Further, if the function $g$ is onto, then each element of $T$ is of the form $g(a)$, for some $a \in S$, so that $g(e)$ is an identity element of the semigroup $T$. □

A homomorphism from a semigroup to itself is called an *endomorphism*. The set of semigroup endomorphisms of a semigroup $S$ is denoted by $\text{End}(S)$.

**Corollary 8.4** *For any semigroup $S$, the set $\text{End}(S)$ is a monoid with respect to the binary operation given by the composition of functions.*

***Proof.*** Since End$(S)$ contains the identity function $i_S : S \to S$, it is nonempty. Also, it follows from Theorem 8.1 that End$(S)$ is closed with respect to binary operation given by the composition of functions. Further, since the composition of the function is always associative, it follows that End$(S)$ is a semigroup. Hence, End$(S)$ is a monoid, with the identity function $i_S : S \to S$ as the identity element. □

**Theorem 8.17** *Let* $(S, *)$, $(T, \triangle)$, $(V, \oplus)$ *be semigroups, and let* $h : S \to T$, $g : T \to V$ *be semigroup homomorphisms. Then we have* $h \circ g : S \to V$ *is a semigroup homomorphism.*

***Proof.*** For $a, b \in S$, we have

$$\begin{aligned}(h \circ g)(a * b) &= h\big(g(a * b)\big)\\ &= h\big(g(a) \triangle g(b)\big)\\ &= h\big(g(a)\big) \oplus h\big(g(b)\big)\\ &= \big(h \circ g\big)(a) \oplus \big(h \circ g\big)(b).\end{aligned}$$

Therefore $h \circ g : S \to V$ is a semigroup homomorphism. □

**Definition 8.16** Let $(M, *, e_M)$ and $(T, \triangle, e_T)$ be two monoids. A function $g : M \to T$ is called a *monoid homomorphism* if it satisfies the condition (8.2.5), for all $a, b \in S$, and also $g\big(e_M\big) = e_T$. In this case, the terms such as epimorphism, monomorphism, and isomorphism have the same meaning as defined previously for a semigroup homomorphism. We say two monoids $(M, *, e_M)$ and $(T, \triangle, e_T)$ are **isomorphic**, and write $M \cong T$, if there is a *monoid isomorphism* $f : S \to T$ such that $f\big(e_M\big) = \big(e_T\big)$.

**Proposition 8.3** *Let* $(M, *, e_M)$ *be a monoid, and* $\triangle$ *be a binary operation on a nonempty set* $T$. *Suppose* $g : M \to T$ *is a surjective function such that* $g\big(e_M\big) = e_T$, *and we have*

$$g\big(a * b\big) = g(a) \triangle g(b), \qquad \textit{for all} \quad a, b \in S.$$

*Then* $(T, \triangle)$ *is a monoid. Also,* $T$ *is commutative if so is* $M$. *Moreover, for any invertible element* $a \in M$, *we have* $g(a) \in T$ *is invertible.*

***Proof.*** As shown in Proposition 8.1, since each element of $T$ is of the form $g(x)$, for some $x \in S$, it follows that the operation $\triangle$ is associative. It thus follows that $(T, \triangle)$ is a semigroup. We also have that

$$g(a) \triangle e_T = g(a) \triangle g(e_M) = g\big(a * e_M\big) = g(a).$$

Similarly, $e_T \triangle g(a) = g(a)$. Therefore, $e_T$ is an identity element of $T$. Hence $(T, \triangle, e_T)$ is a monoid. Further, if the monoid $M$ is commutative, then

$$g(a) \triangle g(b) = g\big(a * b\big) = g\big(b * a\big) = g(b) \triangle g(a)$$

implies that $T$ is commutative. Moreover, if $a \in M$ has inverse $a^{-1}$, then

$$g(a) \triangle g(a^{-1}) = g(a * a^{-1}) = g(e_M) = e_T;$$
$$g(a^{-1}) \triangle g(a) = g(a^{-1} * a) = g(e_M) = e_T.$$

Hence $g(a)$ is invertible, and $\big(g(a)\big)^{-1} = g(a^{-1})$. □

The next theorem proves that every monoid $M$ is embedded in the monoid $\mathrm{End}(M)$.

**Theorem 8.18** *For any monoid $(M, *, e)$, there is a monomorphism $g : S \to \mathrm{End}(S)$.*

***Proof.*** For a fixed $a \in M$, we define $f_a : M \to M$ as $f_a(b) = a * b$, for all $b \in M$, known as the *left translation map*. Now, since

$$\begin{aligned} f_{a*b}(c) = (a * b) * c \qquad\qquad\qquad & = a * (b * c) \\ = f_a(b * c) = f_a\big(f_b(c)\big) & \\ = \big(f_a \circ f_b\big)(c), & \end{aligned}$$

it follows that $f_{a*b} = f_a \circ f_b$. Therefore, we may take $g : M \to \mathrm{End}(M)$ defined by $g(a) = f_a$, for all $a \in M$. Clearly, $g(e) = f_e = i_M$, the identity element of the monoid $\mathrm{End}(M)$. To show that $g$ is a semigroup homomorphism, let $a, b \in G$. We then have

$$g(a * b) = f_{a*b} = f_a \circ f_b = g(a) \circ g(b),$$

which proves $g$ is a semigroup homomorphism from $S$ into $\mathrm{End}(S)$. Notice that, for any $a, b \in S$, we have

$$g(a) = g(b) \;\Rightarrow\; f_a(x) = f_b(x),\ \forall x \in S, \;\Rightarrow\; a * x = b * x \;\Rightarrow\; a = b,$$

by taking $x = e$. Therefore, $g$ is a monomorphism. □

The above theorem tells us that every monoid can be *faithfully represented* by a semigroup of monoid homomorphisms. In general, we often find it interesting to determine when elements of an algebraic structure can be identified with some well-known mathematical objects such as numbers, matrices, mappings of a given set, etc. A fundamental *embedding theorem* proven in the next part of the section states that each group can be embedded into a group of permutations (Theorem 8.20).

**Definition 8.17** Let $(G, *)$ and $(G', \cdot)$ be two groups. A function $\phi : G \to G'$ is called a **group isomorphism** if it is one-one, onto, and we have

$$\phi(a * b) = \phi(a) \cdot \phi(b), \qquad \text{for all } \ a, b \in G.$$

We also say that $G$ and $G'$ are **isomorphic groups** (or algebraically identical), and write $G \cong G'$.

The next theorem plays a crucial role in the subsequent discussion. It says every homomorphism between groups takes the identity element to the identity element and the inverse to the inverse.

**Theorem 8.19** *Let $(G, *)$ and $(G', \cdot)$ be two groups with $e$ and $e'$ being the identity elements, respectively. Let $\phi : G \to G'$ be a homomorphism. Then we have*

1. $\phi(e) = e'$.
2. $\big(\phi(a)\big)^{-1} = \phi\big(a^{-1}\big)$, *for all* $a \in G$.

***Proof.*** To prove (1), notice that as $e * e = e$ and $\phi$ is a homomorphism, we have

$$e' \cdot \phi(e) = \phi(e) = \phi\big(e * e\big) = \phi(e) \cdot \phi(e),$$

which gives $\phi(e) = e'$, by cancellation law. For (2), since we have

$$a * a^{-1} = e = a^{-1} * a, \qquad \text{for any} \quad a \in G,$$

it follows that

$$\begin{aligned} \phi\big(a\big) \cdot \phi\big(a^{-1}\big) &= \phi\big(a * a^{-1}\big) = \phi\big(e\big) = e'; \\ \phi\big(a^{-1}\big) \cdot \phi\big(a\big) &= \phi\big(a^{-1} * a\big) = \phi\big(e\big) = e'. \end{aligned}$$

Hence, we have $\big(\phi(a)\big)^{-1} = \phi\big(a^{-1}\big)$, for all $a \in G$. □

It follows from the above theorem that the two sets

$$\ker\big(f\big) = f^{-1}(e') = \big\{x \in G \mid f(x) = e'\big\}; \tag{8.2.6a}$$
$$\text{Im}\big(f\big) = \big\{y \in G' \mid y = f(x), \text{ for some } x \in G\big\}, \tag{8.2.6b}$$

are respectively subgroups of the groups $G$ and $G'$. In fact, $\ker(f)$ is more interesting group (Theorem 8.43). In general, the constant function $\phi : G \to G$ given by $\phi(a) = e$ is called the *trivial homomorphism*, where $e \in G$ is the identity element. On the other hand, for any group $(G, *)$, the identity function $i_G : G \to G$ give by $i_G(a) = a$ $(a \in G)$ is a *group isomorphism*. Some other illustrations of group isomorphisms are as given in the next example.

**Example 8.11**

(i) Suppose $G$ is the set of integer powers of a *symbol* $x$. That is,

$$G = \big\{x^n \mid n \in \mathbb{Z}\big\},$$

With respect to the binary operation $*$ given by

$$x^n * x^m = x^{n+m}, \qquad \text{for} \quad n, m \in \mathbb{Z},$$

$G$ is the group, with $e := x^0$ as the identity element, and the inverse of $x^n \in G$ is the element $x^{-n} \in G$. It follows easily that the operation $*$ is commutative. In view of the above description of the Abelian group $(G, *)$, it is likely that the group properties of $G$ are similar to the additive group $(\mathbb{Z}, +)$. We now proceed to show exactly that. For, since an arbitrary element of $G$ is of the form $x^n$, we may define a function $\phi : \mathbb{Z} \to G$ as $\phi(n) = x^n$, for $n \in \mathbb{Z}$. We thus have

$$\phi(n+m) = x^{n+m} = x^n * x^m = \phi(n) * \phi(m),$$

for all $n, m \in \mathbb{Z}$. That is, $\phi$ is a homomorphism. In fact, $\phi$ is an isomorphism, because it is a bijective function. Hence, $(G, *)$ is isomorphic to the additive group $(\mathbb{Z}, +)$.

(ii) Suppose both $G$ and $G'$ are the additive group $(\mathbb{R}, +)$. Consider the function $\phi : G \to G'$ given by $\phi(a) = 2a$, for $a \in G$. Clearly, $\phi$ is a homomorphism because

$$\phi(a+b) = 2(a+b) = 2a + 2b = \phi(a) + \phi(b).$$

We also have that

$$\phi(a) = \phi(b) \quad \Rightarrow \quad 2a = 2b \quad \Rightarrow \quad a = b.$$

That is, $\phi$ is an injective function. Moreover, since $\phi(a/2) = a$, for any $a \in G'$, it follows that the function $\phi$ is surjective. Therefore, $\phi$ is an isomorphism. Similarly, when both $G$ and $G'$ are the additive group $(\mathbb{Z}, +)$, it follows that the function $\phi : G \to G'$ given by $\phi(a) = 2a$ is a *monomorphism*, i.e., it is an injective homomorphism. However, in this case, $\phi$ is not an isomorphism because for no element $0 \neq a \in G'$ we have $a/2 \in G$.

(iii) Suppose $G$ is the multiplicative group $(\mathbb{R}^+, \cdot)$ of positive real numbers, and $G'$ be the additive group of real numbers $(\mathbb{R}, +)$. Then the function $\phi : G \to G'$ given by $\phi(a) = \log a$ $(a \in G)$ is a homomorphism:

$$\phi(a \cdot b) = \log(a \cdot b) = \log a + \log b = \phi(a) + \phi(b).$$

We also have that

$$\phi(a) = \phi(b) \quad \Rightarrow \quad \log a = \log b \quad \Rightarrow \quad a = b.$$

That is, the function $\phi$ is one-one. Moreover, in this case, for any $b \in G'$ we have $a := e^b \in G$ such that $\phi(a) = \log(e^b) = b$. Therefore, the function $\phi$ is also onto. Hence, $\phi : (\mathbb{R}^+, \cdot) \to (\mathbb{R}, +)$ is an isomorphism.

**Definition 8.18** Let $(G, *)$ be a group. An **automorphism** of the group $G$ is an isomorphism $\phi : G \to G$. The set of *automorphisms* of a group $G$ is denoted by $\mathrm{Aut}(G)$.

Since the identity homomorphism $i_G \in \mathrm{Aut}(G)$, we have $\mathrm{Aut}(G) \neq \emptyset$.

**Proposition 8.4** *Let $(G, *)$ be a group. For a fixed $a \in G$, let $\phi_a : G \to G$ be the function given by $\phi_a(x) = a * x * a^{-1}$, for $x \in G$. Then $\phi$ is an isomorphism.*

***Proof.*** First, notice that the function $\phi_a$ is onto because we have

$$\phi_a(a^{-1} * x * a) = a * (a^{-1} * x * a) * a^{-1} = x,$$

by associativity of $*$ and identity law. Next, if $\phi_a(x) = \phi_a(y)$, then we have

$$a * x * a^{-1} = a * y * a^{-1} \quad \Rightarrow \quad x = y,$$

by the cancellation law. It thus follows that $\phi_a$ is a bijective function. Finally, for any $x, y \in G$, we have

$$\phi_a(x) * \phi_a(y) = (a * x * a^{-1}) * (a * y * a^{-1}) = a * (x * y) * a^{-1} = \phi_a(x * y).$$

Hence, $\phi_a$ is an isomorphism. □

It follows from this proposition that $\phi_a \in \mathrm{Aut}(G)$. It is called the **inner automorphism** given by the element $a \in G$. The set of *inner automorphisms* of a group $G$ is denoted by $\mathrm{Inn}(G)$. When $G$ is an Abelian group, we have $\mathrm{Inn}(G) = \{i_G\}$, but $\mathrm{Aut}(G)$ contains a non-identity element.

**Proposition 8.5** *Let $(G, *)$ be an Abelian group. Let $\phi : G \to G$ be a function given by $\phi(x) = x^{-1}$, for $x \in G$. Then $\phi$ is an isomorphism.*

***Proof.*** Clearly, the function $\phi$ is one-one and onto. Also, for any $x, y \in G$, we have

$$\phi(x * y) = (x * y)^{-1} = y^{-1} * x^{-1} = x^{-1} * y^{-1} = \phi(x) * \phi(y).$$

Therefore, $\phi$ is a homomorphism, and hence an isomorphism. □

We conclude this section with a proof of the fundamental *embedding theorem* due to Cayley. To prepare for the same, let $(G, *)$ be a group. For a fixed $a \in G$, the function $\pi_a : G \to G$ given by

$$\pi_a(b) = b * a, \qquad \text{for all} \quad b \in G,$$

is called the **right translation** determined by the element $a$. We may write

$$P := \{\pi_a \mid a \in G\}.$$

The central argument of the proof of the *Cayley theorem* is based on the next lemma.

**Lemma 8.1** *The set $P$ is a subgroup of the group of permutations $S_G$ of the set $G$.*

***Proof.*** To show that $P \subseteq S_G$, we need to prove that each *right translation* $\pi_a$ is a bijective function. For, let $\pi_a(b) = \pi_b(c)$, for some $b, c \in G$. Then we have $b * a = c * a$, and so $b = c$, by cancellation law. Therefore, $\pi_a$ is injective. Next, to show that $\pi_a$ is an onto function, let $c \in G$. Then the element $c * a^{-1} \in G$ is such that we have

$$\pi_a(c * a^{-1}) = (c * a^{-1}) * a = c * (a^{-1} * a) = c * e = c.$$

Therefore, each element $c \in G$ has a preimage $c * a^{-1} \in G$. Hence, $\pi_a \in S_G$. Further, for all $a, b, c \in G$, we have

$$\begin{aligned}(\pi_a \circ \pi_b)(c) &= \pi_b(\pi_a(c)) \\ &= \pi_b(c * a) \\ &= (c * a) * b \\ &= c * (a * b) = \pi_{a*b}(c),\end{aligned}$$

which shows that $\pi_a \circ \pi_b = \pi_{a*b}$, and hence $P$ has the closure property with respect to the binary operation $\circ$ given by the composition of functions. The element $\pi_e \in S_G$ satisfies

$$\pi_e * \pi_a = \pi_{e*a} = \pi_a = \pi_{a*e} = \pi_a * \pi_e.$$

That is, $\pi_e \in S_G$ is the identity element. Also, for $\pi_a \in P$, $\pi_{a^{-1}}$ is such that we have

$$\pi_a \circ \pi_{a^{-1}} = \pi_{a*a^{-1}} = \pi_e = \pi_{a^{-1}*a} = \pi_{a^{-1}} \circ \pi_a.$$

It thus follows that $P$ is a subgroup of the permutation group $S_G$. □

**Theorem 8.20 (Cayley's Theorem)** *Every group $G$ is isomorphic to a subgroup of the group of permutation $S_A$, for some set $A$.*

***Proof.*** By above lemma, $P \subset S_G$ is a subgroup. So, we may prove the theorem by taking $A = G$. We claim that the function $\phi : G \to P$ given by $\phi(a) = \pi_a$ is an isomorphism. Now, to show that $\phi$ is a homomorphism, let $a, b \in G$. Then we have

$$\phi(a * b) = \pi_{a*b} = \pi_a \circ \pi_b = \phi(a) \circ \phi(b).$$

Therefore, $\phi : G \to P$ is a homomorphism. Next, to verify that $\phi$ is an injective function, let $\phi(a) = \phi(b)$, for some $a, b \in G$. Then, we have $\pi_a = \pi_b$, and hence

$$\pi_a(e) = \pi_b(e) \;\Rightarrow\; e * a = e * b \;\Rightarrow\; a = b.$$

As $\phi(a) = \pi_a$, we conclude that the function $\phi$ is onto. Hence, $\phi : G \to P$ is an isomorphism. □

An interesting improvement of Theorem 8.20 for a special class of groups is as given below. Notice that the $n$ columns of the identity matrix $I_n \in GL(n, \mathbf{k})$ may be viewed as the unit vectors $e_1, \ldots, e_n \in \mathbf{k}^n$. For $\sigma \in S_n$, the square matrix $P$ of order $n$, with columns given by the vectors $e_{\sigma(1)}, \ldots, e_{\sigma(n)}$, is called the *permutation matrix* determined by the permutation $\sigma$. The inverse of $P$ is given by the transpose matrix $P^t$. It thus follows that the mapping $\sigma \mapsto P : S_n \to GL(n, \mathbf{k})$ is a monomorphism. Therefore, the group $S_n$ itself may be treated as a subgroup of the group $GL(n, \mathbf{k})$. Hence, *every finite group is a matrix group.*

**Remark 8.2** It is in general true that matrices offer a greater freedom for algebraic manipulation of a problem than do the permutations. They also allow us to deal with infinite groups, whereas the symmetric groups simply cannot. In general, an *abstract group* is a capital letter $G$ unless we use the algebraic flexibility of matrices to represent them in a nice manner.

## 8.3 Cyclic Groups

Let $(G, *)$ be a group, and $a \in G$. For $n \in \mathbb{Z}$, we may take

1. $a^0 = e$;
2. $a^{n+1} = a * a^n = a^n * a$, for $n \geq 1$;
3. $a^n = \left(a^{-1}\right)^{-n}$, for $n < 0$.

It then follows that

$$(a^n)^{-1} = a^{-n}, \quad a^n * a^m = a^{n+m}, \quad \text{and} \quad (a^n)^m = a^{nm},$$

for all $m, n \in \mathbb{Z}$. We first prove that $a^n * a^m = a^{n+m}$. There is nothing to prove when $n$ or $m$ is zero. We may thus consider the following two cases:

(i) Case–$I$: When $m > 0$ and $n > 0$. We have

$$\begin{aligned} a^m * a^n &= (a * a * \cdots (m \text{ times})) * (a * a * \cdots (n \text{ times})) \\ &= a * a * \cdots (m + n \text{ times}) \\ &= a^{m+n}. \end{aligned}$$

(ii) Case–$II$: When $m > 0$ and $n < 0$. We also assume that $m > -n$. We have

$$\begin{aligned}
a^m * a^n &= a^m * \left(a^{-1}\right)^{-n} \\
&= \left(a * a * \cdots (m \text{ times})\right) * \left(a^{-1} * a^{-1} * \cdots (-n \text{ times})\right) \\
&= \left(a * a * \cdots (m + n \text{ times})\right) * \left(a * a * \cdots (-n\text{times})\right) \\
&\quad * \left(a^{-1} * a^{-1} * \cdots (-n \text{ times})\right) \\
&= a * a * \cdots (m + n \text{ times}) \\
&= a^{m+n}.
\end{aligned}$$

The other cases are left for the reader as an exercise. Now, since we have

$$a^n * a^{-n} = a^{n-n} = a^0 = e = a^{-n} * a^n,$$

it follows that $(a^n)^{-1} = a^{-n}$. Similarly, it can be shown that $(a^n)^m = a^{nm}$.

**Theorem 8.21** *Let $(G, *)$ be an Abelian group. Then*

$$\left(a * b\right)^n = a^n * b^n, \qquad \textit{for all} \quad a, b \in G \quad \textit{and} \quad n \in \mathbb{Z}.$$

***Proof.*** When $n = 0, 1$, there is nothing to prove. We first assume that $n > 0$. To apply mathematical induction, suppose the assertion holds for $n = m \geq 1$. That is, we assume that $\left(a * b\right)^m = a^m * b^m$. Now, for $n = m + 1$, we have

$$\begin{aligned}
\left(a * b\right)^{m+1} &= \left(a * b\right)^m * \left(a * b\right) = \left(a^m * b^m\right) * \left(a * b\right) \\
&= a^m * \left(b^m * (a * b)\right) = a^m * \left(\left(b^m * a\right) * b\right) \\
&= a^m * \left(\left(a * b^m\right) * b\right) = a^m * \left(a * \left(b^m * b\right)\right) \\
&= a^m * \left(a * b^{m+1}\right) = \left(a^m * a\right) * b^{m+1} \\
&= a^{m+1} * b^{m+1}.
\end{aligned}$$

Therefore, the assertion holds for all positive integers. Notice that, for negative $n$, we may write $n = -m$ $(m > 0)$ so that

$$\begin{aligned}
\left(a * b\right)^n &= \left((a * b)^{-1}\right)^m = \left((b * a)^{-1}\right)^m \\
&= \left(a^{-1} * b^{-1}\right)^m = \left(a^{-1}\right)^m * \left(b^{-1}\right)^m \\
&= a^{-m} * b^{-m} = a^n * b^n.
\end{aligned}$$

Hence the assertion holds for all $n \in \mathbb{Z}$. □

**Theorem 8.22** *A group $(G, *)$ is Abelian if and only if $(a * b)^2 = a^2 * b^2$, for all $a, b \in G$.*

***Proof.*** When $G$ is an Abelian group, we have

$$\begin{aligned}(a*b)^2 &= (a*b)*(a*b) = a*\big(b*(a*b)\big)\\ &= a*\big((a*b)*b\big) = \big(a*\big(a*(b*b)\big)\big)\\ &= a*\big(a*b^2\big) = \big(a*a\big)*b^2\\ &= a^2*b^2.\end{aligned}$$

Conversely, suppose $(a*b)^2 = a^2*b^2$, for all $a, b \in G$. Now, as

$$\big(a*b\big)^2 = a^2*b^2 \quad \Leftrightarrow \quad \big(a*b\big)*\big(a*b\big) = \big(a*a\big)*\big(b*b\big),$$

it follows from the associative law that

$$a*\big(b*(a*b)\big) = a*\big(a*(b*b)\big).$$

Therefore, by the left cancellation law, we obtain

$$b*\big(a*b\big) = a*\big(b*b\big).$$

Again, by using the associative law, we obtain

$$\big(b*a\big)*b = \big(a*b\big)*b$$

which by the right cancellation law gives $b*a = a*b$. This completes the proof. □

**Theorem 8.23** *Let $(G, *)$ be a finite group of order $n$, and $a \in G$. Then there is a positive integer $m$ such that $a^m = e$.*

***Proof.*** Consider the $n+1$ elements given by

$$e = a^0,\ \ a,\ \ a^2,\ \ a^3,\ \ldots,\ \ a^n.$$

Since these are elements of $G$ that only have $n$ elements, we see that $a^i = a^j$, for some $0 \le i < j \le n$. It thus follows that

$$a^{j-i} = a^j*(a^i)^{-1} = a^j*(a^j)^{-1} = e.$$

We may take $m = j - i$ so that $0 < m \le n$. We have thus found a positive integer $m$ such that $a^m = e$. □

**Definition 8.19** Let $G$ be a group, and $g \in G$. We say $g$ is of **infinite order** if, for all integer $n$, the products $g^n$ are distinct elements of the group $G$. Otherwise, $g$ is said to be an element of **finite order**.

Said differently, if $g \in G$ is an element of *finite order*, then we have

$$g^n = g^k, \quad \text{for some integers } k > n.$$

That is, equivalently, we have

$$g^n = e, \quad \text{for some positive integer } \; n.$$

It thus follows from the "well-ordering principle" that there is a *least positive integer*, say $n_0$, such that $g^{n_0} = e$. We thus have The **order** of an element $g \in G$ is thus $n_0$, which we shall write as $o(g) = n_0$. The next theorem is significant.

**Theorem 8.24** *Let $g \in G$ be such that $o(g) = n$. Then*

(i) *$g^r = g^s$ if and only if $n \mid r - s$. In particular, $g^k = 1$ if and only if $n \mid k$.*
(ii) *For any $x \in G$, the* conjugate *$g^x := x^{-1}gx$ of $g$ has the order $n$.*

***Proof.*** By *division algorithm*, we can find positive integers $t$, $q$ such that

$$r - s = nq + t, \qquad \text{where} \;\; 0 \leq t < n,$$

which implies that

$$g^r = g^s \quad \Rightarrow \quad g^{r-s} = e \quad \Rightarrow \quad e = g^{nq+t} = e^q \cdot g^t = g^t.$$

We thus obtain $t = 0$, because $n$ is the least positive integer such that $g^n = e$. Hence, $n \mid r - s$. This proves (i). For (ii), we may first apply induction on $k$ to show that $(x^{-1}gx)^k = x^{-1}g^k x$ holds for each positive integer $k$. Then (ii) follows from (i). This completes the proof □

Clearly, in any group, the identity element is the sole element of order 1.

**Example 8.12**

(i) Consider the group of 4th roots of unity, $U_4 = \{1, -1, i, -i\}$, with respect to multiplication of complex numbers. It follows from the Table 8.1 that we have

$$i^1 = i, \qquad i^2 = -1, \qquad i^3 = -i, \quad \text{and} \quad i^4 = 1;$$
$$(-i)^1 = -i, \qquad (-i)^2 = -1, \qquad (-i)^3 = i, \quad \text{and} \quad (-i)^4 = 1.$$

Therefore, we have $o(i) = o(-i) = 4$. Also, $o(-1) = 2$. Notice that the order of each element of $U_4$ is a divisor of $o(U_4) = 4$.

(ii) Table 8.4 lists the order of elements in the group $(\mathbb{Z}_7^*, \times_7)$.
(iii) It follows from Table 8.3 that in the case of the symmetric group $S_3$ we have

$$o(\tau_1) = o(\tau_2) = o(\tau_3) = 2, \quad \text{and} \quad o(\rho_1) = o(\rho_2) = 3.$$

As shown later, the same holds for any finite group.

**Table 8.4** Order of elements in the group $(\mathbb{Z}_7^*, \times_7)$

| $a$ mod 7 | 1 | 2 | 3 | 4 | 5 | 6 |
|---|---|---|---|---|---|---|
| $\text{ord}_7(a)$ | 1 | 3 | 6 | 3 | 6 | 2 |

(iv) By Example 8.5 (v), for any nonempty set $X$, $(\mathscr{P}(X), \oplus)$ is a group with the empty set $\emptyset$ as the identity element. For each $A \in \mathscr{P}(X)$, we have

$$A \oplus A = (A \setminus A) \cup (A \setminus A) = (A \cup A) \setminus (A \cap A) = \emptyset.$$

Therefore, $A$ an *involution*, i.e., it is an element of order two.

The next theorem proves that $a \in G$ generates a subgroup of order $o(a)$.

**Theorem 8.25** *Let $G$ be a group, and $a \in G$ has order $n \geq 1$. Then the set*

$$H := \{e, a, a^2, a^3, \ldots, a^{n-1}\} \subseteq G \tag{8.3.1}$$

*is a subgroup of order $n$.*

***Proof.*** We may recall that $n$ is the smallest positive integer such that $a^n = e$. Let $(a)$ denote the right side set in (8.3.1). We first show that the elements in $(a)$ are distinct. For, if

$$a^i = a^j \quad \text{for} \quad 0 \leq i,\ j \leq n-1, \ \text{ with } i < j,$$

so that $a^{j-i} = e$, for $0 < j - i \leq n - 1$, then we obtain a contradiction to the fact that $o(a) = n$. Therefore, in particular, we have $|H| = n$. We next show that $H$ is a subgroup of $G$. For, notice that

$$a^n = e \quad \Rightarrow \quad a^{n+1} = a^n * a = e * a = a;$$
$$a^{n+2} = a^n * a^2 = e * a^2 = a^2;$$

We also have that

$$a^{-1} = e * a^{-1} = a^n * a^{-1} = a^{n-1}, \quad a^{-2} = a^{n-2};$$
$$\ldots$$

It thus follows that $(a) = \{a^n \mid n \in \mathbb{Z}\}$. Clearly, $(a) \neq \emptyset$. Also, for $x, y \in (a)$, we have $x = a^\ell,\ y = a^m$, for some integer $\ell, m$, so that

$$x * y^{-1} = a^\ell * (a^m)^{-1} = a^\ell * a^{-m} = a^{l-m} \in (a).$$

This proves that $H := (a)$ is a subgroup of $G$. □

**Definition 8.20** Let $G$ be a group, and $a \in G$. The set $\big(a\big)$ given by

$$\big(a\big) := \{a^n \mid n \in \mathbb{Z}\}$$

is a group, with identity $e = a^0$. It is called the *cyclic subgroup* generated by $a$.

Theorem 8.25 proves that the cyclic subgroup generated by an element in a group is indeed a group. For an Abelian group $G$ with operation denoted by $+$, we write

$$a^n := a + a + \cdots + a = na.$$

Therefore, the cyclic subgroup generated by $a$ in this case is expressed as

$$\big(a\big) = \{na \mid n \in \mathbb{Z}\}.$$

The class of finite groups, known as **cyclic groups**, find application to many problems in computer science such as those related to cryptography, coding theory, and network design. The main idea in general is that the group is known once its "generator" is found. Therefore, *cyclic groups* are determined by a special element it contains. For some classes of groups, cyclic groups act like *building blocks* (see Remark 8.3).

**Definition 8.21** Let $G$ be a group. We say $G$ is a **cyclic group** if there is some $a \in G$ such that every element of the group $G$ is of the form $a^n$ for some integer $n$. In this case, $a$ is called a **generator** of the *cyclic group* $G$. The group $G$ is called a **finite cyclic group** if $o(a) < \infty$. Otherwise, $G$ is called an *infinite cyclic group.*

It follows from Theorem 8.24 that $G$ is a *finite cyclic group* of order $n$ if for some $e \neq a \in G$, $o(a) = n$. Therefore, a group $G$ is a *finite cyclic group* if for some $a \in G$ the order of $G$ equals the order of the cyclic subgroup generated by the element $a$. That is, $G = \big(a\big)$.

**Example 8.13** It follows from Example 8.12 (i) that the group $U_4$ is a cyclic group, with $i$ (and also $-i$) as a generator. We also have that

$$\big(-1\big) = \{1, -1\} \quad \text{and} \quad \big(1\big) = \{1\}.$$

More generally, the group of $n$th roots of unity $U_n$ is a finite cyclic group with $\omega$ as a generator:

$$U_n = \{\omega, \omega^2, \ldots, \omega^{n-1}, \omega_n = 1\}, \quad \text{where} \quad \omega = e^{2\pi i/n}.$$

Also, the group $\big(\mathbb{Z}_n, +_n\big)$ of residue classes modulo $n$ is a finite cyclic group. For the residue class $1 \in \mathbb{Z}_n$, we have

$$1^m = k, \qquad \text{where } 0 \leq k < n \text{ and } m \equiv k \bmod n.$$

We also have that

$$\begin{aligned}(1) &= \{1^m \mid m \in \mathbb{Z}\}\\ &= \{0, 1, \ldots, n-1\}\\ &= \mathbb{Z}_n.\end{aligned}$$

It thus follows that $(\mathbb{Z}_n, +_n)$ is a finite cyclic group, with the residue class 1 as a generator. The group

$$G := \{a \in \mathbb{Z}^* \mid a \text{ is relatively prime to } n\}, \tag{8.3.2}$$

with respect to the operation $\times_n$, as introduced earlier in Example 8.4 (iv), needs further investigation to understand its cycle structure. Also, the group $(\mathbb{Z}, +)$ is an infinite cyclic group, with 1 (and also $-1$) as a generator.

It follows from Theorem 8.29 that $\mathbb{Z}_n$ may be regarded as the "universal" finite cyclic group of order $n$; and, t follows from Theorem 8.30 that the infinite cyclic group $(\mathbb{Z}, +)$ is a "universal" infinite cyclic group. The next theorem proves that if $a$ is a generator for a cyclic group then so is $a^{-1}$.

**Theorem 8.26** *Let $G$ be a group, and $a \in G$. Then we have $(a) = (a^{-1})$.*

***Proof.*** Notice that we have

$$\begin{aligned}(a^{-1}) &= \{(a^{-1})^n \mid n \in \mathbb{Z}\}\\ &= \{a^{-n} \mid n \in \mathbb{Z}\}\\ &= \{a^m \mid m \in \mathbb{Z}\}\\ &= (a).\end{aligned}$$

This completes the proof. □

**Theorem 8.27** *Every cyclic group is Abelian. The converse is not true.*

***Proof.*** Suppose $G$ is a cyclic group, with a generator $a \in G$. That is,

$$G = (a) = \{a^n \mid n \in \mathbb{Z}\}.$$

Therefore, for $x, y \in G$, we can write $x = a^n$ and $y = a^m$, for some $n, m \in \mathbb{Z}$, so that

$$x * y = a^n * a^m = a^{n+m} = a^{m+n} = a^m * a^n = y * x.$$

This proves that $G$ is an Abelian group. For the second part, consider the set $V_4 = \{e, a, b, c\}$. Table 8.5 is the product table of $V_4$ with respect to the binary operation given by the following relations:

**Table 8.5** Product Table for the Klein four group $V_4$

| $\cdot$ | e | a | b | c |
|---|---|---|---|---|
| e | e | a | b | c |
| a | a | e | c | b |
| b | b | c | e | a |
| c | c | b | a | e |

$$a \cdot a = e, \qquad b \cdot b = e, \qquad c \cdot c = e;$$
$$a \cdot b = b \cdot a = c, \qquad b \cdot c = c \cdot b = a, \qquad c \cdot a = a \cdot c = b.$$

It thus follows that $V_4$ is a group, with $e$ as the identity element. Since every element of $V_4$ is an *involution*, by Theorem 8.7, $V_4$ is an Abelian group. This is known as the *Klein four group*. Clearly, it is not a cyclic group, because none of the elements of $V_4$ is of order four. □

The next theorem uses the *division algorithm*.

**Theorem 8.28** *Every subgroup of a cyclic group is a cyclic group.*

***Proof.*** Suppose $G$ is a cyclic group generated by element $a \in G$. That is,

$$G = (a) = \{a^m \mid m \in \mathbb{Z}\}.$$

Let $H$ be a subgroup of $G$. If $H = \{e\}$, then $H = (e)$ is cyclic. We may thus assume that $H \neq \{e\}$. Notice that, as $H \subseteq G$, all the elements of $H$ are powers of $a$. Also, $a^n \in H \Rightarrow a^{-n} \in H$. Therefore, it is possible to find the smallest positive integer $n$ such that $a^n \in H$. We claim that $H = (a^n)$. For, since

$$a^n \in H \quad \Rightarrow \quad (a^n)^k \in H, \qquad \text{for every} \quad k \in \mathbb{Z},$$

it follows that $(a^n) \subseteq H$. To prove the reverse inclusion, let $h = a^m \in H$. By Theorem A.5, we can find integers $q$ and $r$ such that $m = qn + r$, where $0 \leq r < n$. Now, since $H$ is a subgroup, and $a^m,\ a^n \in H$, it follows $[a^n]^{-r} \in H$. We also have that

$$a^r = a^{m-qn} = a^m * (a^n)^{-r} \in H.$$

However, since $0 \leq r < n$ and $a^r \in H$, we conclude $r = 0$; for, otherwise $r < n$ would be the smallest positive integer satisfying the condition $a^r \in H$. Therefore, we obtain

$$m = qn \quad \Rightarrow \quad a^m = (a^n)^q \in H,$$

and so $(a^n) \subseteq H$. Hence, $H = (a^n)$, i.e., it is a cyclic group. □

The next two theorems prove that, up to isomorphism, $(\mathbb{Z}_n, +_n)$ is the only finite cyclic group of order $n$, and $(\mathbb{Z}, +)$ is the only infinite cyclic group.

**Theorem 8.29** *Every finite cyclic group of order n is isomorphic to* $(\mathbb{Z}_n, +_n)$.

***Proof.*** Let $G$ be a cyclic group of order $n$ generated by an element $a \in G$. Then $o(a) = n$, and we have

$$G = (a) = \{e, a, a^2, \cdots, a^{n-1}\}.$$

We may thus define $\phi : G \to \mathbb{Z}_n$ by $\phi(a^\ell) = \ell$. We first show that $\phi$ is a homomorphism. For, we have

$$\phi(a^\ell * a^m) = \phi(a^{\ell+m}) = \phi(a^{\ell+_n m}) = \ell +_n m = \phi(a^\ell) +_n \phi(a^m),$$

which shows that $\phi$ is a homomorphism. Clearly, $\phi$ is onto. Also, $\phi(a^\ell) = 0 \in \mathbb{Z}_n$ implies that $\ell = 0 \bmod n$, i.e., $\ell = nk$, for some integer $k$, and so $a^\ell = e$. That is, $K_\phi = \{e\}$. Hence, the function $\phi$ is an isomorphism. □

**Theorem 8.30** *Any infinite cyclic group is isomorphic to* $(\mathbb{Z}, +)$.

***Proof.*** The proof is similar to one given for Theorem 8.29. Let $G$ is an infinite cyclic group generated by element $a \in G$. We then have

$$G = (a) = \{a^m \mid m \in \mathbb{Z}\}.$$

Therefore, we may consider the mapping $\phi : \mathbb{Z} \to G$ given by $\phi(n) = a^n$, for $n \in \mathbb{Z}$. Now,

$$\phi(n + m) = a^{n+m} = a^n * a^m = \phi(n) * \phi(m),$$

implies that $\phi$ is a homomorphism. Clearly, $\phi$ is onto. Also, $\phi(n) = e$ implies $a^n = e$, and so necessarily $n = 0$; for, if $n \neq 0$, then $G = \{e, a, a^2, \cdots, a^{m-1}\}$ for some $m$, contrary to our assumption. Therefore, $K_\phi = \{e\}$. Hence, $\phi$ is an isomorphism. □

**Definition 8.22** Let $G$ be a finite group. The **exponent** of $G$ is the least integer $n > 1$ such that $g^n = 1$, for all $g \in G$. It is denoted by $e(G)$.

Clearly, $e(G)$ is the *least common multiple* of the numbers $o(g)$, $g \in G$. The next theorem proves that every finite Abelian group $G$ of exponent $e$ contains an element of order $e$. In particular, if $e(G) = |G|$, then $G$ is a cyclic group.

**Theorem 8.31** *Let G be a finite Abelian group of exponent e. Then G contains an element of order e.*

***Proof.*** Using the *fundamental theorem of arithmetic*, we may write

$$e = e(G) = p_1^{n_1} p_2^{n_2} \cdots p_k^{n_k},$$

where $p_i$ are distinct primes, and $n_i \geq 1$. By Definition 8.22, we can find elements $g_i \in G$ such that $o(g_i)$ is divisible by $p_i^{n_i}$, for $i = 1, \ldots, k$. So, a suitable power $x_i$ of the element $g_i$ has the order $p_i^{n_i}$. Take $g = x_1 \cdots x_k$, and suppose $g^m = 1$, where $m \geq 1$. It thus follows that, for

$$q_i = \prod_{j=1,\, j\neq i}^{n} p_j^{n_j}, \quad \text{we have} \quad x_i^{mq_i} = 1.$$

However, $(q_i, p_i^{n_i}) = 1$ also gives $p_i^{n_i} \big| m$. Therefore, $e \big| m$. Hence, it follows from the relation $g^e = 1$ that $o(g) = e$. This completes the proof. □

## 8.4 Cosets and Lagrange's Theorem

The focus of the section is to prove the fundamental *Lagrange's theorem*, which says that the order of a subgroup $H$ of a finite group $(G, *)$ is a divisor of the order of the group. The proof we give here uses the *congruence relation* $\equiv$ as defined below: For $a, b \in G$, we take

$$a \equiv b \quad \Leftrightarrow \quad a^{-1} * b \in H. \tag{8.4.1}$$

The next theorem proves that $\equiv$ is an equivalence relation on $G$. It is called the **congruence relation** on $G$ modulo $H$.

**Theorem 8.32** *Let $(G, *)$ be a group, and $H$ be a subgroup of $G$. The relation* (8.4.1) *on $G$ is an equivalence relation.*

***Proof.*** We need to show that $\equiv$ is reflexive, symmetric, and transitive relation.

(i) Since $H$ is a subgroup, for every $a \in G$, we have $a^{-1} * a = e \in H$. Therefore, $a \equiv a$, proving that $\equiv$ is reflexive.

(ii) Let $a \equiv b$. Then, by (8.4.1), we have $a^{-1} * b \in H$. Therefore, since $H$ is a subgroup, it follows that $(a^{-1} * b)^{-1} \in H$. Also,

$$(a^{-1})^{-1} = a \quad \text{and} \quad \left(a * b\right)^{-1} = b^{-1} * a^{-1}$$

implies that $b^{-1} * a \in H$ so that $b \equiv a$. That is, $\equiv$ is symmetric.

(iii) Finally, to show that $\equiv$ is transitive, let $a \equiv b$ and $b \equiv c$. It then follows from (8.4.1) that

$$a^{-1} * b \in H \quad \text{and} \quad b^{-1} * c \in H.$$

Therefore, since $H$ is a subgroup of $G$, we have

$$\left(a^{-1} * b\right) * \left(b^{-1} * c\right) \in H.$$

Now, since we have

$$\begin{aligned}(a^{-1} * b) * (b^{-1} * c) &= a^{-1} * (b * (b^{-1} * c)) \\ &= a^{-1} * ((b * b^{-1}) * c) \\ &= a^{-1} * (e * c) \\ &= a^{-1} * c,\end{aligned}$$

it follows that $a^{-1} * c \in H$, i.e., $a \equiv c$, by (8.4.1).

Hence, $\equiv$ is an equivalence relation on $G$. □

The equivalence class of $a \in G$ with respect to relation (8.4.1) is given by

$$\begin{aligned}[a] &= \{b \in G \mid a \equiv b\} \\ &= \{b \in G \mid a^{-1} * b \in H\} \\ &= \{b \in G \mid a^{-1} * b = h, h \in H\} \\ &= \{b \in G \mid b = a * h, h \in H\} \\ &= \{a * h \mid h \in H\} \\ &= a * H.\end{aligned}$$

Therefore, the equivalence classes of the group $G$ relative to the subgroup $H$ are given by

$$a * H = \{a * h \mid h \in H\}, \qquad \text{for} \quad a \in G. \tag{8.4.2}$$

Each gives a *left coset* of the subgroup $H$ given by the element $a \in G$.

**Definition 8.23** Let $(G, *)$ be a group, and $H$ be a subgroup of $G$. The **left coset** of $H$ given by an element $a \in G$ is the set $a * H \subset G$ given by (8.4.2).

Notice that the relation $\equiv$ on $G$ given by

$$a \equiv b \qquad \Leftrightarrow \qquad a * b^{-1} \in H. \tag{8.4.3}$$

is also an equivalence relation on $G$. In this case, the equivalence classes of the group $G$ relative to the subgroup $H$ are given by

$$H * a = \{h * a \mid h \in H\}. \tag{8.4.4}$$

This is called the **right coset** of $H$ given by an element $a \in G$. In general, we may write the set of the left cosets of $H$ in $G$ as $(G/H)_\ell$, and the set of right cosets of $H$ in $G$ as $(G/H)_r$. These collections may not be the same in general, but the two are in one-to-one correspondence. For, it follows easily that the mapping $a * H \mapsto H * a : (G/H)_\ell \to (G/H)_r$ is a bijective function, for any $a \in G$.

**Definition 8.24** Let $(G, *)$ be a group, and $H$ be a subgroup of $G$. The number of left cosets (or the right cosets) of $H$ in $G$ is called the **index** of the subgroup $H$ in the group $G$. It is usually denoted by $[G, H]$ (or $i_G(H)$).

For an Abelian group $G$, every left coset is also a right coset, and vice-versa. Accordingly, we may call $a * H = H * a$ simply as the coset of $H$ in $G$ given by the element $a \in G$. Some simple illustrations are given in the next example.

**Example 8.14**

(i) Consider the subgroup $H = U_2 = \{1, -1\}$ of the Abelian group $G = U_4 = \{1, -1, i, -i\}$. The cosets of $H$ in $G$ are given by

$$\begin{aligned} 1\,H &= \big\{1, -1\big\} = H = H1 \\ -1\,H &= \big\{1, -1\big\} = H = H(-1) \\ iH &= \big\{i, -i\big\} = Hi \\ -iH &= \big\{i, -i\big\} = iH = H(-i). \end{aligned}$$

Since $H$ and $iH = Hi$ are the only *distinct cosets*, we have $\big[U_4, U_2\big] = 2$.

(ii) Consider the subgroup $H = \big\{0, 2\big\}$ of the group $G := \mathbb{Z}_4 = \big\{0, 1, 2, 3\big\}$. Then the cosets of $H$ in $\mathbb{Z}_4$ are given by

$$\begin{aligned} 0 +_4 H &= \big\{0, 2\big\} = H = H +_4 0 \\ 1 +_4 H &= \big\{1, 3\big\} = H +_4 1 \\ 2 +_4 H &= \big\{2, 4\big\} = \{0, 2\} = H = H +_4 2 \\ 3 +_4 H &= \big\{3, 1\} = \{1, 3\big\} = H +_4 1 = H +_4 3. \end{aligned}$$

Since $H$ and $1 +_4 H$ are the only *distinct cosets*, we have $[G : H] = 2$. Also, the subgroup $H := 5\mathbb{Z}$ has five *distinct cosets* in the additive group $G := \mathbb{Z}$ given by

$$\begin{aligned} 0 + 5\mathbb{Z} &= \big\{ \ldots, -15, -10, -5, 0, 5, 10, 15, \ldots \big\} = 5 + 5\mathbb{Z} = 10 + 5\mathbb{Z} = \cdots; \\ 1 + 5\mathbb{Z} &= \big\{ \ldots, -14, -9, -4, 1, 6, 11, 16, \ldots \big\} = 6 + 5\mathbb{Z} = 11 + 5\mathbb{Z} = \cdots; \\ 2 + 5\mathbb{Z} &= \big\{ \ldots, -13, -8, -3, 2, 7, 12, 17, \ldots \big\} = 7 + 5\mathbb{Z} = 12 + 5\mathbb{Z} = \cdots; \\ 3 + 5\mathbb{Z} &= \big\{ \ldots, -12, -7, -2, 3, 8, 13, 18, \ldots \big\} = 8 + 5\mathbb{Z} = 13 + 5\mathbb{Z} = \cdots; \\ 4 + 5\mathbb{Z} &= \big\{ \ldots, -11, -6, -1, 4, 9, 14, 19, \ldots \big\} = 9 + 5\mathbb{Z} = 14 + 5\mathbb{Z} = \cdots. \end{aligned}$$

Therefore, in this case, we have $[G : H] = 5$.

(iii) The left cosets of the subgroup $H = \{e, (1\,2)\}$ in the group $S_3$ are as computed in (8.2.2). Also, the right cosets of the subgroup $H$ in $S_3$ are given by

$$\begin{aligned} He = H &= H(1\,2); \\ H(2\,3) = \{(2\,3),\ (1\,3\,2)\} &= H(1\,3\,2); \\ H(1\,2\,3) = \{(1\,2\,3),\ (1\,3)\} &= H(1, 3). \end{aligned} \tag{8.4.5}$$

Therefore, $xH \neq Hx$, in general. However, we have $[G : H] = 3$.

The properties of the *congruence relation* (8.4.1), as stated in the next theorem, are special cases of Theorem 3.8. The reader is encouraged to fill the details.

**Theorem 8.33** *Let G be a group, and H be a subgroup of G. Then we have*

*(i)* $a \in a * H$*, In particular,* $a * H$ *is nonempty;*
*(ii)* $c \in a * H \quad \Rightarrow \quad a * H = c * H$*;*
*(iii) For* $a, c \in G$*, we have* $a * H = c * H$ *or* $a * H \cap c * H = \emptyset$*;*
*(iv) the group G is a disjoint union of all the left cosets of H in G.*

An element $a \in a * H$ is called a **coset representative**. By (ii), we may choose any element of the coset $a * H$ as a *coset representative*.

**Definition 8.25** A **left transversal** (or a *left coset representatives*) of $H$ in $G$ is a set of elements $\{x_i \mid i \geq 1\}$ (with $x_1 = e$) such that G is the disjoint union of the left cosets $x_i * H$. Similarly, a **right transversal** (or a *right coset representatives*) of $H$ in $G$ is a set of elements $\{y_i \mid i \geq 1\}$ (with $y_1 = e$) such that G is the disjoint union of the right cosets $H * y_i$.

By a remarkable theorem of Philip Hall published in 1935 we know that, for a subgroup of *index* $k$ in a finite group $G$, it is always possible to choose elements $t_1, \ldots, t_k \in G$ such that

$$G = \cup_{i=1}^{k} t_i * H = \cup_{i=1}^{k} H * t_i.$$

Although different equivalence classes in general may not have the same cardinality, it follows easily that any two left cosets of a subgroup of a finite group have the same cardinality. Equivalently, we have $|H| = |a * H|$, for all $a \in G$. We are now ready to prove the main theorem of this section. The proof given below uses the decomposition of a group in terms of the *left cosets*, as suggested by Theorem 8.33.

**Theorem 8.34 (Lagrange's Theorem)** *Let G be a finite group, and H be a subgroup of G. Then* $o(H)$ *divides* $o(G)$*.*

***Proof.*** As said above, we can write $G$ as $G = \bigcup_{i \in I} a_i * H$, where the index set $I$ is chosen such that $a_j * H \neq a_k * H$ whenever $j \neq k \in I$. Therefore, $|I| = [G : H]$, and we have

$$o(G) = \sum_{i \in I} o(a_i * H) = [G : H]\, o(H),$$

where the last equality holds because each left coset of $H$ in $G$ has exactly $o(H)$ number of elements. Hence, $o(H)$ divides $o(G)$, and also $o(G)/o(H)$ equals $[G : H]$, the index of the subgroup $H$ in the group $G$. □

**Corollary 8.5** *Every group of prime order is cyclic.*

***Proof.*** Let a group $G$ has the order $p$, where $p$ is a prime number. As $p \geq 2$, there is some $a \in G$ such that $a \neq e$. As the cyclic group $H = \big(a\big)$ generated by $a$ is a subgroup of $G$, it follows from Theorem 8.34 that order of $|H|$ is a divisor of $|G| = p$. Therefore, $o(a) = 1$ or $p$. However, since $H = \big(a\big)$ has more than one element, we conclude $|H| = p$. That is, the group $G$ and the subgroup $H = \big(a\big)$ have the same number of elements. Hence, $G = \big(a\big)$. □

Moreover, it follows from Theorem 8.25 that the order of every element of a finite group is a divisor of the order of the group.

**Corollary 8.6** *Let $G$ be a finite group, and $a \in G$. Then we have $o(a)$ divides $o(G)$. We also have that $a^{o(G)} = e$.*

***Proof.*** It follows from Theorem 8.25 that there is a subgroup $H$ of order $o(a)$. Hence, by Theorem 8.34, $o(a)$ divides $o(G)$. Next, we that $a^{o(G)} = e$. For, let $o(a) = n$. Then, we have $o(G) = nk$, for some $k$. Therefore

$$a^{o(G)} = a^{nk} = (a^n)^k = e^k = e. \square$$

For the next corollary, we may recall that Euler's *totient function* $\varphi(n)$ counts the number of integers $1 \leq a < n$ such that $\gcd(a, n) = 1$.

**Corollary 8.7** (***Euler Theorem***) *If* Euler Theorem *$a$ and $n$ are relatively prime, then $a^{\varphi(n)} \equiv 1 \bmod n$. In particular, if $p$ is a prime and $1 \leq a \leq p - 1$, then $a^{p-1} \equiv 1 \bmod p$ (This is known as **Fermat's Little Theorem**).*

***Proof.*** The proof of the first part follows by applying the second part of Corollary 8.6 to the group $G$ as defined in (8.3.2). The proof of *Fermat's little theorem* follows from the fact that $\varphi(p) = p - 1$ when $p$ is a prime (see (5.2.4)). □

***Proof. (A Second Proof of Fermat's Little Theorem)*** We use a strong form of the principle of induction to prove that $a^p \equiv a \pmod p$, for *any* positive integer $a$. Since $1^p \equiv 1 \bmod p$, the assertion holds trivially for $a = 1$. We may thus assume that the assertion holds for $a = k$. That is, we have $k^p \equiv k \bmod p$. Now, since in all the terms except first and last in the expansion

$$\big(k+1\big)^p = k^p + pk^{p-1} + \frac{p(p-1)}{2}k^{p-2} + \cdots + pk + 1,$$

are divisible by $p$, we obtain

$$\begin{aligned}\big(k+1\big)^p &\equiv \big(k^p + 1\big) \bmod p \\ &\equiv \big(k+1\big) \bmod p.\end{aligned}$$

Therefore, it follows that $(k + 1)^p \equiv (k + 1) \bmod p$. This completes the proof. □

The two theorems stated in Corollary 8.7 play important roles in public key cryptography, as discussed in the last chapter. We will also apply the explicitly formula (5.2.2) of $\varphi(n)$, for a specified value of $n$.

## Double Cosets Decomposition

The equivalence relation we define below plays an important role in solving some problems in theoretical computer science. We start with the relation $R$ defined on a group $G$ that leads to the concept of *double cosets* of a pair of subgroups of the group $G$, say $A$ and $B$. For $x, y \in G$, let

$$x\,R\,y \quad \Leftrightarrow \quad y = a * x * b, \quad \text{for some } a \in A \text{ and } b \in B. \tag{8.4.6}$$

The next theorem proves that $R$ is an equivalence relation on $G$.

**Theorem 8.35** *Let $G$ be a group, and $A$, $B$ be two subgroups of $G$. Then the $R$ on $G$ defined by* (8.4.6) *is an equivalence relation, with the equivalence class of an element $x \in G$ given by*

$$A * x * B = \big\{a * x * b \mid a \in A \text{ and } b \in B\big\}, \tag{8.4.7}$$

*which is called the* **double coset** *of $G$ determined by the pair $(A, B)$.*

***Proof.*** Notice that, as both $A$ and $B$ are subgroups of $G$, we have $x = e * x * e$, where $e \in G$ is the identity element. Therefore, $x\,R\,x$, for all $x \in G$. That is, relation $R$ is reflexive. Next, suppose $x\,R\,y$ so that, for some $a \in A$ and $b \in B$, we have $y = a * x * b$. It then follows that

$$x = a^{-1} * y * b^{-1}, \qquad \text{with } a^{-1} \in A \text{ and } b^{-1} \in B.$$

That is, $y\,R\,x$. Therefore, the relation $R$ is symmetric. To prove that $R$ is transitive, let $x\,R\,y$ and $y\,R\,z$, so that for some $a_1, a_2 \in A$ and $b_1, b_2 \in B$ we have $y = a_1 * x * b_1$ and $z = a_2 * y * b_2$. Then, since

$$\begin{aligned} z &= a_2 * y * b_2 \\ &= a_2 * \big(a_1 * x * b_1\big) * b_2 \\ &= \big(a_2 * a_1\big) * x * \big(b_1 * b_2\big), \end{aligned}$$

with $a_2 * a_1 \in A$ and $b_1 * b_2 \in B$, it follows that $x\,R\,z$. Therefore, the relation $R$ is transitive. Hence, the relation $R$ given by (8.4.7) is an equivalence relation. Finally, for $x \in G$, the equivalence class of $x$ contains elements of $G$ of the form $a * x * b$, with $a \in A$ and $b \in B$. That is, the equivalence $[x]$ is the double coset $A * x * B$. This completes the proof. □

A simple modification of the proof of Theorem 8.33 gives that each double coset is nonempty; two double cosets are either disjoint or identical; and, the union of disjoint double cosets determined by the pair of subgroups $(A, B)$ is $G$. Said differently, the disjoint double cosets form a *partition* of the group $G$.

## *Discrete Logarithm*

A number of important *cryptosystems*, as discussed in the last chapter of this book, apply the *discrete logarithm*. We may recall that the logarithm function of a positive real number $x$, say to the base $a > 0$, is the inverse of the exponential function $a^x$. That is, we must have $x = a^{\log_a x}$. We intend to define an analogous function for modular arithmetic such that it shares most properties of the usual logarithm. As our description on the topic is very crisp, the interested reader may find more details in excellent texts such as [8] or [9].

Let $n > 1$ be a fixed integer, and $a$ be an integer such that $\gcd(a, n) = 1$. By Euler's theorem, we have

$$a^{\varphi(n)} \equiv 1 \pmod{n}, \tag{8.4.8}$$

where as usual $\varphi(n)$ is Euler's totient function. Now, consider the following generalization of (8.4.8):

$$a^m \equiv 1 \pmod{n}. \tag{8.4.9}$$

If the **order** of $a$ mod $n$ is $m$, then we write $m := \text{ord}_n(a)$. We also say $m$ is the **exponent** to which $a$ belongs to mod$n$ or that $m$ is the **length** of the period generated by $a$. Clearly, we have $\text{ord}_n(1) = 1$. For example, it follows from the list of powers of 7 mod 19 given in Table 8.6 that $\text{ord}_{19}(7) = 3$.

Notice that, as long as $a$ and $n$ are relatively prime, there is at least one integer $m$ satisfying (8.4.9), namely, we may take $m = \varphi(n)$. More generally, we say $a$ is a **primitive root** of $n$ if $\gcd(a, n) = 1$, and the least integer $m$ satisfying (8.4.9) is the number $\varphi(n)$. Therefore, all the powers

$$a,\ a^2,\ \ldots,\ a^{\varphi(n)},$$

**Table 8.6** Powers of 7 mod 19

| $m$ | 1 | 2 | 3 | 4 | 5 |
|---|---|---|---|---|---|
| $7^m$ mod 19 | 7 | 11 | 1 | 7 | 11 |

are distinct $(\bmod n)$, and they are relatively prime to $n$. Clearly, not all integers have primitive roots. We know that, for any prime $p > 2$ and positive integer $b$, we have $2, 4, p^b$, and $2p^b$, are the only integers with primitive roots [8].

When $n = p$ is a prime, we have $a^{p-1} \equiv 1 \bmod p$, by *Fermat's little theorem*. That is, $\text{ord}_n(a) \le p - 1$. Notice that the order of $a$ may not be $p - 1$, in general, but it is a divisor of $p - 1$, by Corollary 8.6.

**Theorem 8.36** *Let p be a prime, and an integer a be relatively prime to p. If m is an integer such that* (8.4.9) *holds for* $m = p$, *then* $n := \text{ord}_p(a)$ *divides m. In particular, we have* $n|p - 1$.

***Proof.*** Let $d = \gcd(n, m)$. By Lemma A.1, we can write $d = x\,n + y\,m$, for some integers $x, y$. Since $a^n \equiv a^m \equiv 1 \bmod p$, we obtain

$$a^d = a^{x\,n+y\,m} \equiv 1 \bmod p.$$

It thus follows from the *minimality* of $n$ that $n \le d$. We also have $d|n$, and so $d = n$. Therefore, $n|m$. □

Suppose $a$ is a *primitive root* of $p$, for a prime $p$. For example, with $p = 19$, the primitive roots are 2, 3, 10, 13, 14, and 15. As said above, all the powers

$$a,\ a^2,\ \dots\ ,\ a^{p-1},$$

are distinct $(\bmod n)$, and they are relatively prime to $n$. We also have that, for any integer $b$,

$$b \equiv r \pmod{p}, \qquad \text{for some } r \text{ satisfying } \ 0 \le r \le (p-1). \tag{8.4.10}$$

Therefore, there is unique exponent $k$ such that

$$b \equiv a^k \pmod{p}, \qquad \text{for some } \ 0 \le k \le (p-1). \tag{8.4.11}$$

The exponent $k$ is called the **discrete logarithm** of the number $b$, with the base $a$ mod $p$. We may denote it by $\text{dlog}_{a,p}(b)$. Notice that

$$\begin{aligned} a^0 \bmod p = 1 \bmod p = 1 \quad &\Rightarrow \quad \text{dlog}_{a,p}\left(1\right) = 0; \\ a^1 \bmod p = a \quad &\Rightarrow \quad \text{dlog}_{a,p}\left(a\right) = 1. \end{aligned} \tag{8.4.12}$$

It is easy to verify that the discrete logarithms satisfy most properties of the usual logarithms (Exercise 8.27). In general, the discrete logarithm $\text{dlog}_{a,n}$ exists only if $\gcd(a, n) = 1$. Furthermore, for any given $a, x$, and prime $p$, it is simpler task to compute $b$ from the equation

$$b \equiv a^x \pmod{p}. \tag{8.4.13}$$

On the other hand, given $a$, $y$, and $p$, it is in general very difficult to solve (8.4.13) for $x$, i.e., to compute the discrete logarithm $\text{dlog}_{a,p}(b)$. The level of difficulty in terms of the asymptotically fastest known algorithm appears to be at the same level as getting the prime factorization for the RSA cryptosystem.

## 8.5 Normal Subgroups and Quotient Groups

An equivalence relation $E$ defined on an algebraic structure $(X, *)$ is called a *congruence relation* if it satisfies the *substitution property* with respect to the operation $*$. That is,

$$\big(x_1, x_1'\big),\ \big(x_2, x_2'\big) \in E \quad \Rightarrow \quad \big(x_1 * x_2, x_1' * x_2'\big) \in E. \tag{8.5.1}$$

Recall that each equivalence relation on a set $X$ provides a *partition* of the set $X$ in terms of associated equivalence classes. The *quotient set* of $X$ relative to an equivalence relation $E$ is the set of associated equivalence classes, which is usually denoted by $X/E$.

**Theorem 8.37** *Let $(S, *)$ and $(T, \triangle)$ be semigroups, and $g : S \to T$ be a semigroup homomorphism. Then the relation $R$ on $S$ defined by*

$$(x, y) \in R \quad \Leftrightarrow \quad g(x) = g(y), \ \ \textit{with} \ \ x, y \in S,$$

*is a congruence relation.*

***Proof.*** We first show that $R$ is an equivalence relation. For, we have

1. $R$ is reflexive because $g(x) = g(x) \ \Rightarrow \ (x, x) \in R$;
2. $R$ is symmetric because $(x, y) \in R \ \Rightarrow \ g(x) = g(y)$ or $g(y) = g(x) \ \Rightarrow$ $(y, x) \in R$;
3. $R$ is transitive because $(x, y),\ (y, z) \in R \ \Rightarrow \ g(x) = g(y)$ and $g(y) = g(z)$, and so $g(x) = g(z) \ \Rightarrow \ (x, z) \in R$.

Therefore, $R$ is an equivalence relation. It remains to show that $R$ satisfies the substitution property with respect to $*$. For, let $x_1, x_1', x_2, x_2' \in S$ be such that $(x_1, x_1'), (x_2, x_2') \in R$. We then have $g(x_1) = g(x_1')$ and $g(x_2) = g(x_2')$ so that

$$g\big(x_1 * x_2\big) = g(x_1) \triangle g(x_2) = g(x_1') \triangle g(x_2') = g\big(x_1' * x_2'\big).$$

Therefore, $\big(x_1 * x_2, x_1' * x_2'\big) \in R$, which completes the proof. □

**Theorem 8.38** *Let $R$ be a congruence relation on a semigroup $(S, *)$, and let the operation $\oplus$ on the quotient set $S/R$ be defined by*

$$\big[a\big] \oplus \big[b\big] = \big[a * b\big], \qquad \textit{for} \ \ a, b \in S. \tag{8.5.2}$$

*Then* $(S/R, \oplus)$ *is a semigroup, and the mapping* $a \mapsto [a] : S \to S/R$ *gives a semigroup homomorphism.*[1]

***Proof.*** We first need to verify that the operation $\oplus$ is well defined. For, if $[a] = [a']$ and $[b] = [b']$, then $(a, a') \in R$ and $(b, b') \in R$ implies $(a * b, a' * b') \in R$, i.e., $[a * b] = [a' * b']$. It thus follows that

$$\big[a\big] \oplus \big[b\big] = \big[a * b\big] = \big[a' * b'\big] = \big[a'\big] \oplus \big[b'\big].$$

We next show that $\oplus$ is associative. For, let $[a], [b], [c] \in S/R$. Then we have

$$\begin{aligned}[a] \oplus \big([b] \oplus [c]\big) &= [a] \oplus [b * c] = [a * (b * c)] \\ &= [(a * b) * c] = [a * b] \oplus [c] \\ &= ([a] \oplus [b]) \oplus [c].\end{aligned}$$

Therefore, $(S/R, \oplus)$ is a semigroup. Next, the mapping $g : S \to S/R$ defined by $g(a) = [a]$ satisfies

$$g\big(a * b\big) = \big[a * b\big] = [a] \oplus [b] = g(a) \oplus g(b),$$

which proves that $g$ is a homomorphism from $S$ to $S/R$. □

We next introduce an important property of a subgroup $H$ of a group $(G, *)$ such that the set $(G/H)_\ell$ of all left cosets of $H$ in $G$ is a group with respect to the binary operation $\otimes$ as given below: For $X = a * H,\ Y = b * H \in G/H$, let

$$X \otimes Y = a * H \otimes b * H = \big(a * b\big) * H. \tag{8.5.3}$$

Such a group is called the *quotient group* (or the *factor group*) of the group $G$ by the subgroup $H$ (Definition 8.27). Notice that the operation (8.5.3) is induced by the operation $*$ of the group $G$. We now discuss a special property of the subgroup $H$ that makes $(G/H, \otimes)$ a group. For, since a coset can be represented by any of its elements, it is important to verify that the definition (8.5.3) is independent of the choice for coset representatives of elements in the quotient set $G/H$. That is, if $a * H = c * H$ and $b * H = d * H$, then we must have

$$a * H \otimes b * H = c * H \otimes d * H \quad \text{or, equivalently,} \quad \big(a * b\big) * H = \big(c * d\big) * H. \tag{8.5.4}$$

We first show that the above condition holds when $G$ is an Abelian group.

**Theorem 8.39** *Let* $G$ *be an Abelian group, and* $H$ *a subgroup of* $G$. *Then the binary operation* $\otimes$ *on* $G/H$ *given by* (8.5.3) *is well defined.*

[1] This homomorphism is called the *natural epimorphism* from $S$ onto $S/R$.

***Proof.*** Let $a * H = c * H$ and $b * H = d * H$, for some $a, b, c, d \in G$. As $a \in a * H$ implies that $a \in c * H$, we have $a = c * h_1$, for some $h_1 \in H$. Similarly, $b = d * h_2$, for some $h_2 \in H$. Using the assumption that $G$ is Abelian, we have

$$\begin{aligned} a * b &= \big(c * h_1\big) * \big(d * h_2\big) \\ &= c * \big(h_1 * \big(d * h_2\big)\big) \\ &= c * \big(\big(d * h2\big) * h_1\big) \\ &= c * \big(d * \big(h_2 * h_1\big)\big) \\ &= \big(c * d\big) * \big(h_2 * h_1\big) \\ &= \big(c * d\big) * h, \end{aligned}$$

where $h = h_2 * h_1 \in H$. It thus follows that $a * b \in (c * d) * H$, and so we obtain

$$\big(a * b\big) * H = \big(c * d\big) * H \quad \text{or, equivalently,} \quad a * H \otimes b * H = \big(c * H\big) \otimes (d * H).$$

This completes the proof. □

In general, let $a \in G$ and $h \in H$. Suppose $h * a = a * h_1$, for some $h_1$. With notations as in the above proof, we have

$$\begin{aligned} a * b &= \big(c * h_1\big) * \big(d * h_2\big) \\ &= c * \big(h_1 * \big(d * h_2\big)\big) \\ &= c * \big((d * h2) * h_3\big) \\ &= c * \big(d * (h_2 * h_3)\big) \\ &= \big(c * d\big) * \big(h_2 * h_3\big) \\ &= (c * d) * h, \end{aligned}$$

where $h = h_2 * h_3 \in H$. Once this is proved, we can see that the operation $\otimes$ is well defined. In particular, if the group $H$ is a subgroup of an Abelian group, then $h * a = a * h_1$ with $h = h_1$. In general, as shown in Example 8.14 (iii), we may not have $xH = Hx$, for all $x$ of a group $G$ and subgroup $H$ of $G$. In general, a *left transversal* of a subgroup $H$ in $G$ may not be the same as its *right transversal*. Therefore, the property stated in the next definition holds significance.

**Definition 8.26** Let $G$ be a group, and $N$ be a subgroup of $G$. We say $N$ is a **normal subgroup** of $G$ if we have

$$a * N = N * a, \qquad \text{for every} \quad a \in G. \tag{8.5.5}$$

Every subgroup $N$ of an Abelian group $G$ is a *normal subgroup*, because we have

$$a * N = \big\{a * n \mid n \in N\big\} = \big\{n * a \mid n \in N\big\} = N * a.$$

However, in most interesting situations, we are more keen to study normal subgroups of a non-Abelian group. The next example gives some illustrations.

**Example 8.15** We first recall that the smallest order non-Abelian group is the *symmetric group* $S_3$. The six elements of this group are given by

$$S_3 := \{e, (1\,2), (1\,3), (2\,3), (1\,3\,2), (1\,2\,3)\}.$$

As introduced earlier in (8.1.8), we may use the following notations:

$$\begin{aligned} \tau_1 &= (1\,2), \qquad \tau_2 = (1\,3), \qquad \tau_3 = (2\,3), \\ \rho_1 &= (1\,3\,2), \qquad \rho_2 = (1\,2\,3). \end{aligned}$$

Consider the subgroup $H = \{e,\ (2\,3)\}$ of the group $S_3$. Taking $x = \tau_1 \in S_3$, we have

$$x\,H = \{(1\,2),\ (1\,3\,2)\} \quad \text{and} \quad H\,x = \{(1\,2),\ (1\,2\,3)\}.$$

which implies that $H$ is not a normal subgroup of $S_3$. By Example 8.14 (iii), we also know that the subgroup $H = \{e,\ (1\,2)\}$ is not a normal subgroup of $S_3$. Next, consider the subgroup

$$K := \{e, (1\,3\,2), (1\,2\,3)\} \subset S_3.$$

In this case, it follows from Table 8.3 that we have

$$\begin{aligned} \tau_1\,K &= \{\tau_1, \tau_3, \tau_2\} \quad \text{and} \quad K\,\tau_1 = \{\tau_1, \tau_2, \tau_3\}; \\ \tau_2\,K &= \{\tau_2, \tau_1, \tau_3\} \quad \text{and} \quad K\,\tau_2 = \{\tau_2, \tau_3, \tau_1\}; \\ \tau_3\,K &= \{\tau_3, \tau_2, \tau_1\} \quad \text{and} \quad K\,\tau_3 = \{\tau_3, \tau_1, \tau_2\}. \end{aligned}$$

Therefore, $K$ is a normal subgroup of the group $S_3$.

Theorem 8.39 remains valid for any group $G$ if the condition $h * a = a * h_1, \forall a \in G$ holds for any subgroup $H$ of the group $G$. Clearly, such a condition holds when $H$ is a normal subgroup.

**Theorem 8.40** *Let $G$ be a group, and $N$ a normal subgroup of $G$. Then binary operation $\otimes$ on $G/H$ given by* (8.5.3) *is well defined.*

The next theorem gives a simple characterization of a subgroup to be normal.

**Theorem 8.41** *Let $G$ be a group, and $N$ be a subgroup of $G$. Then the four statements given below are equivalent:*

1. *For each $a \in G$ and $n \in N$, $a * n * a^{-1} \in N$. Or, equivalently, $a * N * a^{-1} \subseteq N$.*
2. *For each $a \in G$, we have $a * N * a^{-1} = N$.*

3. *Each left coset of N is also a right coset of N, in the group G.*
4. *N is a normal subgroup of G.*

***Proof.*** We first show that (1) $\Rightarrow$ (2). Clearly, if $a * n * a^{-1} \in N$ for each $a \in G$ and $n \in N$, then $a * N * a^{-1} \subseteq N$, for all $a \in G$. Now, replacing $a$ by $a^{-1}$, we obtain

$$\begin{aligned} a^{-1} * N * a \subseteq N \quad &\Leftrightarrow \quad \left(a * a^{-1}\right) * N * \left(a * a^{-1}\right) \subseteq a * N * a^{-1} \\ &\Leftrightarrow \quad N = e * N * e \subseteq a * N * a^{-1}. \end{aligned}$$

It thus follows that $a * N * a^{-1} = N$. Next, to show that (2) $\Rightarrow$ (3), let $a * N * a^{-1} = N$, for all $a \in G$. We then have

$$a * N * a^{-1} * a = N * a \quad \Leftrightarrow \quad a * N * e = N * a \quad \Leftrightarrow \quad a * N = N * a.$$

Therefore, we conclude that every left coset of $N$ in the group is also a right coset of $N$ in $G$. Next, to show that (3) $\Rightarrow$ (4), suppose each left coset of $N$ in $G$ is also a right coset of $N$ in $G$. Let $a \in G$, and consider the left coset $a * N$. By our assumption, we have $a * N = N * b$, for some $b \in G$. Therefore, as $a \in a * n = N * b$, we have $N * b = N * a$. It thus follows that $a * N = N * a$. Finally, we show that (4) $\Rightarrow$ (1). For, suppose $a * N = N * a$, for every $a \in G$, so that we have

$$\begin{aligned} a * N * a^{-1} &= \left(a * N\right) * a^{-1} \\ &= \left(N * a\right) * a^{-1} \\ &= N * \left(a * a^{-1}\right) \\ &= N * e = N. \square \end{aligned}$$

**Theorem 8.42** *Let N be a normal subgroup of a group G. Then G/N is a group under the induced operation ⊗.*

***Proof.*** We know that the operation $\otimes$ satisfies the closure property. We now verify the other axioms of a group. Let $X = a * N$, $Y = b * N$, and $Z = c * N$ be arbitrary elements in $G/N$. We first verify that the operation $\otimes$ satisfies the associative law. For, we have

$$\begin{aligned} X \otimes (Y \otimes Z) &= (a * N) \otimes ((b * N) \otimes (c * N)) \\ &= (a * N) \otimes ((b * c) * N) \\ &= (a * (b * c)) * N \\ &= ((a * b) * c) * N \\ &= ((a * b) * N) \otimes (c * N) \\ &= ((a * N) \otimes (b * N)) \otimes (c * N) \\ &= (X \otimes Y) \otimes Z. \end{aligned}$$

Also, the element $N = e * N$ is the identity element, because we have

$$(e * N) \otimes (a * N) = (e * a) * N = a * N;$$
$$(a * N) \otimes (e * N) = (a * e) * N = a * N.$$

Finally, the inverse of $a * N$ is $a^{-1} * N$. For, we have

$$(a * N) \otimes (a^{-1} * N) = (a * a^{-1}) * N = e * N = N;$$
$$(a^{-1} * N) \otimes (a * N) = (a^{-1} * a) * N = e * N = N.\Box$$

**Definition 8.27** The group $G/N$ determined by a normal subgroup $N$ is called the **quotient group** (or the **factor group**) of a group $G$.

When $G$ is a finite group, by Lagrange's theorem, we have $o(G/N) = o(G)/o(N)$.

**Corollary 8.8** *Let $N$ be a subgroup of an Abelian group $G$. Then $G/N$ is an Abelian group under the induced operation $\otimes$.*

***Proof.*** Since $G$ is Abelian, every subgroup of $G$ is normal and therefore $\otimes$ is well defined and is a binary operation on $G/N$. We have just proved that $G/N$ is a group. $G/N$ is Abelian because

$$\begin{aligned}(a * N) \otimes (b * N) &= (a * b) * N \\ &= (b * a) * N \\ &= (b * N) \otimes (a * N).\Box\end{aligned}$$

**Proposition 8.6** *Let $N$ be a normal subgroup of a group $G$, and let $\pi : G \to G/N$ be given by $\pi(a) = a * N$, for all $a \in G$. Then we have $\pi$ is an epimorphism.*

***Proof.*** Notice that, for any element $a * N \in G/N$, we have $\pi(a) = a * N$, and so the function $\pi$ is onto. Also, by the definition of $\otimes$, we have

$$\pi\big(a * b\big) = \big(a * b\big) * N = a * N \otimes b * N = \pi\big(a\big) \otimes \pi\big(b\big).$$

It thus follows that $\pi(a * b) = \pi(a) \otimes \pi(b)$, and hence $\pi$ is a homomorphism. $\Box$

The epimorphism $\pi : G \to G/N$ is also known as the *natural projection*.

**Theorem 8.43** *Let $[G, *]$ and $[G', \cdot]$ be two groups, and $e$, $e'$ be respectively their identity elements. Suppose $\phi : G \to G'$ is a homomorphism. Then we have*

1. *the* kernel *of $\phi$ given by*

$$K_\phi = \big\{a \in G \mid \phi(a) = e'\big\}$$

*is a normal subgroup of $G$.*

2. *For any subgroup $K$ of $G$, $\phi(K)$ is a subgroup of $G'$. Further, if $\phi$ is an isomorphism, then $\phi(K)$ is isomorphic to $K$.*
3. *The preimage of any element $a' \in G'$ is $a * K_\phi$, where $a$ is any preimage of $a'$.*

***Proof.*** We first prove (1). Notice that $K_\phi \neq \emptyset$ because $\phi(e) = e'$ implies that $e \in K_\phi$. Therefore, to prove that $K_\phi$ is a subgroup of $G$, it suffices to show that $a * b^{-1} \in K_\phi$, for all $a, b \in K_\phi$. For, let $a, b \in K_\phi$ so that we have

$$\phi(a * b^{-1}) = \phi(a).\phi(b^{-1}) = \phi(a).[\phi(b)]^{-1} = e' * e'^{-1} = e',$$

which proves that $K_\phi$ is a subgroup of $G$. Finally, let $a \in K_\phi$ and $n \in G$. Then we have

$$\begin{aligned}\phi(n * a * n^{-1}) &= \phi(n).\phi(a).\phi(n^{-1}) \\ &= \phi(n).e'.\phi(n^{-1}) \\ &= \phi(n).\phi(n^{-1}) \\ &= \phi(n * n^{-1}) \\ &= \phi(e) = e'\end{aligned}$$

Therefore, $n * a * n^{-1} \in K_\phi$, for all $a \in K_\phi$ and $n \in G$. Hence, $K_\phi$ is a normal subgroup of $G$. Next, to prove (2), let $K$ be any subgroup of $G$. Let $x, y \in \phi(K)$. Then $x = \phi(a)$, $y = \phi(b)$ for some $a, b \in K$. Since $e \in K$, we have $\phi(e) \in \phi(K)$ and therefore $\phi(K)$ is nonempty. Since $a, b \in K$, $a * b^{-1} \in K$. Now we have

$$\begin{aligned}x.y^{-1} &= \phi(a).[\phi(b)]^{-1} \\ &= \phi(a).\phi(b^{-1}) \\ &= \phi(a * b^{-1}) \in \phi(K).\end{aligned}$$

Therefore, $\phi(K)$ is a subgroup of $G'$. Next, suppose $\phi$ be an isomorphism. We may consider the function $\chi : K \to \phi(K)$ given by $\chi(a) = \phi(a)$, for $a \in K$. Then we have

$$\chi(a * b) = \phi(a * b) = \phi(a).\phi(b) = \chi(a).\chi(b).$$

That is, $\chi$ is a homomorphism. Also, since $\phi$ is one-one, it follows that $\chi$ is also one-one. Clearly, $\chi$ is onto. Therefore $\phi : K \to \phi(K)$ is an isomorphism. Finally, to prove (3), let $a \in G$ be any preimage of $a' \in G'$ so that $\phi(a) = a'$. If $x \in G$ is any other preimage of $g'$, then we have $\phi(x) = a'$, and therefore we have

$$\begin{aligned}\phi(x * a^{-1}) &= \phi(x).\phi(a^{-1}) \\ &= \phi(x).(\phi(a))^{-1} \\ &= a'.a'^{-1} = e'.\end{aligned}$$

Therefore, $x * a^{-1} \in K_\phi$ or $x \in a * K_\phi$, for all $a \in G$. This proves that every preimage of $a'$ is in $a * K_\phi$. Now let $x \in a * K_\phi$. Then $x = a * k$ where $k \in K_\phi$ and

$$\begin{aligned}\phi(x) &= \phi(a * k) \\ &= \phi(a).\phi(k) \\ &= \phi(a).e' \\ &= \phi(a) = a'.\end{aligned}$$

This shows that every element in $a * K$ is a preimage of $a'$. Thus the set of all preimage of $a' \in G'$ is $a * K_\phi$. □

The next theorem is used to verify when a homomorphism is injective.

**Theorem 8.44** *Let $(G, *)$ and $(G', \cdot)$ be groups, with identity elements $e$ and $e'$, respectively. A homomorphism $\phi : G \to G'$ is one-one if and only if $K_\phi = \{e\}$.*

***Proof.*** Clearly, if $\phi$ is one-one, then $K_\phi$ consists of a single element. Such an element must be $e$ because $\phi(e) = e'$. Conversely, suppose $K_\phi = \{e\}$. Now, if $a, b \in G$ are such that $\phi(a) = \phi(b)$, it then follows that

$$\phi(a * b^{-1}) = \phi(a) \cdot \phi(b^{-1}) = \phi(a) \cdot (\phi(b))^{-1} = \phi(b) \cdot (\phi(b))^{-1} = e',$$

which implies $a * b^{-1} \in K_\phi = \{e\}$. Therefore, $a * b^{-1} = e$, i.e., $a = b$. Hence, $\phi$ is a one-one function. □

**Theorem 8.45 (First Fundamental Theorem)** *Let $(G, *)$ and $(G', \cdot)$ be two groups, $\phi : G \to G'$ be an epimorphism, with kernel $K = K_\phi$. Then we have $(G/K, \otimes)$ is isomorphic to the group $G'$.*

***Proof.*** By part (1) of Theorem 8.43, $K$ is a normal subgroup of $G$, $G/K$ is a group under the induced operation $\otimes$. Notice that $\phi : G \to G'$ maps $a$ to $\phi(a)$ and the natural homomorphism $\sigma : G \to G/K$ defined by $\sigma(a) = a * K$ maps $a$ to $a * K$. We need to define a mapping from $G/K$ to $G'$. The natural choice of such function is the function $\chi : G/K \to G'$ given by $\chi(a * K) = \phi(a)$. First of all, this function $\chi$ should be well defined; that is, if $a * K = b * K$, then we must have $\phi(a) = \phi(b)$. But this will follow easily. Note that $a * K = b * K$ implies $a \in b * K$ and hence $a = b * k$, $k \in K$. Therefore

$$\begin{aligned}\phi(a) &= \phi(b * k) \\ &= \phi(b) * \phi(k) \\ &= \phi(b) * e' \\ &= \phi(b).\end{aligned}$$

Thus the function is well defined. Let us now verify that the function $\chi$ is a homomorphism. Since $\phi$ is a homomorphism, we have

$$\begin{aligned}\chi((a * K) \otimes (b * K)) &= \chi((a * b) * K) \\ &= \phi(a * b) \\ &= \phi(a).\phi(b) \\ &= \chi(a * K).\chi(b * K).\end{aligned}$$

Therefore, $\chi$ is a homomorphism from $G/K$ to $G'$. Clearly the mapping $\chi$ is onto; since $\phi$ is onto, any element in $G'$ is of the form $\phi(a)$, $a \in G$ and the preimage of any element $\phi(a)$ in $G'$ under $\chi$ is $a * K$. The proof of the theorem will be complete if we show that $\chi$ is one-to-one. The mapping $\chi : G/K \to G'$ is one-to-one if the kernel of the homomorphism $\chi$ contains the identity element $e * K$. Since

$$\begin{aligned}K_\chi &= \{a * K \mid \chi(a * K) = e'\} \\ &= \{a * K \mid \phi(a) = e'\} \\ &= \{a * K \mid a \in K\} = \{K\},\end{aligned}$$

it follows that $\chi$ is one-to-one as desired. □

In the next example, we give some interesting illustrations.

**Example 8.16**

(i) Let $G$ and $G'$ be the Abelian group $(\mathbb{R}, +)$, and consider the function $\phi : G \to G'$ given by $\phi(a) = 2a$. Then $\phi$ is a homomorphism, with the kernel $K_\phi$ given by

$$K(\phi) = \{a \mid \phi(a) = 0\} = \{a \mid 2a = 0\} = \{0\}.$$

Therefore, $\phi$ is injective. Clearly, $\phi$ is surjective, and hence it is an isomorphism.

(ii) Let $G$ be the Abelian group $(\mathbb{Z}, +)$ and $G'$ be the cyclic group $C_n$ with a generator $a \in G'$. We may write $C_n$ multiplicatively, i.e., the binary operation on $C_n$ is taken as the product. It follows easily that the function $f : G \to G'$ given by $f(r) = a^r$ is a homomorphism. We also have

$$\ker(f) = \{r \in \mathbb{Z} \mid r \equiv 0 \bmod n\} = n\mathbb{Z} \quad \text{and} \quad \mathrm{Im}(f) = G'.$$

It thus follows from Theorem 8.45 that $\mathbb{Z}/n\mathbb{Z} \cong C_n$. Therefore, every cyclic group of order $n$ is isomorphic to $\mathbb{Z}_n$.

(iii) Let $G$ and $G'$ be respectively the Abelian groups $(\mathbb{R}, +)$ and $(\mathbb{R}_{>0}, \cdot)$. Then the *exponential function* $\exp : G \to G'$ is a homomorphism, because we have

$$\exp(a + b) = \exp(a)\exp(b).$$

Clearly, the map exp has the trivial kernel. Is exp an isomorphism?

(iv) Let $G$ be the dihedral group $D_3$ generated by $a$ and $b$, with $a^3 = e$, $b^2 = e$ and $ba = a^{-1}b$, and $G'$ be the cyclic group generated by $c$ such that $c^2 = e$. We may write the binary operations for both these groups multiplicatively. Define a function $f(a^r b^s) = c^s$. It follows easily that $f : G \to G'$ is a homomorphism, $\text{Im}\,(f) = G'$, and $\ker(f)$ is isomorphic to a cyclic group of order three.

(v) Let $G$ be the cyclic group of order six with a generator $a \in G$, and $G'$ be the cyclic group of order two, say generated by $c \in G'$. We may write the binary operations for both these groups multiplicatively. Define a function $f(a^r) = c^r$. It follows easily that $f : G \to G'$ is a homomorphism, $\text{Im}\,(f) = G'$, and $\ker(f)$ is isomorphic to a cyclic group of order three.

(vi) The function $\psi$ introduced in (5.2.6), as given by $\psi(a) = (a, a)$ for $a \in \mathbb{Z}_{mn}^*$, is a homomorphism. We also know that it is a bijective function when $\gcd(m, n) = 1$. It thus follows that we have

$$\psi : \mathbb{Z}_{mn}^* \to \mathbb{Z}_m^* \times \mathbb{Z}_n^*$$

is an isomorphism, whenever $\gcd(m, n) = 1$.

**Remark 8.3** The need to study *solvable, nilpotent, locally finite*, and a host of other important classes of groups, arise naturally when a group is analyzed in terms of its *substructures* and the related *quotient groups*. In an advanced course on group theory, the main objective is to study these types of groups by looking at their *subgroup structure* and the formation of the *quotient groups* by normal subgroups in a central series of the group. Therefore, a group may be studied by understanding its *internal structures*. We may also study it as an *extensions* of a pair of *structurally known* groups, given that one *acts* onto the other in a nice way.

## 8.6 Rings and Fields

The two important algebraic structures introduced here are motivated by the arithmetic properties satisfied by usual addition and multiplication on the standard number sets such as consisting of integers, the rational numbers, or the real numbers. Of course, similar arithmetic properties hold for some other types of sets with two binary operations. For example, we already know how to *sum* and the *multiply* two polynomials or two square matrices. Our main focus in this section is to discuss some important properties of algebraic structures with two binary operations. We may start with the next definition.

**Definition 8.28** A nonempty set $R$ with two binary operations

$$+ : R \times R \to R \quad \text{and} \quad \cdot : R \times R \to R,$$

is called a **ring** if we have

1. $(R, +)$ is an Abelian group with the *additive identity* $0 \in R$, called the **zero element** of the ring. Also, the *additive inverse* of an element $a \in R$ is denoted by $-a \in R$;

2. $(R, \cdot)$ is a semigroup;
3. the *sum* $(+)$ distributes over the *product* $(\cdot)$, i.e., for all $a, b, c \in R$, we have

$$\begin{aligned} a \cdot (b+c) &= a \cdot b + a \cdot c; \\ (b+c) \cdot a &= b \cdot a + c \cdot a. \end{aligned} \tag{8.6.1}$$

We usually write a ring as $(R, +, \cdot)$.

The two binary operations $+$ and $\cdot$ are respectively called the **sum** and the **product** of the ring $R$. We usually omit the notation $\cdot$ while writing the product of two elements of $R$. That is, for $a, b \in R$, we may write $a \cdot b$ simply as $a\,b$. The relations (8.6.1) are the *distributive laws*, which are also known as the **compatibility condition** of the *ring*. In the next theorem, we prove some basic properties of a ring.

**Theorem 8.46** *Let $(R, +, \cdot)$ be a ring. Then, for any $a, b \in S$, we have*

1. $a \cdot 0 = 0 \cdot a = 0$,
2. $a \cdot (-b) = -(a \cdot b)$,
3. $(-a) \cdot b = -(ab)$,
4. $(-a) \cdot (-b) = a \cdot b$.

***Proof.*** To prove (1), let $a \in R$. Then, we have

$$a \cdot 0 + a \cdot 0 = a \cdot (0+0) = a \cdot 0 = 0 + a \cdot 0,$$

so that, by cancellation law, we have $a \cdot 0 = 0$. Similarly, it follows that $0 \cdot a = 0$. To prove (2), let $a, b \in R$. Then

$$a \cdot b + a \cdot (-b) = a \cdot (b + (-b)) = a \cdot 0 = 0$$

and

$$a \cdot (-b) + a \cdot b = a \cdot (-b + b) = a \cdot 0 = 0.$$

We thus have $-(a \cdot b) = a \cdot (-b)$. To prove (3), let $a, b \in R$. Then we have

$$\begin{aligned} a \cdot b + (-a) \cdot b &= (a + (-a)) \cdot b = 0 \cdot b = 0; \\ (-a) \cdot b + a \cdot b &= (-a + a) \cdot b = 0 \cdot b = 0. \end{aligned}$$

It thus follows that $-(a \cdot b) = (-a) \cdot b$. To prove (4), since $(-a) \cdot b = -(a \cdot b) = a \cdot (-b)$, by replacing $b$ by $-b$ we obtain

$$(-a) \cdot (-b) = -(a \cdot (-b)) = a \cdot (-(-b)),$$

or, using the first and last terms, as $-(-b) = b$, we have $(-a) \cdot (-b) = a \cdot b$. This completes the proof. □

Most interesting properties of a ring $(R, +, \cdot)$ are actually statements about the product. For example, we say $R$ is a **commutative ring** if the product $(\cdot)$ is a commutative operation. That is, we have

$$a \cdot b = b \cdot a, \qquad \text{for all} \quad a, b \in R. \tag{8.6.2}$$

Therefore, a ring $(R, +, \cdot)$ is **non-commutative** if we can find a pair of elements $a, b \in R$ such that $a \cdot b \neq b \cdot a$. Some other concepts associated with a ring $(R, +, \cdot)$ are as given below.

(i) An element $0 \neq 1 \in R$ is called a **unit element** (or simply **unity**) if we have

$$1 \cdot a = a \cdot 1 = a, \qquad \text{for all} \quad a \in R. \tag{8.6.3}$$

Said differently, 1 is the identity element of the semigroup $(R, \cdot)$. We thus also call 1 as the *multiplicative identity* of the ring $R$. A ring with a unit element is called a **unital ring**.

(ii) Let $(R, +, \cdot, 1)$ be a *unital ring*. An element $0 \neq u \in R$ is called a **unit** if there is some $v \in R$ such that we have

$$u \cdot v = v \cdot u = 1. \tag{8.6.4}$$

That is, $u$ is an invertible element of the monoid $(R, \cdot, 1)$. We also call $v$ a *multiplicative inverse* of $u$, and write $v = u^{-1}$.

(iii) An element $0 \neq a \in R$ is called a **zero divisor** if there is an element $0 \neq b \in R$ such that we have $a \cdot b = 0$. Notice that then $b$ is also a *zero divisor*.

(iv) A commutative unital ring $(R, +, \cdot, 1)$ is called an **integral domain** if it contains no zero divisor. That is, for all pairs of nonzero elements $a, b$, we have $a \cdot b \neq 0$. Equivalently, for $a, b \in R$, we have

$$a \cdot b = 0 \quad \Rightarrow \quad a = 0 \;\text{ or }\; b = 0. \tag{8.6.5}$$

Examples of some interesting rings are given below.

**Example 8.17**

(i) We know from our school mathematics that, with respect to the usual addition $(+)$ and multiplication $(\cdot)$, the set of integers $\mathbb{Z}$ satisfies all the three conditions of Definition 8.28, and so it is a commutative ring with unity 1. It is called the **integer ring**.

(ii) The sets of rational numbers $\mathbb{Q}$, the real numbers $\mathbb{R}$, and also the set of complex numbers $\mathbb{C}$, are commutative rings with respect to usual addition and multiplication. These are respectively called the *ring of rationals*, *ring of reals*, and the *ring of complex numbers*. Together with the *ring of integers* $(\mathbb{Z}, +, \cdot)$, these are also known as the **number rings**. Clearly, all these rings are integral domains with unity 1.

(iii) Consider the set $\mathbb{Z}_n$ of residue classes modulo $n$ given by

$$\mathbb{Z}_n = \{0, 1, 2, \ldots, n-1\}$$

It is a commutative ring with respect to the operations *addition modulo n* ($+_n$) and the *multiplication modulo n* ($\times_n$). The element $1 \in \mathbb{Z}_n$ is the unity. This is called the **ring of residue classes** modulo **n**. This ring is not an integral domain, in general. For example, as in Table 8.7, we have

$$2 \times_6 3 = 0 \quad \text{and also} \quad 3 \times_6 4 = 0.$$

Therefore, the ring $(\mathbb{Z}_6, +_6, \times_6)$ has 2, 3, and 4 as *zero divisors*.

Interesting applications of the ring $(\mathbb{Z}_n, +_n, \times_n)$ are obtained when $n$ is a prime.

(iv) A **polynomial** $p(x)$ in indeterminate $x$ over a commutative ring $(R, +, \cdot)$ with unity 1 is an expression of the form

$$p(x) = a_0 + a_1x + a_1x^2 + \cdots + a_nx^n, \tag{8.6.6}$$

where $a_i \in R$ and $n \geq 0$. We call $a_i$ the coefficient of $x^i$ in $p(x)$. The **degree** of a *polynomial* $p(x)$, denoted by deg$(p)$, is the largest $k$ such that $a_k \neq 0$; in this case, $a_k$ is called the **leading coefficient** of the polynomial $p(x)$. A polynomial with *leading coefficient* 1 is called a **monic polynomial**. A polynomial of degree zero is $p(x) = a_0$, with $a_0 \neq 0$. On the other hand, a **zero polynomial** $p(x)$, i.e., when all $a_i$'s in (8.6.6) are zero, has the degree $-\infty$. The set of all polynomials in indeterminate $x$ over a *unital ring* $R$ is denoted by $R[x]$. The set $R[x]$ is a commutative unital ring with respect to *coefficient-wise addition* and *multiplication by convolution*, where the constant polynomial $p \equiv 1$ is the unit element. For

$$f(x) = a_0 + a_1x + \cdots + a_mx^m \quad \text{and} \quad g(x) = b_0 + b_1x + \cdots + b_nx^n$$

the addition and multiplication in $R[x]$ are respectively given by

**Table 8.7** Product table for the set $\mathbb{Z}_6$ with respect to operation $\times_6$

| $\times_6$ | 0 | 1 | 2 | 3 | 4 | 5 |
|---|---|---|---|---|---|---|
| 0 | 0 | 0 | 0 | 0 | 0 | 0 |
| 1 | 0 | 1 | 2 | 3 | 4 | 5 |
| 2 | 0 | 2 | 4 | 0 | 2 | 4 |
| 3 | 0 | 3 | 0 | 3 | 0 | 3 |
| 4 | 0 | 4 | 2 | 0 | 4 | 2 |
| 5 | 0 | 5 | 4 | 3 | 2 | 1 |

$$\begin{aligned} f(x) + g(x) &= ((a_0 + b_0) + (a_1 + b_1)x + \cdots ; \\ f(x) \cdot g(x) &= c_0 + c_1 x + \cdots + c_{m+n} x^{m+n}, \end{aligned} \tag{8.6.7}$$

where the coefficients $c_k$ of $x^k$ are given by

$$c_k = \sum_{i=0}^{k} a_i b_{k-i} = a_0 b_k + a_1 b_{k-1} + \cdots + a_k b_0. \tag{8.6.8}$$

We call $(R[x], +, \cdot, 1)$ the **polynomial ring** over the ring $R$. In particular, when $R$ is any of the *number rings* as given in (ii), we come across concepts such as the ring of *integer polynomials*, the ring of *rational polynomials*, the ring of *real polynomials*, and the ring of *complex polynomials*. In applications such as related to coding theory and cryptography, we also use the polynomials with coefficients taken from the *ring of residue classes* modulo $n$, $\mathbb{Z}_n$, especially so when $n$ is a prime.

(v) Let $(R, +, \cdot)$ be a commutative ring with unity 1. For integers $n \geq 1$, let $M_n(R)$ be the set of square matrices of order $n$ over $R$. That is, for $A = (a_{ij}) \in M_n(R)$, we have each $(i, j)$-entry $a_{ij} \in R$. The set $M_n(R)$ is a ring with respect to two binary operations sum $+$ and product $\cdot$ as extended from operations $(+)$ and $(\cdot)$ of the ring $R$. This is called the **n × n matrix ring** over the ring $R$. In general, the *matrix ring* $M_n(R)$ is *non-commutative*, for $n \geq 2$. For example, we have

$$A = \begin{pmatrix} 1 & 2 \\ 3 & 4 \end{pmatrix}, \quad B = \begin{pmatrix} 1 & 0 \\ -1 & 2 \end{pmatrix} \quad \Rightarrow \quad AB \neq BA.$$

In particular, when $R$ is $\mathbb{Z}$, $\mathbb{Q}$, $\mathbb{R}$, or $\mathbb{C}$, we respectively have the *integer matrix ring*, the *rational matrix ring*, the *real matrix ring*, or the *complex matrix ring*. The matrix ring over the ring $(\mathbb{Z}_n, +_n, \times_n)$ is called the *modular ring* modulo $n$.

All the commutative rings $(\mathbb{Q}, +, \cdot)$, $(\mathbb{R}, +, \cdot)$, and $(\mathbb{C}, +, \cdot)$, have the property that every nonzero number is a unit. The same is true for the commutative ring $(\mathbb{Z}_n, +_n, \times_n)$ when $n$ is a prime.

**Definition 8.29** A ring $(R, +, \cdot)$ with unity 1 is called a **skew field** if every nonzero element of $R$ is a unit, i.e., has a (unique) multiplicative inverse. A commutative *skew field* is called a **field**. We denote a *field*[2] as $\mathbb{F}$.

Notice that both $(\mathbb{F}, +)$ and $(\mathbb{F}^*, \cdot)$ are Abelian groups, where $\mathbb{F}^* := \mathbb{F} - \{0\}$. Therefore, a field in general refers to a number system wherein one can perform all four arithmetic operations. In particular, it follows that the sets $\mathbb{Q}$, $\mathbb{R}$, and $\mathbb{C}$ are fields with respect to usual addition and multiplication. The classical notation $\mathbb{K}$ for a field,

[2] The American mathematician *Eliakim H. Moore* (1862–1932) coined the term "field" for the first time in 1893. The German mathematician *Leopold Kronecker* (1823–1891) called a field as a *domain of rationality*, in the context of *rational functions*.

**Table 8.8** Product tables of the field $(\mathbb{Z}_5, +_5, \times_5)$

| $+_5$ | 0 | 1 | 2 | 3 | 4 |
|---|---|---|---|---|---|
| 0 | 0 | 1 | 2 | 3 | 4 |
| 1 | 1 | 2 | 3 | 4 | 0 |
| 2 | 2 | 3 | 4 | 0 | 1 |
| 3 | 3 | 4 | 0 | 1 | 2 |
| 4 | 4 | 0 | 1 | 2 | 3 |

| $\times_5$ | 0 | 1 | 2 | 3 | 4 |
|---|---|---|---|---|---|
| 0 | 0 | 0 | 0 | 0 | 0 |
| 1 | 0 | 1 | 2 | 3 | 4 |
| 2 | 0 | 2 | 4 | 1 | 3 |
| 3 | 0 | 3 | 1 | 4 | 2 |
| 4 | 0 | 4 | 3 | 2 | 1 |

as used by Dedekind in his definition given in 1871, is derived from the German word *Körper* (meaning *body*).

An example of a non-commutative skew field is given below.

**Example 8.18** Let $\mathbb{F}$ denote the field $\mathbb{R}$ or $\mathbb{C}$. With notations as in Example 8.6 (vi), let $\mathscr{Q}$ denote the set of square matrices of order 2 given by

$$\boldsymbol{x} = x_0\, I + x_1\, \mathbf{i} + x_2\, \mathbf{j} + x_3\, \mathbf{k}, \quad \text{with } x_i \in \mathbb{F}.$$

The matrices in $\mathscr{Q}$ are known as *Hamilton's quaternions*. We can see that $\mathscr{Q}$ is closed with respect to the usual addition or multiplication of matrices. Also, since $\mathbf{ij} \neq \mathbf{ji}$, $\mathscr{Q}$ is non-commutative. Now, for

$$\boldsymbol{x}^* = x_0\, I - x_1\, \mathbf{i} - x_2\, \mathbf{j} - x_3\, \mathbf{k},$$

we have

$$\begin{aligned} \boldsymbol{x}\,\boldsymbol{x}^* &= \boldsymbol{x}^*\boldsymbol{x} = x_0^2 + x_1^2 + x_2^2 + x_3^2 \\ &= \|\boldsymbol{x}\|^2, \text{ (say)} \end{aligned}$$

It thus follows that, if $\boldsymbol{x} \neq \mathbf{0}$, then we may take $\boldsymbol{x}^{-1} = \|\boldsymbol{x}\|^{-2}\boldsymbol{x}^*$ so that

$$\boldsymbol{x}^{-1}\boldsymbol{x} = \boldsymbol{x}\,\boldsymbol{x}^{-1} = I.$$

Therefore, $\mathscr{Q}$ is a skew field, which behaves much like the field $\mathbb{F}$.

**Example 8.19** For each prime $p$, the commutative ring $(\mathbb{Z}_p, +_p, \times_p)$ is a field. In particular, the set $\mathbb{Z}_5 = \{0, 1, 2, 3, 4\}$ is a field with respect to $+_5$ and $\times_5$. The *product tables* of $\mathbb{Z}_5$ for the two operations are given by Table 8.8. The associativity of the two operations follows from the fact that the usual addition and multiplication of integers are so. The left side table shows that $(\mathbb{Z}_5, +_5)$ is a group, with additive identity 0, and the *additive inverses* of elements 1, 2, 3, 4 are respectively the elements 4, 3, 2, 1. Also, the right side table shows that $(\mathbb{Z}_5^*, \times_5)$ is a group, with multiplicative identity 1, and the *multiplicative inverses* of elements 1, 2, 3, 4 are respectively the elements 1, 3, 2, 4. The symmetry of the two tables show that

$(\mathbb{Z}_5, +_5)$ and $(\mathbb{Z}_5^*, \times_5)$ are Abelian groups. Finally, the distributivity of $\times_5$ over $+_5$ follows from the fact that the usual multiplication of integers distributes over the addition. Therefore, $(\mathbb{Z}_5, +_5, \times_5, 0, 1)$ is a field.

**Theorem 8.47** *Every field is an integral domain. The Converse is not true. However, every finite integral domain is a field.*

***Proof.*** Let $\mathbb{F}$ be a field with unity 1, and $0 \neq u \in \mathbb{F}$. Since $u$ is a unit, there is some $0 \neq v \in \mathbb{F}$ such that we have $u \cdot v = v \cdot u = 1$. If possible, suppose there is some $0 \neq u' \in \mathbb{F}$ such that $u \cdot u' = 0$. Then, we obtain

$$v \cdot (u \cdot u') = v \cdot 0 = 0 \quad \Rightarrow \quad (v \cdot u) \cdot u' = 0 \quad \Rightarrow \quad 1 \cdot u' = 0,$$

which contradicts our assumption that $u' \neq 0$. Therefore, $u$ cannot be a *zero divisor*. Hence, $\mathbb{F}$ is an integral domain. The set of integers $\mathbb{Z}$ is an example of an *infinite* integral domain that is not a field. Finally, let $R$ be a finite integral domain, which we may write as

$$R = \{a_1 = 0, a_2 = 1, a_3, \dots, a_n\},$$

where 1 is the unit element of $R$. Now, for any $i \neq 1$, we have $a_i a_j \neq 0$, for any $j = 2, \dots, n$, because $R$ is without zero divisors. Therefore, we must have

$$a_i a_j = 1, \qquad \text{for some} \quad j = 2, \dots, n.$$

This shows that every nonzero element in $R$ is a unit. Hence, $R$ is a field. □

For any field $\mathbb{F}$, the polynomial ring $\mathbb{F}[x]$ and the ring of integers $\mathbb{Z}$ share many interesting properties. For example, as (8.6.8) helps show, $\mathbb{F}[x]$ is an integral domain. Also, *division algorithm* holds in the ring $\mathbb{F}[x]$. That is, if $f,\ g \in \mathbb{F}[x]$ are polynomials of degrees $m$ and $n > 0$, respectively, then we can find unique polynomials $q,\ r \in \mathbb{F}[x]$ such that

$$f(x) = g(x)q(x) + r(x), \quad \text{with} \quad 0 \leq \deg(r) < n. \tag{8.6.9}$$

In particular, when $r(x) = 0$, we say $g(x)$ **divides** $f(x)$. Also, if $\mathbb{K}$ is a field containing $\mathbb{F}$, then $a \in \mathbb{K}$ is a **zero** of a polynomial $p \in \mathbb{F}[x]$ if we have $p(a) = 0$. In this case, we also say that $a$ is a **root** of the equation $p(x) = 0$. It is a well-known theorem that every nonzero polynomial $f \in \mathbb{F}[x]$ has at most $\deg(f)$ zeros. Notice that it follows from (8.6.10) that an element $a \in \mathbb{F}$ is a *zero* of a polynomial $p \in \mathbb{F}[x]$ if and only if $(x - a)$ *divides* the polynomial $p$, i.e., $p(x) = (x - a)s(x)$, for some polynomial $s \in \mathbb{F}[x]$.

**Definition 8.30** A polynomial $f \in \mathbb{F}[x]$ of degree at least one is called **irreducible** if, for some polynomials $a,\ b \in \mathbb{F}[x]$,

$$f(x) = a(x)b(x) \quad \Rightarrow \quad \deg(a) = 0 \ \text{or} \ \deg(b) = 0. \tag{8.6.10}$$

Said differently, $f(x)$ is *irreducible* if cannot be expressed as a product of two no-constant polynomials in $\mathbb{F}[x]$. Otherwise, we say $f(x)$ is **reducible** over $\mathbb{F}$.

The property of "irreducibility" of polynomials depends on the underlying field. For example, the polynomial $x^2 + 1$ is irreducible over $\mathbb{R}$, but it is *reducible* over $\mathbb{C}$, because the factorisation $x^2 + 1 = (x + i)(x - i)$ is valid over the polynomial ring $\mathbb{C}[x]$, where $i^2 = -1$. The Eisenstein criterion gives a sufficient condition for a polynomial over $\mathbb{Q}$ to be irreducible [2].

**Definition 8.31** A **subring** of a ring $(R, +, \cdot)$ is a nonempty set $S \subseteq R$ such that $S$ itself is a ring with respect to the induced operations $+$ and $\cdot$. A *subring* $I$ of a commutative ring $(R, +, \cdot)$ is called an **ideal** if we have

$$a \cdot x \in I, \quad \text{for all } a \in R \text{ and } x \in I. \tag{8.6.11}$$

Clearly, with respect to usual addition and multiplication, the integer ring $\mathbb{Z}$ is a subring of the ring of rationals $\mathbb{Q}$, which in turn is a subring of the ring of reals $\mathbb{R}$. It can be shown easily that every ideal $I$ of the integer ring $\mathbb{Z}$ is of the form $I = n\mathbb{Z}$, for $n \geq 0$; and, every ideal $J$ of the *modular ring* $(\mathbb{Z}_n, +_n, \cdot_n)$ is of the form $J = k\mathbb{Z}/n\mathbb{Z}$, for $0 < k \leq n$ such that $k$ divides $n$. More generally, for any ideal $I$ of a commutative ring $(R, +, \cdot)$, we may define a new ring $R/I$ consisting of elements of $R$ *modulo* $I$, i.e., the subsets of $R$ of the form

$$a + I = \{a + x \mid x \in R\}, \quad \text{for } a \in R. \tag{8.6.12}$$

The *addition* and *multiplication* on $R/I$ are defined by

$$\begin{aligned} (a + I) + (b + I) &= (a + b) + I; \\ (a + I) + (b + I) &= (ab) + I, \end{aligned} \tag{8.6.13}$$

for all $a, b \in R$. It follows easily that the two binary operations in (8.6.13) are well defined, and they inherit all the ring properties of Definition 8.28. We call $R/I$ the **quotient ring** of $R$ *modulo* $I$. The element $0 + I = I \in R/I$ is the zero element; and, if 1 is the unit element of $R$, then $1 + I$ is the unit element of the *quotient ring* $R/I$. In particular, with $I = n\mathbb{Z}\,(n \geq 2)$, the quotient ring $\mathbb{Z}/n\mathbb{Z}$ is the modular ring $(\mathbb{Z}_n, +_n, \cdot_n)$.

**Definition 8.32** Let $(R, +, \cdot)$ and $(S, \oplus, \odot)$ be rings. A function $f : R \to S$ is called a **ring homomorphism** (or simply a **homomorphism**) if we have

$$\begin{aligned} f(a + b) &= f(a) \oplus f(b); \\ f(a \cdot b) &= f(a) \odot f(b), \end{aligned} \tag{8.6.14}$$

for all $a, b \in R$. A one-one homomorphism is called a **monomorphism**; and, an onto homomorphism is called an **epimorphism**. Also, a one-one onto homomorphism is called an **isomorphism**. We say two rings $R$ and $S$ are **isomorphic** if there exists an *isomorphism* $f : R \to S$.

Let $f : R \to S$ be a *ring homomorphism*. The **kernel** of $f$ is a subset of $R$ given by

$$\text{Ker}\big(f\big) := \big\{x \in R \mid f(x) = 0\big\}. \tag{8.6.15}$$

It follows from the *ideal condition* (8.6.11) that Ker$(f)$ is an ideal of $R$. Notice that every ring homomorphism from a field $\mathbb{F}$ to a nonzero ring $R$ is a one-one function. For, it follows from the *ideal condition* (8.6.11) that a field cannot have a non-trivial ideal. More generally, we can show that a ring homomorphism $f : R \to S$ is a *monomorphism* if and only if Ker$(f) = \{0\}$. Also, the **image** Im$(f)$ is a subring of $S$ given by

$$\text{Im}\big(f\big) := \big\{y \in S \mid f(x) = y, \text{ for some } x \in R\big\}, \tag{8.6.16}$$

which may not be ideal, in general. Now, for any ideal $I$ of a commutative ring $R$, there is a *natural epimorphism* $\pi_I : R \to R/I$ given by

$$\pi_I\big(a\big) := a + I, \quad \text{for } a \in R, \tag{8.6.17}$$

with Ker$(\pi) = I$. More generally, for any epimorphism $f : R \to S$ with $K =$ Ker$(f)$, there is an isomorphism $\widehat{f} : R/K \to \text{Im}(f)$ given by

$$\widehat{f}\big(a + K\big) := f\big(a\big), \quad \text{for } a \in R, \tag{8.6.18}$$

such that $\widehat{f} \circ \pi_K = f$ holds over $R$. In this case, we say that the *epimorphism* $f$ *factors through the epimorphism* $\pi_K : R \to R/K$.

**Example 8.20** Clearly, the *inclusion function* $f : \mathbb{Z} \to \mathbb{R}$ defined by $f(a) = a$ $(a \in \mathbb{Z})$ is a one-one ring homomorphism. Next, let $\mathbb{F}$ be a field with unity 1. We can see that the function $\theta : \mathbb{Z} \to \mathbb{F}$ defined by

$$\theta\big(n\big) = n \cdot 1 := 1 + 1 + \cdots + 1 \text{ (n terms)} \tag{8.6.19}$$

satisfies the two conditions in (8.6.14). Therefore, $\theta$ is a ring homomorphism with the kernel given by

$$\text{Ker}\big(\theta\big) = \big\{n \in \mathbb{Z} \mid n \cdot 1 = 0\big\}. \tag{8.6.20}$$

**Definition 8.33** The **characteristic** of a commutative ring $R$ with unity 1 is the smallest positive integer $n$ such that

$$n \cdot a = a + \cdots + a \, (n\text{terms}) = 0, \qquad \text{for all } \; a \in R, \tag{8.6.21}$$

We denote the *characteristic* of a ring $R$ by ch$(R)$.

Notice that (8.6.21) is equivalent to the condition

$$n \cdot 1 = 1 + \cdots + 1 \,(n \text{terms}) = 0. \tag{8.6.22}$$

Now, let $\mathbb{F}$ be a field with unity 1. For ring homomorphism $\theta$ given by (8.6.19), the $\mathrm{Ker}(\theta)$ given by (8.6.20) is precisely the ideal generated by $\mathrm{ch}(\mathbb{F})$, i.e., $\mathrm{Ker}(\theta) = \big\langle \mathrm{ch}(\mathbb{F})\big\rangle$, and the image $\mathrm{Im}(\theta)$ is necessarily an integral domain, by the *fundamental theorem of homomorphisms*. It thus follows that the field $\mathbb{F}$ contains the integral domain $\mathbb{Z}$ or the field $\mathbb{Z}_p$, where $p$ is a prime. Therefore, $\mathbb{F}$ contains the field $\mathbb{Q}$ or the field $\mathbb{Z}_p$.

**Definition 8.34** The **prime subfield** of a field $\mathbb{F}$ is the smallest field $\mathbb{P} \subseteq \mathbb{F}$.

Hence, the characteristic of a field $\mathbb{F}$ is 0 if $\mathbb{Q} \subseteq \mathbb{F}$ is the *prime subfield*. Otherwise, $\mathrm{ch}(\mathbb{F}) = p$, for some prime $p$, i.e., the prime subfield $\mathbb{P} \subseteq \mathbb{F}$ has the order $p$. We can show directly that if $n = \mathrm{ch}(\mathbb{F}) \neq 0$ them $n$ is a prime. For, suppose $n = ab$, with $a > 1$ and $b > 1$. By minimality of $n$, we have

$$\begin{aligned} a \cdot 1 &= 1 + \cdots + 1 \,(a \text{ terms}) \neq 0; \\ b \cdot 1 &= 1 + \cdots + 1 \,(b \text{ terms}) \neq 0. \end{aligned}$$

But then, we have

$$(a \cdot 1)(b \cdot 1) = ab \cdot 1 = n \cdot 1 = 0,$$

which is not possible in a field. Therefore, $n$ is a prime.

**Exercises 8**

8.1 Show that the binary operation $*$ on the set $\mathbb{Z}^+$ given by $x * y = x - y$ is not a closed operation, but it is so on the set $\mathbb{Z}$. Is $*$ a commutative operation? Is it an associative operation?

8.2 Let $\mathbb{Z}^+$ denote the set of non-negative integers, and $E := 2\mathbb{N}$ the set of positive even integers. Show that

(a) $\mathbb{Z}^+$ is a monoid with respect to the usual addition $+$;
(b) $\mathbb{Z}^+$ is a monoid with respect to the usual multiplication $\times$;
(c) $E := 2\mathbb{N}$ is a semigroup with respect to the usual addition $+$. Is it a monoid?
(d) $\mathbb{Z}^+$ is a semigroup with respect to the binary operation given by $x * y = \max\{x, y\}$, for $x, y \in \mathbb{Z}^+$. Is $(\mathbb{Z}^+, *)$ a monoid?

8.3 With $S = \{a, b\}$, consider the set $S^S$ of *self-maps* of the set $S$. Show that $(S^S, \circ)$ is a non-commutative semigroup.

8.4 Show that the intersection of two congruence relations on a set is again a congruence relation.

8.5 Let $X$ be a nonempty set, with a binary operation $*$. An element $x \in X$ is called an *idempotent element* for the operation $*$ if we have $x * x = x$. Show that every finite semigroup has an idempotent element.

8.6 Let $*$ be a binary operation on a nonempty set $X$. An element $0_\ell \in X$ ($0_r \in X$) is called a *left zero* (a *right zero*) for $*$ if we have $0_\ell * x = 0_\ell$ ($x * 0_r = 0_r$), for all $x \in X$. If $0_\ell = 0_r = 0$ (say), then 0 is a *zero element* for $*$. That is,

$$0 * x = x * 0 = 0, \qquad \text{for all} \quad x \in X.$$

Moreover, zero element is always unique.

8.7 Let $(S, *)$ be a semigroup. Show that if $a, b \in S$ are left zeros, then so is their product $a * b$.

8.8 Let $(S, *)$ be a semigroup. An element $a \in S$ is called *left cancellable* if

$$a * x = a * y \quad \Rightarrow \quad x = y, \qquad \text{for all} \quad x, y \in S.$$

Show that if $a, b \in S$ are left cancellable then so is the element $a * b$.

8.9 Show that the set of all left invertible elements in a monoid form a submonoid.

8.10 Show that the binary operation $*$ on the set $\mathbb{Z}$ given by

$$a * b = a + b - ab, \qquad \text{for} \quad a, b \in \mathbb{Z},$$

is an associative and commutative. Find the identity element of the semigroup $(\mathbb{Z}, *)$, and also the inverse of an element $a \in \mathbb{Z}$, if it exists.

8.11 Consider the binary operation $*$ on the set $\mathbb{Z}^+$ given by $x * y = x$. Show that $*$ is associative, but not commutative. Does $(\mathbb{Z}^+, *)$ contains an idempotent element? Find the identity element of the semigroup $(\mathbb{Z}^+, *)$. Does it contain any left or right inverse for an element?

8.12 Consider the binary operation $*$ on the set $\mathbb{Z}^+$ given by $x * y := \text{lcm}(x, y)$. Show that $*$ is commutative and associative. Does the semigroup $(\mathbb{Z}^+, *)$ contains an idempotent element? Find its identity element of the semigroup $(\mathbb{Z}^+, *)$. Does it contain any left or right inverse for an element?

8.13 Show that $(\mathbb{Z}_n, +_n)$ is an Abelian group, where $+_n$ denotes the addition modulo $n$.

8.14 Let $\mathbb{Q}^+$ be the set of all positive rational numbers, and let the binary operation $*$ on $\mathbb{Q}^+$ be given by

$$a * b = \frac{ab}{2}, \qquad \text{for} \quad a, b \in \mathbb{Q}^+.$$

Show that $(\mathbb{Q}^+, *)$ is an Abelian group.

8.15 Let $G = \mathbb{R} \setminus \{-1\}$, and consider the operation $*$ on $G$ given by $a * b = a + b + ab$. Show that $(G, *)$ is an Abelian group.

8.16 Show that the set of all polynomials of degree less than or equal to 3 is a group under the addition of polynomials. Is the set of all polynomials of degree 3 a group under the addition of polynomials?

8.17 Show that a group has exactly one idempotent element.

8.18 Let $(S, *)$, $(T, \Delta)$ be two semigroups, and $g : S \to T$ be a semigroup homomorphism. Show that $g(a)$ is an idempotent if so is $a \in S$.

8.19 Let $X^*$ denote the free semigroup generated by a set $X$ with $n$ elements. Show that, for any semigroup $(S, \oplus)$ generated by $n$ elements, there is a homomorphism $g : X^* \to S$.

8.20 Consider the two Abelian groups: $\mathbb{R}$ with respect to addition and the *circle group* $\mathbb{S}^1 := \{z \in \mathbb{C} \mid |z| = 1\}$ with respect to complex multiplications. Prove that the function $\phi : \mathbb{R} \to \mathbb{S}^1$ given by $\phi(x) = e^{ix}$, for $x \in \mathbb{R}$, is a group homomorphism. Also, find the kernel of $\phi$.

8.21 Show that any two cyclic groups of the same order are isomorphic.

8.22 Let $H$ be a subgroup a group $G$ and consider the distinct cosets $a * H$. Show that only one of these $a * H$ is a subgroup of $G$.

8.23 Let $G$ be a group, and $H$ be a subgroup of $G$. Show that there is a bijective correspondence between any two left cosets of $H$ in $G$. Further, when $H$ is finite, each left coset has exactly $o(H)$ number of elements.

8.24 Supply a proof of Theorem 8.33.

8.25 Verify Lagrange's theorem for (i) all subgroups of the group $G = U_4$; and, (ii) all subgroups of the group $\mathbb{Z}_4 = \{0, 1, 2, 3\}$ with respect to the operation $+_4$.

8.26 Suppose two subgroups $H$ and $K$ of a group $G$ are respectively of orders $p$ and $q$, where $p$ and $q$ are relatively prime. Show that $H \cap K = \{e\}$, where $e$ is the identity element of $G$.

8.27 Let $\text{dlog}_{a,p}(b)$ denote the *discrete logarithm* of the number $b$ to the base $a$ mod $p$. Show that

$$\text{dlog}_{a,p}(xy) \equiv \left[\text{dlog}_{a,p}(x) + \text{dlog}_{a,p}(y)\right] \pmod{\varphi(p)};$$
$$\text{dlog}_{a,p}(y^k) \equiv \left[k \times \text{dlog}_{a,p}(y)\right] \pmod{\varphi(p)}$$

8.28 Show that the intersection of normal subgroups of a group is a normal subgroup.

8.29 Let $(G, *)$ be a group. Let $H$ be a subgroup of $G$ such that $x^2 \in H$, for all $x \in G$. Show that $H$ is a normal subgroup, and the quotient group $G/H$ is Abelian.

8.30 Show that the *center* $\zeta(G)$ of a group $(G, *)$ is a normal subgroup.

8.31 Let $p$ and $q$ be two distinct primes. Show that the only homomorphism $f : \mathbb{Z}_p \to \mathbb{Z}_q$ is the trivial homomorphism.

8.32 Show that the quotient group $GL(2, \mathbb{R})/SL(2, \mathbb{R})$ is isomorphic to the multiplicative group of nonzero real numbers $\mathbb{R}^*$.

8.33 Consider $\mathbb{Q}$ as an additive group. Show that, if $r$ and $s$ are nonzero rationals, there is a unique $\alpha \in Aut_{\mathbb{Z}}(\mathbb{Q})$, with $\alpha(r) = s$. Hence, deduce that $Aut_{\mathbb{Z}}(\mathbb{Q})$ is isomorphic to the field of nonzero rationals $\mathbb{Q}^*$.

8.34 Let $G$ and $H$ be groups, and $\phi : G \to H$ a homomorphism with kernel $K$. Show that, if $K = \{e, a\}$ then $x^{-1}ax = a$ for all $x \in G$. Furthermore, show that there is a homomorphism

(a) from the orthogonal group $\mathscr{O}(3, \mathbb{R})$ of $3 \times 3$ real matrices onto a group of order 2 with kernel the special orthogonal group $\mathscr{S}\mathscr{O}(3, \mathbb{R})$;
(b) from the symmetry group $S_3$ to a group of order 2 with a kernel of order 3;
(c) from $\mathscr{O}(3, \mathbb{R})$ onto $\mathscr{S}\mathscr{O}(3, \mathbb{R})$, with kernel of order 2;
(d) from $S_3$ to a group of order 3 with a kernel of order 2.

8.35 Let $C_n$ be the cyclic group with $n$ elements and $D_{2n}$ the dihedral group with $2n$ elements. If $n$ is odd and $f : D_{2n} \to C_n$ is a homomorphism, show that $f(x) = e$, for all $x \in D_{2n}$. What can be said if $n$ is even? Find all the homomorphisms from the cyclic group $C_n$ of order $n$ generated by $a$, say, to $C_m$ the cyclic group generated by $b$, say. If $n = m$, show that there are $\phi(n)$ isomorphisms, where $\phi(n)$ is Euler's *totient function*.

8.36 Show that the function $\theta$ given by

$$\theta \begin{pmatrix} a & b \\ 0 & d \end{pmatrix} = \log ad$$

is a homomorphism from the group $G$ of upper triangular matrices onto the Abelian group $(\mathbb{R}, +)$. What is the kernel $K$ of $\theta$?

8.37 Let $G$ be the group of $3 \times 3$ real matrices of determinant 1 of the form

$$\begin{pmatrix} a & 0 & 0 \\ b & x & y \\ c & z & w \end{pmatrix}.$$

Find a homomorphism from $G$ onto the group $GL_2(\mathbb{R})$, and also find its kernel.

8.38 Prove that the set of all square matrices of order 3 over $\mathbb{Z}$ is a ring with respect to usual matrix addition and multiplication. Is this ring an integral domain? If not, find some zero divisors in this ring.

8.39 Give some more example of rings, fields.

8.40 Let $(S, +, \cdot)$ be a ring. Show that, for $a, b, c \in S$, we have

a. $a \cdot (b - c) = a \cdot b - a \cdot c$,
b. $(a - b) \cdot c = a \cdot c - b \cdot c$.

8.41 Let $(S, +, \cdot)$ be a ring, and $a, b \in S$. Show that

$$(a + b)^2 = a^2 + a \cdot b + b \cdot a + b^2, \qquad \text{where } a^2 = a \cdot a.$$

## References

1. Allenby, R.B.J.: Rings, Fields and Groups: An Introduction to Abstract Algebra. Edward Arnold, London (1983)
2. Dummit, D.S., Foote, R.M.: Abstract Algebra (2/e). Wiley (1999)

3. Wisbauer, R.: Foundations of Module and Ring Theory (1991)
4. Morandi, P.: Fields and Galois Theory. Springer, GTM-167, NY, Inc. (1996)
5. Dym, H.: Linear Algebra in Action. Graduate Studies in Mathematics, vol. 78. AMS (2007)
6. Feller, W.: An Introduction to Probability Theory (2/e), vol. 2. Wiley (1971)
7. Strang, G.: Linear Algebra and Its Applications (4/e). Cengage (2006)
8. Ore, O.: Invitation to Number Theory. MAA, Washington, D.C. (1967)
9. Leveque, W.: Elementary Theory of Numbers. Dover, NY (1990)

# Chapter 9
# Basic Linear Algebra

*Mathematics is not about numbers, equations, computations, or algorithms: it is about understanding.*

**Arthur Cayley** (1821–1895)

This chapter presents some fundamental concepts of Linear algebra that find applications to problems related to *data science, machine learning, natural language processing*, and also *computer vision*. In machine learning, it helps gain insight into certain basic computational tools that one needs to study *loss functions, regularization, covariance matrix, support vector machine*, and the dimensionality reduction technique such as the *principal component analysis*. We also discuss some other applications in later chapters. In general, linear algebra is a study of vector spaces, linear transformations, and related matrix representations. The reader familiar with the core topics on the subject as in [1–3], may benefit from the related interpretations in terms of matrices. We recommend texts such as [4–9], for any missing details and further studies.

## 9.1 Real Euclidean Spaces

We start our discussion with a very special class of vector spaces, known as the *real Euclidean spaces*, which consists of $n$-tuples $\boldsymbol{x} = (x_1, \ldots, x_n)$ of real numbers. Such an $n$-tuple is called an *$n$ - dimensional vector* (or simply an **n-vector**) over the field $\mathbb{R}$. The set of *$n$-vectors* is written as $\mathbb{R}^n$. In actual applications, the elements of this set are usually written as the column vectors $(x_1, \ldots, x_n)^t$. As known for the case

 
V. Ravichandran and A. K. Razdan, *Fundamental Discrete Structures*,
University Texts in the Mathematical Sciences,
https://doi.org/10.1007/978-981-96-0069-4_9

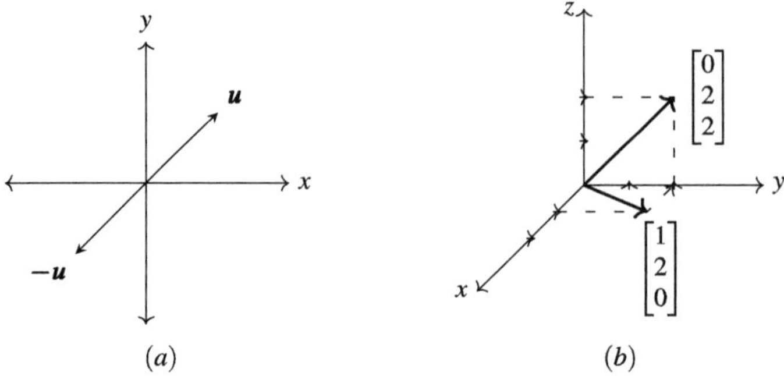


**Fig. 9.1** Vectors in $\mathbb{R}^2$ and $\mathbb{R}^3$

when $n = 2$ or $n = 3$, in geometric terms, a vector $\boldsymbol{v}$ in $\mathbb{R}^n$ with the *Position vector* at the point $\boldsymbol{x} = (x_1, \ldots, x_n)$ is a *directed line segment* with tail at the origin $\mathbf{0} \in \mathbb{R}^n$ and the head at $\boldsymbol{x}$. In particular, a vector $\boldsymbol{u}$ in the *plane* is a directed line segment with the position vector given by an ordered pair $(x_1, x_2) \in \mathbb{R}^2$ (see Fig. 9.1a). Similarly, a triple $(x_1, x_2, x_3) \in \mathbb{R}^3$ such as

$$\boldsymbol{v}_1 = \begin{bmatrix}1 & 2 & 0\end{bmatrix}^t \quad \text{or} \quad \boldsymbol{v}_2 = \begin{bmatrix}0 & 2 & 2\end{bmatrix}^t$$

is the position vector of a vector in the *space* (see Fig. 9.1b).

To model the fundamental laws governing various physical phenomena, we need to work in the space $\mathbb{R}^n$, for $n > 3$. The structural properties of these spaces described below help solve diverse types of problems of the related natural processes. The concepts we introduce below are prototypical for the abstractions we develop in later sections. However, we are not discussing the infinite dimension case in this text.

**The set $\mathbb{R}^n$ is a vector space** For two vectors $\boldsymbol{x}, \boldsymbol{y} \in \mathbb{R}^n$ given by

$$\boldsymbol{x} = \big(x_1, \ldots, x_n\big) \quad \text{and} \quad \boldsymbol{y} = \big(y_1, \ldots, y_n\big), \tag{9.1.1}$$

and a real number $\lambda$, the two *pointwise operations* given by

$$\big(x_1, \ldots, x_n\big) + \big(y_1, \ldots, y_n\big) = (x_1 + y_1, \ldots, x_n + y_n); \tag{9.1.2a}$$

$$\lambda \cdot (x_1, \ldots, x_n) = \big(\lambda\, x_1, \ldots, \lambda\, x_n\big), \tag{9.1.2b}$$

are respectively called the *vector addition* and the *scalar multiplication* on $\mathbb{R}^n$. The set $\mathbb{R}^n$ is an Abelian group with respect to (9.1.2a). The *zero element* is the vector

$$\mathbf{0} = \big(0, 0, \ldots, 0\big) \in \mathbb{R}^n,$$

which is called the $n$-dimensional **zero vector**. Also, the additive inverse of a vector $\boldsymbol{x} \in \mathbb{R}^n$ is the vector $-1 \cdot \boldsymbol{x}$ given by

$$(-1) \cdot \boldsymbol{x} = (-x_1, \ldots, -x_n).$$

We also have $\boldsymbol{x} - \boldsymbol{y} = \boldsymbol{x} + (-\boldsymbol{y})$ and $\boldsymbol{x} - \boldsymbol{x} = \boldsymbol{0}$. Further, the Abelian group $\mathbb{R}^n$ satisfies the following properties with respect to the scalar multiplication (9.1.2b):

(a) $\lambda \cdot (\boldsymbol{x} + \boldsymbol{y}) = \lambda \cdot \boldsymbol{x} + \lambda \cdot \boldsymbol{y}$;
(b) $(\lambda + \mu) \cdot \boldsymbol{x} = \lambda \cdot \boldsymbol{x} + \mu \cdot \boldsymbol{x}$;
(c) $(\lambda\, \mu) \cdot \boldsymbol{x} = \lambda \cdot (\mu \cdot \boldsymbol{x})$;
(d) $1 \cdot \boldsymbol{x} = \boldsymbol{x}$.

Notice that each one of these four axioms is well-known for $\mathbb{R}^2$ and $\mathbb{R}^3$. In general, we say that the set $\mathbb{R}^n$ is a *vector space* over the field $\mathbb{R}$ with respect to the *vector addition* (9.1.2a) and the *scalar multiplication* (9.1.2b).

**Example 9.1** Suppose a restaurant serves $n$ dishes to its customers. If a customer orders $x_j$ quantity of the $j$th dish, then the $n$-tuple $\boldsymbol{x} = (x_1, \ldots, x_n) \in \mathbb{R}^n$ may be called as the *meal vector* of the customer. After the conclusion of the meal, a bill is generated using a function $P_{\boldsymbol{w}} : \mathbb{R}^n \to \mathbb{R}$, where the $j$th coordinate $w_j$ of the vector $\boldsymbol{w} \in \mathbb{R}^n$ is the cost of the $j$th dish. The customer must pay the amount

$$P_{\boldsymbol{w}}(\boldsymbol{x}) = w_1 x_1 + \cdots + w_n x_n.$$

Clearly, it makes little sense to talk about the *angle* between the meal vectors of two customers or the *distance* between them. However, such geometric notions play a significant role in modeling some of the vital real-world problems. We next introduce important geometric concepts such as the **angle** and the **distance** between two vectors of the space $\mathbb{R}^n$, and also the concept of the **length** of a vector in $\mathbb{R}^n$. For, let $\boldsymbol{x}$ and $\boldsymbol{y}$ be vectors as in (9.1.1). The concept of *angle* between two *nonzero vectors* $\boldsymbol{x}$ and $\boldsymbol{y}$ is based on the notion of *inner product* $\langle \boldsymbol{x}, \boldsymbol{y} \rangle$ given by the usual **dot product**:

$$\langle \boldsymbol{x}, \boldsymbol{y} \rangle := \boldsymbol{x} \cdot \boldsymbol{y} = x_1 y_1 + \cdots + x_n y_n; \tag{9.1.3}$$

the concept of *length* uses the notion of **norm** on $\mathbb{R}^n$ given by

$$\|x\| := \sqrt{\boldsymbol{x} \cdot \boldsymbol{x}}; \tag{9.1.4}$$

and, the *distance* between $\boldsymbol{x}$ and $\boldsymbol{y}$ uses notion of **Euclidean metric** given by

$$d_2(\boldsymbol{x}, \boldsymbol{y}) = \|x - y\| := +\sqrt{(x_1 - y_1)^2 + \cdots + (x_n - y_n)^2}, \tag{9.1.5}$$

The following properties of the *dot product* on $\mathbb{R}^n$ are well-known for the cases $n = 2$ and $n = 3$: For all $\boldsymbol{x}, \boldsymbol{x}', \boldsymbol{y} \in \mathbb{R}^n$ and scalar $\lambda \in \mathbb{R}$, we have

(e) (positive) $\boldsymbol{x} \cdot \boldsymbol{x} \geq 0$;
(f) (definiteness) $\boldsymbol{x} \cdot \boldsymbol{x} = 0 \Leftrightarrow \boldsymbol{x} = \boldsymbol{0}$;
(g) (symmetry) $\boldsymbol{x} \cdot \boldsymbol{y} = \boldsymbol{x} \cdot \boldsymbol{y}$;
(h) (linearity) $(\lambda \boldsymbol{x} + \mu \boldsymbol{x}') \cdot \boldsymbol{y} = \lambda(\boldsymbol{x} \cdot \boldsymbol{y}) + \mu(\boldsymbol{x}' \cdot \boldsymbol{y})$.

**Definition 9.1 (Standard Euclidean Space)** The vector space $\mathbb{R}^n$ with the inner product given by (9.1.3) is called the **standard Euclidean space**.

The Euclidean distance of $\boldsymbol{x} \in \mathbb{R}^n$ from the origin $\mathbf{0}$ is called the *magnitude* of a vector $\boldsymbol{x}$. Notice that we have $\|\boldsymbol{x}\| = d_2(\boldsymbol{x}, \mathbf{0})$, which is the *norm* of the vector $\boldsymbol{x}$, as defined in (9.1.4). We also call $\|\boldsymbol{x}\|$ the **length** of the vector $\boldsymbol{x}$. It satisfies the following properties:

(i) (*positiveness*) $\|\boldsymbol{x}\| \geq 0$; and, $\|\boldsymbol{x}\| = 0 \Leftrightarrow \boldsymbol{x} = \mathbf{0}$.
(j) (*absolute homogeneity*) $\|\lambda\, \boldsymbol{x}\| = |\lambda|\, \|\boldsymbol{x}\|$.
(k) (*triangle inequality*) $\|\boldsymbol{x} + \boldsymbol{y}\| \leq \|\boldsymbol{x}\| + \|\boldsymbol{y}\|$.

In particular, a vector $\boldsymbol{x} \in \mathbb{R}^n$ is called a *unit vector* if $\|\boldsymbol{x}\| = 1$. More generally, for any $\mathbf{0} \neq \boldsymbol{x} \in \mathbb{R}^n$, the *unit vector* $\widehat{\boldsymbol{x}} = \boldsymbol{x}/\|\boldsymbol{x}\|$ is called the *normalized unit vector* in the direction of the vector $\boldsymbol{x}$. The set of unit vectors in $\mathbb{R}^n$ is called the *unit sphere* of dimension $(n-1)$. It is denoted by $\mathbb{S}^{n-1}$. So, we have

$$\mathbb{S}^{n-1} := \{\boldsymbol{x} \in \mathbb{R}^n \mid \|\boldsymbol{x}\| = 1\}, \qquad \text{for } n \geq 1. \tag{9.1.6}$$

Notice that $\mathbb{S}^0 \subset \mathbb{R}$ is the discrete set $\{-1, 1\}$; $\mathbb{S}^1 \subset \mathbb{R}^2$ is the usual unit circle in the *plane*, which we also view a boundary of the unit disk in $\mathbb{C}$; and, $\mathbb{S}^2 \subset \mathbb{R}^3$ is the usual unit sphere in the *space*. The next theorem, known as the *Cauchy-Schwartz inequality*, proves an important property of the elements of the unit sphere $\mathbb{S}^{n-1}$. It also finds applications in numerous other situations.

**Theorem 9.1 (Cauchy-Schwarz Inequality)** *For any two vectors $\boldsymbol{x}, \boldsymbol{y} \in \mathbb{R}^n$,*

$$|\langle \boldsymbol{x}, \boldsymbol{y}\rangle| \leq \|\boldsymbol{x}\|\, \|\boldsymbol{y}\|, \tag{9.1.7}$$

*with equality if and only if the two vectors differ by a scalar in $\mathbb{R}$.*

***Proof.*** Since there is nothing to prove if $\boldsymbol{y} = 0$, we may assume that $\boldsymbol{y} \neq \mathbf{0}$. Now, for $\lambda \in \mathbb{R}$, we have

$$\begin{aligned} 0 &\leq \langle \boldsymbol{x} - \lambda \boldsymbol{y}, \boldsymbol{x} - \lambda \boldsymbol{y}\rangle \\ &= \langle \boldsymbol{x}, \boldsymbol{x}\rangle - \lambda\langle \boldsymbol{y}, \boldsymbol{x}\rangle - \lambda\langle \boldsymbol{x}, \boldsymbol{y}\rangle + \lambda^2\langle \boldsymbol{y}, \boldsymbol{y}\rangle \\ &= \|\boldsymbol{x}\|^2 - 2\lambda\langle \boldsymbol{x}, \boldsymbol{y}\rangle + \lambda^2\|\boldsymbol{y}\|^2 \end{aligned}$$

As $\|\boldsymbol{y}\| \neq 0$, we take $\lambda = \langle \boldsymbol{x}, \boldsymbol{y}\rangle/\|\boldsymbol{y}\|^2$ so that we have

$$\begin{aligned} 0 &\leq \|\boldsymbol{x}\|^2 - 2\lambda\langle \boldsymbol{x}, \boldsymbol{y}\rangle + \lambda^2\|\boldsymbol{y}\|^2 \\ &= \|\boldsymbol{x}\|^2 - 2(\langle \boldsymbol{x}, \boldsymbol{y}\rangle/\|\boldsymbol{y}\|^2)\langle \boldsymbol{x}, \boldsymbol{y}\rangle + (\langle \boldsymbol{x}, \boldsymbol{y}\rangle/\|\boldsymbol{y}\|^2)^2\|\boldsymbol{y}\|^2 \\ &= \|\boldsymbol{x}\|^2\|\boldsymbol{y}\|^2 - (\langle \boldsymbol{x}, \boldsymbol{y}\rangle)^2, \end{aligned}$$

where the last equality is obtained by multiplying throughout with $\|y\|^2$. To complete proof of the first part, take square roots on both sides of the inequality

$$(\langle x, y\rangle)^2 \leq \|x\|^2 \|y\|^2.$$

The proof of the second part is left for the reader as an exercise. □

The inequality (9.1.7) motivates the concepts introduced in the next definition.

**Definition 9.2 (Angle)** We say that the **angle** between two *nonzero vectors* $x, y \in \mathbb{R}^n$ is a unique number $\varphi \in [0, \pi]$ if the following relation holds (Fig. 9.2):

$$\cos\varphi = \frac{\langle x, y\rangle}{\|x\| \, \|y\|}. \tag{9.1.8}$$

Also, the **projection of $y$ onto $x$** is the vector

$$\text{proj}_x(y) = \left\langle \frac{x}{\|x\|}, \frac{y}{\|y\|} \right\rangle x, \qquad \text{for} \quad x \neq 0. \tag{9.1.9}$$

We may recall that two vectors $x$ and $y$ in $\mathbb{R}^2$ or $\mathbb{R}^3$ are *perpendicular* if and only if $\cos\varphi = 0$; or, equivalently, $\text{proj}_x(y) = 0$.

**Definition 9.3 (Orthogonality)** Two vectors $x, y \in \mathbb{R}^n$ are said to be *orthogonal* if we have $\langle x, y\rangle = 0$. In this case, we write $x \perp y$.

Since the inner product on $\mathbb{R}^n$ is given by the usual dot product, we can easily construct orthogonal vectors in the Euclidean space $\mathbb{R}^n$. For example, the vectors $x = (1, 1)$ and $y = (-1, 1)$ in $\mathbb{R}^2$ are orthogonal, because their dot product

$$x \cdot y = (1, 1) \cdot (-1, 1) = -1 + 1 = 0.$$

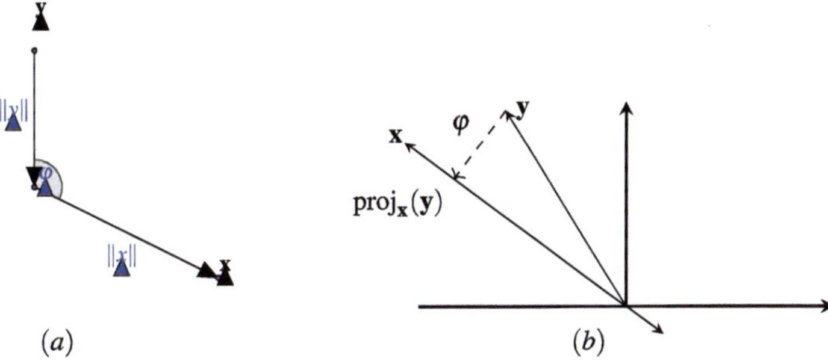


**Fig. 9.2** Angle and projection between vectors

More generally, for a set $U \subseteq \mathbb{R}^n$, the *orthogonal complement* $U^\perp$ of $U$ is the set

$$U^\perp := \{\boldsymbol{x} \in \mathbb{R}^n \mid \langle \boldsymbol{u}, \boldsymbol{x} \rangle = 0, \text{ for all } \boldsymbol{u} \in U\} \tag{9.1.10}$$

Further, the inequality (9.1.7) applied to the identity

$$\|\boldsymbol{x} + \boldsymbol{y}\|^2 = \|\boldsymbol{x}\|^2 + 2\langle \boldsymbol{x}, \boldsymbol{y} \rangle + \|\boldsymbol{y}\|^2, \tag{9.1.11}$$

proves *triangle inequality* for the norm. The identity (9.1.11) also proves

$$\|\boldsymbol{x} + \boldsymbol{y}\|^2 = \|\boldsymbol{x}\|^2 + \|\boldsymbol{y}\|^2 \quad \Leftrightarrow \quad \langle \boldsymbol{x}, \boldsymbol{y} \rangle = 0, \tag{9.1.12}$$

which is a far-reaching generalization of the well-known *Pythagoras theorem*. We will use the next identity repeatedly in the sequel.

$$\|\boldsymbol{x} + \boldsymbol{y}\|^2 - \|\boldsymbol{x} - \boldsymbol{y}\|^2 = 2\|\boldsymbol{x}\|^2 + 2\|\boldsymbol{y}\|^2, \text{ for all } \boldsymbol{x}, \boldsymbol{y} \in \mathbb{R}^n. \tag{9.1.13}$$

This is known as the *parallelogram Law*. A real vector space with a norm wherein (9.1.13) holds is necessarily an inner product space with respect to the inner product given by

$$\langle \boldsymbol{x}, \boldsymbol{y} \rangle = \frac{1}{4}\left(\|\boldsymbol{x} + \boldsymbol{y}\|^2 - \|\boldsymbol{x} - \boldsymbol{y}\|^2\right), \text{ for all } \boldsymbol{x}, \boldsymbol{y} \in \mathbb{R}^n. \tag{9.1.14}$$

This is known as the *polarization identity*.

**Definition 9.4** A **metric space** is a nonempty set $X$ together with a function $d : X \times X \to \mathbb{R}$ such that the following three properties hold, for all $x, y \in X$:

(l) (*positive-definiteness*) $d(x, y) \geq 0$; and, $d(x, y) = 0 \Leftrightarrow x = y$.
(m) (*symmetry*) $d(x, y) = d(y, x)$.
(n) (*triangle inequality*) $d(x, y) \leq d(x, x') + d(x', y)$, for every $x' \in X$.

The function $d$ is called a **metric** on $X$. We write a *metric space* as $(X, d)$.

It is easy to see that the *Euclidean metric* $d_2$ given by (9.1.5) satisfies the above three conditions. Therefore, $(\mathbb{R}^n, d_2)$ is a *metric space*. Notice that we can also view the norm $\|\boldsymbol{x}\|$ of a vector $\boldsymbol{x} \in \mathbb{R}^n$ as the distance of $\boldsymbol{x}$ from the origin $\boldsymbol{0}$, i.e., $\|\boldsymbol{x}\| := d_2(\boldsymbol{x}, \boldsymbol{0})$. In fact, the set $\mathbb{R}^n$ is also a metric space with respect to many other metrics. For example, there is a class of metrics first introduced by the German mathematician *Herman Minkowski* (1864–1909). These are known as **Minkowski's p-metrics**.

**Definition 9.5 (Minkowski Metrics)** For $p > 0$, and vectors $\boldsymbol{x},\ \boldsymbol{y} \in \mathbb{R}^n$ as in (9.1.1), the function $d_p : \mathbb{R}^n \times \mathbb{R}^n \to \mathbb{R}$ given by

$$d_p\big(\boldsymbol{x},\ \boldsymbol{y}\big) := \left(\sum_{k=1}^{n} |x_k - y_k|^p\right)^{1/p} \tag{9.1.15}$$

is a metric on $\mathbb{R}^n$. We also have

$$d_\infty\big(\boldsymbol{x},\ \boldsymbol{y}\big) := \max\big\{|x_1 - y_1|,\ \ldots\ ,|x_n - y_n|\big\} \tag{9.1.16}$$

is a metric on $\mathbb{R}^n$, which is called the *Chebyshev metric*.

Of course, the Minkowski 2-metric is the usual Euclidean metric $d_2$. The *Minkowski metric* $d_1$ is also called the **Manhattan metric** (or the **taxicab metric**). The supervised machine learning technique, known as the $K$-nearest neighbors algorithm (or simply KNN algorithm), uses the *Manhattan metric* $d_1$ or the Euclidean metric $d_2$. KNN algorithm applies to the classification and regression problems, where $K$ refers to the number of nearest neighbors. It is also known as the *lazy learning algorithm*. Furthermore, the metric $\delta$ on the plane $\mathbb{R}^2$ given by

$$\delta\big(\boldsymbol{x}, \boldsymbol{y}\big) = \begin{cases} d_2(\boldsymbol{x}, \boldsymbol{y}), & \text{if } \boldsymbol{x} \text{ and } \boldsymbol{y} \text{ are collinear} \\ d_2(\boldsymbol{x}, \boldsymbol{0}) + d_2(\boldsymbol{y}, \boldsymbol{0}), & \text{if } \boldsymbol{x} \text{ and } \boldsymbol{y} \text{ are not collinear} \end{cases} \tag{9.1.17}$$

is called the **railway metric**. To better understanding the practical aspect of the *railway metric*, suppose the origin $\boldsymbol{0} \in \mathbb{R}^2$ represents the *start point* for all the local trains commuting passengers in a city. To move from a station $\boldsymbol{x}$ to a station $\boldsymbol{y}$, one first takes a train from $\boldsymbol{x}$ to $\boldsymbol{0}$ and then a train from $\boldsymbol{0}$ to $\boldsymbol{y}$, unless the two stations $\boldsymbol{x}$ and $\boldsymbol{y}$ lie on the same line, in which case one would take a direct train. In some cases such as text analytics, we also use *cosine metric*, given in terms of cosine of the angle between the vectors $\boldsymbol{x}$ and $\boldsymbol{y}$ (see Eq. (9.1.8)).

**Remark 9.1** Since the idea of *metric* provides a framework to discuss "nearness" of points in an abstract metric space, it is possible to use various forms of "convergences" of infinite sums. In this latter context, since the following inequalities hold over $\mathbb{R}^n$:

$$d_\infty\big(\boldsymbol{x},\ \boldsymbol{y}\big) \le d_2\big(\boldsymbol{x},\ \boldsymbol{y}\big) \le d_1\big(\boldsymbol{x},\ \boldsymbol{y}\big) \le 2\, d_\infty\big(\boldsymbol{x},\ \boldsymbol{y}\big),$$

it doesn't matter which particular Minkowski metric $d_1$, $d_2$, or $d_\infty$ we use.

## 9.2 Vector Spaces

In all that follows, a reference to a field $\mathbb{F}$ means the set $\mathbb{R}$ or the set $\mathbb{C}$. By a **vector space** over a field $\mathbb{F}$ we mean a nonempty set $V$, together with two operations

$$+ : V \times V \to V \quad \text{and} \quad \cdot : \mathbb{F} \times V \to V,$$

respectively called **vector addition** and **scalar multiplication**, such that $V$ is an Abelian group with respect to the *vector addition* $+$, and the *scalar multiplication* given by $(\lambda, v) \mapsto \lambda \cdot v$, for $\lambda \in \mathbb{F}$ and $v \in V$, is compatible with the vector addition in the sense that the following axioms hold, for all $\lambda, \mu \in \mathbb{F}$ and $v_1, v_2, v \in V$:

$$\begin{aligned} \lambda \cdot (v_1 + v_2) &= \lambda \cdot v_1 + \lambda \cdot v_2; \\ (\lambda + \mu) \cdot u &= \lambda \cdot u + \mu \cdot u; \\ (\lambda \mu) \cdot u &= \lambda \cdot (\mu \cdot u); \\ 1 \cdot v &= v. \end{aligned} \tag{9.2.1}$$

We may also write a *vector space* $V$ over a field $\mathbb{F}$ as $V_{\mathbb{F}}$, especially when the field $\mathbb{F}$ needs emphasis. In general, elements of a vector space are called *vectors*, and elements of a field $\mathbb{F}$ are called *scalars*. The *zero element* $0 \in V$ is called the **zero vector**. Clearly, $V = \{0\}$ is a vector space over every field, with obvious addition and scalar multiplication, which we call the **trivial** vector space. A vector space over the field $\mathbb{R}$ is called a **real vector space**; and, a vector space over the field $\mathbb{C}$ is called a **complex vector space**. The next example gives some simple illustrations of *non-trivial* vector spaces.

**Example 9.2** Each field $\mathbb{F}$ is a vector space over its every subfield. In this case, the vector addition and scalar multiplication are given by the two binary operations on the field $\mathbb{F}$. In particular, the field $\mathbb{C}$ is a real vector space, and the field $\mathbb{R}$ is a vector space over $\mathbb{Q}$. As shown later, these two vector spaces are very different structurally. Next, consider the set $P_n[x]$ of polynomials of degree at most $n$ with coefficients in a field $\mathbb{F}$. We know that $P_n[x]$ is an Abelian group with respect to "coefficient-wise addition". For $p(x) = a_0 + a_1 x + \cdots + a_n x^n \in P_n[x]$ and $\lambda \in \mathbb{F}$, we define the *pointwise scalar multiplication* as

$$\lambda \cdot p(x) = (\lambda a_0) + (\lambda a_1)x + \cdots + (\lambda a_n)x^n. \tag{9.2.2}$$

We can see that this definition satisfies the four *vector space axioms* as specified in (9.2.1). Therefore, $P_n[x]$ is a vector space over the field $\mathbb{F}$. More generally, with respect to pointwise addition and scalar multiplication, the set of functions $f : [a, b] \to \mathbb{F}$ is a vector space over the field $\mathbb{F}$. Also, the Abelian group $M_{m \times n}(\mathbb{F})$ of $m \times n$ matrices is a vector space over the field $\mathbb{F}$ with respect to the scalar multiplication given by

$$\lambda \cdot A = \big(\lambda\, a_{ij}\big), \quad \text{for } \lambda \in \mathbb{F} \text{ and } A = \big(a_{ij}\big) \in M_{m\times n}(\mathbb{F}). \tag{9.2.3}$$

In particular, the set of square matrices $M_n(\mathbb{F})$ of order $n$ is a vector space over the field $\mathbb{F}$. We may recall that the set of positive real numbers $V = \mathbb{R}_{>0}$ is an Abelian group with respect to the "sum" given by $u \oplus v = u\, v$, for $u, v \in V$, where $1 \in \mathbb{R}$ is the additive identity for the operation $\oplus$; and, $1/a$ is the additive inverse of an element $a \in V$. Define the scalar multiplication on $V$ by

$$\lambda \cdot u = u^{\lambda}, \quad \text{for } u \in V \text{ and } \lambda \in \mathbb{R}. \tag{9.2.4}$$

We can see that the above definition satisfies the four axioms given in (9.2.1). Therefore, $V = \mathbb{R}_{>0}$ is a real vector space.

**Example 9.3 (Scalar Vector Space)** Let $\mathbb{F}$ a real or complex field. We generalize the fundamental concepts introduced earlier in Sect. 9.1. The element $\boldsymbol{x} = \big(x_1, \ldots, x_n\big) \in \mathbb{F}^n$ is called an $n$-tuple over the field $\mathbb{F}$. Let $V$ be the set of $n$-tuples, and let $\boldsymbol{u}, \boldsymbol{v} \in V$ be given by

$$\boldsymbol{u} = \big(u_1, \ldots, u_n\big) \quad \text{and} \quad \boldsymbol{v} = \big(v_1, \ldots, v_n\big).$$

With respect to *pointwise* vector addition and scalar multiplication given by

$$\begin{aligned} \big(u_1, \ldots, u_n\big) + \big(v_1, \ldots, v_n\big) &= (u_1 + v_1, \ldots, u_n + v_n); \\ a \cdot (u_1, \ldots, u_n) &= \big(a\, u_1, \ldots, a\, u_n\big), \end{aligned} \tag{9.2.5}$$

the set $V$ is easily seen to be a vector space over $\mathbb{F}$, where $u_i,\ v_i, a \in \mathbb{F}$. In particular, when $\lambda = -1$, the *negative vector* of $\boldsymbol{u}$ is the scalar multiple $(-1)\boldsymbol{u}$ given by

$$(-1)\boldsymbol{u} = \big(-u_1, -u_2, -u_3, \ldots, -u_n\big).$$

We also have $\boldsymbol{u} - \boldsymbol{v} = \boldsymbol{u} + (-\boldsymbol{v})$ and $\boldsymbol{u} - \boldsymbol{u} = \boldsymbol{0}$, where

$$\boldsymbol{0} = \big(0, 0, \ldots, 0\big) \in \mathbb{F}^n$$

is the *zero vector* of the space $\mathbb{F}^n$. This vector space is very significant in various geometry applications. We may recall that, in the special case when $\mathbb{F} = \mathbb{R}$, ordered pair $(x, y) \in \mathbb{R}^2$ is a usual 2-vector; and, triple $(x, y, z) \in \mathbb{R}^3$ is a usual 3-vector. In general, we may call the elements in $\mathbb{F}^n$ as the $n$-dimensional vectors over the field $\mathbb{F}$. In many situations, we may view $\boldsymbol{v} = \big(v_1, \ldots, v_n\big) \in \mathbb{R}^n$ as a *column matrix*.

**Example 9.4** In this next illustration, we give some *non-examples*. First, consider the Abelian group $V = \mathbb{C}$ with respect to usual addition. It is not a vector space with respect to the scalar multiplication given by

$$a \cdot z = x + i(a\, y), \quad \text{for } a \in \mathbb{R} \text{ and } z = x + iy \in \mathbb{C},$$

because for $z = x + iy$ with $x \neq 0$ we have

$$\big(a+b\big)\cdot z = x + i\big(a\,y + b\,y\big) = x + i(a\,y) + i(b\,y) \neq a\cdot z + b\cdot z.$$

This shows that the second condition in (9.2.1) fails. Next, the Abelian group $V = \mathbb{R}^2$ with respect to pointwise addition is not a vector space with respect to the scalar multiplication given by

$$a\cdot(x, y) = (a^2x, y), \quad \text{for } a \in \mathbb{R} \text{ and } \quad (x, y) \in \mathbb{R}^2,$$

because for $z = (1, 0) \in V$ and $a, b \in \mathbb{R}\,(b \neq 0)$ we have

$$\big(a+b\big)\cdot z = (a^2, 0) \neq (a^2, 0) + (b^2, 0) = a\cdot z + b\cdot z.$$

We may also take $a = 1, b = -1$, and $(0, 0) \neq (x, y) = z \in \mathbb{R}^2$, to draw the same conclusion. This shows that the second condition in (9.2.1) fails. Also, the set of nonzero real polynomials is not an Abelian group with respect to the usual addition and scalar multiplication given by (9.2.2) because it does not contain the zero polynomial. So, it cannot be a vector space. Notice that the set of invertible matrices in $V$ is not a vector space, because the sum of two invertible matrices need not be invertible.

## *Subspaces*

Let $V$ be a vector space over a field $\mathbb{F}$, and $W$ be a nonempty subset of $V$. Given two vectors $u, v \in W$ and a scalar $\lambda \in \mathbb{F}$, we know that the sum $u + v \in V$, and so is the scalar product $\lambda\,u$. We ask: *Does $u + v$ and $\lambda\,u$ belong to $W$, for every $u, v \in W$ and $\lambda \in \mathbb{F}$?* The answer to the question is "NO", in general. For a simple illustration, consider the real vector space $V = \mathbb{R}^2$, and let

$$W_1 = \big\{(x_1, 1) \in V \mid x_1 \in \mathbb{R}\big\},$$

Now, for $u = (x_1, 1), v = (y_1, 1) \in W_1$, the sum $u + v = (x_1 + y_1, 2) \notin W_1$. So, the sum of two vectors in $W_1$ may not be a vector in $W_1$, in general. However, in certain cases, both *subspace condition* (9.2.6) may hold. For, let

$$W_2 = \big\{(x_1, 0) \in V \mid x_1 \in \mathbb{R}\big\}.$$

In this case, for any $u = (x_1, 0), v = (y_1, 0)$, both the sum $u + v = (x_1 + y_1, 0) \in W_2$ and the scalar product $\lambda\,u = (\lambda x_1, 0)$ belong to $W_2$. It is easy to verify that the set $W_2$ also satisfies the other properties of a vector space. Therefore, $W_2 \subset V$ is a vector space over the field $\mathbb{F}$ on its own.

**Definition 9.6** Let $V$ be a vector space over a field $\mathbb{F}$. A nonempty set $W \subseteq V$ is called a **subspace** of $V$ if $W$ itself is a vector space over the field $\mathbb{F}$ with respect to the induced vector addition and scalar multiplication.

It follows easily that $W \subseteq V$ is a *subspace* if and only if we have $u + v \in W$ and $\lambda\, u \in W$, for all $u, v \in W$ and $\lambda \in \mathbb{F}$. It thus follows that $\lambda\, u + \mu\, v \in W$, for all $u, v \in W$ and $\lambda,\ \mu \in \mathbb{F}$. Conversely, taking $\lambda = \mu = 1$, we obtain $u + v \in W$; and, taking $\mu = 0$, we obtain $\lambda\, u \in W$. Therefore, $W$ is a subspace of $V$ if and only if

$$\lambda\, u + \mu\, v \in W, \quad \text{for all } u, v \in W \text{ and } \lambda,\ \mu \in \mathbb{F}. \tag{9.2.6}$$

Clearly, $\{0\}$ and $V$ are subspaces of $V$. Any other subspace of $V$ is called a **proper subspace**. To avoid trivialities, a subspace would always mean a *proper subspace*. In particular, consider the subset $W_1 = \{(x, y, 0) \mid x, y \in \mathbb{R}\}$ of the real vector space $V = \mathbb{R}^3$. For $\boldsymbol{u} = (x_1, y_1, 0)$ and $\boldsymbol{v} = (x_2 y_2, 0) \in W_1$ and $\lambda,\ \mu \in \mathbb{R}$, we have

$$\begin{aligned}\lambda\, \boldsymbol{u} + \mu\, \boldsymbol{v} &= \lambda(x_1, y_1, 0) + \mu(x_2, y_2, 0)\\ &= (\lambda x_1 + \mu x_2, \lambda y_1 + \mu y_2, 0) \in W_1.\end{aligned}$$

Therefore, by (9.2.6), $W_1$ is a subspace of $V$. Also, the set

$$W_2 = \big\{(x, y, ax + by) \mid x, y \in \mathbb{R}\big\}$$

is a subspace of $V$. For, let $\boldsymbol{u} = (x_1, y_1, ax_1 + by_1)$, $\boldsymbol{v} = (x_2 y_2, ax_2 + by_2) \in W_2$ and $\lambda,\ \mu \in \mathbb{R}$. Then we have

$$\begin{aligned}\lambda\, \boldsymbol{u} + \mu\, \boldsymbol{v} &= \lambda(x_1, y_1, ax_1 + by_1) + \mu(x_2, y_2, ax_2 + by_2)\\ &= (\lambda x_1 + \mu x_2, \lambda y_1 + \mu y_2, a(\lambda x_1 + \mu x_2) + b(\lambda y_1 + \mu y_2)),\end{aligned}$$

which is again an element of $W_2$. Notice that we just proved that the set of points on a line passing through the origin is a subspace of $\mathbb{R}^3$.

**Example 9.5** Consider the vector space $V = M_n(\mathbb{F})$ of all square matrices of order $n$ over a field $\mathbb{F}$. Its subset $W$ consisting of symmetric matrices is a subspace, because the sum of two symmetric matrices is symmetric, and also a scalar multiple of a symmetric matrix is symmetric. Similarly, the subsets of $V$ consisting of the lower triangular matrices, the upper triangular matrices, and the diagonal matrices, are all subspaces of $V$. However, as said earlier, the subset $GL(n, \mathbb{F})$ of $V$ consisting of invertible matrices is not a subspace, because the sum of two invertible matrices need not be invertible. Also, if $V$ is the vector space of functions $f : [a, b] \to \mathbb{R}$, then the subset $W := C[a, b]$ consisting of continuous real-valued functions defined on the interval $[a, b]$ is a subspace of $V$, because $W$ is closed with respect to pointwise addition and scalar multiplication of functions in $V$.

**Theorem 9.2** *For $A \in M_n(\mathbb{F})$, consider the homogeneous linear system*

$$A\boldsymbol{x} = \boldsymbol{0}, \quad \textit{where} \quad \boldsymbol{x} \in \mathbb{R}^n. \tag{9.2.7}$$

*Then the set of all of the solution of* (9.2.7) *is a subspace of* $\mathbb{R}^n$.

***Proof.*** Suppose $\boldsymbol{x}, \boldsymbol{x}'$ are two solutions of (9.2.7). That is, we have

$$A\boldsymbol{x} = \boldsymbol{0} \quad \text{and} \quad A\boldsymbol{x}' = \boldsymbol{0}.$$

Now, for any $\lambda, \mu \in \mathbb{F}$, we have

$$\begin{aligned} A\big(\lambda\boldsymbol{x} + \mu\boldsymbol{x}'\big) &= \lambda A\boldsymbol{x} + \mu A\boldsymbol{x}' \\ &= \lambda\boldsymbol{0} + \mu\boldsymbol{0} = \boldsymbol{0}, \end{aligned}$$

which shows that $\lambda\boldsymbol{x} + \mu\boldsymbol{x}'$ is also a solution of (9.2.7). It thus follows from (9.2.6) that the claim holds. □

The vector space of all solutions of a homogeneous linear system $A\boldsymbol{x} = \boldsymbol{0}$ is called the *null space* of the matrix $A$. In particular, since $x = -2t,\ y = t,\ z = t$ are the solutions of the linear system

$$x + y + z = 0, \quad y - z = 0,$$

it follows that $W = \big\{(-2t, t, t) \mid t \in \mathbb{R}\}$ is a subspace of the vector space $V = \mathbb{R}^3$. We can also verify this directly. For, let $\boldsymbol{u} = (-2t, t, t),\ \boldsymbol{v} = (-2s, s, s)$, and $\lambda, \mu$ be scalars. Since

$$\lambda\mathbf{u} + \mu\mathbf{v} = (-2(\lambda t + \mu s), \lambda t + \mu s, \lambda t + \mu s)$$

belongs to $W$, it is a subspace.

**Definition 9.7** Let $V$ be a vector space over a field $\mathbb{F}$. For subspaces $U, W \subseteq V$, let

$$U + W := \big\{u + w \mid u \in U \text{ and } w \in W\big\}. \tag{9.2.8}$$

We call $U + W$ the **sum** of subspaces $U$ and $W$ of $V$. Furthermore, we call (9.2.6) a **direct sum** if $U \cap W = \{0\}$. We then write the sum as $U \oplus W$.

The *sum* of subspaces $U$ and $W$ defined by (9.2.8) is easily seen to be a subspace of $V$ (use (9.2.6)). For example, if

$$U = \big\{(x, 0, 0) \mid x \in \mathbb{R}\big\} \quad \text{and} \quad W = \big\{(0, y, 0) \mid y \in \mathbb{R}\big\}$$

then we have

$$U + W = \{(x, y, 0) \mid x,\ y \in \mathbb{R}\}.$$

For $Z = \{(y, y, 0) \mid y \in \mathbb{R}\}$, we have $U + W = U + Z$. Also, the real vector space $M_n(\mathbb{R})$ is a *direct sum* of the two subspaces consisting of symmetric and skew-symmetric matrices. Notice that the concept introduced in Definition 9.7 can be extended to any finite number of subspaces.

**Definition 9.8** Let $V$ be a vector space over a field $\mathbb{F}$. We say $V$ is a **direct sum** of subspaces $U_1, \ldots, U_m$ of $V$ if

$$V := U_1 + \cdots + U_m, \tag{9.2.9}$$

and we have

$$U_j \cap (U_1 + \cdots + \widehat{U}_j + \cdots + U_m) = \{0\}, \quad \text{for all} \quad j = 1, \ldots, m, \tag{9.2.10}$$

where $\widehat{U}_j$ means the subspace $U_j$ is "ejected" from the sum. In this case, we write

$$V := U_1 \oplus \cdots \oplus U_m, \tag{9.2.11}$$

We call the right side of (9.2.9) the **sum** of subspaces $U_1, \ldots, U_m$.

## 9.3 Basis and Dimension

In the majority of cases, we need linear algebra techniques to solve a *system of linear equations* (or simply SLEs). A *linear equation* in unknowns $x_1, \ldots, x_n$ over a field $\mathbb{F}$ is an expression of the form

$$a_{i1}\, x_1 + a_{i2}\, x_2 + \cdots + a_{in}\, x_n = b_i, \quad \text{with } a_{ij} \in \mathbb{F}, \text{ for } 1 \le i \le m. \tag{9.3.1}$$

where $x_j$'s are usually *numbers, vectors, matrices, sequences*, or even *functions*. More generally, expressions involving "finite sums of scalar multiples of vectors" such as on the left side of (9.3.1) are of fundamental importance in many applications. For example, every vector $\boldsymbol{v} = (a_1, a_2, a_3) \in \mathbb{R}^3$ can be written as

$$\mathbf{v} = a_1\mathbf{i} + a_2\mathbf{j} + a_3\mathbf{k},$$

where **i**, **j**, and **k**, are the unit vectors along $x$, $y$, and $z$ axes of the 3-dimensional Euclidean space $\mathbb{R}^3$. Also, by replacing $x_j$'s on the left side of (9.3.1) by formal power $x^j$, the resulting expression is a vector in the vector space $P_n[x]$. We may also recall that the "general solution" of an ordinary LDE with constant coefficients of order $n$ is an expression of the form (9.3.1), where $x_i$'s are *fundamental solutions* of the associated homogeneous LDE, and $b_i$ is a *particular solution*.

In general, as we discover shortly, the concept of *linear combination* of vectors introduced in the next definition is fundamental to various types of applications in the majority of engineering branches, including computer science. For example, in data science, every training data instance can be written as a linear combination.

**Definition 9.9** Let $V$ be a vector space over a field $\mathbb{F}$, and let $v \in V$. We say $v$ is a **linear combination** of vectors $v_1, v_2, \ldots, v_k \in V$ if we have

$$v = \lambda_1 v_1 + \lambda_2 v_2 + \cdots + \lambda_k v_k, \tag{9.3.2}$$

for some scalars $\lambda_1, \lambda_2, \ldots, \lambda_k \in \mathbb{F}$.

For example, with respect to the standard *unit vectors* $\boldsymbol{e}_1, \boldsymbol{e}_2, \ldots, \boldsymbol{e}_n$ along the $n$-axis of the Euclidean space $\mathbb{R}^n$ given by

$$\boldsymbol{e}_1 = \big(1, \ldots, 0\big), \ \ \boldsymbol{e}_2 = \big(0, 1, 0 \ldots, 0\big), \ \ldots \ , \ \boldsymbol{e}_n = (0, \ldots, 0, 1), \tag{9.3.3}$$

every vector $\boldsymbol{v} = \big(v_1, \ldots, v_n\big) \in \mathbb{R}^n$ can be expressed uniquely as

$$\boldsymbol{v} = v_1 \boldsymbol{e}_1 + v_2 \boldsymbol{e}_2 + \cdots + v_n \boldsymbol{e}_n. \tag{9.3.4}$$

Also, for the vectors $\boldsymbol{u} = (1, 0, 0)$, $\boldsymbol{v} = (1, 1, 0)$, and $\boldsymbol{w} = (1, 1, 1)$ in the vector space $V = \mathbb{R}^3$, if $\boldsymbol{x} = (v_1, v_2, v_3) \in \mathbb{R}^3$ is an arbitrary vector then

$$\lambda_1 + \lambda_2 + \lambda_3 = v_1, \lambda_2 + \lambda_3 = v_2, \lambda_3 = v_3$$

implies that

$$\lambda_1 \boldsymbol{u} + \lambda_2 \boldsymbol{v} + \lambda_3 \boldsymbol{w} = (\lambda_1 + \lambda_2 + \lambda_3, \lambda_2 + \lambda_3, \lambda_3) = \lambda_1 \boldsymbol{u} + \lambda_2 \boldsymbol{v} + \lambda_3 \boldsymbol{w}$$

Or, equivalently, if we have

$$\lambda_1 = v_1 - v_2, \quad \lambda_2 = v_2 - v_3, \lambda_3 = v_3,$$

then we have

$$\boldsymbol{x} = (v_1 - v_2)\boldsymbol{u} + (v_2 - v_3)\boldsymbol{v} + v_3 \boldsymbol{w}.$$

Next, consider the vectors $\boldsymbol{x} = (3, -1, 2)$ and $\boldsymbol{y} = (-1, 2, -1)$. Consider the vectors $\mathbf{u}$ and $\mathbf{v}w$. If $\boldsymbol{u} = (7, -4, 5)$ is a linear combination of $\boldsymbol{x}$ and $\boldsymbol{y}$, then for some scalars $\lambda_1$ and $\lambda_2$ we have $\boldsymbol{u} = \lambda_1 \boldsymbol{x} + \lambda_2 \boldsymbol{y}$. Since

$$\lambda_1 \boldsymbol{x} + \lambda_2 \boldsymbol{y} = (3\lambda_1 - \lambda_2, -\lambda_1 + 2\lambda_2, 2\lambda_1 - \lambda_2),$$

we must have

$$3\lambda_1 - \lambda_2 = 7, \quad -\lambda_1 + 2\lambda_2 = -4, \quad 2\lambda_1 - \lambda_2 = 5.$$

The solution of the above linear system is given by $\lambda_1 = 2$ and $\lambda_2 = -1$. We thus obtain $\boldsymbol{u} = 2\boldsymbol{x} - 2\boldsymbol{y}$. On the other hand, if $\boldsymbol{v} = (1, 3, 1) = \lambda_1 \boldsymbol{x} + \lambda_2 \boldsymbol{y}$, for some scalars $\lambda_1$ and $\lambda_2$, then the resulting linear system is

$$3\lambda_1 - \lambda_2 = 1, \quad -\lambda_1 + 2\lambda_2 = 3, \quad 2\lambda_1 - \lambda_2 = 1.$$

This system is inconsistent, and so $\boldsymbol{v}$ is not a linear combination of $\boldsymbol{x}$ and $\boldsymbol{y}$. We start our formal discussion with the next theorem.

**Theorem 9.3** *Let $V$ be a vector space over a field $\mathbb{F}$, and $v_1, v_2, \ldots, v_k \in V$. Let $W$ be the set of linear combinations of these $k$ vectors. Then $W$ is the smallest subspace of $V$ containing the $k$ vectors $v_1, v_2, \ldots, v_k$.*

***Proof.*** Since we can write $v_i \in W$ as

$$v_i = 0\,v_1 + \cdots + 0\,v_{i-1} + 1\,v_i + 0\,v_{i+1} + \cdots + 0\,v_k,$$

it follows that $W \neq \emptyset$. Next, for $u,\ v \in W$, we have

$$u = \lambda_1\,v_1 + \cdots + \lambda_k\,v_k \quad \text{and} \quad v = \mu_1\,v_1 + \cdots + \mu_k\,v_k,$$

for some scalars $\lambda_i, \mu_i$. Let $\lambda,\ \mu \in \mathbb{F}$. Then

$$\begin{aligned}\lambda\,u + \mu\,v &= \lambda\big(\lambda_1\,v_1 + \cdots + \lambda_k\,v_k\big) + \mu\big(\mu_1\,v_1 + \cdots + \mu_k\,v_k\big)\\ &= \big(\lambda\lambda_1 + \mu\mu_1\big)v_1 + \cdots + \big(\lambda\lambda_k + \mu\mu_k\big)v_k,\end{aligned}$$

implies that $\lambda\,u + \mu\,v$ is a linear combination of $v_1, v_2, \ldots, v_k \in V$, and so $\lambda\,u + \mu\,v \in W$. Therefore, $W$ is a subspace of $V$. To complete the proof, we show that any subspace $W'$ of $V$ containing $v_1, v_2, \ldots, v_k$ also contains the subspace $W$. For, recall that every linear combination of elements of a subspace lies in the subspace itself. In particular, since $W'$ is a subspace of $V$ and vectors $v_1, v_2, \ldots, v_k \in W'$, it follows that their linear combination also belongs to $W'$. Hence, $W \subset W'$. □

**Definition 9.10** Let $V$ be a vector space over a field $\mathbb{F}$, and $S = \{v_1, v_2, \ldots, v_k\} \subset V$. The subspace $W$ of the linear combinations of vectors in $S$ is called the **subspace spanned by** the set $S$ (or simply the **span** of $S$). We also say that $S$ spans $W$, and write $W := \mathrm{Span}(S)$ (or $W := \mathrm{Span}\{v_1 \ldots, v_k\}$.

The geometry of the above concept is that the *span* $\mathrm{Span}(S)$ consists of those vectors in $V$ that can be "potentially reached" using only the vectors in the set $S$ by *stretching, shrinking*, or by applying *rotations*, as represented by the *coefficient scalars*. The concept of *basis* introduced later would ensure that there is only one way to do so!

**Example 9.6** The set $\{1, x, x^2, \ldots, x^n\}$ spans the vector space of all polynomials of degree at most $n$ in the indeterminate $x$. Also, by (9.3.4), the vector space $\mathbb{R}^n$ is spanned by the standard unit vectors $\boldsymbol{e}_1, \boldsymbol{e}_2, \ldots, \boldsymbol{e}_n$ given by (9.3.3). In particular, consider $\mathbf{v}_1 = (1, 0, 0)$, $\mathbf{v}_2 = (0, 1, 1)$ and $\mathbf{v}_3 = (1, 0, 1)$ in $\mathbb{R}^3$. Then

$$\lambda_1\mathbf{v}_1 + \lambda_2\mathbf{v}_2 + \lambda_3\mathbf{v}_3 = \left(\lambda_1 + \lambda_3, \lambda_2, \lambda_2 + \lambda_3\right) = \left(v_1, v_2, v_3\right),$$

where $v_1 = \lambda_1 + \lambda_3$, $v_2 = \lambda_2$, $v_3 = \lambda_3 + \lambda_2$; or, equivalently,

$$\lambda_1 = v_1 + v_2 - v_3, \quad \lambda_2 = v_2, \quad \lambda_3 = v_3 - v_2.$$

It thus follows that every vector in $\mathbb{R}^3$ is a linear combination of vectors $\mathbf{v}_1$, $\mathbf{v}_2$, and $\mathbf{v}_3$. Therefore, we have

$$\text{Span}\left\{\mathbf{v}_1, \mathbf{v}_2, \mathbf{v}_3\right\} = \mathbb{R}^3. \tag{9.3.5}$$

Also, consider $\mathbf{v}_1 = (1, 0, 0)$, $\mathbf{v}_2 = (0, 1, 0)$ in $\mathbb{R}^3$. Then

$$\lambda_1\mathbf{v}_1 + \lambda_2\mathbf{v}_2 = (\lambda_1, \lambda_2, 0) \quad \Rightarrow \quad \text{Span}\{\mathbf{v}_1, \mathbf{v}_2\} = \{(x, y, 0) : x, y \in \mathbb{R}\}.$$

Therefore, in this case, $\text{Span}\{\mathbf{v}_1, \mathbf{v}_2\} \neq \mathbb{R}^3$. In fact, $\text{Span}\{\mathbf{v}_1, \mathbf{v}_2\}$ is a plane.

In general, for any $0 \neq v \in V$, $\text{Span}(\{v\}) = \{\lambda\, v \mid \lambda \in \mathbb{F}\}$ is called an **abstract line** in $V$ along the vector $v$. Also, for two non-collinear vectors $v_1, v_2 \in V$, i.e., $v_1 \neq \lambda\, v_2$, for any scalar $\lambda$, the span

$$\text{Span}\left(\{v_1, v_2\}\right) = \left\{\lambda\, v_1 + \mu\, v_2 \mid \lambda, \mu \in \mathbb{F}\right\}$$

is an **abstract plane**, containing vectors $v_1$ and $v_2$. A similar geometric interpretation applies to three non-planar vectors in $V$, and so on.

**Definition 9.11** For a matrix $A = \left(a_{ij}\right) \in M_{m\times n}(\mathbb{F})$, the $m$ vectors $\mathbf{r}_1, \mathbf{r}_2, \ldots, \mathbf{r}_m \in \mathbb{F}^n$ given by the rows of $A$ are called the **row vectors** of $A$. That is, we have

$$\begin{aligned}
\mathbf{r}_1 &= \left(a_{11}\; a_{12}\; a_{13}\; \cdots\; a_{1n}\right), \\
\mathbf{r}_2 &= \left(a_{21}\; a_{22}\; a_{23}\; \cdots\; a_{2n}\right), \\
&\;\;\vdots \\
\mathbf{r}_m &= \left(a_{m1}\; a_{m2}\; a_{m3}\; \cdots\; a_{mn}\right)
\end{aligned}$$

Similarly, the $n$ vectors $\mathbf{c}_1, \mathbf{c}_2, \ldots, \mathbf{c}_n \in \mathbb{F}^m$ given by the columns of $A$ are called the **column vectors** of the matrix $A$. That is, we have

$$\mathbf{c}_1 = \left(a_{11}\ a_{21}\ a_{31}\ \cdots\ a_{m1}\right)^t,$$
$$\mathbf{c}_2 = \left(a_{12}\ a_{22}\ a_{32}\ \cdots\ a_{m2}\right)^t,$$
$$\vdots$$
$$\mathbf{c}_n = \left(a_{1n}\ a_{2n}\ a_{3n}\ \cdots\ a_{mn}\right)^t,$$

where $^t$ denotes the *transpose operation.*

With reference to above definition, the vector space spanned by the row vectors of a matrix $A = \left(a_{ij}\right) \in M_{m\times n}(\mathbb{F})$ is called the **row space** of $A$, and the vector space spanned by the column vectors is the **column space** of $A$. Notice that both row space and the null space of $A$ are subspaces of $\mathbb{F}^n$, whereas the column space is a subspace of the vector space $\mathbb{F}^m$. In particular, for the matrix

$$A = \begin{pmatrix} 1 & 0 & 1 \\ 1 & 1 & 1 \end{pmatrix},$$

we have $\mathbf{r}_1 = (1\ 0\ 1)$ and $\mathbf{r}_2 = (1\ 1\ 1)$ are the row vectors of $A$ while $\mathbf{c}_1 = (1\ 1)^t$ and $\mathbf{c}_2 = (0\ 1)^t$ and $\mathbf{c}_3 = \mathbf{c}_1$ are the column vectors of $A$. The subspace

$$R = \{\lambda\mathbf{r}_1 + \mu\mathbf{r}_2 : \lambda, \mu \in \mathbb{R}\}$$

is the row space of $A$. The subspace

$$C = \{\lambda\mathbf{c}_1 + \mu\mathbf{c}_2 : \lambda, \mu \in \mathbb{R}\}$$

is the column space of $A$. The solutions of the system $A\mathbf{x} = \mathbf{0}$ are given by the equation

$$x + z = 0, \quad x + y + z = 0 \ \text{ or, equivalently, } \ x + z = 0, \quad y = 0.$$

By solving this linear system, we obtain $x = t, y = 0, z = -t$. Thus the set of all solutions of $A\mathbf{x} = \mathbf{0}$ is $N = \{(t\ 0\ -t) : t \in \mathbb{R}\}$. This is clearly a vector subspace of $\mathbb{R}^3$ and is the null space of $A$.

**Definition 9.12** Let $V$ be a vector space over a field $\mathbb{F}$. A subset $S = \{v_1, v_2, \ldots, v_k\}$ of $V$ is said to be **linearly independent** if we have

$$\lambda_1 v_1 + \lambda_2 v_2 + \cdots + \lambda_k v_k = 0 \quad \Rightarrow \quad \lambda_1 = \lambda_2 = \cdots = \lambda_k = 0. \tag{9.3.6}$$

Also, we say $S$ is **linearly dependent** if it is not linearly independent. That is, for some $\lambda_1, \lambda_2, \ldots, \lambda_k \in \mathbb{F}$ (not all zero), we have

$$\lambda_1 v_1 + \lambda_2 v_2 + \cdots + \lambda_k v_k = 0. \tag{9.3.7}$$

Notice that no $v_i$ in $S$ is the zero vector if it is a *linearly independent* set. Also, a singleton set $S = \{v\}$ is *linearly dependent* if and only if for some scalar $\lambda \neq 0$ we have $\lambda\, v = 0$. In turn, this holds if and only if $v = 0$. Therefore, the only singleton set that is linearly dependent consists of the zero vector 0. More generally, any set $S = \{v_1, v_2, \ldots, v_k\}$ with $v_1 = 0$ (say) is necessarily linearly dependent. For, we may take $\lambda_1 = 1$ and $\lambda_2 = \cdots = \lambda_k = 0$, so that we have

$$1 \cdot v_1 + 0 \cdot v_2 + \cdots + 0 \cdot v_k = 0.$$

On the other hand, $S = \{v\}$ is linearly independent if and only if $v \neq 0$. Next, the set $S = \{v_1, v_2\} \subset V$ is linearly dependent if for some scalars $\lambda_1$ and $\lambda_2$ we have

$$\lambda_1\, v_1 + \lambda_2\, v_2 = 0,$$

where at least one of the two scalars $\lambda_1$ and $\lambda_2$ is not zero. WLOG, let $\lambda_1 \neq 0$. We can thus write $v_1 = -\big(\lambda_2/\lambda_1\big)v_2$. Conversely, if a vector $v_1 \in V$ is a *scalar multiple* of a vector $v_2 \in V$, then we have $v_1 = \lambda\, v_2$ or $v_1 - \lambda\, v_2 = 0$, for some scalar $\lambda \neq 0$. It thus follows that the set $S = \{v_1, v_2\}$ is linearly dependent. Therefore, two vectors $v_1,\ v_2 \in V$ are linearly dependent if and only if one is a scalar multiple of the other. Equivalently, the set $S = \{v_1, v_2\}$ is linearly independent if and only if none of the two vectors is scalar multiple of the other vector. For example, the set $S_1 := \{(1, 0, 0), (0, 1, 0)\}$ is linearly independent, while the two vectors $\boldsymbol{u} = (1, 1, 2)$ and $\boldsymbol{v} = (2, 2, 4)$ are linearly dependent, because they *lie on the same line* in $\mathbb{R}^3$. In general, the set $S$ consisting of $n$ *standard unit vectors* given by (9.3.3) in the vector space $\mathbb{R}^n$ is linearly independent, because we have

$$\lambda_1\boldsymbol{e}_1 + \lambda_2\boldsymbol{e}_2 + \cdots + \lambda_n\boldsymbol{e}_n = 0 \quad \Leftrightarrow \quad (\lambda_1, \lambda_2, \ldots, \lambda_n) = \mathbf{0}. \tag{9.3.8}$$

Also, the set $S = \{1, x, x^2, \ldots, x^n\}$ in the vector space $V = P_n[x]$ of polynomials of degree at most $n$ is linearly independent. For, let $\lambda_0, \lambda_1, \lambda_2, \ldots, \lambda_n$ be scalars such that

$$\lambda_0 1 + \lambda_1 x + \lambda_2 x^2 + \cdots + \lambda_n x^n = 0.$$

Since this is true for all $x$, we get $\lambda_0 = 0$ by taking $x = 0$. Now by differentiating the above equation, we get

$$\lambda_1 + 2\lambda_2 x + \cdots + n\lambda_n x^{n-1} = 0.$$

Taking $x = 0$ in this equation, we get $\lambda_1 = 0$. By proceeding this way, we see

$$\lambda_0 = \lambda_1 = \lambda_2 = \ldots = \lambda_n = 0.$$

Therefore, $S$ is a linearly independent set.

**Theorem 9.4** *A subset $S = \{v_1, v_2, \ldots, v_k\}$ of a vector space $V$, with two or more vectors, is linearly dependent if and only if at least one of the vectors in $S$ is a linear combination of other vectors in $S$.*

***Proof.*** Let $S$ be linearly dependent. Then for some scalars $\lambda_1, \lambda_2, \ldots, \lambda_k$ (not all zero) we have

$$\lambda_1 v_1 + \lambda_2 v_2 + \cdots + \lambda_k v_k = 0.$$

Now, if $\lambda_i \neq 0$, then we can write

$$v_i = \frac{-\lambda_1}{\lambda_i} v_1 - \cdots - \frac{\lambda_{i-1}}{\lambda_i} v_{i-1} - \frac{\lambda_{i+1}}{\lambda_1} v_{i+1} - \cdots - \frac{\lambda_k}{\lambda_1} v_k,$$

which shows that $v_i$ is a linear combination of the remaining $k - 1$ vectors in $S$. Conversely, if one of the vectors, say $v_1$, is a linear combination of vectors $v_2, \ldots, v_k \in S$, then for some scalars $\lambda_2, \lambda_3, \ldots, \lambda_k$ we can write

$$v_1 = \lambda_2 v_2 + \cdots + \lambda_k v_k$$

so that it follows that $S$ is linearly dependent. □

For example, consider $\mathbf{v}_1 = (1, 0, 0)$, $\mathbf{v}_2 = (1, 0, 1)$ and $\mathbf{v}_3 = (3, 0, 1)$ in $\mathbb{R}^3$. Since

$$2\mathbf{v}_1 + \mathbf{v}_2 = 2(1, 0, 0) + (1, 0, 1) = (2, 0, 0) + (1, 0, 1) = (3, 0, 1) = \mathbf{v}_3,$$

it follows that the set $S := \{\mathbf{v}_1, \mathbf{v}_2, \mathbf{v}_3\} \subset \mathbb{R}^3$ is linearly dependent. More generally, it follows from (9.3.4) that the standard unit vectors $\boldsymbol{e}_1, \boldsymbol{e}_2, \ldots, \boldsymbol{e}_n$ spans the vector space $\mathbb{F}^n$; and, (9.3.8) implies that these vectors constitute a linearly independent set. That is, every vector $\boldsymbol{v} = (v_1, \ldots, v_n) \in \mathbb{F}^n$ can be expressed uniquely as

$$\boldsymbol{v} = v_1\, \boldsymbol{e}_1 + v_2\, \boldsymbol{e}_2 + \cdots + v_n\, \boldsymbol{e}_n.$$

In this case, we say $S = \{\boldsymbol{e}_1, \ldots, \boldsymbol{e}_n\}$ is a **standard basis** of $\mathbb{F}^n$. The next definition introduces the fundamental concept of *basis* more formally.

**Definition 9.13** Let $V$ be a vector space over a field $\mathbb{F}$. A nonempty set $B \subset V$ is called a **basis** of $V$ if every vector $v \in V$ can be expressed uniquely as

$$\lambda_1 v_1 + \lambda_2 v_2 + \cdots + \lambda_n v_n, \quad \text{for some } \lambda_i \in \mathbb{F} \text{ and } v_i \in B. \tag{9.3.9}$$

Equivalently, a set $B \subset V$ is a *basis* if (i) $B$ spans $V$; and, (ii) $B$ is linearly independent, over the field $\mathbb{F}$. The verification of equivalence is left to the reader as a simple exercise. Therefore, the concepts introduced in our previous discussion are integral to the formulation of Definition 9.13.

**Example 9.7** We may recall that every complex number $z \in \mathbb{C}$ has unique expression as $z = x + i\,y$, for some $x,\ y \in \mathbb{R}$, where $i = \sqrt{-1}$. So, the real vector space

$\mathbb{C}$ has the set $B = \{1, i\}$ as the *standard basis*. Also, the vector space $P_n[x]$ has the set $B = \{1, x, x^2, \ldots, x^n\}$ as the *standard basis*. As shown earlier, it is linearly independent and spans $P_n[x]$. The following four *matrix units* constitute the *standard basis* of the vector space $M_2(\mathbb{F})$:

$$E_1 = \begin{pmatrix} 1 & 0 \\ 0 & 0 \end{pmatrix}, \quad E_2 = \begin{pmatrix} 0 & 1 \\ 0 & 0 \end{pmatrix}, \quad E_3 = \begin{pmatrix} 0 & 0 \\ 1 & 0 \end{pmatrix}, \quad E_4 = \begin{pmatrix} 0 & 0 \\ 0 & 1 \end{pmatrix} \tag{9.3.10}$$

However, for vectors $\mathbf{v}_1 = (1, 0, 0)$, $\mathbf{v}_2 = (0, 1, 1)$ and $\mathbf{v}_3 = (1, 0, 1)$, the set $B = \{\mathbf{v}_1, \mathbf{v}_2, \mathbf{v}_3\}$ spans $\mathbb{R}^3$, by (9.3.5), and it is a linearly independent set. Therefore, $B$ is the (non-standard) basis for $\mathbb{R}^3$.

The next theorem gives an important characterization for a set to be a basis.

**Theorem 9.5** *Let $V$ is a vector space over a field $\mathbb{F}$, with* $\dim(V) = n < \infty$. *We have*

1. *(**Basis Reduction Theorem**) Every basis is a* minimal spanning set *of $V$. That is, if $V =$* Span$(B)$, *with $|B| > n$, then $B$ contains a basis of $V$.*
2. *(**Basis Extension Theorem**) Every* maximal linearly independent *subset of $V$ constitute a basis. That is if $B \subset V$ is linearly independent, and $|B| < n$, then $B$ can be "extended" to form a basis of $V$.*
3. *(**Dimension Uniqueness**) Any two basis of $V$ have the cardinality $n$.*

***Proof.*** Since $|B| > n$ implies it is not a basis for $V$, so the set $B$ is linearly dependent. Therefore, some vector $v \in B$ is a linear combination of other vectors in $B$. Let $B' = B - \{v\}$. Notice that we have $V = \text{Span}(B')$. Now, if $B'$ is linearly independent, then it is a basis for $V$. Otherwise, we can find a vector $v' \in B'$ that is a linear combination of some other vectors in $B'$. Let $B'' = B' - \{v'\} = B - \{v, v'\}$. It still spans $V$. If $B''$ is linearly independent, then it is a basis for $V$. Continuing with this procedure, we ultimately obtain a subset of $B$ that is linearly independent and spans $V$. This proves (1). Next, let $B$ be a linearly independent set in $V$, with $|B| < n$. Then $B$ does not span $V$. We may extend $B$ to another set $B'$ by adding an appropriate vector $v \in V - B$ so that $B'$ is linearly independent. If $B'$ spans $V$, then it is a basis for $V$. Again, if $B'$ does not span $V$, then we may add another vector $v' \in V - B'$ to $B'$ to obtain a set $B''$ so that it is linearly independent. If $B''$ spans $V$, then $B''$ is a basis for $V$. Continuing with this procedure, we ultimately obtain a linearly independent set that spans $V$. This proves (2). Finally, we prove any two bases of the vector space $V$ have the same number of elements. For, let $B$ be a basis for $V$ having $n$ elements. If $B'$ be yet another basis of $V$, with $n'$ elements, then we must have $n' \leq n$, because $B$ is a basis for $V$, and set with more than $n$ elements has to be linearly dependent. Since $B$ is a basis for $V$, so and any set with less than $n$ elements does not span $V$. But, $S'$ spans $V$. Therefore, we must have $n' \geq n$. Hence, we have $n = n'$. This completes the proof. □

By an **ordered basis** of a vector space we usually mean a basis whose elements are ordered by "indexing". A more commonly used term is **frame**. For example, while referring to a sequence of data expressed as *coordinates*.

**Definition 9.14** Let $B$ be an *ordered basis* of a vector space $V$ over a field $\mathbb{F}$ so that each vector $v \in V$ has a unique representation in $\mathbb{F}^n$ given by $(\lambda_1, \ldots, \lambda_n)$ as in (9.3.9). This $n$-tuple is called the **coordinate vector** of $v$ relative to $B$, which is usually denoted by $[v]_B$. That is, we have

$$[v]_B := (\lambda_1, \ldots, \lambda_n). \tag{9.3.11}$$

We often express $[v]_B$ as a column vector so that we may also call it the **matrix of vector** $v$ relative to basis $B$. In this latter case, we also write $[v]_B$ as $M_B(v)$ (or simply $M(v)$, if $B$ is clear from the context).

**Example 9.8** The *coordinate vector* of a vector $\boldsymbol{v} = (v_1, v_2, \ldots, v_n) \in \mathbb{R}^n$ relative to the *standard basis* $\{\boldsymbol{e}_1, \ldots, \boldsymbol{e}_n\}$ (with the natural ordering) is the *point* $(v_1, v_2, \ldots, v_n)$ itself. We next consider the vectors $\boldsymbol{v}_1 = (1, 0, 0)$, $\boldsymbol{v}_2 = (0, 1, 1)$, and $\boldsymbol{v}_3 = (1, 0, 1)$, in $\mathbb{R}^3$, and let $B = \{\boldsymbol{v}_1, \boldsymbol{v}_2, \boldsymbol{v}_3\}$. Then $B$ is a basis of $\mathbb{R}^3$. We order $B$ by indexing, i.e., as $\boldsymbol{v}_1 < \boldsymbol{v}_2 < \boldsymbol{v}_3$. For a vector $\boldsymbol{v} = (v_1, v_2, v_3) \in \mathbb{R}^3$, we have

$$\boldsymbol{v} = (v_1 + v_2 - v_3)\boldsymbol{v}_1 + v_2\boldsymbol{v}_2 + (v_3 - v_2)\boldsymbol{v}_3.$$

Therefore, in this case, we have

$$[\boldsymbol{v}]_B = (v_1 + v_2 - v_3, v_2, v_3 - v_2).$$

Also, for the ordered basis $B = \{1, x, x^2\}$ of the real vector space $P_2[x]$, we have

$$[5 + 3x - x^2]_B = (5, 3, -1).$$

In general, the idea of the coordinate vector of a vector justifies why it is often said that *an ordered basis to a real vector space $V$ is what a coordinate system is to the usual Euclidean space $\mathbb{R}^n$, for some $n$*. The idea of *dimension* introduced in the next definition formalizes the statement. It is fundamental to applications of linear algebra to various engineering branches, including computer science.

**Definition 9.15** Let $V$ be a vector space over a field $\mathbb{F}$, with a *basis* $B$. The **dimension** of $V$, denoted by $\dim_{\mathbb{F}}(V)$ (or simply $\dim(V)$, refers to the number $|B|$. A vector space $V$ is said to be **finite dimensional** if it has a basis with a finite number of elements.[1] Otherwise, it is called **infinite dimensional**.

It follows from the part (3) of Theorem 9.5 that the above definition of the *dimension* of a *finite-dimensional* vector space makes sense. Notice that $\dim(V)$ depends on the underlying field $\mathbb{F}$.

[1] We treat $V = \{0\}$ as a finite-dimensional vector space (even though it has no basis).

**Theorem 9.6** *Every finite-dimensional space has a basis.*

***Proof.*** Since $\text{Span}(V) = V$, the proof follows using part (1) of Theorem 9.5. □

Notice that we can also prove Theorem 9.6 by applying part (2) of Theorem 9.5 to $\emptyset$, which is vacuously linearly independent. For a simple illustration, we show that the set

$$S = \{\boldsymbol{v}_1 = (1, 1, 1),\ \boldsymbol{v}_2 = (1, 0, 1),\ \boldsymbol{v}_3 = (1, -1, -1)\}$$

is a basis of the real vector space $V = \mathbb{R}^3$. As said above, since the dimension of $\mathbb{R}^3$ is 3, it suffices to show that $S$ is linearly independent. Equivalently, we show that the matrix $A$ with columns consisting of the transpose of the row vectors $\boldsymbol{v}_1$, $\boldsymbol{v}_2$, and $\boldsymbol{v}_3$ is nonsingular. This latter assertion holds because $\det(A) \neq 0$. The next example gives some more illustrations.

**Example 9.9** It follows from Example 9.7 that

$$\dim_{\mathbb{R}}(\mathbb{C}) = 2, \quad \dim_{\mathbb{F}}(P_n[x]) = n + 1, \quad \text{and} \quad \dim_{\mathbb{F}}(M_2(\mathbb{F})) = 4.$$

The last equality follows from (9.3.10). In general, we have $\dim(M_{m\times n}(\mathbb{F})) = mn$. Also, it follows from our previous discussion that $\dim(\mathbb{F}^n) = n$. On the other hand, using Theorem 9.12, a simple "cardinality argument" implies that $\dim_{\mathbb{Q}}(\mathbb{R}) = \infty$ (see Exercise 9.22).

Let $V$ be a vector space over a field $\mathbb{F}$ of dimension $n$. Suppose

$$B := \{b_1, b_2, \dots, b_n\} \quad \text{and} \quad C := \{c_1, c_2, \dots, c_n\} \tag{9.3.12}$$

are two ordered bases of $V$. Let $v \in V$. An expression of the coordinate vector $(\lambda_1, \dots, \lambda_n)$ of $v$ relative to $B$ in terms of the coordinate vector $(\mu_1, \dots, \mu_n)$ of $v$ relative to $C$ is given by

$$\lambda_i = \sum_{j=1}^{n} a_{ij}\, \nu_j, \quad \text{for} \quad i = 1, \dots, n. \tag{9.3.13}$$

This is called the *change of basis formula*. To express it in matrix notation we write the coordinate vectors $[v]_B$ and $[v]_C$ as column vectors so that (9.3.13) transforms to $[v]_B = (a_{ij})[v]_C$. The related verifications are left to the reader as an exercise.

**Definition 9.16** A subspace $U$ of a vector space $V$ over a field $\mathbb{F}$ is called a **direct summand** of $V$ if, for some subspace $W$ of $V$, we have $V = U + W$, with $U \cap W = \{0\}$, and every element $v \in V$ can be written as $v = u + w$, for some $u \in U$ and $w \in W$. In this case, we write, $V = U \oplus W$.

It can be shown easily that $V = U \oplus W$ if and only if every element $v \in V$ can be written uniquely as $v = u + w$, for some $u \in U$ and $w \in W$. The next theorem proves a fundamental fact about vector spaces.

**Theorem 9.7** *Let $V$ be a finite-dimensional vector space over a field $\mathbb{F}$, with* $\dim(V) = n$, *and $U$ be a subspace of $V$. Then we have $V = U \oplus W$, for some subspace $W$ of $V$.*

***Proof.*** It should be clear that $\dim(U) = m \leq n$. Let $B_U = \{b_1, \ldots, b_m\}$ be a basis of $U$. In particular, $B_U \subset V$ is a linearly independent set. Therefore, by part (2) of Theorem ??, there are $k = n - m$ vectors $c_1, \ldots, c_k \in V - U$ such that

$$B_V = \{b_1, \ldots, b_m, c_1, \ldots, c_k\}$$

is a basis of $V$. We may take $W := \text{Span}(\{c_1, \ldots, c_k\})$. It then follows easily that $U \cap W = \{0\}$, and every element $v \in V$ can be written as $v = u + w$, for some $u \in U$ and $w \in W$. Therefore, $V = U \oplus W$. □

We conclude our discussion here with the next definition.

**Definition 9.17** A subspace of $\mathbb{R}^n$ of dimension $n - 1$ is called a **hyperplane**, and

$$H^{\perp} := \{\boldsymbol{x} \in \mathbb{R}^n \mid \langle \boldsymbol{y}, \boldsymbol{x} \rangle, \text{ for all } \boldsymbol{y} \in H\}$$

is called the *orthogonal complement* of the *hyperplane $H$*.

For example, each "point" in $\mathbb{R}$ is a *hyperplane* (of dimension 0); every "line" in $\mathbb{R}^2$ is a hyperplane (of dimension 1); every "plane" in $\mathbb{R}^2$ is a hyperplane (of dimension 2); and, so on. See Sect. 12.1, for a detailed discussion.

## 9.4 Linear Transformations

Let $U$ and $V$ be two vector spaces over the same field $\mathbb{F}$. An *additive* and *homogeneous* function $T : U \to V$ is called a *linear transformation*. A formal definition is as given below.

**Definition 9.18** A function $T : U \to V$ is called a **linear transformation** (or a **linear map**) if we have

$$T(\lambda u + \mu v) = \lambda Tu + \mu Tv, \quad \text{for all } u, v \in U \text{ and } \lambda, \mu \in \mathbb{F}. \tag{9.4.1}$$

A linear transformation is thus a linear combination preserving function between vector spaces. As in the above definition, we follow the convention of writing $T(x)$ simply as $Tx$, etc.

**Example 9.10** Clearly, the *zero mapping* $\mathbf{0} : u \mapsto 0\,(\forall u \in U)$, and the identity mapping $I_U : u \mapsto u\,(\forall u \in U)$ are linear transformations. Also, the *derivative function* $D : P[x] \to P[x]$ given by $D(p) = p'$; and, the *multiplication function* $M_x : P[x] \to P[x]$ given by $M_x p(x) = x^2 p(x)$ is a linear map, for $p \in P[x]$.

In general, the set $\mathcal{L}(U, V)$ of linear maps $T : U \to V$ is a vector space over the field $\mathbb{F}$ with respect to addition and scalar multiplication as defined below: For $T, S \in \mathcal{L}(U, V)$, $u \in U$, and $\lambda \in \mathbb{F}$, we have

$$\begin{aligned}(T + S)u &:= Tu + Su;\\ (\lambda T)u &:= \lambda Tu,\end{aligned} \tag{9.4.2}$$

When $U = V$, we write $\mathcal{L}(U, V)$ simply as $\mathcal{L}(V)$. An element of the vector space $\mathcal{L}(V)$ is called an **operator** on $V$.

We first consider the most important particular case when $U = \mathbb{R}^n$ and $V = \mathbb{R}^m$, for $m, n \geq 1$. Recall that an ordered basis to a real vector space is what a coordinate system is to the usual Euclidean space. We thus find that the dynamics of a linear map $T \in \mathcal{L}(\mathbb{R}^n, \mathbb{R}^m)$ is governed by the matrix $M_T$ of $T$ obtained by using the standard bases for the two Euclidean spaces. In general, a function $f : \mathbb{R}^n \to \mathbb{R}^m$ is given by

$$f(\boldsymbol{x}) := (w_1, \ldots, w_m), \quad \text{for} \quad \boldsymbol{x} = (x_1, \ldots, x_n) \in \mathbb{R}^n, \tag{9.4.3}$$

where the vector $\boldsymbol{w} = (w_1, \ldots, w_m) \in \mathbb{R}^m$ is given by

$$\begin{aligned} w_1 &= f_1(x_1, \ldots, x_n)\\ w_2 &= f_2(x_1 \ldots, x_n)\\ &\vdots\\ w_m &= f_m(x_1, \ldots, x_n), \end{aligned} \tag{9.4.4}$$

for some functions $f_i : \mathbb{R}^n \to \mathbb{R}$, with $i = 1, \ldots, m$. For example, with

$$\begin{aligned} w_1 &= f_1(x_1, x_2, x_3) := x_1 + x_1x_2 + x_3\\ w_2 &= f_2(x_1, x_2, x_3) := x_2 + x_3, \end{aligned}$$

we obtain a function $f : \mathbb{R}^3 \to \mathbb{R}^2$ given by

$$f(x_1, x_2, x_3) = (w_1, w_2) = (x_1 + x_1x_2 + x_3, x_2 + x_3).$$

We may also take the following functions for $f_i$'s:

(a) $f : \mathbb{R} \to \mathbb{R}$, given by $f(x) = x^2 + x$;
(b) $f : \mathbb{R}^3 \to \mathbb{R}$, given by $f(x, y, z) = x^2 + xy + z^2$;
(c) $f : \mathbb{R}^n \to \mathbb{R}$, given by $f(x_1, x_2, \ldots, x_n) = x_1 + x_2^2 + x_3^3 + \ldots + x_n^n$.

In particular, when each $f_i$ is a *linear map*, so is the function $f : \mathbb{R}^n \to \mathbb{R}^m$, as in (9.4.3). So, a linear map $T : \mathbb{R}^n \to \mathbb{R}^m$ in general is given by

$$T\boldsymbol{x} = \boldsymbol{w}, \quad \text{with vectors } \boldsymbol{x} \text{ and } \boldsymbol{w} \text{ are as above,} \tag{9.4.5}$$

where we have

$$w_1 = a_{11}x_1 + a_{12}x_2 + \cdots + a_{1n}x_n \tag{9.4.6}$$
$$w_2 = a_{21}x_1 + a_{22}x_2 + \cdots + a_{2n}x_n \tag{9.4.7}$$
$$\vdots \tag{9.4.8}$$
$$w_m = a_{m1}x_1 + a_{m2}x_2 + \cdots + a_{mn}x_n. \tag{9.4.9}$$

Viewing the standard unit vector $\boldsymbol{e}_i \in \mathbb{R}^n$ (defined earlier in (9.3.3)) as the column $n$-vector that has 0's in each entry except the $i$th where it is 1, we have

$$T\boldsymbol{e}_i := \begin{pmatrix} a_{1i} & a_{2i} & \cdots & a_{mi} \end{pmatrix}^t, \quad \text{for} \quad i = 1, \ldots, n, \tag{9.4.10}$$

so that the right side of (9.4.6) can be written as

$$T\begin{pmatrix} x_1 \\ x_2 \\ \vdots \\ x_n \end{pmatrix} = x_1 \begin{pmatrix} a_{11} \\ a_{21} \\ \vdots \\ a_{m1} \end{pmatrix} + \cdots + x_n \begin{pmatrix} a_{1n} \\ a_{2n} \\ \vdots \\ a_{mn} \end{pmatrix}. \tag{9.4.11}$$

It is computationally fruitful to keep track of real numbers $a_{ij}$ in the form of a matrix. Therefore, we may write the SLEs (9.4.6) in matrix form as

$$\begin{pmatrix} w_1 \\ w_2 \\ \vdots \\ w_m \end{pmatrix} = \begin{pmatrix} a_{11} & a_{12} & \cdots & a_{1n} \\ a_{21} & a_{22} & \cdots & a_{2n} \\ \vdots & \vdots & \ddots & \vdots \\ a_{m1} & a_{m2} & \cdots & a_{mn} \end{pmatrix} \begin{pmatrix} x_1 \\ x_2 \\ \vdots \\ x_n \end{pmatrix}. \tag{9.4.12}$$

Further, taking

$$\mathbf{w} = \begin{pmatrix} w_1 \\ w_2 \\ \vdots \\ w_m \end{pmatrix}, \quad \mathbf{x} = \begin{pmatrix} x_1 \\ x_2 \\ \vdots \\ x_m \end{pmatrix}, \quad \text{and} \quad A = \begin{pmatrix} a_{11} & a_{12} & \cdots & a_{1n} \\ a_{21} & a_{22} & \cdots & a_{2n} \\ \vdots & \vdots & \ddots & \vdots \\ a_{m1} & a_{m2} & \cdots & a_{mn} \end{pmatrix}, \tag{9.4.13}$$

we can write (9.4.12) in more compact form as

$$\boldsymbol{w} = T\boldsymbol{x} = A\,\boldsymbol{x}. \tag{9.4.14}$$

**Example 9.11** Consider the function $T : \mathbb{R}^4 \to \mathbb{R}^3$ given by

$$\begin{aligned} 9w_1 &= 2x_1 + x_2 + 4x_3 - 2x_4; \\ w_2 &= 3x_1 + x_2 - x_3 + 2x_4; \\ w_3 &= x_1 + x_2 + x_3 + 3x_4. \end{aligned}$$

It is the linear map $\boldsymbol{w} = T\boldsymbol{x} = A\,\boldsymbol{x}$, where

$$\boldsymbol{w} = \begin{pmatrix} w_1 \\ w_2 \\ w_3 \end{pmatrix} = A\,\boldsymbol{x} = \begin{pmatrix} 2 & 1 & 4 & -2 \\ 3 & 1 & -1 & 2 \\ 1 & 1 & 1 & 3 \end{pmatrix} \begin{pmatrix} x_1 \\ x_2 \\ x_3 \\ x_4 \end{pmatrix}.$$

In particular, for $\boldsymbol{x} = (1, 2, -1, 0)$,

$$\begin{aligned} 9w_1 &= 2 + 2 - 4 - 0 = 0, \\ w_2 &= 3 + 2 + 1 + 0 = 6, \\ w_3 &= 1 + 2 - 1 + 0 = 2, \end{aligned}$$

which we can also compute as

$$\begin{pmatrix} w_1 \\ w_2 \\ w_3 \end{pmatrix} = \begin{pmatrix} 2 & 1 & 4 & -2 \\ 3 & 1 & -1 & 2 \\ 1 & 1 & 1 & 3 \end{pmatrix} \begin{pmatrix} 1 \\ 2 \\ -1 \\ 0 \end{pmatrix} = \begin{pmatrix} 0 \\ 6 \\ 2 \end{pmatrix}.$$

**Definition 9.19 (Standard Matrix)** The matrix $A$ in (9.4.37) obtained using the standard basis of $\mathbb{R}^n$, it is called the **standard matrix** of $T$, which is usually denoted by $M_T$ (or $M(T)$).

Notice that $A$ in Example 9.11 is the *standard matrix* of $T$. On the other hand, we may also treat $T$ as the linear map given by the matrix $A$. More precisely, we say $T := T_A \in \mathcal{L}(\mathbb{R}^n, \mathbb{R}^m)$ is obtained as **multiplication by matrix** $A \in M_{m\times n}(\mathbb{R})$. The proof of the next theorem follows from the above discussion.

**Theorem 9.8** *Every linear map $T : \mathbb{R}^n \to \mathbb{R}^m$ is actually a **matrix transformation** given by $\boldsymbol{x} \mapsto M_T\boldsymbol{x}$, where the action of $T$ on the standard unit vectors $\boldsymbol{e}_i$ determines the structure of the matrix $M_T \in M_{m\times n}(\mathbb{R})$.*

Therefore, to visualize what a linear map $T : \mathbb{R}^n \to \mathbb{R}^m$ do to a vector $\boldsymbol{x} \in \mathbb{R}^n$, we view $T$ as the multiplication by the standard matrix $M_T \in M_n(\mathbb{F})$. Recall that a **reflection** across a hyperplane $H \subset \mathbb{R}^n$ is an *operator* $\theta \in \mathcal{L}(\mathbb{R}^n)$ such that

$$\theta\boldsymbol{v} = \begin{cases} -\boldsymbol{v}, & \text{for } \boldsymbol{v} \in H^{\perp} \\ \boldsymbol{v}, & \text{for } \boldsymbol{v} \in H \end{cases} \tag{9.4.15}$$

We also call $H$ the "plane of reflection" of the *reflection* $\theta$. In particular, it should be clear that the standard matrices of the reflections on $\mathbb{R}^2$ across (i) $x$-axis; (ii) $y$-axis; (iii) the line $y = x$; and, (iv) the line $y = -x$, are respectively given by

$$\begin{pmatrix} 1 & 0 \\ 0 & -1 \end{pmatrix}, \quad \begin{pmatrix} -1 & 0 \\ 0 & 1 \end{pmatrix}, \quad \begin{pmatrix} 0 & 1 \\ 1 & 0 \end{pmatrix}, \quad \begin{pmatrix} 0 & -1 \\ -1 & 0 \end{pmatrix} \tag{9.4.16}$$

Also, the standard matrices of the operators $\rho_1, \rho_2 \in \mathcal{L}(\mathbb{R}^2)$ that rotate every point of $\mathbb{R}^2$ about the origin *(counterclockwise)* (i) through an angle $\pi/4$ radians; and, (ii) through an angle $\phi$ radians, are respectively given by

$$M_{\rho_1} = \frac{1}{2}\begin{pmatrix} \sqrt{2} & -\sqrt{2} \\ \sqrt{2} & \sqrt{2} \end{pmatrix} \quad \text{and} \quad M_{\rho_2} = \begin{pmatrix} \cos\phi & -\sin\phi \\ \sin\phi & \cos\phi \end{pmatrix}. \tag{9.4.17}$$

More generally, let $U$ and $V$ be vector spaces over a field $\mathbb{F}$ respectively of dimensions $n$ and $m$. We fix an ordered basis $B = \{b_1, \ldots, b_n\}$ of $U$ and an ordered basis $C = \{c_1, \ldots, c_m\}$ of $V$. The technique used in the above special case applies *mutatis mutandis*. The $m \times n$ matrix of a linear map $T : U \to V$ with respect to bases $B$ and $C$, denoted by $[T]_B^C := (a_{ij})$, is obtained by using the coefficients appearing in the expression for $Tb_j$ when written as the linear combination of basis vectors in $C$. Said differently, the $i$th column of the matrix $[T]_B^C$ is the *coordinator vector* in $\mathbb{F}^m$ of the vector $Tb_j$, for $j = 1, \ldots, n$. That is,

$$\begin{aligned} Tb_1 &= a_{11}c_1 + a_{21}c_2 + \cdots + a_{m1}c_m; \\ Tb_2 &= a_{12}c_1 + a_{22}c_2 + \cdots + a_{m2}c_m; \\ &\vdots \\ Tb_n &= a_{1n}c_1 + a_{2n}c_2 + \cdots + a_{mn}c_m. \end{aligned} \tag{9.4.18}$$

Sometimes, we also write the matrix $[T]_B^C$ as $M_T[B, C]$ (or simply $M_T$, if the underlying bases are clear from the context). In terms of the *coordinate vector* $[u]_B$ of $u \in U$ and the coordinate vector $[v]_C$ of $v = Tu \in V$ respectively with respect to bases $B$ and $C$, both written as the column vectors, we can write the system (9.4.18) simply as

$$[v]_C = [T]_B^C [u]_B, \qquad \text{for all} \quad u \in U. \tag{9.4.19}$$

Unless stated otherwise, in the special case when $U = \mathbb{F}^n$ and $V = \mathbb{F}^m$, the $m \times n$ matrix $M_T$ of a linear map $T : \mathbb{F}^n \to \mathbb{F}^m$ is the *standard matrix*, where the elements of $\mathbb{F}^n$ and $\mathbb{F}^m$ are written as the column vectors (see Theorem 9.8). We may recall that the $k$th column of the matrix $M_T$ is the *coordinate vector* of $T\boldsymbol{e}_k$, where $\boldsymbol{e}_k$ is the $k$th vector in the standard basis of the scalar vector space $\mathbb{F}^n$.

**Definition 9.20** Let $V$ be vector space over a field $\mathbb{F}$ of dimensions $n$. Let $B = \{e_1, \ldots, e_n\}$ be an ordered basis of $V$. The matrix $[T]_B$ of a linear map $T$ on $V$ with respect to basis $B$ is a square matrix of order $n$ given by

$$[T]_B = \begin{pmatrix} | & | & & | \\ Te_1 & Te_2 & \cdots & Te_n \\ | & | & & | \end{pmatrix}, \tag{9.4.20}$$

where the $k$th column of $[T]_B$ is the coordinate vector of $Te_k$ with respect to the basis $B$, for $k = 1, \ldots, n$

Said differently, $Te_1$ is the vector where the basis vector $e_1$ lands after transformation; $Te_2$ is the vector where the basis vector $e_2$ lands after transformation; and, so on. For $u = a_{11}e_1 + \cdots + a_{1n}e_n \in V$, if

$$v = Tu = \lambda_1 e_1 + \cdots + \lambda_n e_n, \tag{9.4.21}$$

then we have

$$\lambda_j = a_{j1}e_1 + \cdots + a_{jn}e_n, \quad \text{for} \quad j = 1, 2, \ldots, n. \tag{9.4.22}$$

Therefore, the concepts of coordinate vectors and bases go in opposite ways. Depending on our choice for a basis, a vector in one coordinate system is different from what it is in another coordinate system. This is precisely the content of the next theorem.

**Theorem 9.9** **(Change of Basis)** *Let $V$ be vector space over a field $\mathbb{F}$ of dimension $n$, and let $T \in \mathcal{L}(V)$. Let $B = \{e_1, \ldots, e_n\}$ and $B' = \{e'_1, \ldots, e'_n\}$ be two ordered basis of $V$. We may write*

$$A_{old} := [T]_B \qquad \textit{and} \qquad A_{new} := [T]_{B'}. \tag{9.4.23}$$

*Then there is a matrix $P \in GL(n, \mathbb{F})$ such that*

$$A_{new} = P^{-1} A_{old} P. \tag{9.4.24}$$

*The matrix $P$ is called the matrix of* ***change of basis****.*

***Proof.*** We may define the matrix of *change of basis* $P = (p_{ij})$ in a way so that the column vectors of $P$ are the coordinate vectors of "new" basis vector $e'_j$ when expressed as linear combinations in terms of vectors in the "old" basis $B$. That is,

$$e'_j = p_{1j}e_1 + \cdots + p_{nj}e_n, \quad \text{for} \quad j = 1, \ldots, n. \tag{9.4.25}$$

Since both $B$ and $B'$ are basis, $P$ is a nonsingular matrix. For $v \in V$, we have

$$[v]_B = P\,[v]_{B'}, \tag{9.4.26}$$

where $[v]_B = (\lambda_1, \ldots, \lambda_n)^t$ and $[v]_{B'} = (\mu_1, \ldots, \mu_n)^t$ are respectively the coordinate vectors of $v$ with respect to basis $B$ and $B'$. To verify (9.4.26), we have

$$\begin{aligned} v &= \sum_{j=1}^{n} \mu_j\, e'_j \\ &= \sum_{j=1}^{n} \mu_j \left( \sum_{i=1}^{n} p_{ij} e_i \right) \qquad \text{(by 9.4.25)} \\ &= \sum_{i=1}^{n} \left( \sum_{j=1}^{n} p_{ij} \mu_j \right) e_i, \end{aligned}$$

it thus follows from (9.4.22) that

$$\lambda_i = p_{i1}\mu_1 + \cdots + p_{in}\mu_n, \quad \text{for } i = 1, 2, \ldots, n.$$

Next, if $w = Tv$ has the coordinate vectors $[w]_B$ and $[w]_{B'}$ respectively with respect to basis $B$ and $B'$, we obtain

$$\big[w\big]_{B'} = \big[T\big]_{B'}\big[v\big]_{B'} \quad \text{and} \quad \big[w\big]_B = \big[T\big]_B\big[v\big]_B,$$

using (9.4.26) and $[w]_B = P\,[w]_{B'}$. We now have

$$\begin{aligned} \big[w\big]_B &= P\big[w\big]_{B'} = P\,\big(\big[T\big]_{B'}\big[v\big]_{B'}\big) = \big(P\big[T\big]_{B'}\big)\big[v\big]_{B'}; \\ \big[w\big]_B &= \big[T\big]_B\big[v\big]_B = \big[T\big]_B\big(P\big[v\big]_{B'}\big) = \big(\big[T\big]_B P\big)\big[v\big]_{B'}, \end{aligned}$$

which implies $\big[T\big]_B P = P\big[T\big]_{B'}$. □

**Definition 9.21 (Similar Matrices)** Two matrices $A$ and $B$ are called **similar matrices** if $B = P^{-1}AP$, for some nonsingular matrix $P$.

Now, consider the *determinant function* $\det : M_n(\mathbb{F}) \to \mathbb{F}$. Using the multiplicative properties of det, it follows from (9.4.26) that

$$\begin{aligned} \det\big(A_{\text{new}}\big) &= \det\big(P^{-1}\,A_{\text{old}}\,P\big) \\ &= \det\big(P\big)^{-1}\det\big(A_{\text{old}}\big)\det\big(P\big) \\ &= \det\big(A_{\text{old}}\big). \end{aligned} \tag{9.4.27}$$

Therefore, any two matrices of an operator $T$ with respect to different bases have the same *determinant*. Hence the **determinant** of an operator $T$ is defined by

$$\det\big(T\big) = \det\big(M_T\big), \tag{9.4.28}$$

where $M_T$ is the matrix of $T$ with respect to some basis of $V$. In particular, *similar matrices* have the same determinant. A yet another useful function $\text{tr} : M_n(\mathbb{F}) \to \mathbb{F}$, called the **trace function**, is defined by

$$\operatorname{tr}\big(A\big) = a_{11} + \cdots + a_{nn}, \quad \text{for} \quad A = \big(a_{ij}\big) \in M_n(\mathbb{F}). \tag{9.4.29}$$

Clearly, the *trace function* is linear, and we also have

$$\operatorname{tr}\big(A\,B\big) = \operatorname{tr}\big(B\,A\big), \quad \text{for all} \quad A,\ B \in M_n(\mathbb{F}). \tag{9.4.30}$$

In particular, similar matrices have the same trace. The relation (9.5.27) connects the above two concepts.

**Remark 9.2** The determinant $\det(M_T)$ is a measure of "relative change in area/volume" affected by the operator $T$. An operator with zero determinant squishes vectors. Specifically, operators with nonzero determinants have some interesting applications. In majority of applications, we have $V = \mathbb{R}^n$ and $B$ is the standard basis $\{\boldsymbol{e}_1, \ldots, \boldsymbol{e}_n\}$.

We next consider another linear map $S : W \to U$, where $W$ is a vector space over the same field $\mathbb{F}$ of dimension $p$. Let $[S]_D^B \in M_{n\times p}(\mathbb{F})$ be the matrix of $S$ with respect to a fixed basis $D = \{d_1, d_2, \ldots, d_p\}$ given by

$$\big[S\big]_D^B = \begin{pmatrix} b_{11} & b_{12} & \cdots & b_{1p} \\ b_{21} & b_{22} & \cdots & b_{2p} \\ \vdots & \vdots & \ddots & \vdots \\ b_{n1} & b_{n2} & \cdots & b_{np} \end{pmatrix} \tag{9.4.31}$$

It then follows easily that the matrix of the composite linear map $T \circ S : W \to V$ is an $m \times p$ matrix $[T \circ S]_D^C$ given by

$$\big[T \circ S\big]_D^C := \begin{pmatrix} c_{11} & c_{12} & \cdots & c_{1p} \\ c_{21} & c_{22} & \cdots & c_{2p} \\ \vdots & \vdots & \ddots & \vdots \\ c_{m1} & c_{m2} & \cdots & c_{mp} \end{pmatrix} \tag{9.4.32}$$

where we have

$$c_{k\ell} = a_{k1}b_{1\ell} + a_{k2}b_{2\ell} + \cdots + a_{kn}b_{n\ell} \tag{9.4.33}$$

The proof of the next theorem is left for the reader as an exercise.

**Theorem 9.10** *Let $U$, $V$, and $W$ be vector spaces over the same field $\mathbb{F}$, respectively of positive dimensions $n$, $m$, and $p$, and consider the linear maps $S : W \to U$ and $T : U \to V$. Then the matrix representation of the composite linear maps $T \circ S : W \to V$ is the product of the matrix representation of $T$ with the matrix* *representation of S, assuming they are obtained with respect to some fixed basis of respective vector spaces. We thus have*

$$[T \circ S]_D^C = [T]_B^C \cdot [S]_D^B. \tag{9.4.34}$$

It is mainly due to the above association between the composition of linear maps among vector spaces and the usual matrix multiplication of the associated matrices that the fundamental algebraic properties of the **matrix algebra** hold.

## Rank-Nullity Theorem

Let $U$ and $V$ be two vector spaces over a field $\mathbb{F}$, and $T \in \mathcal{L}(U, V)$. Let

$$N(T) := \{u \in U \mid Tu = 0\}; \tag{9.4.35a}$$

$$R(T) := \{v \in V \mid v = Tu, \quad \text{for some } u \in U\} \tag{9.4.35b}$$

These two sets are subspaces of the respective vector spaces (use (9.2.6)). We call $N(T)$ the **null space** of $T$, and $R(T) = \mathrm{Im}(T)$ is called the **range space** of $T$. We also call $N(T)$ the **kernel** of $T$, and may also write it as $\ker(T)$. For example, consider the linear map $T : \mathbb{R}^2 \to \mathbb{R}^3$ given by $(x, y) \mapsto (x, y, x + y)$. We have $N(T) = \{(0, 0)\}$, and $R(T) = \mathrm{Span}\{(1, 0, 1), (0, 1, 1)\}$. Also, for the *derivative map* $D : P_n[x] \to P_{n-1}[x]$ given by $p \mapsto p'$, we have $N(D)$ consists of "constant polynomials", and $R(D) = \mathrm{Span}\{x, x^2, \ldots, x^n\}$.

**Definition 9.22** Let $U$ and $V$ be two vector spaces over a field $\mathbb{F}$, with $\dim(U) < \infty$, and $T \in \mathcal{L}(U, V)$. The **nullity** of $T$ is the dimension of the *null space* $N(T)$; and, the **rank** of $T$ is the dimension of the *range space* $R(T)$.

For the linear map $T(x, y) = (x, y, x + y)$, the *nullity* is 0 and the *rank* is 2. Notice that the sum of these two integers is precisely the dimension of $\mathbb{R}^2$. This is not a coincidence. It is also known as the *conservation of dimensions* formula.

**Theorem 9.11 (Rank-Nullity Theorem)** *Let $U$ and $V$ be two vector spaces over a field $\mathbb{F}$, and $T : U \to V$ be a linear map. If $U$ is finite dimensional, then* $\dim(R(T)) < \infty$ *and we have*

$$\dim(U) = \dim(N(T)) + \dim(R(T)). \tag{9.4.36}$$

***Proof.*** Let $\dim(U) = n$. Since subspace of a finite-dimensional vector space is finite dimensional, we have $N(T)$ is finite dimensional, say $\dim(N(T)) = m$. Let $K = \{b_1, b_2, \ldots, b_m\}$ be a basis of $N(T)$. For $k = n - m$, we find elements $c_1, c_2, \ldots, c_k \in U - N(T)$ such that the set

$$B = \{b_1, b_2, \ldots, b_m, c_1, c_2, \ldots, c_k\}$$

is a basis of $U$. The proof is completed by showing that the set

$$R = \{T(c_1),\ T(c_2), \dots,\ T(c_k)\}$$

is a basis of $R(T)$. For, let $v \in R(T)$ so that we have $Tu = v$, where

$$u = \lambda_1 b_1 + \cdots + \lambda_m b_m + \mu_1 c_1 + \ldots + \mu_k c_k,$$

for some scalars $\lambda_i,\ \mu_i \in \mathbb{F}$. Applying $T$ to the above expression, we obtain

$$v = Tu = \mu_1\, Tc_1 + \ldots + \mu_k\, Tc_k,$$

which proves $\text{Span}(R) = R(T)$. Next, to show that $R$ is linearly independent, let

$$\mu_1\, Tc_1 + \ldots + \mu_k\, Tc_k = 0,$$

for some scalars $\mu_i \in \mathbb{F}$. Using linearity of $T$, this implies $\mu_1\, c_1 + \ldots + \mu_k\, c_k \in N(T)$. So, we can write

$$\mu_1\, c_1 + \ldots + \mu_k\, c_k = \lambda_1\, b_1 + \cdots + \lambda_m\, b_m,$$

for some scalars $\lambda_i \in \mathbb{F}$. However, since $B$ is a basis of $U$, we have $\mu_1 = \cdots = \mu_k = 0$. Hence, $R$ is a basis of $R(T)$. □

Notice that if $U$ and $V$ are finite-dimensional vector spaces over a field $\mathbb{F}$ with $\dim(U) > \dim(V)$, then no linear map $T : U \to V$ is one-one, i.e., $N(T) \neq \{0\}$. In particular, a homogeneous system of linear equation $A\boldsymbol{x} = \mathbf{0}$, with more unknowns than the number of equations, must have a nonzero solution. On the other hand, if $\dim(U) < \dim(V)$, then no linear map $T : U \to V$ is onto, i.e., $R(T) \neq V$. In particular, a non-homogeneous system of linear equation $A\boldsymbol{x} = \boldsymbol{b}$, with more equations than the number of unknowns, has no solution for some choice of vector $\boldsymbol{b}$.

**Definition 9.23** Let $U$ and $V$ be two vector spaces over the same field $\mathbb{F}$. An **isomorphism** $T : U \to V$ is a bijective linear map. Two vector spaces $U$ and $V$ are said to be **isomorphic**, denoted by $U \cong V$, if there is an isomorphism $T : U \to V$.

The next result is a fundamental theorem of finite-dimensional vector spaces. It says a basis to a finite-dimensional vector space is what a coordinate system is to the geometry of a configuration in usual space.

**Theorem 9.12** *Let $V$ be a finite-dimensional vector space over a field $\mathbb{F}$, with $\dim(V) = n$. Then we have $V \cong \mathbb{F}^n$.*

***Proof.*** Let $B = \{b_1, \ldots, b_n\}$ be a basis of $V$, and let $v \in V$. We can write $v$ uniquely as the sum

$$v = \lambda_1 b_1 + \lambda_2 b_2 + \cdots + \lambda_n b_n, \tag{9.4.37}$$

for some $n$-tuple $(\lambda_1, \lambda_2, \ldots, \lambda_n) \in \mathbb{F}^n$. We claim that the function $T : V \to \mathbb{F}^n$ given by $v \mapsto M_B(v)$ is an isomorphism, where

$$M_B(v) := (\lambda_1, \ldots, \lambda_n)^t$$

is the *matrix of vector* $v$. Clearly, $T$ is a linear map. Also, the uniqueness of representation of $v$ in (9.4.37) implies that $T$ is injective. Further, the *natural* linear map $S : \mathbb{F}^n \to V$ given by

$$(\lambda_1, \ldots, \lambda_n)^t \mapsto \lambda_1 b_1 + \cdots + \lambda_n b_n$$

is the inverse of $T$. Hence, $V \cong \mathbb{F}^n$. □

Let $U$ and $V$ be finite-dimensional vector spaces over a field $\mathbb{F}$, with $\dim(U) = n$ and $\dim(V) = m$. Theorem 9.12 provides an isomorphism between the vector spaces $\mathcal{L}(U, V)$ and $M_{m\times n}(\mathbb{F})$. For, let $B_U = \{b_1, \ldots, b_n\}$ and $B_V = \{c_1, \ldots, c_m\}$ be respectively fixed bases of vector space $U$ and $V$. Now, for any $T \in \mathcal{L}(U, V)$, let $\Theta(T) \in M_{m\times n}(\mathbb{F})$ be the matrix representation of $T$ with respect to above two bases. We already noted that the function $\Theta : \mathcal{L}(U, V) \to M_{m\times n}(\mathbb{F})$ is a linear map. To prove that $\Theta$ is injective, let $\Theta(T) = 0$ so that we have $T(b_k) = 0$, for all $k = 1, \ldots, n$. Therefore, $T \equiv 0$. Next, let $A = (a_{ij}) \in M_{m\times n}(\mathbb{F})$. The $T \in \mathcal{L}(U, V)$ defined by

$$Tb_j := \sum_{i=1}^{m} a_{ij} w_i, \quad \text{for} \quad j = 1, \ldots, n,$$

is such that we have $\Theta(T) = A$. Therefore, $\Theta$ is surjective. Hence, we obtained an isomorphism, as asserted. Notice that, since the vector space $M_{m\times n}(\mathbb{F})$ is of dimension $mn$, with a basis given by the **matrix units** $E_{ij} \in M_{m\times n}(\mathbb{F})$. We may recall that the *matrix unit* $E_{ij}$ has all entries 0, except the $(i, j)$-entry that is 1. It thus follows from the isomorphism $\Theta$ that $\dim(\mathcal{L}(U, V)) = mn$.

**Definition 9.24** Let $V$ vector space over a field $\mathbb{F}$. A linear map $T \in \mathcal{L}(V) := \mathcal{L}(V, V)$ is called an **operator** on $V$.

Let $V$ be a vector space over a field $\mathbb{F}$, and $U$, $W$ be two subspaces such that $V = U \oplus W$. Then each $v \in V$ can be written uniquely as $v = u + w$, for some $u \in U$ and $w \in W$. Define $\pi_U, \pi_W : V \to V$ by

$$\pi_U v := u \quad \text{and} \quad \pi_W v = w. \tag{9.4.38}$$

It follows easily that both $\pi_U$ and $\pi_W$ are linear maps, i.e., $\pi_U, \pi_W \in \mathcal{L}(V)$. They are respectively called the **projection maps** onto the subspaces $U$ and $W$. Clearly, $N(\pi_U) = W$ and $R(\pi_U) = U$. Similarly, $N(\pi_W) = U$ and $R(\pi_W) = W$. In particular, we obtain $\dim(V) = \dim(U) + \dim(W)$. This claim also holds when $V$ is a direct sum of a finite number of subspaces as in (9.2.11). In this latter situation, we extend the concept of *projection maps* to obtain $\pi_1, \ldots, \pi_m$ as operators on $V$ such that $\pi_i := \pi_{U_i}$, for $i = 1, \ldots, m$.

**Theorem 9.13** *Suppose* $\dim(V) < \infty$, *and* $T \in \mathcal{L}(V)$. *Then* $T$ *is* invertible *if and only if it is injective if and only if it is surjective.*

***Proof.*** By Theorem 9.11, we have

$$\dim\big(V\big) = \dim\big(N(T)\big) + \dim\big(R(T)\big).$$

Now, if $T$ is injective, then $N(T) = \{0\}$ implies that

$$\dim\big(V\big) = \dim\big(R(T)\big) \quad \text{i.e.,} \quad R\big(T\big) = V.$$

So, $T$ is surjective. Also, if $T$ is surjective, then $R(T) = V$ implies that

$$\dim\big(N(T)\big) = 0 \quad \text{i.e.,} \quad N\big(T\big) = \{0\}.$$

So, $T$ is injective. □

Therefore, every injective or surjective operator on a finite-dimensional vector space $V$ is an *automorphism* on $V$.

**Definition 9.25** Let $V$ vector space over a field $\mathbb{F}$. An invertible operator $T \in \mathcal{L}(V)$ is called an **automorphism** on $V$. The set of automorphisms on $V$ is denoted by $\mathrm{Aut}_{\mathbb{F}}(V)$ (or simply $\mathrm{Aut}(V)$, if the underlying field $\mathbb{F}$ is clear from the context).

For $1 \leq i \leq 9$, consider the operators $T_i \in \mathcal{L}(\mathbb{R}^2)$ given by $T_i(\boldsymbol{x}) = A_i\boldsymbol{x}$, where

$$A_1 = \begin{pmatrix} 1 & 0 \\ 0 & 1 \end{pmatrix}, \quad A_2 = \begin{pmatrix} 0 & -1 \\ -1 & 0 \end{pmatrix}, \quad A_3 = \begin{pmatrix} -1 & 0 \\ 0 & -1 \end{pmatrix},$$
$$A_4 = \begin{pmatrix} 0 & 1 \\ -1 & 0 \end{pmatrix}, \quad A_5 = \begin{pmatrix} -1 & 0 \\ 0 & 1 \end{pmatrix}, \quad A_6 = \begin{pmatrix} 1 & 0 \\ 0 & -1 \end{pmatrix},$$
$$A_7 = \begin{pmatrix} 0 & 1 \\ 1 & 0 \end{pmatrix}, \quad A_8 = \begin{pmatrix} 1 & 0 \\ 0 & 0 \end{pmatrix}, \quad A_9 = \begin{pmatrix} 0 & 0 \\ 0 & 1 \end{pmatrix}$$

We then have that $T_1, \ldots, T_6$ are automorphisms on $\mathbb{R}^2$, because the matrices $A_1, \ldots, A_6$ are nonsingular. However, since $A_7, \ldots, A_9$ are singular, $T_7, \ldots, T_9$ are not automorphisms. In general, the set $\mathrm{Aut}(V)$ constitutes a group with respect to the binary operation *composition*. The next theorem follow from the isomorphism $\Theta : \mathcal{L}(U, V) \to M_{m\times n}(\mathbb{F})$, as a particular case when $U = V$ is of dimension $n$.

**Theorem 9.14** *Let $V$ be an n-dimensional vector space over a field $\mathbb{F}$, and fix a basis $B = \{v_1, \ldots, v_n\}$ of $V$. The group $G = \mathrm{Aut}(V)$ is isomorphic to the group $GL(n, \mathbb{F})$ of nonsingular matrices of order n.*

In geometry, if $C \subset \mathbb{R}^n$ is a configuration, then for every $T \in \mathrm{Aut}\left(\mathbb{R}^n\right)$ the image $T(C) \subset \mathbb{R}^n$ is a *change in perspective* of $C$ [10].

## 9.5 Eigenvectors and Eigenvalues

Let $V$ be an $n$ dimensional *nonzero* vector space over a field $\mathbb{F}$, and $T \in \mathcal{L}(V)$. A *proper subspace* $U$ of $V$ is called an **invariant subspace** relative to $T$ if $Tu \in U$, for every $u \in U$. We also call $U$ a **T-invariant** subspace (or simply an *invariant subspace* of $T$). For example, both $N(T)$ and $R(T)$ are invariant subspaces, in general. In particular, every subspace $P_n[x]$ of the real vector space $P[x]$ is invariant relative to the derivative operator $D : P[x] \to P[x]$. We also have that

$$P[x] = P_0[x] \oplus P_1[x] \oplus P_2[x] \oplus \cdots$$

In general, to find invariant subspaces $U_1, U_2, U_3, \ldots,$ of an operator $T$ on an infinite-dimensional vector space such that

$$V = U_1 \oplus U_2 \oplus U_3 \oplus \cdots, \tag{9.5.1}$$

is an unresolved fundamental problem, known as the *invariant subspace problem*. As discussed later, in some particular cases, it is far more easier to understand the structure of $T$ in terms of the operators $T\big|_{U_j}$, for $j = 1, \ldots, m$, whenever such a decomposition of $V$ is possible. We first consider the simplest case of a 1-dimensional invariant subspaces of an operator $T \in \mathcal{L}(V)$. That is, a subspace of the form

$$\mathbb{F}\, v := \{\lambda\, v \mid \lambda \in \mathbb{F}\}, \quad \text{for} \quad 0 \neq v \in V. \tag{9.5.2}$$

In this case, the subspace $\mathbb{F}\, v$ is $T$-invariant if we have

$$Tv = \lambda\, v, \quad \text{for some} \quad \lambda \in \mathbb{F}. \tag{9.5.3}$$

On the other hand, if some nonzero vector $v \in V$ satisfies the relation (9.5.3), then the 1-dimensional subspace in (9.5.2) is actually an invariant subspace of $T$. Therefore, an operator $T \in \mathcal{L}(V)$ has a 1-dimensional invariant subspace of the form (9.5.2) if and only if it has an eigenvalue, in the sense of the next definition.

**Definition 9.26** **(Eigenvalue and Eigenvector)** A scalar $\lambda \in \mathbb{F}$ is called an **eigenvalue** (or a **characteristic value**) of an operator $T \in \mathcal{L}(V)$ if the relation (9.5.3) holds, for some nonzero vector $v \in V$, which is called an **eigenvector** (or a **characteristic vector**) of $T$ corresponding to *eigenvalue* $\lambda$.

By Theorem 9.8, we can use the matrix of an operator $T \in \mathcal{L}(V)$ with respect to a specified basis of $V$ to visualize what $T$ does to a vector $v \in V$. In particular, for an operator $T \in \mathcal{L}(\mathbb{F}^n)$, the *standard matrix* $M_T$ of $T$ is used. So, the *multiplication operator*: $\boldsymbol{x} \mapsto M_T\boldsymbol{x}$ helps visualize action of the operator $T$ on a vector $\boldsymbol{x} \in \mathbb{F}^n$. An *eigenvector* corresponding to an *eigenvalue* $\lambda$ of $T$ is thus a vector $\boldsymbol{x} \in \mathbb{F}^n$ such that $M_T$ doesn't change the "direction" of the vector $\boldsymbol{x}$. Also, for $\lambda \in \mathbb{R}$, it gives a measure of how much the eigenvector $\boldsymbol{x}$ is "stretched" or "squished" or "rotated" by the operator $T$. For $\lambda \in \mathbb{C}$, the geometry is bit more involved.

**Example 9.12** Consider reflections on $\mathbb{R}^2$ across (i) $x$-axis; (ii) $y$-axis; (iii) the line $y = x$; and, (iv) the line $y = -x$, with the standard matrices respectively given by

$$M_x = \begin{pmatrix} 1 & 0 \\ 0 & -1 \end{pmatrix}, \quad M_y = \begin{pmatrix} -1 & 0 \\ 0 & 1 \end{pmatrix},$$
$$M_{y=x} = \begin{pmatrix} 0 & 1 \\ 1 & 0 \end{pmatrix}, \quad M_{y=-x} = \begin{pmatrix} 0 & -1 \\ -1 & 0 \end{pmatrix}$$

Since $M_x$ doesn't change the direction of the unit vector $\mathbf{i} = (1, 0)$ fixed, and maps the unit vector $\mathbf{j} = (0, 1)$ to the vector $-\mathbf{j}$, it follows that 1 and $-1$ are eigenvalues of the reflections on $\mathbb{R}^2$ across $x$-axis, with eigenvectors $\mathbf{i}$ and $\mathbf{j}$, respectively. Similarly, the reflections on $\mathbb{R}^2$ across $y$-axis has eigenvalues $-1$ and 1 with corresponding eigenvectors $\mathbf{i}$ and $\mathbf{j}$. The reader is encouraged to find the eigenvalues and eigenvectors of the other two reflection matrices. Next, we consider the (counterclockwise) *rotation* $T_\theta$ of the plane $\mathbb{R}^2$ about the origin through an angle $\theta$ radians. In this context, the standard matrix of $T_\theta$ is given by

$$M_\theta = \begin{pmatrix} \cos\theta & -\sin\theta \\ \sin\theta & \cos\theta \end{pmatrix}. \tag{9.5.4}$$

Clearly, $T_\theta$ has no eigenvectors unless $\theta \equiv 0 \pmod{\pi}$: When $\theta \equiv \pi \pmod{2\pi}$, each nonzero vector in $\mathbb{R}^2$ is an eigenvector with eigenvalue $-1$; and, when $\theta \equiv 0 \pmod{2\pi}$, each nonzero vector in $\mathbb{R}^2$ is an eigenvector with eigenvalue 1. On the other hand, if $T_{\pi/2} : \mathbb{C}^2 \to \mathbb{C}^2$ is a rotation about the origin $\mathbf{0} \in \mathbb{C}^2$ through an angle $\pi/2$ radians so that we have

$$T_{\pi/2}\big(z_1, z_2\big) = \big(-z_1, z_2\big), \quad \text{for} \quad z_1, z_2 \in \mathbb{C},$$

then $\pm i$ are two eigenvalues of $T_{\pi/2}$. Moreover, the eigenvectors corresponding to eigenvalue $i$ are vectors of the form $(z, -i\,z)$ $(z \in \mathbb{C})$; and, the eigenvectors corresponding to eigenvalue $-i$ are vectors of the form $(z, i\,z)$ $(z \in \mathbb{C})$.

**Definition 9.27 (Spectrum)** The set of all eigenvalues of an operator $T \in \mathcal{L}(V)$ is called the **spectrum** of $T$, which is usually denoted by $\operatorname{spec}(T)$.

It is easy to see that, for any fixed $\lambda \in \mathrm{spec}(T)$, the set of eigenvectors corresponding to $\lambda$ is a nonzero subspace of $V$. It is called the **eigenspace** of $T$ corresponding to eigenvalue $\lambda$, which is usually denoted by $E_\lambda(T)$ (or simply $E_\lambda$). That is,

$$E_\lambda(T) = \{v \in V \mid Tv = \lambda\, v\}. \tag{9.5.5}$$

**Theorem 9.15** *Let $V$ be an $n$ dimensional* nonzero *vector space over a field $\mathbb{F}$, and $T \in \mathcal{L}(V)$. A scalar $\lambda \in \mathbb{F}$ is an eigenvalue of $T$ if and only if $T - \lambda I$ is not invertible.*

***Proof.*** Since we can rewrite (9.5.3) as

$$(T - \lambda\, I)v = 0, \quad \text{for some} \quad 0 \neq v \in V, \tag{9.5.6}$$

it follows that $(T - \lambda\, I)^{-1}$ exists. For the converse, notice that the set of eigenvectors of $T$ corresponding to the eigenvalue $\lambda$ is the null space $N(T - \lambda\, I)$. □

By Theorem 9.10, we know that

$$T^n = T \circ T \circ \cdots \circ T, \quad \text{for} \quad n \geq 0 \tag{9.5.7}$$

belongs to $\in \mathcal{L}(V)$. We may take $T^0 = I$, the identity map on $V$. If $T$ is an invertible operator, then we can extend the definition (9.5.28) to non-positive integers. Now, for a polynomial $p \in P[x]$ given by

$$p(x) = \lambda_0 + \lambda_1 x + \cdots + \lambda_m x^m, \tag{9.5.8}$$

we define an operator $p(T) \in \mathcal{L}(V)$ by

$$p(T) = \lambda_0\, I + \lambda_1\, T + \cdots + \lambda_m\, T^m, \tag{9.5.9}$$

As a special case, suppose $V$ is a complex vector space. For $0 \neq v \in V$, the set

$$\{v, T(v), T^2(v), \ldots, T^n(v)\}$$

is linearly dependent. So, for complex numbers $z_0, z_1, \ldots, z_n$ (not all zero), we have

$$z_0\, v + z_1\, Tv + \cdots + z_n\, T^n v = 0 \tag{9.5.10}$$

Let $m$ be the largest such that $z_m \neq 0$. Notice that we have $0 < m \leq n$. By the *fundamental theorem of algebra*, we can find $\lambda_j \in \mathbb{C}$ such that

$$z_0 + z_1\, z + \cdots + z_n\, z^n = w(z - \lambda_1)(z - \lambda_2) \cdots (z - \lambda_m) \tag{9.5.11}$$

holds for all $z \in \mathbb{C}$, with $w \neq 0$. It thus follows that

$$\begin{aligned} 0 &= z_0\, v + z_1\, Tv + \cdots + z_n\, T^n v \\ &= \big(z_0\, I + z_1\, T + \cdots + z_n\, T^n\big)v \\ &= w\big[(T - \lambda_1\, I)\big) \cdots \big(T - \lambda_m\, I)\big]v, \end{aligned}$$

which means that $T - \lambda_j I$ is not injective for at least one $j$. Therefore, $T$ has an eigenvalue. We have thus proven the next theorem.

**Theorem 9.16** *Every operator $T \in \mathcal{L}(\mathbb{C}^n)$ has an eigenvector. Moreover, for some complex number $w \neq 0$, we have*

$$\mathbb{C}\, w = \big\{\lambda\, w \mid \lambda \in \mathbb{C}\big\}, \tag{9.5.12}$$

*is invariant relative to $T$.*

What the above theorem demonstrated is the main goal of linear algebra, in general. That is, we need to find a nice basis of $V$ of an operator $T \in \mathcal{L}(V)$ such that the matrix $M_T$ of $T$ with respect to some basis has a reasonably simple structure.

**Definition 9.28** A matrix $A = \big(a_{ij}\big) \in M_n(\mathbb{F})$ is called an **upper triangular matrix** if we have $a_{ij} = 0$, for all $i > j$. We usually write an upper triangular matrix as

$$\begin{pmatrix} \lambda_1 & & & * \\ & \lambda_2 & & \\ & & \ddots & \\ \mathbf{0} & & & \lambda_n \end{pmatrix}, \tag{9.5.13}$$

where $\mathbf{0}$ indicates all below diagonal entries are zero. Similarly, we can define a **lower triangular matrix**.

Notice that an *upper triangular matrix* or a *lower triangular matrix* is nonsingular if and only if all the diagonal entries are nonzero. The next theorem establishes an interesting connection between upper triangular matrices and invariant subspaces of an operator $T$.

**Theorem 9.17** *Suppose $T$ is an operator on an $n$-dimensional vector space $V$ over a field $\mathbb{F}$. Let $B = \{b_1, \ldots, b_n\}$ be a basis of $V$. Then the following are equivalent:*

1. *With respect to $B$, the matrix $M_T$ is upper triangular;*
2. *$Tb_j \in \operatorname{Span}\big(\{b_1, \ldots, b_j\}\big)$, for each $j = 1, \ldots, n$;*
3. *$\operatorname{Span}\big(\{b_1, \ldots, b_j\}\big)$ is invariant under $T$, for each $j = 1, \ldots, n$.*

***Proof.*** The only non-trivial part that needs proof is the implication (2) $\Rightarrow$ (3). For, let $1 \leq j \leq n$ be fixed. Since

$$Tb_1 \in \text{Span}\left(\{b_1\}\right) \subset \text{Span}\left(\{b_1, \ldots, b_j\}\right);$$
$$Tb_2 \in \text{Span}\left(\{b_1, b_2\}\right) \subset \text{Span}\left(\{b_1, \ldots, b_j\}\right);$$
$$\vdots$$
$$Tb_j \in \text{Span}\left(\{b_1, \ldots, b_j\}\right),$$

for any $v \in V$ that is a linear combination of vectors $b_1, \ldots, b_j$ we have $T(v) \in \text{Span}(\{b_1, \ldots, b_j\})$. So, $\text{Span}(\{b_1, \ldots, b_j\})$ is invariant under $T$. □

We see that the condition (1) of Theorem 9.17 holds for complex vector spaces. More generally, suppose it holds for an operator $T \in \mathcal{L}(V)$ such that $M_T$ is of the form as in (9.5.13) with respect to a basis $B = \{b_1, \ldots, b_n\}$ of $V$. Then, for $\lambda \in \mathbb{F}$, the operator $T - \lambda I$ is not invertible if and only if $\lambda = \lambda_j$, for some $j = 1, \ldots, n$. Therefore, the eigenvalues of $T$ are precisely the scalars $\lambda_j$ in the matrix (9.5.13).

**Definition 9.29** Let $V$ be an $n$ dimensional vector space over a field $\mathbb{F}$. We call an operator $T \in \mathcal{L}(V)$ **diagonalisable** if $M_T \in M_n(\mathbb{F})$ is a **diagonal matrix** with respect to some basis of $V$.

Clearly, the matrix $M_T$ of an operator $T \in \mathcal{L}(V)$ is a *diagonal matrix* with respect to a basis $\{b_1, \ldots, b_n\}$ of $V$ if and only if $b_i$'s are eigenvectors of $T$, and the diagonal entries of $M_T$ are precisely the corresponding eigenvalues. Unfortunately, not every linear operator is *diagonalizable*.

**Example 9.13** The matrix of the identity operator $I : V \to V$ with respect to any basis is the identity matrix $I_n$, which has only one eigenvalue. Also, not every linear operator on a complex vector space is diagonalizable. We may consider the typical linear operator $T : \mathbb{C}^2 \to \mathbb{C}^2$ given by $T(x_1e_1 + x_2e_2) = x_2e_1$, where $\{e_1, e_2\}$ is the standard basis for $\mathbb{R}^2$. In general, we can characterize linear operators $T : \mathbb{C}^2 \to \mathbb{C}^2$ in terms of the number of distinct eigenvalues $T$ may have. It is far more complicated to deal with an operator $T : \mathbb{C}^n \to \mathbb{C}^n$. The complete answer is provided by the fundamental theorem classifying operators on finite-dimensional spaces in terms of *Jordan normal forms*.

The next theorem gives a sufficient condition for $T \in \mathcal{L}(V)$ to be diagonalisable.

**Theorem 9.18** *Let $V$ be an $n$ dimensional vector space over a field $\mathbb{F}$, and $T \in \mathcal{L}(V)$. If $T$ has $n$ distinct eigenvalues, then the associated eigenvectors form a basis of $V$, and $M_T$ is a diagonal matrix with respect to this basis.*

***Proof.*** Let $\lambda_1, \ldots, \lambda_n$ be $n$ distinct eigenvalues of $T$, with the associated eigenvectors $v_1, \ldots, v_n$. We know that the set $B = \{v_1, \ldots, v_n\}$ is linearly independent. Therefore, it is a basis, because $n = \dim(V)$. Clearly, as said earlier, $M_T$ is a diagonal matrix with respect to $B$. □

The next theorem gives a generalization of Theorem 9.18.

**Theorem 9.19** *Let $V$ be an $n$-dimensional vector space over a field $\mathbb{F}$, and $T \in \mathcal{L}(V)$ with $m$ distinct eigenvalues $\lambda_1, \ldots, \lambda_m$. Then the following are equivalent:*

1. *$M_T$ is diagonal with respect to some basis of $V$;*
2. *$V$ has a basis consisting of eigenvectors of $T$;*
3. *there exists* 1*-dimensional invariant subspaces $U_1, \ldots, U_m$ of $T$ such that*

$$V = U_1 \oplus U_2 \oplus \cdots \oplus U_m; \tag{9.5.14}$$

4. *$V = N(T - \lambda_1 I) \oplus \cdots \oplus N(T - \lambda_m I)$;*
5. $\dim(V) = \dim\big(N(T - \lambda_1 I)\big) + \cdots + dim\big(N(T - \lambda_m I)\big)$.

***Proof.*** The equivalence of (1) and (2) is known. Next, if (2) holds, then $V$ has a basis consisting of eigenvectors of $T$, say $B = \{v_1, \ldots, v_n\}$. It thus follows that $U_j := \mathbb{F}\, e_j$ satisfies (9.5.14). Therefore, (2) $\Rightarrow$ (3). Conversely, if (3) holds, and we take $0 \neq v_j \in U_j$, then $B = \{v_1, \ldots, v_n\}$ is basis of $V$ consisting of eigenvectors of $T$. Hence, (3) $\Rightarrow$ (2). To complete the proof, it suffices to show that (2) $\Rightarrow$ (4) $\Rightarrow$ (5) $\Rightarrow$ (2). The details are left to the reader as an exercise. □

More generally, fix an operator $T \in \mathcal{L}(V)$. Suppose there is a "nice" decomposition of $V$ of the form

$$V = U_1 \oplus U_2 \oplus \cdots \oplus U_m, \tag{9.5.15}$$

where each $U_j$ is a $T$-invariant subspace of $V$. By Theorem 9.19, we know that $U_j$ are 1-dimensional $T$-invariant subspaces of $V$ if and only if we have

$$V = N\big(T - \lambda_1 I\big) \oplus \cdots \oplus N\big(T - \lambda_m I\big), \tag{9.5.16}$$

where $\lambda_1, \ldots, \lambda_m$ are $m$ distinct eigenvalues of $T$. It is known that a decomposition of $V$ of the form (9.5.16) doesn't hold in general. However, it does for some nice operators $T \in \mathcal{L}(V)$ such as when $T$ is *self-adjoint*. In the next part, our goal is to show that for the case when $V$ is a complex vector space we have

$$V = N\big((T - \lambda_1 I)^n\big) \oplus \cdots \oplus N\big((T - \lambda_m I)^n\big), \tag{9.5.17}$$

The concept we introduce in the next definition helps us to achieve this goal.

**Definition 9.30 (Generalized Eigenvector)** Let $V$ be an $n$-dimensional vector space over a field $\mathbb{F}$, and $T \in \mathcal{L}(V)$. A vector $v \in V$ is called a **generalized eigenvector** corresponding to an eigenvalue $\lambda \in \mathbb{F}$ of $T$ if we have

$$T\big((T - \lambda I)^k\big)v = 0, \quad \text{for some} \quad k \in \mathbb{Z}^+. \tag{9.5.18}$$

Clearly, every eigenvector of $T$ corresponding to an eigenvalue is a *generalized eigenvector*; for, take $k = 1$. On the other hand, the operator $T : \mathbb{C}^3 \to \mathbb{C}^3$ given by

$$T\big(z_1, z_2, z_3\big) = \big(z_2, 0, z_3\big), \quad \text{for} \quad z_1, z_2, z_3 \in \mathbb{C},$$

has the property that

$$T^2\big(z_1, z_2, 0\big) = 0, \quad \text{for all} \quad z_1, z_2 \in \mathbb{C},$$

so that every vector $v \in \mathbb{C}^3$ with the last coordinate zero is a generalized eigenvector of $T$. Notice that none of these vectors is an eigenvector. In fact, we have

$$\mathbb{C}^3 = \big\{(z_1, z_2, 0) \mid z_1, z_2 \in \mathbb{C}\big\} \oplus \big\{(0, 0, z_3) \mid z_3 \in \mathbb{C}\big\},$$

where the first summand on the right side consists of the generalized eigenvectors of the operator $T$ corresponding to the eigenvalue 0, and the second summand consists of the generalized eigenvectors corresponding to the eigenvalue 1. We will prove shortly that the decomposition of the above type is possible for every operator on a finite-dimensional complex vector space.

As before, for any fixed $\lambda \in \text{spec}(T)$, the **generalized eigenspace** of $T$ is the subspace of generalized eigenvectors of $T$ corresponding to eigenvalue $\lambda$. In this case, we denote it by $G_\lambda(T)$ (or simply $G_\lambda$). That is,

$$G_\lambda := \big\{v \in V \mid (T - \lambda\, I)^k v = 0, \text{ for some } k \in \mathbb{Z}^+\big\}. \tag{9.5.19}$$

The **multiplicity** of an eigenvalue $\lambda$ of $T$, denoted by $M(\lambda)$, is the dimension of the subspace $G_\lambda \subset V$. We have that the *multiplicity* of an eigenvalue $\lambda$ of $T$ equals $\dim(N((T - \lambda\, I)^n))$. In fact, as shown later, $M(\lambda)$ is exactly the number of times $\lambda$ appears on the diagonal of the upper triangular matrix representation of $T$, whenever available (Theorem 9.21).

**Remark 9.3** The concepts introduced above are fundamental to many applications in computer science (see Sect. 9.7). A deeper understanding of the related linear algebra techniques is crucial to make good machine learning models such as *PAC*.

We conclude this part with the next important fact about real vector spaces.

**Theorem 9.20** *Every operator $T$ on a finite dimension real vector space $V \neq 0$ has an invariant subspace of dimension* 1 *or* 2.

***Proof.*** Let $\dim(V) = n$. Choose $0 \neq v \in V$. The there exists real numbers $a_0, a_1, \ldots, a_n$ (not all zero) such that we have

$$a_0 v + a_1 T v + a_2 T^2 v + \cdots + a_n T^n v = 0.$$

We can factorize the associated real polynomial

$$p(x) := a_0 + a_1 x + a_2 x^2 + \cdots + a_n x^n$$

into linear and quadratic factors as given below:

$$p(x) = c\big(x - \alpha_1\big) \cdots \big(x - \alpha_m\big)\big(x^+\beta_1 x + \gamma_1\big) \cdots \big(x^+\beta_k x + \gamma_k\big),$$

where $0 \neq c, \alpha_i, \beta_i, \gamma_i \in \mathbb{R}$, and $m + k \geq 1$. It thus follows that

$$\begin{aligned} 0 &= a_0 v + a_1 T v + a_2 T^2 v + \cdots + a_n T^n v \\ &= \big(a_0 I + a_1 T + a_2 T^2 + \cdots + a_n T^n\big) v \\ &= c\big[(T - \alpha_1 I) \cdots \big(T - \alpha_m I\big)\big(T^2 + \beta_1 T + \gamma_1 I\big) \cdots \big(T^2 + \beta_k T + \gamma_k I)\big] v, \end{aligned}$$

which implies that at least one linear factor $T - \alpha_i I$ or a quadratic factor $T^2 + \beta_j T + \gamma_j I$ is not one-one. In the first case, $T$ has an invariant subspace of dimension 1; and, in the second case, it has an invariant subspace of dimension 2. Notice that if

$$T^2 u = -\beta_j T u - \gamma_j u, \quad \text{for some} \quad 0 \neq u \in V,$$

then the subspace $\mathrm{Span}(\{u, Tu\})$ of dimension 1 or 2 is an invariant subspace under $T$. This completes the proof. □

## *Cayley-Hamilton Theorem*

Let $V$ be a complex vector space of dimension $n$, and $T \in \mathcal{L}(V)$. It is known that $V$ can have many different bases with respect to which $M_T$ is upper triangular. However, all must have the eigenvalues of $T$ as the diagonal entries. In general, when $T$ has $m < n$ distinct eigenvalues, the next theorem proves that an eigenvalue $\lambda \in \mathrm{spec}(T) \subset \mathbb{C}$ appears on the diagonal of an upper triangular matrix representation $M_T$ exactly $\dim(N(T - \lambda I)^n)$ times. Notice that this number doesn't depend on our choice of basis for $V$. The proof given below is an adaptation of Theorem 8.10 in [11].

**Theorem 9.21** *Let $V$ be a complex vector space of dimension $n$, and $T \in \mathcal{L}(V)$. For every basis of $V$ such that $M_T$ is upper triangular, a scalar $\lambda \in \mathbb{C}$ appears on the diagonal of $M_T$ exactly* $\dim(N(T - \lambda I)^n)$ *times.*

***Proof.*** Clearly, it suffices to prove the theorem for the case when $\lambda = 0$. The proof is by induction on $n$. Clearly, the assertion holds for $n = 1$. We may thus assume that $n > 1$, and the assertion holds for any complex vector space of dimension $n - 1$. We fix a basis $\{b_1, \ldots, b_n\}$ of $V$ such the related matrix $M_T$ is upper triangular of the form

$$\begin{pmatrix} \lambda_1 & & & & * \\ & \lambda_2 & & & \\ & & \ddots & & \\ & & & \lambda_{n-1} & \\ \mathbf{0} & & & & \lambda_n \end{pmatrix}, \tag{9.5.20}$$

Take $U = \text{Span}(\{b_1, \ldots, b_{n-1}\})$. Clearly, it is a $T$-invariant subspace, and the matrix of the restriction $T\big|_U$ is given by

$$\begin{pmatrix} \lambda_1 & & & * \\ & \lambda_2 & & \\ & & \ddots & \\ \mathbf{0} & & & \lambda_{n-1} \end{pmatrix}, \tag{9.5.21}$$

By our induction hypothesis, 0 appears on the diagonal of the matrix in (9.5.21) exactly $\dim(N((T\big|_U)^{n-1})$ times. However, since $\dim(U) = n - 1$ implies that

$$N\big((T\big|_U)^{n-1}\big) = N\big((T\big|_U)^n\big), \tag{9.5.22}$$

it follows that

$$0 \text{ appears on the diagonal of the matrix in (9.5.21) } \dim(N(T\big|_U)^n) \text{ times.} \tag{9.5.23}$$

To complete the proof, we need to consider the following two cases:

**Case–1**: Let $\lambda_n \neq 0$. In this case, we need to show that

$$N\big(T^n\big) \subset U, \tag{9.5.24}$$

which implies $N(T^n) = N((T\big|_U)^n)$, and so 0 appears on the diagonal of the matrix in (9.5.20) exactly $\dim(N(T^n)$ times, by (9.5.23). Notice that

$$M_{T^n} = M_T^n = \begin{pmatrix} \lambda_1^n & & & & * \\ & \lambda_2^n & & & \\ & & \ddots & & \\ & & & \lambda_{n-1}^n & \\ \mathbf{0} & & & & \lambda_n^n \end{pmatrix}$$

The above gives

$$T^n b_n = u + \lambda_n^n b_n, \quad \text{for some} \quad u \in U.$$

Now, to prove (9.5.24), let $v \in N(T^n)$. We can write

$$v = u' + \alpha\, b_n, \quad \text{where } u' \in U \text{ and } \quad \alpha \in \mathbb{F}.$$

It thus follows that

$$0 = T^n v = T^n u' + \alpha\, T^n b_n = T^n u' + \alpha\, u + \alpha\, \lambda_n^n b_n.$$

Because $T^n u'$ and $\alpha\, u$ are in $U$, and $b_n \notin U$, it follows that $\alpha\, \lambda_n^n = 0$. However, $\alpha\, \lambda_n \neq 0$, so $\alpha = 0$. Therefore, $v = u' \in U$.

**Case–2**: Let $\lambda_n = 0$. In this case, we need to show that

$$\dim\big(N(T^n)\big) = \dim\big(N((T\big|_U)^n)\big) + 1, \tag{9.5.25}$$

which shows 0 appears on the diagonal of the matrix in (9.5.20) exactly $\dim(N(T^n)$ times, by (9.5.23). We have

$$\begin{aligned}\dim\big(N(T^n)\big) &= \dim\big(U \cap N(T^n)\big) + \dim\big(U + N(T^n)\big) - \dim\big(U\big) \\ &= \dim\big(N((T\big|_U)^n)\big) + + \dim\big(U + N(T^n)\big) - \big(n - 1\big).\end{aligned}$$

Suppose we are able to show that $N(T^n) - U \neq \emptyset$, it then follows that

$$n = \dim\big(V\big) \geq \dim\big(U + N(T^n)\big) > \dim\big(U\big) = n - 1,$$

so that we obtain $\dim(U + N(T^n)) = n$. Using this in the previous equation, we get (9.5.25). It thus remains to find a vector in $N(T^n) - U$. For, we may chose $u \in U$ such that $u - b_n \in N((T\big|_U)^n)$. □

**Definition 9.31 (Characteristic Polynomial)** Let $V$ be a complex vector space of dimension $n$, and an operator $T \in \mathcal{L}(V)$ has $m$ distinct eigenvalues $\lambda_1, \ldots, \lambda_m$ with associated multiplicities $d_1, \ldots, d_m$. Then the polynomial

$$\chi\big(M_T\big) := \big(z - \lambda_1\big)^{d_1} \cdots \big(z - \lambda_m\big)^{d_m} \tag{9.5.26}$$

is called the **characteristic polynomial** of $T$.

Since the sum of the multiplicities of all the eigenvalues of $T$ equals $\dim(V)$, the degree of the *characteristic polynomial* $\chi_{\mathbb{C}}(T)$ is $n$. The following interesting identity connects the concepts of the *trace* (9.4.29) and *determinant* (9.4.28) of the matrix $M_T \in M_n(\mathbb{F})$ with the eigenvalues of the matrix $M_T$:

$$\det\big(\lambda\, I - M_T\big) := \lambda^n - \operatorname{tr}\big(M_T\big)\lambda^{n-1} + \cdots + \big(-1\big)^n \det\big(M_T\big). \tag{9.5.27}$$

We thus see that the eigenvalues of $M_T$ are the zeros of the above form of the characteristic polynomial of $M_T$. The next theorem is a fundamental result of linear algebra. It is due to the English mathematician *Arthur Cayley* (1821–1895) and the Irish physicist *William R. Hamilton* (1805–1865).

**Theorem 9.22 (Cayley-Hamilton Theorem)** *Every linear operator on a finite-dimensional complex vector space satisfies its characteristic polynomial.*

***Proof.*** Let $V$ be an $n$-dimensional complex vector space, and $T \in \mathcal{L}(V)$. We have to show that $\chi\big(T\big) = 0$. Let $\{b_1, \ldots, b_n\}$ be a basis of $V$ such that

$$M_T = \begin{pmatrix} \lambda_1 & & & * \\ & \lambda_2 & & \\ & & \ddots & \\ \mathbf{0} & & & \lambda_n \end{pmatrix}, \tag{9.5.28}$$

In view of (9.5.26), it suffices to show that

$$\big[(T - \lambda_1 I)\cdots(T - \lambda_n I)\big]b_k = 0, \quad \text{for} \quad k = 1, \ldots, n. \tag{9.5.29}$$

For $k = 1$, it follows from (9.5.28) that we have $Tb_1 = \lambda_1 b_1$. To complete the proof by induction on $k$, suppose $1 < k \le n$, and that

$$\begin{aligned} 0 &= \big(T - \lambda_1 I\big)b_1 \\ &= \big(T - \lambda_1 I\big)\big(T - \lambda_1 I\big)b_2 \\ &\vdots \\ &= \big(T - \lambda_1 I\big)\cdots\big(T - \lambda_{k-1} I\big)b_{k-1} \end{aligned}$$

Once again, it follows from (9.5.28) that we have

$$\big(T - \lambda_k I\big)b_k \in \text{Span}\,\big(\{b_1, \ldots, b_{k-1}\}\big)$$

Therefore, by our induction hypothesis, we have (9.5.29) holds. □

## 9.6 Inner Product Spaces

In general, an **inner product space** is a vector space $V$ over a real or complex field $\mathbb{F}$ together with a function $\langle\, ,\, \rangle : V \to \mathbb{F}$ such that the following properties hold: For all vectors $u, v, w \in V$ and scalars $\lambda,\ \mu \in \mathbb{F}$, we have

I1 **(conjugate symmetry)** $\langle u,\ v\rangle = \overline{\langle v,\ u\rangle}$;
I2 **(linearity)** $\langle \lambda\, u + \mu\, v,\ w\rangle = \lambda\,\langle u,\ w\rangle + \mu\,\langle v,\ w\rangle$;
I3 **(positive-definiteness)** $u \neq 0 \Rightarrow \langle u,\ u\rangle > 0$.

The function $\langle\,,\rangle$ is called an **inner product** on $V$. We usually write an *inner product space* as a pair $(V, \langle\,,\rangle)$. By $(I1)$, $\langle u,\ u\rangle \in \mathbb{R}$. Taking $\lambda = \mu = 1$ in $(I3)$, we conclude the *inner product* $\langle\,,\rangle$ is **additive** in the first argument; and, taking $\lambda = 1,\ \mu = 0$ in $(I3)$, we conclude **homogeneity** of $\langle\,,\rangle$ in the first argument. Also, it follows from $(I1)$ and $(I2)$ that $\langle\,,\rangle$ is *conjugate-linear* in the second argument, i.e., we have

$$\langle u,\ \lambda\, v\rangle = \overline{\lambda}\,\langle u,\ v\rangle.$$

Similarly, we can deduce some other interesting properties of $\langle\,,\rangle$. In all that follows, we shall use the notation $\langle\,,\rangle$ for an inner product on a real or complex vector space. For example, if $\boldsymbol{u} = (z_1, \ldots, z_n),\ \boldsymbol{v} = (w_1, \ldots, w_n) \in \mathbb{C}^n$, then

$$\langle \boldsymbol{u},\ \boldsymbol{v}\rangle := z_1\overline{w}_1 + \cdots + z_n\overline{w}_n \tag{9.6.1}$$

is an inner product on $\mathbb{C}^n$. Also, on the vector space $P_n[x]$, the function

$$\langle p,\ q\rangle := \int_0^1 p(x)\overline{q(x)}\,dx, \quad \text{for} \quad p, q \in P_n[x], \tag{9.6.2}$$

is an inner product. We shall make use of the following identity repeatedly:

$$\langle u,\ v\rangle = \langle u,\ w\rangle, \ \text{for all}\ u \in V \quad \Rightarrow \quad v = w. \tag{9.6.3}$$

Any finite-dimensional real vector space with an inner product is called an **Euclidean vector space**, and an **Euclidean space** is an *affine space* such that the associated real vector space is an *Euclidean vector space*, where an **affine space** over $\mathbb{R}$ is an abstraction to do affine geometry [12]. We next introduce the concept of "orthogonality" for general inner product spaces.

**Definition 9.32 (Orthogonal Vectors)** Let $(V, \langle\ ,\ \rangle)$ be an *inner product space*, and $u,\ v \in V$. We say $u$ and $v$ are **orthogonal**[2] if $\langle u,\ v\rangle = 0$. In this case, we write $u \perp v$.

It follows from axioms $(I1)$–$(I3)$ that the function $\|\ \| : V \to \mathbf{R}$ given by

$$\|v\| := +\sqrt{\langle v,\ v\rangle}, \qquad \text{for} \quad v \in V, \tag{9.6.4}$$

satisfies the following three properties, for all $u, v \in V$ and $\lambda \in \mathbb{F}$:

N1 (*positive*) $\|v\| \geq 0$; and, $\|v\| = 0 \Leftrightarrow v = 0$.
N2 (*absolute homogeneity*) $\|\lambda\, v\| = |\lambda|\,\|v\|$.
N3 (*triangle inequality*) $\|u + v\| \leq \|u\| + \|v\|$.

We call $\|\ \|$ the **norm** on $V$ induced by the inner product $\langle\ ,\ \rangle$. In geometric terms, we interpret the *norm* $\|v\|$ as the **length** of the vector $v \in V$. In particular, a vector

---

[2] The term is derived from the Greek work *orthogonios* meaning *right-angled*, as in perpendicularity.

$v \in V$ is called a **unit vector** (or a *normalized vector*) if we have $\|v\| = 1$. Recall that we can also interpret $\|v\|$ as the **distance** of $v \in V$ from the origin $0 \in V$.

**Remark 9.4** Most of the facts stated earlier for the real inner product space $\mathbb{R}^n$ hold good for an arbitrary real or complex inner product space $(V, \langle\,,\,\rangle)$. More specifically, the *Cauchy-Schwartz inequality, triangle inequality, Pythagoras theorem,* and *Parallelogram law* hold. It is easy to reformulate the proofs given earlier.

## *Orthonormal Basis*

While studying a configuration in the *standard Euclidean space* $\mathbb{R}^n$, it is convenient to work with a specified orthonormal basis (Definition 9.34). The same idea is used in numerous application areas such as *quantum mechanics, data analysis, machine learning, data compression, signal processing*, and many more. We first describe a recursive algorithm that takes a finite linearly independent set $X$ as input and converts it into an orthonormal set $Y$ such that $\text{Span}(X) = \text{Span}(Y)$. It is known as the *Gram-Schmidt orthogonalization*, which was first applied independently by the Danish mathematician *Jørgen P. Gram* (1850–1916), and the German mathematician *Erhard Schmidt* (1876–1959), in some concrete cases.

To begin with, notice that the standard basis $S = \{\boldsymbol{e}_1, \ldots, \boldsymbol{e}_n\}$ of the Euclidean space $\mathbb{R}^n$ has the following important property: For each vector $\boldsymbol{x} \in \mathbb{R}^n$, we have a unique expression

$$\boldsymbol{x} = \sum_{k=1}^{n} \langle \boldsymbol{x}, \boldsymbol{e}_k \rangle \boldsymbol{e}_k. \tag{9.6.5}$$

This holds because each $\boldsymbol{e}_i \in S$ is a unit vector and $\langle \boldsymbol{e}_j, \boldsymbol{e}_k \rangle = 0$, for all $1 \le j \ne k \le n$.

**Definition 9.33 (Orthonormal Set)** Let $(V, \langle\,,\,\rangle)$ be an inner product space, and $B = \{u_1, \ldots, u_m\} \subset V$. We say $B$ is an **orthonormal set** if we have

$$\langle u_i, u_j \rangle = \delta_{ij}, \quad \text{for all} \quad 1 \le i,\ j \le m. \tag{9.6.6}$$

Notice that every nonzero inner product space contains an orthonormal set: For $0 \ne v \in V$, the vector $u = v/\|v\| \in V$ is such that $\|u\| = 1$. Also, (9.6.6) implies that each $u_i$ is a unit vector, and $u_j \perp u_k$, for all $1 \le j \ne k \le n$, i.e., distinct $u_i$'s are *pairwise orthogonal*. Clearly, the concept of *orthonormality* extends to an infinite linearly independent set if $V$ is infinite dimensional. The next identity follows by repeated applications of the *Pythagoras theorem* (9.1.12):

$$\|\lambda_1 u_1 + \cdots + \lambda_m u_m\| = |\lambda_1|^2 + \cdots + |\lambda_m|^2, \tag{9.6.7}$$

for all scalars $\lambda_i \in \mathbb{F}$. In particular, it follows that every *orthonormal set* is linearly independent. Therefore, with $m = \dim(V)$, it follows that $B$ is a basis of $V$.

**Definition 9.34 (Orthonormal Basis)** Let $(V, \langle\,,\,\rangle)$ be an inner product space. An *orthonormal set* $B \subset V$ is called an **orthonormal basis** if we have $V = \text{Span}(B)$.

For example, the standard basis $S = \{\boldsymbol{e}_1, \ldots, \boldsymbol{e}_n\}$ is an *orthonormal basis* of $\mathbb{R}^n$. The property (9.6.5) holds for any orthonormal basis of a finite dimension inner product space $(V, \langle\,,\,\rangle)$. That is, if $\dim(V) = n$ and $B = \{u_1, \ldots, u_n\}$ is an orthonormal basis of $V$, then for every $v \in V$ we have

$$v = \langle v, u_1\rangle u_1 + \cdots + \langle v, u_n\rangle u_n; \tag{9.6.8a}$$

and so by (9.6.7) we also have

$$\|v\|^2 = |\langle v, u_1\rangle|^2 + \cdots + |\langle v, u_n\rangle|^2. \tag{9.6.8b}$$

We now describe how **Gram-Schmidt procedure** converts a linearly independent set $\{v_1, v_2, \ldots\}$ in to an orthonormal set $\{u_1, u_2, \ldots\}$ such that

$$\text{Span}\left(\{v_1, v_2, \ldots, v_n\}\right) = \text{Span}\left(\{u_1, u_2, \ldots, u_n\}\right), \quad \text{for all} \quad n \geq 1. \tag{9.6.9}$$

To begin with, since $0 \neq v_1 \in V$ implies $u_1 = v_1/\|v_1\|$ is a unit vector, and we have $\text{Span}(\{v_1\}) = \text{Span}(\{u_1\})$. Therefore, (9.6.9) holds for $n = 1$. We next suppose $n > 1$, and that the set $\{v_1, v_2\} \subset V$ is linearly independent. With $u_1$ as above, take $u_2' = v_2 - \langle v_2, u_1\rangle\, u_1$. Since $v_1$ and $v_2$ are linearly independent, we have $u_2' \neq 0$, and

$$\begin{aligned}\langle u_2', u_1\rangle &= \langle v_2, u_1\rangle - \langle v_2, u_1\rangle\langle u_1, u_1\rangle \\ &= \langle v_2, u_1\rangle - \langle v_2, u_1\rangle\, \|u_1\|^2 \\ &= 0, \qquad (\because\ u_1 \text{ is a unit vector})\end{aligned}$$

which implies that $u_2' \perp u_1$. So, for the vector $u_2 = u_2'/\|u_2'\|$, we have

$$\|u_2\| = 1 \quad \text{and} \quad \langle u_1, u_2\rangle = 0.$$

Therefore, $\{u_1, u_2\}$ is an orthonormal set in $V$. Furthermore, since $u_2'$ is the linear combination of $v_1$ and $v_2$, and also $v_2$ is the linear combination of $u_1$ and $u_2$, it follows that $\text{Span}(\{u_1, u_2\}) = \text{Span}(\{u_1, u_2\})$. Proceeding inductively, we find that the nonzero vector $u_n'$ given by

$$u_n' = v_n - \sum_{i=1}^{n-1} \langle v_n, u_i\rangle u_i \tag{9.6.10}$$

is such that $u_n' \perp u_i$, for all $i = 1, \ldots, n-1$, by Exercise 9.54. Therefore, for $u_n = u_n'/\|u_n'\|$, the set $\{u_1, \ldots, u_n\}$ is orthonormal. Finally, since (9.6.10) implies that

$$\mathrm{Span}\left(\{v_1, v_2, \ldots, v_n\}\right) \subset \mathrm{Span}\left(\{u_1, u_2, \ldots, u_n\}\right), \tag{9.6.11}$$

and both these subspaces have the same dimension, we conclude that (9.6.9) holds, for all $n \geq 1$.

**Theorem 9.23** *Let $(V, \langle\,,\,\rangle)$ be a finite-dimensional inner product space. Then $V$ has an orthonormal basis.*

***Proof.*** Let $\dim(V) = n$, and $B = \{v_1, \ldots, b_n\}$ be a basis of $V$. By Gram-Schmidt procedure, we find an orthonormal set $B' = \{u_1, \ldots, u_n\}$ such that (9.6.9) holds. Therefore, $B'$ is an orthonormal basis of $V$. □

More generally, every orthonormal set in a finite-dimensional inner product space can be extended to an orthonormal basis. Also, if an operator $T \in \mathcal{L}(V)$ is such that the matrix representation $M_T$ is upper triangular with respect to some basis, then $M_T$ is also upper triangular with respect to some orthonormal basis of $V$. The next result is known as *Schur's theorem*.

**Theorem 9.24 (Schur's Theorem)** *Let $(V, \langle\,,\,\rangle)$ be a finite-dimensional complex inner product space, and $T \in \mathcal{L}(V)$. Then $M_T$ is an upper triangular matrix with respect to some orthonormal basis of $V$.*

The concept in the next definition will prove to be very useful in the sequel.

**Definition 9.35 (Orthogonal Complement)** Let $(V, \langle\,,\,\rangle)$ be an *inner product space*, and $U \subset V$. The set

$$\left\{v \in V \mid \langle u, v\rangle = 0, \text{ for all } u \in U\right\} \tag{9.6.12}$$

is called the **orthogonal complement** of $U$. It is denoted by $U^\perp$.

It follows easily that the *orthogonal complement* $U^\perp$ of a set $U \subset V$ is a subspace of $V$. In particular, we have $V^\perp = \{0\}$ and $\{0\}^\perp = V$. Also, for two subsets $U_1, U_2 \subset V$, we have $U_1 \subset U_2 \Rightarrow U_1^\perp \supset U_2^\perp$.

**Theorem 9.25** *Let $(V, \langle\,,\,\rangle)$ be a finite-dimensional inner product space, and $U$ be a subspace of $V$. Then we have*

$$V = U \oplus U^\perp, \tag{9.6.13}$$

*where $U^\perp$ is the* orthogonal complement *of $U$.*

***Proof.*** Let $\dim(U) = m$, and $\{u_1, \ldots, u_m\}$ be an orthonormal basis of $U$. For $v \in V$, consider the two elements given by

$$\begin{aligned} u &= \langle v, u_1 \rangle u_1 + \cdots + \langle v, u_m \rangle u_m; \\ w &= v - \langle v, u_1 \rangle u_1 - \cdots - \langle v, u_m \rangle u_m. \end{aligned} \tag{9.6.14}$$

Clearly, we have $u \in U$. Also, for each $j = 1, \ldots, m$,

$$\langle w, u_j \rangle = \langle v, u_j \rangle - \langle v, u_j \rangle = 0$$

implies that $w$ is orthogonal to every vector in the subspace $\mathrm{Span}(\{u_1, \ldots, u_m\})$. So, $w \in U^\perp$. It thus follows that

$$V = U + U^\perp. \tag{9.6.15}$$

Since $v \in U \cap U^\perp \Rightarrow \langle v, v \rangle = 0 \Rightarrow v = 0$, it follows that (9.6.13) holds. □

**Definition 9.36 (Orthogonal Projection)** Let $(V, \langle\ ,\ \rangle)$ be an inner product space, and $U$ be a subspace of $V$. The **orthogonal projection** of $V$ onto $U$, denoted by $\pi_U$, is an operator on $V$ defined by $\pi_U v = u$, where $v = u + w$ is as given by the decomposition (9.6.13).

We can see the *orthogonal projection* $\pi_U$ has the following properties:

$$N(\pi_U) = U^\perp, \quad R(\pi_U) = U, \quad \pi_U^2 = \pi_U.$$

Also, for every $v \in V$, we have

$$v - \pi_U v \in U^\perp \quad \text{and} \quad \|\pi_U v\| \le \|v\|.$$

Moreover, if $\{u_1, \ldots, u_m\}$ is an orthonormal basis of $U$, then

$$\pi_U v = \langle v, u_1 \rangle u_1 + \cdots + \langle v, u_m \rangle u_m, \tag{9.6.16}$$

for every $v \in V$. The *minimization problem*: *Given a finite-dimensional subspace $U$ of an inner product space $V$, and a point $v \in V$, find a $u \in U$ such that $\|v - u\|$ is minimum.* The next theorem tells that $u = \pi_U v$ is the solution, where the right side is usually computed using the formula (9.6.16).

**Theorem 9.26 (Minimization Problem)** *Let $(V, \langle\ ,\ \rangle)$ be an inner product space, $U$ be a finite-dimensional subspace of $V$, and $v \in V$. Then, for every $u \in U$, we have*

$$\|v - \pi_U v\| \le \|v - u\|. \tag{9.6.17}$$

*If the above is an equality, for some $u \in U$, then we have $u = \pi_U v$.*

***Proof.*** For $u \in U$, we have

$$\begin{aligned}\|v - \pi_U v\|^2 &\leq \|v - \pi_U v\|^2 + \|\pi_U v - u\|^2 \\ &= \|(v - \pi_U v) + (\pi_U v - u)\|^2 \\ &= \|v - u\|^2,\end{aligned}$$

where the first equality in above is due to *Pythagoras theorem*, which applies because $v - \pi_U v \in U^\perp$ and $\pi_U v - u \in U$. Therefore, (9.6.17) follows by taking square roots on both sides of the last inequality. Notice that it is an equality if and only if $\|\pi_U v - u\| = 0$. This completes the proof. □

An illustration given in the next example may help to understand the procedure.

**Example 9.14 ([11])** We find a real polynomial $p(x)$ of degree at most 5 over the interval $[-\pi, \pi]$ such that the following integral has the least possible value:

$$\int_{-\pi}^{\pi} |\sin x - p(x)|^2 dx$$

The problem suggests to consider the vector space $V = C[-\pi, \pi]$ of real-valued continuous functions on the interval $[-\pi, \pi]$, with the inner product given by

$$\langle f, g \rangle = \int_{-\pi}^{\pi} f(x)g(x)\, dx, \qquad \text{for} \quad f, g \in V.$$

We have $v(x) = \sin x \in V$, and $U = P_5[x]$. We first apply Gram-Schmidt procedure to the basis $\{1, x, \ldots, x^5\}$ to produce an orthonormal basis $\{u_1, u_2, \ldots, u_6\}$ of $U$. Subsequently, using the above inner product and (9.6.16) (with $m = 6$), we obtain

$$\pi_U v = 0.987862\, x - 0.15527\, x^3 + 0.00564312\, x^5$$

The verification of details is left to the reader as an exercise.

## *Dual Space and Adjoint Operator*

Let $V$ be a vector space over a field $\mathbb{F}$. A function $T : V \to \mathbb{F}$ is called a *functional* on $V$. A **linear functional** is a functional on $V$ that is also a linear map. In all that follows, we will write *linear functional* simply as a functional. The set of functionals on a vector space $V$ is denoted by $V^*$. It is called the **algebraic dual** (or simply **dual**) of the vector space $V$. For example, consider the vector space $P[x]$ of real polynomials over the interval $[0, 1]$. The functional $\varphi : P[x] \to \mathbb{R}$ given by

$$\varphi\, p := \int_0^1 p(x)\, dx, \quad \text{for} \quad p \in P[x],$$

is a functional on $P[x]$. In general, the *dual* $V^*$ of a vector space $V$ over a field $\mathbb{F}$ is itself a vector space over the same field, where the addition and scalar multiplication are given by (9.4.2).

Suppose $\dim(V) = n$, and $B = \{b_1, \ldots, b_n\}$ is a basis of $V$. Now, given a functional $\phi \in V^*$, and a vector $v = \sum_{i=1}^{n} \lambda_i b_i \in V$, we have

$$\phi\, v = \sum_{i=1}^{n} \lambda_i \phi\, b_i = \lambda_1 \mu_1 + \cdots + \lambda_n \mu_n,$$

where $\mu_i = \phi\, b_i \in \mathbb{F}$. Conversely, for a given

$$\boldsymbol{\mu} = \big(\mu_1, \ldots, \mu_n\big) \in \mathbb{F}^n, \tag{9.6.18}$$

there is a unique $\phi \in V^*$ such that

$$\phi\, b_j = \mu_j, \quad \text{for} \quad j = 1, \ldots, n. \tag{9.6.19}$$

We can define such a functional $\phi : V \to \mathbb{F}$ as follows: For $v = \sum_{j=1}^{n} \lambda_j b_i$,

$$\phi\, v := \sum_{j=1}^{n} \lambda_j \mu_j. \tag{9.6.20}$$

For $v = \sum_{j=1}^{n} \alpha_j b_j$, $w = \sum_{j=1}^{n} \beta_j b_j$, and scalars $\lambda,\ \mu \in \mathbb{F}$, we have

$$\begin{aligned}
\phi\big(\lambda v + \mu w\big) &= \sum_{j=1}^{n} \big(\lambda \alpha_j + \mu \beta_j\big) \beta_j \\
&= \lambda \left( \sum_{j=1}^{n} \alpha_j \beta_j \right) + \mu \left( \sum_{j=1}^{n} b_j \beta_j \right) \\
&= \lambda \phi\, v + \mu \phi\, w,
\end{aligned}$$

which shows $\phi \in V^*$. Notice that this $\phi$ satisfies the relation (9.6.19), which implies the functional $\phi$ is necessarily unique. For, if there is a second $\varphi \in V^*$ that satisfies the relation $\varphi\big(b_j\big) = \beta_j$, for $j = 1, \ldots, n$, then

$$\varphi\big|_B = \phi\big|_B \quad \Rightarrow \quad \varphi = \phi \;\text{ on }\; V,$$

because $\operatorname{Span}(B) = V$. In particular, it follows that every functional $\phi$ on $\mathbb{F}^n$ is given by the *dot product*. That is, for each $\boldsymbol{\lambda} = (\lambda_1, \dots, \lambda_n) \in \mathbb{F}^n$, there exists a unique $\boldsymbol{\mu} = (\mu_1, \dots, \mu_n) \in \mathbb{F}^n$ such that

$$\phi \boldsymbol{\lambda} = \boldsymbol{\lambda} \cdot \boldsymbol{\mu} = \lambda_1 \mu_1 + \cdots + \lambda_n \mu_n. \tag{9.6.21}$$

This assertion holds even for some infinite-dimensional inner product spaces such as $\ell_p$ spaces. The next theorem proves that the dual $V^*$ is the same as $V$, up to an isomorphism, when $V$ is finite dimensional.

**Theorem 9.27** *Let $V$ be an $n$-dimensional vector space over a field $\mathbb{F}$. Then we have the dual $V^*$ is isomorphic to $V$.*

***Proof.*** Let $B = \{b_1, \dots, b_n\}$ be a basis of $V$. Taking $\boldsymbol{\mu}$ in (9.6.18) to be the $i$th standard vectors $\boldsymbol{e}_i \in \mathbb{F}^n$, we find a unique functional $\phi_i : V \to \mathbb{F}$ such that we have

$$\phi_i \, b_k = \delta_{ik}, \quad \text{for} \quad i = 1, \dots, n, \tag{9.6.22}$$

by (9.6.19). We thus obtain a unique $n$-tuple $(\phi_1, \dots, \phi_n)$ of elements in $V^*$. To complete the proof, we show that $B' = \{\phi_1, \dots, \phi_n\}$ is a basis of $V^*$. First, we prove $B'$ is linearly independent. For, if $\sum_{i=1}^n \gamma_i \phi_i = 0$ in $V^*$, for some scalars $\gamma_i$, then

$$\sum_{i=1}^n \gamma_i \phi_i = 0 \quad \Rightarrow \quad \left( \sum_{i=1}^n \gamma_i \phi_i \right) b_k = 0$$

for all $k = 1, \dots, n$. So, we must have $\gamma_k = 0$, for all $k = 1, \dots, n$, by (9.6.22). Next, if $\phi \in V^*$ with $\phi \, b_i = \gamma_i$, for $i = 1, \dots, n$, then for $v = \sum_{i=1}^n \beta_i b_i \in V$, we have

$$\begin{aligned} \phi \, v &= \phi \left( \sum_{i=1}^n \beta_i b_i \right) = \sum_{i=1}^n \beta_i \phi \, x_i \\ \Rightarrow \quad \phi \, v &= \sum_{i=1}^n \beta_i \gamma_i = \left( \sum_{i=1}^n \gamma_i \phi_i \right) v. \end{aligned}$$

That is, $\phi = \sum_{i=1}^n \gamma_i \phi_i$, which shows $\operatorname{Span}(B') = V^*$. Therefore, $B'$ is a basis of $V^*$. Hence, the mapping $b_i \mapsto \phi_i$, for $i = 1, \dots, n$, extends to an isomorphism $V \to V^*$. This completes the proof. □

For each $v \in V$, there is a *natural functional* $e_v : V^* \to \mathbb{F}$ given by

$$e_v \varphi := \varphi \, v, \qquad \text{for} \quad \varphi \in V^*, \tag{9.6.23}$$

which is called the **evaluation map** on $V^*$. Therefore, the *double dual* $V^{**}$ of $V$ is non-trivial, unless $V$ is the trivial vector space. Further, for $v \in V$, the function

$\Phi : V \to V^{**}$ given by $\Phi(v) = e_v$ is a linear map. Notice that, when $V$ is finite dimensional, we have $V \cong V^{**}$, by Theorem 9.27.

We now prove that (9.6.21) holds for a finite-dimensional inner product space. Notice that, for any fixed $v_0 \in V$, the mapping $v \mapsto \langle v, v_0 \rangle$ is a functional on $V$, by the linearity of the inner product. The next theorem proves that every element of $V^*$ is essentially of this form only.

**Theorem 9.28 (Representation Theorem)** *Let $(V, \langle\, , \rangle)$ be a finite-dimensional inner product space, and $\varphi \in V^*$. Then there exists a unique $v_0 \in V$ such that*

$$\varphi\, v = \langle v, v_0 \rangle, \qquad \textit{for all} \quad v \in V. \tag{9.6.24}$$

***Proof.*** Let $\dim(V) = n$, and $\{u_1, \ldots, u_n\}$ be an orthonormal basis. By (9.6.8a), for any $v \in V$, we can write

$$v = \langle v, u_1 \rangle u_1 + \cdots + \langle v, u_1 \rangle u_1$$

so that we obtain

$$\begin{aligned} \varphi\, v &= \langle v, u_1 \rangle \varphi(u_1) + \cdots + \langle v, u_n \rangle \varphi(u_n) \\ &= \langle u, \overline{\varphi(u_1)}\, u_1 + \cdots + \overline{\varphi(u_n)}\, u_n \rangle \end{aligned}$$

We may thus take $v_0 = \overline{\varphi(u_1)}\, u_1 + \cdots + \overline{\varphi(u_n)}\, u_n$. Finally, the uniqueness of $v_0$ in the statement of the theorem follows from (9.6.3). □

The concept we introduce next is fundamental to many applications in STEM.

**Definition 9.37 (Adjoint)** Let $V$ and $W$ be two inner product spaces, where the inner product in both cases is denoted by $\langle\, , \rangle$, and let $T \in \mathcal{L}(V, W)$. For a fixed $w \in W$, let $\varphi \in V^*$ be given by $v \mapsto \langle Tv, w \rangle$. By Theorem 9.28, there is a unique vector $T^*w \in V$ such that we have

$$\langle Tv, w \rangle = \langle v, T^*w \rangle, \qquad \text{for all} \quad v \in V. \tag{9.6.25}$$

The linear map $T^* \in \mathcal{L}(W, V)$ is called the **adjoint** of $T$.

The next example illustrates the computation of *adjoint* $T^*$ in a concrete situation.

**Example 9.15** We first consider a linear map $T \in \mathcal{L}(\mathbb{R}^3, \mathbb{R}^2)$ given by

$$T(x, y, z) = (z, 3x - 2y), \qquad \text{for} \quad \boldsymbol{x} = (x, y, z) \in \mathbb{R}^3.$$

Since the adjoint $T^* \in \mathcal{L}(\mathbb{R}^2, \mathbb{R}^3)$ must satisfy (9.6.25), we have

$$\langle T(x, y, z), (u, v) \rangle = \langle (x, y, z), T^*(u, v) \rangle, \qquad \text{for all} \quad \boldsymbol{x} = (u, v) \in \mathbb{R}^2.$$

We thus obtain

$$\begin{aligned}\big\langle (x, y, z), T^*(u, v)\big\rangle &= \big\langle (z, 3x - 2y), (u, v)\big\rangle \\ &= zu + 3xv - 2yv = \big\langle (x, y, z), (3v, -2v, u)\big\rangle,\end{aligned}$$

for all $(x, y, z) \in \mathbb{R}^3$. Therefore, by applying (9.6.3), it follows that

$$T^*\big(u, v\big) = \big(3v, -2v, u\big), \qquad \text{for all} \quad \big(u, v\big) \in \mathbb{R}^2.$$

We may recall that the **conjugate transpose** of a matrix $A = \big(a_{ij}\big) \in M_{m\times n}(\mathbb{F})$ is the matrix $A^* = \big(b_{ij}\big) \in M_{n\times m}(\mathbb{F})$ obtained by interchanging the rows and columns of $A$ followed by taking complex-conjugate of each entry. That is, we have $b_{ij} := \overline{a_{ji}}$, for all $i, j$. For example, the two matrices given below are mutually *complex transpose*:

$$\begin{pmatrix} 1 & 5 \\ 2 + 3i & 7 \\ -4 & -4i \end{pmatrix} \quad \text{and} \quad \begin{pmatrix} 1 & 2 - 3i & -4 \\ 5 & 7 & 4i \end{pmatrix}$$

In particular, over the field $\mathbb{R}$, we recover the concept of the **transpose** of a real matrix. The next theorem proves that, with respect to orthonormal bases, the matrix of a linear map $T \in \mathcal{L}(V, W)$ and the matrix of its adjoint $T^* \in \mathcal{L}(W, V)$ are complex transpose of each other.

**Theorem 9.29** **(Complex Transpose)** *Let $V$ and $W$ be two finite-dimensional inner product spaces, say of dimensions $n$ and $m$, respectively. Let $T \in \mathcal{L}(V, W)$, $B_V := \{v_1, \ldots, v_n\}$ be an orthonormal basis of $V$, and $B_W := \{u_1, \ldots, u_m\}$ be an orthonormal basis of $W$. Then the matrix $M_T \in M_{m\times n}(\mathbb{F})$ of $T$ with respect to $B_V$ and the matrix $M_{T^*} \in M_{n\times m}(\mathbb{F})$ of the adjoint $T^*$ with respect to basis $B_W$ are complex transpose of each other.*

***Proof.*** By (9.6.8a), we have

$$Tv_k = \big\langle Tv_k, u_1\big\rangle u_1 + \cdots + \big\langle Tv_k, u_m\big\rangle u_m, \quad \text{for} \quad k = 1, \ldots, n.$$

We thus have that the $(j, k)$-entry of $M_T$ is given by $\langle Tv_k, u_j\rangle$. By the same reasoning, and using the relation (9.6.25), it follows that the $(j, k)$-entry of $M_{T^*}$ is given by $\langle T^*u_k, v_j\rangle$. Now, since we have

$$\big\langle T^*u_k, v_j\big\rangle = \big\langle u_k, Tv_j\big\rangle = \overline{\big\langle Tv_j, u_k\big\rangle},$$

we conclude that $M_{T^*}$ is the complex transpose of $M_T$. □

## *Special Operators*

We start with the next definition.

**Definition 9.38 (Special Operators)** Let $V$ be an inner product space, and $T \in \mathcal{L}(V)$. We say

1. $T$ is **self-adjoint** if $T = T^*$.
2. $T$ is **normal** if $TT^* = T^*T$.
3. $T$ is **positive semidefinite** (or simply **positive**) if $T$ is self-adjoint and we have $\langle Tv, v\rangle \geq 0$, for all $v \in V$.

Further, an operator $S \in \mathcal{L}(V)$ is called a **square root** of $T$ if we have $S^2 = T$.

Some important observations related to these special operators that we need in the sequel are recorded below. In some cases, proofs are included. First, notice that sum of two self-adjoint operators is a self-adjoint operator, and so is a real multiple of a *self-adjoint operator*. In general, a self-adjoint operator $T$ on a complex inner product space $V$ behaves largely like *complex conjugation*. For example, *every eigenvalue of $T$ is a real number*. For, if we have $Tv = \lambda\, v$ $(v \neq 0)$, then

$$\begin{aligned}\lambda\, \|v\|^2 &= \big\langle \lambda\, v, v\big\rangle = \big\langle Tv, v\big\rangle = \big\langle v, T^*v\big\rangle \\ &= \big\langle v, Tv\big\rangle = \big\langle v, \lambda\, v\big\rangle = \overline{\lambda}\, \|v\|^2,\end{aligned}$$

implies that $\lambda = \overline{\lambda}$, i.e., $\lambda$ is real. Moreover, the condition

$$\big\langle Tv, v\big\rangle \in \mathbb{R}, \qquad \text{for all} \quad v \in V, \tag{9.6.26}$$

characterizes $T$ as a self-adjoint operator. The details are left for the reader as an exercise. Next, every self-adjoint operator is a *normal operator*, but the converse is not true. It is easy to construct such matrix examples. The condition

$$\big\|Tv\big\| = \big\|T^*v\big\|, \qquad \text{for all} \quad v \in V, \tag{9.6.27}$$

characterizes $T$ as a normal operator. For, we have

$$\begin{aligned} T \text{ is normal } &\Leftrightarrow\ T^*T - TT^* = 0 \\ &\Leftrightarrow\ \big\langle (T^*T - TT^*)v, v\big\rangle = 0, \quad \text{for all} \quad v \in V, \\ &\Leftrightarrow\ \big\langle (T^*T)v, v\big\rangle = \big\langle (TT^*)v, v\big\rangle, \quad \text{for all} \quad v \in V, \\ &\Leftrightarrow\ \big\|Tv\big\|^2 = \big\|T^*v\big\|^2, \quad \text{for all} \quad v \in V, \end{aligned}$$

In particular, for a normal operator $T$, we have $N(T) = N(T^*)$, and also

$$Tv = \lambda\, v \quad \Rightarrow \quad T^*v = \overline{\lambda}\, v, \tag{9.6.28}$$

for the same $0 \neq v \in V$ and $\lambda \in \mathbb{F}$. This follows from (9.6.27), and the fact that $(T - \lambda I)$ is also a normal operator. The above observations also imply that eigenvectors of $T$ corresponding to distinct eigenvalues are orthogonal. For, if we have $Tu = \lambda_1 u$ and $Tv = \lambda_2 v$, with $\lambda_1 \neq \lambda_2$ and $u,\ v$ nonzero in $V$, then

$$(\lambda_1 \neq \lambda_2)\langle u, v\rangle = \langle \lambda_1 u, v\rangle - \langle u, \overline{\lambda_2} v\rangle = \langle Tu, v\rangle - \langle u, T^* v\rangle = 0,$$

which implies that $\langle u, v\rangle = 0$, i.e., $u$ and $v$ are orthogonal. The *spectral theorem* proved next is a fundamental fact about the normal operators.

**Theorem 9.30 (Complex Spectral Theorem)** *Let V be a finite-dimensional complex inner product space, and $T \in \mathcal{L}(V)$. Then T is normal if and only if V has an orthonormal basis consisting of eigenvectors of T.*

***Proof.*** If $V$ has an orthonormal basis $B_V$ consisting of eigenvectors of $T$, then by Theorem 9.29 the matrix $M_T$ of $T$ and the matrix $M_{T^*}$ of the adjoint $T^*$ with respect to the same orthonormal basis $B_V$ are complex transpose of each other. Also, since $M_T$ is a diagonal matrix, so is $M_{T^*}$. Therefore, $T$ commutes with $T^*$. To prove the converse, suppose $T$ is a normal operator on $V$. Let $\dim(V) = n$. By Theorem 9.24, the matrix $M_T$ is upper triangular with respect to some orthonormal basis of $V$, say $B = \{v_1, \ldots, v_n\}$. Let

$$M_T = \begin{pmatrix} a_{11} & a_{12} & \cdots & a_{1n} \\ 0 & a_{22} & \cdots & a_{2n} \\ & & \ddots & \vdots \\ 0 & & \cdots & a_{nn} \end{pmatrix} \tag{9.6.29}$$

The proof is completed by showing that all off-diagonal entries of $M_T$ are necessarily zero. For, since

$$\|Tv_1\|^2 = |a_{11}|^2 \quad \text{and} \quad \|T^* v_1\|^2 = |a_{11}|^2 + |a_{12}|^2 + \ldots + |a_{1n}|^2,$$

and also by (9.6.27) we have $\|Tv_1\| = \|T^* v_1\|$, it follows that all entries $a_{12}, \ldots, a_{1n}$ are zero, expect possibly $a_{11}$. Again, since

$$\|Tv_2\|^2 = |a_{22}|^2 \quad \text{and} \quad \|T^* v_2\|^2 = |a_{22}|^2 + |a_{23}|^2 + \ldots + |a_{2n}|^2,$$

and also by (9.6.27) we have $\|Tv_2\| = \|T^* v_2\|$, it follows that all entries $a_{23}, \ldots, a_{2n}$ are zero, expect possibly $a_{22}$. Continuing with the same procedure, it follows that all off-diagonal entries of $M_T$ are zero. Therefore, $M_T$ in (9.6.29) is diagonal. □

We conclude our discussion by proving some interesting facts about the *positive operators*. We know that when $V$ is a complex inner product space, an operator $T \in \mathcal{L}(V)$ is self-adjoint if and only if $\langle Tv, v\rangle \in \mathbb{R}$, for all $v \in V$. In such a situation, we can drop the assumption that $T$ is self-adjoint. Some simple examples of positive operators include the classes of orthogonal projections, and also operators $S \in \mathcal{L}(V)$ of the form $S := T^2 + aT + b$, where $T \in \mathcal{L}(V)$ is a self-adjoint operator and $a^2 < 4b$ in $\mathbb{R}$. To prove the later, let $0 \neq v \in V$. We then have

$$\begin{aligned}
\langle (T^2 + aT + b)v, v\rangle &= \langle T^2 v, v\rangle + a\langle Tv, v\rangle + b\langle v, v\rangle \\
&= \langle Tv, Tv\rangle + a\langle Tv, v\rangle + b\,\|v\|^2 \\
&\geq \|Tv\|^2 - |a|\,\|Tv\|\,\|v\| + b\,\|v\|^2 \\
&= \left(\|Tv\| - \frac{|a|\|v\|}{2}\right)^2 + \left(b - \frac{a^2}{4}\right)\|v\|^2 \\
&> 0,
\end{aligned}$$

where the first inequality follows by using Cauchy-Schwartz inequality (see Remark 9.4). The characterization of positive operators $T \in \mathcal{L}(V)$ given in the next theorem is the main result that we need in the sequel.

**Theorem 9.31** *Let $V$ be an $n$-dimensional inner product space, and $T \in \mathcal{L}(V)$. Then the following are equivalent:*

1. *$T$ is positive;*
2. *$T$ is self-adjoint, and all eigenvalues in the* spectrum $\operatorname{spec}(T)$ *are non-negative;*
3. *$T$ has a positive* square root*;*
4. *$T$ has a self-adjoint square root;*
5. *there exists an operator $S \in \mathcal{L}(V)$ such that $T = S^*S$.*

***Proof.*** First, we prove (1) $\Rightarrow$ (2). For, if $T$ is positive, then $T$ is self-adjoint, by definition. Also, if $0 \neq v \in V$ is an eigenvector corresponding to an eigenvalue $\lambda$ of $T$, then

$$0 \leq \langle Tv, v\rangle = \langle \lambda\, v, v\rangle = \lambda\,\langle v, v\rangle$$

implies that $\lambda \geq 0$. Next, to prove (2) $\Rightarrow$ (3), suppose $T$ is a self-adjoint operator such that all eigenvalues of $T$ are non-negative. By Theorem 9.30, $V$ has an orthonormal basis $\{v_1, \ldots, v_n\}$ consisting of the eigenvectors of $T$. Let $\lambda_1, \ldots, \lambda_n$ be the corresponding eigenvalues. Notice that each $\lambda_i$ is non-negative, by assumption. It then follows that the operator $S \in \mathcal{L}(V)$ defined by

$$Sv_i = \sqrt{\lambda_i}\, v_i, \qquad \text{for} \quad i = 1, \ldots, n,$$

is positive, and satisfies $S^2 = T$. The implication (3) $\Rightarrow$ (4) is trivial. Now, suppose there exists a self-adjoint operator $S \in \mathcal{L}(V)$ such that $T = S^2$. Since $S^* = S$, we have (4) $\Rightarrow$ (5) holds. To conclude the proof, suppose (5) holds. Then the condition $T = S^*S$ implies that $T$ is self-adjoint. Also, since

$$\langle Tv, v\rangle = \langle S^*Sv, v\rangle = \langle Sv, Sv\rangle \geq 0, \quad \text{for} \quad v \in V,$$

we have $T$ is positive. □

The next theorem justifies writing the square root of an operator $T$ as $\sqrt{T}$.

**Theorem 9.32** *Let $V$ be an inner product space. Every positive operator $T \in \mathcal{L}(V)$ has a unique square root.*

***Proof.*** Let $\lambda_1, \ldots, \lambda_m$ denote the distinct eigenvalues of $T$ so that, by Theorem 9.31, we have all $\lambda_j$'s are non-negative. Since $T$ is self-adjoint, by Exercise 9.61, we have

$$V = N(T - \lambda_1 I) \oplus + \cdots \oplus N(T - \lambda_m I), \tag{9.6.30}$$

Suppose $S \in \mathcal{L}(V)$ is a positive square root of $T$, and a scalar $\alpha$ is an eigenvalue of $S$. For $v \in N(S - \alpha I)$, we thus have $Sv = \alpha v$, and so

$$Tv = S^2v = \alpha^2 v \quad \Rightarrow \quad v \in N(T - \alpha^2 I). \tag{9.6.31}$$

Therefore, $\alpha^2 = \lambda_j$, for some $j = 1, \ldots, m$. That is, $\alpha = \sqrt{\lambda_j}$, for some $j$. Moreover, by (9.6.31), we have

$$N(S - \sqrt{\lambda_j}\, I) \subset N(T - \lambda_j I). \tag{9.6.32}$$

It thus follows that $\sqrt{\lambda_1}, \ldots, \sqrt{\lambda_m}$ are the only possible eigenvalues of the operator $S$. Since $S$ is positive, by Exercise 9.61, we can write

$$V = N(T - \sqrt{\lambda_1}\, I) \oplus + \cdots \oplus N(T - \sqrt{\lambda_m}\, I) \tag{9.6.33}$$

Hence, using (9.6.30), (9.6.33), and (9.6.32), we conclude that

$$N(S - \sqrt{\lambda_j}\, I) = N(T - \lambda_j I), \quad \text{for all} \quad j = 1, \ldots, m.$$

Said differently, on the null space $N(T - \lambda_j I)$, the operator $S$ is *multiplication by* $\sqrt{\lambda_j}$. This shows that $T$ uniquely determines $S$. □

We may recall that $GL(n, \mathbb{F})$ is the group of nonsingular matrices of order $n$ over a field $\mathbb{F}$. A matrix $A \in GL(n, \mathbb{F})$ is said to be **orthogonal** if we have $AA^t = I$. The set of orthogonal matrices is denoted by $O(n, \mathbb{F})$ (or simply as $O(n)$). Since $\det(A^t) = \det(A)$, for any matrix $A \in M_n(\mathbb{R})$, it follows that $\det(A) = 1$ or $\det(A) = -1$, for any $A \in O(n)$. Therefore, the set

$$SO(n) = \{A \in O(n) \mid \det(A) = 1\},$$

is a normal subgroup of the group $O(n)$. It is called the group of **special orthogonal matrices** over the field $\mathbb{F}$. We also call it the group of **rotations**, especially when

$\mathbb{F}$ is the field of real numbers. The next theorem characterizes the elements of the groups $O(n)$ and $SO(n)$. As before, our emphasis is to think in terms of operators on an inner product space $V$.

**Definition 9.39** Let $V$ be an inner product space, and $T \in \mathcal{L}(V)$. We say $T$ is an **isometry** on an $V$ if it is a *length preserving operators*, i.e., $\|Tv\| = \|v\|$, for all $v \in V$.

The term *isometry* is derived from the Greek work *isometría* meaning equal measurement. Notice that every *isometry* on a finite-dimensional inner product space is an invertible operator. In particular, if the scalars $\lambda_1, \dots, \lambda_n$ are of unit modulus, and we have $Tv_j = \lambda_j v_j$, for some orthonormal basis $\{v_1, \dots, v_n\}$ of $V$, then (9.6.8a) and (9.6.8b) implies that $T$ is an isometry on $V$.

**Theorem 9.33 (Isometry)** *Let $V$ be an $n$-dimensional inner product space, and $T \in \mathcal{L}(V)$. Then the following are equivalent*

1. *$T$ is an isometry;*
2. *$\langle Tu, Tv\rangle = \langle u, v\rangle$, for all $u, v \in V$;*
3. *$T^*T = I$;*
4. *For each orthonormal set $\{v_1, \dots, v_m\}$ in $V$, we have $\{Tv_1, \dots, Tv_m\}$ is also an orthonormal set;*
5. *there exists an orthonormal basis $\{v_1, \dots, v_n\}$ of $V$ such that $\{Tv_1, \dots, Tv_n\}$ is also orthonormal;*
6. *$T^*$ is an isometry;*
7. *$\langle T^*u, T^*v\rangle = \langle u, v\rangle$, for all $u, v \in V$;*
8. *$TT^* = I$;*
9. *For each orthonormal set $\{v_1, \dots, v_m\}$ in $V$, we have $\{T^*v_1, \dots, T^*v_m\}$ is also an orthonormal set;*
10. *there exists an orthonormal basis $\{v_1, \dots, v_n\}$ of $V$ such that $\{T^*v_1, \dots, T^*v_n\}$ is also orthonormal.*

***Proof.*** The proof is left for the reader as an exercise. □

We can also see that $AA^t = I$, where the matrix $A := M_T$ of an isometry $T$ obtained with respect to some orthonormal basis of $V$. It thus follows that a matrix $A \in O(n)$ if and only if the column vectors of $A$ are orthonormal if and only if the row vectors of $A$ are orthonormal.

We may recall that, for a unit vector $\boldsymbol{u} \in \mathbb{R}^n$, a **reflection** in the hyperplane $H = \boldsymbol{u}^\perp$ is an operator $\theta \in \mathcal{L}(\mathbb{R}^n)$ given by

$$\theta(\boldsymbol{x}) = \boldsymbol{x} - 2\langle \boldsymbol{x}, \boldsymbol{u}\rangle \boldsymbol{u}.$$

Compare this with (9.4.15).

**Definition 9.40 (Householder Matrix)** The **Householder matrix** $H$ of a unit vector $\boldsymbol{u} \in \mathbb{R}^n$, and for reflection in the line $L(\boldsymbol{u})^{\perp}$, is defined by

$$H = I - 2\,\boldsymbol{u}\,\boldsymbol{u}^t. \tag{9.6.34}$$

The *Householder matrix* is a symmetric orthogonal matrix, and we have

$$H\big(\boldsymbol{u}\big) = \boldsymbol{u} - 2\,\boldsymbol{u}\big(\boldsymbol{u}\boldsymbol{u}^t\big) = -\boldsymbol{u}. \tag{9.6.35}$$

Also, if $\boldsymbol{w} \perp \boldsymbol{u}$, then we have

$$H\big(\boldsymbol{w}\big) = \boldsymbol{w} - 2\,\boldsymbol{u}\,\boldsymbol{u}^t\,\boldsymbol{w} = \boldsymbol{w}. \tag{9.6.36}$$

So, the Householder matrix $H$ induces reflection in the plane perpendicular to the line $L(\boldsymbol{u})$.

## 9.7 Some Important Applications

We discuss methods of *matrix decompositions, least square fitting, computations of matrix norms*, and some of their applications such as those related to *data compression* and the *principal component analysis*. We need a *matrix decomposition* to optimize the machine-related calculations. The main point is that, while dealing with *big data matrices*, technical issues such as algorithmic complexity and computational efficiency become critical. A matrix decomposition may also help to understand the geometry of the related data transformations. We recommend [13–15], for further studies. One may learn to implement efficient matrix algorithms by reading any standard text on the subject *numerical linear algebra* [16, 17].

### *Matrix Decompositions*

A **matrix decomposition** takes a matrix as input and returns certain special types of matrices as its factors that may be *triangular matrices, diagonal matrices*, or *orthogonal matrices*. We start with the *lower-upper decomposition* (or simply *LU-decomposition*). It is the simplest form of matrix decomposition, which is commonly applied to solve smaller order square systems of linear equations. It also helps in computing determinants and finding the inverse of a matrix.

**Definition 9.41 (LU Decomposition)** Let $A$ be a square matrix. By an **LU-decomposition** of $A$ we mean a (unique) factorization of the form $A = L\,U$, where $L$ is a lower triangular matrix, and $U$ is an upper triangular matrix. We also say that $A$ is **LU-decomposable**.

An *LU-decomposition* $A = L\,U$ is *unique* when the factor $L$ has all diagonal entries 1. In this case, $L$ is called *unit lower triangular* matrix. Suppose a square matrix $A$ is *LU-decomposable*. We can find elementary matrices $E_1, \ldots, E_n$ such that

$$L = (E_1^{-1} \cdots E_n^{-1}) \quad \text{and} \quad U = E_n \cdots E_1 A. \tag{9.7.1}$$

More precisely, start with the equation $A = IA$. To obtain an LU-decomposition of $A$, provided $(1, 1)$-entry of $A$ is nonzero, we apply a sequence of row operations to $A$, and the corresponding inverse column operations to $I$:

$$\begin{aligned} A = IA &= \big(IE_1^{-1}\big)\big(E_1A\big) \\ &= \big(IE_1^{-1}E_2^{-1}\big)\big(E_2E_1A\big) \\ &\;\vdots \\ &= \big(IE_1^{-1}\cdots E_n^{-1}\big)\big(E_n\cdots E_1A\big) \\ &= L\,U. \end{aligned}$$

This algorithm is known as the *Gaussian elimination with partial pivoting* (or the *Gaussian elimination with row pivoting*). The next example illustrates the procedure.

**Example 9.16** We may write

$$A = \begin{pmatrix} 1 & 1 & 0 \\ 1 & 2 & 1 \\ 2 & 1 & 1 \end{pmatrix} = \begin{pmatrix} 1 & 0 & 0 \\ 0 & 1 & 0 \\ 0 & 0 & 1 \end{pmatrix} \begin{pmatrix} 1 & 1 & 0 \\ 1 & 2 & 1 \\ 2 & 1 & 1 \end{pmatrix}$$

Use the row operations $R_2 \to R_2 - R_1$; $R_3 \to R_3 - 2R_1$ to reduce the second factor, and the corresponding inverse column operations $C_1 \to C_1 + C_2$; $C_1 \to C_1 + 2C_3$ on the first factor. We thus obtain

$$A = \begin{pmatrix} 1 & 0 & 0 \\ 1 & 1 & 0 \\ 2 & 0 & 1 \end{pmatrix} \begin{pmatrix} 1 & 1 & 0 \\ 0 & 1 & 1 \\ 0 & -1 & 1 \end{pmatrix}$$

Now, apply the row operation $R_3 \to R_3 + R_2$ to the second matrix and the inverse column operation $C_2 \to C_2 - C_3$ to the first matrix. We get

$$A = \begin{pmatrix} 1 & 0 & 0 \\ 1 & 1 & 0 \\ 2 & -1 & 1 \end{pmatrix} \begin{pmatrix} 1 & 1 & 0 \\ 0 & 1 & 1 \\ 0 & 0 & 2 \end{pmatrix},$$

which is an *LU-decomposition* of the matrix $A$.

However, not every square matrix is LU-decomposable. For, consider the matrix

$$A = \begin{pmatrix} 0 & 1 \\ 1 & 0 \end{pmatrix}.$$

By comparing $A$ with the product

$$LU = \begin{pmatrix} 1 & 0 \\ c & 1 \end{pmatrix} \begin{pmatrix} a & b \\ 0 & d \end{pmatrix} = \begin{pmatrix} a & b \\ ac & bc + d \end{pmatrix}$$

it follows that $a = 0$. Therefore, $A = L\,U$ holds if and only if all the entries in the first column of $A$ are zero. Hence, the matrix $A$ is not LU-decomposable. Moreover, since $A$ is nonsingular, we conclude that not every nonsingular matrix is LU-decomposable. In general, for $a \neq 0$, we have

$$\begin{pmatrix} a & b \\ c & d \end{pmatrix} = \begin{pmatrix} 1 & 0 \\ c/a & 1 \end{pmatrix} \begin{pmatrix} a & b \\ 0 & (ad - bc)/a \end{pmatrix}. \tag{9.7.2}$$

More generally, every nonsingular matrix $A$ of order $n$ can be factored as

$$A = P\,L\,U \quad \text{or} \quad P^t A = LU,$$

where $P$ is a permutation matrix, $L$ is unit lower triangular matrix, and $U$ an upper triangular nonsingular matrix. Notice that $P^t A$ is the matrix obtained from $A$ by re-ordering the rows. The idea of LU-decompositions extends easily to rectangular matrices, say of the type $m \times n$, such that $L$ is a square unit lower triangular matrix, and $U$ an $m \times n$ rectangular matrix. The LU-decomposition is the standard method of solving a general system of linear equations.

We next define the *Cholesky decomposition* of a square matrix. Recall that a square matrix $A$ of order $n$ is *symmetric* if $A = A^T$; and, *positive semidefinite* if $\boldsymbol{x}^t A \boldsymbol{x} > 0$, for all nonzero $n$-vectors $\boldsymbol{x}$. We may recall that if all eigenvalues of $A$ are non-negative, then it is a positive semidefinite matrix.

**Definition 9.42 (Cholesky Decomposition)** Let $A$ be a square matrix. A **Cholesky decomposition** of $A$ is a factorization of the form $A = L\,L^t$, where $L$ is a lower triangular matrix with positive diagonal entries. We also call $L$ the **square root** of the matrix $A$.

For a square matrix of order two, the iterative formula to find the *square root* $L$ of $A$ is obtained by writing

$$\begin{pmatrix} a_{11} & a_{12} \\ a_{21} & a_{22} \end{pmatrix} = \begin{pmatrix} \ell_{11} & 0 \\ \ell_{12} & \ell_{22} \end{pmatrix} \begin{pmatrix} \ell_{11} & \ell_{12} \\ 0 & \ell_{22} \end{pmatrix} = \begin{pmatrix} \ell_{11}^2 & \ell_{11}\ell_{12} \\ \ell_{11}\ell_{12} & \ell_{12}^2 \ell_{22}^2 \end{pmatrix},$$

Using the symmetry, we compute $\ell_{ij}$'s columnwise as follow:

$$\ell_{11}=\sqrt{a_{11}}, \qquad \ell_{12}=\frac{1}{\ell_{11}}a_{21}, \qquad \ell_{22}=\frac{1}{\sqrt{\ell_{12}}}\sqrt{a_{22}}. \tag{9.7.3}$$

Similarly, for a matrix of order three, the formula to find $L$ is found by writing

$$\begin{aligned}\begin{pmatrix}a_{11}&a_{12}&a_{13}\\a_{21}&a_{22}&a_{23}\\a_{31}&a_{32}&a_{33}\end{pmatrix}&=\begin{pmatrix}\ell_{11}&0&0\\\ell_{21}&\ell_{22}&0\\\ell_{31}&\ell_{32}&\ell_{33}\end{pmatrix}\begin{pmatrix}\ell_{11}&\ell_{21}&\ell_{31}\\0&\ell_{22}&\ell_{32}\\0&0&\ell_{33}\end{pmatrix}\\&=\begin{pmatrix}\ell_{11}^2&\ell_{11}\ell_{21}&\ell_{11}\ell_{31}\\\ell_{21}\ell_{11}&\ell_{21}^2+\ell_{22}^2&\ell_{21}\ell_{31}+\ell_{22}\ell_{32}\\\ell_{31}\ell_{11}&\ell_{31}\ell_{21}+\ell_{32}\ell_{22}&\ell_{31}^2+\ell_{32}^2+\ell_{33}^2\end{pmatrix}\end{aligned}$$

Once again, we compute $\ell_{ij}$'s columnwise:

$$\begin{aligned}\ell_{11}&=\sqrt{a_{11}}, & \ell_{22}&=\sqrt{a_{22}-\ell_{21}^2}\,,\\ \ell_{21}&=\frac{1}{\ell_{11}}a_{21}, & \ell_{32}&=\frac{1}{\ell_{22}}\left(a_{32}-\ell_{31}\ell_{21}\right),\\ \ell_{31}&=\frac{1}{\ell_{11}}a_{31}, & \ell_{33}&=\sqrt{a_{33}-\ell_{31}^2-\ell_{32}^2}\end{aligned} \tag{9.7.4}$$

The next example illustrates the procedure.

**Example 9.17** To find the square root $L$ of a matrix of order three, we write

$$\begin{pmatrix}1&3&5\\3&13&23\\5&23&42\end{pmatrix}=\begin{pmatrix}\ell_{11}^2&\ell_{11}\ell_{21}&\ell_{11}\ell_{31}\\\ell_{21}\ell_{11}&\ell_{21}^2+\ell_{22}^2&\ell_{21}\ell_{31}+\ell_{22}\ell_{32}\\\ell_{31}\ell_{11}&\ell_{31}\ell_{21}+\ell_{32}\ell_{22}&\ell_{31}^2+\ell_{32}^2+\ell_{33}^2\end{pmatrix}$$

By (9.7.4), we have

$$\begin{aligned}\ell_{11}&=1, & \ell_{22}&=\sqrt{13-9}=2,\\ \ell_{21}&=3, & \ell_{32}&=\frac{1}{2}\left(23-15\right)=4,\\ \ell_{31}&=5, & \ell_{33}&=\sqrt{42-25-16}=1\end{aligned}$$

The Cholesky decomposition of the given matrix is thus as follows:

$$\begin{pmatrix}1&3&5\\3&13&23\\5&23&42\end{pmatrix}=\begin{pmatrix}1&0&0\\3&2&0\\5&4&1\end{pmatrix}\begin{pmatrix}1&3&5\\0&2&4\\0&0&1\end{pmatrix}$$

In general, the following iteration formulas hold

$$
\begin{aligned}
\ell_{ii} &= \left(a_{ii} - \sum_{j=1}^{i-1} \ell_{ij}^2\right)^{1/2}, \quad \text{for } i = 1, \dots, n; \\
\ell_{ij} &= \frac{1}{\ell_{jj}}\left(a_{ij} - \sum_{k=1}^{j-1} \ell_{ik}\ell_{jk}\right), \quad \text{for } i = 1, \dots, n \text{ and } j = 1 \dots, i-1.
\end{aligned}
\tag{9.7.5}
$$

The Cholesky decomposition for a positive semidefinite matrix $A$ is nearly twice as efficient as a LU-decomposition. More precisely, no pivoting is needed, and $U = L^t$ implies only half the operations are performed. It finds applications in solving linear least squares for regression, and also in various simulation and optimization techniques.

**Definition 9.43 (QR Decomposition)** A **QR-decomposition** of a square real matrix $A$ is a factorisation of the form $A = QR$, where $Q$ is orthogonal and $R$ is upper triangular. For a complex matrix $A$, the factor $Q$ is a unitary matrix, i.e., $Q^* = Q^{-1}$.

The related *QR-algorithm* is used to solve linear least square problems. The columns of $Q$ consist of orthogonal unit vectors. *QR-decomposition* for a nonsingular matrix is unique, subject to that the diagonal entries of $R$ are positive. We can obtain QR-decomposition by applying the *Gram-Schmidt orthogonalization*, *Householder transformations*, or *Givens rotations*. In general, we apply the Gram-Schmidt procedure to the linearly independent column vectors of $A$, which we may write as $\boldsymbol{a}_1, \dots, \boldsymbol{a}_k$, where $k = \text{rank}(A)$. For now, suppose $A$ is a nonsingular real matrix of order $n$ so that we have

$$
\begin{aligned}
&\boldsymbol{u}_1 = \boldsymbol{a}_1, \qquad \boldsymbol{e}_1 = \frac{\boldsymbol{u}_1}{\|\boldsymbol{u}_1\|}; \\
&\boldsymbol{u}_2 = \boldsymbol{a}_2 - \langle \boldsymbol{a}_2, \boldsymbol{e}_1\rangle \boldsymbol{e}_1, \qquad \boldsymbol{e}_2 = \frac{\boldsymbol{u}_2}{\|\boldsymbol{u}_2\|}; \\
&\ \vdots \\
&\boldsymbol{u}_n = \boldsymbol{a}_n - \langle \boldsymbol{a}_n, \boldsymbol{e}_1\rangle \boldsymbol{e}_1 - \cdots - \langle \boldsymbol{a}_n, \boldsymbol{e}_{n-1}\rangle \boldsymbol{e}_{n-1}, \qquad \boldsymbol{e}_n = \frac{\boldsymbol{u}_n}{\|\boldsymbol{u}_n\|}
\end{aligned}
$$

The QR-decomposition is then given by

$$
A = \left\{ \boldsymbol{e}_1 \,\middle|\, \boldsymbol{e}_2 \,\middle|\, \cdots \,\middle|\, \boldsymbol{e}_n \right\} \left\{ \begin{array}{cccc} \langle \boldsymbol{a}_1, \boldsymbol{e}_1\rangle & \langle \boldsymbol{a}_2, \boldsymbol{e}_1\rangle & \cdots & \langle \boldsymbol{a}_n, \boldsymbol{e}_1\rangle \\ 0 & \langle \boldsymbol{a}_2, \boldsymbol{e}_2\rangle & \cdots & \langle \boldsymbol{a}_n, \boldsymbol{e}_2\rangle \\ \vdots & \vdots & \ddots & \vdots \\ 0 & 0 & \cdots & \langle \boldsymbol{a}_n, \boldsymbol{e}_n\rangle \end{array} \right\}
\tag{9.7.6}
$$

We only need to find the vector $\boldsymbol{e}_1, \dots, \boldsymbol{e}_n$, because the rest of the computation follows directly using the dot products $\langle \boldsymbol{a}_i, \boldsymbol{e}_j\rangle$. The next examples illustrate the procedure.

**Example 9.18 (Gram-Schmidt Orthogonalisation)** Let $\boldsymbol{a}_1 = (1 \quad 1 \quad 0)^t$, $\boldsymbol{a}_2 = (1 \ 0 \ 1)^t$, and $\boldsymbol{a}_3 = (0 \ 1 \ 1)^t$. By Gram-Schmidt procedure, we have

$$
\begin{aligned}
\boldsymbol{u}_1 &= \boldsymbol{a}_1 = (1\ 1\ 0)^t, \\
\boldsymbol{e}_1 &= \frac{\boldsymbol{u}_1}{\|\boldsymbol{u}_1\|} = \frac{1}{\sqrt{2}}(1\ 1\ 0)^t; \\
\boldsymbol{u}_2 &= (1\ 0\ 1)^t - \frac{1}{\sqrt{2}}\big\langle(1\ 0\ 1)^t, (1\ 1\ 0)^t\big\rangle(1\ 1\ 0)^t = \frac{1}{2}(1\ -1\ 1)^t, \\
\boldsymbol{e}_2 &= \frac{\boldsymbol{u}_2}{\|\boldsymbol{u}_2\|} = \frac{1}{\sqrt{6}}(1\ -1\ 2)^t; \\
\boldsymbol{u}_3 &= \boldsymbol{a}_3 - \big\langle\boldsymbol{a}_3, \boldsymbol{e}_1\big\rangle\boldsymbol{e}_1 - \big\langle\boldsymbol{a}_3, \boldsymbol{e}_2\big\rangle\boldsymbol{e}_2 \\
&= (0\ 1\ 1)^t - \frac{1}{2}\big\langle(0\ 1\ 1)^t, (1\ 1\ 0)^t\big\rangle(1\ 1\ 0)^t - \frac{1}{4}\big\langle(0\ 1\ 1)^t, (1\ -1\ 1)^t\big\rangle(-1\ 1\ 1)^t \\
&= \frac{1}{\sqrt{3}}(-1\ 1\ 1)^t, \\
\boldsymbol{e}_3 &= \frac{\boldsymbol{u}_3}{\|\boldsymbol{u}_3\|} = \frac{1}{\sqrt{3}}(-1\ 1\ 1)^t
\end{aligned}
$$

It thus follow from (9.7.6) that

$$
\begin{pmatrix} 1 & 1 & 0 \\ 1 & 0 & 1 \\ 0 & 1 & 1 \end{pmatrix} = \begin{pmatrix} \frac{1}{\sqrt{2}} & \frac{1}{\sqrt{6}} & \frac{-1}{\sqrt{3}} \\ \frac{1}{\sqrt{2}} & \frac{-1}{\sqrt{6}} & \frac{1}{\sqrt{3}} \\ 0 & \frac{2}{\sqrt{6}} & \frac{1}{\sqrt{3}} \end{pmatrix} \begin{pmatrix} \frac{2}{\sqrt{2}} & \frac{1}{\sqrt{2}} & \frac{1}{\sqrt{2}} \\ 0 & \frac{3}{\sqrt{6}} & \frac{1}{\sqrt{6}} \\ 0 & 0 & \frac{2}{\sqrt{3}} \end{pmatrix}
$$

The upper triangular matrix is obtained directly by using the dot products.

More generally, if $A$ is an $m \times n$ rectangular matrix with linearly independent columns, then by applying the Gram-Schmidt procedure we obtain an $m \times n$ matrix with columns as the orthonormal vectors. So, by keeping track of column operation, it is possible to write $A = QR$, where $R$ is an $n \times n$ upper triangular matrix with all diagonal entries positive. We next discuss *eigendecomposition* of a matrix. In this case, factors are determined by the eigenvalues and eigenvectors of the matrix.

**Definition 9.44 (Eigendecomposition)** An **eigendecomposition** of a square matrix $A$ of order $n$ is a factorisation of the form

$$A = B\,\Sigma_n\,B^{-1}, \tag{9.7.7}$$

where the diagonal matrix $\Sigma_n$ has $n$ eigenvalues of $A$ as the diagonal.

If $\lambda_1, \ldots, \lambda_n$ are $n$ eigenvalues of $A$ and $\boldsymbol{b}_1, \ldots, \boldsymbol{b}_n$ are the corresponding eigenvectors, then we can also write (9.7.7) as

$$A\,\boldsymbol{b}_i = \lambda_i\,\boldsymbol{b}_i, \qquad \text{for} \quad i = 1, \ldots, n. \tag{9.7.8}$$

Therefore, the $n$ columns of the matrix $B$ are the $n$ eigenvectors of $A$. For a normal or real symmetric matrix, an eigendecomposition is also known as the **spectral decomposition** (Theorem 9.30). We may recall that a matrix $A$ is normal if $AA^* = A^*A$. Every normal matrix $A$ of order $n$ can be *eigendecomposed* as $A = B\,\Sigma_n\,B^*$. In this case, we can choose the eigenvectors of $A$ to be orthonormal so that we also have $BB^* = I$ (Theorem 9.33). In particular, all unitary, Hermitian, and also skew-Hermitian matrices, are eigendecomposable. We often sort eigenvalues in decreasing order, say $\lambda_1 \geq \lambda_2 \geq \cdots \geq \lambda_n$, and write the product $A = B\,\Sigma_n\,B^t$ as

$$A = \sum_{i=1}^{n} \lambda_i\, \boldsymbol{b}_i\, \boldsymbol{b}_i^t. \tag{9.7.9}$$

For a *diagonalizable* matrix $A$, eigendecomposition always exists. Equivalently, when $A$ has $n$ linearly independent eigenvectors. We know that this latter condition holds when $A$ has $n$ distinct eigenvalues.

**Definition 9.45 (Singular Values Decomposition)** Let $A$ be an $n \times m$ real matrix. The **singular value decomposition** (or simply SVD) of $A$ is a factorisation of the form

$$A = U\,\Sigma\,V^t, \tag{9.7.10}$$

where $U$ and $V$ are orthogonal matrices of orders $n$ and $m$, respectively, and $\Sigma$ is an $n \times m$ rectangular diagonal matrix with non-negative real values as the diagonal. These diagonal values are called the **singular values** of the matrix $A$.

As above, it can be arranged to have the *singular values* in $\Sigma$ placed in descending order. The number of nonzero singular values equals the rank of $A$. The columns of $U$ are called the **left-singular vectors** of $A$, and the columns of $V$ are called the **right-singular vectors**. In geometric terms, the right side of (9.7.11) consists of a rotation, followed by a rescaling transformation, and again a rotation. When $A$ is a complex matrix, $U$ and $V$ are unitary matrices, and in place of $V^t$ we have the conjugate transpose $V^*$ of the matrix $V$. The *singular value decomposition* is a general type of factorization that finds applications in numerous subject areas. In data science, it applies to high dimensional data matrix of the form

$$A = \begin{pmatrix} | & | & & | \\ \mathbf{X}_1 & \mathbf{X}_2 & \cdots & \mathbf{X}_m \\ | & | & & | \end{pmatrix}, \tag{9.7.11}$$

where each $\mathbf{X}_j \in \mathbb{F}^n$ represents data for a *dominating feature* of the dataset. In the majority of cases, the value $n$ is much larger than the value $m$.

**Theorem 9.34** *Every real matrix A admits a decomposition of the form* (9.7.11).

***Proof.*** Suppose $A$ be an $n \times m$ matrix. Since $A^t A$ is symmetric and positive semidefinite, there exists an $m \times m$ orthogonal matrix $V$ whose column vectors are the eigenvectors $\boldsymbol{v}_i$ of the matrix $A^t A$ having non-negative eigenvalues, say $\lambda_1, \ldots, \lambda_m$, by Theorem 9.31. That is,

$$A^t A\, \boldsymbol{v}_i = \lambda_i\, \boldsymbol{v}_i, \quad \text{for} \;\; i = 1, 2, \ldots, m. \tag{9.7.12}$$

Suppose $r := \text{rank}(A)$. We may assume that $\lambda_1 \geq \lambda_2 \geq \cdots \geq \lambda_r > 0$, and $\lambda_j = 0$, for $j = r+1, \ldots, m$. Take $\sigma_i := \sqrt{\lambda_i}$, for $i = 1, 2, \ldots, m$. It thus follows that

$$\mathbf{v}_i^t A^t A\, \boldsymbol{v}_i = \lambda_i\, \boldsymbol{v}_i^t \boldsymbol{v}_i = \lambda_i, \quad \text{for} \;\; i = 1, 2, \ldots, m. \tag{9.7.13}$$

We thus have

$$\left\| A\boldsymbol{v}_i \right\| = \sigma_i, \quad \text{for} \;\; i = 1, 2, \ldots, m. \tag{9.7.14}$$

Therefore, vectors $\boldsymbol{u}_i := A\boldsymbol{v}_i / \sigma_i$ form an orthonormal basis of the space $R(A)$:

$$\boldsymbol{u}_i^t \boldsymbol{u}_i = \frac{(A\boldsymbol{v}_i)^t A\boldsymbol{v}_j}{\sigma_i \sigma_j} = \frac{\lambda_j\, \boldsymbol{v}_i^t \boldsymbol{v}_j}{\sigma_i \sigma_j} = \delta_{ij} \tag{9.7.15}$$

We can add to it an orthonormal basis $\{\boldsymbol{u}_{r+1}, \ldots, \boldsymbol{u}_n\}$ of the null space $N(A^t)$ so that $U = \left[\boldsymbol{u}_1, \boldsymbol{u}_2, \ldots, \boldsymbol{u}_n\right]$ is an orthogonal matrix. Finally, since $A\boldsymbol{v}_i = \sigma_i \boldsymbol{u}_i$ for all $i$, we obtain the desired singular value decomposition as

$$A = U\, \Sigma\, V^t, \quad \text{with} \;\; \Sigma = \text{diag}\left(\sigma_1, \sigma_2, \ldots, \sigma_r, 0, \ldots, 0\right) \tag{9.7.16}$$

This completes the proof. □

We can thus interpret the singular value decomposition of a rectangular matrix $A$ in terms of the eigendecomposition of matrices $A^t A$ and $AA^t$. The *left-singular vectors* of $A$ are precisely the eigenvectors of the square matrix $AA^t$; and, the *right-singular vectors* of $A$ are precisely the eigenvectors of the square matrix $A^t A$. Also, the nonzero singular values of $A$ are obtained by taking the square roots of the eigenvalues of the matrix $A^t A$ or $AA^t$. The *polar decomposition* of a square matrix is a special case of the singular value decomposition.

**Corollary 9.1 (Polar Decomposition)** *If $A$ is a square matrix of order $n$, then there exists an orthogonal matrix $O$ and a positive matrix $P$ such that $A = OP$.*

***Proof.*** Let $A = U\, \Sigma\, V^t$ be a singular value decomposition of $A$. Now, for the orthogonal matrix $O = UV^t$ and positive matrix $P = V\, \Sigma\, V^t$, we have

$$A = UV^t \left(V\, \Sigma\, V^t\right) = O\, P. \tag{9.7.17}$$

This completes the proof. □

The next theorem gives a geometric interpretation of the largest *singular value*.

**Theorem 9.35 (Largest Singular Value)** *Let $A$ be an $n \times m$ real matrix of rank $r$. Then the largest singular values of $A$ are given by*

$$\sigma_1 = \max\left\{\left\|A\mathbf{x}\right\| \mid \mathbf{x} \in \mathbb{S}^{m-1}\right\}, \tag{9.7.18}$$

*where $\mathbb{S}^{m-1} \subset \mathbb{R}^m$ is the* unit sphere.

***Proof.*** Let $\mathbf{v}_1, \mathbf{v}_2, \ldots, \mathbf{v}_m$ be an orthonormal basis of $\mathbb{R}^m$ consisting of eigenvectors of the matrix $A^t A$, with eigenvalues given by

$$\sigma_1 \geq \sigma_2 \geq \cdots \geq \sigma_r > 0 \geq 0 \geq \cdots \geq 0 \tag{9.7.19}$$

We may write

$$\mathbf{x} := c_1\mathbf{v}_1 + c_2\mathbf{v}_2 + \cdots + c_m\mathbf{v}_m, \quad \text{with} \quad c_i \in \mathbb{R}, \tag{9.7.20}$$

so that we obtain

$$\begin{aligned}\left\|A\mathbf{x}\right\|^2 &= \mathbf{x}^t A^t A\mathbf{x} \\ &= \mathbf{x} \cdot \left(c_1\sigma_1^2\mathbf{v}_1 + c_2\sigma_2^2\mathbf{v}_2 + \cdots + c_r\sigma_r^2\mathbf{v}_r\right) \\ &= c_1^2\sigma_1^2 + c_2^2\sigma_2^2 + \cdots + c_r^2\sigma_r^2\end{aligned} \tag{9.7.21}$$

It thus follows from (9.7.21) that

$$\left\|A\mathbf{x}\right\|^2 \leq \sigma_1^2\left(c_1^2 + c_2^2 + \cdots + c_m^2\right) \leq \sigma_1^2, \tag{9.7.22}$$

if we have $\|\mathbf{x}\| = 1$. Notice that the equality in (9.7.22) holds if $\mathbf{x} = \mathbf{v}_1$. This completes the proof. □

We can use singular value decomposition to generalize the concept of the inverse of a matrix. The next example illustrates the procedure.

**Example 9.19** .

The next remark gives a brief on the development of the concept.

**Remark 9.5** The concept of SVD for a square matrix was first introduced in 1870 by the Italian mathematician *Eugenio Beltrami* (1835–1990) and the French mathematician *Camille Jordan* (1838–1922). In 1913, the French engineer and mathematician *Léon Autonne* (1859–1916) found independently such decompositions for complex matrices. In 1936, the American physicist *Carl Eckhart* (1902–1973) and the English mathematician *Grace Young* (1868–1944) extended the concept to rectangular matrices. In 1965, the American mathematician *Gene H Golub* (1932–2007) and the Canadian mathematician and computer scientist *William Kahan* (born 1933) introduced a numerical version of SVD. In 1970, Golub proposed an algorithm to compute the singular values of a rectangular matrix. The interested reader may refer [18], for more details.

In today's world of rapidly evolving communication technologies, we need to have high data transmission rates and enough data storage capacities. In particular, for applications related to an *image processing*, the first step is to read the image in grayscale, which involves the acquisition of a large dataset of numbers arranged as an $m \times n$ matrix that represents values of each pixel. A common strategy to obtain such a real matrix is by using *colorimetry*[3] that is based on the *principle of constant luminance*. It also involves encoding and restructuring to reduce the size of data for efficient storage and transmission (see [19] or [20]). This is the realm of **data compression**. Approximation of a large dataset by a linear polynomial is also a typical case. We introduce below *singular value decomposition* based method of *data compression*.

## *Matrix Norms*

In this part, we consider matrix norms. We start with the next definition.

**Definition 9.46** A **norm** on a vector space $V$ over a field $\mathbb{F}$ is a function $\| \; \| : V \to \mathbf{R}$ such that, for all $u, v \in V$ and $\lambda \in \mathbb{F}$, we have

(1) (positivity) $\|v\| \geq 0$; and, $\|v\| = 0 \Leftrightarrow v = 0$.
(2) (scaling) $\|\lambda\, v\| = |\lambda|\, \|v\|$.
(3) (triangle inequality) $\|u + v\| \leq \|u\| + \|v\|$.

We also call $\|x\|$ the **vector norm** of $x \in V$.

Every inner product $\langle\, ,\, \rangle$ on a vector space $V$ gives a *vector norm* defined by (9.6.4). We also call it the vector norm *induced by the inner product* on $V$. In particular, on the Euclidean space $\mathbb{F}^d$, we have the following four norms: For $\boldsymbol{x} = (x_1, \ldots, x_d)$, let

(a) Manhattan or 1-norm: $\|(x_1, \ldots, x_d)\|_1 := |x_1| + \cdots + |x_d|$.
(b) Euclidean or 2-norm: $\|(x_1, \ldots, x_d)\|_2 := \left(x_1^2 + \cdots + x_d^2\right)^{1/2}$.
(c) Minkowski's $p$-norm: $\|(x_1, \ldots, x_d)\|_p := \left(x_1^p + \cdots + x_d^p\right)^{1/p}$.
(d) Maximum or $\infty$-norm: $\|(x_1, \ldots, x_d)\|_\infty := \max\left\{|x_1|, \ldots, |x_d|\right\}$.

Notice that 1-norm and 2-norm are the special cases of Minkowski's $p$-norm. In numerous subject areas, we also need to consider norms on the vector space

[3] The **colorimetry** is the science of quantifying and describing the human color preception.

$M_{m\times n}(\mathbb{F})$, which are called the *matrix norms* [21]. We introduced below some that find applications in the analysis and design of algorithms for *numerical computations*.

**Definition 9.47 (Matrix Norm)** A norm $|||\ |||$ defined on the vector space $M_{m\times n}(\mathbb{F})$ is called a **matrix norm**. A *matrix norm* $|||\ |||$ is called **sub-multiplicative** if we have

$$|||A\,B||| \leq |||A|||\ |||B|||\,, \qquad \text{for all} \quad A,\ B \in M_{m\times n}(\mathbb{F}). \tag{9.7.23}$$

In what follows, we may assume that $m = n$, i.e., matrices under consideration are square matrix of order $n$. Also, $\mathbb{F}$ is a real or complex field. In particular, when both $A$, $B$ in (9.7.23) are the identity matrix $I$, it follows that we always have $|||I||| \geq 1$. Some vector norms such as those defined above may be used to define *matrix norms* on the vector space $M_n(\mathbb{F})$. The one introduced in the next definition is known as the *column-sum matrix norm*.

**Definition 9.48 (Column-Sum Norm)** For $A = \big(a_{ij}\big)$, the largest **column-sum norm** of $A$ is given by

$$\big\|A\big\|_1 := \max_j \sum_{i=1}^{n} \big|a_{ij}\big|. \tag{9.7.24}$$

It is also known as the 1-norm.

Analogous to the one given in Definition 9.48, the matrix norm introduced in the next definition is called the *row-sum matrix norm*.

**Definition 9.49 (Row-Sum Norm)** For $A = \big(a_{ij}\big)$, the largest **row-sum norm** of $A$ is given by

$$\big\|A\big\|_\infty := \max_i \sum_{j=1}^{n} \big|a_{ij}\big|. \tag{9.7.25}$$

It is also known as the **infinite norm**.

The matrix norm $|||A|||$ of the matrix $A$ given in the next definition is called the *induced norm*. It is derived viewing the matrix $A$ as an operator $\boldsymbol{x} \mapsto A\boldsymbol{x} : \mathbb{F}^n \to \mathbb{F}^n$. We also say that matrix norm $|||A|||$ of $A$ is "subordinate to the vector norm" $\|x\|$.

**Definition 9.50 (Operator Norm)** The **operator norm** of $A$ is defined by

$$|||A||| := \sup_{\boldsymbol{x}\neq\boldsymbol{0}} \frac{\|A\,\boldsymbol{x}\|}{\|\boldsymbol{x}\|} = \sup_{\|x\|=1} \big\{\big\|A\,\boldsymbol{x}\big\| \mid \boldsymbol{x} \in \mathbb{F}^n\big\}. \tag{9.7.26}$$

The *operator norm* (9.7.26) satisfies the *compatible condition*:

$$\big\|A\boldsymbol{x}\big\| \leq |||A|||\ \big\|\boldsymbol{x}\big\|, \qquad \text{for all} \quad \boldsymbol{x} \in \mathbb{F}^n. \tag{9.7.27}$$

We can use the above condition to show that (9.7.23) holds. Therefore, $A \mapsto |||A|||$ is a norm on the vector space $M_n(\mathbb{F})$. We can also define the operator norm as

$$|||A||| := \inf_{x\in\mathbb{F}^n} \{\lambda \in \mathbb{R} \mid \|Ax\| \le \lambda\|x\|\}. \tag{9.7.28}$$

In particular, it follows that $|||I||| = 1$. We cannot use the operator norm (9.7.26) to compute the norm of a matrix. Instead, we use a **subordinate matrix norm** of $A \in M_n(\mathbb{F})$, which is obtained using a specific vector norm $\|\ \|$ on the right side of the definition (9.7.26). In this way, the two types of matrix norms we introduced earlier are called the **induced matrix norms**.

**Theorem 9.36** *For every matrix $A = (a_{ij}) \in M_n(\mathbb{R})$, we have*

1. $\|A\|_\infty = \sup_{\|x\|_\infty=1} \{\|Ax\|_\infty \mid x \in \mathbb{R}^n\} = \max_i \sum_{j=1}^n |a_{ij}|.$
2. $\|A\|_1 = \sup_{\|x\|_1=1} \{\|Ax\|_1 \mid x \in \mathbb{R}^n\} = \max_j \sum_{i=1}^n |a_{ij}|.$
3. $\|A\|_2 = \sup_{\|x\|_2=1} \{\|Ax\|_2 \mid x \in \mathbb{R}^n\} = \sqrt{\rho(A^t A)} = \sqrt{\rho(A\,A^t)}.$

*Furthermore, $\|A^T\|_2 = \|A\|_2$, the norm $\|\ \|_2$ is orthogonally invariant, which means that $\|A\|_2 = \|UAV\|_2$, for all orthogonal matrices $U$, $V$, and if $A$ is a normal matrix, then $\|A\|_2 = \rho(A)$.*

***Proof.*** For a fixed $\mathbf{0} \ne x = (x_i, \dots, x_n) \in \mathbb{R}$, let $N = \|x\|_\infty$ so that we have $|x_j| \le N$, for $j = 1, \dots, n$. We then have

$$|(Ax)_i| = \left|\sum_{j=1}^n a_{ij}x_j\right| \le \sum_{j=1}^n |a_{ij}|\,|x_j| \le N \sum_{j=1}^n |a_{ij}|. \tag{9.7.29}$$

We may take

$$K = \max_{1\le i\le n} \sum_{j=1}^n |a_{ij}|. \tag{9.7.30}$$

It thus follows that

$$\begin{aligned}\frac{\|Ax\|_\infty}{\|x\|_\infty} &= \frac{\max_{1\le i\le n} |(Ax)_i|}{\|x\|_\infty} = \frac{\max_{1\le i\le n} |(Ax)_i|}{N}\\ &\le \frac{\max_{1\le i\le n} \left(N \sum_{j=1}^n |a_{ij}|\right)}{N} = K, \qquad \text{for all} \quad \mathbf{0} \ne x \in \mathbb{F}.\end{aligned}$$

Therefore, $\|A\|_\infty \le K$. To show that $\|A\|_\infty \ge K$, let $m$ be an index $i$ in (9.7.30) such that $K = \sum_{j=1}^n |a_{mj}|$, and define $\mathbf{0} \ne x = (x_i, \dots, x_n)$ so that $x_j$ and $a_{mj}$ have the same sign; and, $|x_j| = 1$, for $j = 1, \ldots, n$. We may take $x_j = 1$ or $-1$ when $a_{mj} = 0$. We have

$$\left|\sum_{j=1}^{n} a_{mj}x_j\right| = \sum_{j=1}^{n} |a_{mj}|\,|x_j|.$$

It thus follows that

$$\begin{aligned}\|A\|_\infty \|\boldsymbol{x}\|_\infty \geq \|A\boldsymbol{x}\|_\infty &= \max_{1\leq i\leq n}\left|\sum_{j=1}^{n} a_{ij}x_j\right| \\ &\geq \left|\sum_{j=1}^{n} a_{mj}x_j\right| = \sum_{j=1}^{n} |a_{mj}|\,|x_j| = K.\end{aligned}$$

Since $\|\boldsymbol{x}\|_\infty = 1$, we obtain

$$\|A\|_\infty \|\boldsymbol{x}\|_\infty \geq \|A\boldsymbol{x}\|_\infty \geq C\,\|\boldsymbol{x}\|_\infty.$$

Hence, $\|A\|_\infty \geq K$. This proves (1). The proof of (2) follows from this using the fact that $\|A\|_1 = \|A^t\|_\infty$. Finally, to prove (3), let $\lambda_1, \ldots, \lambda_n$ be the non-negative eigenvalues of the matrix $A^tA$. By *eigendecomposition*, we can find an orthogonal matrix $Q$ and a diagonal matrix $D$ with $\lambda_i$'s as diagonal values such that $A^tA = Q^tDQ$. Therefore, we obtain

$$\|A\boldsymbol{x}\|_2^2 = (A\boldsymbol{x})^t A\boldsymbol{x} = \boldsymbol{x}^t(A^tA)\boldsymbol{x} = (Q\boldsymbol{x})^t D(Q\boldsymbol{x}).$$

So, by taking $\boldsymbol{y} = Q\boldsymbol{x}$, we have

$$\begin{aligned}\|A\|_2^2 &= \max_{\|\boldsymbol{x}\|_2=1} \|A\boldsymbol{x}\|_2^2 = \max_{\|\boldsymbol{x}\|_2=1} (Q\boldsymbol{x})^t D(Q\boldsymbol{x}) \\ &= \max_{\|\boldsymbol{x}\|_2=1} \boldsymbol{y}^t D\boldsymbol{y} = \max_{\|\boldsymbol{x}\|_2=1} \sum_{i=1}^{n} \lambda_i y_i^2 \\ &\leq \left(\max_{1\leq i\leq n} \lambda_i\right) \max_{\|\boldsymbol{x}\|_2=1} \|\boldsymbol{y}\|_2^2\end{aligned}$$

Since we also have

$$\|\boldsymbol{y}\|_2^2 = (Q\boldsymbol{x})^t Q\boldsymbol{x} = \boldsymbol{x}^t Q^t Q\boldsymbol{x} = \boldsymbol{x}^t I\boldsymbol{x} = \|\boldsymbol{x}\|_2^2 = 1,$$

taking the square root on both sides of the last inequality gives proof of (3). The proof of the remaining assertions is left for the reader as an exercise. □

The norm $\| \ \|_2$ on $M_n(\mathbb{F})$ is often called the **spectral norm**. The following properties hold for any *subordinate matrix norm*:

(a) the matrix and vector norms are compatible, i.e., (9.7.27) holds;
(b) the matrix norm is a *sub-multiplicative* norm, i.e., it satisfies the condition (9.7.23).
(c) $\|I\| = 1$, where $I \in M_n(\mathbb{F})$ is the identity matrix.

The norm introduced in the next definition is not an *induced norm*.

**Definition 9.51 (Frobenius Norm)** The norm on $M_n(\mathbb{F})$ defined by

$$\|A\|_F := \left( \sum_{i,j=1}^{n} |a_{ij}|^2 \right)^{1/2}, \tag{9.7.31}$$

is called the **Frobenius norm**, where $A = \left(a_{ij}\right) \in M_n(\mathbb{F})$.

We may also define the *Frobenius norm* on $M_n(\mathbb{F})$ as

$$\|A\|_F := \sqrt{\operatorname{tr}\left(A^* A\right)} = \sqrt{\operatorname{tr}\left(A\, A^*\right)}, \tag{9.7.32}$$

which is known as the *Hilbert-Schmidt norm* or the *Schur norm*. In fact,

$$\langle A, B \rangle_F := \operatorname{tr}\left(A^* A\right), \tag{9.7.33}$$

is the *Frobenius inner product* on the vector space $M_n(\mathbb{F})$. Moreover, since $\operatorname{tr}(A^* A)$ is the sum of squares of the *singular values* $\sigma_i$'s of $A$, we also have

$$\|A\|_F := \sqrt{\sigma_1^2 + \cdots + \sigma_n^2}, \tag{9.7.34}$$

**Theorem 9.37** *The Frobenius norm* (9.7.32) *is indeed a matrix norm. Also, it is unitarily invariant, and we have*

$$\sqrt{\rho(A^* A)} \le \|A\|_F \le \sqrt{n}\sqrt{\rho(A\, A^*)}, \tag{9.7.35}$$

*for all* $A \in M_n(\mathbb{F})$.

***Proof.*** For the first claim it suffices to show that the Frobenius norm $\|A\|_2$ is sub-multiplicative. For, by Cauchy-Schwartz inequality, we have

$$\begin{aligned}
\|AB\|_F &= \left(\sum_{i=1}^{n}\sum_{j=1}^{n}\left|\sum_{k=1}^{n}|a_{ik}b_{kj}|\right|\right)^{1/2}\\
&\le \left(\sum_{i=1}^{n}\sum_{j=1}^{n}\left(\sum_{k=1}^{n}a_{ik}\right)\left(\sum_{k=1}^{n}b_{kj}\right)\right)^{1/2}\\
&= \left(\sum_{i=1}^{n}\sum_{k=1}^{n}a_{ik}\right)^{1/2}\left(\sum_{k=1}^{n}\sum_{j=1}^{n}b_{kj}\right)^{1/2}\\
&= \|A\|_F\|B\|_F.
\end{aligned}$$

For the second claim we show that, for all unitary matrix $U$, $V$, we have

$$\|A\|_F = \|UA\|_F = \|AV\|_F = \|UAV\|_F.$$

The details are left for the reader as an exercise. □

Notice that the Frobenius norm doesn't satisfy the property that $\|I\| = 1$. We conclude by providing some general results. Recall that the *spectrum* of a positive matrix $B$ has positive eigenvalues (Theorem 9.31), where the maximal one is called the *spectral radius* of the matrix $B$, denoted by $\rho(B)$. In particular, for any matrix $A$, we have $B = A^*A$ or $B = AA^*$ is a positive matrix. So, the *spectral radius* $\rho(A^*A)$ or $\rho(AA^*)$ is well defined.

**Definition 9.52 (Spectral Radius)** The **spectral radius** of $B$, denoted by $\rho(B)$, is given by

$$\rho(B) := \max_{1\le i\le n}|\lambda_i|, \tag{9.7.36}$$

where the scalars $\lambda_i \in \mathbb{F}$ are the eigenvalues of the matrix $B$.

**Theorem 9.38** *For any matrix norm $|||\ |||$, and for any $A \in M_n(\mathbb{F})$, we have*

$$\rho(A) \le |||A|||, \tag{9.7.37}$$

*where $\rho(A)$ is given by* (9.7.36).

***Proof.*** Let $\lambda$ be an eigenvalue of the matrix $A$ with an eigenvector $\boldsymbol{x} \in \mathbb{F}^n$. Then, we have

$$|\lambda|\|\boldsymbol{x}\| = \|\lambda\boldsymbol{x}\| = \|A\boldsymbol{x}\| \le |||A|||\ \|\boldsymbol{x}\|,$$

which proves (9.7.37), because $\boldsymbol{x} \neq \boldsymbol{0}$. □

**Theorem 9.39** *Let* $|||\ |||$ *be a sub-multiplicative matrix norm, and A be such that* $|||A||| < 1$. *Then the matrix* $I - A$ *is nonsingular, and we have*

$$\left\|(I - A)^{-1}\right\| \leq \frac{1}{1 - \|A\|}. \tag{9.7.38}$$

***Proof.*** By Theorem 9.38, for all eigenvalues $\lambda_i$ of $A$, we have $|\lambda_i| \leq |||A||| < 1$. If $\boldsymbol{x}_i$ is an eigenvector associated with the eigenvalue $\lambda_i$, then

$$(I - A)\boldsymbol{x}_i = (1 - \lambda_i)\boldsymbol{x}_i, \quad \text{for} \quad i = 1, \ldots, n,$$

which implies that $1 - \lambda_i \neq 0$ are eigenvalues of the matrix $I - A$, for $i = 1, \ldots, n$. Therefore, $I - A$ is nonsingular. Now,

$$\begin{aligned} 1 = |||I||| &= \left|\left|\left|(I - A)^{-1}(I - A)\right|\right|\right| \\ &= \left|\left|\left|(I - A)^{-1} - (I - A)^{-1}A\right|\right|\right| \\ &\geq \left|\left|\left|(I - A)^{-1}\right|\right|\right| - \left|\left|\left|(I - A)^{-1}A\right|\right|\right| \\ &\geq \left|\left|\left|(I - A)^{-1}\right|\right|\right| - \left|\left|\left|(I - A)^{-1}\right|\right|\right|\, |||A||| \\ &= \left|\left|\left|(I - A)^{-1}\right|\right|\right| \left(1 - |||A|||\right). \end{aligned}$$

Dividing throughout by $(1 - \|A\|)$, the proof follows. □

## *Principal Component Analysis*

We know that, with an increase in the number of features (or variables), the data become more sparse. The related *over-fitting issues* make the training of machine learning models harder. It is thus a major concern in data science to describe a large dataset efficiently using only a small number of features. A *dimensionality reduction technique* such as the *principal component analysis* (or simply PCA) helps reduce the number of features while preserving as much variability as possible. Since the involved distance metrics fail to distinguish between near and far points, PCA is an important tool that needs to be applied to deal with the *curse of dimensionality*.

The PAC algorithms also help us with data visualization, simplification of complex datasets, and improve overall computational efficiency. It is based on the concepts of *eigenvalues* and *eigenvectors*. The eigenvalues are sorted in descending order, and the top $n$ is called the **principle components**. Since PCA is affected by the scale, we first ensure that the mean of each feature of the dataset is *zero*. The stepwise procedure to implement PAC is given below:

(a) Find the *axes* (as in the orthonormal basis sense), accounting for the largest variance in the dataset. These are the *principal components*.

(b) Use *heatmap* of the covariance matrix $C$ (say of order $n$) to get an idea about how the features vary together.
(c) Find the eigendecomposition of $C$. The eigenvectors give the directions of the maximum variance, i.e., the principal components, and the eigenvalues estimate the variance captured by each *principal component*. We write $C = PDP^{-1}$, where the columns of $P$ are the eigenvectors of $C$, and $D$ is the diagonal matrix of associated eigenvalues.
(d) Finally, center the dataset and multiply it by the matrix $P$ to project the original data onto the dimensions provided by the principle components.

Therefore, the *projected dataset* is obtained as

$$X_{\text{proj}} := X \cdot P.$$

The optimal number of dimensions is chosen subject to that the number of components adds up to at least 95% of the explained variance of the dataset. Notice that, since PAC finds a low-dimensional representation of the dataset while keeping a good share of the variance instead of making predictions, it is actually an *unsupervised ML algorithm*.

For example, a dataset about crimes in a city has numerous features including "crime against person—with injuries" and "crime against person—without injuries". Clearly, the places with high rates of the first type of crime must also have high rates of the second type of crime. So, the two features are highly correlated. Therefore, it is possible to reduce the dimensions of the dataset by diminishing the redundancies in the data.

**Exercises 7**

9.1 Check whether the following are real vector spaces:

a. The Abelian group $V = \mathbb{C} = \mathbb{R}^2$ with respect to the scalar multiplication defined by $\alpha \cdot z = (x, \alpha y)$, for $\alpha \in \mathbb{R}$ and $z = (x, y)$.
b. The Abelian group $V = \mathbb{R}_{>0}$ with respect to the addition and scalar multiplication defined by $x + y = xy$ and $\alpha x = x$, for $x, y \in \mathbb{R}_{>0}$ and $\alpha \in \mathbb{R}$.
c. The Abelian group $V = \mathbb{R}^2$ with respect to the scalar multiplication given by $\alpha(x, y) = (\alpha^2 x, \alpha^2 y)$, for $\alpha \in \mathbb{R}$ and $(x, y) \in \mathbb{R}^2$.
d. The Abelian group $V = \mathbb{C}$ with respect to the addition given by $z + w = (x + u + 1, y + v - 1)$, for $z = (x, y)$ and $w = (u, v) \in \mathbb{C}$, and pointwise scalar multiplication.

9.2 Let $V$ be a vector space over a field $\mathbb{F}$. Show that $\emptyset \neq W \subseteq V$ is a subspace if and only if $(a)$ $u + v \in W$, for all $u, v \in W$; and, $(b)$ $\lambda u \in W$, for all $u \in W$ and for all $\lambda \in \mathbb{F}$.

9.3 The following are simple illustrations of the sum of subspaces.

a. Let $V = \mathbb{R}^2$. Show that, for

$$U := \{(x, x) \mid x \in \mathbb{R}\} \quad \text{and} \quad W := \{(x, -x) \mid x \in \mathbb{R}\},$$

we have $V = U + W$.

b. Let $V = \mathbb{R}^4$, and consider the subspaces $U$ and $W$ given by

$$U := \{(x, y, z, w) \in V \mid x + y + z = 0,\ x + 2y - z = 0\};$$
$$W := \{(s, 2s, 3s, t) \mid s,\ t \in \mathbb{R}\},$$

Describe the sum $U + W$.

9.4 The following are simple illustrations of the direct sum of subspaces.

a. Let $V = \mathbb{R}^2$. Show that, for

$$U := \{(x, 2x) \mid x \in \mathbb{R}\} \quad \text{and} \quad W := \{(x, 3x) \mid x \in \mathbb{R}\},$$

we have $V = U \oplus W$.

b. Let $V_n = \mathbb{R}^n$ $(n \geq 2)$, and consider the subspaces $U$ and $W$ given by

$$U_n := \{(a_1, a_2, \ldots, a_n) \in V \mid a_1 + a_2 + \cdots + a_n = 0\};$$
$$W_n := \{(a_1, a_2, \ldots, a_n) \in V \mid -a_1 + a_2 - \cdots + (-1)^n a_n = 0\},$$

Show that $V_n = U_n + W_n$, it is a direct sum only when $n = 2$.

9.5 Prove the second assertion of Theorem 9.1.

9.6 Show that the vectors $\boldsymbol{v}_1 = (1, 0, 0)$, $\boldsymbol{v}_2 = (0, 1, 1)$, and $\boldsymbol{v}_3 = (1, 0, 1)$ in $\mathbb{R}^3$ are linearly independent.

9.7 Show that the vectors $\boldsymbol{v}_1 = (1, 0, 0)$, $\boldsymbol{v}_2 = (0, 1, 1)$, and $\boldsymbol{v}_3 = (1, 1, 1)$ in $\mathbb{R}^3$ are linearly dependent

9.8 Show that a set $S \subset \mathbb{F}^n$ with more than $n$ vectors is linearly dependent.

9.9 Let $V$ be a finite-dimensional vector space over a field $\mathbb{F}$. Show that a set $B \subset V$ is a basis if and only if (i) $B$ spans $V$; and, (ii) $B$ is linearly independent, over the field $\mathbb{F}$.

9.10 Show that the real vector space of $2 \times 2$ symmetric matrices has the dimension 3.

9.11 Consider the vector space $M_2$ of all matrices of order two over a field $\mathbb{F}$. Show that the set $S = \{E_1, E_2, E_3, E_4\}$, where $E_i$ are as in (9.3.10), is a basis. What is the coordinate vector of an arbitrary matrix in $M_2$?

9.12 Let $S$ be a subset of the real vector space $\mathbb{R}^3$ consisting of the three vectors $\boldsymbol{v}_1 = (1, 2, 3)$, $\boldsymbol{v}_2 = (2, 5, 3)$, and $\boldsymbol{v}_3 = (1, 0, 8)$. Show that $S$ is a basis for $\mathbb{R}^3$.

9.13 Let $S$ be a subset of the real vector space $\mathbb{R}^4$ consisting of the four vectors $\boldsymbol{v}_1 = (1, 0, 0, 0)$, $\boldsymbol{v}_2 = (a, 1, 0, 0)$, $\boldsymbol{v}_3 = (b, c, 1, 0)$, and $\boldsymbol{v}_4 = (d, e, f, 1)$. Show that $S$ is a basis for $\mathbb{R}^4$.

9.14 Let $S$ be a subset of the real vector space $\mathbb{R}^3$ consisting of the three vectors $\boldsymbol{v}_1 = (1, 1, 1)$, $\boldsymbol{v}_2 = (1, 1, 0)$, and $\boldsymbol{v}_3 = (1, 0, 0)$. Show that $S$ is a basis for $\mathbb{R}^3$.

9.15 Let $V$ be finite-dimensional vector space over a field $\mathbb{F}$, and $S$ be a basis for $V$ with $n$ vectors. Show that any set in $V$ with more than $n$ elements is linearly dependent.

9.16 Let $V$ be finite-dimensional vector space over a field $\mathbb{F}$, and $S$ be a basis for $V$ with $n$ vectors. Show that any set with less than $n$ elements cannot span $V$.

9.17 Let $V$ be a vector space over a field $\mathbb{F}$, and $S$ be a nonempty subset of $V$. Let $S$ be linearly independent subset of $V$, with $v \notin \text{Span}(S)$. Show that the set $S' = S \cup \{v\}$ is also linearly independent.

9.18 Let $V$ be a vector space over a field $\mathbb{F}$, and $S$ be a nonempty subset of $V$. Show that if $\mathbf{v} \in S$ and $\mathbf{v} \in \text{Span}(S - \{\mathbf{v}\})$ then $\text{Span}(S) = \text{Span}(S - \{\mathbf{v}\})$.

9.18 Give a proof of *change of basis formula* as in (9.3.13).

9.20 Let $V$ be a vector space over a field $\mathbb{F}$ of dimension $n$, and $S \subset V$ has exactly $n$ vectors. Show that $S$ is a basis for $V$ if either $S$ spans $V$ or $S$ is linearly independent.

9.21 Let $V$ be a vector space over a field $\mathbb{F}$, and $U$, $W$ be two subspace of $V$. Show that

$$\dim(U + W) = \dim(U) + \dim(V) - \dim(U \cap W).$$

9.22 Justify that the Abelian group $\mathbb{Q}$ is not a real vector space.

9.23 Consider the real vector space $V = P_3[x]$. Let $D : V \to V$ be the derivative, and $T : V \to V$ be given by $p \mapsto p(x + 1)$. Compute matrices $M_D$ and $M_T$ with respect to the standard basis $\{1, x, x^2, x^3\}$. Also, show that $M_{T \circ D} = M_T \cdot M_D$.

9.24 Prove the relation (9.4.32). More generally, give a proof of Theorem 9.10.

9.25 With notations as explained in the text, prove the relation (9.4.19).

9.26 Show that every subspace of a finite-dimensional vector space, say of dimension $n$, is finite dimensional of dimension at most $n$.

9.27 Let $V$ be a vector space over a field $\mathbb{F}$, and $U$, $W$ be two subspace of $V$. Show that $V = U \oplus W$ if and only if every vector $v \in V$ can be written uniquely as $v = u + w$, for some $u \in U$ and $w \in W$.

9.28 Let $U$ and $V$ be two vector spaces over a field $\mathbb{F}$, with $\dim(U) = m < \infty$, and $T : U \to V$ be a linear map. If $B = \{b_1, b_2, \ldots, b_m\}$ is a basis of $U$, show that

$$R(T) = \text{Span}\left(\left\{T(b_1), T(b_2), \ldots, T(b_m)\right\}\right).$$

9.29 Consider the real vector spaces $U = P_3[x]$ and $V = M_2(\mathbb{R})$. For the matrix

$$A = \begin{pmatrix} 1 & 1 \\ -1 & 1 \end{pmatrix},$$

define $T : U \to V$ as $T(p) = p(A)$. That is, if $p(x) = 1 + 2x - 3x^3$, then $p(A) = I + 2A - 3A^2$. Show that $T$ is a linear map, with

$$N(T) = \text{Span}\left(\{x^3 - 2x + 4, x^2 - 2x + 2\}\right) \text{ and} R(T) = \text{Span}\left(\left\{I_2, \begin{pmatrix} 0 & -1 \\ 1 & 0 \end{pmatrix}\right\}\right)$$

9.30 Let $U$ and $V$ be two finite-dimensional vector spaces over a field $\mathbb{F}$, say of dimensions $n$ and $m$, respectively. Show that every linear map $T : U \to V$ is actually a matrix transformation given by $M_T \in M_{m\times n}(\mathbb{F})$, where the action of $T$ on a given basis of $U$ determines the structure of $M_T$.

9.31 Let $U$ and $V$ be two vector spaces over a field $\mathbb{F}$, and $T : U \to V$ be linear map. Show that $N(T)$ is a subspace of $U$ and $R(T)$ is a subspace of $V$.

9.32 Let $U$ and $V$ be two vector spaces over a field $\mathbb{F}$, and $T : U \to V$ be linear map. Show that $T$ is injective if and only if $N(T) = 0$.

9.33 Let $U$ and $V$ be two vector spaces over a field $\mathbb{F}$, and $T : U \to V$ be linear map. Show that $T$ is an isomorphism if and only if $T$ is bijective. Further, show that $T^{-1}$ is also linear when defined.

9.34 Let $U$ and $V$ be two finite-dimensional vector spaces over a field $\mathbb{F}$. Show that $U$ and $V$ are isomorphic if and only if $\dim(U) = \dim(V)$.

9.35 Show that no eigenvector of a linear map corresponds to two distinct eigenvalues.

9.36 Let $V$ be an $n$ dimensional *nonzero* vector space over a field $\mathbb{F}$, and $T \in \mathcal{L}(V)$. Suppose $\lambda_1, \ldots, \lambda_m$ are distinct eigenvalues of $T$, with nonzero vectors $v_1, \ldots, v_m$ as corresponding eigenvectors. Show that $\{v_1, \ldots, v_m\}$ is linearly independent. In particular, each $T \in \mathcal{L}(V)$ has at most $\dim(V)$ distinct eigenvalues.

9.37 Suppose $V$ is an $n$ dimensional complex vector space, and $T \in \mathcal{L}(V)$. Show that $M_T$ is upper triangular with respect to some basis of $V$.

9.38 Suppose $V$ is an $n$ dimensional vector space over a field $\mathbb{F}$, and $T \in \mathcal{L}(V)$ is such that $M_T$ is upper triangular with respect to a basis of $V$. Show that $T$ is invertible if and only if all diagonal entries of $M_T$ are nonzero.

9.39 Show that exactly one of the following three holds for an operator $T : \mathbb{C}^2 \to \mathbb{C}^2$:

a. $T$ has two distinct eigenvalues $\lambda$ and $\mu$. Further, there is a basis of eigenvectors $e_1, e_2$ of $\mathbb{C}^2$ such that

$$M_T = \begin{pmatrix} \lambda & 0 \\ 0 & \mu \end{pmatrix}.$$

b. $T$ has exactly one eigenvalue $\lambda$, but it is diagonalisable, then $T = \lambda I_2$ so that

$$M_T = \begin{pmatrix} \lambda & 0 \\ 0 & \lambda \end{pmatrix}$$

with respect to any basis.

c. $T$ has exactly one eigenvalue $\lambda$, and it is not diagonalisable, there exists a basis $e_1, e_2$ for $\mathbb{C}^2$ such that

$$M_T = \begin{pmatrix} \lambda & 1 \\ 0 & \lambda \end{pmatrix}.$$

Notice that $e_1$ is an eigenvector with eigenvalue $\lambda$, but $e_2$ is not.

9.40 Show that every operator on an odd dimension real vector space has an eigenvalue.

9.41 Suppose $T \in \mathcal{L}(V)$ is such that $T^2 = T$. Show that $V = N(T) \oplus R(T)$.

9.42 Suppose $V$ is a vector space over a field $\mathbb{F}$, and $T \in \mathcal{L}(V)$ such that $N(T^m) = N(T^{m+1})$, for some $m \in \mathbb{Z}^+$. Show that

$$N(T^0) \subset N(T^1) \subset \cdots \subset N(T^m) = N(T^{m+1}) = \cdots$$

9.43 Suppose $V$ is an $n$ dimensional vector space over a field $\mathbb{F}$, and $T \in \mathcal{L}(V)$. Show that

$$N(T^n) = N(T^{n+1}) = N(T^{n+2}) = \cdots$$

Hence deduce that the relation (9.5.22) holds.

9.44 Suppose $V$ is an $n$ dimensional vector space over a field $\mathbb{F}$, and $T \in \mathcal{L}(V)$. Show that the set of generalized eigenvectors of $T$ corresponding to an eigenvalue $\lambda \in \mathbb{F}$ equals $N\big((T - \lambda I)^n\big)$.

9.45 Suppose $V$ is an $n$ dimensional vector space over a field $\mathbb{F}$, and $T \in \mathcal{L}(V)$. Show that

$$R(T^n) = R(T^{n+1}) = R(T^{n+2}) = \cdots$$

Hence deduce that the relation (9.5.22) holds.

9.46 Suppose $V$ is an $n$ dimensional complex vector space, and $T \in \mathcal{L}(V)$. Show that the sum of multiplicities of all the eigenvalues of $T$ is precisely $n = \dim(V)$.

9.47 (*Structure Theorem for Complex Vector Spaces*) Suppose $V$ is an $n$ dimensional complex vector space, and $T \in \mathcal{L}(V)$. Let $\lambda_1, \ldots, \lambda_m$ be distinct eigenvalues of $T$, and $U_1, \ldots, U_m$ be the corresponding subspaces consisting of generalized eigenvectors of $T$. Show that

a. $V = U_1 \oplus \cdots \oplus U_m$;
b. each $U_j$ is $T$-invariant;
c. each $(T - \lambda_j I)\big|_{U_j}$ is nilpotent.

Recall that an operator $T$ is said to be *nilpotent* if $T^k = 0$, for some $k > 0$.

9.48 Suppose $V$ is an $n$ dimensional complex vector space, and $T \in \mathcal{L}(V)$. Show that $V$ has a basis consisting of generalized eigenvectors of $T$.

9.49 Let $A \in M_n(\mathbb{F})$. Show that TFAE

a. If we choose a basis $\{u_1, \ldots, u_n\}$ of $\mathbb{F}^n$ and define a linear map $T : \mathbb{F}^n \to \mathbb{F}^n$ by

$$Tu_j = a_{1j}u_1 + \cdots + a_{nj}u_n, \quad \text{for} \quad j = 1, \ldots, n,$$

then we can find a basis $\{e_1, \ldots, e_n\}$ for $\mathbb{F}^n$ and $d_i \in \mathbb{F}$ such that $T(e_j) = d_j e_j$.

b. There is a nonsingular matrix $P \in M_n(\mathbb{F})$ such that $P^{-1}AP$ is diagonal. (We may recall that, in this case, we say $A$ is *diagonalizable matrix*)

9.50 With $\theta$ real, diagonalise the following matrix over $\mathbb{C}$:

$$\begin{pmatrix} \cos\theta & -\sin\theta \\ \sin\theta & \cos\theta \end{pmatrix}$$

9.51 Let $(V, \langle\,,\,\rangle)$ be a 2-dimensional real inner product space, and $T \in \mathcal{L}(V)$. Show that the following are equivalent

a. $T$ is normal but not self-adjoint;
b. the matrix $M_T$ with respect to every orthonormal basis of $V$ is of the form

$$\begin{pmatrix} a & -b \\ b & a \end{pmatrix}, \qquad \text{with} \quad b \neq 0;$$

c. the matrix $M_T$ with respect to some orthonormal basis of $V$ is of the form

$$\begin{pmatrix} a & -b \\ b & a \end{pmatrix}, \qquad \text{with} \quad b > 0.$$

9.52 Let $A \in M_2(\mathbb{C})$. Show that exactly one of the following three holds:

a. We can find a nonsingular $2 \times 2$ complex matrix $P$ such that

$$P^{-1}AP = \begin{pmatrix} \lambda & 0 \\ 0 & \mu \end{pmatrix}, \quad \text{with} \ \lambda \neq \mu.$$

b. $A = \lambda I$, for some $\lambda$.
c. We can find a nonsingular $2 \times 2$ complex matrix $P$ such that

$$P^{-1}AP = \begin{pmatrix} \lambda & 1 \\ 0 & \lambda \end{pmatrix}, \quad \text{for some} \ \lambda.$$

9.53 Let $(V, \langle\,,\,\rangle)$ be a finite-dimensional inner product space, and $U$ be a subspace of $V$. Show that $\left(U^{\perp}\right)^{\perp} = U$.

9.54 Let $(V, \langle\,,\,\rangle)$ be a real or complex inner product space, and $S = \{u_1, u_2, \ldots, u_n\}$ be an orthonormal set in $V$. Show that $v - v_n \in S^{\perp}$, for all $v \in V$, where

$$v_n = \langle v, u_1 \rangle u_1 + \cdots + \langle v, u_n \rangle u_n.$$

In particular, $v_n$ is the *orthogonal projection* of $v$ onto the subspace = Span$(S)$.

9.55 Let $V$ and $W$ be two inner product spaces, and let $T \in \mathcal{L}(V, W)$. Show that the mapping $T^* : W \to V$ given by (9.6.25) is a linear map. Also, for $S, T \in \mathcal{L}(W, V)$ and $\lambda \in \mathbb{F}$, show that the following algebraic laws hold:

$$(\lambda S + T)^* = \overline{\lambda} S^* + T^*, \quad (T^*)^* = T, \quad I^* = I, \quad (ST)^* = T^* S^*.$$

9.56 Let $V$ and $W$ be two inner product spaces, $T \in \mathcal{L}(V, W)$, and $T^* \in \mathcal{L}(W, V)$. Show that

a. $N(T^*) = (R(T))^\perp$;
b. $R(T^*) = (N(T))^\perp$;
c. $N(T) = (R(T^*))^\perp$;
d. $R(T) = (N(T^*))^\perp$;

In particular, we have $T$ is injective if and only if $T^*$ is surjective; and, $T$ is surjective if and only if $T^*$ is injective.

9.57 Let $V$ and $W$ be two inner product spaces, $T \in \mathcal{L}(V, W)$, and $T^* \in \mathcal{L}(W, V)$. Show that a scalar $\lambda$ is an eigenvalue of $T$ if and only if $\overline{\lambda}$ is an eigenvalue of $T^*$. Also, show that a subspace $U$ of $V$ is $T$-invariant if and only if $U^\perp$ is $T^*$-invariant.

9.58 Let $V$ be a complex inner product space, and $T \in \mathcal{L}(V)$. Show that

$$\langle Tv, v \rangle = 0, \quad \text{for all } v \in V \quad \Rightarrow \quad T \equiv 0.$$

Hence deduce that $T$ is self-adjoint if and only if $\langle Tv, v \rangle \in \mathbb{R}$, for all $v \in V$.

9.59 Let $V$ be a real inner product space, and $T \in \mathcal{L}(V)$ be self-adjoint. Show that $T$ has an eigenvalue.

9.60 (*Real Spectral Theorem*) Let $V$ be a real inner product space, and $T \in \mathcal{L}(V)$. Show that $V$ has an orthonormal basis consisting of eigenvectors of $T$ if and only if $T$ is self-adjoint.

9.61 Let $V$ be a real or complex inner product space, and $T \in \mathcal{L}(V)$ be respectively a self-adjoint or normal operator. Show that, if $\lambda_1, \ldots, \lambda_m$ are distinct eigenvalues of $T$, then we have

$$V = N(T - \lambda_1 I) \oplus + \cdots \oplus N(T - \lambda_m I)$$

Also, show that each vector in every null space $N(T - \lambda_j I)$ is orthogonal to all vectors in the other subspaces of the above decomposition.

9.62 Let $T \in O(n)$. Show that $T$ is product of $m$ reflections with $0 \le m \le n$. (If $m = 0$, $T = \iota$. Otherwise, we can find reflections $\rho_1, \rho_2, \ldots, \rho_m$ such that $T = \rho_1 \rho_2 \cdots \rho_m$. If $m$ is even, $T \in SO(n)$. If $m$ is odd, $T \notin SO(n)$.

9.63 Let $T \in O(2)$. Show that one of two things must happen.

a. $T \in SO(2)$, and we can find $0 \le \theta < 2\pi$ such that, with respect to any orthonormal basis, $M_T$ has one of the two possible forms

$$\begin{pmatrix} \cos\theta & -\sin\theta \\ \sin\theta & \cos\theta \end{pmatrix} \quad \text{or} \quad \begin{pmatrix} \cos\theta & \sin\theta \\ -\sin\theta & \cos\theta \end{pmatrix}.$$

b. $T \notin SO(2)$, and we can find an orthonormal basis of $\mathbb{R}^2$ with respect to which $M_T$ has the form

$$\begin{pmatrix} -1 & 0 \\ 0 & 1 \end{pmatrix}.$$

9.64 Recover the addition formulae for cos and sin by considering the product of the rotation matrices

$$\begin{pmatrix} \cos\theta & -\sin\theta \\ \sin\theta & \cos\theta \end{pmatrix} \quad \text{and} \quad \begin{pmatrix} \cos\phi & -\sin\phi \\ \sin\phi & \cos\phi \end{pmatrix}$$

9.65 For $T \in O(3)$, show that we can find an orthonormal basis with respect to which $T$ has a matrix

$$\begin{pmatrix} \pm 1 & 0 & 0 \\ 0 & \cos\theta & -\sin\theta \\ 0 & \sin\theta & \cos\theta \end{pmatrix}.$$

Further, if the plus sign is taken, show that $T \in SO(3)$; and, if the minus sign is taken, $T \notin SO(3)$. (It follows from Exercise 9.64 that *every rotation has an axis*. Things are more complicated in higher dimensions. The interested reader can look at Exercise ??.)

9.66 Let $V$ be an inner product space, and $T \in \mathcal{L}(V)$. Show that $T$ is an isometry if and only if there is an orthonormal basis of $V$ consisting of eigenvectors of $T$ such that the modulus of all corresponding eigenvalues is 1.

9.67 With the notation as in Theorem 9.33, show that TFAE

a. There is an orthonormal basis $\{v_1, \ldots, v_n\}$ with respect to which $T$ has a diagonal matrix $D$ with $d_{11} = -1$, $d_{ii} = 1$, for all $2 \le i \le n$.
b. $T^2 = I$.
c. $T \in O(n)$.
d. $\det(M_T) = -1$.
e. If $\|u\| = \|v\|$, then $\boldsymbol{x} = \boldsymbol{y}\,\widehat{u}$, for some unit vector $\widehat{u}$.

9.68 Obtain LU-decomposition for the matrix

$$A = \begin{pmatrix} 2 & 1 & 3 \\ 4 & 3 & 7 \\ 6 & 7 & 17 \end{pmatrix},$$

9.69 Let $A = \left(a_{ij}\right)$ be a nonsingular matrix of order $n$, with $a_{11} = 0$. Show that $A$ is not $LU$-decomposable. Does there exist an $PLU$-composition for $A$, where $P$ is the permutation matrix?

## References


1. Strang, G.: Linear Algebra and Its Applications (4/e). Cengage (2006)
2. Lancaster, P., Tismenetsky, M.: The Theory of Matrices: With Applications (2/e). Academic Press (1985)
3. Dym, H.: Linear Algebra in Action. Graduate Studies in Mathematics, vol. 78. AMS (2007)
4. Bretscher, O.: Linear Algebra with Applications (2/e). Pearson (2008)
5. Lax, P.: Linear Algebra and Its Applications (2/e). Wiley-Interscience (2013)
6. Lay, D.C., Lay, S.R., McDonald, J.J.: Linear Algebra and Its Applications (5/e). Pearson (2014)
7. Yang, Y.: A Concise Text on Advanced Linear Algebra. Cambridge University Press (2014)
8. Friedberg, S. and Insel, A., *Linear Algebra* (5/e), Pearson, 2018
9. Singh, K.: Linear Algebra: Step by Step. OUP (2013)
10. Carrell, J.B.: Groups, Matrices, and Vector Spaces—A Group Theoretic Approach to Linear Algebra. Springer (2017)
11. Axler, S.: Linear Algebra Done Right (3/e). Springer (2015)
12. Berger, M.: Geometry I (2/e). Universitext, Springer (2009)
13. Lay, D.: Linear Algebra and Its Applications (3/e). Addison-Wesley (2003)
14. Shifrin, T., Adams, M.: Linear Algebra: A Geometric Approach. Freeman (2001)
15. Savov, I.: No Bullshit Guide to Linear Algebra (2/e). Minireference Co. (2017)
16. Trefethen, L.N., Bau, D., III.: Numerical Linear Algebra. Society for Industrial and Applied Mathematics, Philadelphia (1997)
17. Conte, S.D., Boor, C.: Elementary Numerical Analysis: An Algorithmic Approach. Classics in Applied Mathematics SIAM (2017)
18. Stewart, G.W.: On the early history of the singular value decomposition. SIAM Rev. **35**(4), 551–566
19. Poynton, C.: Digital Video. Wiley (1996)
20. Poynton, C.: Digital Video and HDTV Algorithm and Interfaces. Kaufmann Publishers, San Francisco (2003)
21. Horn, R.A., Johnson, C.R.: Matrix Analysis (2/e). Cambridge University Press (2013)

# Chapter 10
# Graph Theory

*The origins of graph theory are humble, even frivolous.*

***Norman L. Biggs*** *(1941–1912)*

Graph is a fascinating modern tool of computation. It consists of "nodes" and "links". The main emphasis of any graph-based study is always on how nodes are linked together. The graph theory finds applications in many subject areas such as *multiprocessor systems, data mining, networking, computer vision, statistical database, pattern recognition*, and several others ([1, 2]). A major role it plays in the development of graph algorithms such as used to find the *shortest path, minimum spanning tree, an element in a data, graph planarity*, and so on [3]. The number of practical problems that can be analyzed using a graph algorithm is enormous ([4, 5]). Graph-based concepts presented here concern some fundamental processes and configurations related to computer science. In this chapter, we introduce some important classes of graphs and also discuss some of their interesting properties. The content presented here is motivated by the standard texts [6, 7], and [8]. We recommend the reader [9–11], or [12], for further studies.

## 10.1 A Brief Introduction

Graph theory is a branch of mathematics that dates back to the eighteenth century [13]. The roots can be traced to the 1736 paper of the Swiss mathematician *Leonard Euler* (1707–1783). In this paper, Euler introduced a prototype of a

 
V. Ravichandran and A. K. Razdan, *Fundamental Discrete Structures*,
University Texts in the Mathematical Sciences,
https://doi.org/10.1007/978-981-96-0069-4_10

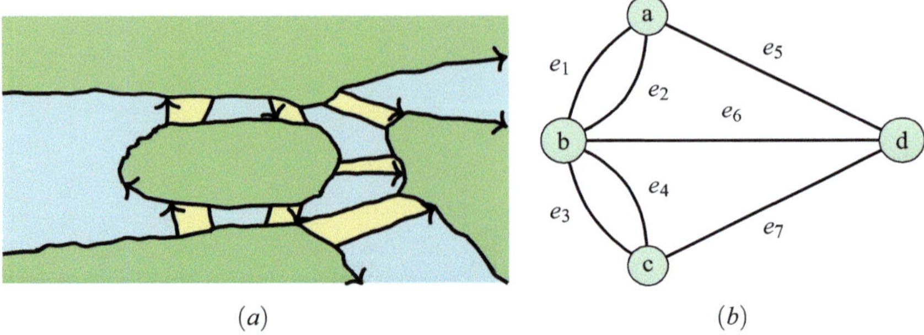


**Fig. 10.1** Königsberg seven bridges, and the related graph model

multigraph to settle the *Königsberg bridges problem* negatively. The context of the problem is the four islands in the eastern Prussian town of Königsberg (now called *Keliningard* in Russia) spread across the river *Pregel* that are connected by the seven bridges (see Fig. 10.1a). A precise statement of the problem is as follows:

**Königsberg Bridges Problem**: *Does there exist a closed walk across the four islands that uses the seven bridges exactly once?*

Euler was quick to realize that the topography of the region holds no significance to a possible solution to the problem. He mainly focused on how the four islands are linked together by seven bridges.[1] He represented the four islands by points $a, b, c, d$ in the plane, and joined a pair of points by an *arc* if the corresponding islands are linked by a bridge. The **graph model** shown in Fig. 10.1b is what we now call a *Multigraph*. In modern terminology, we call the four points $a, b, c, d$ as *nodes* (or *vertices*) of the graph $G$, and the seven arcs $e_1, e_2, e_3, e_4, e_5, e_6, e_7$ as the *links* (or *edges*). In this case, we write $G = (V, E)$, where

$$V = \{a, b, c, d\} \quad \text{and} \quad E = \{e_1, e_2, e_3, e_4, e_5, e_6, e_7\}.$$

Euler solved the Königsberg bridges problem by applying a simple graph theoretic argument. As shown later a walk of the required type exists if and only if there are either 0 or 2 vertices with an odd number of connections (Theorem 10.29). However, since all the vertices are of odd degrees in the graph corresponding to the *Graph model* shown in Fig. 10.1b, there is no walk that covers all seven bridges exactly once. The whole discussion ultimately led to the introduction of an important class of graphs (see Sect. 10.5).

**Definition 10.1** A **graph** $G$ is a pair of sets $(V, E)$, with $V \neq \emptyset$, together with a mapping $\phi$ defined on $E$ that assigns a pair of vertices $u, v \in V$ (not necessarily

[1] As a matter of fact, that is precisely the approach one needs to adopt while applying the *Graph theory* in the study of a practical situation.

distinct) to each element $e \in E$, called the **end verticess** of $e$. The elements of the sets $V$ and $E$ are respectively called **vertices** and **edges** of the graph $G$.

In what follows, we shall view an edge in $E$ as a 2-*subset* of the set of vertices $V$. In this case, the graph $G$ is also called an **undirected graphs**. Therefore, for an *Undirected graph*, connections are specified by 2-subsets of the set of vertices. Such types of graphs play a crucial role in solving various types of problems in computer science. A display of connections between vertices in a particular drawing of a graph is not very significant.

**Remark 10.1** The term "graph" first appeared in a paper of Sylvester published in 1878, and the notation $G = (V, E)$ first appeared in 1894.

## 10.2 Basic Terminology

We introduce some basic terminology and fix notations for use in the sequel. Let $G = (V, E)$ be an undirected graph, and $u \neq v \in V$. We say vertices $u$ and $v$ are *Adjacent vertices* if $e := \{u, v\} \in E$. We also say that the *Incident edge* $e$ has $u$ and $v$ as *End vertices*. We may also write $e$ simply as $uv$. An edge $e \in E$ with the same end vertex $v \in V$ is called a *Loop at* $v$. Also, multiple edges between a pair of distinct vertices are called the *Parallel edges*. Two distinct edges $e_1, e_2 \in E$ are called *Adjacent edges* if $(i)$ $e_1$ and $e_2$ are not parallel edges; and, $(ii)$ they share a vertex. A graph having multiple edges or loops or both is called a *Multigraph*. The term *multigraph* refers to the fact that the edges such as $e_3$ and $e_4$ in Fig. 10.1b are represented by the same 2-subset $\{b, c\} \subset V$. The next example illustrates these concepts.

**Example 10.1** Let $G$ be a graph with vertices $v_1, \ldots, v_7$ and edges $e_1, \ldots, e_{11}$ given by

$$\begin{array}{llll} e_1 = v_1v_2, & e_2 = v_2v_3, & e_3 = v_3v_4, & e_4 = v_3v_5, \\ e_5 = v_2v_5, & e_6 = v_1v_6, & e_7 = v_6v_5, & e_8 = v_5v_5, \\ e_9 = v_5v_7, & e_{10} = v_1v_7, & e_{11} = v_1v_7. & \end{array}$$

A drawing of this multigraph $G$ is shown in Fig. 10.2. The edge $e_8$ is a loop, while the edges $e_{10}$ and $e_{11}$ are parallel edges. Also, the three edges $e_9$, $e_{10}$, and $e_{11}$, are incident on the vertex $v_7$. Said differently, two vertices $v_7$ and $v_5$ are incident to the edge $e_9$. Further, the edges $e_9$, $e_{10}$ are adjacent. However, the edges $e_9$ and $e_6$ are non-adjacent.

**Definition 10.2** A graph without multiple edges and loops is called a **simple graph**.

Notice that each nonempty set $V$ defines a simple graph $G = (V, E)$, where we may take the set $E$ as a collection of 2-element subsets of the set $V$. In general, a

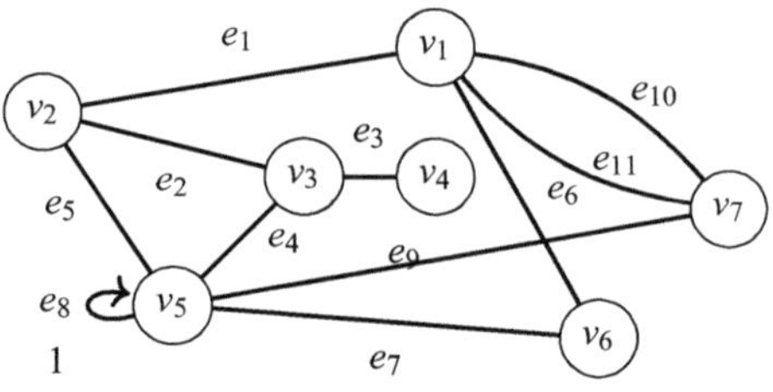


**Fig. 10.2** A drawing of multigraph defined in Example 10.1

graph $G = (V, E)$ is said to be **finite** if both the sets $V$ and $E$ are finite. The number $p = |V|$ is called the **order** and the number $q = |E|$ is called the **size**, of the graph $G$; in this case, we may also write the graph $G$ as $(p, q)$-*graph*. It follows that, in particular, a simple graph with $n$ vertices can have at most $\binom{n}{2} = n(n-1)/2$ edges. Therefore, the maximum number of simple graphs with $n$ vertices is $\leq 2^{n(n-1)/2}$. In this book, we are mainly dealing with *finite graphs* having some nice properties.

**Example 10.2** If cities across the globe are taken as *nodes* and the available air routes among them as the (directed) *connections*, then the "graph model" so obtained gives an air transport *flow chart*. Moreover, if each connection between two cities is assigned the travel cost as a *weight*, then this model may help find an "optimal route". The same model may also be used to identify alternative best possible routes. In addition, it would be easier to see when the two cities are weakly connected in the sense that the stopping services in certain sectors will affect going from one city to another.

## *Degree Sequence*

**Definition 10.3** Let $G = (V, E)$ be a graph, and $v \in V$. The **degree** of vertex $v$, denoted by $\deg_G(v)$ (or simply $\deg(v)$), is the number of edges incident to vertex $v$, subject to that the degree of a loop is counted twice. An **isolated vertex** $v \in V$ is the one that has $\deg(v) = 0$, and a vertex $v \in V$ is called a **pendant vertex** if $\deg(v) = 1$.

Notice that, for a simple graph of order $n$, the degree of each vertex is at most $n - 1$, and there are two or more vertices of the same degree.

**Example 10.3** For the graph $G = (V, E)$ in Example 10.1, as drawn in Fig. 10.2, we have

$$\deg(v_1) = 4, \quad \deg(v_2) = 3, \quad \deg(v_3) = 3, \quad \deg(v_4) = 1,$$
$$\deg(v_5) = 6, \quad \deg(v_6) = 2, \quad \deg(v_7) = 3.$$

Notice that we also have $\sum_{v \in V} \deg(v) = 22 = 2e$, where $e = |E| = 11$.

The next theorem is popularly known as the *First Theorem of Graph Theorem.*

**Theorem 10.1** (Euler, 1736) *Let $G = (V, E)$ be a simple graph. Then the sum of degrees of all vertices of G is even. That is,* $\sum_{v \in V} \deg(v) = 2\,|E|$.

***Proof*** The proof follows from the simple fact that each edge $e = uv \in E$ is counted twice while computing the degree-sum, and so contributes 2 to the left side of the asserted equality. □

In particular, if a graph $G$ has 10 vertices, where 4 have the degree 4 and the rest 6 are each of degree 5, then the number of edges $G$ has is given by

$$\begin{aligned} |E| &= \frac{1}{2}\Big[\big(\deg(v_1) + \cdots + \deg(v_4)\big) + \big(\deg(v_5) + \cdots + \deg(v_{10})\big)\Big] \\ &= \frac{1}{2}\Big[4 \times 4 + 6 \times 5\Big] = 23. \end{aligned}$$

In a graph, a vertex of odd degree is called an *odd vertex*, and a vertex of even degree is called an *even vertex*. It follows from the next corollary that, in a party involving 12 persons, the number of persons who shook hands with all present in the party must be even.

**Corollary 10.1** (Handshaking Lemma) *In any graph $G = (V, E)$, the number of* odd vertices *is even.*

***Proof*** Consider the partition of the set of vertices $V$ given by

$$V_1 = \big\{v \in V : \ \deg(v) \text{ is odd}\big\} \quad \text{and} \quad V_2 = \big\{v \in V : \ \deg(v) \text{ is even}\big\}.$$

It then follows from Theorem 10.1 that

$$2\,|E| = \sum_{v \in V} \deg(v) = \sum_{v \in V_1} \deg(v) + \sum_{v \in V_2} \deg(v),$$

and so $|V_1|$ must be odd because the left side is even. More specifically, let $v_1, \ldots, v_k$ be all the vertices in $V$ of odd degree, say $\deg(v_i) = 2\ell_i + 1$ for $i = 1, \ldots, k$. It follows from Theorem 10.1 that

$$\deg(v_1) + \deg(v_2) + \cdots + \deg(v_k) + \sum_{v \in V - \{v_1, v_2, \ldots, v_k\}} \deg(v) = 2\,|E|.$$

Therefore, as $\deg(v)$ is even for $v \in V \setminus \{v_1, v_2, \ldots, v_k\}$, we have

$$(2\ell_1 + 1) + (2\ell_2 + 1) + \cdots + (2\ell_k + 1)$$

is an even number, i.e., $k =$ an even number $- 2(\ell_1 + \cdots + \ell_k)$ is an even number, which proves that the number of vertices of odd degree in any graph is even. □

**Definition 10.4** Let $G = (V, E)$ be a finite graph. Suppose the vertices of $G$ are labeled such that $V = \{v_1, \ldots, v_n\}$, with $d_1 \geq d_2 \geq \cdots \geq d_n$, where $d_i = \deg(v_i)$. Then the list $(d_1, \ldots, d_n)$ is called the **degree sequence** of the graph $G$. In general, we write

$$\Delta(G) := \max\{d_i : 1 \leq i \leq n\} = d_1;$$
$$\delta(G) := \min\{d_i : 1 \leq i \leq n\} = d_n.$$

We say a list $(d_1, \ldots, d_n)$ of non-increasing non-negative integers is **realizable** if it is the *degree sequence* of some graph.

**Theorem 10.2** *A list $(d_1, \ldots, d_n)$ of non-increasing non-negative integers is realizable if and only if the sum $d_1 + \cdots + d_n$ is even.*

***Proof*** It follows directly from Theorem 10.1 that the condition is necessary. To prove that it is sufficient as well, suppose for a list $(d_1, \ldots, d_n)$ of non-increasing non-negative integers the sum $d_1 + \cdots + d_n$ is even. We have to construct a graph $G = (V, E)$ having $(d_1, \ldots, d_n)$ as the *degree sequence*. Consider $V = V_1 \bigcup V_2$, where

$$V_1 = \{v_i \in V : \deg(v_i) = d_i \text{ is odd}\};$$
$$V_2 = \{v_j \in V : \deg(v_j) = d_j \text{ is even}\}.$$

Each pair of distinct elements in $V_1$ contribute an edge to the set $E$, and for each element in $V_2$ take $\lfloor (d_j/2) \rfloor$ number of loops at $v_j$. □

Notice that the graph constructed in the above proof is not a simple graph. In general, not every list $(d_1, \ldots, d_n)$ of non-increasing non-negative integers is *realizable*. For example, without loops, the sequence (2, 0, 0) is not *realizable*. However, it can shown that there exists a loop-less graphs (possibly with multiple edges) with degree sequence $(d_1, \ldots, d_n)$ of non-increasing non-negative integers if and only if the sum $d_1 + \cdots + d_n$ is even, and $d_1 \leq d_2 + \cdots + d_n$.

**Definition 10.5** A list $(d_1, \ldots, d_n)$ of non-negative integers is called a **graphic sequence** if it is realizable as the degree sequence of a simple graph.

For example, (1, 1, 0) is a graphic sequence because it is the degree sequence of a simple graph consisting of the line $P_2$ and an isolated vertex. Also, (2, 2, 1, 1) is a graphic sequence because it is the degree sequence of a simple graph obtained as the *sum* of $P_2$ and an isolated vertex. Notice that Theorem 10.1 and Corollary 10.1 impose natural restrictions on non-negative integers $d_i$ so that $(d_1, \ldots, d_n)$ is a graphic sequence. For example, (3, 3, 2, 1, 1, 1) cannot be a graphic sequence because the degree-sum is not even. Alternatively, we can use handshaking lemma to arrive at the same conclusion, because the number of odd integers it contains is not even.

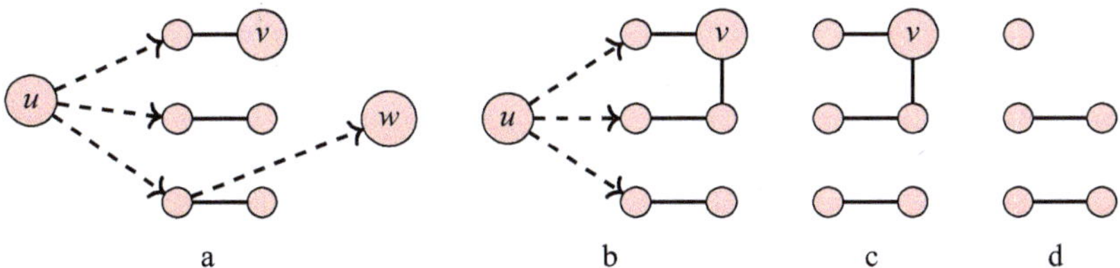


**Fig. 10.3** Construction of a simple graph for a graphic sequence

**Example 10.4** Here, we verify whether the list (3, 3, 3, 3, 3, 2, 2, 1) is a graphic sequence. Notice that it does satisfy the conditions of Theorem 10.1 and Corollary 10.1. Clearly, the vertex set $V$ of a simple graph $G = (V, E)$ that we seek must contain a vertex $u$ of degree 3 that has three neighbors each of degree 3. By Theorem 10.3, such a simple graph exists if and only if the list (2, 2, 2, 3, 2, 2, 1) is a graphic sequence. Once again, by the same result, the latter list is a graphic sequence if and only if the list (1, 1, 1, 2, 2, 1) is a graphic. We continue applying the same *reduction process* until we arrive at a sequence that is realizable trivially. For example, the next reduction yields the list (1, 0, 1, 1, 1), which we may write as (1, 1, 1, 1, 0). The drawings of the related simple graphs (obtained in the reverse order) are as shown in Fig. 10.3.

**Theorem 10.3** (Havel (1955), Hakimi (1962)) *For $n \geq 2$, a list $s = \left(1 \leq d_1, \ldots, d_n\right)$ of non-increasing non-negative integers is graphic if and only if the list given by*

$$s_1 = \left(d_2 - 1, d_3 - 1, \ldots, d_{d_1+1} - 1, d_{d_1+2}, \ldots, d_n\right)$$

*is a degree sequence of a simple graph of order $n - 1$.*

***Proof*** First, suppose $G_1 = (V_1, E_1)$ is a simple graph with degree sequence $s_1$. We may label the vertices of $G_1$ as

$$v_2, \ldots, v_{d_1+1}, v_{d_1+2}, \ldots, v_n$$

such that $\deg\left(v_i\right) = d_i$, for $i = 2, \ldots, n$. Take $V = V_1 \bigcup \{v_1\}$ such that the graph $G = (V, E)$, with

$$E = E \bigcup \left\{v_1 v_2, \ldots, v_1 v_{d_1+1}\right\},$$

is a simple graph with the degree sequence $s$. To prove the converse, suppose $s$ is a graphic sequence. We choose a simple graph $G = (V, E)$, with the degree sequence $s$, such that $V = \left\{v_1, \ldots, v_n\right\}$ has the property that $\deg\left(v_i\right) = d_i$ for $i = 1, \ldots, n$, and the sum of the degrees of the vertices adjacent to $v_1$ is maximum. It then follows that $v_1$ is adjacent to vertices having degrees $d_2, \ldots, d_{d_1+1}$. For, if not, we can find vertices $v_r, v_s \in V$ with $d_r > d_s$ such that $v_1$ is adjacent to $v_s$, but $v_1$ is not adjacent to $v_r$. However, as $d_r > d_s$, there is a vertex $u \in V$ such that $u$ is adjacent to $v_r$, but

$u$ is not adjacent to $v_s$. Now, remove the edges $v_1v_s$ and $v_ru$, and add the edges $v_1v_r$ and $v_su$, so that the new graph $G'$ so obtained though has the degree sequence $s$, but the sum of the degrees of the vertices in $G'$ adjacent to $v_1$ is larger than that in $G$, which contradicts the maximality property of $G$. Therefore, $v_1$ is adjacent to vertices having degrees $d_2, \ldots, d_{d_1+1}$. Hence, the graph $G_1 = G - v_1$ of order $n - 1$ has the degree sequence $s_1$. □

**Remark 10.2** In 1973, D. J. Kleitman and D. L. Wang extended Theorem 10.3 in their paper published in Volume-6 of *Discrete Mathematics*. Some other necessary and sufficient conditions for a list $(d_1, \ldots, d_n)$ to be a *graphic sequence* are also known. In 1991, G. Sierksma and H. Hoogeveen listed seven such characterizations in their paper published in Volume-15 of *Journal of Graph Theory*. More recently, in 2008, A. Tripathi and H. Tyagi gave two new characterizations in their paper published in Volume-156 of *Discrete Applied Mathematics*.

We can represent a loop-less graph $G$ as the *incidence matrix*, denoted by $I(G)$, which reflects how edges are incident to vertices. We can also view a graph without parallel edges as the *adjacency matrix*, denoted by $A(G)$, which reflects adjacency among vertex. A **labeled graph** is a simple graph having vertices labeled.

**Definition 10.6** (*Incidence Matrix*) Let a graph $G = (V, E)$ has order $n$ and size $m$. Suppose vertices are labeled as $v_1, \ldots, v_n$ and the edges as $e_1, \ldots, e_m$. The **incidence matrix** of the graph $G$ is the $n \times m$ matrix $I(G) = (a_{ij})$, with entries $a_{ij}$ given by

$$a_{ij} = \begin{cases} 1, & \text{if } j\text{th edge} e_j \text{is incident with} i \text{th vertex} v_i \\ 0, & \text{otherwise.} \end{cases}$$

As every edge is incident on two (different) vertices, each column of the matrix $I(G)$ has exactly two 1s. Further, the sum of each row equals the degree of the corresponding vertex. The total number of 1s in the matrix $A$ equals twice the number of edges. Identical columns correspond to parallel edges. An isolated vertex is presented by a row all of whose entries are zero.

**Definition 10.7** (*Adjacency Matrix*) Let a graph $G = (V, E)$ has order $n$ and size $m$. Suppose vertices are labeled as $v_1, \ldots, v_n$. The **adjacency matrix** of the graph $G$ is the square matrix $A(G) = (a_{ij})$ of order $n$, with entries $a_{ij}$ given by

$$a_{ij} = \begin{cases} 1, & \text{if the vextex} v_i \text{ is adjacent to vertex} v_j \\ 0, & \text{otherwise.} \end{cases}$$

For example, the incidence matrix $I(G)$ and the adjacency matrix $A(G)$ of the graph $G$ shown in Fig. 10.4 are as given below:

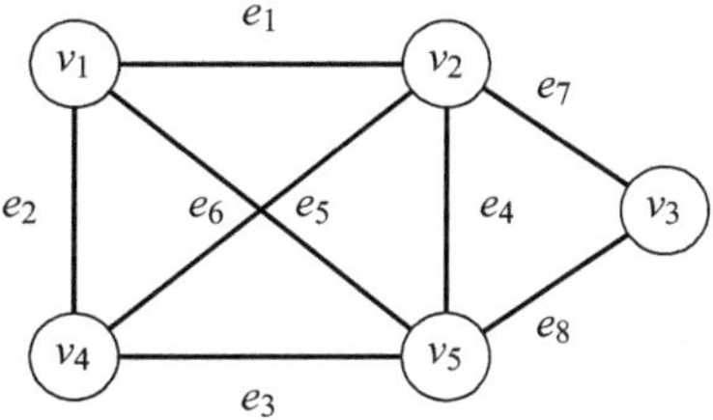


**Fig. 10.4** A graph whose incidence matrix is given below

$$I(G) = \begin{array}{c} \\ v_1 \\ v_2 \\ v_3 \\ v_4 \\ v_5 \end{array} \begin{array}{c} \begin{array}{cccccccc} e_1 & e_2 & e_3 & e_4 & e_5 & e_6 & e_7 & e_8 \end{array} \\ \begin{pmatrix} 1 & 1 & 0 & 0 & 1 & 0 & 0 & 0 \\ 1 & 0 & 0 & 1 & 0 & 1 & 1 & 0 \\ 0 & 0 & 0 & 0 & 1 & 0 & 1 & 1 \\ 0 & 1 & 1 & 1 & 0 & 1 & 0 & 0 \\ 0 & 0 & 1 & 0 & 0 & 0 & 0 & 1 \end{pmatrix} \end{array};$$

$$A(G) = \begin{array}{c} \\ v_1 \\ v_2 \\ v_3 \\ v_4 \\ v_5 \end{array} \begin{array}{c} \begin{array}{ccccc} v_1 & v_2 & v_3 & v_4 & v_5 \end{array} \\ \begin{pmatrix} 0 & 1 & 0 & 1 & 1 \\ 1 & 0 & 1 & 1 & 1 \\ 0 & 1 & 0 & 0 & 1 \\ 1 & 1 & 0 & 0 & 1 \\ 1 & 1 & 1 & 1 & 0 \end{pmatrix} \end{array}$$

Notice that the adjacency matrix of a complete graph $K_n$ is $J - I$, where $J$ is a square matrix of order $n$ with all entries equal to 1 and $I$ is the $n \times n$ identity matrix.

**Definition 10.8** (*Graph Isomorphism*) Two simple graphs $G_1 = (V_1, E_1)$ and $G_2 = (V_2, E_2)$ are said to be isomorphic if there is a bijection $\theta : V_1 \to V_2$ such that, for every pair $u, v \in V_1$, we have

$$e = uv \in E_1 \qquad \Longleftrightarrow \qquad f = \theta(u)\theta(v) \in E_2.$$

Therefore, $\theta$ induces an *incidence preserving* bijection between the edge sets $E_1$ and $E_2$.

That is, two vertices in $G_1$ are incident on an edge in $G_1$ if and only if their images under $\theta$ are incident vertices with the corresponding edge in $G_2$. It can be shown that two graphs $G_1$ and $G_2$ (without self-loops) are *isomorphic* if the incidence matrix of one, say $I(G_1)$, can be obtained by permuting the rows and columns of the incidence matrix of the graph $I(G_2)$.

**Definition 10.9** (*Induced & Spanning Subgraphs*) Let $G = (V, E)$ be a graph, and $V' \subset V$. By $E[V']$ we denote the set of all edges $e \in E$ that have their end vertices

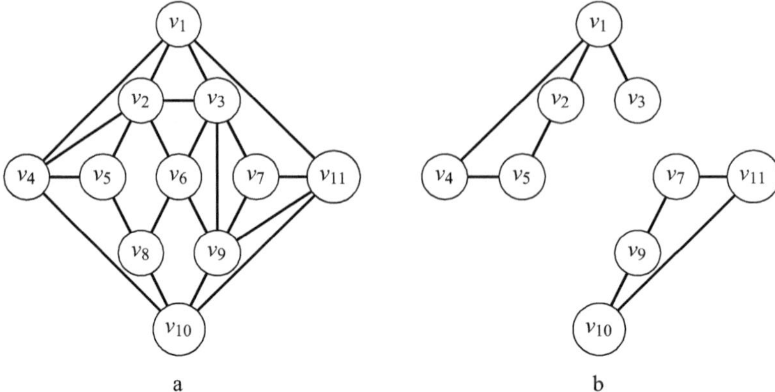


**Fig. 10.5** A graph and a subgraph

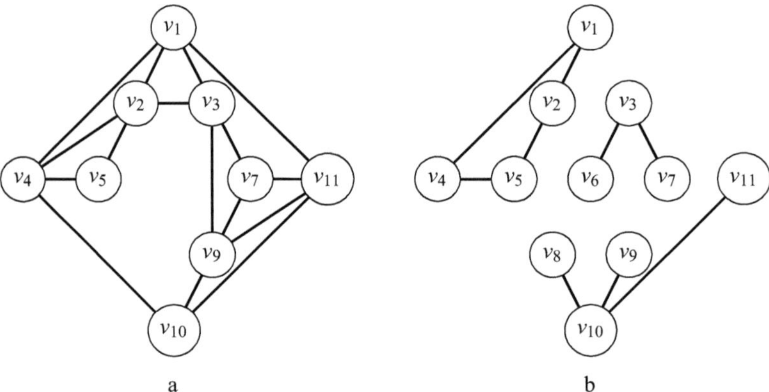


**Fig. 10.6** An induced subgraph and a spanning subgraph

in the set $V'$, and the graph $G\left[V'\right] = \left(V', E[V']\right)$ is called the **induced subgraph** on $V'$. Each graph of the form $\left(V', E'\right)$, with $V' \subset V$ and $E' \subset E[V']$ is called a **subgraph** of $G$, and a subgraph with $V = V'$ is called a **spanning subgraph**.

Therefore, a graph $H$ is a subgraph of a graph $G$ if all the vertices and all the edges of $H$ are in $G$ and each edge in $H$ has the same end vertices in $H$ as in $G$. Some examples are as shown in Figs. 10.5 and 10.6.

In terms of applications of graph theory in designing and analysis of networks, certain subgraphs of the associated *graph model* are *paths* and *cycles*. It is desirable to have every node (vertex) of the network *reachable* from any other node (vertex). Therefore, the main focus in most such situations is to find a path between nodes that is optimal subject to certain specified constraints. We start with the next definition.

**Definition 10.10** (*Walk*) Let $G = (V, E)$ be a graph. By a **walk** $w$ in $G$ we mean a finite alternating sequence of vertices and edges such that each edge is incident with

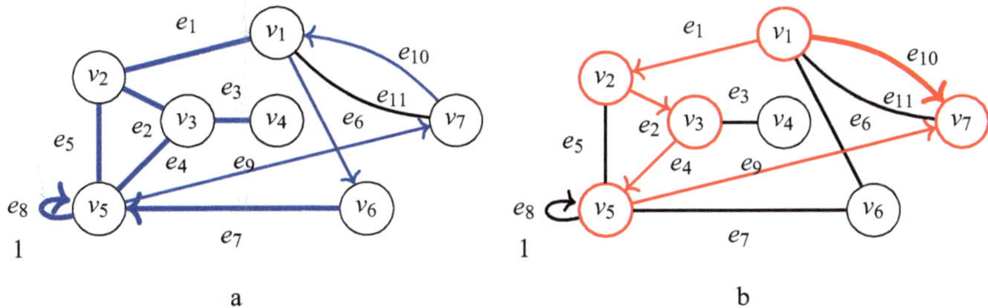


**Fig. 10.7** Some examples of walk, trail, circuit, and a path

the vertices preceding and following it. That is, for some $v_0, v_1, \ldots, v_n \in V$, we have

$$w = v_0 e_1 v_1 e_2 \cdots v_{n-1} e_n v_n \qquad \text{where} \quad e_i = v_{i-1} v_i, \quad \text{for} \quad i = 1, \ldots, n.$$

When $G$ is a simple graph, we also write the above walk as the list $(v_0, v_1, \ldots, v_n)$ of vertices or as the list $(e_1, \ldots, e_n)$ of edges; two vertices $v_0$ and $v_n$ are called the *terminal vertices* of the walk $w$, and $v_1, \ldots, v_{n-1}$ are called its *internal vertices*. A walk with the same terminal vertex is called a *closed walk*; otherwise, it is called an *open walk*.

**Definition 10.11** (*Trail*) A walk $w = (e_1, \ldots, e_n)$ is called a **trail** if no two edges in $w$ appears more than once. A trail with the same terminal vertex is called a **circuit**. The *length of a trail* is the number of edges the trail traverses.

**Example 10.5** In the left side graph shown in Fig. 10.7, we have

$$\begin{aligned} &v_1 e_1 v_2 e_2 v_3 e_4 v_5 e_8 v_5 e_5 v_2 e_2 v_3 e_3 v_4 && \text{is an } \textit{walk} \text{ from } v_1 \textit{ to } v_4; \\ &v_1 e_1 v_2 e_2 v_3 e_4 v_5 e_8 v_5 e_5 v_2 && \text{is a } \textit{trail} \text{ from } v_1 \textit{ to } v_2; \\ &v_6 e_7 v_5 e_8 v_5 e_9 v_7 e_{10} v_1 e_6 v_6 && \text{is a } \textit{circuit} \text{ based at } v_6. \end{aligned}$$

Similarly, we can write other possible walks, trials, and circuits, for various $u, v \in V$.

**Definition 10.12** (*Path*) Let $G = (V, E)$ be a simple graph, and consider two distinct vertices $u, v \in V$. An open trail $P = (u, v_1, \ldots, v_{n-1}, v)$ connecting $u$ and $v$ is called a **path** (or a *simple path*) if no internal vertex $v_i \in V$ (and hence no edge) appears more than once. It is also known as a *u-v path*. A closed path is called a **cycle**.

**Example 10.6** In the right side graph of Fig. 10.7, we have

$$\begin{aligned} &v_1 e_1 v_2 e_2 v_3 e_4 v_5 e_9 v_7 && \text{is an } \textit{open path} \text{ of length 4 from } v_1 \text{ to } v_7; \\ &v_1 e_1 v_2 e_2 v_3 e_4 v_5 e_9 v_7 e_{10} v_1 && \text{is a } \textit{cycle} \text{ of length 5.} \end{aligned}$$

Similarly, we can write other possible walks, trials, and circuits, for various $u, v \in V$.

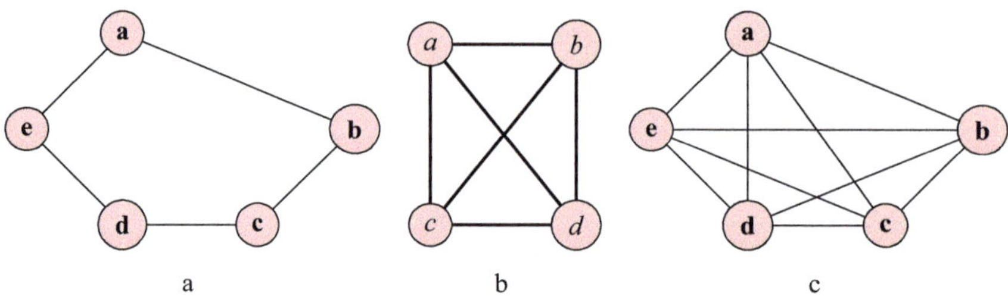


**Fig. 10.8** The cycle $C_5$, and complete graphs $K_4$, $K_5$

As said earlier, the *length* of a path is the number of edges traversed by the path. We will write a $u$-$v$ path of length $n \geq 1$ as $P_n$. Notice that $P_2$ is a *line*. And, for $n \geq 3$, a *cycle of length n* (or simply an **n-cycle**) is a graph with both the order and the size equal to $n$. It is denoted by $C_n$. We can visualize $n$ vertices of $C_n$ as points on a circle such that two vertices are adjacent if and only if they are consecutive points. Therefore, $C_n$ has exactly $n$ edges. In particular, $C_3$ is a *triangle*, $C_4$ is a *square*, and a 5-cycle $C_5$ is as shown on the left side of Fig. 10.8.

In some applications, we also use the notation $u = \sigma(P)$ for the *start vertex* of a path $P = (u, v_1, \ldots, v_{n-1}, v)$, and $v = \tau(P)$ for the end vertex. Therefore, the path $P$ is *closed* if $\sigma(P) = \tau(P)$. Clearly, every walk $w$ with $\sigma(w) \neq \tau(w)$ contains a path from $u$ to $v$. For, if the vertex $v_i = v_j$ $(i < j)$ appears as internal vertices in the walk $w$, we can remove all vertices $v_{i+1}, \ldots, v_{j-1}$ from the walk. In particular, any closed walk of odd length $> 1$ contains a cycle. What does a closed walk not containing a cycle look like?

**Definition 10.13** (*Complete Graph*) A **complete graph** of degree $n$, denoted by $K_n$, is the graph on $n$ vertices every pair of vertices of $K_n$ are adjacent.

For example, the complete graphs $K_4$ and $K_5$ are respectively shown in Fig. 10.8b and c. The degree of each vertex of the *Complete graph* $K_n$ is $n - 1$, and so its size is $\binom{n}{2}$. In general, a graph on $n$ vertices may have at most $n(n-1)/2$ edges. The concept introduced in Definition 10.13 was first defined by the German mathematician *August F. Möbius* (1790–1868) in 1840.

**Definition 10.14** (*Connected Graph*) A graph $G$ is called a **connected graph** if every pair of vertices $u,\ v \in V(G)$ can be connected by a $u$-$v$ path. Otherwise, $G$ is said to be *Disconnected*.

The relation "is connected to" on the set of vertices $V(G)$ is an equivalence relation. First, the relation is *reflexive*, because each vertex $v \in V(G)$ is connected to itself by the *trivial path* $(v)$ (of length zero). Also, it is *symmetric*, because for every $u$-$v$ path, there is a reverse $v$-$u$ path. Finally, given a $u$-$v$ path and a $v$-$w$ path, the concatenation of two paths gives a $u$-$w$ path. Therefore, the relation "connected" is *transitive*. The subgraph induced by an equivalence class of vertices with respect to the relation "is connected to" is called a *Component* of the graph $G$.

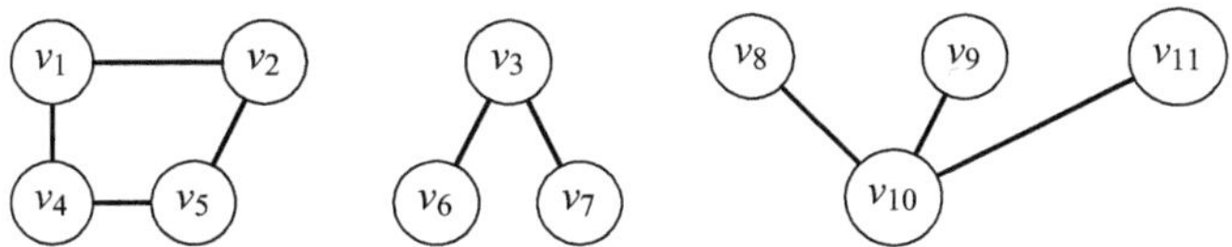


**Fig. 10.9** A graph with 11 vertices and 3 components

**Definition 10.15** (*Component*) Let $G$ be a graph. A **component** $C$ of $G$ is a maximal connected subgraph, where $C$ is *maximal* in the sense that it is not contained in any larger connected subgraph of $G$.

In particular, $G$ is a *Connected graph* if and only if it is the only component of itself. In general, a graph may have two or more components. It follows easily that, for a graph $G$ of order $n$ with $k$ components, we have

$$\left|E(G)\right| \leq \frac{(n-k)(n-k+1)}{2}. \tag{10.2.1}$$

For a graph $G$ of order $n$, size $e$, and with $k$ components, the **rank** $r$ and **nullity** $\mu$ of $G$ is given by

$$r = n - k \qquad \text{and} \qquad \mu = e - n + k. \tag{10.2.2}$$

Notice that the graph consisting of a complete graph of degree $n-k+1$, and $k-1$ isolated vertices, has $k$ components, $n$ vertices, and precisely $(n-k+1)(n-k)/2$ edges. As an example, consider the graph shown in Fig. 10.9, with $n = 11$ vertices and $k = 3$ components.

**Remark 10.3** For any two vertices $u, v \in V(G)$ lying in the same component of a graph $G$, let $d = d(u, v)$ denotes the length of a shortest path between $u$ and $v$. Notice that such a path always exists. The length function $d : V(G) \times V(G) \to \mathbb{R}$ as defined above is called the **distance** between the vertices $u$ and $v$. In the case of a connected graph, it is indeed a metric function in the usual sense.

**Definition 10.16** (*Graph Complement*) Let $G = (V, E)$ be an undirected simple graph. The **complement** of $G$, denoted by $\overline{G}$, is the graph with the vertex set $V$ and two vertices are adjacent in $\overline{G}$ if and only if they are not adjacent in $G$. We say $G$ is **self-complementary** if it is isomorphic to $\overline{G}$.

This definition implies that if $V(G) = n$ and $E(G) = m$ then we have

$$E\big(\overline{G}\big) = E\big(K_n\big) - E\big(G\big) = \frac{n(n-1)}{2} - m.$$

In particular, if $G$ is *Self-complementary* so that $E\big(\overline{G}\big) = m$, then we have $4m = n(n-1)$. It thus follows that $n = 4k$ or $n = 4k + 1$, for any self-complementary

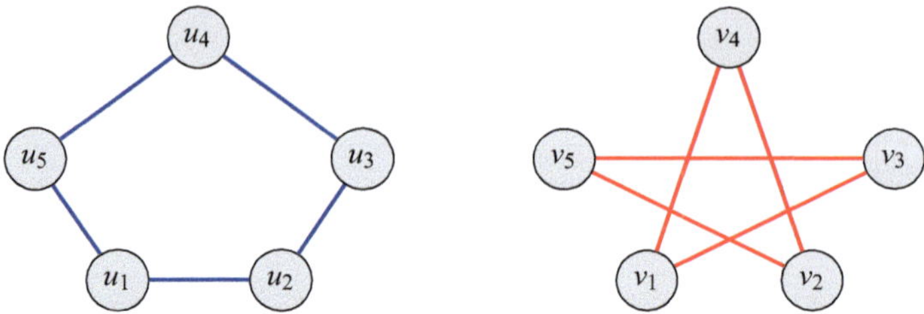


**Fig. 10.10** $C_5$ is self-complementary

graph. It is easy to construct a self-complementary graph with $4k$ and $4k + 1$ vertices, for any $k$. Clearly, cycle $C_4$ and $C_5$ are *Self-complementary* graphs.
For example, the third graph in Fig. 10.10 on five vertices is self-complementary, which is also known as *bull*. Notice hat if $G$ has $n$ vertices and $e$ edges then its complement $\overline{G}$ has $n$ vertices and $n(n-1)/2 - e$ edges. If $G$ is self-complementary then

$$e = \frac{n(n-1)}{2} - e \quad \text{or} \quad 4e = n(n-1).$$

Therefore, either $n = 4k$ or $n = 4k + 1$. It is simple to construct a self-complementary graph with $4k$ and $4k + 1$ vertices, for any $k$.

**Theorem 10.4** *If an undirected simple graph $G = (V, E)$ is disconnected then $\overline{G}$ is connected.*

***Proof*** The proof is left for the reader as an exercise. □

A typical practical problem that motivates the concept introduced in the next definition asks to find the minimum number of weekly *time periods* required to schedule meetings of committees in a way so that no two committees with common members have a meeting during the same time period. For the associated graph model, we may take a vertex for each committee, and treat to vertices adjacent if the corresponding committees have some common members. Then, to solve the above problem, we assign *time periods* to vertices in such a manner that endpoints of each edge receive different *time periods*.

**Definition 10.17** (*Stable Set & Clique*) Let $G = (V, E)$ be a graph. A set of vertices $W \subset V$ is called an **independent set** (or a **stable set**) if all pairs of vertices in $W$ are non-adjacent. A set of vertices $W \subset V$ is called a **clique** if all pairs of vertices in $W$ are adjacent. The **clique number** of a graph $G$, denoted by $\omega(G)$, is the maximal of the clique numbers of $G$.

If the *Meetings scheduling problem* has the graph model shown in Fig. 10.11, then we need to assign three time periods, one each of the vertices in the *independent set* $\{v_1, v_3, v_6\}$. Further, the members in a *Clique* must not have the same meeting time period. Clearly, in this case, the minimum number of time periods required is 3.

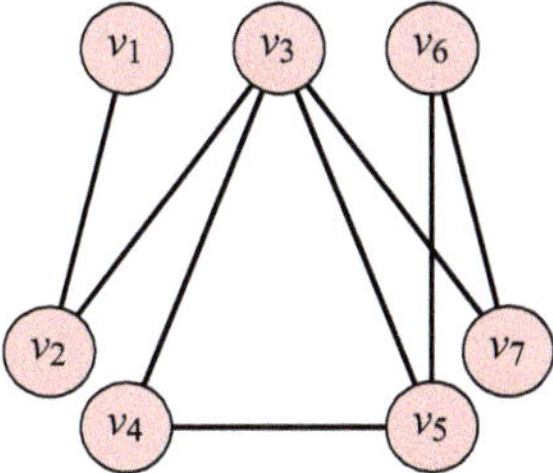


**Fig. 10.11** A graph model of meetings schedule with three time periods

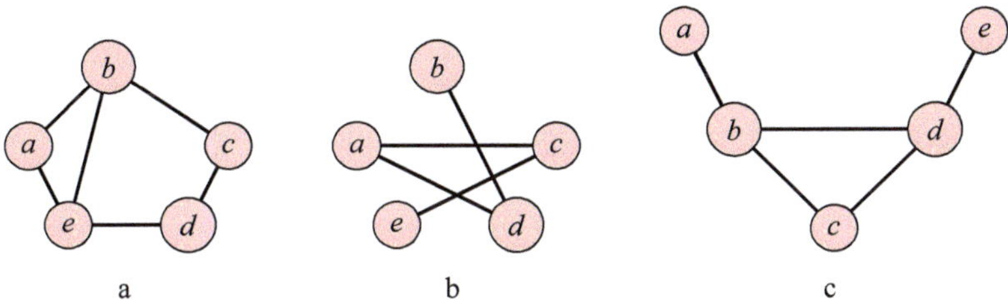


**Fig. 10.12** A graph $G$, its complement $\overline{G}$, and the *bull*

The graph in Fig. 10.12a has the *clique number* three. The next theorem proves that every graph on six vertices has a clique of size three or its complement has a clique of size three.

**Theorem 10.5** *A graph $G$ of order six contains $K_3$ or so does the complement $\overline{G}$.*

***Proof*** We show that at least six people are to be invited to a party to make sure either three of them are mutual friends or three are pairwise strangers. Let the vertices of a graph $G$ of order 6 represent the people invited to the party, and for each pair of friends, there be an edge between the corresponding vertices. We thus need to show that either $G$ or $\overline{G}$ contains $K_3$. Fix a vertex $v$ of $G$. Then $v$ is adjacent to the other five vertices in $G$ or $\overline{G}$. Therefore, there are three or more vertices adjacent to $v$ in $G$ or in $\overline{G}$. We first consider the case when $v$ is adjacent to vertices $u_1, u_2, u_3$ of $G$. If any two of these three vertices are adjacent, then $G$ contains the triangle $K_3$. If not $u_1, u_2, u_3$ are the vertices of a triangle in $\overline{G}$. Other cases can be proved similarly. □

Figure 10.13 shows a 6-vertex triangle-free graph $G$ with $\overline{G}$ containing two triangles. It is next shown that we need to invite 18 people to a party to guarantee that there are either four mutual friends or four mutual strangers. We can picture the situation using a graph. For, let $G$ be a graph of order 18 with vertices representing the people at the party, and there is an edge between two vertices if the corresponding pair of people are friends. It thus remains to show that $G$ or $\overline{G}$ contains $K_4$.

**Theorem 10.6** *A graph $G$ with at least 18 vertices contains $K_4$ or so does the complement $\overline{G}$.*

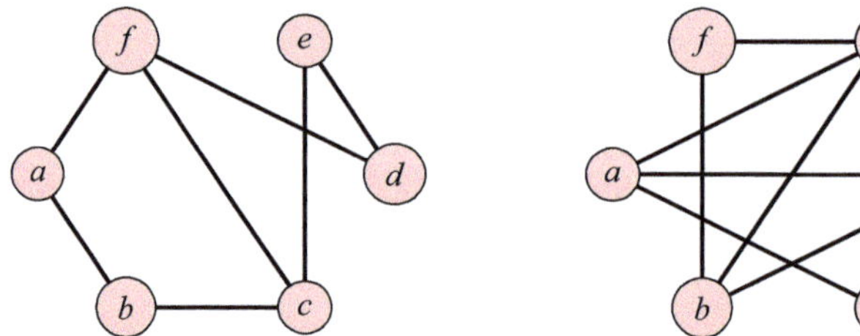


**Fig. 10.13** A 6-vertex triangle-free graph $G$ with $\overline{G}$ containing two triangles

***Proof*** Fix a vertex of $G$, say $v$. Let $F$ be the set of vertices of $G$ that are adjacent to $v$, and $S$ be the set of vertices of $G$ that are non-adjacent to $v$. Notice that vertices in $S$ are adjacent in the graph $\overline{G}$. By pigeonhole principle, we have $|F| \geq 9$ or $|S| \geq 9$. That is, there are nine or more vertices adjacent to $v$ in $G$ or in $\overline{G}$. If $|S| \geq 9$ then it contains $K_4$ or three are three independent vertices. In the former case, we are done. In the latter case, adjoining $v$ to the three independent vertices shows the existence of $K_4$ in $\overline{G}$. If $|F| \geq 9$, we use the fact that $R(4, 3) = R(3, 4)$ to claim that either there is $K_3$ in $F$ or four independent vertices. In the latter case, we have a $K_4$ in $\overline{G}$. In the former case, adjoining $v$ to $K_3$ in $F$ shows the existence of $K_4$ in $G$. □

A story about this theorem goes as follows [14]: In 1950, during a study related to friendship between children, the Hungarian philosopher and sociologist *Sándor Szalai* (1912–1983) observed that among any group of about twenty children there are four children who are mutual friends or else four children no two of whom are friends. Subsequently, a brief discussion with the three eminent Hungarian mathematicians Paul Erdős, Paul Turán, and Vera Sós, helped him understand that it is a mathematical proposition rather than a sociological conclusion. This is indeed a special case of Ramsey's theory, as discussed in the last section of the chapter.

## 10.3 Important Classes of Graphs

Recall that $P_n$ $(n \geq 2)$ and $C_n$ $(n \geq 3)$ respectively denote the path and cycle of length $n$. Clearly, the degree of each vertex of $C_n$ is two. We next introduce another important class of graphs, known as the *Bipartite graphs*. A typical practical problem that can be modeled by these types of graphs is about assignment of $m$ jobs to $n$ people, where it is assumed that not everybody has the skill set to do all $m$ jobs. The concept introduced in the next definition was first given by Möbius.

**Definition 10.18** (*Bipartite Graph*) A graph $G = (V, E)$ is called a **bipartite graph** if there exists mutually disjoint independent sets $V_1$ and $V_2$ such that $V = V_1 \cup V_2$, and for each edge $e = uv \in E$ we have $u \in V_1$ and $v \in V_2$. Further, the bipartite graph $G$ is said to be **complete** if each vertex in $V_1$ is adjacent to every vertex of $V_2$.

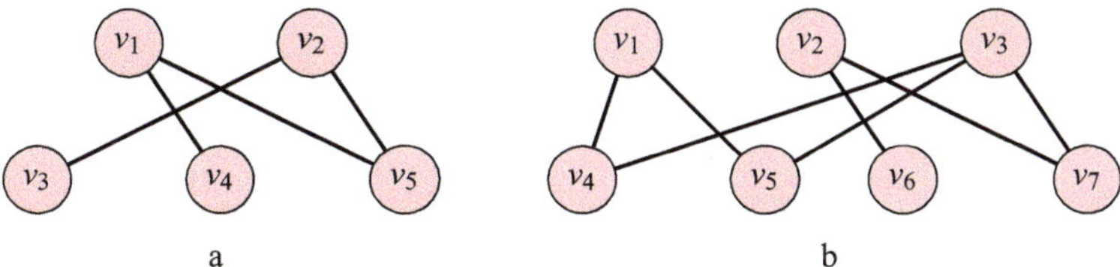


**Fig. 10.14** A (2, 3)-bipartite graph and a (3, 4)-bipartite graph

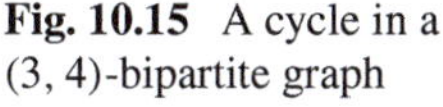

**Fig. 10.15** A cycle in a (3, 4)-bipartite graph

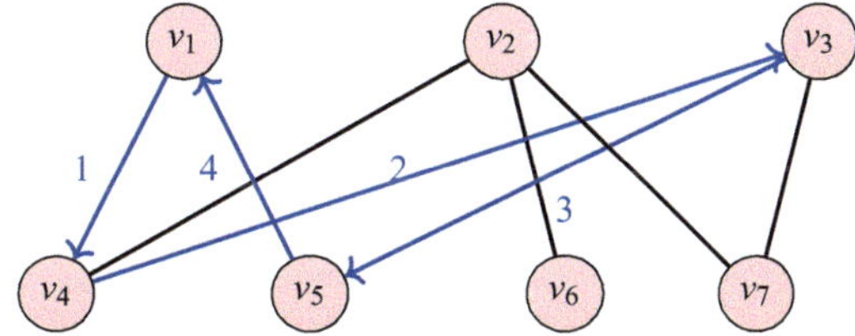


When $|V_1| = m$ and $|V_2| = n$, the complete bipartite $G$ is written as $K_{m,n}$. For example, the two drawings shown in Fig. 10.14 are respectively examples of bipartite graphs of sizes 5 and 7. On the other hand, the middle drawing in Fig. 10.16 is an example of the complete bipartite graph $K_{3,3}$.

**Theorem 10.7** *A graph $G = (V, E)$ is bipartite if and only if it contains no odd cycle.*

***Proof*** We first assume that $(V_1, V_2)$ is a bipartition of $V$. Consider a cycle $C$ in $G$. WLOG, suppose the starting vertex of $C$ is in the vertex set $V_1$. Then $C$ has its second vertex in $V_2$, the third vertex in $V_1$, and so on. As the last vertex of cycles (which is the same as the first vertex in $C$) is in $V_1$, we must have that the cycle $C$ is of even length. (See Fig. 10.15.) Conversely, suppose a graph $G = (V, E)$ contains no odd cycle. For a fixed arbitrary vertex $v \in V$, write $V_1$ for the set of vertices that are at an odd distance from the vertex $v$, and take $V_2 = V - V_1$. We claim that $(V_1, V_2)$ is a bipartition of the set $V$. Notice that it is enough to show both $V_1$ and $V_2$ are independent sets. We prove $V_1$ is an independent set, and the argument applies to prove the independence of $V_2$. For, suppose there is an edge between two vertices $u, u' \in V_1$. Then putting together the two paths of odd length from the vertex $v$ to the vertices $u$ and $u'$ with the edge connecting $u$ and $u'$ produces a circuit of odd length, which cannot be the case. Therefore, no two vertices in $V_1$ are adjacent. $\square$

Notice that $K_{m,n}$ has $m + n$ vertices and $mn$ edges. The graphs $K_{m,n-m}$ with $n$ vertices have $mn - m^2$ edges. Since $1 \leq m \leq n - 1$ the maximum of $mn - m^2$ is $n^2/4$. Therefore, every bipartite with $n$ vertices can have atmost $\lfloor n^2/4 \rfloor$ edges. The graph $K_{\lfloor n/2 \rfloor, \{n/2\}}$ has exactly edges. Note that $\lfloor r \rfloor$ is the greatest integer not exceeding the real number $r$ and $\{r\} = -\lfloor -r \rfloor$. Further, it follows from Theorem 10.7 that the complete bipartite graphs $K_{m,n}$ contains no *triangle* $K_3$. We use the fact to prove our next theorem.

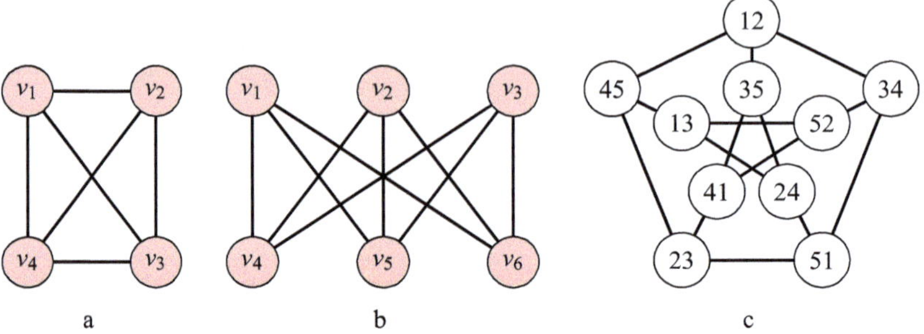


**Fig. 10.16** $K_4$, $K_{3,3}$, and Petersen graph, are cubic graphs

**Theorem 10.8** *Among all n vertices graph without a triangle, the maximum number of edges is* $\lfloor n^2/4 \rfloor$.

***Proof*** The result is true for $n = 1, 2, 3$. More generally, we consider the two cases:
**Case-I**: When $n$ is even. Assume that the result is true for any even $n \le 2m$. Now let $G$ be a graph with $n = 2m + 2$ vertices with no triangles. If $G$ is totally disconnected, $G$ has no triangles and the number of edges is $0 \le n^2/4$. So assume that $G$ is not totally disconnected. Therefore there are adjacent vertices $u$ and $v$ in $G$. Now the subgraph $G' = G - \{u, v\}$ obtained by removing the vertices $u$ and $v$ is a graph with $2m$ vertices and no triangles. Hence $G'$ has atmost $4m^2/4 = m^2$ edges. If any vertex $w$ in $G$ is adjacent to both $u$ and $v$ in $G$ then $G$ has a triangle. Therefore no point of $G$ can be adjacent to both $u$ and $v$ in $G$. Thus if $u$ is adjacent to $k$ vertices in $G'$ then $v$ can be adjacent to $2m - k$ vertices in $G'$. Then $G$ has at most the following number of edges:

$$m^2 + k + (2m - k) + 1 = m^2 + 2m + 1 = (m + 1)^2 = n^2/4 = \lfloor n^2/4 \rfloor.$$

This proves the result for even $n$.
**Case-II**: When $n$ is odd. Assume that the result is true for any odd $n \le 2m + 1$. Let $G$ be a graph with $2m + 3$ vertices which is not totally disconnected. Let $u$ and $v$ be adjacent vertices in $G$. Consider as in Case-I. Since $G'$ has $2m + 1$ points no $K_3$, it has $\lfloor \frac{4m^2+4m+1}{4} \rfloor = m^2 + m$ edges. Now $G$ can have the following number of edges:

$$m^2 + m + k + (2m + 1 - k) + 1 = m^2 + 3m + 2 = \lfloor \frac{(2m + 3)^2}{4} \rfloor.$$

This proves the result. □

Notice that $K_{\lfloor n/2 \rfloor, \{n/2\}}$ has maximum possible edges.

**Definition 10.19** (*Regular Graph*) A graph $G = (V, E)$ is called **regular** of degree $r$ (or simply **r -regular**) if $\deg(v) = r$, for all vertex $v \in V$. A regular graph of degree 3 is called a *cubic graph*.

Clearly, the cycle $C_n$ is regular of degree 2, and the complete graph $K_n$ is a regular graph of degree $n - 1$. Further, the complete bipartite graph $K_{m,n}$ is regular if and only if $m = n$; in this latter case, $K_{n,n}$ is regular of degree $n$. By Theorem 10.1, for a regular graph $G$ of degree $r$ with $n$ vertices and $e$ edges, we have $e = nr/2$. Therefore, the complete graph $K_n$ has exactly $n(n - 1)/2$ edges. By *handshaking lemma*, every cubic graph has an even number of vertices. Figure 10.16 shows drawings of three important cubic graphs.

The drawing shown in Fig. 10.16c is the famous **Petersen graph**. It has ten vertices given by the 2-element subsets of the set $\{1, 2, 3, 4, 5\}$, and two vertices are adjacent if and only if the corresponding 2-element sets are disjoint. Notice that the *Petersen graph* has two disjoint 5-cycles given by

$$12,\ 34,\ 51,\ 23,\ 45 \quad \text{and} \quad 13,\ 52,\ 41,\ 35,\ 24,$$

together with edges that pair up vertices on the two such cycles. Notice that each pair of non-adjacent vertices in this graph has exactly *one common neighbor*. Since the shortest cycle in the Petersen graph is $C_5$, its girth is 5.

**Definition 10.20** (*Girth*) The **girth** of a connected graph $G$ is the smallest integer $g$ such that $G$ contains the cycle $C_g$. Said differently, the *Girth* $g$ is the length of the shortest cycle contained in $G$.

The *breadth-first search* (or simply BFS) algorithm explores the vertices of a graph in breadth-ward motion, visiting all the neighbors of a vertex before moving on to the next level of vertices. For smaller order graphs, we can use the BFS algorithm to find the girth of a connected graph $G$. The main idea is to find, for each vertex, the length of the shortest cycle it is part of. Notice that for bipartite graph search ends when the shortest cycle found is of even length. It is interesting to find the girth of a planar graph. The concept has numerous real-world applications. In transportation networks, the girth of a graph can help identify potential traffic bottlenecks or areas with a higher likelihood of congestion. In social network analysis, the girth can provide insights into the structure of social networks and help identify communities or cliques. In computer networks, the girth can be used to analyze the resilience and robustness of the network topology.

**Definition 10.21** An $r$-regular spanning subgraph of a graph is called an **r -factor**. When the edge set of a graph can be divided into $r$-factors, we say the graph admits an **r -factorization** (or a *decomposition*). In particular, *factorization* (or *resolution*) refers to 1-factorization.

## 10.4 Trees and Rooted Trees

The concept introduced in Definition 10.22 was first given by the German physicist *Gustav Kirchhoff* (1824–1887) in 1845. He was mainly interested in computing currents in electric circuits. Much later, *Arthur Cayley* (1821–1895) also started working on this class of graphs. However, the ideas he introduced ultimately became the standard terminologies of graph theory.

**Definition 10.22** (*Tree*) A graph without any circuit is called a **forest**. A **tree** is a connected forest.

We usually write a tree as $T$. Notice that the girth of a forest is infinite. A *point* $K_1$ is the only tree on a single vertex, and the *line* $K_2$ is a tree having only two pendant vertices. All trees other than these two are called non-trivial trees. For example, some non-trivial trees on 4, 5, and 6 vertices are respectively as shown in Fig. 10.17 through Fig. 10.19. We know that the number of edges in a connected graph on $n$ vertices has at least $n - 1$ edges. Notice that every tree $T$ is a bipartite graph. Fix a vertex $u \in V(T)$. As bipartition we may take $(V_1, V_2)$, where

$$V_1 := \{v \in V(T) \mid d_T(u, v) \text{ is even}\};$$
$$V_2 := \{v \in V(T) \mid d_T(u, v) \text{ is odd}\}.$$

More generally, by Theorem 10.7, we know that a graph is bipartite if and only if it contains no odd cycle. The next theorem proves that a tree on $n$ vertices has exactly $n - 1$ edges.

**Theorem 10.9** *A tree $T$ of order $n$ has $e = n - 1$ edges.*

***Proof*** This proof uses induction on $n$. Clearly, the claim holds for trees with 1, 2, 3, 4, and 5 vertices (see Figs. 10.17 and 10.18). Let $T$ be a tree of order $n$ and the size $e$. We may assume that the claim holds for every tree of order $n - 1$, with $n \leq 5$. By Lemma 10.1, $T$ has a pendant vertex, say $v$. Let $T' := T - e$. Clearly, $T'$ is a tree with $n - 1$ vertices and $e - 1$ edges. Therefore, by induction hypothesis, we have

$$e - 1 = n - 1 - 1 \qquad \text{i.e.,} \qquad e = n - 1.$$

This completes the proof. □

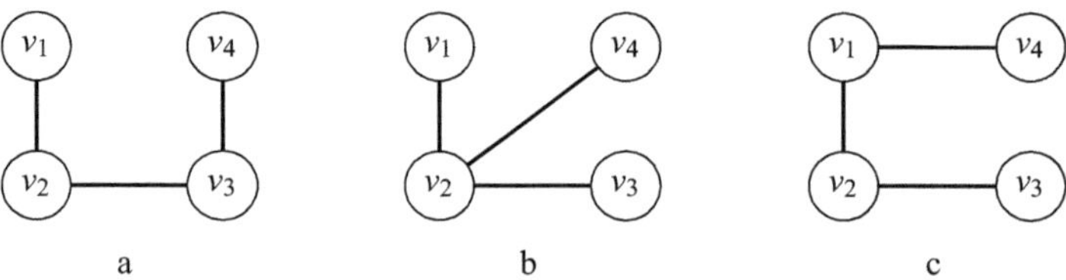


**Fig. 10.17** Two *non-isomorphic* trees on 4 vertices

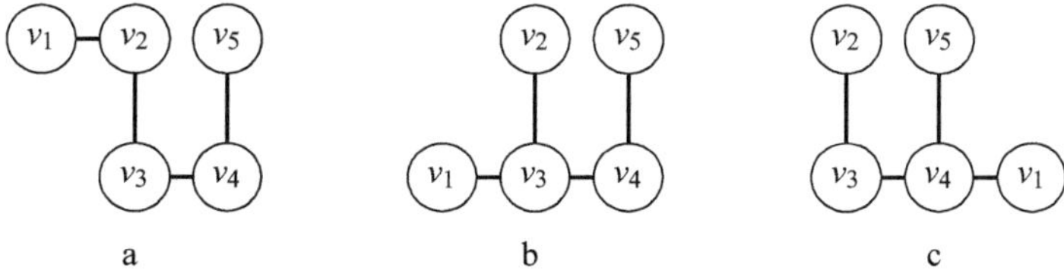


**Fig. 10.18** Three *non-isomorphic* trees on 5 vertices

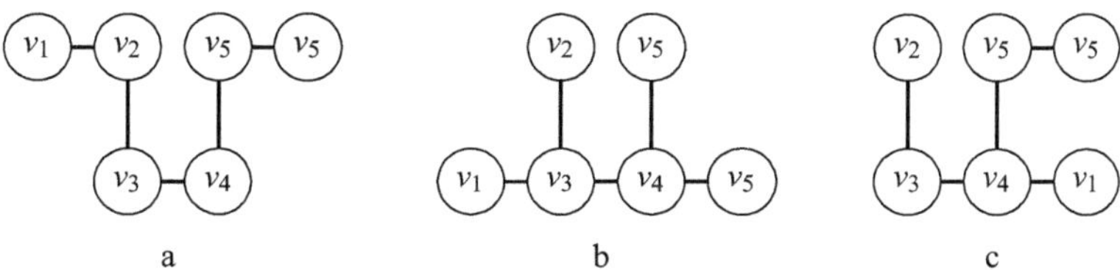


**Fig. 10.19** Three *non-isomorphic* trees on 6 vertices

More generally, a **subtree** of a connected graph $G$ is a subgraph without a cycle, i.e., it is a tree. The converse of Theorem 10.9 holds for connected graphs.

**Theorem 10.10** *A connected graph G of order n and size $e = n - 1$ is a tree.*

***Proof*** For a contradiction, suppose $G$ is not a tree, and let $C$ be a circuit in $G$. By removing an edge from $C$, we obtain a graph $G'$ that is still connected. If $G'$ is not a tree, we repeat the above construction to obtain a graph $G''$ of size one less than the size of the graph $G'$. After repeating this process finitely many times, we end up with a tree $T$ with $n$ vertices and $e - k$ edges, for some integer $k \geq 1$. But then, by Theorem 10.9, we must have

$$e - k = n - 1 \quad \text{i.e.,} \quad e = n - 1 + k.$$

This contradicts the assumption $e = n - 1$, because $k \geq 1$. Therefore, $G$ is a tree. □

**Theorem 10.11** *An acyclic graph G of order n and the size $e = n - 1$ is a tree.*

***Proof*** For a contradiction, suppose $G$ is not a tree. Therefore, $G$ is not connected. Let $G$ has $k$ components, with $k \geq 2$, of the types $(n_i, e_i)$ for $i = 1, \ldots, k$. We then have

$$n_1 + n_2 + \cdots + n_k = n \quad \text{and} \quad e_1 + e_2 + \cdots + e_n = e.$$

Also, for each $i$, we have $e_i = n_i - 1$, because each component is a tree. It thus follows that

$$e = e_1 + e_2 + \cdots + e_n = (n_1 - 1) + (n_2 - 1) + \cdots + (n_k - 1) = n - k.$$

This contradicts the assumption $e = n - 1$, because $k \geq 2$. Therefore, $G$ is a tree. □

A graph $G$ is said to be **minimally connected** if removal of a single edge disconnects $G$. The next theorem proves that this property characterizes trees.

**Theorem 10.12** *A graph is a tree if and only if it is minimally connected.*

***Proof*** Consider a tree $T$ and any edge $(u, v)$ in $T$. If the removal of the edge $(u, v)$ does not disconnect the graph, then there is a path connecting $u$ and $v$ in the graph obtained by removing the edge $(u, v)$ from $T$. This path together with the edge $(u, v)$ forms a circuit in $T$, a contradiction. Conversely, assume that the graph $T$ is minimally connected. Thus it is connected. If there is any circuit in $T$, the removal of any one of the edges in the circuit does not disconnect the graph and this is a contradiction. Thus the graph cannot have circuits. Since $T$ is connected and has no circuits, $T$ is a tree. □

**Theorem 10.13** *A graph $G$ is a tree if and only if for every pair of distinct vertices $u, v \in V(G)$ there is a unique path between $u$ and $v$.*

***Proof*** Let $G$ be a tree and there are two paths between a pair of vertices $u$ and $v$ in $G$. Then the union of these two paths contains a circuit and this is a contradiction. To see this, let $P_1$ and $P_2$ be two paths beginning at $u$ and ending at $v$. Then the first vertex of $P_1$ and $P_2$ are the same. The successive vertices can also be the same. Find the vertex $v_1$ at which the two paths branch out. Then by moving along $P_1$, find the next vertex $v_2$ that is also in $P_2$. Then the portion of the two paths $P_1$ and $P_2$ between $v_1$ and $v_2$ is a required circuit. Conversely, assume that every pair of vertices in a graph $G$ is connected by exactly one path. Then it is clear that the graph $G$ is connected. If there is a circuit in $G$, then it is clear that there are at least two paths between any two vertices in the circuit. Thus the graph has no circuit and this shows that $G$ is a tree. □

In the next theorem, we summarize the results obtained above. Any of these five fundamental characterizations may be used to define a tree.

**Theorem 10.14** *Let $G$ be graph with $n$ vertices and $e$ edges. Then the following are equivalent:*

1. *$G$ is a tree.*
2. *$G$ is connected and is circuitless.*
3. *$G$ is connected and $e = n - 1$.*
4. *$G$ has no circuit and $e = n - 1$.*
5. *There is exactly one path between every pair of vertices in $G$.*
6. *$G$ is a minimally connected graph.*

**Corollary 10.2** *A non-trivial tree has at least two pendant vertices.*

***Proof*** Since a tree of order $n$ has $e = n - 1$ edges, by degree-sum formula we have

$$\sum \deg(v) = 2e = 2(n-1).$$

Now, if $\deg(v) \geq 2$ for all vertices excepting the two $v_1$ and $v_2$ (say), then

$$\deg(v_1) + \deg(v_2) + \sum_{v \neq v_1, v_2} \deg(v) = 2(n-1).$$

It thus follows that we have

$$\begin{aligned}\deg(v_1) + \deg(v_2) &= 2n - 2 - \sum_{v \neq v_1, v_2} \deg(v) \\ &\leq 2n - 2 - (n-2)2 = 2.\end{aligned}$$

Therefore, since $\deg(v_1) \geq 1$, $\deg(v_2) \geq 1$, we obtain $\deg(v_1) = \deg(v_2) = 1$. Hence, $v_1$ and $v_2$ are pendant vertices. □

The **eccentricity** of a vertex $u$ of a graph $G$, denoted by $E(u)$, is given by

$$E(u) = \max\{d(u, v) \mid v \in V(G)\}, \tag{10.4.1}$$

where $d(u, v)$ denotes the length of the shortest path between vertices $u$ and $v$.

**Definition 10.23** The **center** of a tree $T$ is a vertex $c$ such that

$$E(c) = \min\{E(u) \mid u \in V(T)\}, \tag{10.4.2}$$

where $E(u)$ is as defined in (10.4.2).

We prove that every tree has one or two adjacent vertices as the *center*.

**Theorem 10.15** *Every tree $T$ has one or two adjacent centers.*

***Proof*** If $T$ is trivial, then the single vertex is the center. Also, if $T$ has two vertices, then both are centers of $T$. Suppose $T$ is a non-trivial tree that is not the line $K_2$. Then by Corollary 10.2 $T$ has at least two pendant vertices. The removal of these pendant vertices from the tree $T$ uniformly reduces the eccentricities of vertices by one in the so-obtained new tree $T'$. An example illustrating this reduction in eccentricities is given in Fig. 10.15. It thus follows that the center(s) of $T'$ is the same as the center(s) of the tree $T$. The process of deleting all the pendant vertices successively produces either a trivial tree or the line $K_2$, which proves the assertion. □

**Definition 10.24** (*Spanning Tree*) A **spanning tree** $T$ of a connected graph $G$ is a circuit-free spanning subgraph of $G$. Such a sparse subgraph contains all vertices of $G$.

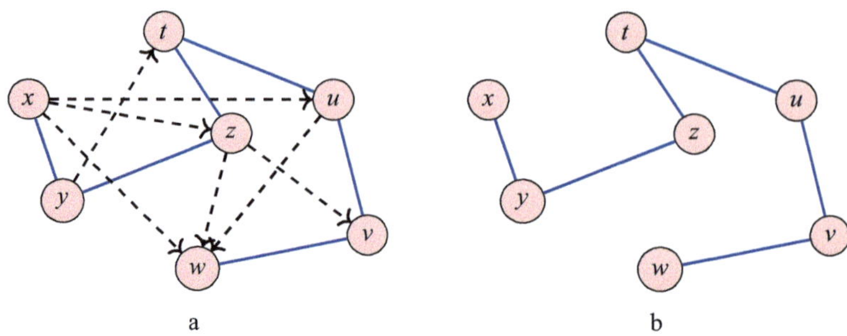


**Fig. 10.20** A spanning tree of a graph (in blue)

For any fixed vertex of the complete graph $K_n$ $(n \geq 2)$, we obtain the complete bipartite graph $K_{1,n-1}$, which is a *Spanning tree* of the graph $K_n$. In particular, all sixteen spanning trees of the graph $K_4$ are shown in Fig. 10.23. The next theorem proves that every connected graph has a spanning tree. The main idea of the proof given below is to "break" all circuits of the given graph. This is illustrated by the spanning tree (in blue) shown in Fig. 10.20b obtained from the graph shown in Fig. 10.20a by systematically "breaking all circuits" while moving along vertices $x$ to $w$.

**Theorem 10.16** *Every connected graph has a spanning tree.*

***Proof*** Let $G$ be the connected graph. If $G$ is circuit-free, then $G$ is its own spanning tree. Next, suppose $C$ is a circuit in $G$. For any edge $e$ on $C$, the subgraph $G' = G - e$ stays connected, because there is at least one more path that connects the endpoints of the edge $e$. Now, if $G'$ is a tree, then once again we are done. Otherwise, we may continue the process by deleting one edge from a circuit $C'$ of the graph $G'$. Of course, this process must terminate after a finite number of steps, and so ultimately we obtain a circuit-free subgraph $T$ of the graph $G$. This is a spanning tree of the graph $G$. □

If $G$ is a graph model of a practical problem, then it is desirable to know how to find "nice" spanning trees that have some additional properties. For example, such is the case for algorithm design problems related to *transportation, routing in network flow, combinatorial optimization, dynamic programming, project management, bioinformatics*, etc. ([2, 15]). The terms introduced in the next two definitions would prove helpful in the sequel.

**Definition 10.25** (*Branch & Chord*) Let $G$ be a connected graph, and $T$ be a spanning tree of $G$. Each edge on $T$ is called a **branch** of $T$, and an edge not on $T$ is called a **chord**.

Suppose the order of a connected graph $G$ is $n$, and $T$ be a spanning tree of $G$. Then $T$ also has $n$ vertices, and the number of edges on $T$ is $n - 1$, by Theorem 10.9. Therefore, with respect to the tree $T$, the number of branches of $G$ is $n - 1$, and so

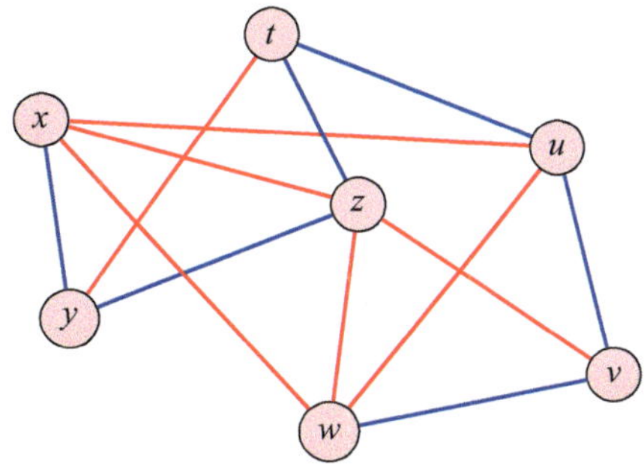


**Fig. 10.21** Branches (blue) and chords (red) of a spanning tree

the number of chords is $e - (n - 1) = e - n + 1$, where $e$ is the size of the graph $G$. For example, with respect to the spanning tree $T$ (in blue) shown in the connected graph $G$ as in Fig. 10.21, the blue edges are branches and the red edges are chords.

**Definition 10.26** (*Fundamental Circuit*) Let $G$ be a connected graph, and $T$ be a spanning tree of $G$. A unique cycle obtained by adding a chord to $T$ is called a **fundamental circuit** of the tree $T$. More generally, a *Fundamental circuit* in a graph is a cycle that cannot be decomposed into smaller cycles.

It is known that every graph may be expressed as a combination of fundamental circuits. This result helps simplify problems related to complex graphs because it suffices to analyze the constituent fundamental circuits individually. The following steps may be used to determine the fundamental circuits in a graph $G$:

(a) Select a spanning tree $T$ of $G$;
(b) Add a chord of $T$ to form a fundamental circuit;
(c) Repeat $(b)$ to obtain all fundamental circuits.

For example, the addition of any red edge to the spanning tree (in blue) shown in Fig. 10.21 gives a fundamental circuit.

Consider the connected graph $G$ shown in Fig. 10.20a, with a spanning tree $T$ (in blue) as in Fig. 10.20b. We illustrate below how to obtain more spanning trees of the graph $G$. The fundamental circuit created from $T$ by adding the chord $xw$ (in red) is shown in Fig. 10.22a. We next delete a branch from the fundamental circuit $T + xw$ to obtain a new spanning tree, say $T_1 = T + xw - xy$, as shown in Fig. 10.22b. Of course, we can generate more spanning trees by deleting some other branch, instead of $xy$. A similar procedure applied to other fundamental circuits of $T$ gives some more spanning trees. The next theorem gives a characterization of a tree in terms of fundamental circuits.

**Theorem 10.17** *Let $G$ be a connected graph. Then $G$ is a tree if and only if every pair of vertices of $G$ is part of a fundamental circuit of $G$.*

**Theorem 10.18** *The number of fundamental circuits in a connected graph equals the number of independent cycles in the graph.*

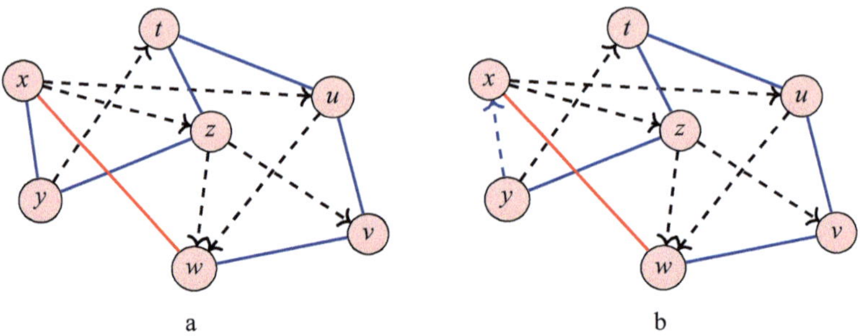


**Fig. 10.22** Fundamental circuit obtained by an adding chord (in red) to spanning tree

Let $s(G)$ denote the number of spanning trees in a connected graph $G$. In 1889, Cayley obtained an elegant formula specifying the number of spanning trees in the complete graph $K_n$ [16]. He proved $s(K_n) = n^{n-2}$, which is known as the **spanning tree formula**. It is the same as the number of *Labeled trees* on $n$ vertices.

**Definition 10.27** (*Graph Labeling*) A **labeling** of a graph $G$ is a function $\lambda$ defined on the vertex set $V(G)$ such that $\lambda(v) \in \{1, 2, \ldots, m\}$ is a **label** of the vertex $v \in V(G)$.

**Theorem 10.19** (Cayley, 1889) *The total number of spanning trees in the complete graph $K_n$ is $n^{n-2}$.*

***Proof*** The proof uses the concept of *Branching forest* [17], which is a digraph with each component a branching. □

Alternatively, one may use the *Prüfer coding* to prove Cayley's *Spanning tree formula* [7]. A **Prüfer sequence** is a string of length $n - 2$ over the alphabet

$$\Sigma = \left[n\right] := \left\{1, 2, \ldots, n\right\}.$$

That is, a *Prüfer sequence* is an ordered list $(x_1, x_2, \ldots, x_{n-2})$, where $x_i \in \left[n\right]$, with repetitions allowed. Clearly, there are $n^{n-2}$ such types of sequences. The main idea of the proof is to show that the set of labeled trees $\mathscr{T}_n$ on $n$ vertices is in bijective correspondence with the set of Prüfer sequence $\mathscr{S}_{n-2}$ of length $n - 2$. The underlying bijective function is called the **Prüfer coding**:

For $n = 2$, we have $\mathscr{T}_2$ contains $K_2$, and $\mathscr{S}_0$ contain the empty string $\epsilon$. Next, for $n = 3$, we have $\mathscr{T}_3$ contains three double lines obtained by deleting an edge from $K_3$, and $\mathscr{S}_1 = \{(1), (2), (3)\}$. Also, for $n = 4$, we have $\mathscr{T}_4$ contains the sixteen spanning trees of the graph $K_4$ obtained by deleting three edges to break circuits, and $\mathscr{S}_2$ contains the Prüfer sequences written as labels of the spanning trees (Fig. 10.23). In general, for a given spanning tree $T$ of $K_n$, a unique Prüfer sequence of length $n - 2$ is defined as follows: Find the leaf with the lowest label. The first term of the Prüfer sequence associated with $T$ is the label of the (unique) neighbor of this leaf. Delete the leaf from $T$, and continue repeating the same operation until $T$ is left with two vertices only. Since the Prüfer coding deletes $n - 2$

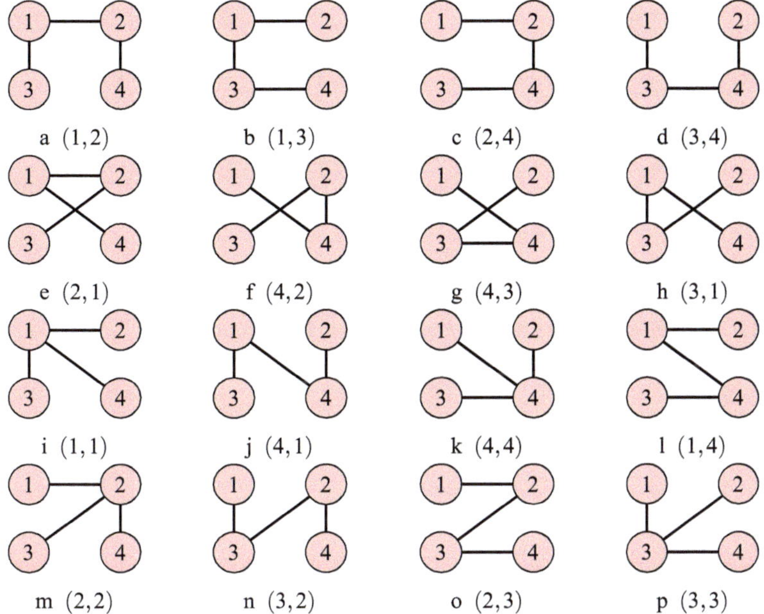


**Fig. 10.23** All sixteen spanning trees of $K_4$

vertices, the sequence so obtained is of length $n-2$. Conversely, it is possible to construct the spanning tree associated with a given Prüfer sequence of the set $[n]$. The underlying procedure is illustrated in Example 10.7.

For example, the lowest label leaf in Fig. 10.23h is 2, so the first term of the associated Prüfer sequence is 3, which is the label of the (unique) neighbor of 2. Deleting the vertex 2 from the tree. Now, the lowest label leaf is 3 with a unique neighbor of 1. so the second term of the Prüfer sequence is 1. Therefore, the Prüfer sequence in this case is obtained as (3, 1). Similarly, the Prüfer sequence associated with the spanning tree shown in Fig. 10.23a is given by (1, 2). And, so on.

We next consider the graph shown in Fig. 10.24a. The node 1 is the leaf with the lowest label, and so we append to the desired list the label of its unique neighbor. Therefore, the first term of the Prüfer sequence is 2. To repeat the procedure, we remove the node 1. Next node 2 is the leaf with the lowest label, and so we append to the Prüfer sequence the label of its unique neighbor that is 3. To repeat the procedure, we remove the node 2. We ultimately obtain (2, 3, 4, 5, 7) as the Prüfer sequence associated with the spanning tree shown in Fig. 10.24b (in blue). However, not every spanning tree of the graph shown in Fig. 10.24a defines a Prüfer sequence (why?).

**Example 10.7** We give a construction of the spanning tree associated with a given Prüfer sequence, say (1, 1, 4, 6, 5) of the set $[7]$. Notice that the tree associated with this sequence has $5+2=7$ nodes. The first number in the given list is 1, and the lowest number not included in the list is 2, so we connect 1 to 2 (Fig. 10.25a). Delete

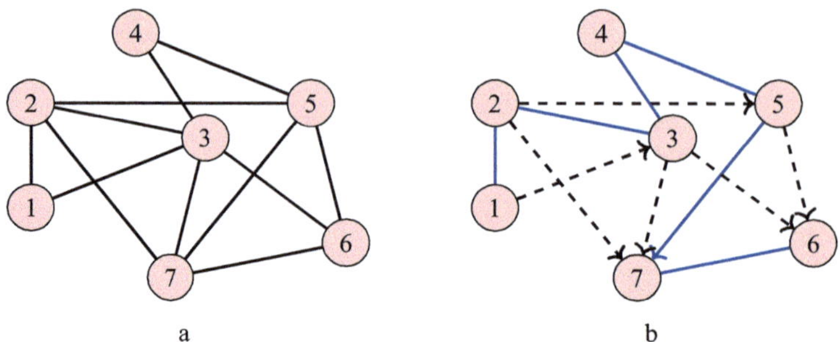


**Fig. 10.24** The Prüfer sequence for spanning tree is (1, 3, 4, 7, 6)

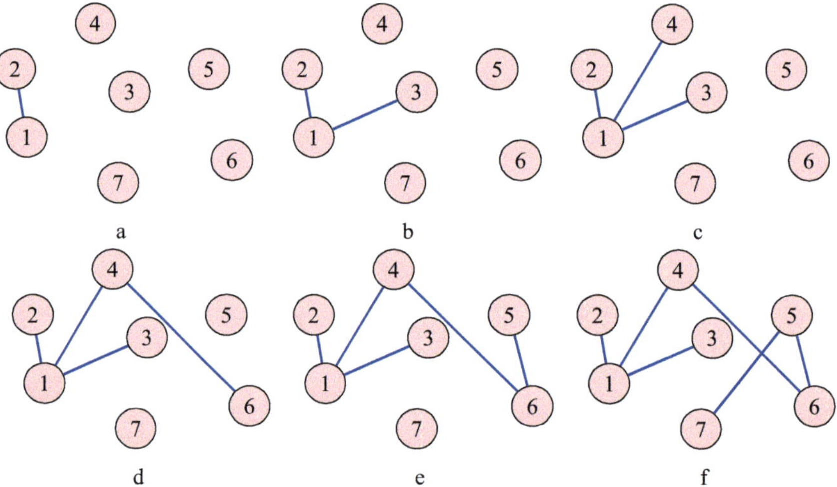


**Fig. 10.25** Spanning trees for the Prüfer sequence (1, 1, 4, 6, 5)

the term 1, and append 2 at the back of the sequence to obtain (1, 4, 6, 5, 2). The first number in the new sequence is again 1, and the lowest number not included in the list is 3, so we connect 1 to 3 (Fig. 10.25b). Delete the term 1, and append 3 at the back of the sequence to obtain (4, 6, 5, 2, 3). The first number in the new list is now 4, and the lowest number not included in the list is 1, so we connect 4 to 1 (Fig. 10.25c). Delete the term 4, and append 1 at the back of the sequence to obtain (6, 5, 2, 3, 1). The first number in the new sequence is 6, and the lowest number not included in the list is now 4, so we connect 6 to 4 (Fig. 10.25d). Delete the term 6, and append 4 at the back of the list to obtain (5, 2, 3, 1, 4). The first number in the new sequence is now 5, and the lowest number not included in the list is 6, so we connect 5 to 6 (Fig. 10.25e). Delete the term 5, and append 6 at the back of the sequence to obtain (2, 3, 1, 4, 6). We have iterated all the way through the given list, and since 5 and 7 are the two missing numbers, so we connect the two (Fig. 10.25f). We are done!

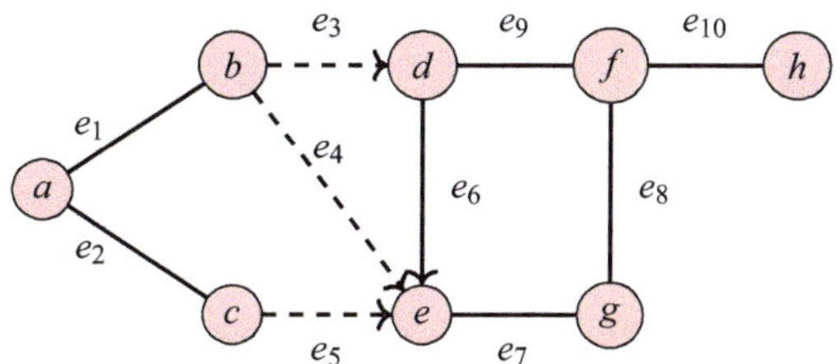


**Fig. 10.26** Example of a cut-set in the graph (indicated by dashed lines)

**Remark 10.4** The "trees" were applied by Staudt and Kirchhoff in their independent work, both published in 1847. In 1857, Cayley used trees to study (cycle-free) saturated hydrocarbon of the type $C_m H_n$, where $C$ and $H$ atoms are the vertices and $C - H$ bonds define the adjacency between the atoms [18]. The technique so developed by him, and also some important theorems published by *George Pólya* during 1935 and 1937, ultimately led to the introduction of a new discipline known as the **enumerative graph theory**. The generalization of results obtained above was given by the Dutch mathematician *Nicolas de Bruijn* (1918–2012) in 1959.

## Cut-Set and Cut-Vertex

We start with the next definition.

**Definition 10.28** Let $G$ be a connected graph. A **cut-set** of $G$ is a *minimal set* $S \subset E$ such that the subgraph $G - S$ is disconnected. The number $|S|$ is called the **edge connectivity** of $G$. In particular, a *Cut-set* consisting of a single edge is called a **cut-edge** (or a **bridge**).

For example, the set of $(n - 1)$ edges incident on a particular vertex of the complete graph $K_n$ is a cut-set. Therefore, there are $n(n - 1)/2$ distinct cut-sets of the graph $K_n$, each of size $n - 1$. Also, every edge of a tree is a *bridge*. In the graph in Fig. 10.26, the sets $\{e_3, e_4, e_5\}$, $\{e_1, e_2\}$, $\{e_7, e_9, \}$, and $\{e_{10}\}$ are some example of cut-sets. We next prove some interesting properties of fundamental circuits and cut-sets of a graph.

**Theorem 10.20** *Every fundamental circuit in a connected graph G contains at least one cut-set. Conversely, every cut-set contains at least one fundamental circuit.*

**Theorem 10.21** *In a connected graph, every cut-set contains at least one branch of every spanning tree of the graph. Conversely, any minimal set of edges of a connected graph G containing at least one branch of every spanning tree of G is a cut-set.*

***Proof*** Let $G$ be a connected graph, and $S \subset E(G)$ be a cut-set. If possible, suppose $T$ is a spanning tree of the graph $G$ such that $S \cap E(T) = \emptyset$. Therefore, we have $T \subset G - S$, where the subgraph $G - S$ is connected. This contradiction proves the assertion. Conversely, let $S \subset E(G)$ be a minimal set such that $S \cap E(T) \neq \emptyset$, for every spanning tree of $G$. Therefore, $G - S$ is a subgraph of $G$ that cannot contain wholly any spanning tree of $G$, and so it is disconnected. It follows by the minimality of the set $S$ with this property that $G - S + e$ has at least one spanning tree of $G$, for any $e \in E(G)$. In particular, $G - S + e$ is connected. This proves that $S$ is a cut-set. □

Consider the cut-set $S$ shown by the dotted lines in Fig. 10.26. Notice that every circuit $C$ in this graph such as $ae_2ce_5ee_4be_1a$ or $be_4ee_6de_3b$ has the property that $|E(C) \cap S|$ is even. The next theorem proves that this holds in general.

**Theorem 10.22** *Every fundamental circuit $C$ in a connected graph $G = (V, E)$ has an even number of edges in common with any cut-set $S \subset E$.*

***Proof*** Consider the subgraph $G - S$. Suppose $V(G - S)$ is the disjoint union of the set of vertices $V_1, V_2 \subset V$. If all vertices of circuit $C$ belong to exactly one of the two sets $V_1$ and $V_2$, then we have $|E(C) \cap S| = |\emptyset| = 0$, which is an even number. Next, assume that the vertices on circuit $C$ belong to both $V_1$ and $V_2$, in parts. Therefore, to traverse $C$, we need to traverse back and forth between the vertices in $V_1$ and $V_2$, and so $|C(E)|$ must be even. We also have that each edge in $S$ has one endpoint in $V_1$ and the other in $V_2$, whereas no other edge in $E$ has this property. Therefore, once again $|E(C) \cap S|$ is even. □

The following steps may be used to find cut-sets in a connected graph $G$:

(a) Select a spanning tree $T$ of $G$;
(b) A chord of $T$ is part of a cut-set $S$ if its removal disconnects the graph;
(c) Repeat $(b)$ to obtain all edges in $S$.

**Definition 10.29** Let $G$ be a connected graph, and $T$ a spanning tree of $G$. A cut-set containing exactly one branch of $T$, and possibly some chords of $T$, is called a **fundamental cut-set** relative to $T$.

For example, consider the spanning trees $T_1$ and $T_2$ (in blue) of the two graphs shown in Fig. 10.27. Then the set $\{e_1, e_2\}$ is a *Fundamental cut-set* of the graph in Fig. 10.27a relative to $T_1$, and $\{de\}$ is a *fundamental cut-set* of the graph in Fig. 10.27b relative to $T_2$.

**Theorem 10.23** *Let $G$ be a connected graph, and $T$ be a spanning tree of $G$. A chord of $T$ that determines a fundamental circuit $C$ occurs in every fundamental cut-set associated with the branches on $C$, and in no other cut-set.*

***Proof*** Let $c$ be a chord with respect to $T$, and let $C = \{c, b_1, b_2, \ldots, b_k\}$, where $b_i$ are $k$ branches of $T$. Recall that every branch of $T$ has a fundamental cut-set associated with it. Let $S_1 = \{b_1, c_1, c_2, \ldots, c_\ell\}$ be the fundamental cut-set associated with $b_1$. Notice that $b_1$ is common to $C$ and $S_1$. It thus follows from Theorem 10.22 that there must be at least one more edge common to $C$ and $S_1$. However, among the remaining edges common to $C$ and $S_1$, only a chord may belong to both. Therefore, the chord $c = c_i$, for some $1 \leq i \leq k$. Hence, $c$ lies in the fundamental cut-set $S_1$

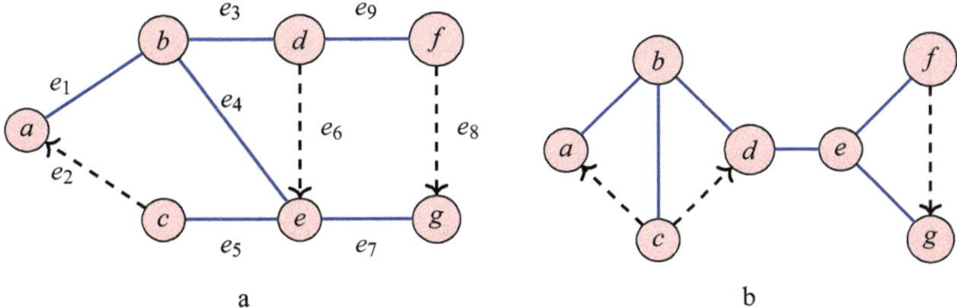


**Fig. 10.27** Fundamental cut-sets relative to a spanning tree

corresponding to the branch $b_1$ of $C$. The same argument holds for fundamental cut-sets associated with branches $b_2, \ldots, b_k$. This proves the chord $c$ lies in every fundamental cut-set associated with branches in $C$. Moreover, if the chord $c$ belongs to any other fundamental cut-set $S'$ in $T$, besides ones associated with $b_2, \ldots, b_k$, then we can find a single edge common to $S'$ and $C$, because none of the branches in $C$ are in $S'$, which contradicts Theorem 10.22. We thus conclude that $c$ belongs only in the fundamental cut-set associated with the branches on $C$, and in no other cut-set. □

**Theorem 10.24** *Let $G$ be a connected graph, and $T$ be a spanning tree of $G$. A branch of $T$ that determines a fundamental cut-set $S$ is contained in every fundamental circuit associated with the chord in cut-set $S$, and in no other.*

***Proof*** Let $b$ be a branch of $T$, and let $S = \{b, c_1, c_2, \ldots, c_r\}$, where $c_j$ are $r$ chords of the spanning tree $T$. Recall that every chord of $T$ has a fundamental circuit associated with it. Let $C_1 = \{c_1, b_1, b_2, \ldots, b_s\}$ be the fundamental circuit associated with $c_1$. Notice that $c_1$ is common to $S$ and $C_1$. It thus follows from Theorem 10.22 that there must be at least one more edge common to $S$ and $C_1$. However, among the remaining edges common to $S$ and $C_1$, only a branch may belong to both. Therefore, the branch $b = b_i$, for some $1 \leq i \leq s$. Hence, $b$ lies in the fundamental circuit $C_1$ corresponding to the branch $b_1$ in the cut-set $S$. The same argument holds for fundamental circuits associated with chords $c_2, \ldots, c_r$. This proves the branch $b$ lies in every fundamental circuit associated with chords in $S$. Moreover, if the branch $b$ belongs to any other fundamental circuit $C'$ in $T$, besides ones associated with $c_2, \ldots, c_r$, then we can find a single edge common to $C'$ and $S$, because none of the chords in $S$ are in $C'$, which contradicts Theorem 10.22. We thus conclude that $b$ belongs only in the fundamental circuit associated with the chords in the fundamental cut-set $S$, and in no other. □

**Theorem 10.25** *The number of cut-sets equals the number of vertices in the graph minus one. The sum of the sizes of all the cut-sets equals to the sum of the sizes of all the edges in the graph.*

**Remark 10.5** The concept of cut-set is used to identify weak spots in a communication network. For, all cut-sets of the corresponding graph model are listed, and we find the one that has the smallest number of edges. Since there are a large number of cut-sets in general, a suitable technique is needed to generate all relevant cut-sets. More generally, cut-sets are used in network design and optimization problems to identify critical points in the network where failures or congestion may occur and to optimize the placement of resources such as routers, switches, and links. By analyzing the cut-sets of a network, we can determine the minimum number of resources needed to ensure connectivity and reliability. (Selvakumar, R., and Gupta, N., *Fundamental Circuits and Cut-sets used in Cryptography*, Journal of Discrete Mathematical Sciences and Cryptography, 15(4), 2012, 287–301).

**Definition 10.30** (*Vertex Connectivity*) The **vertex connectivity** of a connected graph $G = (V, E)$ is the smallest number of vertices that need to be removed (along with all incident edges) to disconnect the graph. The graph $G$ is called **k-connected** if this number is $k$.

A **1-connected** graph is called a **separable graph**. In this case, the related vertex $v \in V$ is called a **cut-vertex**. That is, $v$ is a *cut-vertex* of a *separable graph* $G$ if the subgraph $G - v$ is disconnected. For example, $d$ is a cut-vertex of the separable graph shown in Fig. 10.28a. Also, both $d$ and $e$ are cut-vertices of the separable graph shown in Fig. 10.28b. Notice that the edge $de$ of the graph in Fig. 10.28b is a *bridge*.

The characterizations given in the next two theorems are used in the sequel.

**Theorem 10.26** *Let* $G = (V, E)$ *a connected graph. Then the following are equivalent:*

1. *A vertex* $v \in V$ *is a cut-vertex of* $G$.
2. *There exist vertices* $u, w \in V - \{v\}$ *such that* $v$ *lies on every* $u$-$w$ *path.*
3. *There exits a partition* $(U, W)$ *of the set* $V \setminus \{v\}$ *such that, for any vertices* $u \in U$ *and* $w \in W$, *the vertex* $v$ *lies on every* $u - w$ *path.*

***Proof*** We first prove (1) $\Rightarrow$ (2). For, suppose the subgraph $G - v$ has $k$ components, say $C_1, C_2, \ldots, C_k$. We may choose vertices $u, w \in G - v$ such that $u \in V(C_i)$ and $w \in V(C_j)$, for some $1 \leq u \neq w \leq k$. There is no path between $u$ and $w$ as they belong to distinct components. However, there is at least one path between $u$ and $w$ in the graph $G$. Since there is no path between $u$ and $w$ after removal of the vertex $v$

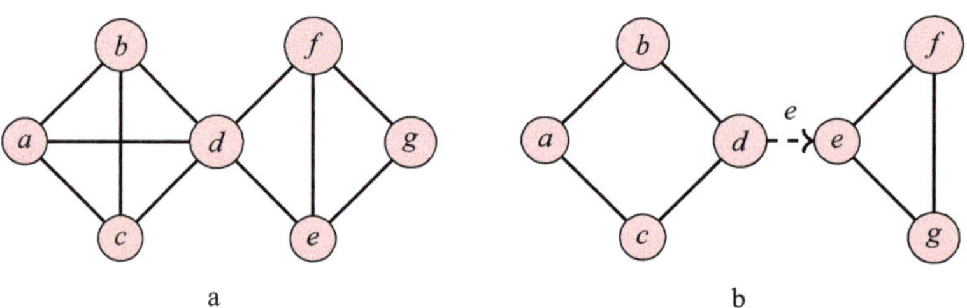


**Fig. 10.28** A graph with a cut-vertex $d$ and a cut-edge $e_5$

from $G$, every path connecting $u$ and $w$ passes through $v$. We next prove (2) $\Rightarrow$ (1). For, suppose $u, w \in V$ are such that all paths $P_1, P_2, \ldots, P_n$ between $u$ and $w$ that pass through $v$. Therefore, in the subgraph $G - v$ there is no path connecting vertices $u$ and $w$, and so $v$ is a cut-vertex of $G$. □

**Theorem 10.27** *Let $G = (V, E)$ a connected graph. Then the following are equivalent:*

1. *An edge $e \in E$ is a bridge of $G$.*
2. *The edge $e$ is not on any cycle in $G$.*
3. *There exist $u, w \in V$ such that $e$ is on every $u$-$w$ path in $G$.*
4. *There exits a partition $(U, W)$ of the set $V$ such that, for any vertices $u \in U$ and $w \in W$, the edge $e$ is on every $u$-$w$ path in $G$.*

***Proof*** Since deletion of an edge $e$ on a cycle in $G$ will not disconnect the graph $G$, it follows that $e$ is not a bridge of $G$. This proves (1) $\Rightarrow$ (2). □

## 10.5 Euler and Hamiltonian Graphs

In the paper we mentioned earlier in the first section, Euler argued that each vertex of a connected graph $G$ must have an even degree if it contains a trail that includes all edges. For, each time a vertex is visited along an edge we need a different edge to continue with the tour. As that is not true for the graph model of *Kö* nigsberg seven bridges problem, in particular, it is not possible to walk through all edges of the right side graph in Fig. 10.8 exactly once, and return to the start point. Therefore, a connected graph having an odd degree vertex cannot have an Eulerian circuit. Therefore, the walk is impossible. We also say that the graph is *non-Eulerian*.

**Definition 10.31** (*Euler Graph*) A walk in a graph $G$ that contains all edges exactly once is called an **Eulerian trail**, and a closed Eulerian trail is called an **Eulerian circuit**. An **Euler graph** is a connected graph with an Eulerian circuit.

**Theorem 10.28** *A simple connected graph $G$ has an Eulerian circuit if and only if the degree of every vertex in $G$ is even.*

***Proof*** First, suppose $G$ is an Euler graph. Then it has an Eulerian circuit, say $C$. Since no two edges in $C$ are the same, every vertex that appears in $C$ is an end vertex for an even number of edges. In particular, the first vertex (which is the same as the last vertex) in $C$ is an end vertex of exactly two edges. Therefore, every vertex of $G$ is of even degree. Conversely, suppose $G$ is a connected graph such that $\deg(v)$ is even, for every vertex $v$ of $G$. Starting from a vertex $v$, we traverse as many edges of $G$ as possible until the vertex $v$ is met again. Of course, it is possible to do so because the degree of every vertex of the graph $G$ is even. Let the closed walk obtained this way be denoted by $\Gamma$. If $\Gamma$ contains all the edges of $G$, then $\Gamma$ is the required Eulerian line. If $\Gamma$ does not contain all the edges of $G$, consider the graph $G' = G - \Gamma$ obtained by removing from $G$ all the edges in $\Gamma$. Clearly, the vertices of $G'$ are of even degree and there is at least one vertex $v'$ that is common to both $G'$ and $\Gamma$. Starting at $v'$, traverse the edges of $G'$ as much as possible and let $\Gamma_1$ be the closed walk obtained this way. Then combining $\Gamma$ and $\Gamma_1$ by traversing from $v$ to $v'$ in $\Gamma$, followed by $\Gamma_1$ and finally the portion of $\Gamma$ from $v'$ to $v$, we get new walk $\Gamma'$ that contain more edges than $\Gamma$. If this walk contains all the edges of $G$, we have the required Eulerian line. Otherwise, this process can be continued to get an Eulerian line. Note that the process will terminate as there are only a finite number of edges. □

The first formal proof of the converse part of the above theorem was first given by *Carl Hierholzer* (1811–1874), published posthumously in 1873. In general, an open walk that contains all the edges of a graph is called an **unicursal line** (or an open **Eulerian line**). Also, a graph having an *unicursal line* is called an **unicursal graph**. A graph is *unicursal graph* if and only if there are exactly two vertices of odd degree. Also, a graph is Euler if and only if it is a union of edge-disjoint circuits. Therefore, an *Euler graph* is the one that contains a cycle that goes through each edge exactly once. For example, the graph on the left side of Fig. 10.29 has an Eulerian trail, and the graph on the left side is an Euler graph. In general, as we discover shortly, the *complete graph* $K_n$ is Euler precisely when $n$ is odd, and every *cycle* $C_n$ is an Euler graph (Fig. 10.30).

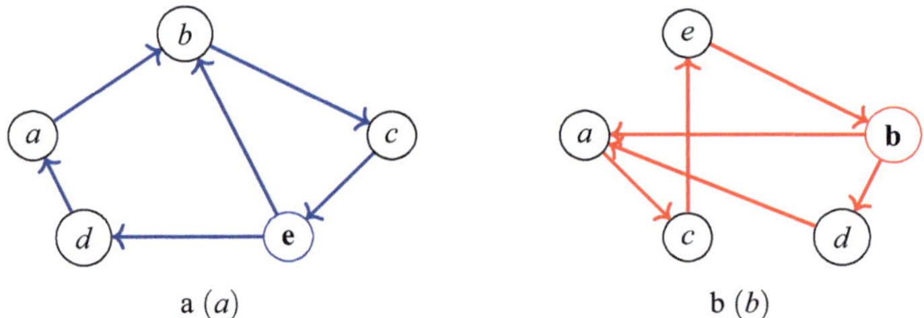


**Fig. 10.29** Two graphs with Eulerian trail (starting from the bold vertex)

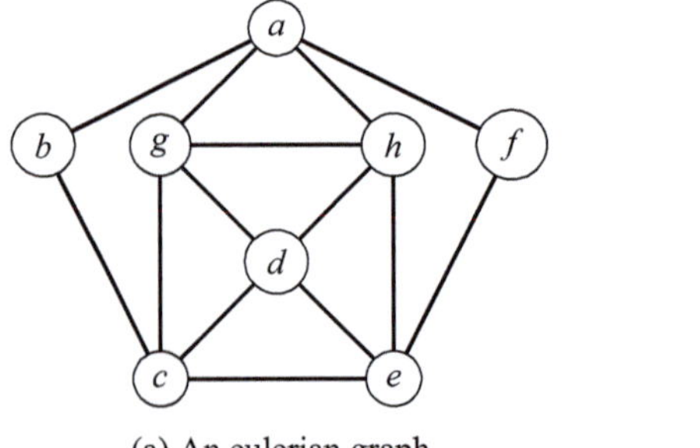

(a) An eulerian graph.

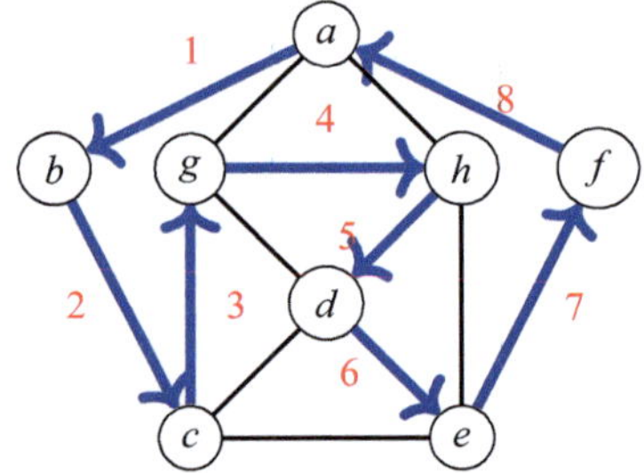

(b) A Hamiltonian cycle (in blue).

**Fig. 10.30** An Eulerian graph with a Hamiltonian cycle

**Theorem 10.29** (Hierholzer, 1873) *A connected graph is Euler if and only if exactly zero or two vertices have odd degree.*

***Proof*** We first assume that a graph $G = (V, E)$ is Euler with an Eulerian circuit $C$, say starting with a vertex $v \in V$. As $v$ is also the last vertex of $C$, it is an even vertex. Moreover, as all the edges in $C$ are distinct, it follows that the degree of all other vertices on $C$ is of even degree. Therefore, each vertex of $G$ has an even degree. Conversely, suppose $G = (V, E)$ is a simple connected graph such that every vertex of $G$ is of even degree. Starting from a vertex $v \in V$, we walk along as many distinct edges as possible. Now, as the degree of every vertex is even, the walk ends only at $v$. Let the closed walk so obtained be denoted by $W$. If $W$ contains all the edges of $G$, then we are done. Otherwise, consider the graph $G' = G - W$, obtained by removing from $G$ all the edges in $W$. Clearly, the vertices of $G'$ are of even degree, and there is at least one vertex $v' \in V$ that is common to both $G'$ and $W$. Therefore, starting at $v'$, we traverse as many distinct edges of $G'$ as possible and write $W_1$ for the closed walk so obtained. Then combining $\Gamma$ and $\Gamma_1$ by traversing from $v$ to $v'$ in $\Gamma$, followed by $\Gamma_1$ and finally the portion of $\Gamma$ from $v'$ to $v$, we get new walk $\Gamma'$ that contain more edges than $\Gamma$. If this walk contains all the edges of $G$, we have the required Eulerian line. Otherwise, this process can be continued to get an Eulerian line. Note that the process will terminate as there are only a finite number of edges. □

The next theorem proves that a graph $G = (V, E)$ is Euler if and only if it is a union of edge-disjoint cycles, i. e., there exists a family of cycles $\mathscr{F}$ in $G$ such that we have

$$C, C' \in \mathscr{F} \quad \Rightarrow \quad E(C) \cap E(C') \neq \emptyset, \quad \text{and} \quad E(G) = \bigcup_{C \in \mathscr{F}} E(C). \tag{10.5.1}$$

More generally, a graph is called a (edge) *decomposable* if it satisfies the above condition for an arbitrary family of induced subgraphs. In particular, when all subgraphs in the family are cycles, we obtain a **cycle decomposition** of the graph. Clearly, in general, a graph may have more than one decomposition. For example,

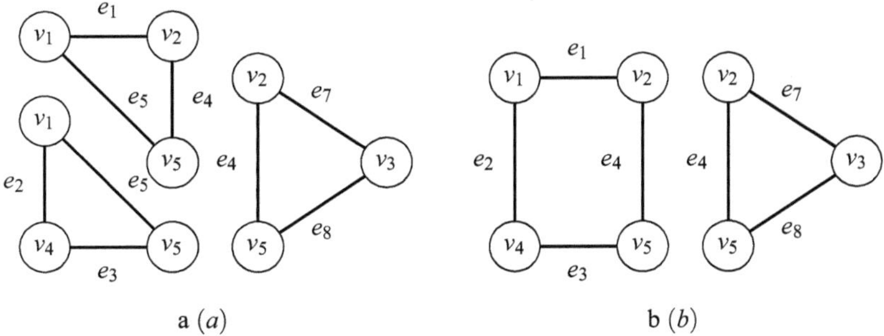


**Fig. 10.31** Two cycle decompositions of the graph as in Fig. 10.4

Fig. 10.31 shows two cycle decompositions of the graph given in Fig. 10.4. The next lemma plays an important role in part of the proof of the next theorem.

**Lemma 10.1** *Let* $G = (V, E)$ *be a simple graph such that* $\deg(v) \geq 2$, *for all* $v \in V$. *Then* $G$ *contains a cycle.*

***Proof*** Starting with a vertex $u = u_0 \in V$, by the given degree condition and since $|V| < \infty$, it is possible to construct a path $p$ of maximal length in $G$, say $p = u_0u_1 \cdots u_{n-1}u_n$. Now, as $\deg(u_n) \geq 2$, there is some vertex $v \in V$, with $v \neq v_{n-1}$. Therefore, by the maximality of the length of $p$, we must have that $v = u_i$, for some $0 \leq i < n-1$. Hence, we obtain a cycle, as asserted. □

**Theorem 10.30** *Show that a graph* $G$ *admits a cycle decomposition if and only if it is an* even graph.

***Proof*** First, suppose $\mathscr{F}$ is a family of cycles in $G$ satisfying the condition (10.5.1), and let $v \in V(G)$. If $\deg(v) = 0$, there is nothing to prove. Otherwise, suppose there are $k \geq 1$ cycles in $\mathscr{F}$ that contains $v$. As each cycle is a 2-regular graph, we must have $\deg(v) = 2k$, and we are done proving the necessary part of the theorem. Conversely, suppose $G$ is an *even graph*. To complete the proof, we use induction on $m = |E(G)|$. If $m = 0$, each vertex of $G$ is an isolated vertex, and so $G$ has a trivial decomposition. We may thus take an even graph $G$, with $m = |E(G)|$, and assume that the assertion holds for any such type of graph with the number of edges $< m$. Now, let $S$ denote the set of vertices of the graph $G$ such that $\deg(v) > 0$, for $v \in S$, and $H$ be the subgraph of $G$ spanned by the set $S$. As $H$ is also an even graph, by Lemma 10.1, it contains a cycle, say $C$. It then follows from our inductive hypothesis that the graph $G' = G - E(C)$ has a cycle decomposition, which together with $C$ provides a cycle decomposition of the graph $G$. □

## *Hamiltonian Graphs*

The concept introduced in the next definition is attributed to the game that William Rowan Hamilton called *Traveler's Dodecahedron* (see Fig. 10.32). In this game, which is also known as *icosian game* (or Hamilton's puzzle), one player specifies a 5-vertex path and the other is challenged to extend the same to a spanning cycle, which is now known as a *Hamiltonian cycle*. However, Thomas Kirkman (1806–1895) was the first to conduct a mathematical study of such types of graphs in 1856. Until 1970, interest in Hamilton cycles was mainly due to its relationship with *Four Color Theorem.*

**Definition 10.32** (*Hamiltonian Graph*) A *Hamiltonian path* (or a *traceable path*) in a graph $G = (V, E)$ is a path that traverses every vertex exactly once. A closed Hamiltonian path is called a *Hamiltonian cycle*. A graph $G$ containing a Hamiltonian cycle is called a **Hamiltonian graph**.

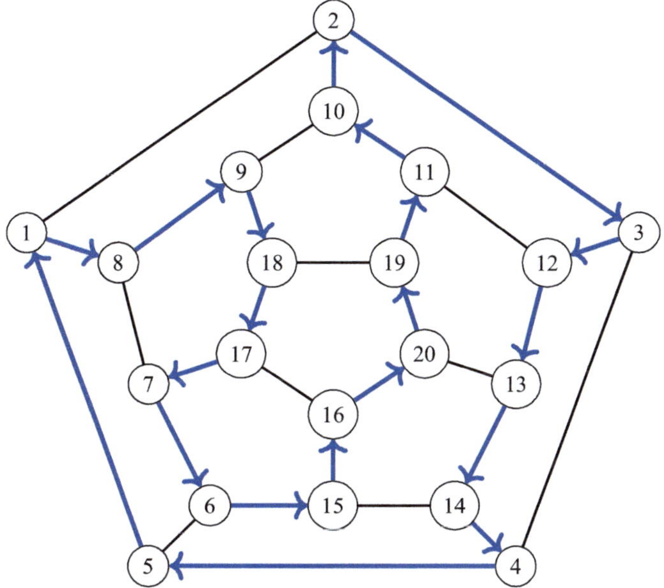


**Fig. 10.32** Hamilton's Dodecahedron with a Hamiltonian cycle

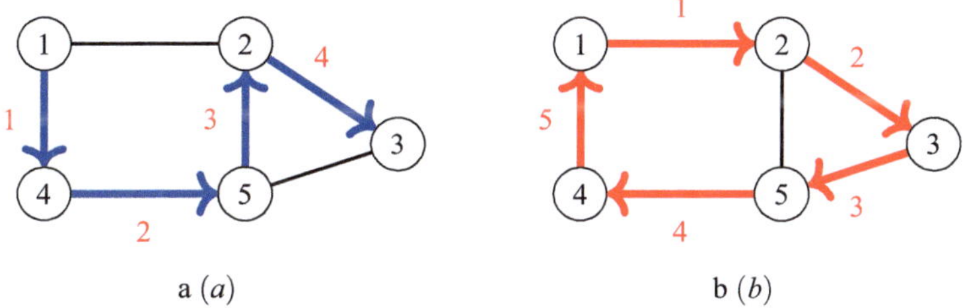


**Fig. 10.33** A Hamiltonian graph with a Hamiltonian path

Notice that loops and parallel edges hold no significance because a multigraph is Hamiltonian if and only if the simple graph obtained by keeping one copy of each non-loop edge is Hamiltonian. To date, no easily testable characterization is known for Hamiltonian graphs.

**Example 10.8** The right side graph in Fig. 10.33 is Hamiltonian, which contains a Hamiltonian path, as shown in the left side graph. Clearly, for $n \geq 3$, the graphs $C_n$ is Hamiltonian, and so is $K_n$ because it contains $C_n$ as a spanning subgraph. More generally, any graph on $n$ vertices is Hamiltonian if it contains a Hamiltonian graph on $n$ vertices as a spanning subgraph. Further, since a spanning cycle in a complete bipartite graph $K_{m,n}$ visits the two partite sets alternatively, there cannot be any Hamiltonian cycle unless $m = n$ (see Fig. 10.34).

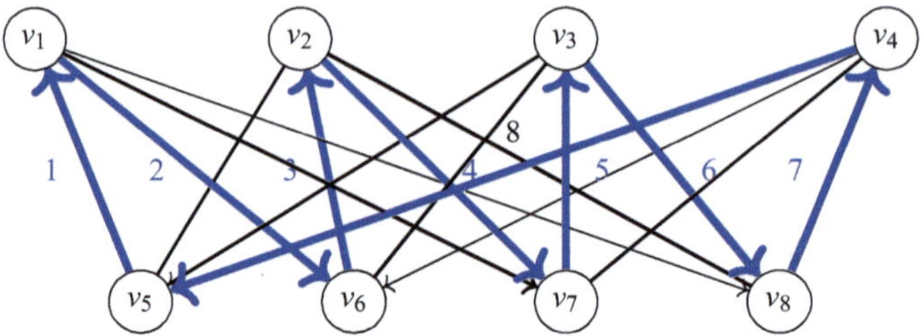


**Fig. 10.34** The bipartite graph $K_{4,4}$ with a Hamiltonian cycle

Though there is no simple necessary and sufficient condition for a Hamiltonian graph, the following theorem, due to Dirac, gives a sufficient condition for the existence of a Hamiltonian circuit.

**Theorem 10.31** *If the degree of every vertex in a simple graph $G$ with $n(\geq 3)$ vertices is at least $n/2$, then $G$ has a Hamiltonian circuit.*

***Proof*** We prove the theorem by contradiction. Suppose the degree of every vertex of a graph $G = (V, E)$ with $n$ vertices is at least $n/2$ and has no Hamiltonian circuit. We further assume that $G$ is a maximal non-Hamiltonian graph in the sense that any addition of edge makes it Hamiltonian. Clearly, $G$ is not the complete graph $K_n$ and therefore there are non-adjacent vertices in $G$. Let $u$ and $v$ be non-adjacent in $G$. The addition of the edge $(u, v)$ to the graph $G$ makes it Hamiltonian. Call the new graph as $G'$. Thus there is a Hamiltonian circuit in $G'$ and hence (by removing the edge $(u, v)$ from the circuit) we have a Hamiltonian path beginning at $u$ and ending at $v$ in $G$. Let $v_1, v_2, \ldots, v_n$ be the Hamiltonian path where $v_1 = u$ and $v_n = v$. Now consider two sets which we define as below:

$$S = \{v_i \in V | (u, v_{i+1}) \in E\} \qquad T = \{v_i \in V | (v_i, v) \in E\}.$$

Note that $S$ consists of vertices $v_i$ if the next vertex $v_{i+1}$ in the Hamiltonian path is incident with $u$. The set $V$ consists of all the vertices incident on $v$. It is clear that $v \notin S$ and since $G$ is simple, $v \notin T$. Thus $v \notin S \cup T$ and hence $|S \cup T| < n$. Let us now show that $S \cap T = \emptyset$. If $S \cap T \neq \emptyset$, then there is an element $v_j \in S \cap T$. Then $v_1, v_2, \ldots, v_j, v_n, v_{n-1}, \ldots, v_{j+1}, v_1$ is a Hamiltonian circuit in $G$ (see Fig. ??) and it is a contradiction to our assumption. Thus $S \cap T = \emptyset$. Thus it follows that

$$d(u) + d(v) = |S| + |T| = |S \cap T| < n.$$

But by our assumption, we have $d(u) \geq n/2$ and $d(v) \geq n/2$, and hence

$$d(u) + d(v) \geq n.$$

Thus we have a contradiction. □

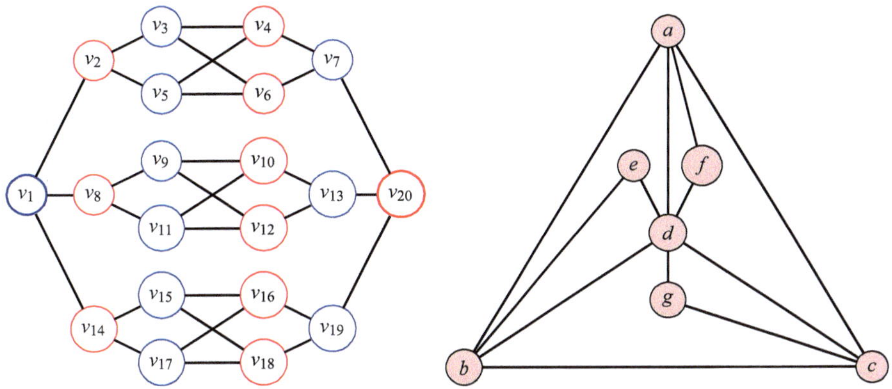


**Fig. 10.35** A cubic bipartite graphs and a non-Hamiltonian graph

**Theorem 10.32** *If a graph $G$ contains a Hamiltonian cycle then, for each $\emptyset \neq S \subset V(G)$, the number of components of the graph $G - S$ is at most $|S|$.*

***Proof*** Since a Hamiltonian cycle in $G$ while leaving a component of $G - S$ for the last time can visit only to some vertex in $S$, and also the arrival in $S$ must use distinct vertices of the set $S$, we conclude that $S$ must have at least as many vertices as the graph $G - S$ has the number of components. □

It thus follows that if the number of components of the graph $G - S$ is more than $|S|$, for some $\emptyset \neq S \subset V(G)$, then $G$ is not Hamiltonian. In particular, a graph with a cut-vertex cannot be Hamiltonian. The graph in Fig. 10.35a is a cubic bipartite having $= 10$ vertices in each partition (as displayed in two different colors), that fails the condition of Theorem 10.32, and hence it is the non-Hamiltonian graph. On the other hand, the graph in Fig. 10.35b is non-Hamiltonian, but it satisfies the condition of Theorem 10.32, which shows that the condition is not sufficient to give the existence of a Hamiltonian cycle. Notice that to get a spanning cycle in the latter case all edges incident to vertices of degree 2 have to be used. However, it is not possible to do so unless the central vertex has three edges incident to it. Indeed, it does contain Hamiltonian paths.

**Remark 10.6** The condition stated in Theorem 10.32 guarantees that $G$ is 2-*connected*,[2] but not the existence of a Hamiltonian cycle.

We next prove some necessary conditions for the existence of a Hamiltonian cycle in a graph. The next theorem is due to Gabriel Dirac.

**Theorem 10.33** (Dirac, 1952) *If $G$ is a graph of order $n := n(G) \geq 3$ such that $\delta(G) \geq n/2$, then $G$ contains a Hamiltonian cycle.*

[2] Recall that a graph is **2-connected** if deletion of a single vertex leaves at most one component.

***Proof*** Notice that the condition $n \geq 3$ is natural because $\delta(K_2) = n(K_2)/2$, but $K_2$ is not Hamiltonian. The proof given here uses *extremality argument*, i.e., we assume there exists a maximal non-Hamiltonian graph $G$ of order $n$ with a minimum degree of a vertex $\geq n/2$, where *maximality* of $G$ is taken in the sense that adding an edge joining two non-adjacent vertices would give a Hamiltonian cycle. Said differently, if $u, v \in V(G)$ are non-adjacent vertices such that $u = v_1, v_2, \ldots, v_{n-1}, v_n = v$ is spanning cycle in the graph $G + e$ containing the new edge $e = uv$, then it suffices to prove the existence of a spanning cycle in $G$ (avoiding the edge $e$). First, notice that if a neighbor $v_{i+1}$ of $u$ directly follows a neighbor $v_i$ of $v$ on the path then $u, v_{i+1}, v_{i+2}, \ldots, v, v_i, v_{i-1}, \ldots, v$ is a spanning cycle. It thus remains to show that there is a common index in the sets $I$ and $J$ given by

$$I = \{i \,:\, u \text{ is adjacent to } v_{i+1}\} \qquad \text{and} \qquad J = \{j \,:\, v \text{ is adjacent to } v_i\}$$

Summing the sizes of the above two sets, we obtain

$$|I \cup J| + |I \cap J| = |I| + |J| = \deg(u) + \deg(v) \geq n.$$

As neither of the two index sets contain $n$, we conclude that $|I \cup J| < n$, and hence $|I \cap J| \geq 1$. We have thus obtained a contradiction by finding a spanning cycle in $G$, which proves no maximal non-Hamiltonian graph exists that satisfies the condition of the theorem. □

In 1960, Oystein Ore noticed that the condition $\delta(G) \geq n(G)/2$ is used only to ensure that the condition

$$\deg(u) + \deg(v) \geq n, \tag{10.5.2}$$

holds for all non-adjacent vertices $u \neq v \in V(G)$. Furthermore, in the above proof, it is enough to assume that $G + uv$ is Hamiltonian. Therefore, the next theorem is a simple modification of Dirac's result.

**Theorem 10.34** (Ore, 1960) *Let $G$ be a graph of order $n := n(G) \geq 3$ such that* (10.5.2) *holds for some pair of non-adjacent vertices $u, v$. Then $G$ contains a Hamiltonian cycle if and only if the graph $G + uv$ is Hamiltonian.*

***Proof*** On the contrary, suppose three exists a non-Hamiltonian graph of order $n := n(G) \geq 3$ such that (10.5.2) holds. Notice that it may be the case that $G$ may still be non-hamiltonian even when a few edges are added. Clearly, we obtain a Hamiltonian graph by adding all missing edges, namely $K_n$. Let $H$ be a non-Hamiltonian graph obtained from $G$ by adding some edges, and it is maximal such. That is, by adding an additional edge, $H$ turns out to be Hamiltonian. Further, we have that distinct non-adjacent vertices in $H$ satisfy the condition (10.5.2). Now, as $H \neq K_n$, it must contain a pair of distinct non-adjacent vertices, say $x, y \in V(H)$. Then, by maximality assumption, we have that $H + xy$ is Hamiltonian. Moreover, every Hamiltonian cycle in $H + xy$ must contain the edge $xy$. Therefore, $H$ contains a

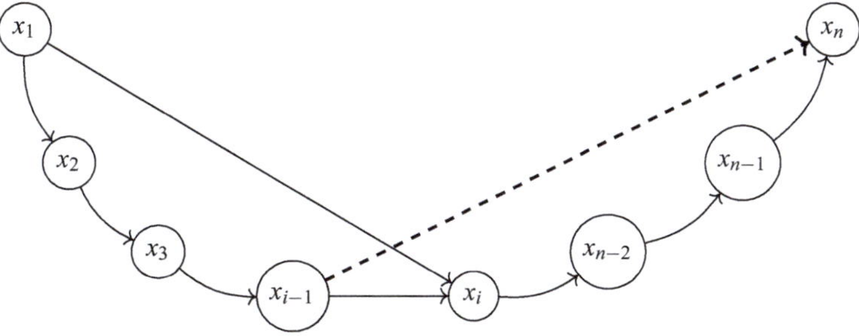


**Fig. 10.36** The path $p$ referred to in Theorem 10.34

Hamiltonian path $p = \left(x = x_1, x_2, \ldots, x_n = y\right)$. It follows that whenever $x_1x_i$ is an edge of $H$, for $2 \geq i \geq n$, then $x_{i-1}x_n$ is *not an edge* of $H$ (see Fig. 10.36). For, otherwise,

$$\left(x = x_1, x_i, x_{i+1}, \ldots, x_n, x_{i-1}, x_{i-2}, \ldots, x_1\right)$$

is a Hamiltonian cycle in $H$, which cannot be the case. Therefore, for each vertex in the set $\left\{x_2, \ldots, x_n\right\}$ that is adjacent to $x_1$, there is a vertex in the set $\left\{x_1, \ldots, x_{n-1}\right\}$ that is not adjacent to $x_n$. It thus follows that

$$\deg\left(x_n\right) \leq (n-1) - \deg\left(x_1\right) \quad \Rightarrow \quad \deg_H\left(x\right) + \deg_H\left(y\right) \leq n-1,$$

which gives the desired contradiction. Hence the proof is complete. □

**Remark 10.7** Let $n = 2m + 1 \geq 3$, and consider the graph $G$ obtained by identifying a vertex of a copy of the complete graph $K_{m+1}$ with a vertex of a second copy of $K_{m+1}$. That is, we have $G = K_1 \oplus 2K_m$. A drawing of $G$ for $n = 7$ is shown in Fig. 10.37. Notice that, as $G$ contains a cut-vertex, it follows from Theorem 10.32 that $G$ is a non-Hamiltonian graph. Further, if $u, v$ are any two non-adjacent vertices of $G$, we have

$$\deg\left(u\right) = \deg\left(v\right) = m \quad \Rightarrow \quad \deg\left(u\right) + \deg\left(v\right) = 2m = n-1.$$

**Fig. 10.37** The graph $G = K_1 \oplus 2K_3$ shows Ore's bound is sharp

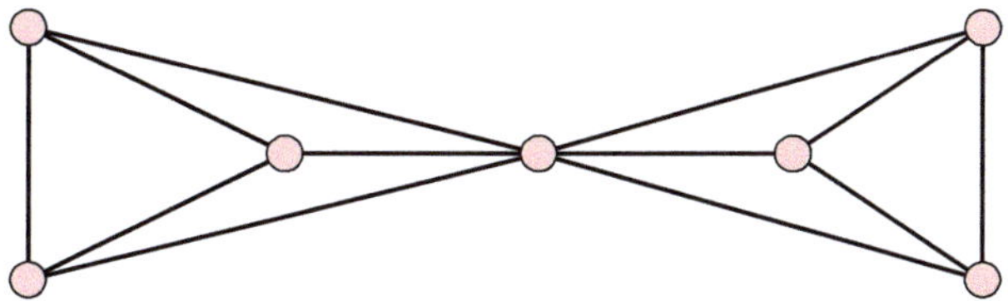

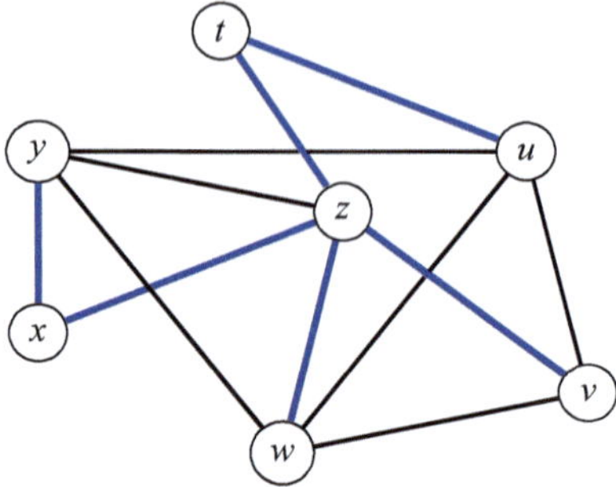


**Fig. 10.38** A non-Hamiltonian graph

Therefore, the lower bound specified by Theorem 10.34 is indeed sharp. Moreover, Dirac's assertion can be deduced from Ore's theorem. For, suppose $G$ is a graph of order $n := n(G) \geq 3$ such that $\delta(G) \geq n/2$. If possible, suppose $G$ is non-Hamiltonian. In particular, $G \neq K_n$, for $n \geq 3$. Therefore, we can find non-adjacent vertices $u, v \in G$. But then, we have

$$\deg(u) \geq \frac{n}{2}, \ \deg(v) \geq \frac{n}{2} \quad \Rightarrow \quad \deg(u) + \deg(v) \geq \frac{n}{2} + \frac{n}{2} = n.$$

Hence, Theorem 10.34 applies to give a contradiction.

Indeed, a Hamiltonian graph $G$ of order $n := n(G) \geq 3$ contains $C := C_n$, and possibly some additional edges joining non-consecutive vertices of $C$. Also, as $C$ is a connected 2-regular subgraph of order $n$, every proper subgraph of $C$ is a path or a (disjoint) union of paths. In particular, it follows that $C$ contains no cycle of order $< n$ as a subgraph, and certainly, no subgraph of $C$ has a vertex of degree 3 or more. Moreover, if $G$ contains a vertex $v$ of degree 2 then both edges incident to $v$ must lie on $C$.

**Example 10.9** The graph $G$ shown in Fig. 10.38 is a non-Hamiltonian graph. Suppose to contrary that it contains a Hamiltonian cycle $C$. By what we said above, as $C$ contains the vertex $t$ (of degree 2), it must also contain edges $tu$ and $tz$. For the same reason, $C$ also contains edges $xy$ and $xz$, and so do the edges $vw$ and $vz$. But then, we must have that the vertex $z$ is incident to three edges on the cycle $C$, which is simply not possible. Therefore, $G$ is a non-Hamiltonian graph. Notice that, however, this graph satisfies the condition of Theorem 10.32.

**Theorem 10.35** (Bondy-Chvátal) *Let $u$, $v$ be two non-adjacent vertices in a graph $G$ of order $n := n(G) \geq 2$ such that* $\deg(u) + \deg(v) \geq n$. *Then the graph $G + uv$ is Hamiltonian if and only if so is the graph $G$.*

***Proof*** Clearly, $G$ is Hamiltonian implies that $G + uv$ is so. To prove the converse, suppose $G + uv$ is Hamiltonian. If possible, suppose $G$ is non-Hamiltonian so that every Hamiltonian cycle in the graph $G + uv$ must contain the edge $uv$. □

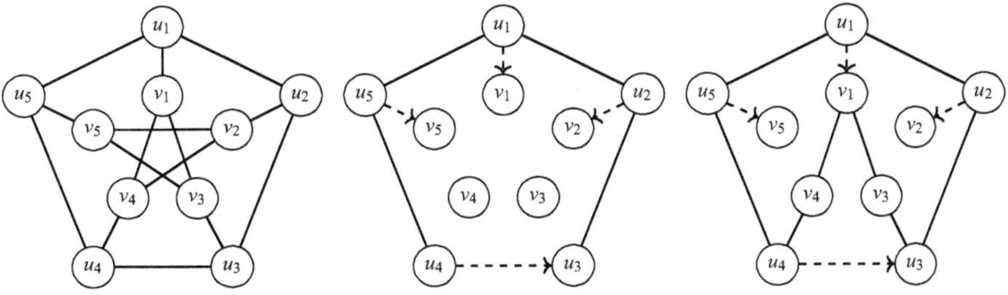


**Fig. 10.39** The Petersen graph is non-Hamiltonian

**Example 10.10** We show the Petersen graph $G$ as shown on left side of Fig. 10.39 is non-Hamiltonian. On the contrary, suppose $G$ contains a Hamiltonian cycle $C$ that necessarily contains 10 edges. Notice that $G$ is a cubic graph, and two of the three edges incident to each vertex of $G$ must belong to the cycle $C$. In general, $C$ may contain all, some, or none of the five edges $u_i v_i$ $(1 \leq i \leq 5)$, and so at least five edges of $C$ belong

$$\text{either to} \quad C' = \big(u_1,\, u_2,\, u_3,\, u_4,\, u_5,\, u_1\big) \qquad \text{or to} \quad C'' = \big(v_1,\, v_3,\, v_5,\, v_2,\, v_4,\, v_1\big)$$

Therefore, either $C'$ contains at least three edges of $C$ or $C''$ contains at least three edges of $C$. We may assume WLOG that $C'$ contains at least three edges of $C$. Notice that all five edges of $C'$ cannot belong to $C$ because no cycle contains a smaller cycle as a subgraph. Suppose that $C$ contains exactly four edges of $C'$, say the edges $u_4u_5,\ u_5u_1,\ u_1u_2,\ u_2u_3$ (as shown in the right side two drawings), where the dashed edges of $G$ cannot belong to $C$. However, the cycle $C$ must then contain the edges $u_4v_4,\ u_3v_3$ as well as $v_1v_3,\ v_1v_4$ (see the right side drawing shown above). The conclusion that $C$ contains an 8-cycle is impossible. The only situation that remains to be investigated is when $C$ contains exactly three edges of $C'$. In this case, we need to explore two possibilities:

1. The three edges of $C'$ on the cycle $C$ are consecutive on $C'$.
2. These three edges are not consecutive on $C'$.

The above two possibilities are respectively depicted in Fig. 10.40. The situation in the left side drawing is impossible as $u_1v_1$ is the only edge incident to $u_1$ that could lie on $C$. Similarly, the situation in the right side drawing is impossible as $C$ would have to contain the smaller cycle $\big(u_4,\ v_4,\ v_1,\ v_3,\ u_3,\ u_4\big)$. Hence, $G$ is non-Hamiltonian.

**Definition 10.33** The **line graph** of a graph $G$, denoted by $L(G)$, is the graph that has the edge set $E(G)$ as the set of vertices, and $e_1, e_2 \in E(G)$ are *adjacent* in $L(G)$ if $e_1$ and $e_2$ share a common end vertex in $G$.

Our main interest in the *line graph* $L(G)$ of a graph $G$ is due to the reason that problems involving edges of $G$ can be phrased as a problem about vertices in $L(G)$. In general, the latter types of problems are comparatively more difficult. Notice that

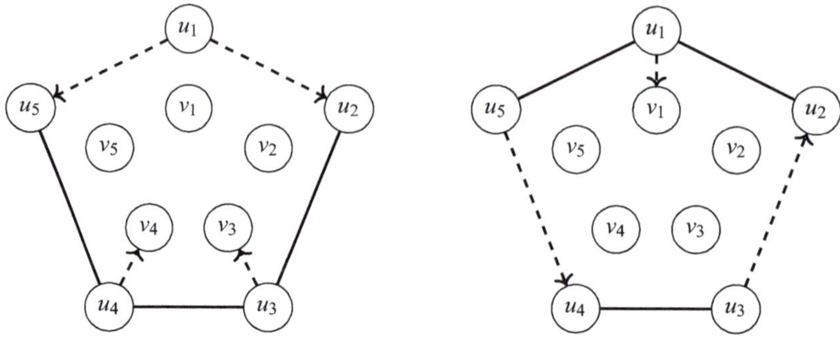


**Fig. 10.40** The case of $C$ containing exactly three edges

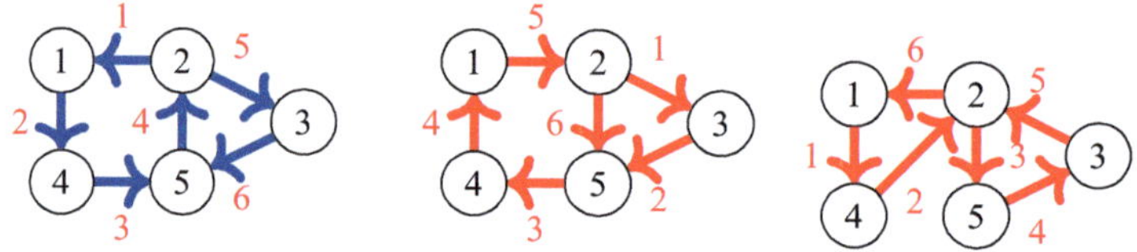

**Fig. 10.41** An example of an Euler graph that is also Hamiltonian

each Eulerian circuit in a graph $G$ yields a Hamiltonian cycle in $L(G)$. However, the converse may not hold in general. For example, there exists 2-connected non-Euler graph whose line graph is Hamiltonian (Fig. 10.41).

## 10.6 Planar Graphs

By a drawing of a graph $G = (V, E)$ we mean a function $f$ defined on $V \cup E$ such that, for each $v \in V$, $f(v)$ is a point in the plane, and it assigns to each edge $e = uv \in E$ a polygonal curve connecting *distinct* points $f(u)$ and $f(v)$. For $e, e' \in E$, a point in $f(e) \cap f(e')$ that is not a common endpoint of the curve $f(e)$ and $f(e')$ is called a **crossing**.

**Definition 10.34** (*Planar Graph*) A graph is said to be **planar** if it has a drawing such that no two edges cross each other.

Therefore, a graph is *non-planar* if all its drawings have some crossing. Notice that a plane graph is a particular planar embedding[3] of a planar graph. Such an embedding of a planar graph $G$ is called a plane graph or plane representation of $G$. For example, the complete graph $K_4$ and the complete bipartite graph $K_{2,n}$, for any $n \geq 2$, are planar graphs. See Fig. 10.42 for the plane representations of these two planar graphs. It also follows by induction that every tree is a planar graph.

[3] The drawing of a graph on the plane without any intersections is known as the embedding.

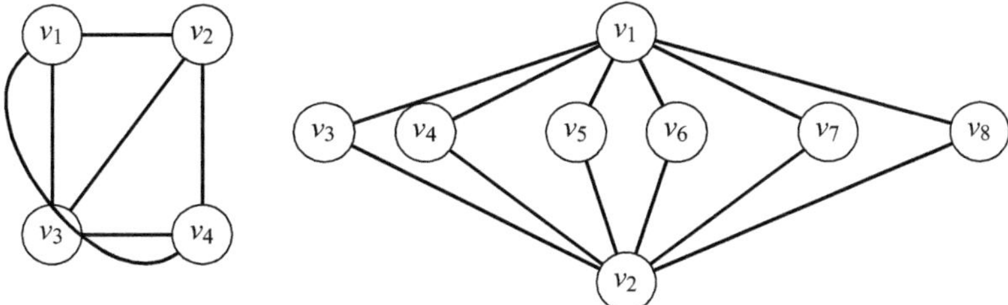


**Fig. 10.42** The graphs $K_4$ and $K_{2,6}$ are planar

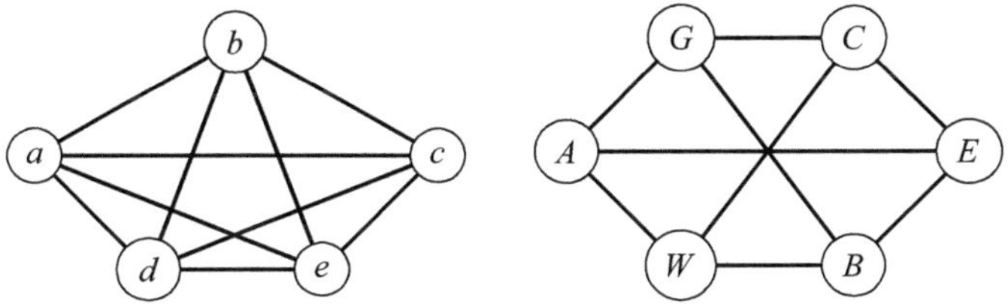


**Fig. 10.43** The graphs $K_5$ and $K_{3,3}$ are non-planar

Let $C$ be a spanning cycle in a graph. Notice that $C$ can be drawn as a *closed curve*, provided the drawing has no *crossings*. Further, chords of $C$ must be drawn inside or outside this curve. Clearly, two chords conflict if their endpoints on $C$ occur in alternating order. Therefore, when two chords conflict, we can draw only one inside $C$ and one outside $C$. In particular, consider the drawings of the graphs $K_5$ and $K_{3,3}$ as shown in Fig. 10.43. A 6-cycle in $K_{3,3}$ has three pairwise conflicting chords. We can put at most one inside and one outside, so it is not possible to complete the embedding. When $C$ is a 5-cycle in $K_5$, at most two chords can go inside or outside. Since there are five chords, again it is not possible to complete the embeddings. Hence, both $K_5$ and $K_{3,3}$ are non-planar.

A simple physical interpretation of non-planarity of $K_{3,3}$ is usually described in terms of making available to three houses, say represented by vertices $A$, $B$ and $C$, the three basic amenities such as *water,gas*, and *electricity*, say represented by vertices $W$, $G$ and $E$ (see Fig. 10.45).

Any planar graph divides the plane into several regions, whereby a *region* (or faces) we mean an open set $U \subset \mathbb{R}^2$ that contains a polygonal $u, v$-*curve*, for every pair of elements $u, v \in U$. Notice that the faces of a planar graph are (pairwise disjoint) *maximal regions* of the plane that contains no point used in the embedding. The following theorem relates the number of regions, vertices, and edges. And, two points $u, v \in \mathbb{R}^2$ lying in no edge of a graph $G$ are in the same region if and only if there is no polygonal $u, v$-*curve* that crosses no edges. Every finite planar graph $G$ has one *unbounded region*, which is called the **outer face** of the graph. Moreover, each cycle in $G$ is viewed as an embedded simple closed curve, with some faces lying inside it and some outside.

**Theorem 10.36** (Euler Formula) *If a connected planar graph has v vertices, e edges and f faces, then* $v - e + f = 2$.

***Proof*** We can prove the assertion by induction on the number of vertices or the number of edges. We first apply induction on the number of edges. Clearly, the assertion holds for any graphs with $e = 0, 1, 2$. So, we may assume that the assertion holds for any graph with $e - 1$ edges, with $e \geq 3$. Let $G$ be a graph with $e$ edges. If there is no circuit in $G$, then $G$ is a tree so that $f = 1$ and $e = v - 1$. Therefore, in this case, we have

$$v - e + f = v - v + 1 + 1 = 2.$$

We next assume that $G$ has a circuit containing an edge $e_1$ of $G$. As the removal of an edge from a circuit reduces the number of faces by one, the graph $G - e_1$ has $n$ vertices, $e - 1$ edges, and $f - 1$ faces. It thus follows that

$$f - 1 = (e - 1) - v + 2 \qquad \text{or} \qquad v - e + f = 2.$$

Alternatively, notice that a connected planar graph on a single vertex is a bouquet of loops, each viewed as a closed curve in the embedding. In particular, when $e = 0$, we have $f = 1$, and so the assertion holds trivially. In general, each added loop passes through a face, and cuts it into two faces (inside and outside). Therefore, each such augmentation increases the edge count and the face count by 1. Hence, once again, the formula holds. We thus have the base step to apply induction on the number of vertices. So, suppose $v > 1$. As $G$ is connected, it contains an edge that is not a loop. By *contracting* such an edge (see Fig. 10.44), we obtain a connected planar graph $G'$

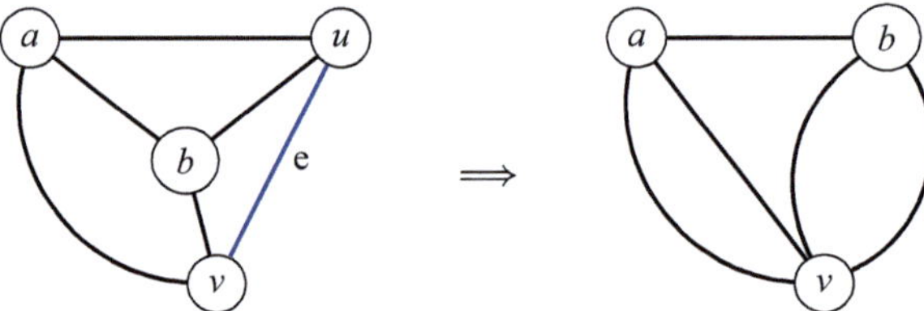


**Fig. 10.44** Graph obtained from the contraction of the edge $e = uv$

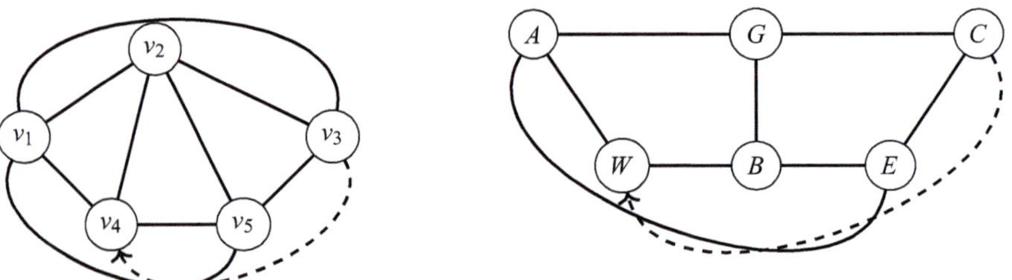


**Fig. 10.45** The graphs $K_5$ and $K_{3,3}$ non-planar

with $v' = v - 1$ vertices, $e' = e - 1$ edges, and $f' = f$ faces, because the number of faces remains the same by shortening the boundaries. Hence, by induction, we have

$$v - e + f = \left(v' + 1\right) - \left(e' + 1\right) + f' = v' - e' + f' = 2.$$

□

The following corollary helps prove certain graphs are non-planar.

**Corollary 10.3** *For any planar graph G,with v vertices, $e > 2$ edges, and f faces, we have $e \leq 3v - 6$. Further, if G is* triangle-free, *i.e., every face is bounded by four or more edges, then $e \leq 2v - 4$.*

***Proof*** It can be assumed that $G$ is connected; otherwise, we may add edges. The central idea is to dispose off the number $f$ from Euler's formula. For, we count the number of edges in all the faces. Notice that, in this counting, every edge that belongs to two faces will be counted once for each face, while those that belong to a single face (for example, as in the case of edges of a tree) will be counted twice for the same face. Therefore, the total number of edges counted this way is $2e$. Also, since every face in a simple graph is bounded by at least 3 edges, we must have $2e \geq 3f$ or $e \geq 3f/2$. Hence, by using the formula $v - e + f = 2$, we have

$$2e \geq 3f \quad \Rightarrow \quad 2e \geq 6 - 3v + 3e, \quad \text{i.e.,} \quad e \leq 3v - 6.$$

Next, if every face is bounded by at least four edges, then we have $2e \geq 4f$, i.e., $e \geq 2f$. Therefore, in this case, we have

$$e \geq 2e - 2v + 4 \quad \text{i.e.,} \quad e \leq 2v - 4.$$

□

**Corollary 10.4** *$K_5$, $K_{3,3}$ are non-planar.*

***Proof*** We first show that $K_5$ is non-planar. It has $n = 5$ vertices and $e = 10$ edges. If it is planar, every face must have at least 3 edges. Hence

$$10 = e < 3n - 6 = 3 \times 5 - 6 = 9$$

must hold which is impossible. If the graph $K_{3,3}$ is planar, then every face must be bounded by at least four edges ($K_{m,n}$ has no triangle!). Since $e = 9$, $n = 6$, the inequality $e \leq 2n - 4$ becomes $9 \leq 12 - 4 = 8$ which is again impossible. □

By above corollary, the complete graph $K_5$ and the complete bipartite $K_{3,3}$ are non-planar. In fact, by Theorem 10.37, any graph containing one of these two latter graphs as a subgraph is also non-planar.

**Corollary 10.5** *If G is simple and planar, then there is a vertex of degree 5 or less in G.*

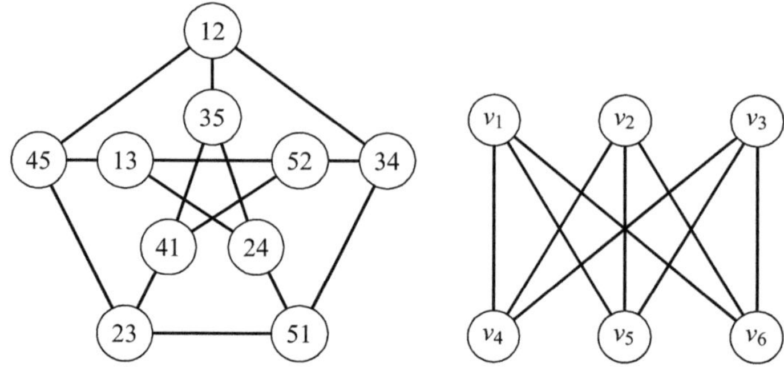


**Fig. 10.46** Peterson graph and the graph $K_{3,3}$

***Proof*** The components of a planar graph are planar and therefore it is enough to prove the corollary for a simple, connected planar graph. If there is no vertex of degree 5 or less in $G$, then $dv \geq 6$ for all vertices. Since $\sum dv = 2e$, it follows that $2e \geq 6n$ or $e \geq 3n$ and this contradicts the fact that $e \leq 3n - 6$ for a simple connected planar graph. □

Two edges are in series if the edges are incident on a common vertex of degree 2. Two graphs are *homeomorphic* of one can be obtained from the other by replacing an edge by two edges in series or by replacing two edges in series by an edge.

**Theorem 10.37** (Kuratowski's Theorem) *A graph is planar if and only if it has no subgraph homeomorphic to* $K_5$ *or* $K_{3,3}$.

***Proof*** For a proof see [19]. □

An elementary contraction of a graph $G$ is obtained by merging two adjacent vertices $u$ and $v$, that is, by removing $u$ and $v$ and adding a vertex $w$ adjacent to those vertices to which $u$ and $v$ are adjacent. A graph $G$ is contractible to a graph $H$ if $H$ can be obtained from $G$ by a sequence of elementary contractions. The complete bipartite graph $K_{3,3}$ is homomorphic to a subgraph of the Peterson graph. Also, the Peterson graph can be contracted to $K_5$.

**Theorem 10.38** *A graph is planar if and only if it has no subgraph contractible to* $K_5$ *or* $K_{3,3}$.

***Proof*** For a proof see [19]. □

## 10.7 Graph Coloring

We start our discussion here with one of the most famous and stimulating problems in graph theory, namely, the **graph coloring conjecture**. In 1852, while working on his school project, *Francis Guthrie* contemplated that four colors suffice to color any map of a country in a manner so that no two adjacent states have the same color. He asked *Frederick Guthrie*, his elder sibling pursuing mathematics at the University College London, about the mathematical validity of the "puzzle". In turn, Frederick took the following problem to his teacher *Augustus de Morgan*[4]: *What would be the least number of colors needed to color states of a country maps such as shown in Fig.* 10.47 *in a way so that no two neighboring states are assigned the same color?* It is assumed that the boundaries are shared by three or more states, and none of them is a *point boundary.*

**Theorem 10.39** *No more than four colors are needed to color any map.*

The above *four color theorem* remained unsolved for more than a century. Several renowned mathematicians attempted to prove this theorem: *Aurther Cayley*, in 1878; *Alfred Kempe* (1849–1922), in 1879; *Peter Tait* (1831–1901), in 1880; and many others. Kempe's proof was accepted to be true until *Percy Heawood* (1861–1955) in 1890 produced a map with 18 regions to refute the claim. In fact, it is now known that a map with nine regions suffices to expose flaws in Kempe's proof. Also, Tait's proof was refuted by the Danish mathematician *Julius Petersen* (1839–1910). A computer-assisted proof of the conjecture published by the German mathematician *Heinrich Heesch* (1906–1995) in 1969 was found to be incomplete. However, in this paper, Heesch developed certain fundamental notions such as the concept of *discharging*, which ultimately helped *Kenneth Appel* (1932–2013) and *Wolfgang Haken* (1928–2022) to formulate a computer-assisted proof of the conjecture in 1976. More specifically, the proof of the theorem took a comprehensive analysis of properties of 1936 discrete cases by computer. More recently, considering only 633 discrete cases, the

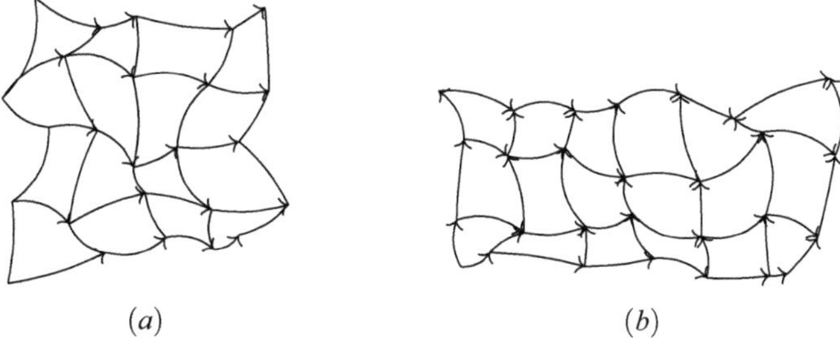


**Fig. 10.47** Maps of two hypothetical countries

[4] We find the first written record of so-called *four color conjecture* during the same time in a letter of De Morgan addressed to Hamilton.

proof was simplified by N. Robertson, D. P. Sanders, P. D. Seymour, and R. Thomas ([20–22]).

**Remark 10.8** The study and generalization of the conjecture by Tait, Heawood, Ramsey, and Hadwiger led to the study of the problem of *coloring of graphs embedded on surfaces* (of arbitrary *genus*). In fact, Tait's reformulation generated a new class of problems such *factorization problems*, studied by Petersen and König. The works of Ramsey on colorations, and more specifically the results obtained by Turán in 1941, have led to the introduction of a new branch of graph theory, which is now known as the *extremal graph theory*.

**Definition 10.35** (*Dual Graph*) The **dual graph** of a planar graph $G = (V, E)$ is the graph $G' = (V', E')$ such that it has a vertex for each region of $G$, and *adjacency* among vertices is defined in terms of common boundaries of the corresponding regions, which is assumed not to be a *point boundary*. That is, $e' = u'v' \in E'$ if the regions corresponding to vertices $u'$ and $v'$ share a common boundary.

**Remark 10.9** The double dual of a connected planar graph is the graph itself. While forming $G'$, we are not sure there is only one edge connecting two vertices. We might have $G'$ to be a multigraph. It is an interesting fact that a connected planar graph is Eulerian if and only if its dual graph is bipartite. Also, a Hamiltonian cycle in a planar graph G corresponds to a partition of the vertices of the dual graph into two subsets (the interior and exterior of the cycle) whose induced subgraphs are both trees.

Since states divide a country map into *regions*, we may see a map as a partition of the plane into connected regions, where the exterior part is called the *unbounded region*. The *dual graph* $G'$ is a model of the *map coloring problem* such that we are interested in finding the minimum number of colors that can be assigned to vertices in a way so that no two adjacent vertices get the same color. Therefore, the "map coloring problem" is equivalent to the "vertex coloring problem" for the planar graph $G'$.

**Definition 10.36** (*Vertex Coloring*) A **vertex coloring** of a graph $G$ is a function $f$ that assigns to each vertex $v \in V(G)$ a color in a set $C = \{c_1, c_2, \ldots, c_m\}$ in a way so that no two adjacent vertices have been assigned the same color.

The concept introduced in the above definition is one of the most important concepts in graph theory. It is used in many real-time applications in computer science. There are various coloring methods that one can use as per need. In *list coloring problem*, each vertex has a list of available colors and we have to find a coloring where the color of each vertex is taken from the list of available colors. This list coloring can be used to model situations where a job can be processed only in certain time slots or can be processed only by certain machines.

**Definition 10.37** (*Chromatic Number*) The least number of colors needed to *vertex color* a connected graph $G$ is called the **chromatic number** of $G$. It is denoted by $\chi(G)$ or $\chi_0(G)$.

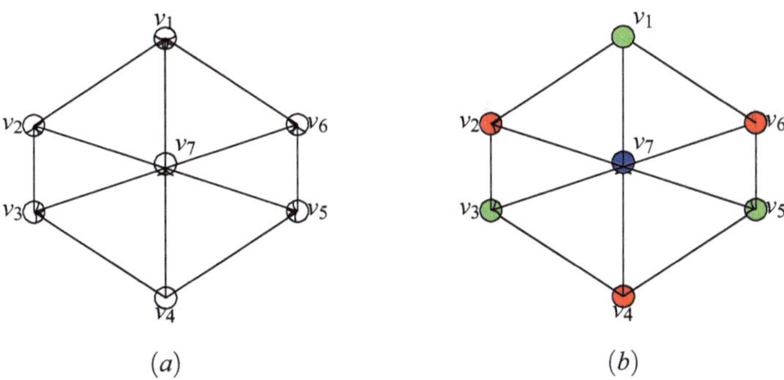


**Fig. 10.48** A 3-colorable graph

The *chromatic number* of a disconnected graph is taken as the largest among the chromatic numbers of its components. When $\chi(G) = k$, we also say $G$ is **k-colorable graph**. Notice that the concept of a chromatic number of a graph doesn't depend on self-loops and parallel edges. Therefore, the graph model of a map coloring problem may be assumed to be an undirected simple graph.

**Example 10.11** For $n \geq 3$, the cycle $C_n$ is of chromatic number 2 or 3. It is 2 when $n$ is even; and, it is 3 when $n$ is odd. Therefore, for the wheel $W_n$, we have $\chi(W_n) = 3$ or 4. In particular, the *vertex coloring* of the graph $G$ with a drawing shown in Fig. 10.48a as given in Fig. 10.48b confirms $\chi(G) = 3$. Notice that a color assigned to the vertex $v_7$ in Fig. 10.50a must be different from the colors assigned to the vertices $v_1, \ldots, v_6$ on the cycle $C_6$, which has the chromatic number 2. Notice that a cycle of even length is bipartite and the chromatic number of any bipartite graph is also 2. For, given a bipartition $(V_1, V_2)$, we only need 2 colors respectively for vertices in $V_1$ and $V_2$. Theorem 10.40 proves that the converse holds. In particular, every non-trivial tree is 2-chromatic. We also have that $\chi(K_n) = n$.

The next theorem proves that every 2-chromatic graph is bipartite.

**Theorem 10.40** *A 2-chromatic graph $G = (V, E)$ is a bipartite.*

***Proof*** Suppose a graph $G$ is 2-colorable, say with colors $c_1$ and $c_2$. Let $V_1$ be the set of vertices with color $c_1$, and $V_2$ be the set of vertices with color $c_2$. By the definition of coloring, $V_1$ and $V_2$ are independent sets, and each edge in $G$ is between a vertex in $V_1$ and a vertex in $V_2$. Also, we have $V = V_1 \cup V_2$. Therefore, $G$ is a bipartite graph. □

In view of Definition 10.36, the problem asks: *Does every connected planar graph has chromatic number* $\leq 4$? It is the next theorem that proves any physical map drawn in the plane can have its regions colored with four colors in such a way that any two regions having a common border have different colors.

**Theorem 10.41** (Four Color Theorem) *If $G$ is a simple planar graph, then the chromatic number $c(G)$ of $G$ satisfies $c(G) \leq 4$.*

The Four Color Problem remained unsolved for more than a century. However, one can easily prove by induction that $\chi(G) \leq 6$ by using the fact that any planar graph has a vertex of degree 5 or less. One has to consider the coloring of $G - v$ where $v$ is a vertex of degree five or less.

**Theorem 10.42** *Every planar graph $G$ is 6-colorable.*

***Proof*** We use induction on $n = |V(G)|$ to prove the claim. Notice that there is nothing to prove for $n \leq 6$. We may thus assume that the claim holds for any planar graph with fewer than $n$ vertices. Now, let $G$ be a planar graph of order $n > 6$. We know that there is at least one vertex $v \in V(G)$ such that $\deg(v) \leq 5$, by Corollary 10.5. Consider the vertex-deleted subgraph $G' = G - v$ obtained by removing from $G$ the vertex $v$ and all edges incident on $v$. Then the planar graph $G' = G - v$ is 6-colorable, by our induction hypothesis. We now proceed to show that $G$ itself is 6-colorable. For, consider the coloring of vertices of $G$ (other than the vertex $v$) that is the same as in the subgraph $G'$. We can assign the vertex $v$ a color that is not assigned to any of the 5 vertices adjacent to $v$. This proves $G$ is 6-colorable. □

Kempe's attempted proof of Theorem 10.41, published in 1879, contains many interesting ideas. The concept of *Kempe chains* introduced therein ultimately helped Heawood to prove that planar graphs can be colored using five colors.

**Theorem 10.43** (Heawood, 1890) *Every planar graph $G$ is 5-colorable.*

***Proof*** We use induction on $n = |V(G)|$ to prove the claim. Notice that there is nothing to prove for $n \leq 5$. We may thus assume that the claim holds for any planar graph with fewer than $n$ vertices. Now, let $G$ be a planar graph with $n$ vertices. We know that there is at least one vertex $v \in V(G)$ such that $\deg(v) \leq 5$, by Corollary 10.5. Consider the vertex-deleted subgraph $G' = G - v$ obtained by removing from $G$ the vertex $v$ and all edges incident on $v$. Then the planar graph $G' = G - v$ is 5-colorable, by our induction hypothesis. We now proceed to show that $G$ itself is 5-colorable. Consider the coloring of vertices of $G$ (other than the vertex $v$) by 5 colors, as in the subgraph $G'$. It remains to color the vertex $v$. Notice that, if $\deg(v) < 5$, then it is possible to color $v$ by the fifth color that is not assigned to any of the four vertices adjacent to $v$. Also, when $\deg(v) = 5$ and the five vertices adjacent to $v$ are not assigned all five colors, we have a color available for $v$. Finally, the only possibility we are left with is when $\deg(v) = 5$, and all available five colors $c_1, c_2, \ldots, c_5$ are assigned respectively to the five vertices $v_1, v_2, \ldots, v_5$ adjacent to $v$. Let these vertices be arranged around $v$ as in Fig. 10.49a. We assert that, among these five vertices, there are two vertices $v_i$ and $v_j$ such that there is no path from $v_i$ to $v_j$ that used the colors $c_i$ and $c_j$ only. For example, if there is a path $P$ from $v_1$ to $v_3$ that uses the colors $c_1$ and $c_3$, then any path from $v_2$ to $v_4$ that uses only $c_2$ and $c_4$ would have to cross the path $P$, and so the assertion holds (see Fig. 10.49b). We thus have two vertices $v_i$ and $v_j$ such that there is no path between them that is colored only with $c_i$ and $c_j$. Now, exchange the colors $c_i$ and $c_j$ all along the paths from $v_i$ with alternating colors $c_i$ and $c_j$. Notice that such an exchange of colors doesn't affect

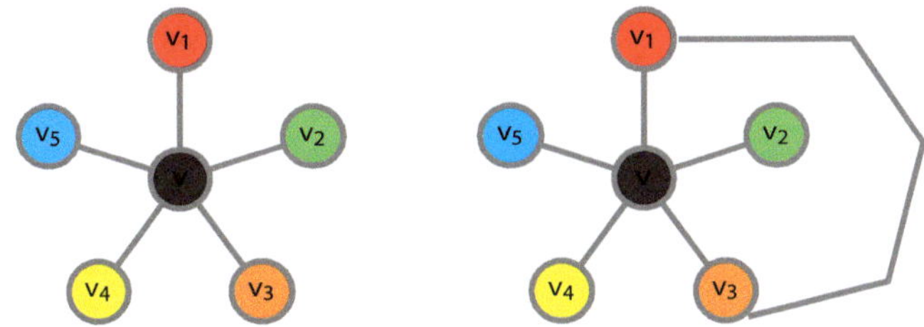


**Fig. 10.49** A vertex of degree five

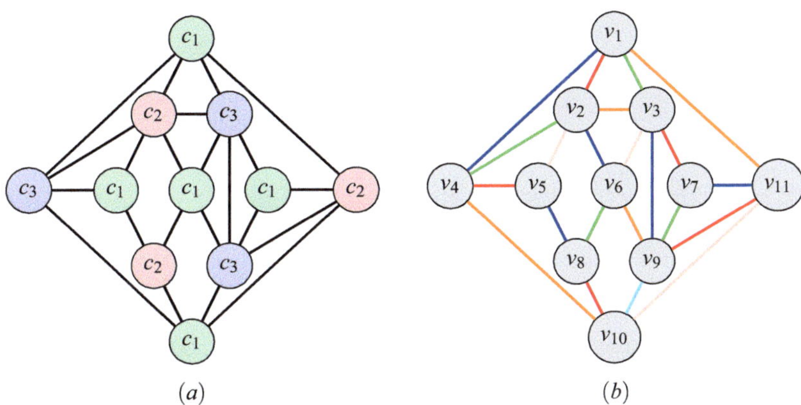


**Fig. 10.50** A graph $G$ with $\chi(G) = 3$ and $\chi'(G) = 6$

the color assigned to $v_j$. Hence, the color $c_i$ is available to color the vertex $v$. This completes the proof. □

**Definition 10.38** Let $G$ be a graph, and $C = \{c_1, c_2, \ldots, c_m\}$ be a set of colors. An **edge coloring** of $G$ is a function $f$ that assigns to each edge in $E(G)$ a color in $C$ in a way so that no two adjacent edges have the same color. The smallest value of $m$ that suffices to *edge color* a graph $G$ in this manner is called the **chromatic index** of $G$. It is denoted by $\chi'(G)$ or $\chi_1(G)$.

We say $G$ is **k-colored graph** when $\chi'(G) = k$. In particular, the term *monochromatic* is used when $k = 1$.

**Example 10.12** The graph in Fig. 10.50a has the chromatic number three, whereas its *chromatic index* appears to be six (see Fig. 10.50b). Does there exist an alternative *edge coloring* that uses a lesser number of colors? Next, in view of colors assigned to the modified *Peterson graph* shown in Fig. 10.51a, we have $\chi(G) = 4$. In fact, it is not possible to color this graph using three or less colors. Notice that the maximal independent set of $G$ consists of three vertices, because the cycles $1, 2, 3, 4, 5$ and $6, 8, 10, 7, 9$ cannot contain more than two independent vertices. If the chromatic

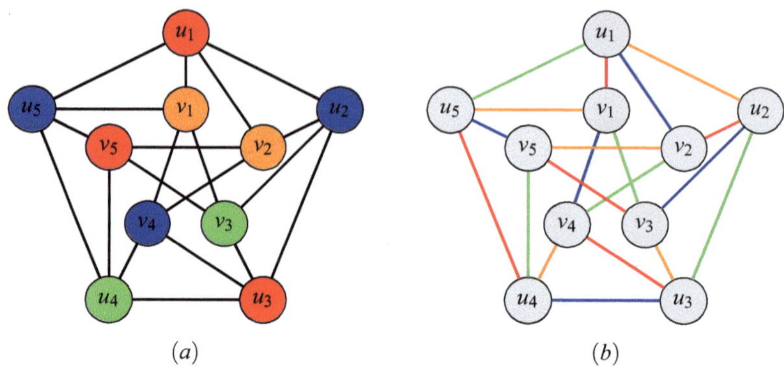


**Fig. 10.51** A modified Petersen graph with $\chi(G) = \chi'(G) = 4$

number of the graph were three, then all 10 vertices of the graph could be partitioned into three (monochromatic) parts. At least one of the parts must have at least four vertices each. But then the graph has 4 independent vertices, which is impossible. Also, in view of edge coloring shown in Fig. 10.51b and a, we have $\chi'(G) = 4$.

It is interesting to know how the coloring of a graph is related to its edge coloring. The next definition implies that the chromatic index of a graph $G$ is the same as the chromatic number of the *line graph* $L(G)$ of the graph $G$.

## 10.8 Ramsey Numbers

The concept introduced here is part of a whole new discipline in the field of combinatorics, known as the *Ramsey theory*. The theory in general studies enumerative problems associated with minimal inescapable nice configurations in sufficiently large systems ([23, 24]. In most contexts, the goal is to show that "*complete disorder is impossible*". The first significant result of this type is due to the Dutch mathematician *Van der Waerden* (1903–1996) about *arithmetic progressions* [25].

**Theorem 10.44** (Van der Waerden's Theorem, 1927) *For any pair of positive integers $p$, $q$, we can find a smallest positive integer $n := W(p, q)$ such that in every partition of the set $\left[n\right] = \{1, 2, \ldots, n\}$ into $p$ subsets there is a partition set containing an arithmetic sequence of $q$ terms.*

The integers $W(p, q)$ are known as the **Waerden numbers**. The reader is referred to lecture notes [26], for a proof of the above theorem, and loads of other important information about the *Waerden numbers* $W(p, q)$ for various values of $p$ and $q$. It is known that $W(2, 3) = 9$, $W(3, 3) = 27$, $W(3, 4) = 293$, etc. ([27, 28]). For example, the set $\{1, 3, 4, 6, 9\}$ of the partition $\big\{\{1, 3, 4, 6, 9\big\} \cup \big\{2, 5, 7, 8\big\}$ of the set $[9]$ contains the arithmetic sequence 3, 6, 9. However, the property fails for the partition $\{\{1, 4, 5, 8\} \cup \{2, 3, 6, 7\}$ of the set $[8]$.

More generally, a theorem due to the German mathematician *Richard Rado* (1906–1989) says that a homogeneous system of linear equations over integers is "partition regular" precisely when the coefficient integer matrix satisfies the so-called "column condition". Other special cases of Rado's theorem are Folkman's theorem and Schur's theorem. There are diverse applications of such types of theorems [25]. The pigeonhole principle plays a significant role in most cases (Theorem 2.15).

Notice that, in terms of $p$-coloring of the set $[n]$, Theorem 10.46 says that the **Waerden number** $W(p, q)$ is the smallest integer such that there exists a "monochromatic" arithmetic progression of length $q$. Our emphasis here is the *Ramsey numbers* that may be seen as representing the boundaries of disorder in the context of "graph Ramsey theory" [29]). In all that follows we are dealing with arbitrary edge coloring of complete graphs. In 1930, the British mathematician *Frank Ramsey* (1903–1930) discovered a theorem that when interpreted in terms of graphs is the statement of our next theorem [30].

**Theorem 10.45** (Ramsey Theorem, 1930) *For any k positive integers $p_1, \ldots, p_k$, there exists a smallest positive integer $n := R(p_1, \ldots, p_k)$ such that every k-coloring of edges of $K_n$ contains a monochromatic clique of size $p_i$, for some i.*

As said above, Theorem 10.45 had been known earlier, though in some other contexts. Recall that a set of $r$ vertices $V'$ of a graph $G$ is called a *clique* of size $r$ if all vertices in $V'$ are mutually adjacent. That is, the subgraph induced by $V'$ is the complete graph $K_r$. On the other hand, a set of $s$ vertices $V'$ of a graph $G$ is called an *independent set* of size $s$ if no two vertices in $V'$ are adjacent.

**Definition 10.39** The integers $n := R(p_1, \ldots, p_k)$ in Theorem 10.45 are known as the **k-color Ramsey numbers** (or simply **Ramsey numbers**).

The Ramsey numbers $R(p_1, \ldots, p_k)$ are most difficult quantities to compute. The computation even for small values of $p_1, \ldots, p_k$ gets messy with an increase in the size of the clique [31]. In this text, we restrict our discussion to the case when $k = 2$. That is, we are looking for a monochromatic clique in graphs $K_n$, say edge colored in *red* and *blue* colors. Figure 10.52 shows three *red-blue* colorings of $K_4$ and $K_5$. Notice that only Fig. 10.52a contains a monochromatic triangle, whereas all in Fig. 10.52d through Fig. 10.52f have it.

We start our discussion with the historical puzzle problem: *What is the least number of people to be invited to a party to guarantee that there are three mutual friends or three mutual strangers?* We know that the answer is six (Theorem 10.5). We now formulate the problem using the concept of edge coloring of a graph.[5] Let the six people be represented by the vertices of the graph $K_6$, and each of the 15 edges represents the relation "friend" or "stranger". We may color an edge *blue* if the

[5] Recall that a graph $G$ is said to be **k-colored** if each edge of $G$ is assigned a color among given $k$ colors.

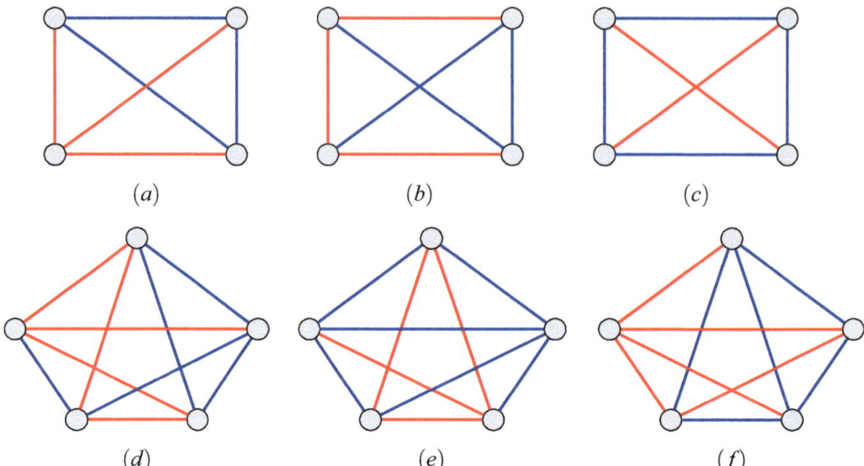


**Fig. 10.52** Three different 2-colorings of $K_4$ and $K_5$

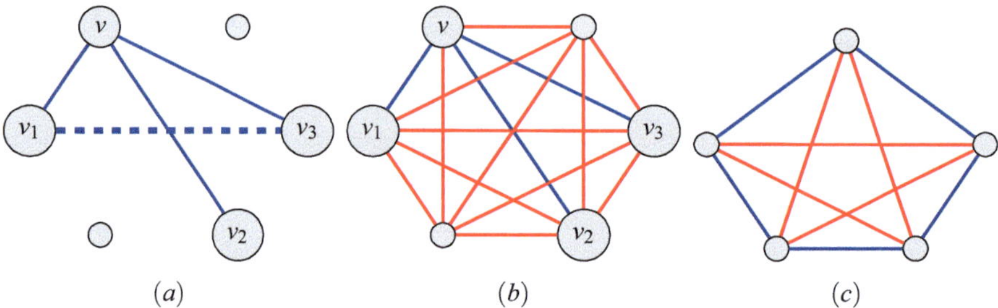


**Fig. 10.53** Pictures for the proof of $R(3, 3, ) = 6$

two incident vertices represent friends and as *red*, otherwise. Then the next theorem answers the question in the puzzle problem.

**Theorem 10.46** *Every* red-blue *coloring of the graph $K_6$ contains a red or a blue triangle.*

***Proof*** Consider a *red-blue* coloring of the graph $K_6$. Fix a vertex $v$ of $K_6$. The five edges incident to $v$ are colored red or blue. Let $R$ be the set of vertices adjacent to $v$ via a red edge, and $B$ be the set of vertices adjacent to $v$ via a blue edge. By the *pigeonhole principle*, at least three of these five vertices adjacent to $v$ are connected by the same color edges. That is, either $|R| \geq 3$ or $|B| \geq 3$. For brevity, suppose $|B| \geq 3$, and let $v_1,\ v_2,\ v_3 \in B$. If any one of the three edges $v_1v_2,\ v_1v_3$, or $v_2v_3$ is blue, then $K_6$ contains a blue triangle (Fig. 10.53a). Otherwise, it already contains red triangles (Fig. 10.53b). □

Notice that the number 6 is the smallest with said property. For, consider a *red-blue* coloring of the graph $K_5$ as shown in Fig. 10.53c. Since every triangle in this coloring has two blue edges and one red or two red edges and one blue edge, it follows that the number of people to be invited has to be more than 5. Therefore, it follows from Theorem 10.46 that the final answer of the puzzle problem is 6. More generally, the integers $n := R(p, q)$ in the next theorem is called the **2-color Ramsey number**.[6]

**Theorem 10.47** *For all positive integers $p$, $q$, there exists a smallest integer $n := R(p, q)$ such that every red-blue edge coloring of $K_n$ contains a monochromatic clique of size $p$ or $q$. We also have*

$$R(p, q) \leq \binom{p+q-2}{p-1}. \tag{10.8.1}$$

***Proof*** A proof can be written using double induction on $p + q$, and argument similar to the proof of Theorem 10.46 ([23], Theorem 3.4). Since $K_1$ is a point, we can assume that $p, q \geq 2$. So, taking $p + q = 4$ as the base case, the assertion follows from the equality $R(2, 2) = 2$, which holds because there is a monochromatic "line" $K_2$ regardless of which particular color is used. Also, since $R$ is symmetric, i.e., $R(p, q) = R(q, p)$, and $R(p, 2) = p$, for all $p$, it can be assumed that $p > 2, q > 2$. Notice that both $R(p - 1, q)$ and $R(p, q - 1)$ are finite, by induction hypothesis. Therefore, proof completes once we are done proving the following recurrence relation:

$$R(p, q) \leq R(p - 1, q) + R(p, q - 1) \tag{10.8.2}$$

For, let $m = R(p - 1, q) + R(p, q - 1)$, and $G$ be a graph of order $m$. Fix a vertex $v$ of $G$. We may consider the following two cases:

**Case-I** Suppose the vertex $v$ has at least $R(p, q - 1)$ neighbors. Applying induction on the number of neighbors of $v$ implies that either they contain an independent set of size $p$, or a clique of size $q - 1$. Next, we extend the clique by adding $v$. Therefore, $G$ contains an independent set of size $p$ or a clique of size $q$.

**Case-II** Suppose the vertex $v$ has at least $R(p - 1, q)$ non-neighbors. Applying induction on the number of non-neighbors of $v$ implies that either they contain an independent set of size $p - 1$, or a clique of size $q$. We can extend the independent set by adding vertices $v$. Once again, we are done.

An explicit upper bound as in (10.8.1) is obtained as follows: Clearly, the inequality holds for the base case when $p = q = 1$, because every graph contains a clique or an independent set of size 1. The inductive step follows from (10.8.2) as given below

[6] More generally, 2-color *Ramsey numbers* $n = R(p, q)$ is the smallest integer such that any graph $G$ of order $\geq n$ contain a *clique* of order $p$ or it has an *independent set* of order $q$.

$$\begin{aligned} R(p,q) &\le R(p-1,q) + R(p,q-1) \\ &\le \binom{p+q-3}{p-2} + \binom{p+q-3}{p-1} \\ &= \binom{p+q-2}{p-1}, \end{aligned}$$

by using the Pascal formula (Theorem 5.13). □

An explicit formula for the Ramsey numbers $R(p, q)$ may never be a possibility. However, it is relatively easy to compute simple instances of these Ramsey numbers. The first non-trivial case corresponds to the "puzzle problem" for the value $p = q = 3$. We have shown that $R(3, 3) = 6$ (Theorem 10.46). Notice that (10.8.1) implies that

$$R(3,4) \le 10, \quad R(3,5) \le 15, \quad \text{and} \quad R(4,4) \le 20.$$

It was proven in 1955 that

$$R(3,4) = 9, \quad R(3,5) = 14, \quad \text{and} \quad R(4,4) = 18. \qquad ([123])$$

In some special cases such as $R(3, 4)$, we know why the difference of 1 appears between above mentioned upper bounds and the corresponding exact values. For, if both Ramsey numbers on the right side of (10.8.2) are even integers, then it can be shown that

$$R(p,q) \le R(p-1,q) + R(p,q-1) - 1. \tag{10.8.3}$$

We show that $R(3, 4) = 9$. It follows from (10.8.3) that

$$R(3,4) \le R(2,4) + R(3,3) - 1 = 4 + 6 - 1 = 9.$$

It thus suffices to bi-color $K_8$ in red and blue such that it neither contains a red $K_3$ nor a blue $K_4$. For, consider the bi-coloring of $K_8$ as shown in Fig. 10.54a. Also, Fig. 10.54b illustrates why it doesn't contain any blue $K_4$.

We can use Theorem 10.6 to show that $R(4, 4) = 18$. We also know that $R(3, 6) = 18$, $R(3, 7) = 23$, $R(3, 8) = 28$, $R(3, 9) = 36$, and $R(4, 5) = 25$ ([33–35]). The exact value of $R(5, 5)$ is as yet unknown, but it lies between 43 and 48 [36]. The following quote is attributed to prolific Hungarian mathematicians *Paul Erdős* (1913–1996) [37]:

> Suppose aliens invade the earth and threaten to obliterate it in a year's time unless human beings can find the Ramsey number $R(5, 5)$. We could marshal the world's best minds and fastest computers, and within a year we could probably calculate the value. If the aliens demanded the Ramsey number $R(6, 6)$, however, we would have no choice but to launch a pre-emptive attack.

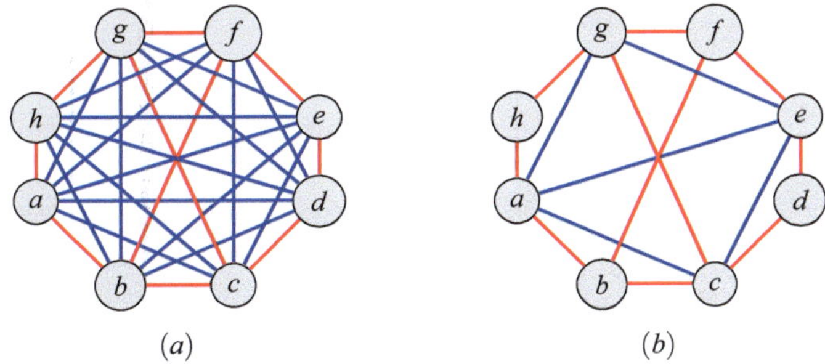


**Fig. 10.54** A red-blue coloring of $K_8$ that proves $R(3, 4) > 8$

While looking for collections of five mutual friends or mutual strangers for the "conjectured" $R(5, 5) = 43$, there are a total of 903 pairwise relationships among forty-three people. That is, we have to search through $2^{903} \sim 10^{271}$ different configurations to look for collections of 5 mutual friends or mutual strangers. By the way, for reference, the universe has about $10^{80}$ atoms in it! For sure, we are in much bigger trouble in the case of six friends or strangers. In general, it follows from (10.8.1) that for the **diagonal Ramsey number** $R(p, p)$, we have

$$R(p, p) \leq \binom{2p-2}{p-1}. \qquad (10.8.4)$$

The lower bound on $R(p, p)$ as in the next theorem is due to Erdős [38].

**Theorem 10.48** (Erdős, 1947) *For any positive integers $p$, $n$ such that*

$$\binom{n}{p} 2^{1-\binom{p}{2}} < 1,$$

*we have $R(p, p) > n$. In particular, for $p \geq 3$, we have $R(p, p) > 2^{\frac{p}{2}}$.*

The lower bound obtained above shows that $R(p, p)$ grow exponentially in $p$. For some Ramsey theorists, the research goal over the years has been to obtain good asymptotics for $R(p, p)$. The upper bound in (10.8.1) obtained by Erdős and *George Szekeres* (1911–2005) in 1935 is sharper than what Ramsey obtained in his original paper [39]. In particular, it follows that $R(3, t) \leq \binom{t+1}{2}$. In 1957, Erdős proved $R(3, t) > t^{t+c_1}$. In 1961, he used probabilistic arguments to prove that

$$R(s, t) > \binom{s+t-2}{s-1}^{c_2}, \qquad \text{for all} \quad s > 3,$$

which shows that the upper bound in (10.8.1) is not very far from being the best possible [40]. The above can also be expressed as $R(3, t) = \Omega\big(t^2/(\log t)^2\big)$. In 1968,

**Table 10.1** Some known values/bounds of Ramsey numbers [45]

| $p, q$ | 3 | 4 | 5 | 6 | 7 | 8 | 9 | 10 |
|---|---|---|---|---|---|---|---|---|
| 3 | 6 | | | | | | | |
| 4 | 9 | 18 | | | | | | |
| 5 | 14 | 25 | 43–48 | | | | | |
| 6 | 18 | 36–40 | 58–85 | 102–161 | | | | |
| 7 | 23 | 49–58 | 80–133 | 115–273 | 205–497 | | | |
| 8 | 28 | 59–79 | 101–194 | 134–427 | 219–840 | 282–1532 | | |
| 9 | 36 | 133–282 | 183–656 | 204–949 | 292–2134 | 343–4432 | 565- | |
| 10 | 40–42 | 92–136 | 149–381 | 204–949 | 292–2134 | 343–4432 | 581- | 798–23556 |

Graver and Yackel proven that $R(3, t) = O\big(t^2 \log\log t/\log t\big)$. Both these bounds were improved in [41] as $R(3, t) = O\big(t^2/\log t\big)$. In 1995, Kim proven that $R(3, t) = \Omega\big(t^2/\log t\big)$ [42]. The next theorem is a huge improvement over the bound obtained earlier in Theorem 10.48.

**Theorem 10.49** $R(3, t) = \Theta\big(t^2/\log t\big)$.

It is now known that $R(p, p) \leq (4 - \epsilon)^p$, for some constant $\epsilon > 0$ [43]. In certain situations, these limits do help compute some less trivial Ramsey numbers $R(p, p)$. The current research focus is on computing some multicolor Ramsey numbers, and how to obtain asymptotics for the Ramsey numbers (see Table 10.1). More generally, as a significant improvement over the known upper bound on the off-diagonal Ramsey number $r(4, t)$, Sam Mattheus and Jacques Verstraete used finite geometry to show that $r(4, t) \geq c\big(t^3/\log^4 t\big)$ [44]. We also know that the best known asymptotic lower bounds is $c'\big(t^{5/2}/\log^2 t\big)$.

## Exercises 10

10.1 Use the principle of induction to prove Corollary 10.1.

10.2 For $k \geq 1$, construct a self-complementary graph with $4k$ and $4k + 1$ vertices.

10.3 For $n \geq 2$, construct a cubic graph with $2n$ vertices.

10.4 Let G be a graph with $v$ vertices and $e$ edges all of whose vertices have degree $m$ or $n$. If $G$ has $p_m > 0$ vertices of degree m and $P_n > 0$ vertices of degree $n$, find $P_m$ and $P_n$ in terms of $v, e, m, n$. (Hint: Solve $nP_n + mP_m = 2e$, $P_n + P_m = v$.)

10.5 Prove that two graphs (without self-loops) are isomorphic if the incidence matrix of one of them can be obtained by permuting the rows and columns of the incidence matrix of the second graph.

10.6 Let $X$ be the adjacency matrix of a graph $G$ with $n$ vertices, and

$$Y = X + X^2 + X^3 + \cdots + X^{n-1}.$$

Show that the graph $G$ is disconnected if and only if at least one of the entries in the matrix $Y$ is zero.

10.7 Show that a simple graph with $n$ vertices and more than $(n-1)(n-2)/2$ edges should be connected.

10.8 Show that a graph $G$ is disconnected if and only if its vertex set $V$ can be partitioned into two nonempty, disjoint subsets $V_1$ and $V_2$ such that there exists no edge in $G$ whose one end vertex is in subset $V_1$ and the other in subset $V_2$.

10.9 In any graph with exactly two vertices of odd degree, show that there is a path joining these two vertices.

10.10 Prove that a connected graph $G$ remains connected after removing an edge $e$ from $G$, if and only if the edge $e$ is in some circuit in $G$.

10.11 In a connected graph $G$ with exactly $2k$ odd vertices, there exist $k$ edge-disjoint subgraphs such that they together contain all edges of $G$ and that each is a unicursal line.

10.12 Let $G$ be a simplex graph with $n$ vertices satisfying the inequality $d(v_i) + d(v_j) \geq n-1$ for every pair of distinct vertices $v_i, v_j$ of $G$. Show that $G$ is connected and has a Hamiltonian path. (A Hamiltonian path is a path that covers every vertex exactly once.)

10.13 What is the number of edge-disjoint Hamiltonian circuits in the complete graph $K_n$?

10.14 Define the distance $d(u, v)$ between two vertices $u, v$ of a connected graph $G$ as the number of edges in the shortest path between the vertices $u$ and $v$. Show that this distance in a connected graph is a metric.

10.15 A graph with each of its vertices assigned a name (label) is called a labeled graph. Show that there are $n^{n-2}$ labeled trees on $n$ vertices.

10.16 Show that every edge in a tree is a cut-edge. What can you say about the vertices of a tree?

10.17 Show that $e = k(n-2)/(k-2)$ in every simple planar graph, with $n$ vertices and $e$ edges, all of whose regions are bounded by exactly $k$ edges. [Hint: $fk = 2e$]

10.18 For any graph $G$, show that $c(G) \leq 1 + \max d(v)$ where the maximum is taken over all the vertices in the graph $G$. When can this bound be improved to $c(G) \leq \max d(v)$?

10.19 Let a graph of order $n$ has $k$ components. Show that $|E(G)| \leq (n-k)(n-k+1)/2$.

10.20 Let $G$ be a graph of order $n$. Suppose $\deg(v) \geq (n-1)/2$, for each vertex $v \in V(G)$. Show that $G$ is a connected graph.

10.21 Show that a graph $G = (V, E)$ is connected if and only if, for every partition $V = V_1 \cup V_2$, there exists an edge $e = uv$, with $u \in V_1$ and $v \in V_2$.

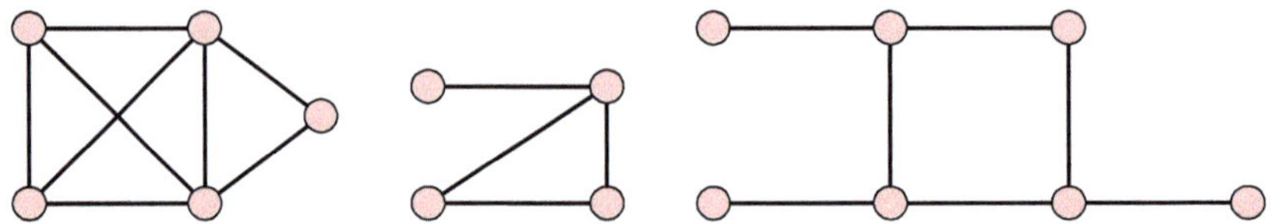

**Fig. 10.55** Graphs for Q. 10.28

10.22 Let $G$ be a graph with $k$ components. What is the minimum number of edges that one needs to add to $G$ so that the extended graph is connected?

10.23 Make drawings of all possible graphs on four vertices. In general, how many graphs are possible on $n$ vertices? How many of them are with exactly $m$ edges, with $m \leq \lceil n^2/2 \rceil$?

10.24 Let $G = (V, E)$ be a bipartite graph, with $|V| = n$. Prove that $|E| \leq \lfloor n^2/4 \rfloor$.

10.25 Show that if a graph $G$ is disconnected then $\overline{G}$ is connected.

10.26 Give an example of an Euler graph that is also Hamiltonian.

10.27 Differentiate between an Euler and a Hamiltonian graph, by using examples. Show that each Hamiltonian path is a spanning tree.

10.28 Which of the three graphs shown in Fig. 10.55 have a Hamiltonian cycle or at least a Hamiltonian path:

10.29 Is it necessary for a Hamiltonian graph to have an Eulerian trail? Justify.

10.30 Suppose that a connected graph $G = (V, E)$ is of order 11 and size 53. Explain why $G$ cannot be Euler. Further, show that $G$ is Hamiltonian.

10.31 Prove that, for any connected planar graph $G = (V, E)$ with $v = |V|$ and $e = |E|$, we have $v - e - r = 2$, where $r$ is the number of regions formed by the edges of $G$.

10.32 Explain edge and vertex coloring, and also how the two concepts are related.

10.33 Draw a planar graph $G = (V, E)$ of order 6 such that $\deg(v) \geq 3$, for all $v \in V$.

10.34 Show that in a connected planar graph with 6 vertices and 12 edges, each of the region is bounded by 3 edges.

10.35 Given the in-order and the post-order of a binary tree $T$ as given below:

In-order : BEHFACDGI and Post-order : HFEABIGDC

Find the tree $T$, and also its pre-order.

10.36 Find the Prüfer sequence for the two trees shown in Fig. 10.57. Also, describe the reverse process in each case.

10.37 Determine which trees have Prüfer sequence that have distinct values in all positions.

10.38 Let $G$ be a connected graph, and $T$ spanning tree $T$ of $G$. Show that the complement $\overline{T}$ contains an edge of any given cycle $C$ in $G$.

10.39 Let $G = (V, E)$ be a connected graph, and $v \in V$ be a vertex of smallest degree. Show that the *edge connectivity* of $G$ cannot exceed $\deg(v)$.

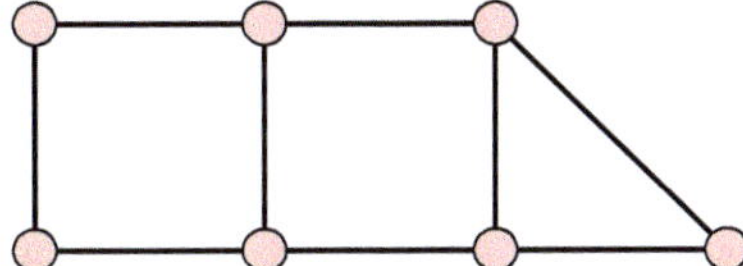

**Fig. 10.56** A Hamiltonian graph without an Eulerian trail

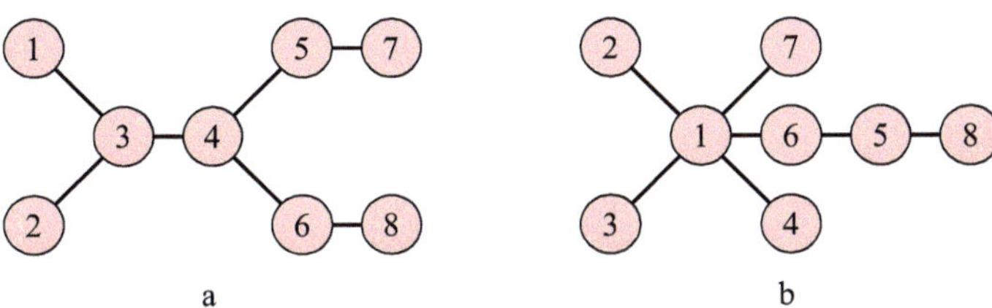


**Fig. 10.57** Two spanning trees for Q. 10.37

10.40 Let $G = (V, E)$ be a connected graph, with edge connectivity $\alpha$. Show that the *vertex connectivity* of $G$ is $\leq \alpha$.
10.41 Let $G = (V, E)$ be a connected graph, with $|V| = n$ and $|E| = m$. Show that the maximum vertex connectivity of $G$ is $\lfloor 2m/n \rfloor$.
10.42 Prove that a non-trivial tree $T$ always contains at least two vertices of degree one.
10.43 Show that a regular binary tree has an odd number of vertices.
10.44 Show that a connected graph with $n$ vertices must have at least $n - 1$ edges.
10.45 Prove that the maximum number of vertices in a binary tree of height $h$ is $2^{h+1}$, for $h \geq 0$.
10.46 Show that if all the internal vertices of a tree have the same degree then it has a total odd number of vertices. Otherwise, give a counter-example to refute the assertion.
10.47 Define what is a binary tree. Further, draw the same for the words: VIREO, WASBLER, EGRET, GROSBEAK, NUTHATCH, and KINGFISHER.
10.48 Define a binary tree, and discuss two important applications of binary trees. Suppose a binary tree with 11 nodes has respectively the *in-order* and *pre-order* given by ABDHIEJKCFG and HDIBJEKAFCG. Draw the binary tree.
10.49 Construct the binary tree for the sequence 19, 25, 6, 9, 11, 21, 47, 8, 4, 37, 77.
10.50 Write the pre-order and the post-order traversal for the binary tree shown in Fig. 10.58.
10.51 Construct the binary tree having the *in-order* and *pre-order* respectively given by DGBEIHJACF and ABDGEHIJCF. Also, find the *post-order* traversal for the same.
10.52 Show that $R(p, 2) = p$, for all $p \geq 2$. Hence deduce that $R(3, 3) = 6$.
10.53 Give a proof of the inequality (10.8.3).
10.54 Show that $R(3, 5) = 14$.

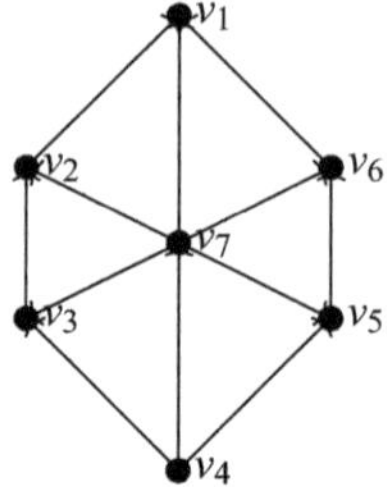


**Fig. 10.58** Binary tree $T$ in Q. 10.51

10.55 Use Theorem 10.6 to show that $R(4, 4) = 18$.

10.56 Show that, for any five plane points in general position, some four of them form a convex quadrilateral.

## References


1. Aho, A.V., Hopcroft, J.E., Ullman, D.: The Design and Analysis of Computer Algorithms. Pearson Education (1974)
2. Ahuja, R.K., Magnanti, T.L., Orlin, J.B.: Network Flows—Theory, Algorithms, and Applications. Prentice Hall, Englewood Cliffs, New Jersey (1993)
3. Cormen, T.H., Leiserson, C.E., Rivest, R.L., Stein, C.: Introduction to Algorithms (3/e). MIT Press (2009)
4. Korte, B., Vygen, J.: Combinatorial Optimization: Theory and Algorithms, Algorithms and Combinatorics, vol. 21. Springer, Berlin Heidelberg, New York (2012)
5. Papadimitriou, C., Steiglitz, K.: Combinatorial Optimization: Algorithms and Complexity. Prentice-Hall (1982)
6. Bollobás, B.: Graph Theory: An Introductory Course. Cambridge University Press, New York (1979)
7. Bondy, J.A., Murthy, U.S.R.: Graph Theory with Applications. North Holland, New York (1976)
8. West, D.B.: Introduction to Graph Theory. Prentice Hall, Upper Saddle River (1995)
9. Chartrand, G., Lesniak, L.: Graphs & Digraphs (3/e). Chapman & Hall, New York (1996)
10. Chartrand, G., Zhang, P.: Introduction to Graph Theory. McGraw-Hill, New York (2005)
11. Gross, A., Jonathan, L., Tucker, T.W.: Topological Graph Theory. John Wiley & Sons Inc, New York (1987)
12. Wilson, R.J.: Introduction to Graph Theory. Oliver and Boyd, Edinburgh (1972)
13. Biggs, N.L., Lloyd, E.K., Wilson, R.J.: Graph Theory 1736–1936. Clarendon Press, Oxford (1976)
14. Fox, J.: Lecture Notes in Combinatorics. https://math.mit.edu/~fox/MAT307-lecture01.pdf
15. Wu, B.Y., Chao, K-M.: Spanning Trees and Optimization Problems. Chapman and Hall/CRC (2004)
16. Cayley, A.: A theorem on trees. Quart. J. Math. **23**, 376–378 (1889)
17. Pitman, J.: Coalescent random forests. J. Comb. Theory (Ser. A) **85**, 165–193 (1999)
18. Cayley, A.: On the theory of the analytical forms called trees. Philos. Mag. 4th series, **13**, 172–176 (1857)
19. Harary, F.: Graph Theory. Narosa Publishing House, New Delhi (1988)
20. Wilson, R. J.: Four Colours Suffice. Allen Lane Science (2002)

21. Fritsch, G., Fritsch, R., Peschke, L.L.: The Four-Color Theorem: History, Topological Foundations, and Idea of Proof. Springer-Verlag, New York Inc. (1998)
22. Gonthier, G.: Formal proof—the four-color theorem. AMS Not. **55**(11), 1382–1393 (2008)
23. Robertson, A.: Fundamentals of Ramsey Theory. Taylor & Francis, CRC Press (2021)
24. Soifer, A. (ed.): Ramsey Theory: Yesterday, Today, and Tomorrow. Birkhäuser, Springer (2011)
25. Graham, R., Rothschild, B., Spencer, J.H.: Ramsey Theory (2/e). John Wiley and Sons, NY (1990)
26. Tao, T.: The ergodic and combinatorial approaches to Szemérdi's theorem. arXiv:math/0604456
27. Khinchin, A.: Three Pearls of Number Theory. Dover, NY (1998)
28. Kouril, M.: Computing the van der Waerden number $W(3, 4) = 293$. Integers **12**, 1–13 (2012)
29. Xu, X., Liang, M., Luo, H.: Ramsey Theory. Walter de Gruyter (2018)
30. Ramsey, F.P.: On a problem of formal logic. Proc. LMS **30**, 264–286 (1930)
31. Conlon, D., Fox, J., Sudakov, B.: Recent developments in graph Ramsey theory. arXiv:1501.02474v3
32. Greenwood, R.E., Gleason, A.M.: Combinatorial relations and chromatic graphs. Can. J. Math. **7**, 1–7 (1955)
33. Graver, J.E., Yackel, J.: Some graph theoretic results associated with Ramsey's theorem. J. Comb. Theory **4**(2), 125–175 (1968)
34. Grinstead, C.M., Roberts, S.M.: On the Ramsey numbers $r(3, 8)$ and $r(3, 9)$. J. Comb. Theory Ser. B **33**(1), 27–51 (1982)
35. McKay, B.D., Radziszowski, S.P.: $R(4, 5) = 25$. J. Graph Theory **19**(3), 309–322 (1995)
36. Angeltveit, V., McKay, B.D.: $R(5, 5) \leq 48$. arXiv:1703.08768v2
37. Graham, R., Spencer, J.: Ramsey theory. Sci. Am. Mag. **263**(1), 112 (1990)
38. Erdős, P.: Some remarks on the theory of graphs. Bull. Am. Math. Soc. **53**(4), 292–294 (1947)
39. Erdős, P., Szekeres, G.: A combinatorial problem in geometry. Compos. Math. **2**, 463–470 (1935)
40. Erdős, P.: Graph theory and probability. Can. J. Math. **13**, 346–352 (1961)
41. Ajtai, M., Komlós, J., Szemerédi, E.: A note on Ramsey numbers. J. Comb. Theory Ser. A **29**(3), 354–360 (1980)
42. Kim, J.H.: The Ramsey number r(3, t) has order of magnitude $t^2/\log t$. Random Struct. & Algorithms **7**(3), 173–207 (1995)
43. Campos, M., Griffiths, S., Morris, R., Sahasrabudhe, J.: An exponential improvement for diagonal Ramsey. arXiv:2303.09521v1
44. Mattheus, S., Verstraete, J.: The asymptotics of R(4,t). arXiv:2306.04007
45. Radziszowski, S.: Small Ramsey numbers. Electron. J. Comb. DSI(16), (2021)

# Chapter 11
# Encoding and Encryption

*Information is the resolution of uncertainty.*
**Claude E. Shannon** (1916–2001)

We introduce some basic concepts concerning two fundamental aspects of information theory, namely, the *error control coding* and the *cryptography*. The information is first "encoded" in a way so that the receiver can "decode" the transmitted message, i.e., the receiver can detect and correct the transmission errors. The overall reliability of a communication system depends on how appropriate is the chosen "data transmission scheme". The core idea due to *Richard W. Hamming* (1915–1998) is to create redundancy in the transmitted message while encoding. On the other hand, the purpose of "encryption" is to secure the data, communicated over public channels or in a computer system with shared resources. Every cryptographer must assume that the attacker knows the system. In this chapter, the topics presented are largely based on some algebraic concepts discussed earlier in Chap. 8. We also use some enumerative techniques discussed earlier in Chap. 5. For number theory-related simple results used here, the reader may see Appendix A.1.

 
V. Ravichandran and A. K. Razdan, *Fundamental Discrete Structures*,
University Texts in the Mathematical Sciences,
https://doi.org/10.1007/978-981-96-0069-4_11

## 11.1 Finite Fields

In this section, we discuss basic concepts of finite fields inasmuch as we need in the sequel. Our main emphasis is on the construction of some finite fields, as needed in a later section on cryptography. As a matter of fact, a majority of *cryptographic protocols* are based on finite fields. We start with the next definition.

**Definition 11.1** A field with finite number of elements is called a **finite field**. We usually write a *Finite field* with $q$ elements as $\mathbb{F}_q$.

It follows easily that the set $\mathbb{Z}_p$ is a field with respect to operations $+_p$ (addition modulo $p$) and $\times$ (multiplication modulo $p$). We may write this finite field as $\mathbb{F}_p$. In particular, as shown in Example 8.19, $\mathbb{F}_5$ is a finite field. We may recall that the characteristic of a finite field is a prime, say $p$. Notice that the Abelian group

$$\mathbb{F}_q := \mathbb{Z}_p \oplus \cdots \oplus \mathbb{Z}_p \ (n \text{ terms})$$

is a vector space over the field $\mathbb{Z}_p$, where the scalar multiplication is given by

$$a \cdot v = a\, v, \qquad \text{for } \ v \in \mathbb{F}_q \ \text{ and } \ a \in \mathbb{F}_p.$$

The concept introduced in the next definition is needed in the sequel.

**Definition 11.2** Let $\mathbb{F}$ and $\mathbb{K}$ be two fields with $\mathbb{F} \subset \mathbb{K}$. Then $\mathbb{K}$ is a vector space over $\mathbb{F}$, say of dimension $n < \infty$. In this case, we say $\mathbb{K}$ is **finite field extension** of the field $\mathbb{F}$ of degree $n$. We write the dimension $n$ of $\mathbb{K}$ over the field $\mathbb{F}$ as $[\mathbb{F} : \mathbb{K}]$. Similarly, we say $\mathbb{K}$ is **infinite field extension** of a field $\mathbb{F}$ if $[\mathbb{F} : \mathbb{K}] = \infty$.

Therefore, the field $\mathbb{F}_q$ contains $q = p^n$ elements, for some positive integer $n$.We also call $\mathbb{F}_q$ a **Galois field** in characteristic $p$, in honor of its discoverer. We may thus also write $\mathbb{F}_q$ as **GF**$(q)$. More generally, a vector space over the field $\mathbb{F}_q$ is called a **modular vector space** in characteristic $p$.

**Example 11.1** The field $\mathbb{C}$ is a *Finite field extension* over the field $\mathbb{R}$ of degree two, because $\{1, i\} \subset \mathbb{C}$ is a basis over $\mathbb{R}$. Recall that each element $z \in \mathbb{C}$ can be written uniquely as $z = x + i\, y$, for some $x,\ y \in \mathbb{R}$. On the other hand, $\mathbb{R}$ is an *infinite field extension* over the field $\mathbb{Q}$. This follows directly by using the usual *Cardinality argument*.

Every finite field $\mathbb{F}$ contains $q = p^n$ elements, for some positive integer $n$.

**Theorem 11.1** (E.H. Moore, 1893) *For any prime $p$ and $n \geq 1$, there is unique finite field with $q = p^n$ elements. More precisely, up to isomorphism, every finite field is the minimal splitting field of the polynomial $x^q - x \in \mathbb{F}_p\big[x\big]$.*

Though Theorem 11.1 guarantees finite fields abound in nature but we need to explicitly construct finite fields, in actual applications. Apart from the finite fields $\mathbb{F}_p$, this

**Table 11.1** Product tables of the field $\mathbb{F}_4 = \{0, 1, \omega, \bar{\omega}\}$

| $+$ | $0$ | $1$ | $\omega$ | $\bar{\omega}$ |
|---|---|---|---|---|
| $0$ | $0$ | $1$ | $\omega$ | $\bar{\omega}$ |
| $1$ | $1$ | $0$ | $\bar{\omega}$ | $\omega$ |
| $\omega$ | $\omega$ | $\bar{\omega}$ | $0$ | $1$ |
| $\bar{\omega}$ | $\bar{\omega}$ | $\omega$ | $1$ | $0$ |

| $\cdot$ | $0$ | $1$ | $\omega$ | $\bar{\omega}$ |
|---|---|---|---|---|
| $0$ | $0$ | $0$ | $0$ | $0$ |
| $1$ | $0$ | $1$ | $\omega$ | $\bar{\omega}$ |
| $\omega$ | $0$ | $\omega$ | $\bar{\omega}$ | $1$ |
| $\bar{\omega}$ | $0$ | $\bar{\omega}$ | $1$ | $\omega$ |

is a difficult task, in general. However, we demonstrate below some methods of establishing the existence of $\mathbb{F}_q$ "by actual construction".

For example, one may use the irreducibility property of polynomials over a field to construct a lower order finite field such as $\mathbb{F}_2$, by way of a *Quotient construction* given by $\mathbb{Z}[i]/(i+1)$. Or, more generally, as the quotient $\mathbb{Z}[i]/(p) \cong \mathbb{F}_{p^2}$, where $p > 2$ is a prime such that $p \equiv 3 \bmod 4$. We may recall that the usual *irreducibility test*[1] works nicely for degree two and three polynomials over a finite field having a small number of elements. We also know by a celebrated theorem of commutative algebra that if $\mathfrak{m}$ is a maximal ideal of the ring $R = \mathbb{Z}[x_1, \ldots, x_n]$ then the quotient $R/\mathfrak{m}$ is a finite field.

A finite field with $q = 2^2$ elements has elements $0, 1, \omega, \bar{\omega}$, and the two binary operations have the product tables given by Table 11.1. Notice that we have $x + x = 0$, for all $x \in \mathbb{F}_4$, and also the following fundamental equations hold:

$$\bar{\omega} = \omega^2 + 1 \quad \text{and} \quad \omega^3 = \bar{\omega}^3 = 1.$$

The next example gives a construction of the finite field $\mathbb{F}_9$.

**Example 11.2** By Theorem 11.1, $\mathbb{F}_9$ is the minimal splitting field of the polynomial $x^9 - x \in \mathbb{Z}_3[x]$, and so its degree is 2. It thus remains to find an irreducible factor of the polynomial $x^9 - x$ of degree 2. Indeed, any irreducible factor of it over $\mathbb{Z}_3$ has degree $\leq 2$. We thus have

$$x^9 - x = x(x-1)(x+1)(x^2+1)(x^2+x-1)(x^2-x-1).$$

Let $\alpha$ be a root of $x^2 + 1$, then the elements of $\mathbb{F}_9$ can be written as $a + \alpha\, b$, where $a, b = 0, 1, 2$. In fact, since $\alpha^2 = -1$, we may treat the elements of $\mathbb{F}_9$ as the "complex numbers" over the field $\mathbb{Z}_3$.

The result proven in the next theorem is a fundamental fact about the structure of the group $\mathbb{F}^* = \mathbb{F}\setminus\{0\}$ of a finite field $\mathbb{F}$.

---

[1] Over a finite field, a quadratic or cubic polynomial $f(x)$ is irreducible if and only if it has a zero there. For example, $f(x) = x^2 + x + 1$, over $\mathbb{Z}_2$; and, $f(x) = x^2 + 2x + 2$, over $\mathbb{Z}_3$, are irreducible.

**Theorem 11.2** *Let $\mathbb{F}$ be a field. Every finite subgroup of the group $G = \mathbb{F}^*$ is cyclic.*

***Proof*** For any $a \in G$, we have $a^e = 1$, where $e = e(G)$ is the exponent of the group $G$. So, $a$ is a root of the polynomial $X^e - 1$ over the field $\mathbb{F}$. We know that $X^e - 1$ cannot have more than $e$ roots in the field $\mathbb{F}$. It thus follows that $|G| \leq e$. However, since we also have $e \leq |G|$ (always), it follows that $e = |G|$. Hence, $G$ is a cyclic group, by Theorem 8.31. □

*An alternative of Theorem* 11.2 The group by $G$ is a finite Abelian group. By the structure theorem for these types of groups, we have

$$G \cong \mathbb{Z}/m_1\mathbb{Z} \oplus \cdots \oplus \mathbb{Z}/m_r\mathbb{Z}, \quad \text{where} \quad m_1 \big| m_2 \big| \cdots \big| m_r = e(G)$$

To prove that $G$ is cyclic, we show that $r = 1$. If possible, suppose $r > 1$, and take a prime $p$ dividing $m_1$. We thus have the cyclic group $\mathbb{Z}/m_1\mathbb{Z}$ has $p$ elements of order $p$. Similarly, $\mathbb{Z}/m_2\mathbb{Z}$ has $p$ elements of order $p$, and so $G$ has at least $p^2$ elements of order $p$. However, all elements of order $p$ in $G$ are roots of the polynomial $X^p - 1$, which cannot have more than $p$ roots over the field $K$. This contradiction shows that $r = 1$. □

**Example 11.3** In case of cyclic group $\mathbb{Z}_{37}^*$, it is easy to find a generator of order 36 such as 2. It is however harder to find a generator for the cyclic group $\mathbb{Z}_{71}^*$. The smallest generator of order 70 is 7. Of course, the Abelian group of order 70 is cyclic. The cyclic group $\mathbb{Z}_{16}^*$, usually described as the quotient

$$\mathbb{Z}_2[x]/\langle x^4 + x^3 + x^2 + x + 1\rangle,$$

has coset containing $x$ that has the order 5.

**Definition 11.3** A **primitive element** of a finite field $\mathbb{F}_q$ is a generator of the cyclic group $\mathbb{F}_q^*$.

For example, 2 is a *Primitive element* of $\mathbb{F}_{11}$, whereas 4 is not. Also, $2 + \sqrt{2}$ is a primitive element of the field $\mathbb{F}_{25}$. It is a difficult problem to find a primitive element of the finite field $\mathbb{F}_q$. No general procedure is known.

**Remark 11.1** In Example 11.2, $\zeta = \alpha + 1$ is a primitive element, but $\alpha$ is not. Operating with powers of $\zeta$ simplifies multiplication, but addition involves more work. The **Zech logarithm** is a function, denoted by $Z(n)$, that is defined on the set $\{0, 1, \ldots, q-1\}$ by the relation $\zeta^{Z(n)} = \zeta^n + 1$. The formula $\zeta^a + \zeta^b = \zeta^{Z(a-b)+b}$ facilitates addition.

## 11.2 Error Control Coding

Modern tools of the algebraic coding theory are used to deal with inescapable aspects of any communication system: **noise**. The hardware implementations of present-day smart machines such as *optical devices, telecom equipments*, and *scanners*, are based on algebraic codes. In this section, we introduce some basic error detection and error control coding schemes. Our presentation is largely based on the expositions given in [1, 2]. The interested reader may also refer [3–8], for deeper insight.

### *Some Preliminaries*

For a *Communication system*, the *channel* refers to a medium over which information is transmitted from a source to a receiver. The transmission may be physical or over a multiplexed medium. The focus of error control coding is to transmit the information reliably and efficiently over a *noisy channel*. Every communication system is developed and engineered considering the fact that "noise" will always be part of the channel, and it would corrupt transmitted data. We prefer to transmit over a *binary symmetric channel* (see Definition 11.13). Also, the bandwidth of a channel refers to its *transmission capacity*. As depicted in Fig. 11.1, a theoretical model of a communication system has the following five components:

(a) a source or a data;
(b) a transmitter;
(c) a channel such as a cellular or a storage device;
(d) a receiver;
(e) a destination or a memory.

We can also view a *storage device* such as a disk drive as a communication system, where data to be stored as memory is the source, the storage device is the channel, and a prospective user of the stored information is the receiver. It is reasonable to expect that the transmitted data may get corrupted during transmission due to "noise" in the channel.

In most realistic situations, a noisy channel has a lower maximum data rate. A *transmission error* (or simply an **error**) is a distortion in the data transmitted

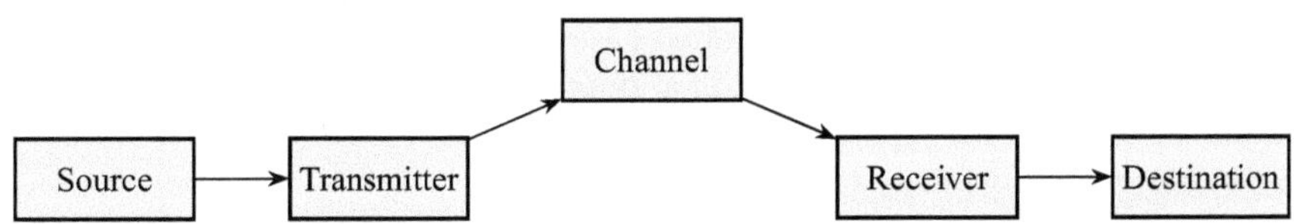


**Fig. 11.1** A theoretical model of a communication system

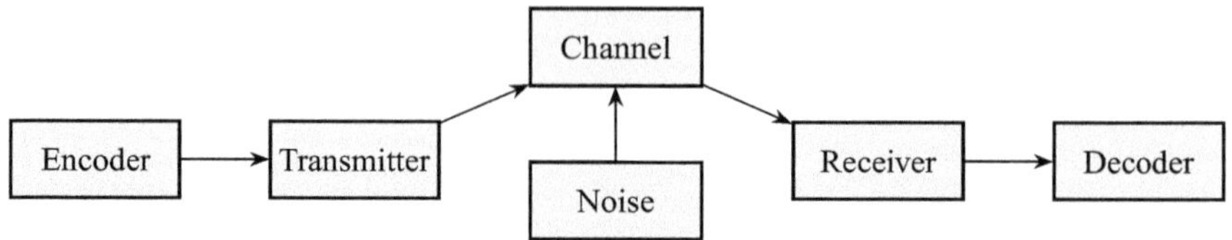


**Fig. 11.2** A realistic model of a communication system

over a channel. The overall reliability of a communication system depends on how appropriate is the "data transmission scheme". The main idea is that the information must be "encoded" before transmission in a way so that the receiver can "decode" the original message, i.e., the receiver can detect and correct any number of transmission errors. It is important that the receiver knows the shape of what to expect. So, to improve the reliability of the transmitted data, we implement an efficient **encoder** at the source and also a related **decoder** at the receiver's end. Figure 11.2 shows a more realistic model of communication over a noisy channel.

**Definition 11.4** (*Encoding*) Let $\mathscr{A}$ be an alphabet of a language such as English, and suppose $c : \mathscr{A} \to \Sigma^*$ is a **coding function**. A function $c^* : \mathscr{A}^* \to \Sigma^*$ given by

$$c^*\big(a_1 a_2 \cdots a_n\big) := c\big(a_1\big)c\big(a_2\big) \cdots c\big(a_n\big), \tag{11.2.1}$$

is called an **encoding function** (or simply **encoding**). A procedure used to retrieve the original message from an encoded message is called **decoding**.

The main purpose of encoding is to create redundancy in the transmitted message. A typical example of a single-digit redundancy is **ISBN code** used by publishers, which can detect any single error and also any double error that occurs due to the transposition of two digits. In a ten-digit *ISBN code*, "message" consisting of the nine-digit $a_1 \cdots a_9$ is encoded as a ten-digit codeword. The last digit $a_{10}$ is called a *check-digit*, which is chosen in a way so the following relation holds

$$1 \cdot a_1 + 2 \cdot a_2 + \cdots + 10 \cdot a_{10} = 0 \pmod{11}.$$

When $a_{10} = 10$ we use the symbol $X$. Check-digit schemes are also used in other situations such as on airplane tickets, bank numbers on cheques, credit card numbers, etc. In general, the codewords in $\Sigma^*$ for which $c^*$ in (11.2.1) is *injective* are called *decodable*. Our emphasis in most cases is on the collection of codewords $C := c\big(\mathscr{A}\big)$. Therefore, a **codeword** is a message that is transformed into another "language" in a manner so that both the sender and the receiver understand it, and these can be unambiguously transformed back.

**Definition 11.5** (*String*) Let $\Sigma$ be a nonempty finite set of symbols. A *string* (or a *word*) $x$ over *alphabet* $\Sigma$ is a "juxtaposed" sequence of symbols $a_1, a_2, \ldots, a_n \in \Sigma$ (or ) of the form

$$x := a_1 a_2 \cdots a_n. \tag{11.2.2}$$

The *length* of a string $x$, denoted by $\ell(x)$, is the number of the symbols $x$ contains.

For $\Sigma = \{0, 1\}$, we call $x$ an *n-bit string*; and, for $\Sigma = \{a, b, c, \ldots\}$, we call $s$ an *n-character string*. For example, $aaaa$, $abc$, and $aabc$, are respectively character strings of length 4, 3, and 4. In general, $\Sigma^n$ denotes the set of strings of length $n$. We write $\Sigma^*$ for the set of strings of arbitrary length $n \geq 0$.

**Definition 11.6** (*Code*) A string that contains a message as a part of it is called a **codeword**. A **code** $C$ is a collection of fixed or variable length *codewords*. When all the codewords in a *code* $C$ over an alphabet $\Sigma$ have the length $n$, we call $\Sigma^n$ the *ambient space* of the "encoding scheme" (Definition 11.4). A collection of an $n$-bit strings is called a **binary code** $C$ of length $n$.

An *error control coding* refers to an algorithm that expresses a sequence of numbers such that any errors which are introduced can be detected and corrected based on the remaining numbers.

**Example 11.4** (*Repetition Code*) The **repetition code** is a simple illustration of a *binary code*, which is about transmitting a string 11111 instead of a 5-bit single bit 1 (say, for a "yes"); and, a 5-bit string 00000 instead of a single bit 0 (say, for a "no"). In this case, under certain mild assumptions about the channel, the recipient uses **majority vote** to detect the possible errors. That is, the received 5-bit string is decoded by "voting" in favor of the bit that occurs most frequently. So, if there have occurred 0, 1, or 2 errors, then we will still decode the received 5-bit string as the intended message.

In general, there has been a long history of how various types of strings of symbols as codes have been used for the purpose of secrecy of back-and-forth communications among states.

**Example 11.5** (*Moorse Code*) In 1830, the American painter *Samuel Morse* (1791–1872) created **Moorse codes** using "dots" and "dashes", as applied more predominantly in telegraph systems of communication. These codes are based on 2-symbol alphabet $\Sigma = \{\bullet, -\}$. Each English letter and numeric digit is assigned a *codeword* consisting of strings over $\Sigma$. Some examples of *Moorse codes* are as given below:

$$\begin{array}{lll} A \mapsto \bullet - ; & B \mapsto - \bullet \bullet\bullet ; & C \mapsto - \bullet - \bullet ; \\ D \mapsto - \bullet \bullet ; & E \mapsto \bullet ; & F \mapsto \bullet \bullet - \bullet ; \\ O \mapsto - - - ; & S \mapsto \bullet \bullet \bullet ; & 7 \mapsto - \bullet \bullet\bullet \end{array} \tag{11.2.3}$$

Now, if A7ESB is the message sent, then the encoding in Morse code is given by

$$A7ESB \mapsto \bullet - - \bullet\bullet\bullet\bullet\bullet\bullet\bullet - \bullet\bullet\bullet$$

For decoding, the mapping (11.2.3) is reversed to create the original message. Notice that the Morse codes are of **variable length**.

After the advent of electronic devices such as computers, better encoding schemes evolved, where a code consists of a number of *code points*, each corresponding to a letter of a language, a numeral, or some other character. In more precise terms, a **code point** is a unique position in a quantized $n$-dimensional space that has a semantic meaning.

**Example 11.6** (*ASCII Code*) In various devices, including computers, the 7-bit *ASCII codes* represent the text, where ASCII stands for "*American Standard Code for Information Interchange*". Modern computer systems use "*Unicode*", with millions of code points, but the first 128 are still the same as the ASCII codes designed originally. The codewords in ASCII are of **fixed length**. Notice that the types of codewords described in this illustration may not be *decodable* even when the *coding function* is injective.

## *Correcting and Detecting Errors*

An **error** refers to a change in one or more bits of sent message that occurred during transmission over a noisy channel. The number of errors is calculated in terms of the "distance" between the transmitted message and the one received by the recipient. The standard difference between the two is also referred as the **error vector**. A change in a single bit is called a *single error*, and two or more changes in the transmitted data are called a *burst error*. The main emphasis is always on developing tools needed to assess what could possibly be the message sent; or, equivalently, to decide which *error vector* most likely occurred. In many cases, **nearest neighbor decoding** is taken as a preferred scheme. It is based on the idea of *minimum distance*. For example, the *majority vote* scheme followed in the case of the binary *repetition code* uses the *nearest neighbor decoding*. Said differently, in this case, the recipient decodes the received $n$-bit string as the "closest codeword".

**Remark 11.2** More generally, *maximum-likelihood decoding* refers to a technique that one uses to determine the most likely transmitted message in a communication system, based on the received signal and statistical models of noise and interference. The underlying methods use a maximum-likelihood estimator to compute the probability for each transmitted message and then select the one with the highest probability. It is also referred to as the decoding algorithm that minimizes the probability of decoding an incorrect codeword when all codewords are equally likely to be transmitted.

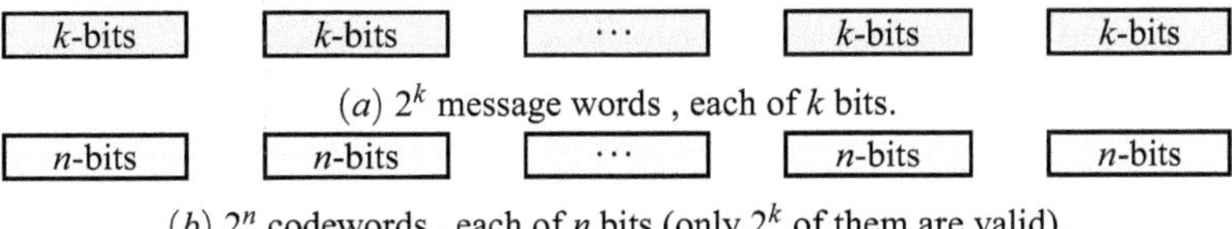


**Fig. 11.3** Message words

While transmitting data over a noisy channel, the *source coding* ensures efficiency and a good scheme for *channel coding* addresses the reliability issues. The pioneering work on source and channel coding is due to the American mathematician and electrical engineer *Claude E. Shannon* ([9, 10]). His *channel coding theorem* predicts the existence of an efficient encoding technique that guarantees a reliable transmission, provided the transmission rate doesn't exceed the channel capacity. It is thus possible to design an efficient and reliable communication system so that transmission errors can be detected and corrected. For this, we need to apply a random encoding scheme so that the probability of *error* is lowest. The description given below is based on the celebrated paper of the American mathematician *Richard Hamming* [11].

The key idea is to send extra bits with the original message, which is called the *redundancy*. In *block coding*, the original message is divided into "blocks", say of $k$ bits, which we may call *message words* (see Fig. 11.3a). For $n > k$, we assign $r = n - k$ redundant bits randomly to each block to make the total length $n = k + r$ (see Fig. 11.3b).

**Definition 11.7** (*Redundancy*) The $n$-bit blocks are called **codewords**, and the additional $r$ bits is the **redundancy** of the code, which is also known as the **parity check** (or **parity digits**).

We use *parity check* to detect errors, and also correct some of these, in certain special cases. In all that follows, we describe how redundancy guarantees that the *encoded message* remains legible to the receiver. It also helps devise various types of techniques for *error detection* and *error correction*. The involved process is known as the *error control coding*, which is also known as the *channel coding* ([1, 12]).

**Example 11.7** We first illustrate how to use parity check to construct a single *error-detection code*. For, add an extra 1 to a bit string of length $n - 1$ so that the total number of 1s in each codeword is even. This is called *even parity coding*. Similarly, *odd parity coding* converts a bit string of length $n - 1$ into a bit string of length $n$ by adding an extra 1 so that the total number of 1s in each codeword is odd. The table below shows a code with four 5-bit strings and the associated parity checks

| Message | Even parity check | Odd parity check |
|---|---|---|
| 00000 | 00000 0 | 00000 1 |
| 00010 | 00010 1 | 00010 0 |
| 01101 | 01101 1 | 01101 0 |
| 11000 | 11000 0 | 11000 1 |

Addition of a bit as parity check helps detect all single errors. However, it cannot detect multiple errors or correct errors. More generally, we may write encoded messages as vectors over the alphabet $\Sigma = \mathbb{F}_q$, the finite field with $q$ elements. A message and a codeword are respectively (row) vectors in the vector spaces $\mathbb{F}_q^k$ and $\mathbb{F}_q^n$ (see Definition 11.10). In the next definition, we have $q = 2$.

**Definition 11.8** (*Hamming Distance*) Let $x, y \in \{0, 1\}^n$ be two strings as in

$$x = x_1 x_2 \cdots x_n \qquad \text{and} \qquad y = y_1 y_2 \cdots y_n. \tag{11.2.4}$$

The **Hamming distance** between $x$ and $y$, denoted by $H(x, y)$, is defined by

$$H(x, y) = \sum_{i=1}^{n} \big(x_i \oplus y_i\big). \tag{11.2.5}$$

Equivalently, the *Hamming distance* $H(x, y)$ is the number of positions where the two bit strings $x$ and $y$ differ, i.e., $H(x, y) := \big\{i \mid x_i \neq y_i\big\}$. The number $\big(1/n\big)H(x, y)$ is called the **relative** Hamming distance.

It follows easily that $H$ is a **metric** on the set $S$. Therefore, the set $S$ is a metric space with respect to *distance function* $H$. Given a string $x \in \{0, 1\}^n$, the **Hamming ball** at $x$ of radius $r > 0$ is the set

$$B_H(x, r) := \big\{y \in\in \{0, 1\}^n \mid H\big(x, y\big) \leq r\big\}. \tag{11.2.6}$$

Also, the **minimum distance** of a *code* $C$, denoted by $d$, is the least nonzero Hamming distance among codewords in the set $C$. That is,

$$d := \min\big\{H(x, y) \mid x, y \in C, \ \text{with } x \neq y\big\}. \tag{11.2.7}$$

For example, let $S := \mathbb{F}_2^6$, and consider the code $C := \{c_1, c_2, c_3, c_4\} \subset S$, where

$$\begin{aligned} c_1 &= 1\,1\,1\,0\,0\,0; & c_2 &= 1\,0\,1\,1\,1\,1; \\ c_3 &= 1\,1\,0\,0\,1\,1; & c_4 &= 1\,0\,1\,0\,1\,1. \end{aligned} \tag{11.2.8}$$

In view of (11.2.5), we have

$$\begin{aligned} H(c_1, c_2) &= 4, & H(c_1, c_3) &= 3, \\ H(c_1, c_4) &= 3, & H(c_2, c_3) &= 3, \\ H(c_2, c_4) &= 1, & H(c_3, c_4) &= 2. \end{aligned}$$

Therefore, the minimum distance among codewords in $C$ is 1. Also, notice that if a received codeword is $c = 1\,1\,1\,1\,1\,1$ then we have

$$H\big(c, c_1\big) = 3, \quad H\big(c, c_2\big) = 1, \quad H\big(c, c_3\big) = 2, \quad H\big(c, c_4\big) = 2.$$

Therefore, $c_2$ is the unique codeword with minimum distance 1.

**Definition 11.9** (*Hamming Weight* Let $x \in \mathbb{F}_2^n$ be a bit string given by $x = x_1 x_2 \cdots x_n$. The **Hamming weight** of $x$, denoted by $\mathrm{wt}(x)$, is defined as the number of 1s in the string $x$.

For example, the *Hamming weight* of the codewords in (11.2.8) are given by

$$\begin{aligned} \mathrm{wt}\big(1\,1\,1\,0\,0\,0\big) &= 3; & \mathrm{wt}\big(1\,0\,1\,1\,1\,1\big) &= 5; \\ \mathrm{wt}\big(1\,1\,0\,0\,1\,1\big) &= 4; & \mathrm{wt}\big(1\,0\,1\,0\,1\,1\big) &= 4. \end{aligned}$$

In general, we have $H(x, y) = \mathrm{wt}(x + y)$ and $\mathrm{wt}(x) = H(x, 0)$. Therefore, the minimum distance $d$ of a code $\mathscr{C}$ is the same as the minimum weight of the codewords in $\mathscr{C} - \{\mathbf{0}\}$. In general, when $d = 1$, the code cannot detect all transmission errors; when $d = 2$, the code can detect, but cannot correct all single errors; and, when $d \geq 3$, the *maximum-likelihood decoding* can correct all the single errors. That is, assuming transmission errors are rare and occur independently in each bit, we look for the codeword that is most likely transmitted. While discussing the error-detection and error-correction capabilities of a code, we always refer to the Hamming distance and Hamming weight. More specifically, the next theorem proves that we must choose either to correct $t$ errors or detect $2t$ errors.

**Theorem 11.3** *A code $\mathscr{C}$ can correct $k$ or less errors if and only if the minimum distance $d$ between any two codewords is at least $2t + 1$. Consequently, the same code detects $2t$ or fewer errors.*

***Proof*** We first show that if $\mathscr{C}$ can correct $t$ or fewer errors, then $d$ must be at least $2t + 1$. On the contrary, suppose there is at least one pair $u, v \in \mathscr{C}$ such that $H(u, u) < 2t + 1$. It may also be assumed that $H(u, v) \geq 2t + 1$. For, otherwise, one cannot even detect $t$ errors. Consider a codeword $u'$ that differs from $u$ in exactly $t$ digits. These $t$ digits are chosen to be any subset of those digits in which $u$ and $v$ differ from one another. This means $H(u, u') = t$. Obviously, by our choice, $H(u', v) \leq t$, because $u$ and $v$ only differ in at most $2t$ digits. Therefore, $u'$ cannot be decoded with certainty, since codeword $v$ is at least as close to $u'$ as $u$ is. This is a contradiction. To prove the converse, assume that $d = 2t + 1$. It thus follows that the code can correct all sets of $t$ errors. Let $u$ be a codeword, and let $u'$ be a received erroneous word that has no more than $t$ error digits. If a decoding rule correctly decodes $u'$ as $u$, then we know that $u'$ is nearer to the codeword $u$ than any other codeword $v$. From the property of the Hamming distance, we thus obtain

$$H(u, u') + H(u', v) \geq H(u, v).$$

Since $H(u, v) \geq 2t + 1$ and $H(u, u') \leq t$, we conclude $H(u', v) > t + 1$. This implies that every codeword $v$ is farther away from $u'$ than $u$. Therefore, $u'$ can be decoded correctly. □

Alternatively, suppose a transmitted codeword $u$ is received as $v$, and that at most $t$ errors have occurred in transmission. Then, by definition, we have $H(u, v) \leq t$. If $w$ is any codeword other than $u$, then $w - u$ is a nonzero codeword. Assume that

$$\begin{aligned} 2t + 1 \leq \mathrm{wt}\left(w - u\right) &= H\left(w, u\right) \\ &\leq H\left(w, v\right) + H\left(v, u\right) \\ &\leq H\left(w, v\right) + t, \end{aligned}$$

so that it follows that $t + 1 \leq H\left(w, v\right)$. Therefore, the codeword closest to $v$ is $u$. Hence, $v$ is correctly decoded as $u$. Said differently, it is possible not to decode a received message at all. In particular, if $\mathrm{wt}(\mathscr{C}) = 4$, then the codeword $\mathscr{C}$ would correct any single error and detect any two errors or detect any three errors.

**Theorem 11.4** *A code $\mathscr{C}$ can detect $k$ or less errors if and only if the minimum distance $d$ between any two codewords in $\mathscr{C}$ is at least $k + 1$.*

***Proof*** We cannot detect a combination of errors if and only if that particular combination transforms a codeword $\mathbf{u}$ into some codeword $\mathbf{c}\,\mathbf{v}$. With $d \geq k + 1$, it would take a combination of at least $k + 1$ errors to change codeword $\mathbf{u}$ into codeword $\mathbf{v}$. Hence, all combinations of $k$ or fewer errors can be detected. □

In general, there are three parameters that decide which code is "good", namely, the *length* of a code; the *number of codewords* a code contains; and, the *Hamming distance* between pair of codewords in a code. Sometimes, we also use alternate distance metrics to evaluate non-binary codes [13].

## *Linear Codes*

Among many different kinds of codes, the most commonly applied is the *linear code*. We may recall that $\mathbb{F}_q$ denotes the finite field with $q$ elements. In general, an **(n, k)-code** $C$ over $\mathbb{F}_q$ is a subset of $\mathbb{F}_q^n$. We call $n$ the *length* and $k$ the *size* of the codewords in $C$. Said differently, an $(n, k)$-*code* is a string over the alphabet $\Sigma = \mathbb{F}_q$, which contains both the message and redundancy. So, a message and a codeword are respectively vectors in the vector spaces $\mathbb{F}_q^k$ and $\mathbb{F}_q^n$ over $\mathbb{F}_q$. In what follows, we may also consider some other representations of a codeword such as in the case of *cyclic codes*. We introduce below some basic concepts related to linear codes.

**Definition 11.10** (*Linear Codes*) An **(n, k)-linear code** $\mathscr{C}$ over the finite field $\mathbb{F}_q$ is a $k$-dimensional subspace of the $n$-dimensional vector space

$$\mathbb{F}_q^n = \mathbb{F}_q \oplus \cdots \oplus \mathbb{F}_q \quad (n \text{ copies}).$$

We also say $\mathscr{C}$ is a **q-ary code**.

Notice that the number of $q$-*ary codes* $\mathscr{C}$ contains is $q^k$. The most commonly used $q$-ary codes are for the case when $q = 2^m$, which are also called the *binary codes*. For example, $\mathscr{C} = \{0000,\ 0101,\ 1010,\ 1111\}$ is a (4, 2) binary code, where the first two digits of each string are the numbers 0, 1, 2, 3, and the last two bits represents the *redundancy*. In the case of linear codes, there are well-defined algorithms for error detection. The related efficiency is based on the linear code's Hamming weight. It may ultimately come down to a choice between detection and correction. In general, it follows from Theorems 11.4 and 11.3 that a code that has minimum distance $d$ can be used to either detect up to $d - 1$ errors or correct up to $\lfloor (d-1)/2 \rfloor$ errors.

**Definition 11.11** We say $\mathscr{C}$ is an $(n, k, d)$-code if each codeword in $\mathscr{C}$ has the length $n$, the size of the set $\mathscr{C}$ is $k$, and the minimum distance between every pair of codewords in $\mathscr{C}$ is $d$.

For example, $\mathscr{C} = \{0000, 0101, 1010, 1111\}$ is a (4, 4, 2)-linear code. In general, a useful way to generate a linear code is by using a *generator matrix*.

**Definition 11.12** Let $\mathscr{C}$ be an $(n, k)$-linear code over a field $\mathbb{F}_q$. The **generator matrix** for $\mathscr{C}$ is a matrix $G \in M_{k\times n}(\mathbb{F}_q)$ such that $k$-rows of this matrix form a basis of $\mathscr{C}$. On the other hand, a **parity check matrix** for $\mathscr{C}$ is a matrix $H \in M_{(n-k)\times n}(\mathbb{F}_q)$ such that we have

$$\mathscr{C} = \{\mathbf{x} \in \mathbb{F}_q^n \mid H \cdot \mathbf{x}^t = \mathbf{0}\}. \tag{11.2.9}$$

A solution $\mathbf{x}$ of the equation $H \cdot \mathbf{x}^t = \mathbf{0}$ is called a **codeword generated** by the parity check matrix $H$.

Notice that $\mathscr{C}$ in (11.2.9) is the *null space* of the linear transformation : $\mathbb{F}_q^n \to \mathbb{F}_q^{n-k}$ given by $\mathbf{x} \mapsto H \cdot \mathbf{x}^t$. Since it has rank precisely $k$, it follows that the rows of the *parity check matrix H* are linearly independent. Also, for any $k$ linearly independent columns of a *generator matrix G* of an $(n, k)$-linear code $\mathscr{C}$, the associated codewords form $\mathscr{C}$ and the remaining $m = n - k$ is called the **redundancy** of $\mathscr{C}$.

**Example 11.8** The purpose here is to find all codewords of (7, 4) binary linear code with a generator matrix $G$ given by

$$G = \left[ I_4 \middle| A \right] = \begin{pmatrix} 1\,0\,0\,0\,1\,1\,1 \\ 0\,1\,0\,0\,1\,0\,1 \\ 0\,0\,1\,0\,1\,1\,0 \\ 0\,0\,0\,1\,0\,1\,1 \end{pmatrix}$$

We may view the binary numbers representing 0, 1, 2, ..., 15 as our message vector $m_i$, and multiply them by $G$ to form the $i$th row of C. All of the individual code words are

$$\mathscr{C} = \begin{pmatrix} m_0 \cdot G \\ \hline \vdots \\ \hline m_{15} \cdot G \end{pmatrix} = \begin{pmatrix} 0\,0\,0\,0\,0\,0\,0 \\ 1\,0\,0\,0\,1\,1\,1 \\ 0\,1\,0\,0\,1\,0\,1 \\ 0\,0\,1\,0\,1\,1\,0 \\ 0\,0\,0\,1\,0\,1\,1 \\ 1\,1\,0\,0\,0\,1\,0 \\ 1\,0\,1\,0\,0\,0\,1 \\ 1\,0\,0\,1\,1\,0\,0 \\ 0\,1\,1\,0\,0\,1\,1 \\ 0\,1\,0\,1\,1\,1\,0 \\ 0\,0\,1\,1\,1\,0\,1 \\ 0\,1\,1\,1\,0\,0\,0 \\ 1\,0\,1\,1\,0\,1\,1 \\ 1\,1\,0\,1\,0\,0\,1 \\ 1\,1\,1\,0\,1\,0\,0 \\ 1\,1\,1\,1\,1\,1\,1 \end{pmatrix},$$

where the parity check matrix is given by

$$H = \left[ -A \middle| I_3 \right] = \begin{pmatrix} 1\,1\,1\,0\,1\,0\,0 \\ 1\,0\,1\,1\,0\,1\,0 \\ 1\,1\,0\,1\,0\,0\,1 \end{pmatrix}$$

The weight of this code is 3, so it is able to correct any single error.

**Example 11.9** Consider two matrices $H_1 \in M_{2\times3}(\mathbb{Z}_2)$ and $H_2 \in M_{1\times3}(\mathbb{Z}_2)$ given by

$$H_1 = \begin{pmatrix} 1\,1\,0 \\ 1\,0\,1 \end{pmatrix} \quad \text{and} \quad H_2 = \begin{pmatrix} 1\ 0\ 1 \end{pmatrix}$$

and let $\mathscr{C} = \mathbb{F}_2^3$. That is, $\mathscr{C}$ consists of the following eight codewords:

$$\begin{aligned} &\mathbf{c}_1 = 0\,0\,0; \quad &\mathbf{c}_2 = 0\,0\,1; \quad &\mathbf{c}_3 = 0\,1\,0; \quad &\mathbf{c}_4 = 0\,1\,1; \\ &\mathbf{c}_5 = 1\,0\,0; \quad &\mathbf{c}_6 = 1\,0\,1; \quad &\mathbf{c}_7 = 1\,1\,0; \quad &\mathbf{c}_8 = 1\,1\,1. \end{aligned} \tag{11.2.10}$$

To find the solutions of $H_1 \cdot \mathbf{c}_i^t = \mathbf{0}$ and $H_2 \cdot \mathbf{c}_i^t = 0$, we have

| $c_i$ | $H_1 \cdot \mathbf{c}_i^t$ | codeword ? |
|---|---|---|
| 000 | 00 | Yes |
| 001 | 01 | No |
| 010 | 10 | No |
| 011 | 11 | No |
| 100 | 11 | No |
| 101 | 01 | No |
| 110 | 01 | No |
| 111 | 00 | Yes |

| $c_i$ | $H_2 \cdot \mathbf{c}_i^t$ | codeword ? |
|---|---|---|
| 000 | 0 | Yes |
| 001 | 1 | No |
| 010 | 0 | Yes |
| 011 | 1 | No |
| 100 | 1 | No |
| 101 | 0 | Yes |
| 110 | 1 | No |
| 111 | 0 | Yes |

Since $d = 3$ in the case of $H_1$, we have $\mathscr{C}$ as a single error-correcting code or a doubly error-detecting code. Also, $d = 1$ in the case of $H_2$ implies that $\mathscr{C}$ is not even a single error-detecting code.

**Theorem 11.5** *A parity check matrix* $H = \big(h_{ij}\big) \in M_{n\times k}(\mathbb{F}_2)$ *generates a codeword of weight* $q$ *if and only if there are* $q$ *columns of* $H$ *such that their* $k$*-tuple sum is zero.*

***Proof*** We first assume that, in a codeword $\mathbf{c}$ generated by $H$, the components $c_{i1}, c_{i2}, \ldots, c_{iq}$ are 1s and the remaining components are 0s, so that $\mathrm{wt}(\mathbf{c}) = q$. It then follows from $H \cdot \mathbf{c}^t = 0$ that

$$h_{i1} \oplus h_{i2} \oplus \cdots \oplus h_{iq} = 0.$$

Conversely, if $q$ distinct columns $h_{i1}, h_{i2}, \ldots, h_{iq}$ of $H$ are such that last equation holds, then we can choose codeword $\mathbf{c}$ wherein $c_{i1}, c_{i2}, \ldots, c_{iq}$ are 1s and the remaining components are 0s. This $n$-tuple will satisfy $H \cdot \mathbf{c}^t = 0$, implying that $\mathbf{c}$ is a codeword of weight $q$ generated by $H$. □

In the case when $\mathscr{C}$ consists of the first $k$ columns, it has a unique generator matrix given by $\big[I_k \mid A\big]$ (standard form), where $A$ is a $m \times n$ matrix such that $H = \big[-A^t \mid I_{n-k}\big]$ is a parity check matrix for the linear code $\mathscr{C}$. Therefore, $H$ is also a generator matrix of a code, denoted by $\mathscr{C}^{\perp}$. It is called the **dual code** (or **orthogonal code**) of the code $\mathscr{C}$. Notice that it is an $(n, n-k)$-linear code. It follows from (11.2.9) that we can also define the *dual code* $\mathscr{C}^{\perp}$ as

$$\mathscr{C}^{\perp} = \big\{\mathbf{x} \in \mathbb{F}_q^n \mid \mathbf{x} \cdot \mathbf{c} = \mathbf{0}, \quad \text{for all } \mathbf{c} \in \mathscr{C}\big\}, \tag{11.2.11}$$

where $\mathbf{x} \cdot \mathbf{y}$ denotes the usual *inner product* of vectors $\mathbf{x}, \mathbf{y} \in \mathbb{F}_q^n$. It is easy to verify that if $G$ and $H$ are, respectively, generator matrix and parity check matrices of a code $\mathscr{C}$, then $H$ and $G$ are respectively generator matrix and parity check matrices of the dual code $\mathscr{C}^{\perp}$. We say a code $\mathscr{C}$ is **self-orthogonal** provided $\mathscr{C} \subseteq \mathscr{C}^{\perp}$; and, it is called **self-dual** when $\mathscr{C} = \mathscr{C}^{\perp}$. Notice that a *self-dual* code has even length and dimension $n/2$.

**Remark 11.3** Code Division Multiple Access (or simply CDMA) is a cell phone standard, which works on orthogonal codes. The core idea is that each cell user is assigned a unique a "*spreading sequence*" so that the network identifies and routes traffic based on these sequences. This is how the network prevents cross-talks among users who are using the same space in a finite frequency band.

**Definition 11.13** (*Symmetric Channel*) A channel is called a $q$-ary **symmetric channel** if the following two conditions hold:

(a) Every transmitted symbol has the same probability $p$ ($< 1/2$) of being received in error; and, the same probability $1 - p$ of being received correctly. We also call $p$ as the *symbol error probability*.
(b) When a symbol is received in error, each of the $q - 1$ possible errors is equally likely

We call it **binary symmetric channel** (or simply BSC) when $q = 2$.

For a $q$-ary code, over a *symmetric channel*, the nearest neighbor decoding scheme maximizes the chances of a decoder to correct errors. This is so mainly because the error vectors of lower weight will occur with higher probability than error vectors of higher weight. A most pertinent question to ask is: *Given a $q$-ary code, what is the probability that $e$ errors occur in a codeword of length $n$?* We first consider some special cases. If no error occurs while transmitting a $q$-ary codeword of length $n$ over a symmetric channel, then the desired probability is $(1 - p)^n$. Next, if a single error occurs, then the probability that the first or the second or the third, ..., symbol is received in error and the rest $n - 1$ are received correctly is given by $p(1 - p)^{n-1}$. It thus follows that

$$np\left(1 - p\right)^{n-1}$$

is the probability of receiving a $q$-ary codeword of length $n$ with exactly one error. Similarly, the probability of receiving a $q$-ary codeword of length $n$ with exactly two errors is given by

$$\binom{n}{2}p^2\left(1 - p\right)^{n-2}.$$

Recall that, over a BSC(p), the 5-bit repetition code can correct up to two errors.

## 11.3 Cryptography

In this section, we briefly introduce three important types of fundamental *cryptosystems*, namely, the *symmetric key cryptosystem, asymmetric key cryptosystem*, and also *hash functions* based cryptosystems. The emphasis is on describing how some mathematical tools are used to design encryption and decryption algorithms. At an elementary level, simple number theoretic cryptosystems may help to safeguard the transmission of data by encrypting the messages. The target audience is assumed to consist of students pursuing an undergraduate course related to network security or some other similar course based on cryptology. The presentation given here is largely based on the expositions given in [14, 15]. The interested reader may also see [16–20], for more details.

## *Cryptosystems*

The prefix "crypt" means *hidden* and the suffix "graphy" means *writing*. So, the main purpose of cryptography is to secure the data, communicated over public channels or in computer systems with shared resources. In the latter case, we also need to safeguard the systems from any unauthorized user or a cyber attacker. An "attack" refers to a sort of malicious activity that attempts to collect, disrupt, deny, degrade, or destroy a cryptosystem or the information available on these systems. A *passive attack* refers to eavesdropping on, or monitoring of, the transmission, whereas by an *active attack* we mean replay, masquerade, modification of messages, and also the denial of service. The aim is to ensure that only the legitimate receiver has access to the transmitted message.

In general, every *cryptosystem* is designed to transform an "unencrypted message", also known as **plaintext**, to an "encrypted message", also known as the **ciphertext**. The **encryption** refers to the process of converting a *plaintext* into a *ciphertext*, and the **decryption** refers to the reverse process. As shown in Fig. 11.4a, *encryption $E$* uses a *secret key $k$* to convert a plaintext $m$ into a ciphertext $c$. Symbolically, we write $c =: E(k, m)$. On the other hand, as shown in Fig. 11.4b, **decryption** $D$ uses the same or different *secret key $k'$* to convert the *ciphertext $c$* back into the *plaintext $m$*. Symbolically, we write $m =: D(k', c)$. In most cases, the size of $c$ is more than that of $m$.

A cryptosystem transforms a plaintext to a ciphertext so that only the legitimate receiver has access to the transmitted message. A formal definition of a cryptosystem is given below.

**Definition 11.14** A **cryptosystem** refers to a collection of algorithms and protocols implemented on a computer system to secure storage and protect online transfer of

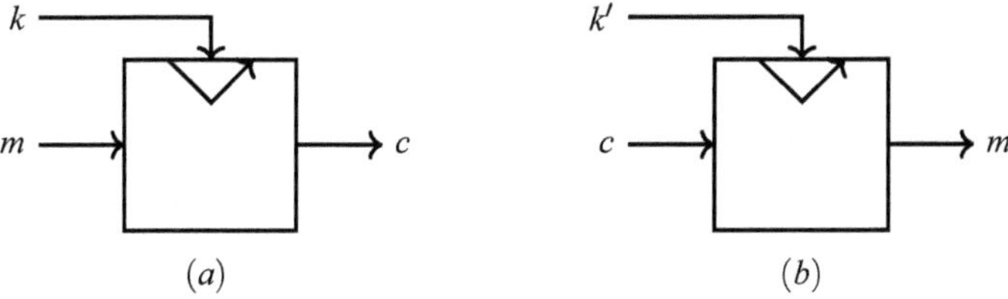


**Fig. 11.4** Encryption and Decryption

information, where a **cryptographic protocol** is the theoretical and practical aspects of provable security over a public channel.

A cryptosystem provides ways how to gain control over the information disseminated in a computer system. While at Bell Labs, Shannon proved that any secure cryptosystem must have all the key features of the cryptographic one-time pad, namely, the secret key has to be random and at least as large as the plaintext. He first wrote this in 1945 as a classified research entitled "*A Mathematical Theory of Cryptography—Case 20878*", which was later published in 1949 as "*Communication Theory of Secrecy Systems*". A *cipher* makes the message unreadable, except for the intended receiver. Based on whether a secretly shared single key or a pair of public keys is used, the two types of *reversible* cryptosystems are $(a)$ and $(b)$ as given below:

(a) In the case of a **symmetric key cryptosystem**, both the sender and the receiver use the same *secret key* to encrypt and decrypt messages. Though it is a faster and simpler encryption system, to exchange the secret key in a secure manner is a serious limitation. Some popular *symmetric key cryptosystems* are such as the *data encryption system* (or simply DES) and the *advanced encryption system* (or simply AES). We will discuss these in the next section.
(b) In the case of a **asymmetric key cryptosystem** (or **public key cryptography**), two different but related keys are used to encrypt and decrypt a message. A pair of secret numbers chosen by each end-user constitutes the private key, and the public key is produced from it by applying a one-way function. Subsequently, a common key is generated by the two users doing the same calculation, but using their own private key and also the public key of the other user. For encryption, the receiver's public key is used, and, for decryption, the receiver's private key is used. Even though the public key is known to everyone, only the intended receiver can decode the sent message, mainly because he alone knows his private key. The *public key cryptography* is used by the *secure sockets layer* (or simply SSL) and *transport layer security* (or simply TLS) protocols to encrypt data sent between the web server and the client. One of the oldest and most widely used *asymmetric key cryptosystem* is the *RSA algorithm*.
(c) On the other hand, an *irreversible* cryptosystem doesn't involve any specific key. It only includes hash algorithms and message authentication codes. We may recall that a *hash function* is a deterministic transformation $H$ that returns a fixed-length string $h := H(k)$, called the **hash value**, for each input key $k$ that may be a string of any length. Since hash functions are one way, strongly collision-free, and easy to compute, the *hash value* so generated for a plaintext makes it impossible to decode the ciphertext. Most operating systems use hash algorithms to encrypt passwords, which are hashed while logging in, and compared to previously stored hash values, so that even when a hacker has access to the database the passwords cannot be read. Some well-known hash functions are such as MD2, MDS, and SHA.

The two fundamental assumptions that each cryptographer makes are that "perfect secrecy" is a myth and every system is vulnerable. Therefore, two major considerations are: *What needs to be protected, and from whom*? A cryptographer thus needs a fair assessment of the asset to be protected, and also the level of risk involved. A majority of cryptosystems are designed to generate *cryptographic key, digital signatures*, in *maintaining passwords*, do necessary data protection and privacy legislation related *verifications*, shield web browsing from *eavesdropping*, and to *safeguard* the online transactions of confidential information. On the other hand, it is the job of a computer engineer to write software consisting of opaque cryptographic libraries, which are specifically designed to meet the information security concerns of the end-users.

The three key objectives that each *cryptosystem* must fulfill are as given below:

(i) The *confidentiality* assures only the intended recipient is able to access and process the sent messages.
(ii) The *integrity* assures that no part of the sent messages is modified in any manner during transmission.
(iii) The *availability* assures promptness of working of the system, and service is not denied to any legitimate users.
(iv) The *authentication* assures the possibility to confirm the identities of sender and receiver, and also the origin/destination of the sent messages.

The first three put together are known as the **CIA triad**, which are fundamental security objectives for both data information and computing services. A firm grasp of basic concepts presented in the next part of the section helps better understand the technical details as given in the subsequent parts. The process of *encryption* makes the transmitted information incomprehensible, and so ensures *confidentiality*. The two classical ciphers described in the next two examples use an *encryption scheme* (or simply a **cipher**) that is based on "shifting" letters of an alphabet by a certain number of letters to the right or left. Since they predate computers, no wonder letters of the alphabet were used as symbols.

**Example 11.10** (*Caesar Cipher*) The **Caesar cipher** described in this illustration is named after *Julius Caesar*. According to the Roman historian Suetonius, Caesar used it to encrypt military communications. In the case of the *Caesar cipher*, the encryption shifts each letter to 3 places right, subject to that the letter is wrapped back to $A$ after reaching the letter $Z$. Clearly, to decrypt, we simply have to shift the letters 3 places to the left. During the time of Crassus and Cicero, these ciphers may have worked fine, knowing that the attackers were not too smart (see Fig. 11.5). Of course, in place of 3, one can use any other number significantly smaller than 26.

The first meaningful improvement of the Caesar ciphers was proposed by the Italian cryptologist Giovan Battista Bellaso in 1553. The reference to the name "Vigenére" is a historical error. The term gained popularity mainly due to the fact that it was used during the American Civil War, and also by the Swiss Army. The French cryptologist Blaise de Vigenére (1523–1596) invented a different cipher, during the same period.

**Example 11.11** (*Vigenére Ciphers*) In this case, encryption follows a *pattern* to shift letter by three positions defined by a *key* consisting of a block of three letters, say CSL. Since $C$ is down two letters after $A$, $S$ is down eighteen letters and $L$ is down

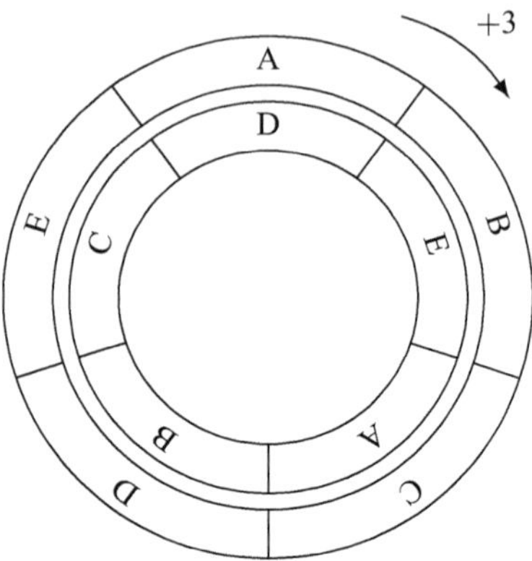


**Fig. 11.5** Modulo 3 Caesar cipher

eleven letters, accordingly in the plaintext the letters are shifted using the values 2, 18, 11. The same pattern is used until the whole of the plaintext is encrypted. It is though more secure than Caesar ciphers, yet the code is easy to break. First figure out the length of the *key* used, followed by using *frequency analysis*. The latter is based on the fact that letters in languages are unevenly distributed. For example, the letter *E* in English is the most frequent one. Therefore, if *X* is most frequent in the encrypted message, then it is likely that the plaintext has *E* at this position.

In 1863, the German cryptologist and archeologist *Friedrich Kasiski* (1805–1881) published a book on cryptography "*Cryptography and the Art of Decryption*" that has an account of a general procedure to decipher Vigenére Ciphers, which now known as *Kasiski examination* [21]. During World War II, the Germans used the cipher device "*Enigma*" to produce nearly unbreakable codes. This was one of the main reasons why Alan Turing designed his "computing device".

A **substitution cipher** in more abstract terms is a *permutation* of letters so that each shift has a unique inverse. Clearly, not every permutation is secure. For that purpose, first, *cipher's permutation need be specified by a key*. For example, in the case of a Vigenére cipher, we must know the key to figure out which one of the twenty-six permutations was used. Second, *different keys should give different permutations*. For example, in the case of a Vigenére cipher, each letter of the key determines a substitution: There are twenty-six distinct letters and as many distinct permutations. Third, loosely speaking, the *permutation must look random*. A permutation that satisfies these three criteria is called a **secure permutation**.

In addition, our *encryption scheme* has to have a **mode of operation** (an algorithm) to support messages of any length. The Caesar ciphers have a trivial *mode of operation* (or simply mode), whereas Vigenére ciphers have a more complex mode because letters at different positions undergo different permutations. That is, if it uses the key of *n* letters, then every *n* consecutive letter needs *n* different permutations. However, since every *n*th letter of the message uses the same permutation, patterns in the ciphertext can be used to decrypt the message by applying frequency analysis. Therefore, we need both a secure permutation and a secure mode.

**Remark 11.4** Since "randomness" makes all involved operations unpredictable, it is a crucial aspect of any cryptosystem. Most operating systems use pseudorandom number generators to ensure randomness. It is done by generating bits randomly. We begin our discussion with a simple pen-and-paper or strong *randomized encryption* in terms of ciphers. In the case of the latter, randomness describes how a pseudorandom generator is applied to construct secure systems.

The popular instant messaging application *WhatsApp* uses symmetric *end-to-end encryption* to guarantee the secrecy of messages exchanged between two users. The term "symmetry" refers to the fact that the same key is used to encrypt and decrypt messages, which is generated using the method first introduced by Whitfield Diffie and Martin Hellman in 1976. In this case, the encryption and decryption happen on the user's phone such that the encryption keys are renewed with each message. That is why absolutely no one other than the intended recipients is able to read the message. In general, a real challenge is to design of one-way algorithms, i.e., they are easy to execute, but to find the starting point from the result is computationally impossible. The process uses advanced mathematical concepts such as elliptic curves, the discrete elliptic logarithm, and modular arithmetic.

The *block ciphers* process messages block per block, with special emphasis on *Advanced Encryption Standard* (or simply AES). On the other hand *stream ciphers* presents ciphers that produce a stream of random-looking bits that are XORed with messages to be encrypted. The most ubiquitous crypto building blocks, known as the *hash functions*, are the only algorithms that don't work with a secret key. The *keyed hashing* is obtained when a hash function is combined with a secret key. It is used to authenticate messages. An *authenticated encryption* such as AES-GCM, is about some algorithms that can both encrypt and authenticate a message.

**Remark 11.5** In 1976, *Whitfield Diffie* and *Martin Hellman* published a number theory based public key cryptosystem. They used exponentiation of a number modulo a prime to encrypt a message that does not allow eavesdroppers to read the ciphertext even when they have access to the encryption key. The idea of *extended asymmetric cryptography* to the notion of key agreement is wherein two parties establish a secret value using only non-secret values.

However, of all public key cryptosystems, the most popular is known is the RSA algorithm, named after the three modern-day cryptographers and computer scientist *Ron Rivest* (born 1947), the Israeli cryptographer and inventor *Adi Shamir* (born 1952), and the American computer scientist *Leonard Adleman* (born 1945). In what follows, we briefly discuss fundamentals related to *public key encryption* and show that RSA leverages the factoring problem in order to build secure encryption and signature schemes with a simple arithmetic operation.

## *Public-Key Encryption: RSA*

Suppose Bob needs to receive the secure message. He chooses two large primes $p$ and $q$, and takes $n = pq$. He also chooses an integer $e < n$ that is relatively prime to $(p-1)(q-1)$. Everybody can use integers $e$ and $n$ as public keys to encrypt messages. However, one must know the primes $p$ and $q$ to decrypt the message. So, if these two primes are large enough (say each containing 150 digits), then no known factorization methods would do the task within any reasonable time. Let us see how the encryption works. We transform plaintext into a sequence of numbers, say using the scheme

$$a \mapsto 01,\ b \mapsto 02,\ c \mapsto 03,\ \ldots$$

with a few extra numbers for blank spaces, commas, etc. Therefore, the ciphertext is a sequence of integers $t < n$ (we can always break the plaintext into smaller pieces before converting it to ciphertext). Now, Alice encrypts each integer $t$ as $c \equiv t^e \bmod n$ and transmits a sequence of $c$'s to Bob. He can decrypt the ciphertext as follows: Since factors $p$ and $q$ are known to Bob, with $m = (p-1)(q-1)$, he applies the Euclidean algorithm on the pair $(e, m)$ to find an integer $d$ such that we have $de \equiv 1 \bmod m$. This is followed by computing $c^d \bmod n$. The result is

$$c^d \equiv \left(t^e\right)^d = t^{de} \bmod n,$$

which gives $c^d \equiv t \bmod n$, by using $de \equiv 1 \bmod m = \varphi(n)$ and Theorem 8.7. Hence, Bob recovered the plaintext. Now, suppose Celia is eavesdropping. She knows the pair $(n, e)$ and also the ciphertext $c$. To decrypt, she will have to find the inverse $d$ of $e \bmod m$, which requires knowledge of factors $p$ and $q$. Our choice of primes $p$ and $q$ in the next example is only for illustration purposes.

**Example 11.12** Let $n = 1073$, with $p = 29$ and $q = 37$, and $e = (p-1)(q-1) = 1008$. Suppose Bob picks the pair (1073, 25) as the public key, and Alice's message for Bob is "*miss piggy*". The corresponding string of integers is given below.

| | m | i | s | s | | p | i | g | g | y |
|---|---|---|---|---|---|---|---|---|---|---|
| t | 13 | 9 | 19 | 19 | 27 | 16 | 9 | 7 | 7 | 25 |

The ciphertext is created by computing $c \equiv t^{25} \bmod n$, for each $t$ in the above table. For example, it starts with $13^{25} \equiv 671 \bmod 1073$, and ultimately gives a string of $c$'s as given in the next table. Bob receives this string. Since Bob is aware of the prime factorization of $n$, he computes the inverse of 25 mod $e$ as follows: Apply the extended algorithm to the private key (25, 1008) to find $1 = 25 \cdot 121 - 1008 \cdot 3$, which gives $d = 121$. Finally, he decrypts Alice's message using the relation $c^{121} \equiv t \bmod n$. We must highlight a problem that is specific to this example. For, letter-for-letter encryption of the message with repetitive letters is easy to break by analyzing the frequency of occurrence of such letters. However, no such problem arises when more realistic key sizes of about 200 digits are used, because then we are encrypting the plaintext in blocks of about 100 letters. Indeed, in this case, there are negligible chance of having coinciding blocks of 100 letters inside a message.

| t | 13 | 9 | 19 | 19 | 27 | 16 | 9 | 7 | 7 | 25 |
|---|---|---|---|---|---|---|---|---|---|---|
| c | 671 | 312 | 901 | 091 | 656 | 1011 | 312 | 922 | 922 | 546 |

In the case of *WhatsApp*, once the common key is available, the encryption and the decryption of the messages take place on the device itself. The AES symmetric cryptographic system—a version of the Rijndael algorithm for 256-bit keys developed in 1998 by the Belgian cryptographers *Vincent Rijmen* (born 1970) and *Joan Daemen* (born 1965)—is used to do so. It is unbreakable with the present computation capacity. The application users generate and renew the common encryption and decryption key with every message which further increases the reliability of the communication system. More specifically, for each batch of messages, a pair of private-public keys are generated on the handset; and, the common root key is computed with their private key and the recipient's public key. From this, a different subkey is generated automatically to encrypt each message. The message header includes the public key. The recipient computes the root key to obtain the subkeys, and ultimately decrypt the messages. The application does the same for their responses, and when the person who started the communication receives them, they renew the private-public keys and replicate the process as many times as necessary. So, the root key and the message subkeys are renewed with each batch of messages.

**Remark 11.6** In 1985, the American mathematician *Neal Koblitz* (born 1948) and the American writer *Victor Miller* (born 1940), independently proposed using Diffie-Hellman methods based on the points on an elliptic curve. Certain algebraic properties of these types of curves have made them apt for cryptographic algorithms and protocols. For example, *WhatsApp* uses the *Montgomery Curve25519* elliptic curve, introduced by the American mathematician and cryptologist *Daniel J. Bernstein* (born 1971) in 2005, whose equation is given by $y^2 = x^3 + 486662x^2 + x$. In general, the advantage of *elliptic curve cryptography* (or simply ECC) is that the keys are relatively small (only 256 bits) and easy to execute. The related mathematical operations are based on the *modular arithmetic*. A sum is defined on the points of the elliptic curve. The fundamental *elliptic curve discrete logarithm* problem (or simply ECDLP) is a special case of the "discrete logarithm problem".

## Exercises 11

11.1 Let $f(x)$ be a monic irreducible polynomial over a field $\mathbb{F}$. Show that $K = \mathbb{F}/\langle f(x)\rangle$ is a field, and the dimension of $K$ over $\mathbb{F}$ is precisely $\deg(f)$.

11.2 Show that $\mathbb{F}_2[x]/\langle x^2 + x + 1\rangle$ is a finite field of order 4.

11.3 Show that, for any odd prime $p$, there are exactly $(p+1)/2$ elements $x \in \mathbb{F}_p$ with $x = y^2$ for some $y \in \mathbb{F}_p$. Hence, conclude that some $x \in \mathbb{F}_p$ is not a **square** in the above sense.

11.4 Describe the elements of the field $\mathbb{F} = \mathbb{Q}(\xi_3)$, where $\xi_3$ is a primitive cube root of unity. Also, write $\mathbb{F}$ as $\mathbb{Q}[x]/\langle f(x)\rangle$, for an appropriate polynomial $f(x)$. What is the dimension of $\mathbb{F}$ over the field $\mathbb{Q}$.

11.5 How many monic degree $d$ irreducible polynomials are there over a field $\mathbb{F}_p$?

11.6 Let $q = p^m$, for some positive integer $m$. How many monic degree $d$ irreducible polynomials are there over a field $\mathbb{F}_q$?

11.7 Let $q = p^m$, for some positive integer $m$. Show that $\mathbb{F}_q^*$ is a cyclic group of order $q - 1$.

11.8 Show that $-1$ is a square modulo an odd prime $p$ if and only if $p = 1 \mod 4$. Also, show that there is some $x \neq 1 \mod p$ so that $x^3 = 1 \mod p$ if and only if $p = 1 \mod 3$.

11.9 Give a proof of Theorem 11.1.

11.10 List the orders of the elements in the field $GF(81)$, and also determine the number of elements that have each allowed order.

11.11 Construct a representation of the field $GF(32)$ using the primitive polynomial $x^5 + x^3 + 1$.

11.12 Let $x, y \in \{0,1\}^n$ be two strings as in (11.2.4). Show that

a. (Positivity) $H(x, y) \geq 0$;
b. $H(x, y) = 0$ if and only if $x = y$;
c. (Symmetry) $H(x, y) = H(y, x)$;
d. (Triangle Inequality) $H(x, z) \leq H(x, y) + H(y, z)$.

11.13 Show that, for any matrix $H \in M_{n\times k}(\mathbb{F}_2)$, the set $C$ given by

$$C = \left\{x \mid x \cdot H = \mathbf{0}_k \mod 2\right\}$$

is a group code with respect to the binary operation $\oplus$.

11.14 Show that the minimum distance $d$ of a linear code $\mathscr{C}$ equals $\mathrm{wt}^*(\mathscr{C})$, the lowest weight nonzero codeword.

11.15 Let $n$ be the length of each codeword in a code $\mathscr{C}$, $k$ be the number of codewords, and $d$ be the minimum Hamming distance between codewords in $\mathscr{C}$. For each of the following codes, find the triple $(n, k, d)$:

a. $\mathscr{C} = \{111, 100, 001, 010\}$.
b. $\mathscr{C} = \{00000, 01111, 10100, 11011\}$.

11.16 Assuming 20-bit strings are used in a company's new $(n, 20, 3)$ error-correcting code, find the least value of $n$ that will permit the code to be used for single bit error correction.

11.17 For any $(n, k)$-block code with minimum Hamming distance $\geq 2t + 1$ between codewords, show that

$$2^{n-k} \geq 1 + \binom{n}{1} + \binom{n}{2} + \cdots + + \binom{n}{t}.$$

11.18 Consider the binary code $\mathscr{C} = \{00100, 10010, 01001, 11111\}$.

a. What is the minimum distance of this code?
b. What is the maximum weight for which the detection of all error patterns is guaranteed?
c. What is the maximum weight for which the correction of all error patterns is guaranteed?
d. Is this code linear? Justify your answer.

11.19 Find the length, dimension, and minimum distance for the binary linear code defined by the parity check matrix

$$H = \begin{pmatrix} 1\,0\,1\,0\,0\,0\,1 \\ 1\,0\,0\,1\,1\,0\,0 \\ 0\,0\,0\,1\,0\,1\,1 \\ 1\,1\,0\,1\,0\,0\,1 \end{pmatrix}$$

Also, find a generator matrix for this code.

## References

1. Lin, S., Costello, D.J.: Error Control Coding: Fundamentals and Applications. Prentice-Hall, Inc., Englewood Cliffs, New Jersey (1983)
2. Huffman, W.C., Pless, V.S.: Fundamentals of Error-Correcting Codes. Cambridge University Press (2003)
3. MacWilliams, F.J., Sloane, N.J.: The Theory of Error-Correcting Codes. North-Holland, Amsterdam (1977)
4. Berlekamp, E.R.: Algebraic Coding Theory. World Scientific Publishing (2015)
5. Peterson, W.W., Weldon, E.J., Tr.: Error-Correcting Code (2/e). MIT Press, Cambridge, MA (1972)
6. Thompson, T.M.: From Error-Correcting Codes through Sphere Packings to Simple Groups. Carus Mathematical Monographs, vol. 21, MAA (1983)
7. Wakerly, J.F.: Error Detecting Codes. Self-Checking Circuits and Applications. Elsevier Science Ltd (1978)
8. Pless, V.S.: Introduction to the Theory of Error-Correcting Codes (3/e). Wiley, New York (1998)
9. Shannon, C.E.: A Mathematical theory of communication. Bell Syst. Tech. J. **27**, 379–423; 623–656 (1948)
10. Shannon, C.E.: Communications in the presence of noise. Proc. IEEE **86**(2), 447–458 (1998)
11. Hamming, R.W.: Error Detecting and Error Correcting Codes. Bell Syst. Tech. J. **29**, 147–160 (1950)
12. Shu, L.: Error Control Coding. Pearson India (2011)

13. van Lint, J.H.: Introduction to Coding Theory (3/e). Springer-Verlag, Berlin (1999)
14. Stallings, W.: Cryptography and Network Security: Principles and Practice (8/e). Pearson Education (2022)
15. Stinson, D.R., Paterson, M.B.: Cryptography: Theory and Practice (4/e). CRC Press (2019)
16. Paar, C., Pelzl, J.: Understanding Cryptography: A Textbook for Students and Practitioners. Springer (2009)
17. Schneier, B.: Applied Cryptography. John Wiley & Sons (2007)
18. Ferguson, N., Schneier, B., Kohno, T.: Cryptography Engineering: Design. Wiley, Principles and Practical Applications (2010)
19. Aumasson, J.-P.: Serious Cryptography: A Practical Introduction to Modern Encryption. No Starch Press (2017)
20. Menezes, A.J., Katz, J., Oorschot, P.C., Vanstone, A.: Handbook of Applied Cryptography. CRC Press (1996)
21. Singh, S.: The Code Book: The Science of Secrecy from Ancient Egypt to Quantum Cryptography. Fourth Estate and Doubleday (1999)

# Chapter 12
# Discrete Optimization

> *Mathematics is not a careful March down a well-cleared highway, but a journey into a strange wilderness, where the explorers often get lost. Rigour should be a signal to the historian that the maps have been made, and the real explorers have gone elsewhere.*
>
> ***W. S. Anglin*** *(1854–1912)*

This chapter introduces some basic concepts of discrete optimization. The emphasis is on providing mathematical description of some optimization algorithms. Convex sets and convex functions play an important role in many different types of optimization problems. We discuss hyperplane separation theorem and the fundamental Farkas lemma of polyhedral combinatorics, Petersen theorem on maximum matchings, and also König-Egerváry theorem. We conclude the chapter with the classical Magner theorem and the max-flow min-cut theorem. In this chapter, we don't dwell on data structures, implementation details, or more sophisticated approaches. A graduate pursuing an advance course in computer science, operation research, or some other related subject area is the intended audience for this chapter. Our presentation is based on the texts [1–3], and also exposition given in the classical three volumes of Schrijver [4]. The reader may refer [5–11], for further details.



V. Ravichandran and A. K. Razdan, *Fundamental Discrete Structures*,
University Texts in the Mathematical Sciences,
https://doi.org/10.1007/978-981-96-0069-4_12

## *Introduction*

A typical *combinatorial problem* involves finding a grouping, ordering, or a set of objects, subject to certain constraints that are usually expressed in terms of available resources. A solution satisfying all the given constraints is called a *feasible solution*. In general, a combinatorial problem may have a large number of feasible solutions about which we only know a descriptive specification. For example, the set of *spanning trees* of a complete graph of order $n$ contains $n^{n-2}$ elements (Theorem 10.19). Also, the set of traveling-salesperson tours $S$ through $n$ points in the Euclidean space $\mathbb{R}^N$ has $(n-1)!/2$ equivalence classes; for instance, notice that to complete a tour of certain fixed total length, it takes any of the $n$ points as the initial point, and ordering of these points may be reversed. The emphasis in most cases is on developing some key mathematical concepts that provide a practically efficient way to avoid enumeration of all possible feasible solutions. We restrict to basic questions such as given below:

(a) How fast can a maximum matching be found in a graph?
(b) What is the best way to traverse a network to get from one point to another as cheaply as possible?
(c) If a network has capacities on arc flows, how can we send as much flow as possible between two points in the network while honoring the arc flow capacities?
(d) If we incur a cost per unit flow on a network with arc capacities and we need to send units of a good that reside at one or more points in the network to one or more other points, how can we send the material at minimum possible cost? In fact, the *minimum-cost flow* problem is one of the most fundamental network flow problems.
(e) Why is the maximum flow equal to the minimum cut?
(f) How to find disjoint flow paths?

For example, to find a *maximum-weight* spanning tree is as difficult as sorting the $\binom{n}{2}$-edge weights. Also, it is notoriously difficult to find a shortest traveling-salesperson tour that visits each point exactly once. This is the famous *traveling-salesperson problem* (or simply TSP). Some other similar types of optimization problems are such as *knapsack problem, job-shop scheduling problem* and the *integer linear programming problem* (or simply ILPP). These types of problems arise frequently in theoretical subject domains such as statistics, physics, and chemistry, and also while dealing with industrial problems concerning *manufacturing, distribution, design, and routing in telecommunication network; crew scheduling in airlines*, etc. A bounded feasible solution that maximizes or minimizes an objective function is called an **optimal solution**. A feasible solution that is not optimal is called an *approximate solution*. An *optimization algorithm* is the one that finds an *optimal solution* to a "minimization problem" or a "maximization problem", subject to certain constraints.

A *minimization problem* (or a *maximization problem*) asks to obtain the minima (or the maxima) of a real-valued *objective function* $f$ defined on the set $S$ consists of all *feasible solutions* to the problem, whenever there exists one. More precisely, once the existence is proven, we need to find an element $s^* \in S$ such that

$$f(s^*) \leq f(s) \quad \text{or} \quad f(s) \leq f(s^*), \quad \text{for all} \quad s \in S.$$

We also say that "$f$ attains a minimum (or a maximum)" on $S$ at the point $s^*$.

When $f$ is a continuous function, and $S$ is a compact set, the existence of minima (or maxima) is guaranteed by some well-known theorems of analysis. Unlike continuous optimization, an optimal solution of a combinatorial problem involves integers or discrete structures.

## 12.1 Polyhedral Combinatorics

In this section, we prove the *Farkas lemma* due to the Hungarian mathematician *Gyula Farkas* (1847–1930) (Theorem 12.4). It is a solvability statement (of the alternative) for convex inequalities, which plays a key role in solving many different types of optimization problems. We begin with an introduction to combinatorial optimization and the theory of polytopes. We first prove *hyperplane separation theorem*. We need some mathematical facts concerning polytopes and linear programming. Recall that an $n$-vector is an element of the vector space $\mathbb{R}^n$. For $n$-vectors $\boldsymbol{x}_1, \ldots, \boldsymbol{x}_d$, the expressions of the form

$$\lambda_1 \boldsymbol{x}_1 + \cdots + \lambda_d \boldsymbol{x}_d; \tag{12.1.1a}$$

$$\lambda_1 \boldsymbol{x}_1 + \cdots + \lambda_d \boldsymbol{x}_d, \quad \text{with} \sum_{k=1}^{d} \lambda_k = 1; \tag{12.1.1b}$$

$$\lambda_1 \boldsymbol{x}_1 + \cdots + \lambda_d \boldsymbol{x}_d, \quad \text{with } \lambda_k \geq 0 \text{ such that} \sum_{k=1}^{d} \lambda_k = 1 \tag{12.1.1c}$$

are, respectively, called a *linear combination*, an *affine combination*, and a *convex combination*, with coefficients $\lambda_1, \ldots, \lambda_d \in \mathbb{R}$. The next definition generalizes these simple concepts.

**Definition 12.1** The set of vectors of the form (12.1.1a) is called the **linear hull** of the $n$-vectors $\boldsymbol{x}_1, \ldots, \boldsymbol{x}_d$; the set of vectors of the form (12.1.1b) is called the **affine hull**; and the set of vectors of the form (12.1.1c) is called the **convex hull.**

Clearly, the *linear hull* contains the origin $\mathbf{0} \in \mathbb{R}^n$. For two $n$-vectors $\boldsymbol{x}_1 \neq \boldsymbol{x}_2$, the linear hull is the plane $\mathbb{R}^2$; their *affine hull* is the line through the two points given by

$$\boldsymbol{x} := \boldsymbol{x}_1 + \lambda(\boldsymbol{x}_2 - \boldsymbol{x}_1), \quad \text{where} \quad -\infty < \lambda < \infty, \tag{12.1.2}$$

and their *convex hull* is the line segment $[\boldsymbol{x}_1, \boldsymbol{x}_2]$. That is, we have

$$[\boldsymbol{x}_1, \boldsymbol{x}_2] := \{\boldsymbol{x}_1 + \lambda(\boldsymbol{x}_2 - \boldsymbol{x}_1) \mid 0 \leq \lambda \leq 1\}. \tag{12.1.3}$$

Notice that (12.1.2) is the vector equation of the line through points $\boldsymbol{x}_1$ and $\boldsymbol{x}_2$. Also, when $0 < \lambda < 1$ in (12.1.3), we have an *open segment*, which we write as $(\boldsymbol{x}_1, \boldsymbol{x}_2)$. If $\mathbf{y} \in (\boldsymbol{x}_1, \boldsymbol{x}_2)$ such that $\mathbf{y} = \lambda \boldsymbol{x}_1 + \nu \boldsymbol{x}_2$ (with $\lambda + \nu = 1$), then we have

$$\frac{\|\boldsymbol{y} - \boldsymbol{x}_1\|}{\|\boldsymbol{y} - \boldsymbol{x}_2\|} = \frac{\nu}{\lambda}. \tag{12.1.4}$$

Similarly, one may define a *left-open segment* and a *right-open segment*. Also, for any three linearly independent vectors in $\mathbb{R}^3$, the linear hull is the whole space; the affine hull is the plane containing the three vectors; and the convex hull is the triangle with the three points as vertices (Fig. 12.1).

**Remark 12.1** We can extend the above three concepts for arbitrary sets $X \subseteq \mathbb{R}^n$. The *convex hull* of $X$, denoted by con. hull$(X)$, is defined as the smallest convex set in $\mathbb{R}^n$ containing $X$. It is easy to see that we have

$$\text{con. hull}\,(X) = \left\{\sum_{k=1}^{t} \lambda_k \boldsymbol{x}_k \mid \boldsymbol{x}_1, \ldots, \boldsymbol{x}_t \in X, \ \text{ with } \sum_{k=1}^{t} \lambda_k = 1 \ \text{ and } \lambda_k \geq 0\right\}. \tag{12.1.5}$$

Similarly, one may define the *affine hull* of $X$, denoted by aff. hull$(X)$. The related verifications are left for the reader as exercises.

**Definition 12.2** The **convex hull** $P$ of a finite set $X = \{\boldsymbol{x}_1, \ldots, \boldsymbol{x}_t\} \subset \mathbb{R}^n$ is called a **polytope** determined by $X$. More specifically, the *polytope* $P$ determined by $X$ is given by

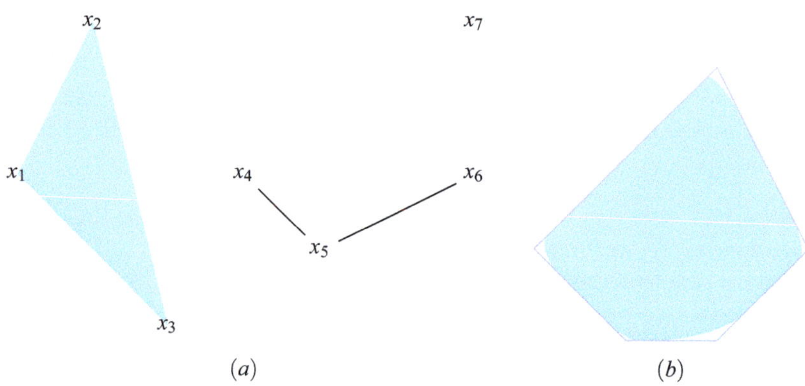


**Fig. 12.1** A regular and non-regular hulls

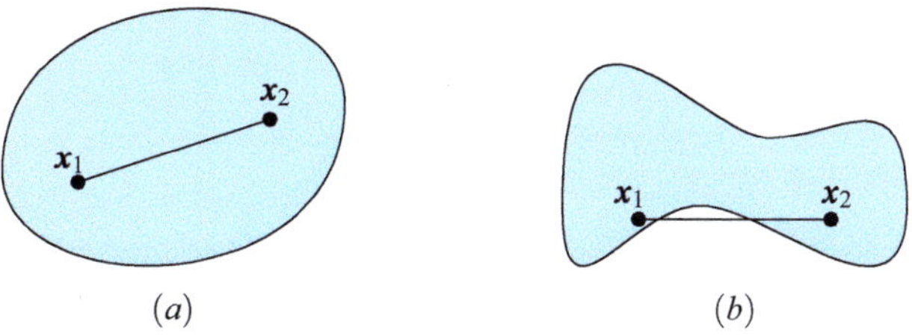


**Fig. 12.2** A convex and non-convex sets

$$P = \left\{ \sum_{k=1}^{t} \lambda_k \boldsymbol{x}_k \mid \lambda_k \geq 0 \text{ and } \sum_{k=1}^{t} \lambda_k = 1 \right\}. \tag{12.1.6}$$

As expected, $P$ is a sharp, corned structure. It is easy to show that a polytope is a compact convex set.

**Definition 12.3** A set $C \subseteq \mathbb{R}^n$ is said to be a **convex set** if, for all $\boldsymbol{x}_1, \boldsymbol{x}_2 \in C$, we have $[\boldsymbol{x}_1, \boldsymbol{x}_2] \subseteq C$.

The set in $(a)$ of Fig. 12.2 is convex, whereas the set in $(b)$ is not. It follows from Definition 12.3 that the *nonempty* intersection of any collection of convex sets is again a convex set. We can also show that, if $\boldsymbol{x}_1$ and $\boldsymbol{x}_2$ are interior points of a convex set $C$, then the line segment $[\boldsymbol{x}_1, \boldsymbol{x}_2]$ is also contained in the interior of the set $C$. Therefore, interior of a convex set is either $\emptyset$ or is itself convex.

**Definition 12.4** A set $\mathscr{H}_n \subset \mathbb{R}^n$ is called a **hyperplane**, with an **offset** $a \in \mathbb{R}$, if for some *nonzero vector* $\mathbf{c} = (c_1, \ldots, c_n) \in \mathbb{R}^n$ we have

$$\mathscr{H}_n = \{\boldsymbol{x} \in \mathbb{R}^n \mid \mathbf{c}^T \boldsymbol{x} = a\}, \tag{12.1.7}$$

where $\mathbf{c}^T \boldsymbol{x}$ is the usual *dot product* and $\mathbf{c}$ is normal to the hyperplane $\mathscr{H}_n$.

For example, a one-dimensional hyperplane is a *point* on the real line; a two-dimensional hyperplane is a *line* in the plane; a three-dimensional hyperplane is a *plane* in the space $\mathbb{R}^3$; and so on. Notice that a line in $\mathbb{R}^3$ is not a hyperplane, because it does not *separate* the space into two parts; indeed, the complement of a line $L \subset \mathbb{R}^3$ is a connected set. In general, each hyperplane $\mathscr{H}_n$ may be seen as the intersection of two (closed) halfspaces given by

$$\mathscr{H}_n^- = \{\boldsymbol{x} \in \mathbb{R}^n \mid \mathbf{c}^T \boldsymbol{x} \leq a\}; \tag{12.1.8a}$$

$$\mathscr{H}_n^+ = \{\boldsymbol{x} \in \mathbb{R}^n \mid \mathbf{c}^T \boldsymbol{x} \geq a\}, \tag{12.1.8b}$$

where $\mathscr{H}_n^-$ is called the *lower halfspace*, and $\mathscr{H}_n^+$ is called the *upper halfspace*. Accordingly, a point $\boldsymbol{x} \in \mathscr{H}_n^-$ is called a *negative point* if we have $\mathbf{c}^T \boldsymbol{x} - a < 0$, and a point $\boldsymbol{x} \in \mathscr{H}_n^+$ is called a *positive point* if we have $\mathbf{c}^T \boldsymbol{x} - a > 0$ (see Fig. 12.3).

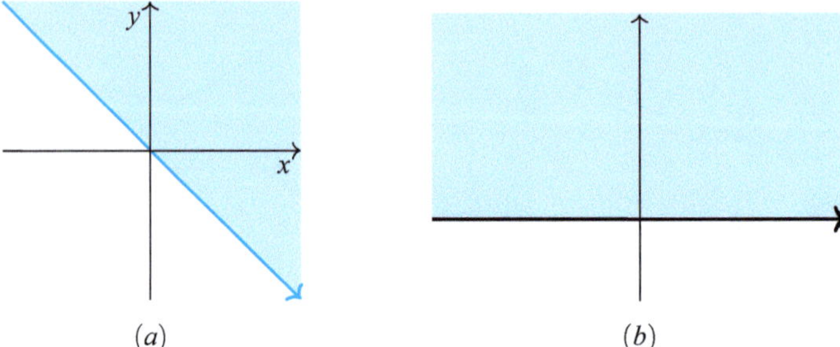


**Fig. 12.3** Two halfspaces in $\mathbb{R}^2$

**Remark 12.2** We can also interpret a *hyperplane* as the level surface of a functional. These play a significant role in classification algorithms such as *support vector machines* (or simply SVMs): *A hyperplane is a "decision boundary" that differentiates the two classes in an SVM*. It also finds applications in linear regression to *separate data points* belonging to different classes, and in *clustering algorithms* to identify clusters of data points in the input space.

Notice that each halfspace is a closed convex set. A basic property of a closed convex set $C$ is that any point $\boldsymbol{x} \notin C$ can be separated from $C$ by a hyperplane $\mathscr{H}_n$. Said differently, if $\boldsymbol{x}$ and $C$ lie separately in two components of the set $\mathbb{R}^n \setminus \mathscr{H}_n$, we say that $\mathscr{H}_n$ separates $\boldsymbol{x}$ and $C$ (Fig. 12.4).

**Theorem 12.1** *Let $C \subset \mathbb{R}^n$ be a closed convex set, and $\boldsymbol{x} \notin C$. There is a hyperplane separation $\boldsymbol{x}$ and $C$.*

***Proof*** We may assume that $C \neq \emptyset$. Otherwise, there is nothing to prove. Let $\boldsymbol{y} \in C$ be a vector that is nearest to $\boldsymbol{x}$. That is, $\|\boldsymbol{x} - \boldsymbol{y}\|$ is minimum. Since $C \neq \emptyset$, there

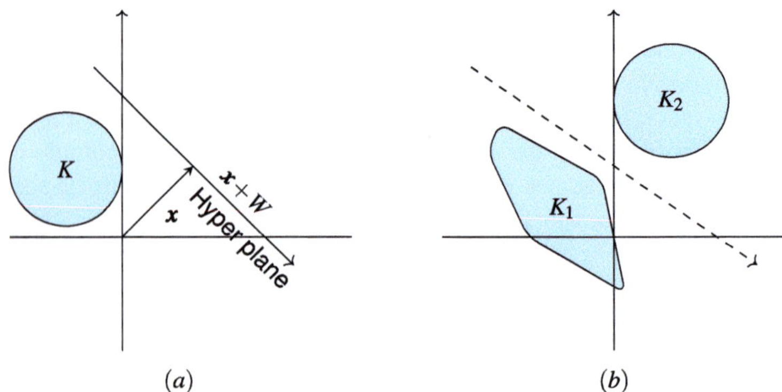


**Fig. 12.4** Hyperplane separation closed convex sets

exists an $r > 0$ such that $\bar{B}(\boldsymbol{x}, r) \cap C \neq \emptyset$. Then $\mathbf{y}$ is the one that minimizes the continuous function $\|\boldsymbol{x} - \mathbf{y}\|$ over the compact set $\bar{B}(\boldsymbol{x}, r) \cap C$. We may take

$$\boldsymbol{c} := \boldsymbol{x} - \boldsymbol{y} \quad \text{and} \quad a := \frac{1}{2}\left(\|\boldsymbol{x}\|^2 - \|\mathbf{y}\|^2\right). \tag{12.1.9}$$

We show that the following two properties hold:

(a) $\mathbf{c}^T \boldsymbol{x} > a$;
(b) $\mathbf{c}^T \mathbf{z} < a$, for each $\mathbf{z} \in C$.

First, since we have

$$\mathbf{c}^T \boldsymbol{x} = (\boldsymbol{x} - \mathbf{y})^T \boldsymbol{x} > (\boldsymbol{x} - \mathbf{y})^T \boldsymbol{x} - \frac{1}{2}\|\boldsymbol{x} - \mathbf{y}\|^2 = a,$$

$(a)$ holds. Also, if $(b)$ is not true, we can find $\mathbf{z} \in C$ such that $\mathbf{c}^T \mathbf{z} \geq a$. Since

$$\mathbf{c}^T \mathbf{y} < \mathbf{c}^T \mathbf{y} + \frac{1}{2}\|\mathbf{c}\|^2 = a,$$

it follows that $\mathbf{c}^T(\mathbf{z} - \mathbf{y}) > 0$. Therefore, for some $\lambda$, with $0 < \lambda \leq 1$, and

$$\lambda < \frac{2\mathbf{c}^T(\mathbf{z} - \mathbf{y})}{\|\mathbf{z} - \mathbf{y}\|^2}, \tag{12.1.10}$$

we may take

$$\mathbf{w} := \lambda \mathbf{z} + (1 - \lambda)\mathbf{y} \tag{12.1.11}$$

so that we have $\mathbf{w} \in C$, and

$$\begin{aligned} \|\mathbf{w} - \boldsymbol{x}\|^2 &= \|\lambda(\mathbf{z} - \mathbf{y}) + (\mathbf{y} - \boldsymbol{x})\|^2 = \|\lambda(\mathbf{z} - \mathbf{y}) - \mathbf{c}\|^2 \\ &= \lambda^2\|\mathbf{z} - \mathbf{y}\|^2 - 2\lambda\mathbf{c}^T(\mathbf{z} - \mathbf{y}) + \|\mathbf{c}\|^2 \\ &< \|\mathbf{c}\|^2 = \|\mathbf{y} - \boldsymbol{x}\|^2. \end{aligned} \tag{12.1.12}$$

Notice that $<$ in the last equation follows from (12.1.10). Since (12.1.12) contradicts the fact that $\mathbf{y}$ is a point in $C$ nearest to $\boldsymbol{x}$, the proof is complete. □

The above theorem provides a characterization of closed convex sets. It also implies that if $C$ is a closed convex set and $\boldsymbol{x} \notin C$, then there exists a halfspace $\mathscr{H}_n^+$ such that $C \subseteq \mathscr{H}_n^+$ and $\boldsymbol{x} \notin \mathscr{H}_n^+$.

**Theorem 12.2** (Hyperplane Separation Theorem) *Let $C, D \subset \mathbb{R}^n$ be two mutually disjoint convex sets. There is a hyperplane $\mathbf{c}^T \boldsymbol{x} - a = 0$ separating $C$ and $D$. That is,*

$$\mathbf{c}^T \boldsymbol{x} - a \geq 0, \ \forall \boldsymbol{x} \in C \quad \textit{and} \quad \mathbf{c}^T \boldsymbol{x} - a \leq 0, \ \forall \boldsymbol{x} \in D. \tag{12.1.13}$$

***Proof*** We give a proof for the case when $C$ and $D$ are closed sets, and one of the two is compact. In this case, strict inequalities in (12.1.13) hold. In view of our assumptions on $C$ and $D$, we can find $\boldsymbol{a} \in C$ and $\boldsymbol{b} \in D$ such that $\|\boldsymbol{a} - \boldsymbol{b}\| = d(C, D)$. Now, we may take the perpendicular bisector between $\boldsymbol{a}$ and $\boldsymbol{b}$ that gives the separating hyperplane. For instance, if it is not, then we can find a point of $C$ (or $D$) on the hyperplane. We thus can construct a point closer to $\boldsymbol{b}$ (or $\boldsymbol{a}$) from the set $C$ (or $D$). □

We know that every closed convex set is the intersection of halfspaces, possibly infinitely many. The next definition introduces an interesting class of closed convex sets, known as the *polyhedra.*

**Definition 12.5** A **polyhedron** is the intersection of finite number of halfspaces.

Said differently, a set $P \subseteq \mathbb{R}^n$ is a *polyhedron* if there is a matrix $A \in M_{m \times n}$, and a vector $\mathbf{b} \in \mathbb{R}^m$, such that

$$P = \{\boldsymbol{x} \in \mathbb{R}^n \mid A\boldsymbol{x} \leq \mathbf{b}\}. \tag{12.1.14}$$

An **extreme point** (or a **vertex**) of a polyhedron is any point that is not a convex combination of any two points of the polyhedron. More generally, according to *Krein-Milman theorem*, every convex, compact nonempty set $C \subseteq \mathbb{R}^n$ can be expressed as the convex hull of its extreme points. In particular, for a feasible region that is a bounded polytope, the optimal value of a linear objective function would always lie at a vertex. The proof is left for the reader as an exercise. Notice that $A\boldsymbol{x} \leq \mathbf{b}$ is equivalent to the following $m$ inequalities:

$$\mathbf{a}_1^T x_1 \leq b_1, \quad \ldots, \quad \mathbf{a}_m^T x_m \leq b_m, \tag{12.1.15}$$

where $\mathbf{a}_1^T, \ldots, \mathbf{a}_m^T$ are the row vectors of the matrix $A$. In the case when $A$ has zero rows, we take $P = \mathbb{R}^n$. The next theorem is fundamental to the theory.

**Theorem 12.3** (Minkowski-Weyl) *Every convex polytope is a polyhedron.*

The terminologies introduced next are needed in the sequel. A point is called the **interior point** of the polyhedron $P$ if it satisfies $m$ inequalities in (12.1.15), but in strict sense. A polyhedron with an *interior point* is called a **full-dimensional polyhedron**. The minimal number of halfspaces needed to describe a *full-dimensional polyhedron* is its essential inequalities. A **facet** of a polyhedron is a set of its points that satisfies an essential inequality as an equality. A **face** $F$ of a polyhedron is a subset of the form:

$$F := \{\boldsymbol{x} \mid \boldsymbol{x} \text{ satisfies a subset of the essential inequalities as equalities}\}.$$

The dimension of a *face* $F$ refers to the dimension of the affine hull of the face. Notice that a vertex of a polyhedron is a unique solution of $m$ linearly independent halfspaces, i.e., a point that satisfies $m$ linearly independent essential inequalities as equalities. The dimension of a face equals

m *minus* the number of equations satisfied.

In particular, a vertex has the dimension 0, and a facet has the dimension $m-1$.

**Definition 12.6** A polyhedron is called a **regular polytope** if all vertices have the same *degree* and all facets have the same number of edges.

Let $f_k$ denote the number of faces of a *regular polytope* of dimension $k \geq 0$. Then the following generalized *Euler formula* holds:

$$\sum_{k=0}^{m-1} (-1)^k f_k = 1 - (-1)^m. \tag{12.1.16}$$

For $m = 3$, we obtain the well-known Euler formula for planar graphs:

$$f_0 - f_1 + f_2 = 2. \tag{12.1.17}$$

It follows from the formula (12.1.17) that, up to a symmetry transformation, *there are only five regular polytopes in dimension three*. For instance, if the degree of each vertex is $v$, and each facet has $e$ edges, then we have

$$f_1 = \frac{e\, f_2}{2} \quad \text{and} \quad f_0 = \frac{2\,e}{v}. \tag{12.1.18}$$

Substituting this in (12.1.17), we obtain

$$f_2 = \frac{4\,v}{2e - (e-2)v}, \quad \text{with} \quad e > 3. \tag{12.1.19}$$

Now, when $e = 3$, we have

$$f_2 = \frac{4\,v}{6 - v} \quad \Rightarrow \quad 1 \leq v \leq 5.$$

As such, the only value $v$ may take is 3, 4, or 5. Also, it follows from (12.1.19) that the only value $f_2$ may take is 4, 8, or 20, because it is an integer. Therefore, in this case, the descriptions $(f_0, f_1, f_2)$ for regular polytopes on the basis of above values are as follows:

$$(4, 6, 4), \quad (6, 12, 18), \quad \text{and} \quad (12, 30, 20).$$

Next, when $e = 4$, we have

$$f_2 = \frac{4\,v}{8 - 2v} \quad \Rightarrow \quad (8, 12, 16).$$

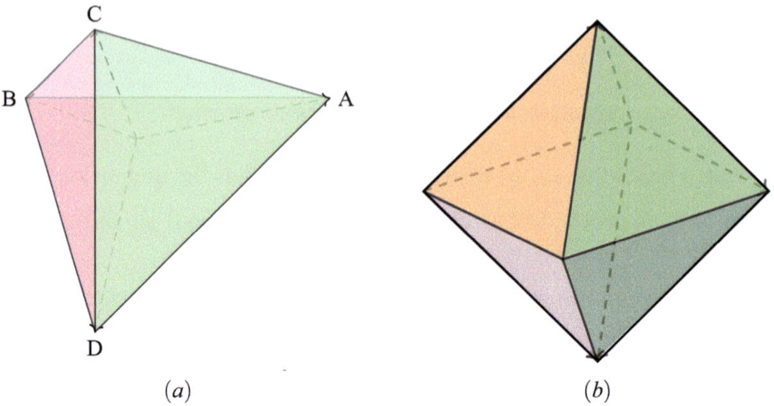


**Fig. 12.5** Regular polytopes with descriptions (4, 6, 4) and (6, 12, 18)

And, for $e = 5$, we have

$$f_2 = \frac{4\,v}{10 - 3v} \quad \Rightarrow \quad \big(20, 30, 12\big).$$

Hence, the only regular polytopes in three dimension have the descriptions as given below:

$$\big(4, 6, 4\big), \ \ \big(8, 12, 16\big), \ \ \big(6, 12, 18\big), \ \ \big(12, 30, 20\big), \ \ \text{and} \ \ \big(20, 30, 12\big). \qquad (12.1.20)$$

These correspond, respectively, to well-known five *platonic solids*, namely, the *tetrahedron, cube, octahedron, dodecahedron*, and the *icosahedron* (see Fig. 12.5). Pythagoras knew of at least three of them and Theaetetus described all five. Much later, Euclid's "Elements" built up geometry and number theory using deductive reasoning. Notice that the cube and the regular octahedron share the same symmetry, as do the regular dodecahedron and icosahedron. Indeed, *symmetry groups* are sometimes named after regular polytopes, for example, the tetrahedral and icosahedral symmetries.

**Definition 12.7** A set $C \subseteq \mathbb{R}^n$ is called a **convex cone** if we have

$$\lambda\,\boldsymbol{x} + \mu\,\mathbf{y} \in C, \quad \text{for all} \ \ \boldsymbol{x},\ \mathbf{y} \in C, \ \ \text{and positive } \lambda,\ \mu \in \mathbb{R}. \qquad (12.1.21)$$

Notice that *convex cone* is a special type of a convex set. For any $X \subset \mathbb{R}^n$, the smallest cone containing $X$, denoted by cone $X$, is the set given by

$$\text{cone}\,\big(X\big) = \left\{ \sum_{k=1}^{t} \lambda_k\,\boldsymbol{x}_k \mid \text{for some } \ \boldsymbol{x}_1, \ldots, \boldsymbol{x}_t \in X, \ \ \text{and non-negative } \lambda_1, \ldots, \lambda_t \right\}. \qquad (12.1.22)$$

The related details are left for the reader as an exercise. A cone $C$ is said to be **finitely generated** if $C = \text{cone}(X)$, for some finite set $X \subset \mathbb{R}^n$.

### *Farkas Lemma*

Let $A$ be an $m \times n$ matrix, and let $\boldsymbol{b} \in \mathbb{R}^n$. Consider the following two systems of linear inequalities:

$$\textbf{System-I:} \qquad A\boldsymbol{x} \le \mathbf{0}, \quad \langle \mathbf{b}, \boldsymbol{x} \rangle > 0, \quad \boldsymbol{x} \in \mathbb{R}^n; \tag{12.1.23a}$$

$$\textbf{System-II:} \qquad A^T\mathbf{y} = \mathbf{b}, \quad \mathbf{y} \ge \mathbf{0}, \quad \mathbf{y} \in \mathbb{R}^m. \tag{12.1.23b}$$

The fundamental solvability theorem related to alternatives for the above two systems is popularly known as the *Farkas lemma* (Theorem 12.4). The content presented in the previous section covers most of the preliminaries needed to prove it. We shall use it to solve some "optimization problems" in the later part of the section. We first consider a simple particular case of the systems given in (12.1.23).

Let $A$ be a square matrix of order two, with row vectors $\mathbf{a}_1$ and $\mathbf{a}_2$. Any vector $\boldsymbol{x}$ in the shaded region satisfies $A\boldsymbol{x} \le 0$. It thus follows that if a vector $\mathbf{b}$ doesn't lie in the acute angle determined by vectors $\mathbf{a}_1$ and $\mathbf{a}_2$, then (12.1.23a) has a solution. It is instructive to draw the hyperplane $\langle \mathbf{b}, \boldsymbol{x} \rangle = 0$, for such a $\mathbf{b}$. On the other hand, for the system equivalent to (12.1.23b) given by $\mathbf{y}^T A = \mathbf{b}^T$, $\mathbf{y}^T \ge \mathbf{0}$, to have a solution the vector $\mathbf{b}$ must lie in the angle determined by vectors $\mathbf{a}_1$ and $\mathbf{a}_2$. This is not possible, because $\mathbf{b}$ is pre-specified, and so its position is fixed. Therefore, exactly one of the two systems has a solution.

**Theorem 12.4** (Farkas Lemma, 1901) *Let $A$ be an $m \times n$ matrix, and let $\mathbf{b} \in \mathbb{R}^n$. Then exactly one of systems of linear inequalities in (12.1.23) has a solution.*

***Proof*** For proof, see [12, p. 64]. □

## 12.2 Matchings and Covers

In this section, we introduce matchings in graphs. We also discuss some applications of matching to other combinatorial-optimization problems. Let $V$ be a nonempty set, and $E$ be the set consisting of 2-subsets of the set $V$, so that $G = (V, E)$ is an undirected simple graph. In all that follows, we refer $G$ simply as a graph. We also recall that a set of vertices $C$ is called a *stable set* (or an *independent set*) of $G$ if we have $e \not\subseteq C$, for every edge $e \in E$.

**Definition 12.8** A stable set is said to be **maximum** if it is not part of any larger stable set of $G$. A stable set is said to be **maximal** if it cannot be extended to a larger stable set.

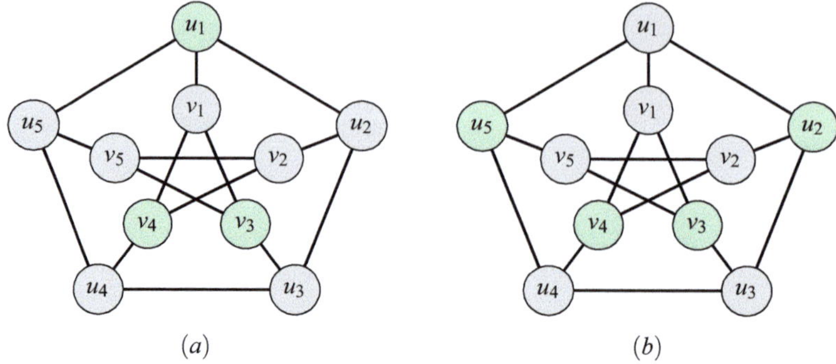


**Fig. 12.6** A maximal and maximum stable sets in the *Petersen graph*

Notice that every *maximum stable set* is also a *maximal stable set*, but the converse may not hold. The stable set of green vertices in Fig. 12.6($a$) is maximal, whereas the stable set of green vertices in Fig. 12.6b is maximum. The next definition introduces a closely related concept.

**Definition 12.9** Let $G = (V, E)$ be a graph. A set of vertices $W \subset V$ is a **vertex cover** (or simply a **covering**) of $G$ if we have $e \cap W \neq \emptyset$, for every edge $e \in E$. A *covering* is said to be a **minimum covering** if no other covering of $G$ has fewer vertices.

In general, for *stable set* any set of vertices $C$ of a graph $G$, we have

$$C \text{ is a} stableset \iff W := V - C \text{ is a covering.} \tag{12.2.1}$$

Therefore, a set of vertices is a *vertex cover* if it includes at least one end vertex of every edge. For example, the blue vertices of the graph shown in Fig. 12.7($a$) are

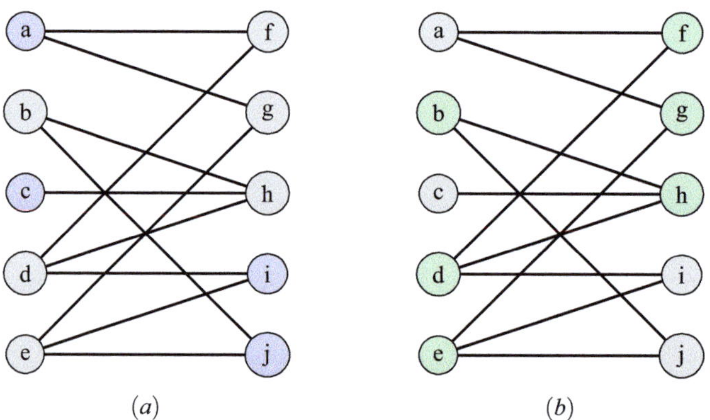


**Fig. 12.7** A stable set (in blue) and a covering (in green) of a bipartite graph

stable, and so the green vertices of the graph shown in Fig. 12.7b form a covering of the underlying bipartite graph, by (12.2.1). Analogously, a set $F \subset E$ is called an **edge covering** of graph $G$ if for each vertex $v \in V$ there is an edge $e \in F$ satisfying $v \in e$. That is, edges in $F$ meet all the vertices of the graph. Therefore, a graph $G$ with an isolated vertex doesn't have an *edge cover*. The concept of a *matching* of a graph introduced in the next definition is closely related to the above notions.

**Definition 12.10** Let $G = (V, E)$ be a graph. A set $M \subset E$ is called a **matching** if no two edges in the set $M$ are adjacent, i.e., $e \cap e' = \emptyset$, for all $e \neq e' \in M$. In a weighted graph, a maximum-weight *matching* is a matching such that the sum of edge weights is maximum.

Equivalently, a set $M \subset E$ is a *matching* if no two edges in $M$ lie on the same path of length $\geq 2$. For a graph $G$, we write

$$\begin{aligned} \alpha(G) &:= \max\{|C| \mid C \text{ is a stable set}\}; \\ \tau(G) &:= \min\{|W| \mid W \text{ is a covering}\}; \\ \nu(G) &:= \max\{|M| \mid M \text{ is a matching}\}; \\ \rho(G) &:= \min\{|F| \mid F \text{ is an edge covering}\}. \end{aligned} \tag{12.2.2}$$

The first three of above numbers are, respectively, called the *stability number*, *covering number*, and the *matching number*, of the graph $G$. Now, since at least one endpoint of every edge $e \in M$ appears in a cover, each vertex cover must contain at least as many elements as any matching set of the graph $G$. The following two inequalities thus hold:

$$\nu(G) \leq \tau(G) \quad \text{and} \quad \alpha(G) \leq \rho(G). \tag{12.2.3}$$

The graph $G = K_3$ shows that strict inequalities are possible. For a proof of the second inequality, consider a matching $M$ of graph $G$. By definition, every covering includes vertices that at least touch edges in $M$. Also, since the edges are disjoint, each vertex can cover at most one edge in $M$. The Hungarian mathematician *Tibor Gallai* (1912–1992) proved in 1959 that equality of one of two inequalities in (12.2.3) implies the equality of the other.

**Theorem 12.5** (Gallai's Theorem) *For any graph $G = (V, E)$ with no isolated vertex, we have*

$$\alpha(G) + \tau(G) = |V| = \nu(G) + \rho(G). \tag{12.2.4}$$

***Proof*** The first equality follows directly from (12.2.1). To prove the second equality, let $M$ be a matching of size $\nu(G)$. For each of the $|V| - 2|M|$ vertices $v$ missed by $M$, add to $M$ an edge covering $v$. Accordingly, we obtain an edge cover of size

$$|M| + (|V| - 2|M|) = |V| - |M|.$$

Therefore, $\rho(G) \leq |V| - \nu(G)$. Second, let $F$ be an edge cover of size $\rho(G)$. For each $v \in V$ delete from $F$, $d_F(v) - 1$ edges incident with $v$. We obtain a matching of size at least

$$|F| - \sum_{v \in V} (d_F(v) - 1) = |F| - (2\,F| - |V|) = |V| - |F|.$$

Hence, $\nu(G) \geq |V| - \rho(G)$. □

The above proof also shows that the smallest edge covering can be derived from a maximum matching, and vice versa. Recall that the *clique number* of a graph $G$ is the maximal size clique in $G$, which is denoted by $\omega(G)$. We also have $\omega(G) = \alpha(\overline{G})$, where $\overline{G}$ is the complement of the graph $G$ (Definition 10.16).

**Definition 12.11** Let $G = (V, E)$ be a graph. A matching $M \subset E$ is called a **maximum matching** if no other matching of $G$ has more edges. That is, a *maximum matching* $M$ contains maximum number of edges.

The problem of "maximum cardinality matching" is about finding a matching of maximum cardinality.

**Definition 12.12** Let $M$ be a matching of a graph $G = (V, E)$. We say $M$ **covers** a vertex $v \in V$ if it is an end vertex of an edge $e \in M$, i.e., $v \cap e \neq \emptyset$, for some $e \in E$. In this case, we also say $v$ is **matched** with respect to $M$. A set of vertices $S \subseteq V$ is said to be covered if $M$ *covers* every vertex $v \in S$. A vertex $v \in V$ is called **M-free** (or *unmatched*) if $M$ doesn't cover $v$.

For example, the matching $M$ (in blue) shown in Fig. 12.8a covers the set of vertices $S = \{u, v, w, x, y, z\}$, but the vertex $t$ is $M$-free. Also, the matching $M$ (in blue) shown in Fig. 12.8b covers the set of vertices $S = \{t, u, v, w, y, z\}$, but the vertex $x$ is $M$-free.

**Definition 12.13** Let $G = (V, E)$ be a graph. A matching of $G$ that is not contained in any larger matching of $G$ is called a **maximum matching**. A *maximum matching* $M$ of $G$ that covers every vertex $v \in V$ is called a **complete matching** (or a **perfect matching**).

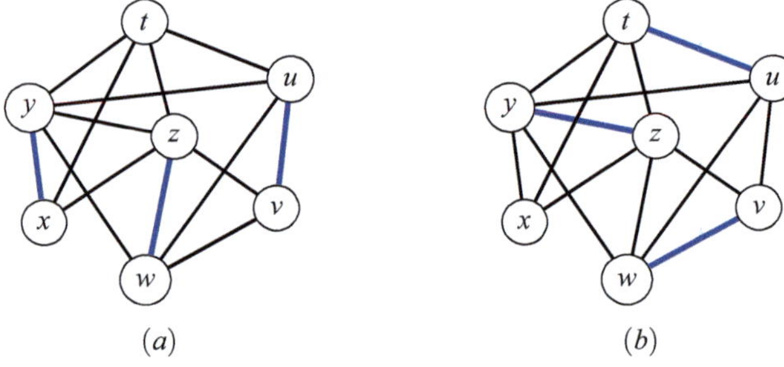


**Fig. 12.8** Two maximum matchings in a graph

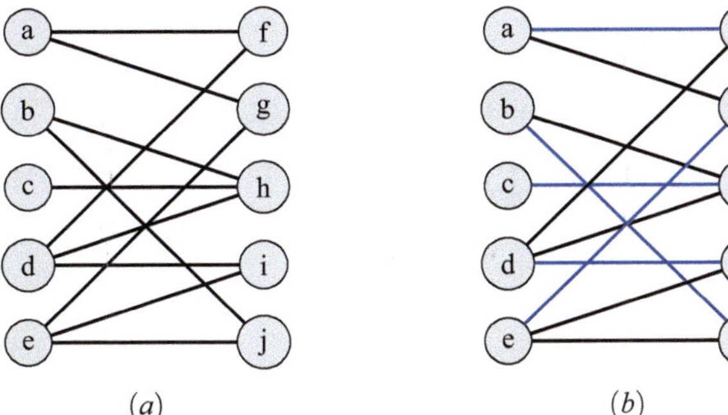


**Fig. 12.9** A bipartite graph and a complete matching (in blue)

Said differently, a *complete matching* of a graph is a maximum matching $M$ that covers all vertices. Therefore, we have $|M| = \big(1/2\big)|V|$. Our aim here is to develop a fast algorithm[1] to find a maximum matching in an $(n, m)$-graph, where $n$ is the *size* of the input graph.

In general, not every graph has a *complete matching*. A complete matching $M$ of a graph of order $n$ has $n/2$ edges, and so it is a *maximum cardinality matching*. The matchings shown in Fig. 12.8 are maximum cardinality matching. Also, the graphs shown in Fig. 12.9b display some complete matchings. Some applications of *perfect matching* in chemistry include *storing information, estimating bond lengths, estimating resonance energy*, and many more.

## *Augmenting Paths*

In all that follows in this part, we describe a "fast" algorithm for finding a maximum matching of a graph $G$. The concept introduced in the next definition is fundamental to our subsequent discussion.

**Definition 12.14** Let $G = (V, E)$ be a graph, and let $M$ be a matching. A path (or a circuit) in $G$ whose edges alternately belong to $M$ and $E - M$ is called an **M-alternating path** (or an **M-alternating circuit**). An **M-augmenting path** in $G$ is an *M-alternating path* between $M$-free vertices.

Notice that the length of an *M-augmenting path* is odd. If $G$ contains an $M$-augmenting path $P$, then we can find a matching $M'$ such that we have

$$\big|M'\big| = \big|M\big| + 1. \tag{12.2.5}$$

[1] By *fast* we mean the run time of algorithm is bounded by a fixed polynomial $f(n)$.

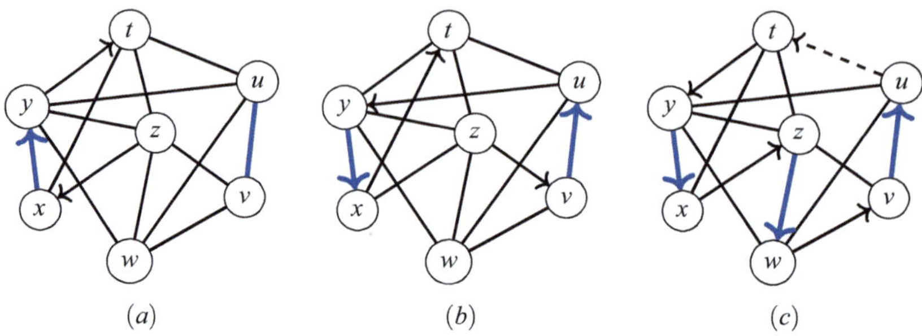


**Fig. 12.10** Three augmenting paths relative to matching $M$ (in blue)

For instance, take $M'$ as the *symmetric difference*

$$M' = M \oplus P := \big(M - P\big) \cup \big(P - M\big) = \big(M \cup P\big) - \big(M \cap P\big). \qquad (12.2.6)$$

Notice that $M'$ reverses the roles of the edges in the path $P$. That is, we first delete the edges of $P$ from the set $M \cap P$ and then add to it the remaining edges on $P$. Therefore, the augmentation process *increases the cardinality of the given matching by one*. For example, the graph in Fig. 12.10a has an $M$-augmenting path $P$ (with arrows) such that we have $M \oplus P = \{zx, yt, uv\}$, and the graph in Fig. 12.10b has an $M$-augmenting path $P$ (with arrows) such that we have $M \oplus P = \{xt, vz, uy\}$. However, for the matching $M$ in Fig. 12.10c (shown in blue), the graph has no $M$-augmenting path.

In this part of the section, we discuss *augmenting paths*. The main idea is to *augment* a non-complete matching $M$ of a graph $G$ to a greater cardinality or to conclude that $M$ is already maximum. One way is to find a path $P$ in the graph $G$ that starts and ends at unmatched vertices, and moves alternately along unmatched edges and matched edges. We then consider the set of edges $M'$ obtained by deleting the edges in $M \cap P$ and adding the rest of the edges on $P$, i.e., we consider the symmetric difference $M \oplus P$. It follows easily that $M'$ is also a matching that has one more edge than the matching $M$. Such a path is called an **augmenting path**. The next theorem due to the Danish mathematician *Julius Petersen* (1839–1910) says that a matching $M$ of greater cardinality is obtained by repeating the above procedure when $M$ has no augmenting paths.

**Theorem 12.6** *(*Petersen, 1891*) Let $G = (V, E)$ be a graph, and let $M$ be a matching. Then $M$ is maximum-cardinality matching if and only if there is no $M$-augmenting path in $G$.*

***Proof*** If $G$ has no $M$-augmenting path, then the matching $M$ is necessarily maximum cardinality, by (12.2.5). To prove the converse, suppose a matching $M$ doesn't contain an augmenting path. We show that if $M'$ is a maximum matching then we necessarily have $|M| = |M'|$. For instance, consider the *symmetric difference* $M \oplus M'$. Since $M$ and $M'$ both induce subgraphs of maximum degree one, it follows that $M \oplus M'$

induce subgraph of maximum degree two. Such a subgraph may consist only of disjoint paths and/or cycles. Also, these paths and cycles contain edges that are alternately in matchings $M$ and $M'$. First, consider the case when there are cycles in the induced subgraph, which must contain an even number of edges. Otherwise, there must be some vertex that is adjacent to two edges in either $M$ or $M'$, which contradicts the fact that both $M$ and $M'$ are matchings. It thus follows that such cycles contain an equal number of edges from $M$ and $M'$. We next consider the case when there are paths in the induced subgraph. Suppose $P$ is one such path of *odd length*. The additional edge $P$ contains is from $M$ or $M'$. In the former case, it contradicts the maximality of $M'$, because $P$ is then an $M'$-augmenting path in $G$. In the latter case, it contradicts the maximality of $M$, for the same reason. Therefore, the paths and cycles induced by $M \oplus M'$ contain an equal number of edges from $M$ and $M'$. Finally, since edges that are not induced by $M \oplus M'$ are either in $M$ and $M'$ or in neither of them, it follows that $M$ and $M'$ are of equal cardinality. Hence, $M$ is a maximum matching. □

We can thus find a maximum-cardinality matching $M$ in $G$ by repeatedly extending an $M$-augmenting path. The related algorithm can be implemented on a bipartite graph $G = (U, V)$ using the method due to van der Waerden [13] and *Dénes König* [14]. For a given matching $M \subset E(G)$ of $G = (U, V)$, construct the bipartite digraph $G_M := (U, V, A)$, where

$$A := \{\bar{e} \mid e \in M\} \cup \{f \mid f \in E(G) - M\}.$$

That is, all edges in $M$ are (reverse) oriented from $V$ to $U$, and the rest in $E - M$ from $U$ to $V$. We then have that every dipath in $G_M$ starting from an $M$-free vertex in $U$ and ending in an $M$-free vertex in $V$ corresponds to an $M$-augmenting path in $G$. With $|U| \le |V|$, we find at most $|U|$ such types of paths. Since each dipath can be identified in at most $O(|A|)$ time by *breadth-first search*, the time complexity of this algorithm is $O(|U||A|)$. This run time can be improved to $O(|U|^{1/2}|A|)$ if, instead of repeatedly augmenting the matching $M$ along a single $M$-augmenting path, we augment $M$ simultaneously along a collection of disjoint $M$-augmenting paths. Since such paths can be found in $O(|A|)$ time, by reasoning on the lengths of the alternating paths, it follows that the algorithm needs only $O(|U|^{1/2})$ iterations. Therefore, the total time complexity of the algorithm is $O(|U|^{1/2}|A|)$ [15].

## *Bipartite Matching Algorithm*

Let $G = (V_1, V_2)$ be a *bipartite graph*, and $W \subset V_1 \cup V_2$. The set of vertices adjacent to the vertices in $W$ is denoted by $N(W)$. That is, we have

$$N(W) := \{v \in V_1 \cup V_2 \mid uv \in E, \ \text{for some } u \in W\}. \tag{12.2.7}$$

Notice that, for any $W \subseteq V_1$, we have $N(W) \subseteq V_2$.

**Definition 12.15** For a bipartite graph $G = (V_1, V_2)$, we say $M \subset E(G)$ is a **matching from $\mathbf{V_1}$ to $\mathbf{V_2}$** if it covers every vertex $v \in V_1$.

If $G = (V_1, V_2)$ contains a complete matching, then it is a maximum cardinality matching. The next theorem is also known as Hall's **marriage theorem**. It characterizes bipartite graphs to have a *complete matching*.

**Theorem 12.7** (Frobenius-Hall) *Given a bipartite graph $G = (V_1, V_2)$, there exists a complete matching $M$ from $V_1$ to $V_2$ if and only if*

$$\left|N(A)\right| \geq \left|A\right|, \qquad \textit{for all} \quad A \subseteq V_1. \tag{12.2.8}$$

***Proof*** We first suppose that $M$ is a complete matching from $V_1$ to $V_2$, and $A \subseteq V_1$. Then for every vertex $v \in A$ there is an edge $e \in M$, with $v \in V_1$ as one end vertex. Therefore, there are at least as many vertices in $V_2$ that are neighbors of vertices in $V_1$ as there are vertices in $V_1$. Hence, the inequality (12.2.7) holds. Conversely, suppose the inequality given in the statement holds. We shall use *strong form of mathematical induction* on $\left|V_1\right|$ to find a complete matching from $V_1$ to $V_2$. Now, if $\left|V_1\right| = 1$, then we have $V_1 = \{v_0\}$, so that

$$\left|N(v_0)\right| \geq \left|\{v_0\}\right| = 1 \quad \Rightarrow \quad V_2 = \{w_0\},$$

and the edge $\{v_0w_0\}$ forms a complete matching. Our *inductive hypothesis* in this case is as follows: Suppose for any bipartite graph $G = (V, E)$, with a bipartition $(V_1, V_2)$ such that $|V_1| = j \leq k$ $(k > 0)$, there exists a complete matching from $V_1$ to $V_2$ whenever the inequality (12.2.7) holds. Now, let $G' = (v', E')$ be a bipartite graph with a bipartition $(X, Y)$ such that $|X| = k + 1$. We shall use the method of "proof by cases" to complete the inductive step.

**Case-1 :** Suppose that for all integers $j$, with $1 \leq j \leq k$, the vertices in every subset of $j$ vertices in $X$ are adjacent to at least $j + 1$ vertices in $Y$. Selecting a vertex $v \in X$, by (12.2.7), we can find a vertex $w \in Y$ such that $w \in N(\{v\})$. Delete all edges from the graph $G'$ that are adjacent to these two vertices. We thus obtain the bipartite graph $H$ with bipartition $(X - \{v\}, Y - \{w\})$. Since $|X - \{v\}| = k$, the induction hypothesis tells us that there is a complete matching from $X - \{v\}$ to $Y - \{w\}$. Adding to this the edge from $vw$ we obtain a complete matching from $X$ to $Y$.

**Case-2 :** Suppose that for some integer $j$, with $1 \leq j \leq k$, there is a set $X_1$ of $j$ vertices such that there are exactly $j$ neighbors of these vertices in $X$, which we may write as $Y_1$. Once again, by our induction hypothesis, there is a complete matching from $X_1$ to $Y_1$. Removing these $2j$ vertices, and also the incident edges, we obtain a bipartite graph $K$ with bipartition $(X - X_1, Y - Y_1)$. Notice that, if the graph $K$ does not satisfy the inequality

$$\left|N(A)\right| \geq \left|A\right|, \quad \text{for all} \quad A \subseteq X - X_1,$$

then we can find a set of $t$ vertices in $X - X_1$, where $1 \leq t \leq k + 1 - j$, such that these have fewer than $t$ vertices in $Y - Y_1$ as neighbors. This implies that the set of $j + t$ vertices of $X$ consisting of these $t$ vertices together with the $j$ vertices we removed earlier from $X$ has fewer than $j + t$ neighbors in $Y$. This contradiction completes the proof. □

The next theorem due to the Hungarian mathematician *Dénes König*[2] (1884–1944) published in 1931 is equivalent to the classical Theorem 12.7. In the same year, another Hungarian mathematician *Jenő Egerváry* (1891–1958) published a generalization for the weighted graphs. Recall that a set of vertices of a graph $G$ is called a **covering** (or a ) if it includes at least one end vertex of every edge. Also, a is called a **minimum covering** if no other covering has fewer vertices.

**Theorem 12.8** (König-Egerváry Theorem) *For a bipartite graph $G$, the size of a maximum matching in $G$ is equal to the size of a minimum vertex cover of the graph.*

***Proof*** For a proof see [16]. □

## 12.3 Menger's Theorem, Flows, and Circulations

In this section, we discuss the maximum-flow problem for digraphs and related cut problems. Although the topic seems less intricate than those of the two previous sections, we discuss the seminal method of Edmonds and Karp that is used to produce an efficient algorithm. Our focus in this part is to prove an important result due to the Austrian-American mathematician and economist *Karl Menger* (1902–1985). In 1927, he proved that the size of the minimum *uv-vertex cut* for any pair of distinct non-adjacent vertices $u$ and $v$ in a (finite or infinite, undirected or directed) graph equals the maximum number of pairwise disjoint $uv$-paths.

### *Menger's Theorem*

We restrict our discussion to (finite) undirected graphs. Of course, we need some preparation.

**Definition 12.16** Let $G = (V, E)$ be an undirected graph. For two not necessarily disjoint sets $A, B \subseteq V$ (possibly empty), an **AB-vertex separator** (or simply **AB-separator**) is a set of vertices such that the vertex-deleted subgraph so obtained contains no path from $A$ to $B$. Each vertex in $A \cap B$ is treated as a path of length zero from $A$ to $B$.

---

[2] Dénes König wrote the first textbook on graph theory.

In particular, for a pair of non-adjacent vertices $u \neq v \in V$, a **uv-vertex-cut** is a set $S \subset V - \{u, v\}$ such that the vertices $u$ and $v$ belong to different components of the subgraph $G - S$. In this case, we also say that $S$ **separates vertices** $u$ and $v$.

**Definition 12.17** The minimum size of a vertex-cut separating non-adjacent vertices $u$ and $v$, denoted by $c(u, v)$, is called the **local cut** of $G$. The function $c(u, v)$ is undefined when $u = v$ or $u$ and $v$ are adjacent.

A vertex-cut separating some pair of non-adjacent vertices of $G$ is a vertex cut of $G$, and one with $k$ elements is a **k-vertex cut**. Similarly, we can define **ef-edge cut** for two non-adjacent edges $e, f \in E$. Also, a **uv-edge separator** is a set of edges that when removed separate $u$ and $v$. For example, the complete graph $K_n$ has no vertex cuts, nor does any multigraph whose underlying simple graph is complete.

**Definition 12.18** Let $G = (V, E)$ be an undirected graph, and $\pi_1, \pi_2$ be two paths in $G$. We say $\pi_1$ and $\pi_2$ are **disjoint paths** if they are vertex disjoint, i.e., they don't share any vertex. Also, paths $\pi_1$ and $\pi_2$ are called **internally disjoint** (or **independent paths**) if they don't share any vertex, other than possibly their endpoints.

**Theorem 12.9** (Vertex Version) *Let $u$ and $v$ be two (distinct) non-adjacent vertices in an undirected graph $G = (V, E)$. The minimum number of vertices in a uv-separator equals the maximum number of internally disjoint uv-paths in $G$.*

The statement of the analogous **edge version** of the above theorem may be obtained by replacing "vertex cut" with "edge cut", and making other appropriate changes. The two versions of *Menger's theorem* are special cases of *max-flow min-cut theorem* that we prove in the next part (Theorem 12.10). Many different proofs of Theorem 12.9 are known [17–19].

## *Flows and Networks*

A **flow network** (or simply a **flow**) is a pair $(D, f)$, where $D = (V, A)$ is a digraph and $f : A \to \mathbb{R}$ is a non-negative function such that each arc $a \in A$ receives a . In a pictorial representation of a *flow network* $(D, f)$ such as in Fig. 12.11, the arcs $a \in A$ are labeled with non-negative capacities $f(a)$, and so we may write a flow simply as $D$. A node $s \in V$ that has only out-flow is called a **source node**, and a node $t \in V$ that has only in-flow is called a **sink node**. Every flow network $D$ satisfies the fundamental (conservation) condition: *The amount of in-flow and out-flow at each vertex $v \in V - \{s, t\}$ of the flow $D$ must be the same.*

Symbolically, we write

$$f\big(d^{in}(v)\big) = f\big(d^{out}(v)\big), \qquad \text{for each} \quad v \in V - \{s, t\}. \tag{12.3.1}$$

We may call $f$ an **s-t−flow** or (or a **flow from s to t**). Since it specifies the amount of flow that is allowable through an arc in $A$, the function $f$ is also known as the

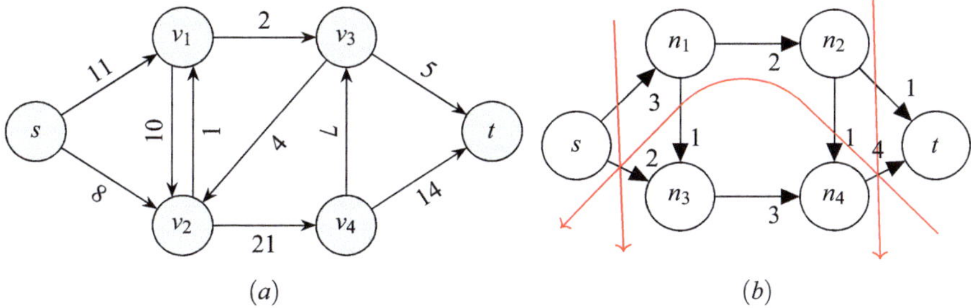


**Fig. 12.11** A pictorial representation of a flow network

**capacity function** of the flow $D$. Let $d(a)$ stand for the *demand* and $c(a)$ for the *capacity* of an arc $a \in A$. We say $f$ is a **feasible flow** in a network if

$$0 \leq d(a) \leq f(a) \leq c(a), \qquad \text{for each} \quad a \in A. \tag{12.3.2}$$

The **value** of an *s-t–flow* $f$ is the net amount of flow leaving *source node* $s$. That is, we have

$$\text{value}\,(f) := f\big(d^{out}(s)\big) - f\big(d^{in}(s)\big). \tag{12.3.3}$$

For a specified **weight function** (or a **cost function**) $w : A \to \mathbb{R}$, and a dipath (or a circuit) $P$ of a flow $(D, f)$, we may write $w(P) = \sum_{a \in P} w(a)$. The **weight** of an *s-t–flow* $f$ is defined as

$$\text{weight}\,(f) := \sum_{a \in A} w(a) f(a). \tag{12.3.4}$$

We say a *feasible flow* $f$ is **minimum-weight flow** if weight $(f) \leq$ weight $(g)$, for any feasible flow $g$ in the flow network $D$. The *minimum-weight flow* problem is to find a minimum-weight $s$-$t$ – flow in $D$.

Some basic concepts of the *flow theory* we introduce here are helpful in solving problems concerning traffic in a computer network, circulation with demands, fluids in pipes, currents in an electrical circuit, or anything similar in which something travels through a network of nodes. Our purpose here is to prove *max-flow min-cut* theorem, and discuss some of its applications. We start with the next definition.

**Definition 12.19** Let $D$ be a flow with a source node $s$ and a *sink node* $t$. A **capacity cut** of $D$ is a partition $\{S, T\}$ of the vertices of $D$ such that $s \in S$ and $t \in T$. The capacity of a *capacity cut* is the sum total of all the capacities of arcs from $S$ to $T$.

The next theorem is due to the American mathematicians *Lester R. Ford Jr.* (1927–2017) and *Delbert R. Fulkerson* (1924–1976) [20]. It is known as the *max-flow min-cut theorem*. For a historical account see [21].

**Theorem 12.10** (Max Flow Min Cut) *The maximum of a* feasible flow *on a digraph $D$ equals the minimum capacity cut of the flow $D$. Moreover, if the capacities of $D$ are integers, then there is a maximal flow with integer values.*

Many different proofs of Theorem 12.10 are known [22–24].

## Exercises 12

12.1. Give a proof of the equality in (12.1.6).

12.2. Show that convex hull of a finite set is always a compact set.[*HINT*: Show that it is the image under a continuous function of a compact set.]

12.3. Show that every polytope is a convex, compact set.

12.4. Show that every halfspace is a convex set.

12.5. Show that a set $C \subseteq \mathbb{R}^n$ is closed convex if and only if $C$ is the intersection of a collection of halfspaces.

12.6. Let $A$ be an $m \times n$ matrix, and let

$$C := \{\mathbf{w} \in \mathbb{R}^m \mid \mathbf{w} = A\boldsymbol{x}, \text{ where } \boldsymbol{x} \in \mathbb{R}^n, \text{ with } \boldsymbol{x} \geq \mathbf{0}\}.$$

Show that $C$ is a closed convex cone.

12.7. Let $f : \mathbb{R}^n \to \mathbb{R}$ be a functional given by $f = \boldsymbol{a}^t\boldsymbol{x}$, and $P$ is a bounded polytope. Show that there always exists an extreme point of $P$ that attains the minimum/maximum value of $f$ on $P$.

12.8. Show that any closed convex set is the intersection of all the half spaces which contain it.

12.9. Let $G = (V, E)$ be a bipartite graph, with a bipartition given by $(V_1, V_2)$, and suppose that the inequality $\deg(u) \geq \deg(v)$ holds, for all $u \in V_1$ and $v \in V_2$. Prove that $V_1$ has a complete matching into $V_2$.

12.10. Show that every regular bipartite graph has a complete matching.

12.11. For $S = [n] := \{1, 2, \ldots, n\}$, let $V_k$ be the set of all subsets of $S$ of cardinality $k$ so that we have $|V_k| = \binom{n}{k}$. Let $k < n/2$, and consider the bipartite graph $G_k := (V_k, V_{k+1})$, where

$$E := \{(u, v) \mid u \in V_k,\ v \in V_{k+1}, \text{ and } u \subset v\}.$$

Prove that the maximum matching in $G_k$ has size $V_k$. Also, prove *Sperner's lemma*: *The maximum number of subsets of $S$ such that no subset is contained into another is* $\binom{n}{\lfloor n/2 \rfloor}$.

# References


1. Cook, W., Cunningham, W., Pulleyblank, W., Schrijver, A.: Combinatorial Optimization. Wiley (1997)
2. Korte, B., Vygen, J.: Combinatorial Optimization: Theory and Algorithms, Algorithms and Combinatorics, Volume-21. Springer, Berlin Heidelberg, New York (2012)
3. Lee, J.: A First Course in Combinatorial Optimization. Cambridge University Press (2004)
4. Schrijver, A.: Combinatorial Optimization - Polyhedra and Efficiency. Springer, Berlin (2003)
5. Nemhauser, G., Wolsey, L.: Integer and Combinatorial Optimization. Wiley (1988)
6. Papadimitriou, C., Steiglitz, K.: Combinatorial Optimization: Algorithms and Complexity. Prentice-Hall (1982)
7. Lovász, L., Plummer, M.P.: Matching Theory, North-Holland Mathematics Studies. North-Holland, Amsterdam (1986)
8. Bertsimas, D., Tsitsiklis, J.N.: Introduction to Linear Optimization, Athena Scientific (1997)
9. Ahuja, R.K., Magnanti, T.L., Orlin, J.B.: Network Flows - Theory, Algorithms, and Applications. Prentice Hall, Englewood Cliffs, New Jersey (1993)
10. Rossi, F., van Beek, P., Walsh, T. (eds.): Handbook of Constraint Programming. Elsevier (2006)
11. Lawler, E.L.: Combinatorial Optimization: Networks and Matroids. Holt, Rinehart and Winston, New York (1976)
12. Berkovitz, L.: Convexity and Optimisation in $\mathbb{R}^n$. Wiley, New York (2002)
13. van der Waerden, B.L.: Ein Satz über Klasseneinteilungen von endlichen Mengen. Abhandlungen aus dem mathematischen Seminar der Hamburgischen Universität **5**, 185–188 (1927)
14. König, D.: Graphok és matrixok. Matematikai és Fizikai Lapok **38**, 116–119 (1931)
15. Hopcroft, J.E., Karp, R.M.: An $n^{5/2}$ algorithm for maximum matchings in bipartite graphs. SIAM J. Comput. **2**(4), 225–231 (1973)
16. Bondy, J.A., Murthy, U.S.R.: Graph Theory with Applications. North Holland, New York (1976)
17. Diestel, R.: Graph Theory (5/e). GTM-173, Springer (2017)
18. Göring, F.: Short proof of Menger's theorem. Discret. Math. **219**(1–3), 295–296 (2000)
19. Aharoni, R.: Menger's theorem for graphs containing no infinite paths. Eur. J. Comb. **4**(3), 201–204 (1983)
20. Ford Jr., L.R., and Fulkerson, D.R., *Maximal flow through a network*, Canadian Journal of Mathematics, Vol-8, 399–404, 1956
21. Ford Jr., L.R., Fulkerson, D.R.: Flows in Networks, p. 1. Princeton University Press (1962)
22. Elias, P., Feinstein, A., Shannon, C.E.: A note on the maximum flow through a network. IRE Trans. Inf. Theory **2**(4), 117–119 (1956)
23. Dantzig, G., Fulkerson, D.R.: On the Max-Flow MinCut theorem of networks–in linear inequalities. Annals of Mathematics Studies, vol. 38. Princeton, New Jersey (1956)
24. Ford Jr., L.R., Fulkerson, D.R.: A simple algorithm for finding the maximum network flows and an application to the Hitchcock problem. Can. J. Math. **9**, 210–18 (1957)

# Appendix
# Supplements

*Mathematics directs the flow of the universe, lurks behind its shapes and curves, holds the reins of everything from tiny atoms to the biggest stars.*

**Edward Frenkel**, In: *Love and Math—The Heart of Hidden Reality*

## A.1 Natural Numbers

In this appendix, we recall some important facts of elementary number theory that are used in various chapters of the main text. We don't have any knowledge who first used the **natural numbers** $1, 2, 3, \ldots$ as mathematical mean to "count objects". However, the existence and properties of the set $\mathbb{N}$ of natural numbers *Natural numbers* are based on the following five "axioms":

(i) $\mathbb{N}$ is a nonempty set.
(ii) Each $n \in \mathbb{N}$ has a *unique* successor, say $\hat{n}$.
(iii) There exists an $n^* \in \mathbb{N}$ that is not the successor of any $n \in \mathbb{N}$.
(iv) For $n_1 \neq n_2 \in \mathbb{N},\ \hat{n}_1 \neq \hat{n}_2$.
(v) The only subset $K \subseteq \mathbb{N}$ that contains $n^*$, and the successors of all elements in $K$, is the set of natural numbers $\mathbb{N}$.

These are known as Peano's axioms *Peano's axioms*, named after the Italian mathematician Giuseppe Peano *Giuseppe Peano* (1858–1932). They first appeared in the pamphlet "*Arithmetices principia, nova methodo exposita*" published in 1889. Peano modified the third axiom in a latter work to include 0 as the first element $n^*$. Brahmagupta *Brahmagupta* (598–668 AD) used the symbol 0 in 628 AD and gave rules to compute with zero[1] in his work *Brāhmasphuṭasiddh*ānta *Brāhmasphuṭasiddhānta* (the

[1] There is an interesting connection between "sifer" and "cyber". The Sanskrit word "shunyam" or the Hindi equivalent "shunya" was translated to Arabic as al-sifer. Fibonacci called it as "cifra" and "cipher" came from it. The word "zero" originated from the Medieval Latin word "zephirum".



V. Ravichandran and A. K. Razdan, *Fundamental Discrete Structures*, University Texts in the Mathematical Sciences,
https://doi.org/10.1007/978-981-96-0069-4

Opening of the Universe) such as $n + 0 = n - 0 = n$ and $n \cdot 0 = 0$. He defined the number zero as the result of subtracting a number from itself. The set consisting of $0, 1, 2, 3, \ldots$ is denoted by $\mathbb{N}_0$, which are known as *non-negative integers*. A construction of $\mathbb{N}_0$ in terms of *ordinal numbers* is due to von Neumann who defined these as follows:

$$0 = \emptyset, \quad 1 = \{\emptyset\}, \quad 2 = \{\emptyset, \{\emptyset\}\}, \quad \text{and so on.}$$

Brahmagupta also gave arithmetical rules in terms of the positive and negative integers such as the product of a negative integer and a positive integer is a negative integer. The symbol $\mathbb{Z}$ (the first letter of the German word "Zahlen" for numbers) denotes the set of all integers $\ldots, -3, -2, -1, 0, 1, 2, 3, \ldots$. According to the Prussian mathematician and logician Leopold Kronecker *Leopold Kronecker* (1823–1891): *God made the integers; all the rest is the work of man.*

***Well-Ordering Principle***

The second axiom provides a "natural order" to the set $\mathbb{N}$. The third axiom says 1 is the least element of the set $\mathbb{N}$. In general, the **well-ordering principle** (or simply WOP) says that *every nonempty subset of the set* $\mathbb{N}$ *has a least element*. As illustrated shortly, this is an important property of the set $\mathbb{N}$. We first show that the well-ordering principle *Well-ordering principle* is equivalent to the principle of induction. Notice that the *fifth axiom* is precisely the principle of induction *Principle of induction*, which we discussed in Chap. 2. For the proof given below, we recall that a set $A \subset \mathbb{R}^+$ is called an inductive set *Inductive set* if $1 \in S$, and $x \in S \;\Rightarrow\; x + 1 \in S$. We thus can reformulate the fifth axiom as "*the set* $\mathbb{N}$ *is the smallest inductive subset of* $\mathbb{R}^+$".

**Theorem A.1** $WOP \;\Leftrightarrow\; POI.$

***Proof*** First, suppose WOP holds for $\mathbb{N}$. Let $S \subseteq \mathbb{N}$ be an inductive set. If $S \neq \mathbb{N}$, then $\mathbb{N} \setminus S$ is a nonempty subset of $\mathbb{N}$, and so $\mathbb{N} \setminus S$ has the least element, say $\ell$, by well-ordering principle. Now, since $1 \in S$, we have $\ell > 1$. But then, $\ell - 1$ is a natural number in $S$, and so we have $\ell \in S$, by inductive property of $S$, which is a contradiction. Hence, $S = \mathbb{N}$. This proves POI. Conversely, suppose $\mathbb{N}$ is the smallest inductive set, and let $S$ be a nonempty subset of $\mathbb{N}$. If possible, suppose $S$ does not have the least element. Then, since $1 \in \mathbb{N}$ is the least element, we have $S \neq \mathbb{N}$. That is, $1 \in K = \mathbb{N} \setminus S$. We claim that $K$ is an inductive set. For instance, let $k \in K$. We proceed to show that $k + 1 \in K$. Again, let us assume for a while that $k + 1 \notin K$. Then, $k + 1 \in S$ implies that $n_0 < k + 1$, for some $n_0 \in S$, because $S$ has no least element. So, if $n_0 - 1 \in S$, we conclude from the previous inequality that $k \in S$, which is not the case by our choice of $k$. But then, we obtain $n_0 - 1 \in \mathbb{N} \setminus S = K$. This, in turn, implies that $n_0 \leq k$, using the relation $n_0 - 1 < k$, so $n_0 \in K$, which is a contradiction because $n_0 \in S$. Hence, we must have that $k + 1 \in K$. That is, $K$ is an inductive set. Therefore, $K = \mathbb{N}$ by POI, i.e., $S = \emptyset$, a contradiction to the fact that $S \neq \emptyset$. Hence, $S$ has the least element. □

**Lemma A.1** (Principle of Mathematical Induction) *For* $n \in \mathbb{N}$, *let* $p(n)$ *be a conjectured proposition about* $n$. *Suppose*

1. *Base Step: $p(1)$ is true and*
2. *Induction Hypothesis: for $k \in \mathbb{N}$, $p(k) \Rightarrow p(k+1)$.*

*Then $p(n)$ holds for all $n \in \mathbb{N}$.*

***Proof*** Let $A = \{n \in \mathbb{N} : p(n) \text{ holds}\}$. By (1), we have $1 \in A$. Also, by (2), $k \in A \Rightarrow k+1 \in A$. Therefore, by Peano's *fifth axiom*, we must have that $A = \mathbb{N}$. That is, $p(n)$ holds for all $n \in \mathbb{N}$. □

**Lemma A.2 (Principle of Mathematical Induction)** *For any fixed $m \in \mathbb{Z}$, let each $p(n)$ be a proposition about an integer $n \geq m$. If*

1. *$p(m)$ is true;*
2. *for an integer $k \geq m$, $p(k) \Rightarrow p(k+1)$,*

*then $p(n)$ holds for all integers $n \geq m$.*

***Proof*** Apply Lemma A.1 to $q(n) = p(n-m+1)$, $n \in \mathbb{N}$. □

We can also use the method of contradiction *Method of contradiction* to prove the weak form of POI.

**Theorem A.2** *If $p(n)$ is a proposition involving natural number $n$ such that*

1. *$p(1)$ is true and*
2. *$p(k) \Rightarrow p(k+1)$, for $k \geq 1$.*

*Then we have $p(n)$ is true, for all $n \geq 1$.*

***Proof*** For a contradiction, suppose $n_0$ is the first natural number such that $p(n_0)$ is not true. We may consider the set $S \subset \mathbb{N}$ consisting of all natural number $m$ such that $p(m)$ is not true. As $n_0 \in S$, it is a nonempty subset of $\mathbb{N}$. Therefore, by *well-ordering principle*, $S$ contains a least natural number, say $m_0$. Notice that, by (1), we have $m_0 \geq 2$, and so $m_0 - 1 \in \mathbb{N} \setminus S$. That is, $p(m_0 - 1)$ is true. But then, by (2), $p(m_0)$ is true, which contradicts our choice of $m_0$. Hence, $p(n)$ is true for all $n \geq 1$. □

***Divisibility***

The concept of **divisibility** as defined below is very useful in describing various important structures introduced in the main text.

**Definition A.1** A nonzero integer $b$ is said to **divide** an integer $a$ if we can write $a = bk$, for some integer $k$. In this case, we also say that $a$ is a **multiple** of $b$, and write $b \mid a$.

It follows directly from this definition that $a \mid 0$, for all $a \neq 0$, and $a \mid 1$ implies $a = \pm 1$. More generally, $a \mid b$ and $b \mid a$ implies $a = \pm b$, which says "divides" is antisymmetric, as proven in the next theorem.

**Theorem A.3** *For any three integers $a$, $b$, and $c$, we have*

$$a \mid b \text{ and } b \mid a \Rightarrow a = b.$$

***Proof*** It follows from the definition that we have

$$b = k_1\, a \quad \text{and} \quad a = k_2\, b, \qquad \text{for some} \quad k_1, k_2 \in \mathbb{N}.$$

We thus have

$$a = \big(k_1 k_2\big) a \;\; \text{i.e.,} \;\; k_1 k_2 = 1 \;\; \text{i.e.,} \;\; k_1 = k_2 = 1,$$

and so $a = b$. □

Also, "divides" has the transitivity property, as proven in the next theorem.

**Theorem A.4** *For any three integers $a$, $b$, and $c$, we have*

$$a \,\big|\, b \;\; \textit{and} \;\; b \,\big|\, c \;\;\Rightarrow\;\; a \,\big|\, c.$$

***Proof*** By the given assumption, we can write

$$b = ak_1 \quad \text{and} \quad c = bk_2, \qquad \text{for some} \quad k_1, k_2 \in \mathbb{Z}.$$

Therefore, it follows that

$$c = bk_2 = a\big(k_1 k_2\big) = ak', \qquad \text{where} \quad k' = k_1 k_2 \in \mathbb{Z}.$$

Hence, we obtain $a|c$, as asserted. □

The divisibility also satisfies the following interesting property:

$$c \,\big|\, a \;\; \text{and} \;\; c \,\big|\, b \;\;\Rightarrow\;\; c \,\big|\, x\, a + y\, b, \tag{A.1.1}$$

for every pair of integers $(x, y)$. The next theorem is known as the division algorithm *Division algorithm*. The proof given below uses the well-ordering principle *Well-ordering principle*.

**Theorem A.5 (Division Algorithm)** *For any two integers $a$ and $b$, with $b > 0$, there exist unique integers $q$ and $r$ such that*

$$a = b\, q + r, \qquad \textit{where} \;\; 0 \le r < b \;\; \textit{and} \;\; q = \big\lfloor a/d \big\rfloor, \tag{A.1.2}$$

*where $\lfloor x \rfloor$ is the largest integer $\le x$.*

***Proof*** More generally, consider the set

$$S := \big\{a + b\, x \mid x \in \mathbb{Z}\big\} \cap \mathbb{N}.$$

Since $b \in S$, it follows by *well-ordering principle* that it contains its least element, say $n_0$. We may write $n_0 = a + b\, x_0$, for $x_0 \in \mathbb{Z}$. For a contradiction, suppose $n_0 \ge b$. But then, we have

$$n_0 - b \geq 0 \quad \Rightarrow \quad n_0 - b = a + b\big(x_0 - 1\big) \in S,$$

which contradicts the minimality of $n_0$. Therefore, we conclude that $0 \leq n_0 < b$. Hence, we may take $q = -x_0$ and $r = n_0$. The uniqueness follows easily. □

The remainder $r$ is also referred to as a **residue**. In particular, when $r = 0$, we have $a = b\,q$. This implies that $b$ "divides" $a$. We also say that $b$ is a "divisor" of $a$.

**Definition A.2** Let $a$ and $b$ be two natural numbers. The **least common multiple** of $a$ and $b$, denoted by $\text{lcm}\big(a, b\big)$, is the smallest positive integer $m$ that is multiple of both $a$ and $b$.

The next theorem proves an interested property of the operator lcm.

**Theorem A.6** *For any $a, b, c \in \mathbb{N}$, we have*

$$lcm\big(ac, bc\big) = lcm\big(a, b\big) \cdot c.$$

***Proof*** We may write

$$m := \text{lcm}\big(ac, bc\big) \quad \text{and} \quad n := \text{lcm}\big(a, b\big) \cdot c.$$

By definition of least common multiple *Least common multiple*, we have

$$\text{lcm}\big(a, b\big) = ax = by, \qquad \text{for some} \quad x, y \in \mathbb{N}.$$

It thus follows that

$$n = \text{lcm}\big(a, b\big) \cdot c = acx = bcy.$$

That is, $n$ is a multiple of both $ac$ and $bc$. On again, by definition, we obtain $m \leq n$. To prove the reverse inequality, since $m$ is the least that is multiple of both $ac$ and $bc$, we can find $x, y \in \mathbb{N}$ so that we can write $m = acx = bcy$. But then, we have $m/c$ is a multiple of both $a$ and $b$, which implies that

$$\text{lcm}\big(a, b\big) \leq \big(m/c\big) \quad \text{or} \quad n = \text{lcm}\big(a, b\big) \cdot c \leq m.$$

Hence, $m = n$, as asserted. □

**Definition A.3** Let $a$ and $b$ be two positive integers. The **greatest common divisor** of $a$ and $b$, denoted by $\gcd(a, b)$, is the largest integer that divides both of them. We say integers $a$ and $b$ are **relatively prime** (or **coprime integers**) if we have $\gcd(a, b) = 1$.

Said differently, $\gcd(a, b)$ is the greatest common divisor *Greatest common divisor* of $a$ and $b$ if $d$ is any divisor of $a$ and $b$, then $d$ also divides $\gcd(a, b)$. We may take $\gcd(0, 0) = 0$ and, for any nonzero integer $a$, we have $\gcd(a, 0) = |a|$, because all nonzero integers divide 0. Since $\gcd(|a|, |b|) = \gcd(a, b)$, we always assume that $a \geq b > 0$. The next theorem is known as the Euclidean algorithm *Euclidean algorithm*. It gives an efficient way to compute the *greatest common divisor* of two positive integers. A proof may be obtained by repetitive use of the division algorithm.

**Theorem A.7 (Euclidean Algorithm)** *Consider two integers $r_0 \geq r_1 > 0$. Suppose $(r_0, r_1, \ldots, r_n)$ and $(q_1, q_2, \ldots, q_n)$ are two finite sequences satisfying the following system of equations:*

$$\begin{aligned}
r_0 &= r_1 \cdot q_1 + r_2, & 0 \leq r_2 < b; \\
r_1 &= r_2 \cdot q_2 + r_3, & 0 \leq r_3 < r_2; \\
r_2 &= r_3 \cdot q_3 + r_4, & 0 \leq r_4 < r_3; \\
&\vdots \\
r_{n-2} &= r_{n-1} \cdot q_{n-1} + r_n, & 0 \leq r_n < r_{n-1}; \\
r_{n-1} &= r_n \cdot q_n + 0.
\end{aligned} \tag{A.1.3}$$

*Then we have* $\gcd(r_0, r_1) = r_n$.

Notice that if $r_2 = 0$ then we have $r_1 = \gcd(r_0, r_1)$. Otherwise, by (A.1.1), we have $d|r_2$, because $d|r_0$ and $d|r_1$ together imply that $d|r_0 - r_1 \cdot q_1 = r_2$. Therefore, we obtain $d = \gcd(r_1, r_2)$. Hence, the algorithm (A.1.3) defines the sequence $\langle r_k \rangle$ recursively as given below:

$$r_{k-1} = q_k \, r_k + r_{k+1}, \qquad \text{where} \quad 0 \leq r_{k+1} < r_k. \tag{A.1.4}$$

At each stage of recursion, we have $d = \gcd(r_k, r_{k+1})$. Notice that the inequality $0 \leq r_{k+1} < r_k$ implies that $\langle r_k \rangle$ is a decreasing sequence of non-negative integers. Therefore, the algorithm (A.1.4) must terminate.

**Example A.1** With $r_0 = 1914$ and $r_1 = 899$, we have

$$\begin{aligned}
1914 &= 899 \cdot 2 + 116; \\
899 &= 116 \cdot 7 + 87; \\
116 &= 87 \cdot 1 + 29; \\
87 &= 29 \cdot 3 + 0.
\end{aligned} \tag{A.1.5}$$

Therefore, $\gcd(1914, 899) = 29$. In practice, it is more convenient to write the related computations in the tabular form, with the table headings as *dividend, divisor, quotient*, and the *remainder*. In particular, each line of the computations (A.1.5) in tabular form may be written as given in Table A.1.

The next corollary, due to the French mathematician Étienne Bézout *Étienne Bézout* (1730–1783), helps prove many important results of elementary number theory such as Euclid's lemma *Euclid's lemma* (Theorem A.9) and the Chinese remainder theorem *Chinese remainder theorem* (see Remark 5.1).

**Table A.1** Tabular form of Euclidean algorithm ((A.1.5))

| Dividend | Divisor | Quotient | Remainder |
|---|---|---|---|
| 1914 | 899 | 2 | 116 |
| 899 | 116 | 7 | 87 |
| 116 | 87 | 1 | 29 |
| 87 | 29 | 3 | 0 |

**Corollary A.1** (Bézout Identity Bézout Identity, 1779) *For any two nonzero integers a and b, there exist an integer d such that*

1. $d = \gcd(a, b)$ *and*
2. *for some integers* $x, y$, *we have* $d = a\,x + b\,y$.

*The integers x and y are called* ***Bézout coefficients***.

***Proof*** Consider the set $S$ given by

$$S := \{a\,x + b\,y \mid x,\ y \in \mathbb{Z},\ \text{ with } a\,x + b\,y > 0\}. \tag{A.1.6}$$

Since $a$ or $-a$ belongs to $S$, we have $S \neq \emptyset$. Therefore, by *well-ordering principle*, $S$ contains least element, say $d$. Let $d = a\,x_0 + b\,y_0$, with $x_0,\ y_0 \in \mathbb{Z}$. It remains to show that $d = \text{GCD}(a, b)$. For instance, by Theorem A.5, we can find integers $q, r$ such that

$$a = d\,q + r, \qquad \text{with } \ 0 \leq r < d.$$

However, if $r \neq 0$, then $r = a - d\,q \in S$ contradicts our choice of $d \in S$. So, we must have $r = 0$ so that $d \big| a$. Similarly, it can be shown that $d \big| b$. Finally, for any integer $c$,

$$c \ \text{divides both } a \text{ and } b \quad \Rightarrow \quad c \ \text{divides } \ a\,x_0 + b\,y_0 = d,$$

which completes the proof. □

It follows from Theorem A.1 that we have

$$\gcd\big(a, b\big) = 1 \quad \Leftrightarrow \quad a\,x + b\,y = 1, \tag{A.1.7}$$

for some integers $x$ and $y$. With reference to Example A.1, it follows from Corollary A.1 that 29 must be of the form $x \cdot 1914 + y \cdot 899$, for some integers $x$ and $y$. A difficult way to compute such a pair of Bézout coefficients $(x, y)$ is by adopting the *back-substitution* approach. That is, we may treat each $r_k$ as a variable, and work backwards through Eq. (A.1.5). For example, it follows from the two equations

$$29 = 116 - 1 \cdot 87 \quad \text{and} \quad 87 = 116 - 7 \cdot 116$$

that we have

$$29 = (-1) \cdot 899 + 8 \cdot 116.$$

We thus obtain from

$$116 = 1914 + (-2) \cdot 899 \quad \text{that} \quad 29 = 8 \cdot 1914 + (-17) \cdot 899.$$

Therefore, a pair of Bézout coefficients in this case is given by $(8, -17)$.

More generally, consider two integers $a$ and $b$, with $d = \gcd(a, b)$. We next describe an extension of the Euclidean algorithm that finds applications in cryptography. It is known as the extended Euclidean algorithm *Extended Euclidean algorithm*. The main purpose of this algorithm is to compute a pair of Bézout coefficients *Bézout coefficients* $(x, y)$ to express $d$ as a linear combination of integers $a$ and $b$. Notice that such an expression of $d$ is not unique. For instance, if we have a pair of Bézout coefficients $(x, y)$, then the other pairs of Bézout coefficients are given by

$$\Big(x - k\lfloor b/d \rfloor,\ y + k\lfloor a/d \rfloor\Big), \quad \text{for arbitrary integer} \quad k.$$

The procedure given below is based on backward recurrence *Backward recurrence* [1].

**Extended Euclidean Algorithm** We may recall that $d$ is only defined when at least one of two integers $a$ and $b$ is nonzero. Now, if $a \neq 0$ and $b = 0$, then we have $d = a$, and so we may write $a = 1 \cdot a + 0 \cdot 0$. Therefore, there is nothing to prove if one of the two integers $a$ and $b$ is zero. Moreover, since $d = (|a|, |b|)$, it can be assumed that both $a$ and $b$ are positive integers. We may thus take $a \geq b > 0$. Taking $r_0 = a$ and $r_1 = b$ in (A.1.3), we can write

$$\begin{aligned}
a &= b \cdot q_1 + r_2, &\quad 0 \leq r_2 < b;\\
b &= r_2 \cdot q_2 + r_3, &\quad 0 \leq r_3 < r_2;\\
r_2 &= r_3 \cdot q_3 + r_4, &\quad 0 \leq r_4 < r_3;\\
&\vdots\\
r_{n-2} &= r_{n-1} \cdot q_{n-1} + r_n, &\quad 0 \leq r_n < r_{n-1};\\
r_{n-1} &= r_n \cdot q_n + 0
\end{aligned} \tag{A.1.8}$$

so that we have $d = \gcd(a, b) = r_n$. We define a sequence $\langle s_k \rangle_{k=0}^{n}$ recursively working backwards as follows: Let $s_n := 0,\ s_{n-1} := 1$, and

$$s_{k-1} := q_k\, s_k + s_{k+1}, \qquad \text{for} \quad k = n-1, \ldots, 2, 1. \tag{A.1.9}$$

We proceed to show that

$$r_n = \big(-1\big)^{n+k+1} r_{k-1}\, s_k + \big(-1\big)^{n+k} r_k\, s_{k-1}, \qquad \text{for} \quad k = 1, 2, \ldots, n. \quad \text{(A.1.10)}$$

We shall use the downward induction *Downward induction*. For $k = n$, we have

$$\begin{aligned} r_n &= \big(-1\big)^{2n+1} r_{n-1}\, s_n + \big(-1\big)^{2n} r_n\, s_{n-1} \\ &= \big(-1\big)^{2n+1} r_{n-1} \cdot 0 + \big(-1\big)^{2n} r_n \cdot 1. \end{aligned}$$

Next, suppose $1 < k < n$, and (A.1.10) holds for $k + 1$. That is, we have

$$r_n = \big(-1\big)^{n+k+2} r_k\, s_{k+1} + \big(-1\big)^{n+k+1} r_{k+1}\, s_k. \qquad \text{(A.1.11)}$$

We show that (A.1.10) also holds for $k$. For instance, substituting $s_{k+1} = s_{k-1} - q_k s_k$ in (A.1.11) and simplifying, we obtain

$$\begin{aligned} r_n &= \big(-1\big)^{n+k+2} r_k \Big[s_{k-1} - q_k s_k\Big] + \big(-1\big)^{n+k+1} r_{k+1}\, s_k \\ &= \big(-1\big)^{n+k+2} r_k\, s_{k-1} + \big(-1\big)^{n+k+1} r_k q_k s_k + \big(-1\big)^{n+k+1} r_{k+1}\, s_k \\ &= \big(-1\big)^{n+k} r_k\, s_{k-1} + \Big[r_k q_k + r_{k+1}\Big]\big(-1\big)^{n+k+1} s_k \\ &= \big(-1\big)^{n+k} r_k\, s_{k-1} + \big(-1\big)^{n+k+1} r_{k-1} s_k. \end{aligned}$$

Therefore, (A.1.10) holds for all $1 \le k \le n$. In particular, for $k = 1$, we have

$$\begin{aligned} r_n &= \big(-1\big)^{n+2} r_0\, s_1 + \big(-1\big)^{n+1} r_1\, s_0 \\ &= \Big[\big(-1\big)^n s_1\Big] r_0 + \Big[\big(-1\big)^{n+1} s_0\Big] r_1. \end{aligned}$$

Hence, we have $x = \big(-1\big)^n s_1$ and $y = \big(-1\big)^{n+1} s_0$.

**Example A.2** Let $a = 187$ and $b = 102$. First, we use Theorem A.7 to find $d := \gcd(a, b)$. The related computations in tabular form are given by

| Dividend | Divisor | Quotient | Remainder | **s-value** |
|---|---|---|---|---|
| 187 | 102 | 1 | 85 | 2 |
| 102 | 85 | 1 | 17 | 1 |
| 85 | 17 | 5 | 0 | 1 |

It thus follows that $d = 17$. Now, we proceed to obtain an expression

$$d = x \cdot 187 + y \cdot 102.$$

By (A.1.9), we have

$$\begin{aligned} s_3 &= 0, \qquad s_2 = 1; \\ s_1 &= q_2 \cdot s_2 + s_3 = 1 \cdot 1 + 0 = 1; \\ s_0 &= q_1 \cdot s_1 + s_2 = 1 \cdot 1 + 1 = 2. \end{aligned}$$

These *s-values*, as shown in the last column of the above table (bottom to top), may be computed directly from the tabulated values using the rule as given below:

$$\big(\text{next } s\big) = \big(\text{last } q\big) \cdot \big(\text{last } s\big) + \big(\text{next-to-last } s\big).$$

Hence, we may take

$$x = \big(-1\big)^3 \cdot 1 = -1 \quad \text{and} \quad y = \big(-1\big)^4 \cdot 2 = 2. \tag{A.1.12}$$

Indeed, we have $(-1) \cdot 187 + 2 \cdot 102 = 17$.

**Example A.3** Let $a = 246$ and $b = 194$. First, we use Theorem A.7 to find $d := \gcd(a, b)$. The related computations in tabular form are given by

| Dividend | Divisor | Quotient | Remainder | **s-value** |
|---|---|---|---|---|
| 246 | 194 | 1 | 52 | 52 |
| 194 | 52 | 3 | 38 | 41 |
| 52 | 38 | 1 | 14 | 11 |
| 38 | 14 | 2 | 10 | 8 |
| 14 | 10 | 1 | 4 | 3 |
| 10 | 4 | 2 | 2 | 2 |
| 4 | 2 | 2 | 0 | 1 |

It thus follows that $d = 2$. Now, we proceed to obtain the expression:

$$d = x \cdot 246 + y \cdot 194.$$

By (A.1.9), we have

$$\begin{aligned} s_7 &= 0, \qquad s_6 = 1; \\ s_5 &= q_6 \cdot s_6 + s_7 = 2 \cdot 1 + 0 = 2; \\ s_4 &= q_5 \cdot s_5 + s_6 = 1 \cdot 2 + 1 = 3; \\ s_3 &= q_4 \cdot s_4 + s_5 = 2 \cdot 3 + 2 = 8; \\ s_2 &= q_3 \cdot s_3 + s_4 = 1 \cdot 8 + 3 = 11; \\ s_1 &= q_2 \cdot s_2 + s_3 = 3 \cdot 11 + 8 = 41; \\ s_0 &= q_1 \cdot s_1 + s_2 = 1 \cdot 41 + 11 = 52. \end{aligned}$$

These *s-values*, as shown in the last column of the above table (bottom to top), may be computed directly from the tabulated values using the rule (A.1.12). Hence, we may take

$$x = (-1)^7 \cdot 41 = -41 \quad \text{and} \quad y = (-1)^8 \cdot 52 = 52.$$

Indeed, we have $(-41) \cdot 246 + 52 \cdot 194 = 2$.

**Remark A.1** Though the above-illustrated extended Euclidean algorithm is best for *manual computations*, but it is not *machine efficient*, mainly because it requires storage for all the related quotients obtained in the Euclidean algorithm. However, there are other algorithms that are *machine efficient* [2].

***Prime Numbers***

An integer $n \geq 2$ is called a **prime number** (or simply a **prime**) if the only divisors of $n$ are 1 and $n$ itself. We know that *there are infinitely many primes*. Euclid's proof of this fundamental fact given in 300 *BC* is a true classic of the number theory. In fact, Euclid's argument given in Theorem 2.12 may be used to generate new prime number *Prime number*.[2] A simple description of a scheme is as given below: Start with $p_1 = 2$, and define inductively $p_{k+1}$ to be the least prime not chosen earlier that divides $m + 1$, where $m$ runs over all the divisors of the product $p_1 p_2 \cdots p_k$. It is an interesting theorem that the sequence of primes $\langle p_n \rangle$ constructed in this manner exhausts all primes. The reader may refer to excellent texts [3] or [4], for more on prime numbers.

An integer is called a **composite number** if it not a prime number. The next theorem proves a fundamental fact about a composite number *Composite number* $n > 1$. It says that set $\mathbb{N}$ has the prime numbers as its building blocks (multiplicatively).

**Theorem A.8 (Fundamental Theorem of Arithmetic)** *Every integer $a > 1$ has a unique expression as a finite product of prime numbers. That is, we have*

$$a = p_1^{k_1} p_2^{k_2} \cdots p_t^{k_t}, \tag{A.1.13}$$

*where each $p_i$ is a prime divisor of $a$, and the integers $k_i \geq 0$, for $i = 1, \ldots, t$. The uniqueness is subject to the arrangement $p_1 < p_2 < \cdots < p_t$.*

We have given a proof of the fundamental theorem of arithmetic *Fundamental theorem of arithmetic* by the principle of induction in the main text (Theorem 8.45). We can also use Euclid's lemma *Euclid's lemma* to prove its existence part.

**Theorem A.9 (Euclid's Lemma)** *For any two integers $a$ and $b$, and a prime number $p$, $p \mid ab$ implies $p \mid a$ or $p \mid b$.*

***Proof*** We prove it for the case when $a = b$. The general case follows on similar lines. If $p$ does not divide $a$, then $a = p_1^{m_1} p_2^{m_2} \cdots p_k^{m_k}$ where none of these $p_i$'s are

[2] The Mersenne number $2^{82589933} - 1$ having 24862048 digits is the largest known prime.

equal to $p$. Then $a^2 = p_1^{2m_1} p_2^{2m_2} \cdots p_k^{2m_k}$ and $p$ is not a factor of $a^2$ which is not possible as $p$ divides $a^2$. □

In what follows, it is more appropriate to write (A.1.13) as

$$a = \prod_{p \in P} p^{a_p}, \qquad \text{where each} \quad a_p \geq 0, \tag{A.1.14}$$

and $P$ denotes the set of all primes. Of course, $a_p = 0$ for almost all $p \in P$.

**Definition A.4** A representation of an integer $a > 1$ given in (A.1.13) or (A.1.14) is called the **prime decomposition** (or **prime factorization**) of the integer $a$. We say $a$ is **square free** if there are no repeated prime factors in the prime decomposition *Prime decomposition* of $a$.

We may also represent an integer $a > 1$ as a list of nonzero exponents. For example,

The integer 15 is the list $[a_3 = 1,\ a_5 = 1]$;
The integer 28 is the list $[a_2 = 2,\ a_7 = 1]$;
The integer 42 is the list $[a_2 = 1,\ a_3 = 1,\ a_7 = 1]$;
The integer 84 is the list $[a_2 = 2,\ a_3 = 1,\ a_7 = 1]$.

Notice that 15 is a square-free *Square-free* integer, and so is 42. The multiplication of two integers in this terminology corresponds to the addition of the exponents. That is, if we have

$$a = \prod_{p \in P} p^{a_p} \quad \text{and} \quad b = \prod_{p \in P} p^{b_p}, \tag{A.1.15a}$$

then the product $c := ab$ is expressed as

$$c = \prod_{p \in P} p^{c_p}, \qquad \text{where} \quad c_p = a_p + b_p, \tag{A.1.15b}$$

for all $p \in P$. For example, with $a = 15$ and $b = 84$, we have

$$c = 15 \cdot 84 = \left(3^1 \times 5^1\right) \cdot \left(2^2 \times 3^1 \times 7^1\right) = 2^2 \times 3^2 \times 5^1 \times 7^1.$$

Notice that it follows from (A.1.15a) that

$$a \mid b \quad \Rightarrow \qquad a_p \leq b_p, \quad \text{for all} \quad p. \tag{A.1.16}$$

Therefore, the greatest common divisor $d$ of two positive integers $a$ and $b$ with prime factorization given by (A.1.15a) may be expressed as

$$d = \prod_{p \in P} p^{d_p}, \qquad \text{where} \quad d_p = \min\left(a_p, b_p\right), \tag{A.1.17}$$

for all $p \in P$. However, this is not a practically viable method to compute $d = \gcd(a, b)$, because determination of the prime factors of a large number is a far more difficult problem. Notice that in the same way the least common multiple $\ell$ of two positive integers $a$ and $b$ with prime factorization given by (A.1.15a) may be expressed as

$$\ell = \prod_{p \in P} p^{\ell_p}, \qquad \text{where} \quad \ell_p = \max\left(a_p, b_p\right). \tag{A.1.18}$$

***Primality Testing***

In various cryptographic system, it is important to determine whether a given large number is a prime. The related algorithm is called **primality testing**. One such important probabilistic algorithm is known as the **Miller-Rabin algorithm**. Though it does not perform a primality testing *primality testing*, but it gives a number that is almost certainly a prime [2]. A deterministic algorithm to test the primality of a large number was first developed in 2002, which is now known as the AKS algorithm [5]. In any such investigation, it is crucial to know how many numbers an algorithm rejects before a prime number is found. We know that, on average, primes near a positive integer $n$ are spaced one in every $\ln(n)$ integers. This is called the *prime number theorem.* Of course, since even integers can be dropped from the list, we only need to test $0.5\ \ln(n)$ integers before a prime is found. In general, we need to understand the *distribution of primes* for better algorithms of primality testing.

Theorem 8.36 has some interesting applications to prime divisors of the numbers of the form $M_n := 2^n - 1$, known as the, *Mersenne primes*[3] and also to prime divisors of the numbers of the form $F_n := 2^{2^n} + 1$, known as the.*Fermat numbers*[4]. For example, $M_{11} = 2047 = 23 \cdot 89$, and

$$F_5 = 4294967297 = 641 \cdot 6700417.$$

The above was used by Euler to disprove Fermat conjecture for such types of numbers. The following two simple observations will be useful in the sequel:

(a) If $p$ is an odd prime, and if $q|M_p$, then $q \equiv 1 \bmod 2p$.
(b) If $q$ divides $F_n$, then $q \equiv 1 \bmod 2^{n+1}$.

***Proof*** Notice that it suffices to prove $(a)$ for prime values of $q$. We may thus assume that $q|2^p - 1$ so that $2^p \equiv 1 \bmod q$. By Theorem 8.36, we have $\text{ord}_p(2)|p$, and so

[3] These types of prime numbers are used in relatively new public key cryptosystem based on elliptical curves, and also in random number generation. The former exploits the fact that reduction modulo of a Mersenne prime does not modify the hamming weight of the output.

[4] Fermat conjecture that these integers are primes attracted attention after Gauss proved that a regular $p$-gon is *constructible* when $p > 2$ is a Fermat prime. The first (almost) complete proof of the converse was given by the French mathematician *Piérre Wantzel* (1814–1848) in a paper published in 1837. The prime factorization of $F_n$ for some $n$ needs further research.

$p = \text{ord}_p(a)$, because $p$ is a prime. Again, it suffices to prove $(b)$ for prime divisors of $q$. Now, $q | F_n$ implies that $2^{2^n} + 1 \equiv 1 \bmod q$, and so

$$2^{2^n} \equiv -1 \bmod q \quad \text{and} \quad 2^{2^{n+1}} \equiv 1 \bmod q.$$

We show that $2^{n+1} = \text{ord}_q(2)$. Indeed, by Theorem 8.36, we have the order divides $2^{n+1}$, and hence is a power of 2. Notice that $2^{n+1}$ is the smallest power of 2 that has this property. On the other hand, by Fermat little theorem, we have $2^{q-1} \equiv 1 \bmod q$, and so $2^{n+1} | (q-1)$, by Theorem 8.36. This completes the proof. □

## A.2 Algorithms

Very roughly, an **algorithm** refers to a finite set of instructions typically applied to complete a computational task on an electronic or mechanical device. Instructions must be well defined in the sense that they provide a precise description of the procedure to be followed by the machine. These instructions include arithmetic operations, mathematical functions, conditional statements, and loops executed in a specific order. Every *Algorithm* takes specified *inputs* to produce one or more *outputs*. We study theoretical aspects of time and space efficiency of an algorithm in the complexity theory. We can write an algorithm as a pseudocode in a natural language, a programming language, or even as a hardware design. Some of the most important types of algorithms are as listed below:

(a) A *Brute force algorithm* is the simplest that can solve most problems, but very *inefficiently*. Some applications include linked lists, queues, stacks, hash maps, and recursion.
(b) A *Greedy algorithm* optimizes at a point, without considering the future steps of the procedure. In some cases, such as problems related to *selection and topological sort, coin change, knapsack*, and *job scheduling*, the approach proves very useful. However, *dynamic programming* approach is preferred for a more reliable solution. Some applications include Bellman-Ford algorithm, matrix multiplication, subset sum, and knapsack problems.
(c) A *Recursive algorithm* uses *recursion functions*. That is, it calls itself to obtain output for the current input by using the returned values for smaller (or simpler) input values. We thus need to have a proper scheme to manage memory. It also requires more computation in comparison to *iterative algorithms*. Some simple applications include sequential search, checking set membership, finding the value of a recursively defined function, computing powers of a number, determination whether a word is a palindrome, and Towers of Hanoi problem.
(d) A *Backtracking algorithm* is a form of recursive algorithm that provides an improvement to brute force algorithm. Some applications include generating binary strings, solving problems related to $n$-queens, knapsack, and graph coloring.

(e) A *Divide-and-conquer algorithm* optimizes by "dividing" the main problem into sub-problems, and then "conquer" it by combining the solutions of such sub-problems. It is one of the most extensively used algorithms. Some examples include binary search, merge, and quick sort.
(f) A *Hashing algorithm* uses some mathematical function to manipulate data to make it unreadable (by chopping and mixing). It is primarily used for *authentication*. Some examples include message digest algorithm such as **MD5**, secure hash algorithm such as **SHA-1** and **SHA-2**, **NTLM**, and **LANMAN**.
(g) A *Randomized algorithm* depends only on inputs provided, and the outputs produced in the preceding parts of computation. The randomness in this case is involved mainly due to the fact that algorithm makes some random choices at certain points of its execution. Some applications include randomized quick sort and Kager's algorithm. However, such types of algorithms are also used in situations when search involves non-random goals.

In what follows, we discuss some *Searching algorithm* and *Sorting algorithm*.

*Searching Algorithm* Suppose we need to find the maximum value in a finite sequence of integers, say $a_1, a_2, \dots, a_n$, which is the input in this case. In a typical **search algorithm**, we set $a_1$ as the *temporary maximum*. Compare it with $a_2$, and if it is larger than $a_1$, we set $a_2$ as the new temporary maximum; otherwise, repeat the previous step if there are more integers in the sequence. Stop when there are no integers left in the sequence to compare with. The temporary maximum obtained at this point is the largest integer in the sequence, which is our output in this case.

In general, searching is the process of finding a particular item in the given list. A *Linear search algorithm* is a sequential searching algorithm that traverses an unordered list to compare each item of the list with the one whose location is to be found. When the match is found, algorithm returns "*location*" as output; otherwise, it returns "*null*". On the other hand, a *binary search algorithm* is a type of divide-and-conquer algorithm such that a sorted list is divided into two halves and then item is compared with the middle item of the list. When the match is found, algorithm returns the "*location*" of the middle item; otherwise, the procedure continues with searching into either of the halves depending upon the output obtained through previous comparisons. Two ways to implement the binary search algorithm are *iterative method* and *recursive method*. The latter one follows the divide-conquer approach.

In the case of a *computational problem*, an algorithm takes a value (or a set of values) as input, and produces a value (or a set of values) as an output. While solving such types of problems, the *input size* such as an array size for *sorting* depends on the problem being studied. An *algorithm design* must ensure **correctness** and **generality**. An algorithm is said to be **correct**[5] if it halts with the correct output, for every admissible input instance. Only a *correct algorithm* solves a given computational problem or executes task in the specified way. Further, we say an algorithm is **general** when it applies to a larger set of *input instances* to produce a *correct output*. However,

[5] There are situations such as finding large primes when an incorrect algorithm may also prove useful, if we somehow control their *error rate*.

some problems are too complicated to be solved efficiently without compromising *generality* or *correctness*.

For the above illustration of *searching algorithm*, a pseudocode is as given by
**Procedure** max $(a_1, a_2, \ldots, a_n$ : integers$)$
max $= a_1$
**for** $i = 2$ to $n$
**if** max $< a_i$ **then** max $= a_i$
**output** *max*

Therefore, the number of steps involved are given by

$$1 + (n - 1) + (n - 1) + 1 = 2n.$$

Recall that the run time an algorithm takes to solve a computation problem depends on the number of steps it uses to complete the task. It is assumed that each execution of the $i$th line of our pseudocode takes time $c_i$, where $c_i$ is a constant. The longest run time for any input of size $n$ is known as the *worst-case* run time. That is, a statement that takes $c_i$ steps to execute and is executed $n$ times will contribute $c_i\, n$ to the total run time. The maximum number of steps an algorithm uses to solve a given problem of a specified input size is known as the **worst-case complexity**. As shown later, for example, the worst-case time complexity of linear search is $O(n)$, and of the binary search is $O(logn)$.

As the next illustration, consider the problem to determine if a given list of integers $A = \langle a_1, a_2, \ldots, a_n\rangle$ contain a value, say $v = 5$. The *input* is the given list of integers and a value $v$; and, the *output* is an index $i$ such that $v = A[i]$ or "*nil*" if $v$ is not in the list $A$. In this case, a pseudocode for this *search algorithm* is given by
**Procedure** search$(a_1, a_2, \ldots, a_n$ : integers$)$
**for** $i = 1$ to $n$
**if** $a_i = 5$ **then output** True
**output** False

In this case, the *worst-case complexity* is given by

$$n + n + 1 = 2n + 1.$$

While describing running time of an algorithm, we ignore constants and consider only the input size that is usually very large. In particular, for a linear search, the worst-case complexity is given by $2n + 2$, and for a binary search, the worst-case complexity is given by $3 + 3\log(n) + 2$. The related algorithms are as follows:

Algorithm for a **linear search** involving integers $a_1, a_2, \ldots, a_n$:
**Procedure** linear search$(x : a_1, a_2, \ldots, a_n$ : integers$)$
**While** $(i \leq n$ and $x \neq a_i)$
i=i+1
**if** $i \leq n$ **then** location $= i$
**output** location

Recall that when we have a sorted sequence $A$ it is possible to check its midpoint against a value $v$, not considering half of the sequence. A typical example is that of binary search algorithm that repeats this procedure, halving the size of the remaining portion of the sequence each time. A pseudocode for binary search algorithm is as given below:

A **binary search** involving integers $a_1 \leq a_2 \leq \cdots \leq a_n$:
**Procedure** binary search($x$ : integer, $a_1, a_2, \ldots, a_n$ : increasing integers)
$i = 1$
$j = n$
$m = \lfloor (i + j)/2 \rfloor$
**While** ($a_m \neq x$ and $i \leq j$)
**begin**
$m = \lfloor (i + j)/2 \rfloor$
**if** $x > a_m$ **then** $i == m + 1$ **else** $j = m - 1$
**end**
**if** $x = a_m$ **then** location $i == m$ **else** location $= 0$
**output** location

The preliminary aspect of *algorithm analysis* is to predict the resources required such as memory, communication bandwidth, and also logic gates. There are mainly two important parameters that determine usefulness of an algorithm: (a) the *algorithm analysis* needed to run the algorithm and (b) the *space requirements* to hold the data, program variables, and so on. In general, the *performance parameters* of a computer program depend on a number of factors such as the computer that is being used, the way the data are represented, and how the program is translated into machine instructions. It is though possible to make precise estimates of the execution time of a program by taking into consideration the above factors, far more useful information can be obtained by analyzing the time of the underlying algorithm.

*Insertion Sorting Algorithm* Suppose we need to **insertion sort** a finite sequence of numbers, say $a_1, a_2, \ldots, a_n$, which is our input in this case. The required output is a permutation of the given sequence. For sorting a small number of elements, *insertion sort* is an efficient algorithm. It takes as a parameter an *array* $A[1, \ldots, n]$ containing the given sequence of length $n$ that needs to be sorted. The array is virtually split into two parts: A *sorted values* and an *unsorted values*. Subsequently, values from the unsorted part are picked and placed at the correct position in the sorted part, always doing comparisons from *right-to-left*. A nice analogy may be given by considering how a player of bridge or gin rummy sort a hand. Starting with an empty left hand, with the cards face down on the table, we remove one card at a time from the table and insert it into the correct position in the left hand. To find the correct position for a card, we compare it with each of the cards already in the hand, from right-to-left. In this way, the input numbers are *sorted in place*. That is, the numbers are rearranged within the array $A$, with at most a constant number of them stored outside the array at any point in time. The input array $A$ contains the sorted output sequence when *insertion sort* is finished.

```
InsertionSort(A)
n = A.length
    for j=2 to n
    key = A[j]
        i = j−1
        while(i>0) and (A[i] > key)
            A[i+1] = A[i]
            i = i−1
        A[i+1] = key;
```

*Merge Sorting Algorithm* The **merge sort** is a kind of recursive algorithm that divides an array into smaller sub-arrays, sorting each sub-array separately, and ultimately merging the sorted sub-arrays back together to form the final sorted array (see Fig. A.1). More specifically, algorithm divides the array into two halves, sort each half, and then merge the sorted halves back together. The same process is repeated until the entire array is sorted. To improve the overall performance, merge sort is often used in conjunction with other algorithms such as *quicksort*. For example, with reference to the merge sort on the array $A = \{39, 27, 43, 3, 9, 82, 10\}$, the lengths of the sorted sequences being merged increase as the algorithm progresses from bottom to top. In general, many programs use *sorting* as an intermediate step. For this reason, a large number of good sorting algorithms have been developed. The appropriateness of a sorting algorithm depends on the number of items to be sorted, the extent to which the items are already somewhat sorted, and the kind of storage device used in the process.

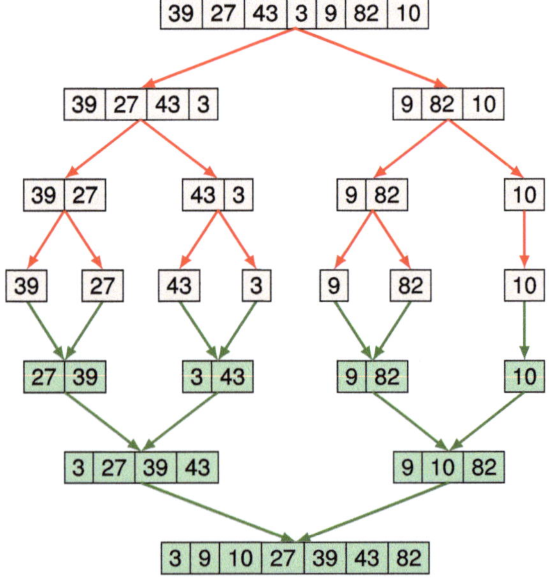


**Fig. A.1** Merge sort algorithm

In general, even analysis of simple algorithm could be challenging. The mathematical tools required may include discrete combinatorics, elementary probability theory, algebraic dexterity, and the ability to identify the most significant terms in a formula. Because the behavior of an algorithm may be different for each possible input, we need a means for summarizing that behavior in simple, easily understood formulas. The *Time complexity* is defined as the amount of time taken by an algorithm to run, as a function $f(n)$ of the *input size* $n$. It measures the time taken to execute each statement of code in an algorithm. It is not about examining the total execution time of an algorithm. For example, **merge sort** has a time complexity of $O(n \log n)$ so that it can sort large arrays relatively quickly. Also, it is *stable sort* in the sense that the order of elements with equal values is preserved during the sort.

## References

1. Glasby, S.P.: Extended Euclid's algorithm via backward recurrence relations. Mathematics Mag. **72**(3), 228–230 (1999)
2. Stallings, W.: Cryptography and Network Security: Principles and Practice (8/e). Pearson Education (2022)
3. Crandall, R., Pomerance, C.: Prime Numbers: A Computational Perspective. Springer (2001)
4. Ribenbiom, P.: The New Book of Prime Number Records. Springer-Verlag, New York (1996)
5. Agrawal, M., Kayal, N., Saxena, N.: PRIMES is in P. IIT Kanpur. Ann, Math (2004)

# Index




V. Ravichandran and A. K. Razdan, *Fundamental Discrete Structures*, University Texts in the Mathematical Sciences,
https://doi.org/10.1007/978-981-96-0069-4

MIX
Papier aus verantwortungsvollen Quellen
Paper from responsible sources
FSC® C105338

If you have any concerns about our products,
you can contact us on
**ProductSafety@springernature.com**

In case Publisher is established outside the EU,
the EU authorized representative is:
**Springer Nature Customer Service Center GmbH**
**Europaplatz 3, 69115 Heidelberg, Germany**

Printed by Libri Plureos GmbH
in Hamburg, Germany